New Results in Numerical and Experimental Fluid Mechanics V

Contributions to the 14th STAB/DGLR Symposium Bremen, Germany 2004

Hans-Josef Rath
Carsten Holze
Hans-Joachim Heinemann
Rolf Henke
Heinz Hönlinger
(Editors)

Springer

Prof. Dr. Hans-Josef Rath
Carsten Holze
ZARM - University of Bremen
Am Fallturm/Hochschulring
D-28359 Bremen

Rolf Henke
Airbus Deutschland GmbH
Senior Project Manager
Hünefeldstraße 1-5
D-28199 Bremen

Dr. Hans-Joachim Heinemann
Co-Managing Editor
„Aerospace Science and Technology, AST"
DLR
Bunsenstraße 10
D-37073 Göttingen

Professor Dr. Heinz Hönlinger
DLR - AE
Bunsenstraße 10
D-37073 Göttingen

ISBN 978-3-642-06995-6 **e-ISBN 978-3-540-33287-9**

Springer is a part of Springer Science+Business Media
springer.com

Softcover reprint of the hardcover 1st edition 2006

Cover-Design: deblik, Berlin

NNFM Editor Addresses

Prof. Dr. Ernst Heinrich Hirschel
(General editor)
Herzog-Heinrich-Weg 6
D-85604 Zorneding
Germany
E-mail: e.h.hirschel@t-online.de

Prof. Dr. Kozo Fujii
Space Transportation Research Division
The Institute of Space
and Astronautical Science
3-1-1, Yoshinodai, Sagamihara
Kanagawa, 229-8510
Japan
E-mail: fujii@flab.eng.isas.jaxa.jp

Dr. Werner Haase
Höhenkirchener Str. 19d
D-85662 Hohenbrunn
Germany
E-mail: werner@haa.se

Prof. Dr. Bram van Leer
Department of Aerospace Engineering
The University of Michigan
Ann Arbor, MI 48109-2140
USA
E-mail: bram@engin.umich.edu

Prof. Dr. Michael A. Leschziner
Imperial College of Science
Technology and Medicine
Aeronautics Department
Prince Consort Road
London SW7 2BY
U. K.
E-mail: mike.leschziner@ic.ac.uk

Prof. Dr. Maurizio Pandolfi
Politecnico di Torino
Dipartimento di Ingegneria
Aeronautica e Spaziale
Corso Duca degli Abruzzi, 24
I-10129 Torino
Italy
E-mail: pandolfi@polito.it

Prof. Dr. Jacques Periaux
Dassault Aviation
78, Quai Marcel Dassault
F-92552 St. Cloud Cedex
France
E-mail: jperiaux@free.fr

Prof. Dr. Arthur Rizzi
Department of Aeronautics
KTH Royal Institute of Technology
Teknikringen 8
S-10044 Stockholm
Sweden
E-mail: rizzi@kth.se

Dr. Bernard Roux
L3M – IMT La Jetée
Technopole de Chateau-Gombert
F-13451 Marseille Cedex 20
France
E-mail: broux@l3m.univ-mrs.fr

Prof. Dr. Yurii I. Shokin
Siberian Branch of the
Russian Academy of Sciences
Institute of Computational
Technologies
Ac. Lavrentyeva Ave. 6
630090 Novosibirsk
Russia
E-mail: shokin@ict.nsc.ru

FOREWORD

This volume contains the paper presented at the 14th DGLR/STAB-Symposium held at the ZARM, Universität Bremen, Germany, November, 16 to 18, 2004. STAB is the German Aerospace Aerodynamics Association, founded towards the end of the 1970's, whereas DGLR is the German Society for Aeronautics and Astronautics (Deutsche Gesellschaft für Luft- und Raumfahrt - Lilienthal Oberth e.V.).

The mission of STAB is to foster development and acceptance of the discipline "Aerodynamics" in Germany. One of its general guidelines is to concentrate resources and know-how in the involved institutions and to avoid duplication in research work as much as possible. Nowadays, this is more necessary than ever. The experience made in the past makes it easier now, to obtain new knowledge for solving today's and tomorrow's problems. STAB unites German scientists and engineers from universities, research-establishments and industry doing research and project work in numerical and experimental fluid mechanics and aerodynamics for aerospace and other applications. This has always been the basis of numerous common research activities sponsored by different funding agencies.

Since 1986 the symposium has taken place at different locations in Germany every two years. In between STAB workshops regularly take place at the DLR in Göttingen. The changing meeting places were established as focal points in Germany's Aerospace Fluid Mechanics Community for a continuous exchange of scientific results and their discussion. Moreover, they are a forum where new research activities can be presented, often resulting in new commonly organised research and technology projects.

It is the fifths time now that the contributions to the Symposium are published after being subjected to a peer review. The material highlights the key items of integrated research and development based on fruitful collaboration of industry, research establishments and universities. Some of the contributions still present results from the "Luftfahrtforschungsprogramm der Bundesregierung (German Aeronautical Research Programme)". Some of the papers report on work sponsored by the Deutsche Forschungsgemeinschaft (DFG, German Research Council) in some of their Priority Programs (Verbundschwerpunkt-Programm) as well as in their Collaborative Research Centres (Sonderforschungsbereiche). Other articles are sponsored by the European Community and are therefore results of cooperation among different organisations. The main areas include numerical simulation and mathematics, aeroelasticity, small and large aspect ratio wings in the context of leading-edge vortices, wake vortices, high lift systems and propulsion integration, and new developments in wind tunnel facilities and measurement techniques. Therefore, this volume gives an almost complete review of the ongoing aerodynamics research work in Germany. The order of the papers in this book corresponds closely to that of the sessions of the Symposium.

The Review-Board, partly identical with the Program-Committee, consisted of A. Altminus (München), G. Ashcroft (Köln), J. Ballmann (Aachen), R. Behr (München), Chr. Breitsamter (München), A. D'Alascio (München), J. Delfs (Braunschweig), G. Eitelberg (Emmeloord), M. Fischer (Bremen), R. Friedrich (München), R. Grundmann (Dresden), A. Gülhan (Köln), H. Hansen (Bremen) , S. Haverkamp (Aachen), St. Hein (Göttingen), P. Hennig (Unterschleißheim), F. Holzäpfel (Weßling), H. Hönlinger (Göttingen), G. Koppenwallner (Katlenburg-Lindau), W. Kordulla (Noordwijk), H. Körner (Braunschweig), E. Krämer (Stuttgart), H.-P. Kreplin (Göttingen), N. Kroll (Braunschweig), D. Kröner (Freiburg), J. Longo (Braunschweig), Th. Lutz (Stuttgart), F. Menter (Otterfing), E. Meyer (München), C. Naumann (Stuttgart), G. Neuwerth (Aachen), W. Nitsche (Berlin), H. Olivier (Aachen), R. Radespiel (Braunschweig), H.-J. Rath (Bremen), U. Rist (Stuttgart), H. Rosemann (Göttingen), R. Schnell (Köln), G. Schrauf (Bremen), W. Schröder (Aachen), D. Schwamborn (Göttingen), J. Sesterhenn (München), E. Steinhardt (München), Chr. Stemmer (Dreseden), P. Thiede (Bremen), J. Thorbeck (Berlin), C. Tropea (Darmstadt), R. Voß (Göttingen), C. Wagner (Göttingen), C. Weiland (München), C. Weishäupl (München), and W. Würz (Stuttgart). Nevertheless, the authors sign responsible for the contents of their contributions.

The editors are grateful to Prof. Dr. E. H. Hirschel as the General Editor of the "Notes on Numerical Fluid Mechanics and Multidisciplinary Design" and to the Springer-Verlag for the opportunity to publish the results of the Symposium.

H.-J. Rath, Bremen
C. Holze, Bremen
H.-J. Heinemann, Göttingen
R. Henke, Bremen
H. Hönlinger, Göttingen

October 2005

Contents (continued)

Page

Laminar Flow Control and Transition

Active Flow Control

Contents (continued)

Contents (continued)

Contents (continued)

High Reynolds-Number Windtunnel Testing for the Design of Airbus High-Lift Wings

Daniel Reckzeh[1], Heinz Hansen[2],
AIRBUS, Aerodynamic Design & Data Domain,
D- 28183 Bremen, Germany

Summary

Present aircraft design methods must be continuously improved. This is due to environmental and future transport market requirements. Promising results are expected from development of advanced high lift systems for civil transport aircraft. At present the validation of the numerical design of a high lift / low speed system is done with subscale aircraft models. The final performance of the new aircraft is derived by extrapolation. This extrapolation to a full-scale aircraft poses a high risk in time and money. The accurate prediction of low speed / high lift performance early in the design process will, therefore, be a key asset in the development of competitive future aircraft. Consequently the Airbus approach for future design verification concentrates on earlier design verification by the use of high-Re facilities, especially the ETW. The experience gained in the qualification and development phase of low-speed testing in ETW via R&T programmes (as EUROLIFT) will be exploited in future in aircraft development work by early use of high-Re benchmark testing in close combination to the design work with advanced CFD-tools.

Drivers for high-Reynolds-number windtunnel testing for design verification

Within the scope of constant pressure to improve aircraft products concerning costs, performance, reliability and emissions, the development and production of advanced high lift systems for new or modified aircraft will play an important role in the future [1]. Design methods for such high lift systems have to be continuously improved due to these environmental and future transport market demands. A crucial prerequisite for the design of efficient and competitive aircraft is a comprehensive understanding of the flow physics of such systems, as well as the ability to optimise these systems in terms of more efficient, yet simpler designs. This, in combination with a high accuracy of flight performance prediction in an early stage of development will provide a strong contribution to the competitiveness of European aircraft manufacturers. Accurate flight performance prediction is a challenging task, this is due to the fact that most of the high lift testing to date has been done at sub-scale conditions [2]. Field performance and handling qualities for the aircraft are then derived by extrapolation. Many of these

---

[1] Head of high-lift devices group, Daniel.Reckzeh@airbus.com
[2] Senior research engineer, high-lift devices group, Heinz.Hansen@airbus.com

scaling effects strongly depend on the Reynolds number as the characteristic parameter between subscale and flight conditions [2, 3, 4]. These scaling effects can introduce an element of risk to the aircraft programme, particularly for large wings, which are designed for high subsonic Mach numbers. This could possibly require expensive design modifications to be made during flight tests. On the other hand the high lift design procedure includes extensive parametric studies at the project design stage, where time is an important factor. Optimisation and refinement are required and wind tunnel tests at reasonably high Reynolds numbers can result in a drastic increase in development costs, if undertaken solely in the wind tunnel. Therefore the combination of experimental and suitable computational investigations is increasingly necessary. The current situation of the aforementioned areas of interest can be summarised as follows:

- Many of the high lift flow phenomena, especially for 3D configurations and at high Re numbers are not fully understood [2, 8].
- Most of the high lift testing until to date has been done at low or moderate Re numbers (Re<7 Mio.).
- Significant adverse scale effects can sometimes be found in flight-testing, but no detailed measurements are available for the detailed understanding of these effects.
- Only a very limited number of tests are available in Europe at flight Re numbers and free flight conditions.
- Scale effects on stall data depend ultimately on some details of the local 3D geometry and the spanwise stall behaviour can be strongly affected by the local Re number.

The need for advanced complex CFD-methods in combination to high-Re testing

Rapid viscous-inviscid interaction codes are a standard industrial tool in the development of high lift systems. However, the extremely complex high lift flows and the need for improved prediction accuracy have led to increased effort of introducing Reynolds averaged Navier-Stokes (RANS) codes as a tool for high lift application. For complex flows with strong separation - even in 2D- an improvement of accuracy for industrial applications in terms of prediction of maximum lift, stall behaviour and for take-off configurations the accurate drag is needed in addition [5]. Common research projects have provided 2D RANS codes for industrial applications. Verification has been accomplished for 3D pilot applications and is currently extended [6]. The challenge of grid generation could be tackled by using unstructured grid approaches, but reliable prediction of all involved flow phenomena requires considerable future effort. The computational results have to be improved for cases with strong separated flows. Previous research activities [2, 7] have shown the importance of the accurate prediction of transition on the high lift elements. The accuracy of computations will have to be driven forward by improved modelling of the main flow phenomena and therefore basic research experiments with detailed flow measurements -especially in 3D- are needed to provide the code developer with the necessary information.

Contributions of the EUROLIFT R&T programme

The role of EUROLIFT [9] can be understood as a qualification phase for the approach to test high-lift design solutions derived from design assessment with complex CFD methods. The main tasks of EUROLIFT were:

- Preparation of an experimental database with detailed flow field information providing numerical code developers with a comprehensive set of information with step-by-step increase of geometric and flow complexity.
- Assessments of state-of-the-art European high lift codes with a clear path to industrial application by using this experimental database.
- Improvement of numerical tools in fields with clear shortcomings and derivation of common European guidelines for the further improvement and development of high lift tools with respect to time effective pre-optimisation and the high accuracy required for industrial applications.
- Extensive use of the unique capabilities of the European cryogenic Transonic Windtunnel (ETW) for high Reynolds number testing at low speed/high lift to bridge the gap between sub-scale testing and flight conditions.
- Application of the improved knowledge for testing of an advanced high lift system up to flight Reynolds numbers and demonstration of the improvement potential.

As far as possible existing results and experiences of different national and European research activities have been used. The CFD tools developed on a national basis have been brought into the EUROLIFT community. A common set of well-defined experiments has given the unique possibility to compare most of the existing high lift CFD tools within Europe. This will lead to a standardised European guideline for fast grid generation and accurate transition and turbulence modelling tools. The integration and application of these tools in the industrial high lift design process has been significantly accelerated.

With the availability of the ETW test facility it is possible to achieve flight Reynolds numbers. Within EUROLIFT this tunnel has been used for the first time for low speed testing of half span models in high lift configuration up to flight conditions. All partners within EUROLIFT could benefit from the assessment and exploitation of this unique European testing facility. Each partner was involved in this process and gains the common knowledge about the advantages of such a tunnel.

EUROLIFT ETW-High-Re verfication procedure of an advanced high-lift system

Within EUROLIFT the European high Re test facility ETW and its unique possibility to use both - pressurized and cryogenic conditions- in combination with cryogenic half model test technique, it was possible for the first time to perform 3D low speed high lift measurements up to flight Re numbers in a wind tunnel [14].

Therefore the approach taken with the KH3Y validation model was testing of selected configurations and high lift element settings over the complete range of Re numbers starting with low Re numbers typical for smaller wind tunnels up to real flight Re numbers. Before these tests with the validation model could be performed it must be assured that this new low speed high lift testing with a newly designed half model balance could be performed in the ETW with a high quality standard. Therefore a pre-testing for the verification of ETW for high lift, high Re testing with half span models was performed. This was an important milestone for the following tests in ETW. For this verification process an existing cryogenic model of Airbus-D (Fig.1) was used as an example of a modern transport aircraft. This model has been extensively tested in the LSWT in Bremen and in the cryogenic non-pressurised tunnel in Cologne (KKK). Based on this information the results from the ETW could be cross checked in the appropriate Ma/Re limits under cryogenic conditions [13, 15]. By use of global flight test results of this aircraft (max. lift, stall behaviour) an additional comparison at flight Re numbers between the new ETW results and flight was possible. This comparison demonstrated a good agreement of the ETW results with other test facilities and the existing flight test results. A detailed overview about these tests has been given in [15]. After this verification phase corresponding to the low Re number validation test in the LSWT a high Re number ETW test for selected configurations has been performed and could provide a database for the assessment of CFD codes up to flight Re numbers. Several configurations have been tested in the ETW up to flight Re numbers: a clean configuration with and without a modified leading edge (behaviour of leading edge stall with Re), a take-off and landing configuration, and a landing configuration with different flap settings. Due to the cryogenic conditions no detailed flow field measurements could be performed but for selected points the boundary layer transition has been observed by I.R detection technique. For some interesting areas and conditions in addition to this the deformation of the flap and the wing bending as a function of the changeable stagnation pressure was measured by a mini CCD camera system.

In a last step the validation model KH3Y (Fig.2) was used in a more realistic complex aircraft configuration as an example of an typical industrial application. Use was made of a so-called multifunctional flap system (Fig.3). With this system an extended fowler flap will replace the aileron. The ‘full-span’ fowler is equipped with a small splitflap (“Gurney-flap” or “mini-TED”) or a camber tabs over the total fowler flap span. These tab elements can be deflected differentially in span-wise direction and used for increased maximum lift, optimising the lift distribution in take-off, roll- and glide path control, as well as for the improvement of cruise performance. This new high lift system was developed and extensively tested by Airbus-D with a large full span model. Corresponding to the KH3Y model geometry at low Re numbers (Re=3 Mio.) within the national German high-lift technology programme HAK [12] was build by Airbus-D within EUROLIFT. In high lift configuration at high deflection angles of the camber tab the effectiveness of the tab is limited by separation at low Re numbers. Tested in the last phase of EUROLIFT tests in ETW the main objective therefore was the assessment of such a system over the complete range of Re numbers from sub-scale testing up to flight

Re numbers by use of the ETW test facility. These measurements helped to improve the understanding of scaling effects and to evaluate design criteria for such a system as a function of Re number. As an example of the significant impact of the Re number on the obtained performance the development of $C_{L,max}$ vs Re is shown in Fig.4. The development of maximum increment lift over Reynolds-number is given for advanced different configurations (extended flap, splitflap and differentiated tab deflections) relative to the reference single slotted flap & aileron layout. The benefit from the advanced flap systems is significantly higher at high Re conditions and also the relation between the layout changes. This example indicates very clearly that a design selection at low Re conditions could have let to a non-optimised solution at flight conditions.

Experience from the A380 design work

The targeted approach with establishing a high-Re 'benchmark' very early in the design process could not yet be applied for A380 [16]. Nevertheless a large amount of tests at 'medium-to-high-Re conditions' was already conducted during the development work. Windtunnels as DNW-KKK, Onera F1 and Qinetiq Q5m provided already Reynolds-numbers up to Re=12 Mio. These tests revealed significant impact of the Reynolds-number on an optimum solution for various design features as the

- Maximum lift & drag performance of various leading edge layouts (slat, droop-nose device, sealed slat, beret-basque-pylon)
- Spanwise combination of slat angles and maximum slat angles
- Slat-setting (lift optimised for landing, drag optimised for take-off)
- Flap-setting (lift optimised for landing, drag optimised for take-off)
- Extension and profile of the unprotected inner wing leading edge
- Winglet efficiency on take off drag

Nevertheless, due to the qualification work for the ETW application for low-speed testing in EUROLIFT which was conducted in parallel to the A380 development process, an ETW check-out campaign for performance risk-mitigation could be finally included in the A380 development work. In this approach a cryogenic halfmodel with the specific wing twist of the landing configuration was manufactured and tested in ETW. It gave very valuable contributions concerning the maximum lift level and the wing separation behaviour under flight conditions, as well as insight in the impact of aeroelastic distortion effects on the high-lift performance with the help of measurements of the deformed wing shape under different windtunnel pressure conditions.

Requirements to high-Re windtunnel facilities in industrial application

From the extensive experience in windtunnel testing for design verification at Airbus following top-line requirements for high-Re facilities can be formulated.

- Flexibility
    - Fast configuration changes, low testing time per configuration

  - Suited for significantly different configurations (e.g. A380, A400M)
  - Engine simulation (high importance esp. for propeller-configurations)
- Quality of results & experimental techniques
  - Good repeatability (short-term & long term)
  - Reliable online results in standardized format
  - Good flow & separation visualisation (tufts, oil, accenaphtene, PSP)
  - Transition visualisation (Infrared, TSP)
  - Deformation measurement to provide information on tested model wing shapes under load (to rule-out "pseudo Re-effects")
- Compatibility to other windtunnels in the 'verification chain'
  - Identical models to be used
  - Identical data procedures & correction methods
- Reasonable relation between cost & effort relative to amount of results & quality
  - For several tasks many variations at low/med Re may be more reasonable than little at high Re-conditions to provide the required amount of data under given time & budget constraints.

Especially the last point can be considered of key importance as the cost of high-Re testing severely conflicts with the usual budget requirements of industrial development programmes.

Conclusions

It is obvious that for further improvement of high-lift configurations' performance by application of advanced system solutions or also simply by better exploitation of state of the art systems by reduction of performance margins to cover the uncertainties in tunnel to flight scaling the experimental verification at flight conditions has to play a central role. The ETW is the only facility where these conditions can be reached. After recent R&T work -especially in EUROLIFT- the ETW can be considered now as qualified for low speed testing and a first application in an aircraft programme has been taken place with an A380 high-lift campaign. Nevertheless, further work is necessary to establish the ETW as a 'robust & known' part in the Airbus windtunnel verification chain. Recent R&T programmes as IHK/HICON, EUROLIFT 2 and FLIRET shall serve this purpose.

In combination to early high-Re testing to provide benchmarks for a new configurations' performance capabilities advanced CFD-tools are required to conduct the detailed design work. It has to be pronounced that these codes have to become more mature, robust and user-friendly than today's CFD-suites, even if impressive developments have been taken place in the last years which led to the establishment of 2D and 3D RANS-codes as integrated analysis tools in the design process and already first cases where design decisions were solely taken based on CFD-results.

The future approach has to include a proper mix of high-Re benchmark testing, CFD-based design assessment and low/medium-Re testing to capture larger amounts of configuration variations and also testing techniques, which are not (yet) applicable at high-Re in ETW (e.g. power simulation).

References

[1] Flaig, A.; R. Hilbig; High-Lift Design for large civil Aircraft; High-Lift System Aerodynamics, CP-515; AGARD, Sept.1993, pp. 31/1-12.

[2] Haines, A.B; Scale Effects on Aircraft and Weapon Aerodynamic; AGARD AG-323, July 1994

[3] Termes P., *et al.* ; Reynolds and Mach Number effects and the 2D-3D correlation based on measurements and computed results for the Garteur take-off configuration; CEAS Forum on High-Lift and Separation Control, University of Bath (UK), March 1995

[4] Fiddes, S.P., Kirby, D.A., Woodward, D.S., Peckham, D.H.; Investigations into the Effects of Scale and Compressibility on Lift and Drag in the RAE 5m Pressurised Low-Speed Wind Tunnel; Aeronautical Journal, Vol. 89, March 1985

[5] Lindblad, I.A.A., de Cock, K.M.J.; CFD prediction of Maximum Lift of a 2D High Lift Configuration; AIAA-99-3180;

[6] Becker, K., Kroll, N., Rossow, C.C., Thiele, F; MEGAFLOW - A numerical Flow Simulation System; ICAS Paper 98-2.7.4, ICAS Congress 1998; Melbourne, September 1998

[7] Arnal, D., Casalis, G., Reneaux, J., Cousteix, J. ; Laminar-Turbulent Transition in Subsonic Boundary Layers : Research and Applications in France; AIAA Paper 97-1905

[8] Delery, J.M., Separation in Two and Three-Dimensional Flows: Still a Subject of Mystery ?; VKI, The Jean J. Ginoux Lecture, 1998

[9] Thiede, P.; EUROLIFT – Advanced High Lift Aerodynamics for Transport Aircraft; AIR&SPACE EUROPE, Vol. 3 No 3+4, 2001

[10] Dargel, G., Schnieder, H.; GARTEUR High Lift Action Group (AG 08) Final Report; GARTEUR TP 043, Nov.1989

[11] Thibert, J.J.; The Garteur High Lift Research Programme; High-Lift System Aerodynamics, CP-515, AGARD, Sept.1993

[12] Hansen, H.; Überblick über das Technologieprogramm Hochauftriebskonzepte (HAK); DGLR Jahrestagung, Bremen 1998

[13] Quest, J.; ETW – High Quality Test Performance in cryogenic environment; AIAA-00-2206

[14] Wright, M.C.N.; Development of the Half Model Testing; Capability at ETW; ICAS 2000

[15] Quest, J.; Hansen, H.; Mesuro, G.G.; First Measurements on an Airbus High Lift; Configuration at ETW up to Flight Reynolds Number; AIAA-2002-0423

[16] Reckzeh, D; Aerodynamc Design of the High-Lift Wing for a Megaliner Aircraft; Aerospace Science & Technology; Oct 2002

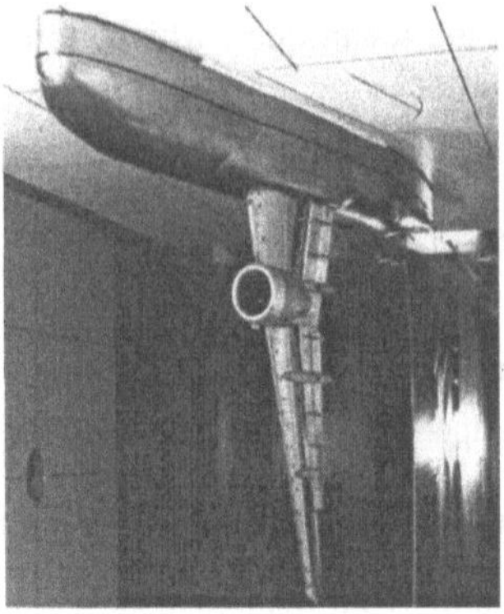

Figure 1: K3DY Airbus model

Figure 2: KH3Y Eurolift model

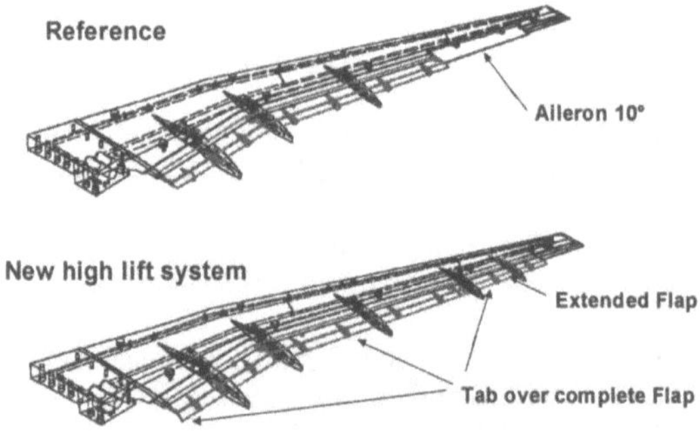

Figure 3: New Flap/Tab system on model KH3Y for test in ETW
Reference: Single Slotted Flaps and drooped aileron
New high-lift system: Extended Single Slotted Flaps with Tabs

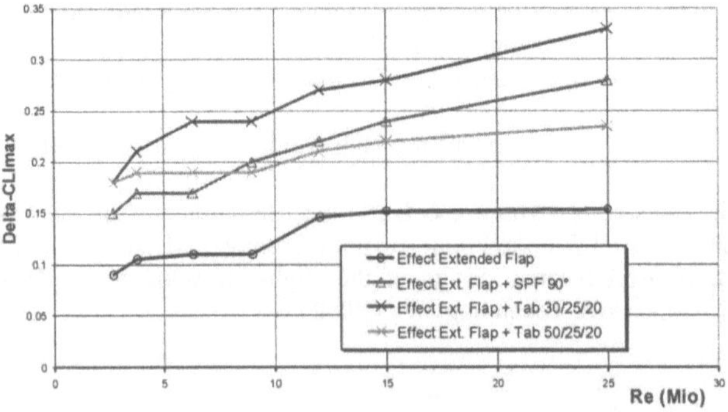

Figure 4: ETW-Results for maximum lift dependency vs. Reynolds-number of different flap system layouts relative to reference single slotted flap (SPF=splitflap, Tab=spanwise differentiated tabs on single slotted flap) (Reference for Reynolds-number: mean wing chord)

Aerodynamic Analysis of Flows with Low Mach- and Reynolds-Number under Consideration and Forecast of Transition on the Example of a Glider

S. Melber-Wilkending[1], G. Schrauf[2], and M. Rakowitz[1]

[1] German Aerospace Center (DLR), Institute of Aerodynamics and Flow Technology
Lilienthalplatz 7, D-38108 Braunschweig, Germany
[2] AIRBUS Deutschland GmbH, D-28183 Bremen, Germany
EMail: Stefan.Melber@DLR.de, Geza.Schrauf@AIRBUS.com, Mark Rakowitz@DLR.de

Summary

For the simulation of low Reynolds- and Mach number flows, a boundary layer code in combination with a database method is coupled with an unstructured RANS code. The technique is validated in 2D using a glider airfoil and in 3D for a glider, showing good agreement with experiments and a well-validated 2D design code.

1 Introduction

The aim of the study at hand is to extend the range of applicability of the DLR TAU-Code [1] for flows at low Reynolds- and Mach numbers. This flow regime is characterised by significant portions of laminar and transitional flow. Therefore, the numerical simulation of this flow regime must include the modelling of transition. The modified TAU-Code can then be used for a wide range of simulations in which laminar flow and transition are of importance, e.g. general aviation, wind turbines and the simulation of models in the wind tunnel. Afterwards TAU will also be used for design tasks like the numerical optimisation of a glider winglet and the wing-root fairing optimisation of the new SB15 glider. This paper describes the development of a process-chain for this purpose, using the boundary layer code COCO [3] and a data base method [4] in combination with TAU. Validation results are presented for a 2D glider airfoil case and in 3D for the simulation of complete speed polars of the glider "Std. Cirrus".

2 Geometry / Grids

For validation in 2D the HQ17 [5] glider airfoil by Horstmann and Quast is used. Four grids are generated with the hybrid unstructured grid generator Centaur [6] with a thickness of the boundary layer (35 layers) which is adapted to four Reynolds-numbers ($Re = 0.7, 1.0, 1.5, 2.0 \cdot 10^6$). The hybrid 2D grid for $Re = 1.5 \cdot 10^6$ is depicted in fig. 1 (left). A 15m "Std. Cirrus" glider by Schempp-Hirth is used for

3D validation. The geometry used here is equivalent to the first version of this glider (1969) with a wing twist of $-0.75°$. The computational geometry is build using CATIA V.5 based on original drawings, the airfoil coordinates and measurements of an original "Std. Cirrus" glider. The geometry includes the wing with wing tip, the fuselage and the wing-root fairing. The empennage is not included. As only symmetrical flow cases are investigated, the half-geometry is meshed using Centaur. The hybrid mesh contains $5.5 \cdot 10^6$ points overall with 20 prism layers for the resolution of the boundary layer. The volume between the outer prism surface and the farfield is filled with tetrahedra. The geometry and grid in the wing-root area are shown in fig. 1 (right).

3 Flow Solution Method

3.1 DLR TAU-Code

The solution of the Reynolds-averaged Navier-Stokes equations (RANS) is carried out using the hybrid unstructured DLR TAU-code [1]. For the closure of the Reynolds-averaged equations the k-ω-SST turbulence model of Menter [7] is used, which combines robustness with the applicability for partly detached flows. Due to the low Mach-numbers and the resulting stiffness of the RANS equations, low Mach-number preconditioning is used [8]. Finally, the central JST-scheme [9] in combination with 80% matrix dissipation [10] assures numerical flow solutions with low numerical dissipation.

3.2 Transition Prediction Method

The concept of the presented transition prediction method is to couple a RANS flow solver, a boundary layer code and a database method for Tollmien-Schlichting waves to predict transition. The main advantage of this method is the usage of "industrial grid densities" for the RANS flow solver, which means a typical grid density (e.g. $\approx 6 \cdot 10^6$ points for a glider) is adequate to get the parameters of the laminar boundary layer from the boundary layer code. To calculate these parameters directly in the RANS flow solver an extremely fine grid (e.g. $\approx 30 \cdot 10^6$ points for a glider) would be needed to achive sufficient resolution of the boundary layer. For complex 3D-configurations this demand would lead to an extensive number of grid points. For this reason the shown coupling of a RANS flow solver and a boundary layer code allows the usage of transition prediction for complex configurations with only small additional costs. The transition prediction method (boundary layer code and data base) is used as an external program. This allows a flexible coupling with any structured or unstructured RANS flow solver (e.g. FLOWer, TAU at DLR) and independent development of the transition prediction method.

In this paragraph the details of the complete process-chain are described: First a RANS flow solver is used to compute a surface pressure distribution for a given configuration based on a fully turbulent flow. Cutting this pressure distribution in planes with a tool of Wilhelm [2], a 2D-flow can be extracted. This method supposes a plane aligned with the streamlines at the upper end of the boundary layer. This situation is found on typical stretched wings whereas e.g. on the wing/fuselage junction this assumption is invalid and the shown method cannot be used there. Based on the pressure distribution in cuts through the flow field, a 2D coordinate system is derived, the pressure distribution is transformed in this coordinate system and splitted at the stagnation point. Two filters for the reduction of local noise and local point clustering provoked by the cutting of unstructured grids are applied on the pressure distribution. Then a boundary layer code - described below - simulates the laminar boundary layer starting from the stagnation points in these cuts.

Two kinds of transition types must be treated in such a 2D-cut: First, at a point of the pressure distribution where laminar boundary layer is inhibited, e.g. on laminar separation bubbles. At such a point the boundary layer code cannot converge to a solution and transition must occur. The second kind of transition is driven by Tollmien-Schlichting waves. This type of transition is detected with a database method (see below) which uses the boundary layer parameters from the boundary layer code and gives an N_{TS}-factor. If $N_{TS} > N_{crit}$, transition will occur. The point of transition is the point which is closer to the stagnation point in a 2D-cut. Afterwards, the transition point is transformed back into the coordinate system of the flow solver. After some iterations the transition points are converged to the final location.

COCO Boundary Layer Code The boundary layer code COCO [3] computes the velocity and temperature profiles of a compressible laminar boundary layer along a swept and conical wing. It uses a new transformation that maps a thickening boundary layer onto a rectangular region, which facilitates the computation of streamwise derivatives needed for non-local stability calculations. The equations are expressed in terms of four coefficients: the acceleration of the inviscid flow driving the boundary layer, the conicity of the wing, the viscosity, and the boundary layer suction.

The equations are discretized with a fourth-order compact scheme ("Mehrstellenverfahren") and the resulting system is solved by a Newton method. One of the features of COCO is its consistent computation of all streamwise derivatives. The user can choose between several possibilities and the chosen streamwise discretisation is then used to solve the differential equation for the velocities at the boundary layer edge, to compute the coefficient of acceleration, to solve the parabolic boundary layer equations, and to compute the streamwise derivatives needed for non-local stability computations. As with every direct method, the boundary layer equations become singular at separation and the Newton iteration fails to converge. This behavior can be used as an indicator for separation.

Data Base Method for Tollmien-Schlichting Waves For a given station in a two-dimensional, incompressible boundary layer, one can present the local spatial amplification rates with the help of level lines in an (F,Re)-diagram, where F is the (non-dimensional) reduced frequency $F = 2\pi f\ \nu/u_e^2$ and Re the Reynolds number $Re = u_e\,\delta_1/\nu$. If we consider a fixed reduced frequency F, the negative value of the spatial amplification rate σ becomes a function of the Reynolds number Re alone. This function can be approximated by two half-parabolas ($i = 0 : Re_0 \leq Re \leq Re_M; i = 1 : Re_M \leq Re \leq Re_1$) with zeros at Re_0 and Re_1, and the maximum σ_M at Re_M:

$$\frac{\sigma}{\sigma_M} = 1 - \left[\frac{Re - Re_M}{Re_i - Re_M}\right]^2, \; i = 1,2 \tag{1}$$

Arnal [11] showed that the four parameters defining the two half-parabolas can be expressed as functions in terms of the reduced frequency F, the shape parameter H_{12}, and the local Mach number M_e at the boundary layer edge. Thus, having performed a boundary layer calculation, we can, for a given frequency, compute the approximate local amplification rates for a Tollmien-Schlichting wave with this frequency and integrate these rates to obtain the N_{TS}-factor. A version has been implemented into the boundary layer code COCO [4].

4 Results

4.1 Airfoil HQ17

In this section the validation of the presented transition prediction method based on the airfoil HQ17 is presented. For comparison, measurements in two different low-speed wind tunnels [5] and a simulation with XFoil [12], a well validated analysis and design program for low-Re number flows, are used. Polars are simulated for four Reynolds numbers: $Re = 0.7, 1.0, 1.5, 2.0 \cdot 10^6$.

As an example in figure 2 for $Re = 1.5 \cdot 10^6$ a comparison is shown. On the right side the transition positions on the airfoil with XFoil and COCO are compared. As XFoil predicts a transition position on the end of the laminar separation bubble, whereas COCO predicts the begin of the bubble, the results of XFoil are modified with the bubble length. On the bottom side XFoil gives a transition position of about 5% behind that of COCO, which leads to a decreased drag, especially at higher angles of attack. The measurements and the TAU-results show a good agreement, whereas the XFoil results at higher angles of attack have an increased lift.

4.2 Complete Glider Configuration

As an example of the numerical simulation in combination with transition prediction of a flow with low Reynolds- and Mach-number, a glider configuration is considered. The aim is to calculate the speed-polar of a "Std. Cirrus" for the following free stream conditions: $H_{flight} = 1000\,m, T_\infty = 281.65\,K, p_\infty = 89875\,Pa$ and $\rho_\infty = 1.112\,kg/m^3$. To get a speed-polar, for every given speed of the flight envelope a corresponding lift coefficient is needed, which can be calculated from the assumption of aerodynamic lift equals weight of the aircraft:

$$c_A = \frac{2\,g\,m_{Glider}}{\rho\,V_\infty^2\,A_{Glider}} \tag{2}$$

and with the parameters of the glider $m_{Glider} = 316.7\ kg, A_{Glider} = 10.04\ m^2$ and $l_\mu = 0.67\ m$ the lift coefficients can be calculated. The angle of attack is calculated iteratively by the flow solver by matching the simulated lift coefficient with the given one. These parameters and the Reynolds-number can be found in table 1.

In figure 3 the streamlines and the simulated transition lines on the upper- and lower surface of the wing for $V_\infty = 100\ km/h$ and $V_\infty = 220\ km/h$ are depicted. At $V_\infty = 100\ km/h$ on the upper side a laminar separation bubble can be found on most of the wing. The transition is located nearly on the thickest point of the wing. At $V_\infty = 220\ km/h$ the transition line on the upper side has moved downstream because of the reduced angle of attack ($\alpha = -5.06^o$ instead of $\alpha = 0.288^o$, compare table 1) and leads to a reduced boundary layer load. The laminar separation bubble has disappeared at this speed. On the lower side the transition line has moved upstream compared to $V_\infty = 100\ km/h$. Especially on the outer wing this leads to a transition location near the leading edge of the wing.

In figure 4 (left) the flow field in the area of the wing/fuselage junction at a speed of $V_\infty = 90\ km/h$ is depicted. On the surface, streamlines are shown which demonstrate the laminar separation bubble on the upper wing surface. The transition occurs above this laminar separation bubble. Furthermore, on the trailing edge of the wing close to the fuselage a flow separation can be found which is noticeable in flight on a real "Std. Cirrus" as a rumbling noise at lower flow speeds. The horseshoe vortex around the wing on the fuselage can also be found in the figure. In the vorticity cuts this vortex is shown on the upper andon the lower side of the wing. Finally, a fuselage-vortex on the upper side of the fuselage (red) due to the stagnation line there is depicted.

The speed-polar of the Std. Cirrus is depicted in figure 4 (right). Because in the simulation the horizontal and vertical tail is not present, the viscous-, profile- and induced drag of the tail is estimated and added to the simulated speed polars. In comparison with flight measurements from the Idaflieg [13] the fully turbulent simulation has an increased drag, and sink-speed respectively. At $V \leq 90\ km/h$ in the simulation a flow separation on the upper wing takes place which is responsible for the drop-down of the speed polar. The main reason for this effect seems to be an over-prediction of flow separation due to the turbulence model. The same effect can be found for the simulation with transition. At $V > 90\ km/h$ the simulated speed polar has a reduced sink velocity compared to the measurement. Possible reasons can be a too small separation bubble drag because of a too coarse grid in this area and the missing induction of the horizontal tail on the wing, which would lead to an increased induced drag of the wing.

5 Conclusion / Further Work

In the presented paper a combination of a boundary layer code, a data base method and a RANS flow-solver is shown to be appropriate for the simulation of flows at low Mach- and Reynolds number under consideration of the laminar/turbulent transition. For the validation of this process-chain the airfoil HQ17 is used, which gives results in agreement with measurements in low Reynolds-number wind-tunnels and an analysis and design system for low Reynolds-number airfoils. The method is also applied for a glider configuration which shows good agreement with measurements.

The presented process-chain will be used for the optimisation of the wing / fuselage junction of a new glider (Akaflieg Braunschweig, SB15) and as a basis for the numerical optimisation of winglets for gliders. Furthermore, trimmed polars with complete gliders can be calculated (including the tail), the results of wind-tunnel measurements can be improved using transition in corresponding numerical simulations. Also, e.g. the efficiency of wind turbines can be improved using the laminar/turbulent transition for design.

Acknowledgments: The authors would like to thank R. Wilhelm (DLR) for the pressure distribution extraction tool and J. Himisch (Technical University of Braunschweig) for the assistance in the post-processing of the numerical simulations.

References

[1] Kroll, N.; Rossow, C.-C.; Becker, K.; Thiele, F.: *MEGAFLOW - A Numerical Flow Simulation System.* 21st ICAS congress, 1998, Melbourne, ICAS-98-2.7.4, 1998.

[2] Wilhelm, R.: *Ein inverses Verfahren zum aerodynamischen Entwurf von Triebwerksgondeln.* Forschungbericht 2004-30, DLR Braunschweig, 2004.

[3] Schrauf, G.: *COCO – A Program to Compute Velocity and Temperature Profiles for local and Nonlocal Stability Analysis of Compressible, Conical Boundary Layers with Suction.* ZARM Technik Report, November 1998.

[4] Schrauf, G.: *Implementation of Data Base Methods into a Boundary Layer Code.* EUROLIFT Technical Report TR 2.6-1, January 2003.

[5] Althaus, D.: *Windkanalmessungen am Profil HQ17/14.38 und Vergleich mit Messungen der University of Technology Delft sowie DFVLR.* Institut für Aerodynamik und Gasdynamik, Universität Stuttgart, Juli 1984.

[6] Kallinderis, Y.: *Hybrid Grids and Their Applications.* Handbook of Grid Generation, CRC Press, Boca Raton / London / New York / Washington, D.C., 1999.

[7] Menter, F.R.: *Zonal Two Equation k-ω Turbulence Models for Aerodynamic Flows.* AIAA-Paper 93-2906, 1993.

[8] Melber, S.; Heinrich, R.: *Implementierung einer Praekonditionierungstechnik für Strömungen kleiner Mach-Zahlen im DLR TAU-Code und Anwendung im Hochauftrieb.* 13th AG STAB/DGLR Symposium München, 12.-14. November 2002.

[9] Jameson,A. ; Schmidt, W.; Turkel, E.: *Numerical Solutions of the Euler Equations by Finite-Volume Methods using Runge-Kutta Time-Stepping Schemes.* AIAA Paper 81-1259, 1981.

[10] Turkel, E.: *Improving the Accuracy of Central Difference Schemes.* 11th International Conference on Numerical Methods in Fluid Dynamics, Lecture Notes in Physics, Vol. 323, pp. 586-591, 1988.

[11] Arnal, D.: *Transition Prediction in Transonic Flow.* IUTAM Symposium Transonicum III, DFVLR-AVA Göttingen, May 1988.

[12] Drela, M.: *XFOIL: An Analysis and Design System for Low Reynolds Number Airfoils.* Conference on Low Reynolds Number Airfoil Aerodynamics, University of Notre Dame, June 1989.

[13] *Vorläufige Ergebnisse der Flugleistungs- und Flugeigenschaftsvermessungen 1974 in Aalen-Elchingen.* DFVLR / Idaflieg, 1974.

Table 1 Input for speed-polar, Std. Cirrus.

$V_\infty[km/h]$	70	75	80	90	100	120	150	220
$V_\infty[m/s]$	19.4	20.8	22.2	25.0	27.8	33.3	41.7	61.1
$c_A[-]$	1.472	1.282	1.127	0.890	0.721	0.501	0.321	0.149
$Re[10^6]$	0.824	0.883	0.941	1.06	1.18	1.41	1.77	2.59
$\alpha[^o]$	14.5	8.68	7.58	2.16	0.288	-1.80	-3.54	-5.06

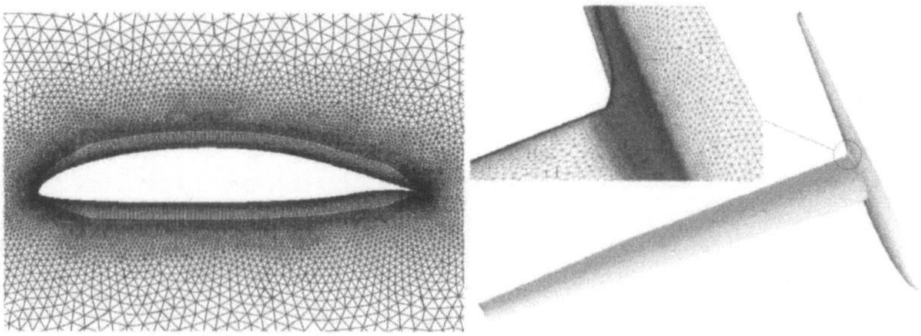

Figure 1 left: Airfoil HQ17, $Re = 1.5 \cdot 10^6$, right: Std. Cirrus and surface grid at wing/fuselage junction.

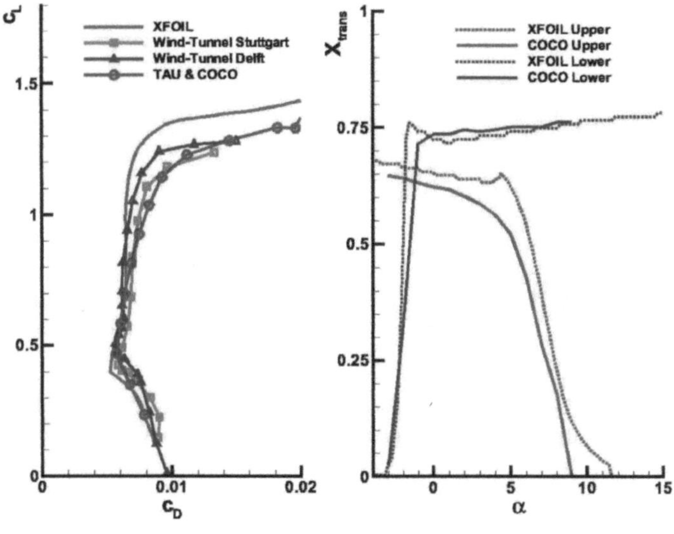

Figure 2 Airfoil HQ17, $Re = 1.5 \cdot 10^6$, comparison of measurement and simulation, polar and transition position.

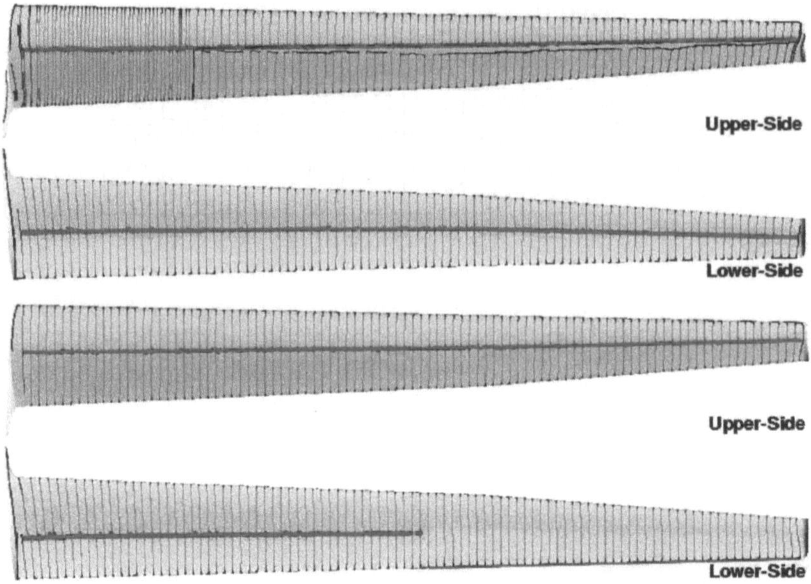

Figure 3 Streamlines on the wing surface, transition line (red) on upper and lower side, velocity: $V = 100\ km/h$ (both top), $V = 220\ km/h$ (both bottom), Std. Cirrus.

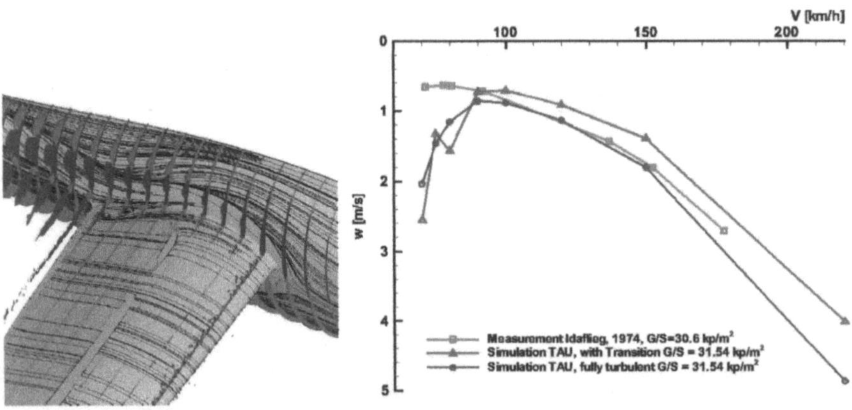

Figure 4 right: Flow fi eld of Std. Cirrus in the area of wing/fuselage junction, $V_\infty = 90\ km/h$, surface stream-lines, vorticity in cuts, left:Speed-Polar of Std. Cirrus.

Analysis of PIV Flow Field Measurements behind the ALVAST-Model in High-Lift Configuration

H. Vollmers, W. Puffert-Meissner, and A. Schröder

DLR Institute of Aerodynamics and Flow Technology
Bunsenstr.10, 37079 Göttingen, Germany
Andreas.Schroeder@dlr.de

Summary

Flow field measurements with a stereoscopic PIV-system were performed behind the DLR-ALVAST half-model in a high lift configuration with and without an ultra-high bypass ratio engine simulator in the DNW-NWB Low-Speed Wind Tunnel Braunschweig at Mach numbers of 0.18 and 0.22 ($\approx$ 62 m/s and $\approx$ 76 m/s). In the plane behind the engine, the vortex position and strength strongly depended on thrust and angle of attack.

1 Introduction

An improved understanding of the viscous and induced drag associated with engine installation was the main objective of the DLR-NLR pre-competitive research co-operation project named Low Speed Propulsion Airframe Integration. One prior test programme with the turbine powered CRUF simulator was carried out in the DNW-LST. As important supplement to these tests, flow field measurements using Particle Image Velocimetry (PIV) were carried out with emphasis on the spatial development of the fan jet flow [2].

During these wind tunnel investigations, flow field measurements in the wake of the ALVAST [3] half model equipped with the high lift wing were carried out with and without CRUF (Counter Rotating Ultra-high bypass Fan) simulator in the DNW-NWB wind tunnel. The flow field measurements behind the wing and the simulator were made using a five-hole rake and a stereoscopic PIV-system. Balance measurements and surface pressure measurements on the wing were performed as well [1]. Results from the PIV measurements will be reported and discussed with respect to variations of the velocity fields and their topology.

2 Test Setup

For the present tests the closed test section running at atmospheric pressure was used with the under floor half-model balance. The longitudinal slots of the test section were kept closed except for small openings for introducing the laser light to illuminate a light sheet for PIV.

The DLR-ALVAST model is a generic wind tunnel model with a span of 3.428 meters, representing a modern transport aircraft similar to an Airbus A320 by scale 1:10. The modular concept of the model allows its individual assembly for different wind tunnel tests as full-span or half-span model either with cruise wing(s) or high lift wing(s). The high lift wings are equipped with inboard/outboard ailerons and with slats and flaps adjustable in a take-off and a landing position. Engine simulators can be installed on both full and half model configurations.

For the tests in DNW-NWB a half-model configuration was assembled with the high lift wing with slats and flaps adjusted in take-off position. The wing is equipped with pressure measurement instrumentation distributed along nine spanwise sections with a total of 617 pressure tappings.

The CRUF simulator unit has a two-stage counter rotating fan of 10" (254 mm) diameter with 8 blades per stage. This fan is driven by a four stage air turbine powered by pressurized air. Drive air was supplied through the balance by a pipe system equipped with three flexible air bridges, so that balance forces due to throughflowing high pressure air was minimized. A gear-box distributes the turbine power on the two counter rotating fan shafts without any RPM reduction. In the fan duct static pressure orifices and rakes with total pressure and temperature probes are installed.

2.1 PIV Setup and Data Processing

For seeding Di-2-Ethylhexyl-Sebacat (DEHS) was used as liquid. The generated particles are small droplets with a mean size of about 1 micro-meter. The seeding was introduced in the settling chamber. One particle generator was sufficient for seeding directly through its hose without a rake. The particles were illuminated by two Nd:YAG lasers of type Quantel Brilliant B. These light sources were pulsed with two independent oscillators with a repetition rate of 10 Hz. The pulse energy at $\lambda = 532$ nm was 2 x 320 mJ. As the PCO cameras utilised for thist test are limited to about 3 Hz, double images have been recorded each 0.3 sec.

To obtain three components of the velocity in the light sheet a stereo set-up was used. Two cameras with 1024 x 1280 pixels and a dynamic range of 12 bits were installed outside the tunnel and thus outside of the tunnel flow viewing through the outlets of the removed breezer flaps into the tunnel.

The measurement areas had dimension of about 20 cm x 20 cm, which results into $\approx$ 50 pixel per centimetre. The distance to the cameras was about 3 to 3.3 m. The opening angle between both cameras was $\approx 50°$. Lenses of 135 mm focus length with maximum aperture of 2.8 were used. The time delay was chosen around 10 μsec. Remotely focusing of the lenses was achieved with a focusing system at the Scheimpflug adapters required for recording of stereo images.

In order to relate the position of both images and the coordinate system of the tunnel, calibration images of a planar table in the plane of the light sheet with a regular grid were taken. From these images the coefficients of the transformation have been evaluated, that maps from pixel values of both cameras into the plane of the light sheet and the pixel displacements into velocity components. This transformation is a general linear projective transformation, which may be inverted using matrix inversion or solving that linear equation. The coefficients of the mapping are determined by an optimization of minimal deviation from given pixel positions using point-wise relations obtained from the calibration images. An example of overlay of a dewarped pair of calibration images from both stereo cameras is given in Fig. 1. The calibration had to be repeated for each camera position (c.f. Fig. 2).

The displacement of particles was evaluated separately for each camera using Gaussian fit for correlation peaks [4]. The size of the interrogation windows were 32 pixels by 32 pixels which corresponds to about 6 mm length of side. This is the area of the flow field which is averaged during evaluation. The images were scanned with step size of 16 pixels in both directions. The three velocity components were evaluated at given points in space (c.f. Fig. 2) using the mapping of the image of each camera into space, the inverted mapping, and the spatial position of the sensor. The local displacement on the transformed points on each camera was obtained by linear interpolation on a triangular grid for the displacement fields of each camera.

The reflections of the laser light sheet from the walls of the tunnel produced reflections on the model which were seen by the cameras. This was unavoidable because spraying sensitive parts like pressure hole and moving parts with black paint was not permitted. Due to this bright background no particles could be distinguished by the cameras in certain areas. As no displacements could be evaluated, these regions were masked and excluded from further analysis.The excluded areas reduced information considerably. As both cameras suffer from the reflections, the omitted areas in the combined stereo results are the union of both regions. The thrust with N_{isa} = 11 050/min was found as a limit for application of PIV because the jet produced condensation in the tunnel for higher thrust.

As the light sheet was fixed to the tunnel also the reference system is fixed to the tunnel. The x coordinate is measured downstream from the centre of the turn table to the light sheet. The y coordinate measures the distance from the centre line of the tunnel and z the normal to the bottom including the peniche, i.e. the span is measured by z.

3 Results

As depicted in Fig. 2, several areas in two sections of the flow field behind the ALVAST-model have been investigated with and without the engine simulator at several conditions, i.e. variations of Mach Number (Ma = 0.18, 0.22, i.e. V $\approx$ 62 m/s, $\approx$ 76 m/s), angle of attack (α = 4°, 12°, 16°), and thrust (N_{isa} = 3 700/min, 10 500/min, 11 050/min). About 100 instantaneous fields were measured for each variation of these parameters. They reveal considerable variations in the velocity fields. Results from the averaged velocity fields are shown in Figs 4 to 8. They include views from back and from top to indicate the geometrical relations of model and regions of measurements. From direct comparisons of the vector fields in these figures the dependence of the structures within the velocity field on thrust and angle of attack can be derived. Especially the changes of the position of the vortices and the topology of the flow field are revealed. In Figs 4 to 6 the position of the vortex moves to the engine with growing thrust. In Figs 7 the same is observed for increasing angle of attack. In contrast, the lateral position of the vortex originating from the outer edge of the outer flap is almost fixed, Fig. 8. Further analysis was applied to investigate the location and strength of vortices by methods described in [5]. Examples are given in Figs 3 and 9. The vortex roll-up is continuing and not only one vortex but also strong vorticity layers are generated by the wing and are observed in the velocity fields. The discrepancies between both applied methods for evaluation of vortex parameters become apparent. While for the method MOV the distribution of vorticity within the observation area is used, the approximation FAM represents the best fit for the velocity field. In spite of the complexity of the flow fields, the variation of vortex position and strength is obvious even for fixed configurations and parameters. Fig. 3 evaluated from fields measured behind the engine indicates much stronger fluctuations in position and strength than the case presented in Fig. 9, which was measured behind the outboard edge of the flap.

4 Conclusions

The forces and the flow fields behind the ALVAST model of DLR have been investigated with emphasis on the effects of thrust of the engine simulator CRUF. Three-component velocity fields were obtained by measurements with PIV in areas of planes behind the wing. The results reveal changes of vortex strength and position for different configurations of thrust and angle of attack as well as for each particular configuration. These variations should be considered as variations of initial conditions for the vortices in the wake behind an aeroplane. Though the change of vortex positions is only within 5% of span with an overlay of $\approx 0.5\%$ of fluctuations they may induce changes in vortex interaction and onset of instabilities.

References

[1] A. Agocs, J. Kompenhans, W. Puffert-Meissner, M. Raffel, H. Richard, A. Schröder, H. Vollmers: "Test Report of Force and Flow Field Measurements on the ALVAST Half-Model in High Lift Configuration in the DNW-NWB Wind Tunnel Phase III", DLR-IB 22415-2002 A16, 2002

[2] G.H. Hegen, W. Puffert-Meissner, L. Dieterle, H. Vollmers: "Flow Field Investigations on a Wing Installed Counter Rotating Ultra-High-Bypass Fan Engine Simulator in the Low Speed Wind Tunnel DNW-LST". 7th European Propulsion Forum on Aspects of Engine/Airframe Integration, Pau, March 1999

[3] Kiock, R.: "The ALVAST Model of DLR". DLR-IB 129-96/22, 1996

[4] Raffel, M., Willert, C. E. & Kompenhans, J.: "Particle Image Velocimetry - A Practical Guide". Springer-Verlag, Berlin, 1998.

[5] H. Vollmers: "Detection of vortices and quantitative evaluation of their main parameters from experimental velocity data". Meas. Sci. Technol. **12**, 2001, pp. 1-9.

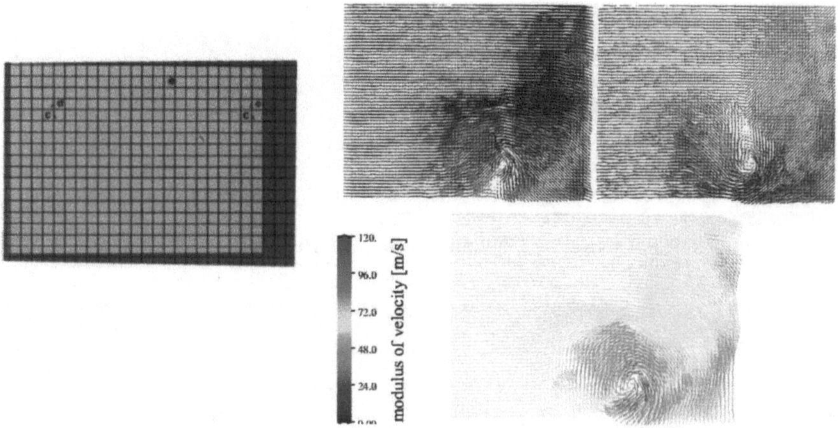

Figure 1 The mapping of a pair of dewarped calibration images from both stereo cameras into one image (left side) demonstrates perfect agreement. The upper vector fields (right side) show the view of both cameras. In the lower part the fields are combined and the vectors are within the plane of the light sheet while the gray values indicate the component normal to the light sheet.

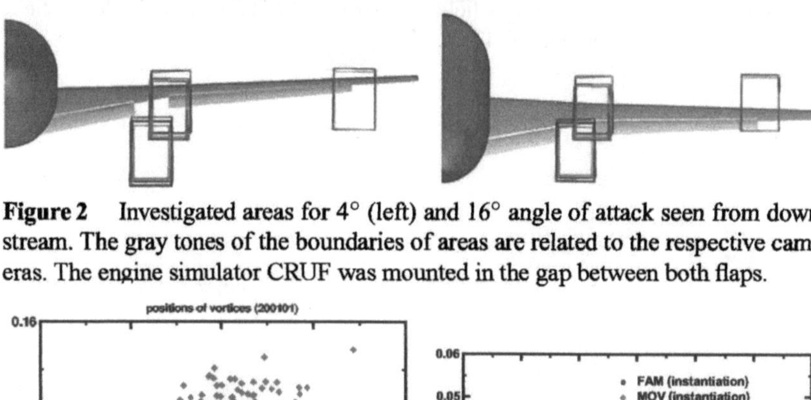

Figure 2 Investigated areas for 4° (left) and 16° angle of attack seen from down stream. The gray tones of the boundaries of areas are related to the respective cameras. The engine simulator CRUF was mounted in the gap between both flaps.

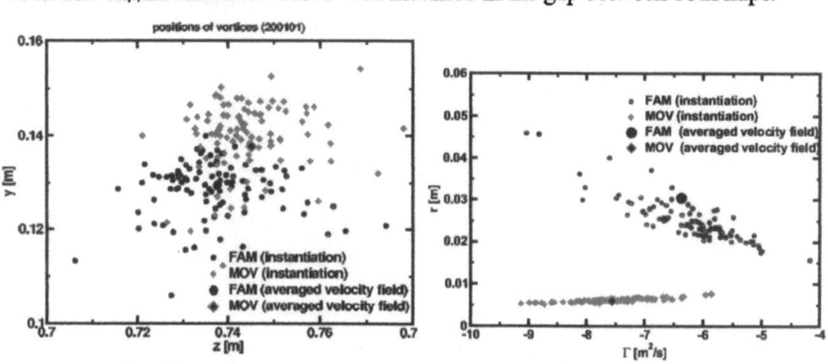

Figure 3 Position of vortices and relation between radius r and circulation Γ obtained via approximation with Oseen Vortex (FAM) and from moments of vorticity (MOV) as described in Ref. [5]. The averaged velocity field measured at position x = 0.696 is depicted in the middle of Fig. 6 (Ma = 0.18, α = 12°, N_{cruf} = 3 700/min).

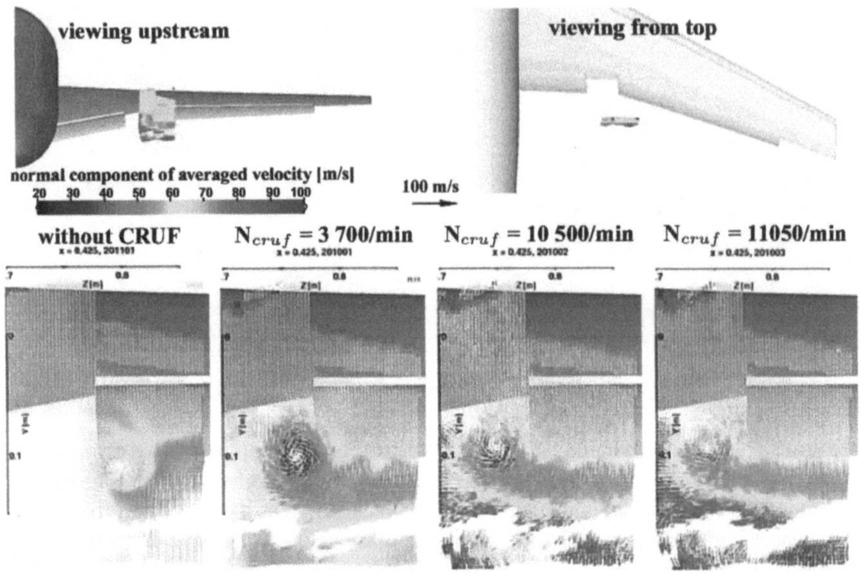

Figure 4 Dependence of flow field on thrust of CRUF. Averaged velocity fields at Ma = 0.18, $\alpha = 12°$, and x = 0.425.

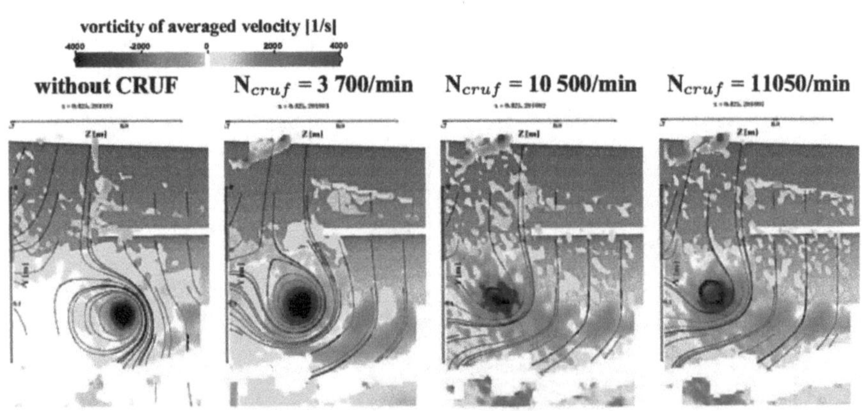

Figure 5 Dependence of vorticity and sectional stream line topology on thrust of CRUF for the averaged velocity fields shown in Fig. 4.

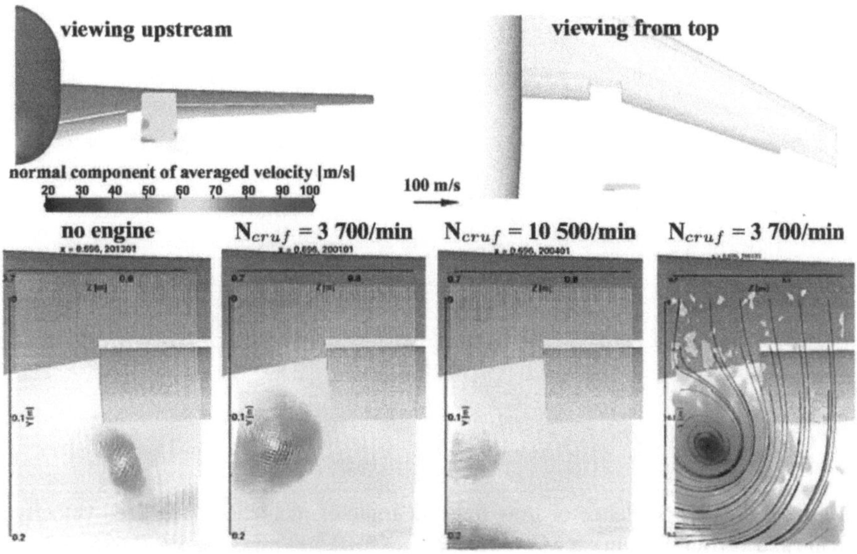

Figure 6 Dependence of flow field on thrust of CRUF. Averaged velocity fields and vorticity (right) at Ma = 0.18, $\alpha = 12°$, and x = 0.696.

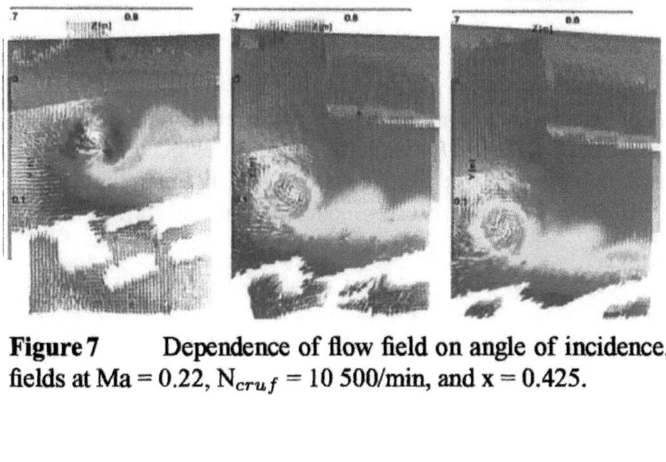

Figure 7 Dependence of flow field on angle of incidence. Averaged velocity fields at Ma = 0.22, N_{cruf} = 10 500/min, and x = 0.425.

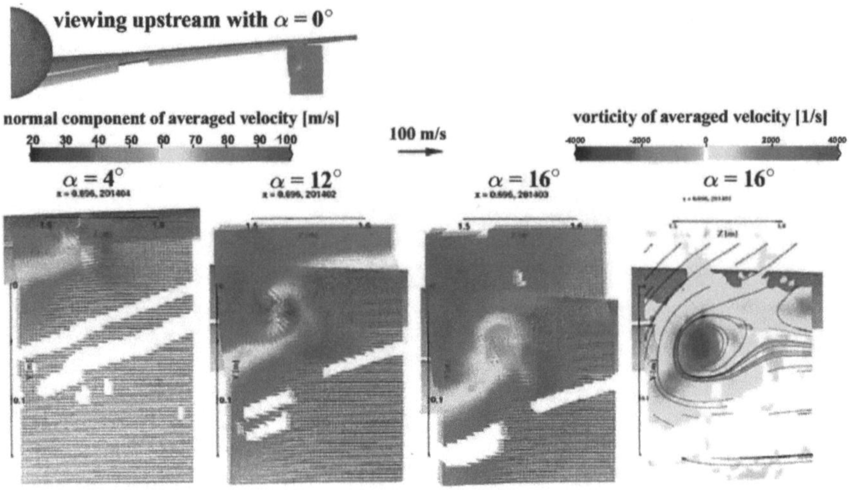

Figure 8 Dependence of flow field on angle of incidence. Averaged velocity fields and vorticity (right) at Ma = 0.22, N_{cruf} = 10 500/min, and x = 0.696.

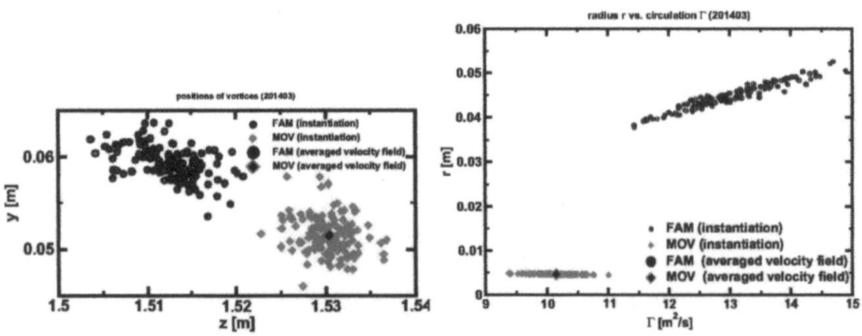

Figure 9 Position of vortices and relation between radius and circulation like in Fig. 3. The averaged velocity field and its vorticity of the vortex generated by the outer edge of the outer flap at position x = 0.696 are depicted on the right of Fig. 8 (Ma = 0.22, N_{cruf} = 10 500/min, $\alpha = 16°$). The fluctuation is much stronger than in Fig. 3

Effect of Differential Flap Settings on the Wake Vortex Evolution of Large Transport Aircraft

C. Bellastrada, C. Breitsamter

Lehrstuhl für Fluidmechanik - Technische Universität München
Boltzmannstrasse 15 - 85748 Garching - Germany
claubell@flm.mw.tum.de

Summary

Wind tunnel investigations have been performed to study the effect of different span loadings, obtained by means of differential flap settings, on the structure of the wake vortex behind a half-model of a four-engine large transport aircraft. Advanced hot-wire anemometry technique has been used to measure the wake vortex velocities at five cross stations up to 4.5 span distances downstream of the model. Results obtained are analysed to identify possible benefits in terms of alleviation of the wake vortex hazard.

1 Introduction

The rolled up trailing vortices shed by landing or starting aircraft pose a serious danger factor for any other aircraft that may encounter them. This implied in the past the introduction of special separation rules to ensure safe operations for flying conditions with respect to Instrument Flight Rules. Nowadays, these separation standards lead to a limitation of the airport efficiency and are generally considered too strict. A reduction of the separation distances, without decreasing the current safety level, would improve the exploitation of the most crowded airports. This is possible only if it is either ensured that the wake vortices have left a safety corridor around the glide path or if their strength is weakened to a tolerable level. Previous studies have shown that by modifying the wingspan loading, the typical rotational velocities within the originated wake can be reduced significantly, [1] and [2]. Changes in the wingspan loading of a flying aircraft can be obtained by means of differential flap setting (DFS), where many flap segments are available. The DFS can also act on the roll up process and on the stability of the wake by producing co–rotating or counter–rotating multiple vortices, thus facilitating the onset of instability mechanisms, [3] and [7].

2 The Experimental Setup

2.1 Half Model and Wind Tunnel

The tests have been performed using a 1:19.25 scale half–model of a typical Large Transport Aircraft, (Fig. 1). The half–span is 1.491 m, the wing mean aerodynamic chord is 0.3569 m. The wing is equipped with sets of fully deployable slats, ailerons and flaps to perform the investigation of approach, landing and taking–off configurations. On the wing two through–flow–nacelles are installed. The model is also provided with a horizontal stabiliser, which can be set on the desired angle with steps of 1 deg. Landing gear, vertical tailplane and blast simulation devices are not available. The measurements were done in the FLM long–range–test section wind tunnel C. The closed test section has a height of 1.8 m, a width of 2.7 m and a length of 21 m, which permits to investigate the wake up to 4.5 spans downstream of the model. The maximum attainable velocity is 30 m/s with a turbulence level less than 0.5 %. The model is positioned on the tunnel floor, slightly downstream of the nozzle exit section, and the wing tip points toward the ceiling of the wind tunnel. The test section is equipped with a three–degree–of–freedom traversing system, which enables to locate a mounted probe in any place of the test section.

2.2 Measurement Technique

Multiple–sensor hot–wire probes have been used to measure the instantaneous flowfield velocities. The sensors consist of three 5–μm–diam. platinum–plated wires having a length of 1.25 mm. The hot–wire probe was operated by a multi–channel constant-temperature anemometer. The output signals of the anemometer bridges have been low–pass filtered at 820 Hz before digitization (16–bit) and amplified for optimal signal level. The sampling rate for each channel was set to 3000 Hz (Nyquist frequency: 1500 Hz) to ensure that all significant flowfield phenomena are detected. The sampling time was 6.4 s resulting in a sample block of 19200 values for each survey point. Anemometer output signals are converted into time–dependent velocity components u, v, and w using a look–up table previously obtained from the velocity and angle dependent calibration of the hot–wire probe.

2.3 Measurement Conditions

The aircraft half–model has been tested in a typical approach configuration (baseline), and in four different DFS configurations. The high lift device settings for the baseline configuration are: (i) inboard slat at 19.6°, (ii) mid-board slat at 23.0°, (iii) outboard slat at 23.0°, (iv) both inboard and outboard flaps at 26°. The horizontal tailplane is fixed at −6.0° and the ailerons are set at 5°. The angle of attack is 7.0° giving a lift coefficient of

1.44 (the target c_L being 1.4). The DFS include two inboard loading 26°/8° and 32°/8°, and two outboard loading 8°/26° and 8°/32° configurations. In this paper only the two DFS cases involving the 32° deflection are considered, the other two configurations are discussed in [6]. The experiments have been done for all configurations at a freestream reference velocity of $U_\infty = 25$ m/s, corresponding to a Reynolds number of $Re = 0.520 \cdot 10^6$ (based on the wing mean aerodynamic chord). Based on the data from previous wind tunnel force measurements, different angles of attack have been set to obtain the same lift coefficient $c_L = 1.44$ for all the DFS configurations. Transition strips are fixed to the wing nose area to achieve turbulent boundary layer avoiding undesirable laminar separation. The surface flow has been monitored by using tufts, which has proven that attached flow is present on wing and control surfaces. The tufts visualization has also revealed local flow separation in the region of the outboard flap for the DFS 8°/32° configuration. The wake vortex has been measured at five cross–flow planes ranging from 0.37 up to 4.5 wingspan distances downstream of the wing reference point, located at the winglet tip trailing edge. The survey points are closely spaced in the wake area with a relative (with respect to the wing span) grid resolution of 0.005 in spanwise (y^*axis) and 0.0034 (z^*axis) in vertical direction based on the wing span. Outside these main areas a relative grid resolution of 0.03 in spanwise and 0.02 in vertical direction was used.

3 Results and Discussion

3.1 The Evolution of the Wake

The normalized axial vorticity $\xi = \overline{\omega}_x \cdot (b/2)/U_\infty$ is used to identify significant vortical structures and to define their location on each cross flow plane. The coordinates y and z are normalized using the wing half span $b/2$. Figure 2 shows the vorticity distribution of the baseline configuration at $x/b = 0.37$. The wake, in its near field, consists basically of six major vortices: the wing tip vortex (WTV), the outboard flap vortex (OFV), the outboard nacelle vortex (ONV), the inboard nacelle vortex (INV), the counter rotating horizontal tail tip vortex (HTV) and the wing fuselage junction vortex (WFV). High vorticity peaks are shown by the WTV, the HTV and by the two similar vortices attributed to the rolled up shear layer of the two engine nacelles, the INV and the ONV. The former has already diffused much of its vorticity at $x/b = 1.0$, while the latter is merging with the OFV, (Fig.3). After the merging process, almost completed at $x/b = 1.0$, the wake rolls up forming, at the most downstream station, a four vortex system: the stronger WTV and OFV–ONV and the weaker INV and HTV, (Fig. 4). The setting of differential flap deflection introduces additional vortices, which are labeled as DFS+ if co–rotating and DFS- if counter–rotating with respect to the WTV and OFV. The DFS+ vortex is clearly detectable in the wake near field for the 32°/8° configuration (Fig. 5), while no well defined regions of concentrated negative vorticity have been identified. The DFS+ vortex is located near the INV, which is displaced upward and slightly inboard with respect to the baseline configuration. Consequently, the position of both the OFV and the ONV is also changed. The OFV has, as expected, a lower peak vorticity because of the minor flap deflection. At one wingspan downstream the development of the wake is characterized by the double merging process between the INV and the DFS+, and between the OFV and the ONV vortex. Since these merging processes have been completed, the wake consists of a progressively rolling up four vortex system. At $x/b = 4.5$ the OFV–ONV has already begun to merge with the dominant WTV, (Fig. 7). Two regions of concentrated vorticity of opposite sign - the DFS+ and the DFS- vortex - are clearly identified in the vorticity plot at $x/b = 0.37$ for the $8°/32°$ configuration, (Fig. 8). They are attributed to the different setting of the inboard and the outboard flap. In this area the INV is not detectable as a well defined vortical structure. Compared with the baseline and the inboard loading configuration the wake of the DFS $8°/32°$ behaves differently in the near field. The DFS+ vortex is orbiting clockwise around the vortex resulting from the merger of the DFS- vortex with the HTV then it dissipates. Conversely, the whole wake rotates counter–clockwise during its roll up process. The ONV and the OFV start immediately to merge together. The latter has a quite low vorticity peak in spite of the high deflection angle of the outboard flap. This is due to the effect of local flow separation around the outer edge of the outboard flap, as confirmed by tufts visualization. The WTV experiences the effects of the modified wingspan loading but does not show appreciable difference, both in terms of vorticity peak and vorticity distribution, between the two DFS configurations. At one wingspan downstream, (Fig. 9), the merging between the OFV and the ONV is not far from being completed, thus generating a single vortex with the same vorticity peak level as the merged WTV. The final outcome at $x/b = 4.5$ is a three vortex wake, where the WTV and the OFV–ONV begin to merge. The peak vorticity of the OFV–ONV system is only slightly changed, while it is sensitively decreased for the WTV, (Fig. 10). Even with the vorticity feeding mechanism from the vortex sheet and the merging with the DFS- the vorticity level of the HTV remains still lower than for the baseline configuration. It appears difficult to build a multiple vortex system with such a vorticity contents that can supply the best suited circulation ratio $\Gamma_1/\Gamma_2 = 0.5$ needed to promote the most effective instability mechanisms observed by [5] and [7]. Taking into account the effects of the DFS on the viscous core radius it is possible to quantify the amount of the diffusion of the wake. According to $\frac{r^*_{c,DFS} - r^*_{c,Baseline}}{r^*_{c,Baseline}}$, the viscous core radius r_c is normalized by using the wing half span $b/2$. Analyzing the $32°/8°$ DFS configuration at $x/b = 4.5$ yields a decrease of the viscous core radius of the WTV and OFV–ONV of 32% and 34%, respectively. The $8°/32°$ configuration still produces a significant reduction of the WTV viscous core radius of about 25% and a minor decrease of 3.5% for the OFV–ONV. A parameter, which permits to judge the benefit brought by the DFS, is the strength of the wake $G^* = \frac{\Gamma}{\Gamma_0^*}$, where Γ_0^* is the initial circulation for an elliptical loaded wing, [4]. Since the wake has not completely rolled up yet, the factor G^* has been evaluated at plane $x/b = 4.5$ (corresponding to a wake non–dimensional lifetime τ of approximately 0.1, [4]) for the WTV and the OFV–ONV only, over circular areas of radius $r/(b/2) = 0.1$ and 0.2, respectively. The 32°/8° configuration shows a strong decrease of about 60% in the OFV–ONV strength $(\frac{G^*_{DFS}-G^*_{Baseline}}{G^*_{Baseline}})$, as expected from the limited flap deflection. The strength of the WTV also decreases by about 14%. Opposite results are found for the 8°/32° configuration: the WTV appears to be slightly stronger (+7%), while the OFV–ONV strenght is reduced by 17%. This behaviour (a stronger vortex was expected) can be explained with the local separation observed in the outer section of the outboard flap. The vortex strength alleviation provided by the DFS is confirmed by the calculation of the induced rolling moment coefficient (see [4]) for a Cessna S-172 following aircraft type with a $C_{L\alpha}$ lift gradient of 2π, a wing span $b_f = 0.2b$, where b is the wingspan of the generating aircraft, and flying at a velocity of 25 m/s. The maximum value of the induced rolling moment coefficient (−0.192) has been found for the baseline configuration in the region close to the OFV–ONV vortex. Both DFS configurations exhibit lower rolling moment coefficient maxima: −0.154 near the WTV for the 32°/8° (here the OFV–ONV is stretched and wrapped around the stronger WTV due to the merging process), and −0.166 and −0.156 for the 8°/32° configuration in the region near the WTV and the OFV–ONV, respectively.

3.2 Conclusions and Outlook

The effects of two DFS configurations on the wake of a large transport aircraft have been experimentally investigated. Deploying the flap segments according to the DFS concept changes the wing loading and introduces additional vortices (DFS+ and DFS-), shed at the inner and/or outer flap side edges. These additional vortices are clearly detectable in the near field wake of the DFS 8°/32° configuration, while only the DFS+ vortex has been identified for DFS 32°/8°. This vortex merges soon with the INV, which has not been observed in the wake of the outboard loading configuration. In the latter case, the DFS+ vortex and the one resulting from the merger of DFS- with the HTV appear orbiting clockwise around each other, while the whole wake is rolling up counter–clockwise. Then the vorticity of DFS+ is reduced within two wingspan distances and the wake consists of three vortices: the WTV, the OFV and the counter–rotating HTV. After the INV-DFS+ and the OFV–ONV merging, the DFS 32°/8° four vortex wake system progresses in its rolling up process, which is characterized at $x/b = 4.5$ by the merging between the OFV–ONV and the dominant WTV. The analysis of the single vortex strength at the most downstream location reveals weaker vortices -

the WTV and OFV–ONV - for the two DFS than for the baseline configuration. These vortices also present a significant decrease in their viscous core in comparison with that observed for the baseline. The calculation of the induced rolling moment on a following aircraft has delivered lower maxima for the two DFS configurations. It must be noted that a wake vortex encounter, if the separation rules are fulfilled, occurs at distances much larger than x/b = 4.5. Nevertheless, the results obtained seem to be promising for having less potentially hazardous wake vortices in the far wake.

References

[1] Ciffone D.L.: Vortex Interactions in Multiple Vortex Wakes behind Aircraft, J. Aircraft, Vol. **14**, No. **5**, 1977, pp. 440–446.
[2] Corsiglia V.R., Rossow V.J., Ciffone D.L.: Experimental Study of the Effect of Span Loading on Aircraft Wakes, J. Aircraft, Vol. **13**, No. **12**, 1976, pp. 968–973.
[3] Coustols E., Stumpf E., Jacquin L., Moens F., Vollmers H., Gerz T.: "Minimised Wake": A Collaborative Research Programme on Aircraft Wake Vortices, AIAA Paper **2003–938** .
[4] de Bruin A.C., Elsenaar A.: A Method to Characterise the Strength and Downstream Decay of Aircraft Wakes. NLR Rept. **NLR–RC–2003**, Jan., 2003.
[5] Fabre D., Jacquin L., Loof A.: Optimal Perturbations in a Four–Vortex Aircraft Wake in Counter–Rotating Configuration, J. Fluid Mech. Vol. **451**, 2002, pp. 319 328.
[6] Rawlinson A.: Transport Aircraft Wake Vortex Modification by Differential Flap Setting, Master Thesis, Rept. **FLM–2004/12**, TU München, May 2004.
[7] Rennich S.C., Lele S.K.: Method for Accelerating the Destruction of Aircraft Vortices, J. Aircraft, Vol. **36**, No. **2**, 1999, pp. 398 404.

Fig. 1: The half-model in the wind tunnel test section.

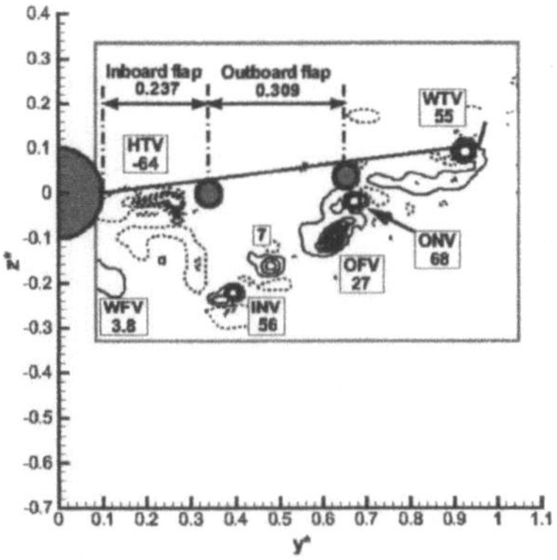

Fig. 2: Baseline. The axial non-dimensional vorticity at $x/b = 0.37$.

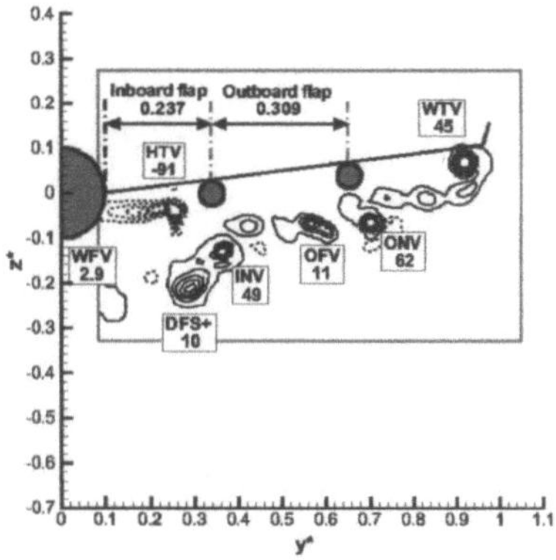

Fig. 5: DFS-32°/8°. The axial non-dimensional vorticity at $x/b = 0.37$.

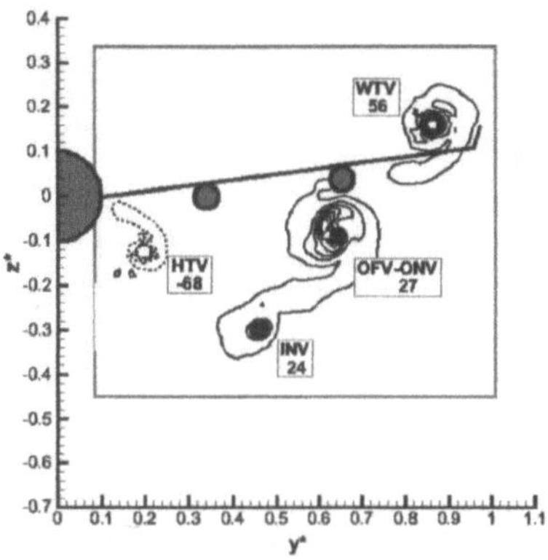

Fig. 3: Baseline. The axial non dimensional vorticity at $x/b = 1.0$.

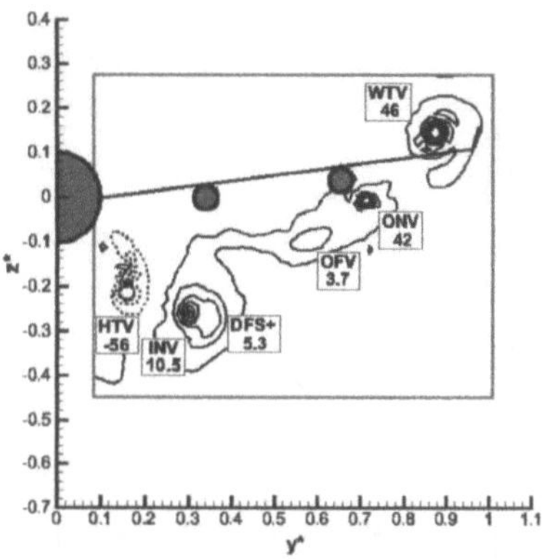

Fig. 6: DFS-32°/8°. The axial non-dimensional vorticity at $x/b = 1.0$.

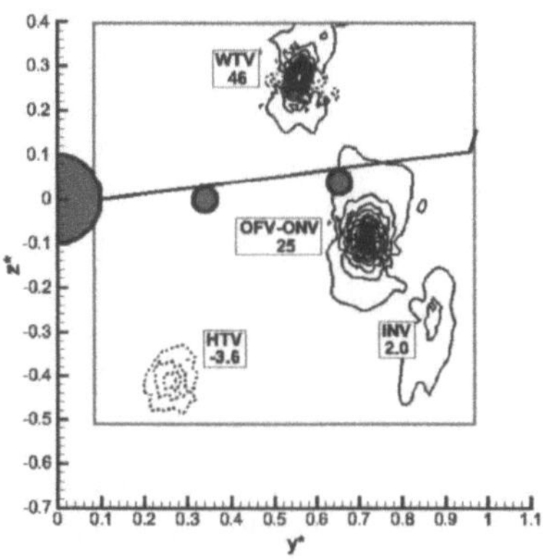

Fig. 4: Baseline. The axial non-dimensional vorticity at $x/b = 4.5$.

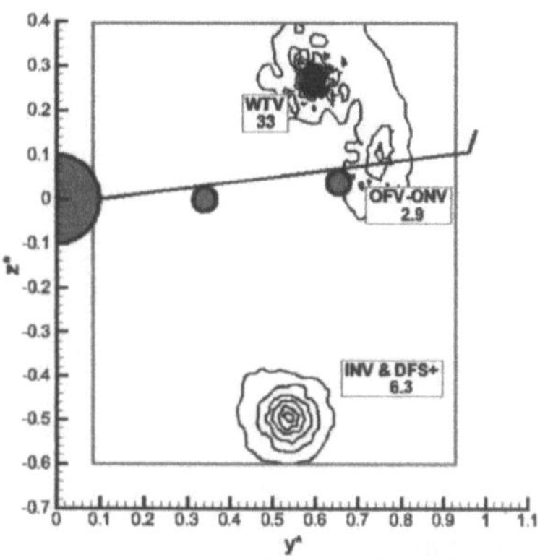

Fig. 7: DFS-32°/8°. The axial non-dimensional vorticity at $x/b = 4.5$.

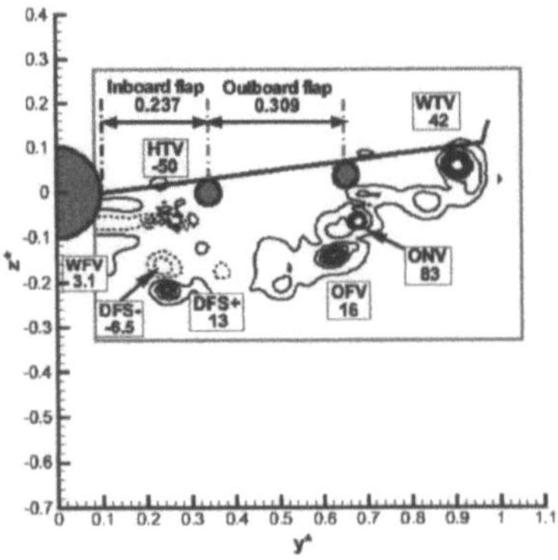

Fig. 8: DFS-8°/32°. The axial non-dimensional vorticity at $x/b = 0.37$.

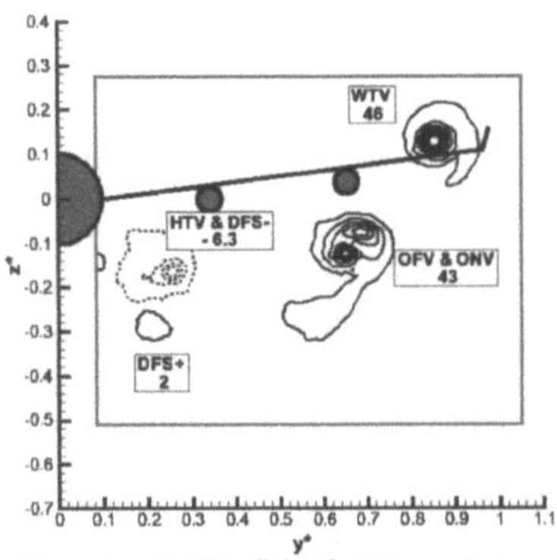

Fig. 9: DFS-8°/32°. The axial non-dimensional vorticity at $x/b = 1.0$.

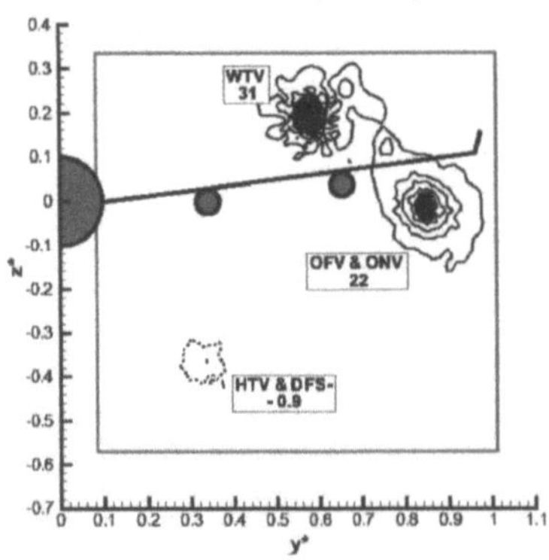

Fig. 10: DFS-8°/32°. The axial non-dimensional vorticity at $x/b = 4.5$.

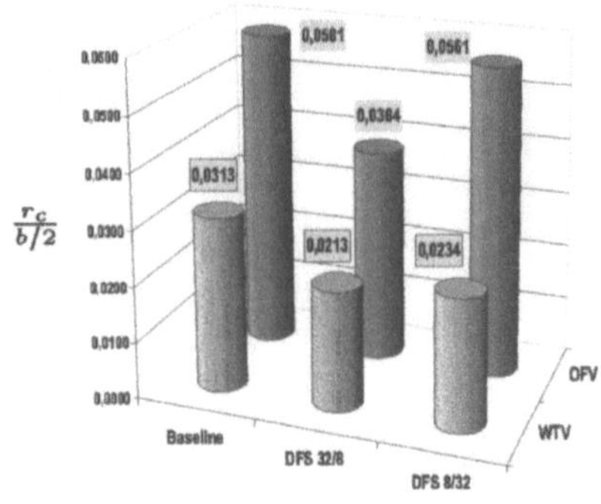

Fig. 11: The normalized viscous core radius of the WTV and OFV at $x/b = 4.5$.

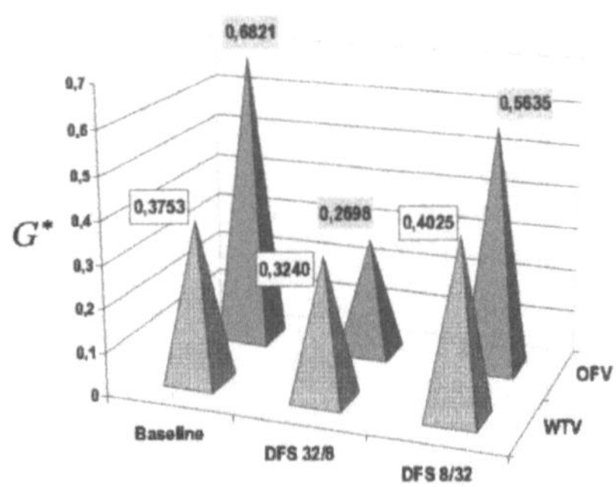

Fig. 12: The vortex strength G^* of the WTV and OFV at $x/b = 4.5$.

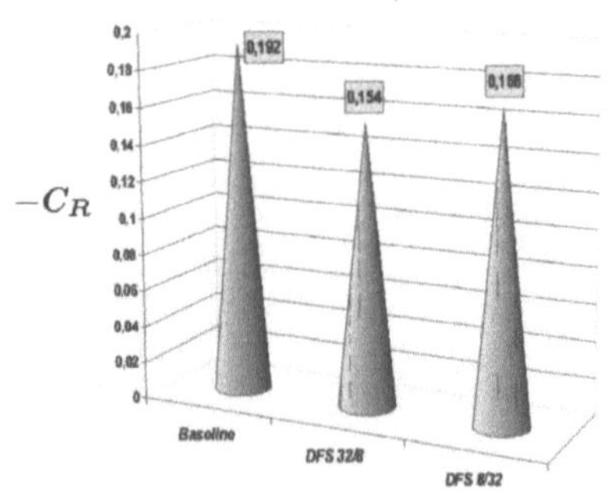

Fig. 13: The maximum induced rolling moment coefficient at $x/b = 4.5$.

Investigations on the Influence of Fins on the Extended Nearfield of a Wing in High-Lift Configuration

Sebastian Kauertz, Günther Neuwerth and Robert Schöll

Institut für Luft- und Raumfahrt, RWTH Aachen, Wüllnerstr. 7, 52056 Aachen, Germany
kauertz@ilr.rwth-aachen.de

Summary

This work presents results obtained from PIV measurements done in the wake of a swept and tapered transport-type aircraft wing model. The measurements are conducted in a water tunnel up to 4 spans behind the model. A fin is mounted on the upper side of the wing just in front of the outer flap edge. The influence of this fin on the structure of the vortex wake and the hazard posed to a following aircraft is investigated. The fin alters the structure of the vortices which may lead to an accelerated decay further downstream. In the investigated cases the fin also stimulates the meandering of the vortices, which alters the dynamic effects when a following aircraft enters the wake.

1 Introduction

Large transport aircraft produce strong vortices which persist behind the aircraft for a very long time. The induced velocities cause forces and moments on a following aircraft, which can be dangerous, especially during landing approach where the altitude is low. This makes it necessary to introduce minimum separation distances between aircraft, which means a following aircraft has to wait until the vortex wake no longer poses a hazard. This limits the capacity of large airports, so it is desirable to find some means of alleviating the vortices in order to shorten the separation distances. One approach, proposed by Rossow [1, 2], is the use of fins, small wings which are mounted on the wing perpendicular to the surface, producing a wake themselves which interacts with the vortices of the main wing. In previous works at the Institut für Luft- und Raumfahrt two special cases were investigated with a fin mounted directly in front of the outer flap edge of a model wing in high-lift configuration, and with two different angles with respect to the flow. Measurements were done up to 0.8 spans behind the wing in a wind tunnel so far, and should be extended up to 4 spans in a water tunnel. Results of these recent measurements are presented in the following.

2 Model and Configuration

The wing model used is a half model and was already used for the previous wind tunnel investigations done by Özger [3]. The model was built according to specifications defined in the Collaborative Research Center SFB 401 of the DFG (Deutsche Forschungsgemeinschaft). It is a swept and tapered wing with a BAC3-11/RES/30/21 airfoil optimized for transonic flight. It is equipped with high-lift devices consisting of two single-slotted flaps and three slats, extended with a constant angle δ_{Flap} and δ_{Slat} respectively. Geometric parameters of the wing model are

$b/2 = 0.675\,\mathrm{m}$	$c = 0.173\,\mathrm{m}$	$\phi = 34°$
$\Lambda = 8.8$	$\delta_{Flap} = 20°$	$\delta_{Slat} = 25°$

with the semispan $b/2$, reference chord c, leading edge sweep ϕ and aspect ratio Λ. Figure 1 shows the geometry of the model. The fin used for the investigations has a NACA0019 profile and a height and chord length of both 0.054 m, which corresponds to 0.04 spans. It can be mounted on the upper surface of the wing just in front of the outer flap edge, with a variable angle to the flow. In the presented results two fixed angles of the fin of $\varepsilon_F = +20°$ and $\varepsilon_F = -20°$ with respect to the main flow were used. In addition the model in high lift configuration without fin was investigated for comparison.

Figure 2 shows local streamlines obtained from wind tunnel experiments at the position where the fin is mounted [3]. The flow around the fin has an inboard component resulting in an angle of incidence of about 15° to the main flow direction due to induced velocities from the wingtip and flap-edge vortices. This gives an effective angle of attack for the fin which is different from its incidence angle to the main flow. The effective angle of attack of the fin is then $\alpha_F \approx 35°$ for the case $\varepsilon_F = +20°$ and thus produces mainly a turbulent wake which mixes with the flap vortex. For the case $\varepsilon_F = -20°$ the angle of attack is $\alpha_F \approx -5°$ and a small concentrated vortex is formed by the fin. This vortex is counter-rotating to the flap-edge vortex.

3 Experimental Setup

Experiments are conducted in a circulating water tunnel of the Institut für Luft- und Raumfahrt, Aachen. The test section of the tunnel has a cross section of $1.5 \times 1.0\,\mathrm{m}$ with a length of 6 m. With the present model measurements up to a distance of 4 spans behind the model are possible. Three Component Particle Image Velocimetry is used to determine the three velocity components in several planes perpendicular to the main flow. Six component force measurements are conducted with a strain gauge balance prior to the PIV measurements to determine lift curves for the investigated configurations. All experiments have been carried out at a freestream velocity of $U_\infty = 1.1$ m/s. The resulting Reynolds number based on the reference chord c is $Re = 190.000$.

3.1 Force measurements

Previous investigations in a wind tunnel with the same model [4] have shown that the fin has an effect on the lift curve of the wing. Especially the fin with $\varepsilon_F = +20°$ results in a decreased lift gradient. The experiments should be conducted with the same overall lift coefficient C_L in all cases. In order to determine the angles of attack for the chosen lift coefficient of $C_L = 1.33$ it was necessary to make force measurements with the model in the water tunnel. This gave an angle of attack of $\alpha_{wing} = 7.0°$ for the cases without fin and $\varepsilon_F = -20°$, and $\alpha_{wing} = 8.2°$ for $\varepsilon_F = +20°$.

3.2 Particle Image Velocimetry

For the measurements in the wake of the model wing Three Component Particle Image Velocimetry (3C-PIV) was used. Several planes perpendicular to the main flow direction have been investigated. The PIV system consists of a pulsed Nd:YAG laser which generates a light sheet in the test section, two CCD cameras which capture the light sheet and a PC for collecting and evaluating data. The water is seeded with small polyamide particles of 55 μm average diameter and a density of 1.02 g/cm^3. By illuminating the particles twice in a distance of a few milliseconds, their motion between the two shots and thus their velocities can be determined. This way the whole field of view of the cameras can be measured at the same time without disturbing the flow. A good solution was found to suppress refraction and distortion occuring when the cameras look into the water-filled test section at an oblique angle. A transparent box, filled with water, is mounted in front of the cameras directly at the glass windows of the test section. The pictures can be substantially improved this way.

As the field of view of the used cameras covers only an area of about 280 x 180 mm, the wake of the wing model has to be divided into several subfields which are measured one after another. After evaluation of the velocity fields they can be assembled into one whole field. This procedure is allowable as the flow is supposed to be stationary over most of the flow field. Recordings for each subfield are made at a frequency of 4 Hz (limited by cameras) over a time of 5 seconds, so that for each field 20 recordings are available. The measurement period of 5 s was considered sufficient as the meandering of the vortices, which was found to be the lowest-frequency instationarity, has a frequency which is still above 1 Hz. At 4 spans downstream of the model where noticeable meandering of the vortices occurs, the flow field differs considerably between the 20 recordings due to the instationary motion of the vortices. It can be seen though that the physical effect of induced rolling moments differs only slightly (cp. ch. 4).

4 Results

The PIV measurements provide velocity distributions containing the vortices of the wing. Downstream positions of $x/b = 0.0, 0.4, 0.8, 2.0$ and 4.0 have been investigated. At each position measurements without fin and with the fin in two positions have been made as described in section 2. From the velocity distributions vorticity has been calculated as well as induced rolling moments on a following wing using strip theory. The induced rolling moments have been computed according to

$$c_{l,ind} = \frac{2}{b_F^2} \int_{-b_F/2}^{b_F/2} 2\pi \, arctan\left(\frac{v_z}{U_\infty}\right) y' dy', \tag{1}$$

with b_F being the span of the following wing and v_z/u_∞ the induced angle of attack on the wing. The lift slope of the following wing has been assumed to be 2π to simplify computations.

Figure 3 qualitatively shows a velocity and vorticity distribution in the plane directly behind the wing tip trailing edge with the fin at position $\varepsilon_F = -20°$. Plotted are instantaneous velocities, i.e. without time averaging. Besides the flap and tip vortices a concentrated fin vortex can be seen. The flap vortex had a longer time to roll up than the tip vortex, as the distance to the wing trailing edge is larger due to wing sweep.

In Figure 4 the distribution for the case $\varepsilon_F = +20°$ is shown. The flow field around the flap vortex is completely different from the former case, the separated flow around the fin introduces high turbulence into the flap vortex.

The maximum induced rolling moments $C_{L,max}$ in the five investigated planes are plotted in Figure 5. These rolling moments have been computed from velocity fields obtained from the 20 recordings. Presented in the figure are the mean of these values and their standard derivation as error margins. The following wing has a span of 20% of the span of the generating model in this case. The maximum standard derivation is in the order of $\sigma = 0.025$ for $\varepsilon_F = -20°$ and $\varepsilon_F = 0°$ and occurs at the farthest downstream position. For $\varepsilon_F = +20°$ it is in the order of $\sigma = 0.06$ and occurs at the nearest position where there is still a lot of turbulence produced by the fin. It can be seen in the near field up to 1 span behind the wing that the fin strongly influences the structure and hence the induced rolling moments in the wake. This was already discovered by Özger [3]. The new results presented here show that in the extended near field the effect nearly vanishes, and no significant influence on the induced rolling moment can be seen. Although the core radius of the stronger flap vortex is enlarged by the fin with $\varepsilon_F = +20°$ which produces a turbulent wake (cp. Figure 6), it is still about an order of magnitude smaller than the span of the following wing of 270 mm. The maximum induced rolling moment occurs when the center of the following wing lies in the center of the strongest vortex, in the present cases the flap vortex. Even if tangential velocities decrease with a larger core radius, they are only shifted to a longer lever arm. Thus the overall rolling moment integrated along the wingspan does not decrease.

Regarding the velocity fields at 4 spans distance to the wing model, a considerable meandering of the vortex centers can be observed. Figure 7 shows positions of the vortex centers determined from the 20 recordings acquired for the case $\varepsilon_F = +20°$ together with the time-averaged vorticity field. The center is defined here as the location of maximum vorticity in the vortex. Mainly the flap vortex is influenced by the fin, where the maximum induced rolling moment occurs as well. It can be seen that the flap vortex moves within a circumference of about 60 mm, corresponding to 0.04 spans. Taking into account the roll damping of typical transport aircraft, it is possible that dynamic effects due to the meandering of the vortices relative to the following aircraft can decrease the peak rolling moments in the practical case. Especially when the vortex motion intensifies further downstream, this can also lead to an accelerated breakdown of the vortex system, which will be investigated in towing tank experiments in the next time.

5 Conclusion

The presented results show that a fin in the two investigated cases has no direct influence on the induced rolling moments in the near wake of the generating aircraft. However, there are no results yet for the far field of this configuration. The altered structure of the wake vortex system may lead to an accelerated decay at distances from about 10 spans and up. Dynamic effects due to the meandering of the vortex centres can lead to alleviated induced rolling moments as well. In the future research into the far-field development of the wake of this configuration will be conducted in towing tank experiments.

Acknowledgements

This work was funded by the DFG (Deutsche Forschungsgemeinschaft) in the Collaborative Research Center SFB 401 "Modulation of Flow and Fluid-Structure Interaction at Airplane Wings" at the RWTH Aachen. Mentioned windtunnel results were taken from work done by Özger [3] at the Institut für Luft- und Raumfahrt.

References

[1] V. J. Rossow: "Effect of Wing Fins on Lift-Generated Wakes". J. Aircraft 15, Nr. 3, 1978, pp. 160-167.

[2] V. J. Rossow: "Lift-generated vortex wakes of subsonic transport aircraft". Prog. in Aerosp. Sciences, Vol. 35, 1999, pp. 507-660.

[3] E. Özger: "Abschwächung des Wirbelnachlaufs von Flugzeugen mit Hilfe von Finnen". VDI Verlag Düsseldorf 2001 (Dissertation RWTH Aachen, Fortschritt-Berichte VDI, Reihe 7, Nr. 422)

[4] E. Özger: "On the Structure and Attenuation of an Aircraft Wake". J. Aircraft 38, Nr. 5, 2001, pp. 878-887

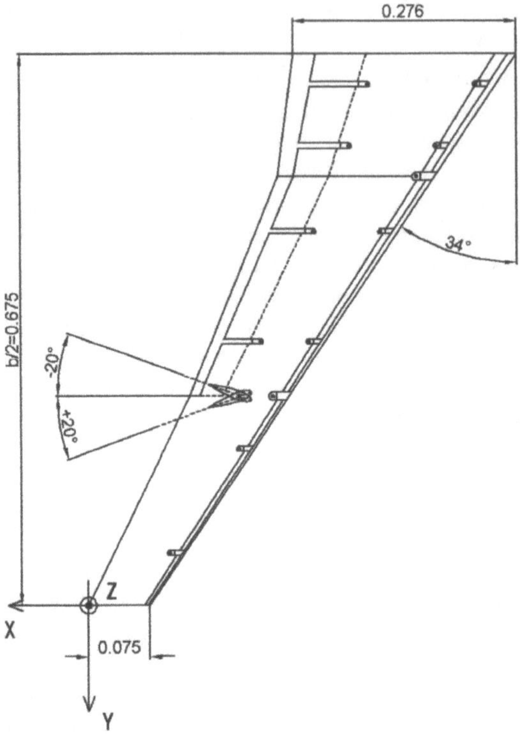

Figure 1 Dimensions of the wing model (lengths in m)

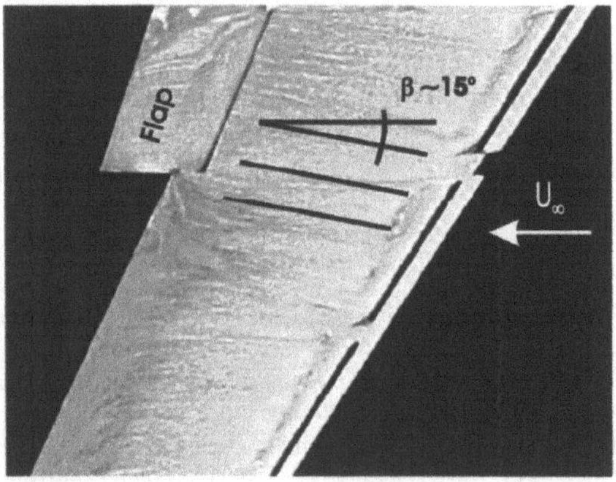

Figure 2 Local streamlines at fin position

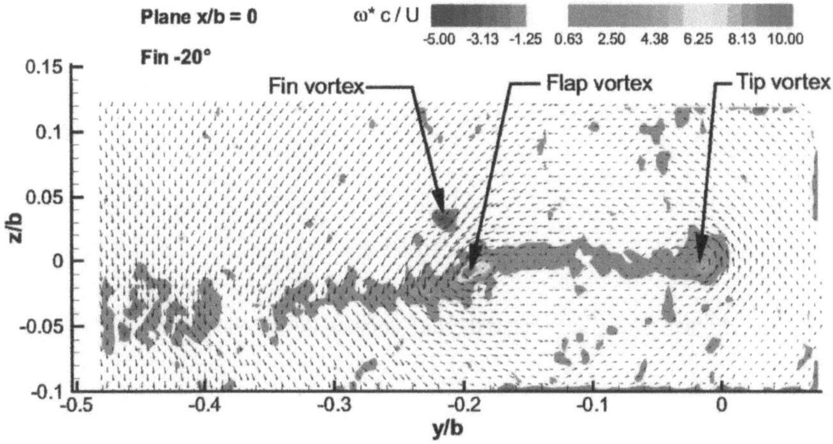

Figure 3 Velocity and vorticity distribution at x/b = 0.0, $\varepsilon_F = -20°$

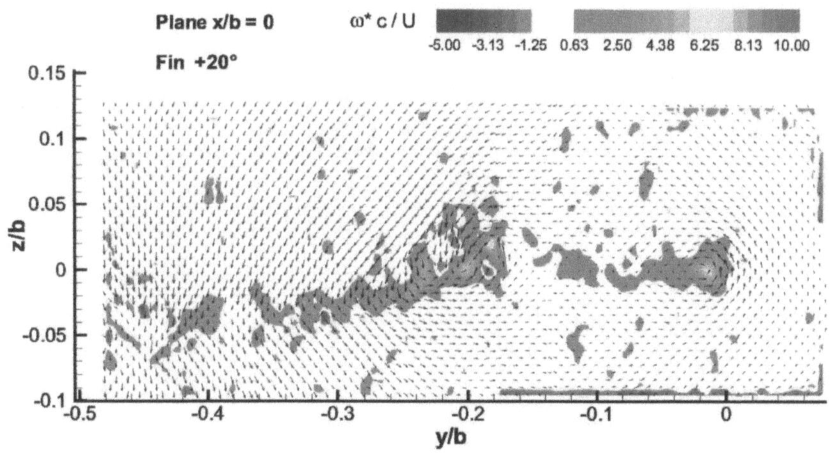

Figure 4 Velocity and vorticity distribution at x/b = 0.0, $\varepsilon_F = +20°$

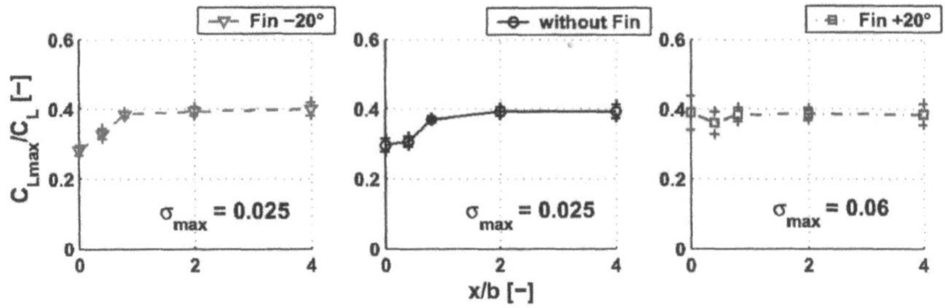

Figure 5 Maximun induced rolling momemt on a following wing

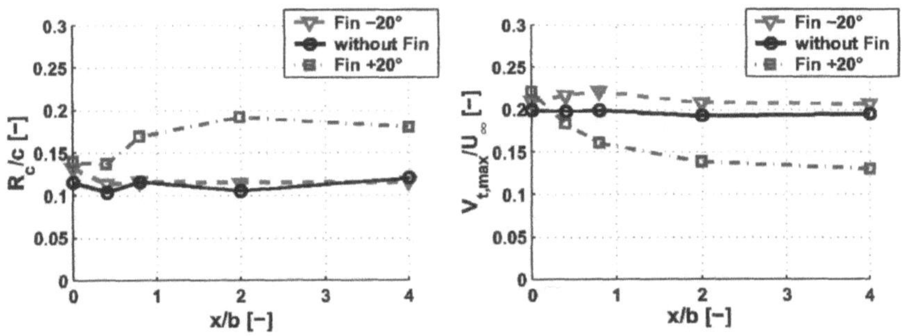

Figure 6 Core radius and maximum tangential velocity of flap vortex

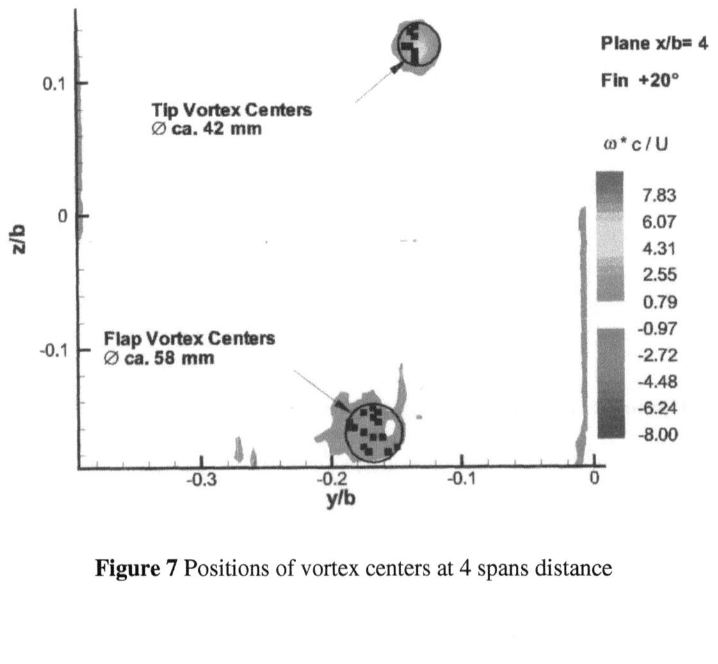

Figure 7 Positions of vortex centers at 4 spans distance

Application of the PSP technique in low speed wind tunnels

U. HENNE
DLR Göttingen, Institut für Aerodynamik und Strömungstechnik,
Bunsenstr. 10, D-37073 Göttingen, Germany
e-mail:ulrich.henne@dlr.de

Summary

Results of pressure field measurements on the A400m half model in the low speed wind tunnel of Airbus Bremen, Germany, are presented. The model is equipped with two propeller simulators driven by pressurized air. The great challenge was to compensate the influence of the warm and cold air pipes which are guided inside the wing, creating large temperature differences on the models surface. The PSP (pressure sensitive paint) measurements were carried out with the mobile DLR PSP system using the DLR02 paint formulation. The influence of propellers with different thrust levels on the pressure distribution was determined successfully. An assessment of the accuracy of the results can be performed by comparison of PSP data with conventional pressure tap data.

1 Introduction

The measurements were conducted to evaluate the feasibility and accuracy of measurements using DLR's Pressure Sensitive Paint (PSP) system in the low speed wind tunnel (LSWT) of Airbus Bremen on a model with high temperature variations on its surface. The main goal from the aerodynamic point of view was to visualize the influence of the propeller at different thrust levels on the flow.
The PSP technique is based on the deactivation of photochemical-excited organic molecules (luminophores), by oxygen molecules [2,3]. The fluorescence intensity reduces with increasing pressure according to the Stern-Volmer relation :

$$I_0/I = 1 + K_q PO_2 , \qquad (1)$$

where I_0 is the photoluminescence in the absence of oxygen (vacuum), I the detected photoluminescence, K_q the quenching constant, and PO_2 is the partial pressure of oxygen. K_q is a function of temperature. Assuming a constant oxygen concentration in air and including second order terms, a more useful equation to calculate the static pressure values p from a measured intensity distribution I for such paint formulations can be derived:

$$p = A(T) + B(T)(I_{ref}/I) + C(T)(I_{ref}/I)^2, \qquad (2)$$

where p is the local static pressure, A, B, and C are temperature dependent calibration coefficients of the pressure sensitive paint formulation, which can be determined in laboratory by varying the pressure level for different temperatures in a calibration chamber using a small paint sample. I_{ref} is the corresponding intensity value for a constant reference pressure. The reference conditions are taken into account by using the corresponding calibration coefficients for these conditions. For the current wind tunnel test the pyrene based paint, DLR02, developed in close collaboration with the University of Hohenheim [1], was used. It is a two color paint which contains a pressure insensitive luminophore which is used for normalization of intensity I (and I_{ref}) to avoid measurement of absolute intensities. While being excited at the same wavelength, this component emits at other wavelengths than the pressure sensitive component and can thus be distinguished by appropriate filters.

Temperature dependence with respect to the pressure of 0.4 %/K for DLR02 paint has been determined in an external calibration chamber. In the case of 100 kPa base pressure this leads to a temperature induced error in pressure of about 40 Pa/K. This means that a temperature deviation of 1 K changes the PSP pressure result by approximately 40 Pa. Since temperature differences of more than 10 K on the model surface are expected, direct measurements of the models surface temperature distribution are required and are achieved by means of thermography technique with an infrared (IR) camera. The paint's IR emission coefficient of approximately 0.95 allows using it as an ideal infrared coating and thus enables the possibility of simultaneous IR and PSP measurements.

2 Experimental Setup

Figure 1 shows a sketch of the experimental setup including an image of the A400m half model, used in this wind tunnel test. The flaps are mounted in landing configuration. The model is equipped with propeller simulators which can be operated at different thrust levels [5]. Alternatively the propellers can be replaced by a simple cap ("dummy spinner"). The inner propeller rotates clockwise whereas the outer one rotates counterclockwise, as indicated by the black arrows. Both are driven by pressurized air heated up to more than 40 °C which cools down after expansion down to -10 °C, depending on the thrust level. The pressurized and expanded air is guided through channels inside the main wing. This explains the high temperature differences on the wing surface. Six sections with 126 pressure taps can be used for comparison with PSP results.

The PSP system is also sketched in Figure 1. It consists of a flash lamp, two CCD cameras with PCs and two workstations. The flash lamp with high intensity output at frequencies up to 40 Hz for direct illumination was already used successfully in TSP measurements at ETW [4]. For adaptation to the current experiment only appropriate filter selection was required to get emission in the region of 320-360 nm. The pressure sensitive emission is measured by the first CCD camera (PSP1), whereas the luminescence of the reference component of the paint is detected by the second camera (PSP2). Filters attached to each camera discriminate both signals. The lamp as well as all cameras were mounted to the side wall of the wind tunnel. The cameras used here are 16 bit water cooled cameras with 1024x1024 pixels on a chip cooled by a Peltier element down to -26 °C to reduce background

and noise. A software for image acquisition with these cameras, triggering and communication with the control program (see below) has been developed and has firstly been used in an industrial wind tunnel. Since in this version only one camera can be controlled by one PC, two PCs are required. Triggering of cameras and flash lamp is synchronized by PC1. Additionally, the IR system of DLR Braunschweig is installed, which is normally used for transition detection [7,6]. It consists of a camera with spatial resolution of 320x240 pixels, a temperature resolution of 20 mK, and a control PC (PC3). Comparison of absolute temperature data from the IR system with results from a PT100, glued to the surface of the wing near the fuselage, show a difference below 0.1 K. In the course of this measurement campaign the software was adapted in such a way that it could be integrated to the PSP data acquisition system, described in the following.
A control program running on the main workstation (WS1) with Solaris as operating system is the main part of the PSP data acquisition system. The workstation is connected via two network cards externally to the wind tunnel network and on the other side to the internal PSP network with all image acquisition PCs (PC1-PC3). The internal high speed network has two main advantages: Firstly, it saves time, because only the second network adapter of the workstation WS1 and not all network components need to be adapted to the wind tunnel network. Furthermore the high data traffic for image saving does not affect the wind tunnel network. On the workstation with connection to the wind tunnel a program is running whose main function is to handle the communication between wind tunnel and image acquisition. After receiving a ready signal from the wind tunnel side, the control program automatically starts image acquisition. When all images are acquired, they are stored into a standard directory structure, required for automated data reduction. After measurement of a series of data points, e.g. an alpha sweep, on the second workstation (WS2) the data reduction starts and delivers preliminary results, showing the flow topology on a 3D surface grid, a few minutes after end of sweep. These results are called preliminary, because they are not corrected with respect to the temperature distribution and absolute pressure calibration. The centralized data acquisition and automated data reduction are parts of DLR's ToPas software.

3 Productivity

PSP often is stated to be a complex and time-consuming measurement technique. Therefore in the following the time schedule for the measurement campaign is presented briefly. In order to save time the model was coated with PSP inside the wind tunnel's test section. The paint must be applied to the model in two layers with drying time of about 12h each. Thus the overall time for PSP coating took two days. During the same time the PSP equipment was mounted, the computers were installed in the control room and a general test of all components and network connections was performed. After fine adjustment and a test series, the measurements started on the third day. Including measurement of reference images without running the wind tunnel and averaging over 4 images in each case, one data point required approximately three minutes. After finishing the measurements the whole PSP system was removed from the wind tunnel in less than half a day. During that time also the paint was removed from the model within 2 hours. Afterwards the model could be used for standard measurements without any impact from the PSP coating.

4 Data Reduction

The goal of the PSP measurement is to obtain the pressure distribution on the model surface, which is delivered at each node of a structured 3D surface grid of the model. For a complete evaluation of the pressure distribution for one angle of attack from one point of view eight images are required. For each PSP camera a "wind off" (at fixed reference pressure) and "wind on" image plus the corresponding IR images for these two cases. A dark image for background subtraction is also needed.

The background subtracted images and the temperature field from the IR images must be aligned to the surface grid. Markers are applied at well defined coordinates on the model surface. After finding markers in the images the correct alignment parameters (model and camera position as well as lens properties and perspective parameters) are calculated by a least square fit of the difference between calculated and known marker positions. Then according to these parameters the pixel values can be assigned to the corresponding node. Now the pressure on each node can be calculated using the calibration from equation (2). For calculation of absolute pressure values an offset is added, which is determined by comparison of seven PSI values with the PSP results at the location of the corresponding pressure taps. Data reduction using DLR's software ToPas is performed by scripting methods with a very high degree of automation and reproducibility.

5 Results and Discussion

All measurements have been performed at velocity U_0 = 60m/s. Four different configurations are compared by sweeping the angle of attack from α=-3.5° up to 14° with 10 steps in between. In the following, configurations are denoted as *DS* ("dummy spinner", see experimental setup), *CT-0*, *CT+0* (see below), and *CT1* (tangential force coefficient C_T=0.1). Since running the propeller simulators exactly without thrust (C_T=0) is difficult to adjust for all α, there are two configurations, one with C_T slightly below 0 (*CT-0*) and another with C_T slightly above 0 *(CT+0)*. The part of the model shown in all following presentation figures is limited by the region visible to the IR camera at all angles of attack since only there the required temperature information is available. Even though the IR lens with shortest focal length available was used, it was not possible to get a full model view for the IR image. Thus, only three sections of pressure taps are available for comparison with PSP data. Since some of the data is confidential, the absolute values of C_P are omitted.

Figure 2 shows the effect of temperature correction by comparing a preliminary PSP result without temperature correction (Figure 2a) with the result after applying the correction using the temperature field obtained from the IR image (Figure 2b). Besides the real pressure distribution some structures can be recognized in Figure 2a which definitely do not originate from the pressure distribution. These structures are not any more visible in the final result, as shown in Figure 2b.

In order to check the accuracy of the PSP results they are compared with PSI data measured simultaneously. In Figure 3 the pressure distribution for configuration *DS* is shown. The black dots indicate the positions of the pressure taps on the model. For each section of pressure taps the diagram shows the pressure values of both pressure measurement techniques. Again the absolute scale of the pressure axis is not given due to confidentiality reasons, but to get an impression of the accuracy a difference of 5 kPa is depicted in Figure 3. The agreement is quite good on the main wing, whereas larger differences are found at the flaps and near the leading edge, as expected due to the viewing angle of more than 30° to these surfaces. A standard deviation between PSI and PSP of 240 Pa is calculated, considering all pressure taps on the main wing and all data points of this measurement campaign. For configurations without thrust and moderate α the standard deviation was about 150 Pa. These values are comparable with the results of the following error estimation.

The main possible sources of errors are reflections of excitation or luminescence light, paint damages, misalignment, and noise of the CCD signal. In the following some error sources will be discussed in more detail, excluding systematic errors like electronic instabilities of cameras and temperature effects. Reflection of luminescence light from the walls was minimized by covering the walls with nonreflecting black foil. Reflections of the excitation light from the fuselage were only found to be a problem in the area between fuselage and inner propeller, which was not visible for the IR system and thus has been excluded from the results. The DLR02 paint is quite sensitive to contamination with oil or solvents. The surface of the nacelles was covered with sticky aluminum tape, which is known to contain glue even at its surface that can affect the luminophores of the pressure sensitive paint. Some holes have been coated with aluminum tape also leading to paint damage and increased errors at some locations which nearly disappeared by using a median filter (see below) but are still possible error sources which cannot be reliably estimated. At the leading edge some dust could be found after the measurement which could affect the intensity ratio of pressure sensitive signal and reference signal. But the affected area is anyhow excluded from the results due to bad optical access. The accuracy of the alignment can be estimated from the mean deviation of 0.4 pixel between the marker positions calculated from 3D position and the positions found in the intensity images. Since one pixel corresponds approximately to 1 mm the spatial accuracy of the pressure distribution is about 0.4 mm. The highest pressure gradients (500 Pa per pixel) can be found at the flaps and near the leading edge leading to possible differences of 200 Pa in the PSP-PSI comparison. The pressure gradients on the main wing are smaller and accordingly their contribution to the error. This effect is more important for higher angles of attack and higher thrust because the pressure gradients (and also intensity gradients) increase, which is in accordance to the mean PSP-PSI differences changing from 150 Pa (see above) up to 400 Pa in these cases. Even though in principle the use of a two color paint should compensate intensity variations over the model, misalignment between pressure sensitive and reference image leads to a significant error if intensity gradients (mostly resulting from inhomogeneous illumination) are as high as in the current measurements. Whereas on the main wing the intensity gradients are smaller than 0.5 %/pixel, on the flaps values of up to 2.5 %/pixel can be found. According to the calibration parameters of the DLR02 paint a relative change in intensity of 1 % changes the pressure result by about 1500 Pa. This leads to possible errors in the resulting pressure data due to misalignment (0.4 pixel) and intensity gradient of up to 300 Pa and 1500 Pa, respectively, and is thus the major error source. To obtain the influence of the noise of the CCD sensor the intensity ratio of the "wind off" images (average of 4 images and background subtracted) was examined. A relative standard deviation of 0.3 % for both cameras was found leading to a statistical error of less than 0.5 % for each pixel in the resulting data. This value is reduced by filtering the images with a 5x5 median filter to 0.1% contributing approximately 150 Pa to the pressure data error.

The comparison of different configurations with mounted propeller is shown in Figure 4. All images are acquired at α=8.5°. As expected the pressure distributions are very much the same for *CT-0* and *CT+0*. In comparison with configuration *DS* there is lower pressure visible at the flaps and the structure between the outer nacelle and the leading edge of the wing has changed. Especially the reduced pressure at the flaps can be understood in terms of increased mass flow beneath the wing induced by the running propellers even with thrust values C_T about 0. This higher mass flow leads to higher flow velocities through the gap between flaps and wing, and thus to lower pressures on the flaps. With higher thrust (C_T=0.1) this effect is much stronger, visualized by the blue colored flaps in Figure 4c. Note that the pressure scale for Figure 4c is shifted towards lower pressures with respect to Figures 4a and 4b in order to show the above described structures. Thus points with the same color represent a lower pressure in Figure 4c than in Figures 4a or 4b. Using this scale, two "dips", areas of higher pressure, can be seen on the flaps at higher thrust levels. They are just located where the fairings are mounted and can be interpreted as blocking of the flow by the fairings. The structure in the middle of the wing between the nacelles must originate from higher impact pressure due to higher mass flow according to the rotation direction of the propellers.

In Figure 5 for three angles of attack α the C_P values of configuration *DS* are subtracted from configuration *CT-0*. This means that where the pressure for case *CT-0* is higher than for *DS* the value ΔC_P is positive (red colored) and where it is lower than for case *DS* the difference is negative (green or blue colored). For small or even negative α (Figure 5a) there is no significant deviation visible whereas for increasing α (Figures 5b and 5c) mainly the pressure at the flaps decreases remarkably when running the propeller simulators without thrust (*CT-0*) in comparison to the results for not mounted propellers (see discussion for Figure 4). Also on the main wing lower pressure is visible which is in agreement to the observation that on-set of separation is shifted to higher angles of attack with running propellers even if the thrust level is below zero, which means that at $\alpha=0$ the flow is not accelerated by the propellers. This can be explained by the fact, that for higher α the running propellers change the flow direction so that the effective angle between wing and flow is reduced. The structure between the outer nacelle and the leading edge of the wing mentioned above is much better visible in the difference representation. Especially for $\alpha=12.4°$ there is a trace with lower pressure in case *DS* which may be induced by a vortex which vanishes with running propellers.

6 Conclusion

The influence of propeller simulators on the pressure distribution on the wing of the A400m model in the low speed wind tunnel of Airbus Bremen (LSWT) was measured successfully by means of the mobile DLR PSP system. The main challenge was to correct for the temperature dependence of the paint for the low speed application of PSP with temperature differences of about 10 K on the observed model. The temperature distribution on the wing was measured by an IR camera which was integrated into the data acquisition system in the course of the current measurement. The temperature correction using data from the IR image was implemented successfully into the ToPas software, which was used for data reduction. Thus, PSP has proven to be an excellent tool as well for qualitative as for quantitative pressure detection on model's surface in a short time. Also it is shown that PSP is able to obtain pressure distributions on model surfaces with high temperature disturbances.

Acknowledgements

I like to thank Dr. Klaus de Groot from DLR Braunschweig for his IR measurement support during the tests. Special thanks for the permission to publish this paper to Airbus Deutschland and to the crew of LSWT for the good team work during the measurements, by name Mr. Klaus Muthreich and Mr. Peter May.

References

[1] F. Döring. *Synthese von Pyrenfarbstoffen und Untersuchung ihrer O_2-druckabhängigen Lumineszenz für die Anwendung als Pressure Sensitive Paint. Dissertation University Göttingen, Germany, DLR-FB 2004-18,* 2004.

[2] R.H. Engler, et al. *Description and assessment of a new optical pressure measurement system (OPMS) demonstrated in the high speed wind tunnel of DLR in Göttingen,* DLR-FB 92-24, 1992.

[3] R.H. Engler and Chr. Klein. *First Results Using the New DLR PSP System – Intensity and Lifetime Measurements,* Conference "Wind Tunnels and Wind Tunnel Test Techniques", Cambridge UK, ISBN 185768 048 0, 1997.

[4] U. Fey, et al. *Transition Detection by Temperature Sensitive Paint at Cryogenic Temperatures in the European Transonic Wind Tunnel (ETW),*

Proceedings of 20th ICIASF, Göttingen, Germany, ISBN 0-7803-8149-1, 2003.

[5] M. Huhnd, *Discretion of Air Motors DA 40/15000 for use on Wind Tunnel Model FLA 5+10,* Laboratory Instrumentation Note, Airbus Deutschland GmbH, M72D04024103, 2004.

[6] J. Meyer, et al. *System Layout and Instrumentation of a Laminar Flow System for the DLR Do228 Test Vehicle*, RaeS Aerospace Aerodynamics Research Conference, London, 2003.

[7] A. Quast, *Detection of Transition by Infrared Image Technique*, IEEE publication CH2449-7/87, ICIASF 1987, 1987.

Figures

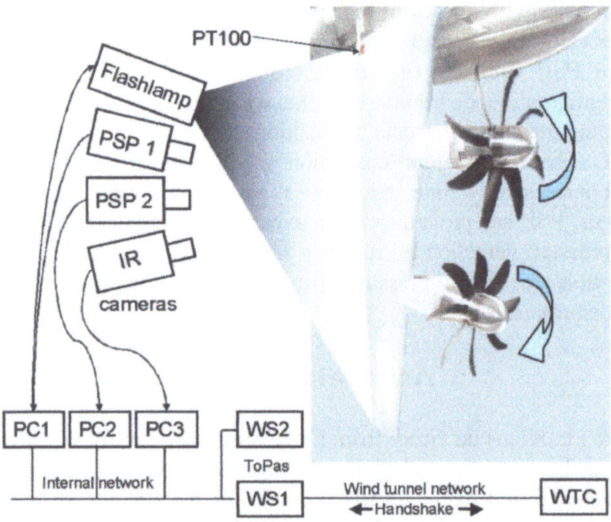

Figure 1 Sketch of experimental setup

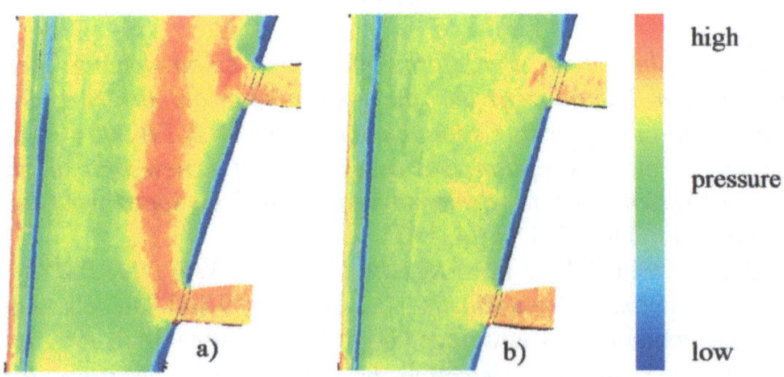

Figure 2 PSP results at C_T=0.1 : a) constant temperature and b) correction with temperature field obtained from IR image

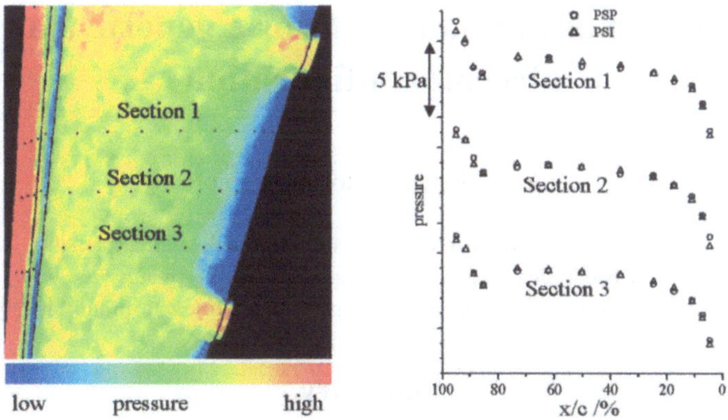

Figure 3 Comparison of PSP and PSI at α=8.5° without propeller simulators (*DS*)

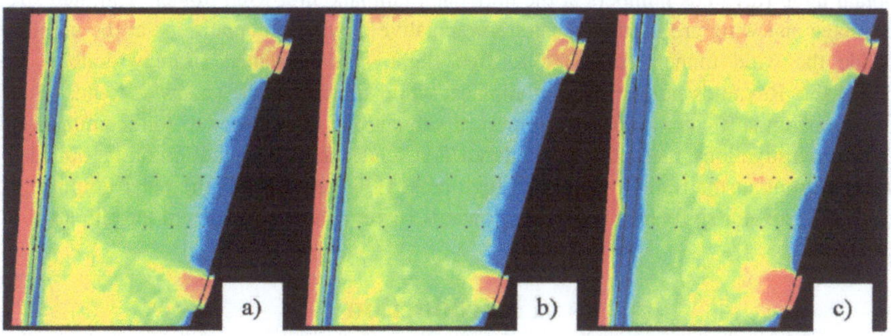

Figure 4 PSP results at α=8.5° with a) *CT-0*, b) *CT+0* (scale from Figure 3), and c) C_T=0.1 (with scale shifted towards lower C_P values)

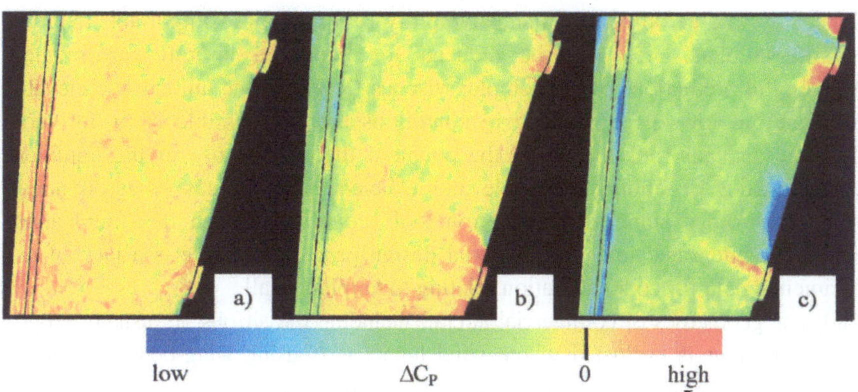

Figure 5 C_P difference between *CT-0* and no propellers (*DS*) at a) α=-3.5°, b) α=8.5° and c) α=12.4°

Investigation of tailplane stall tor a generic transport aircraft configuration

Arne Grote and Rolf Radespiel

Institute of Fluid Mechanics, Braunschweig Technical University
Bienroder Weg 3, 38106 Braunschweig. Germany
Internet: www. tu-braunschweig.de/ism, E-mail: arne.grote@tu-braunschweig.de

Summary

The design of a generic wind tunnel model for tailplane stall investigations to setup an experimental database for code validation is presented. The configuration is optimised to obtain large Reynolds numbers at the horizontal tailplane in a wind tunnel of limited size. Fuselage and wing are used to create representative downwash conditions at the tailplane inside the wind tunnel, compared to free flight. The presented strategy doubles the achievable Reynolds number of the tailplane, while simulating the spanwise trends of the downwash in sufficient accuracy. Numerical simulations preliminary to forthcoming wind tunnel investigations, using the unstructured TAU code, show a separation of the boundary 1ayer starting at the trailing edge with high cross flow velocities at the outer tailplane. A deflected elevator shifts flow separation towards 1ower incidence angles of the tailplane.

1 Introduction

The aerodynamic design of horizontal tailplanes (HTPs) for modyrn transport aircraft is driven by the demand for longitudinal stability and reduced drag at cruise condition, and manoeuvrability at low speed, while avoiding tail stall. Especially, the aerodynamic manoeuvrability becomes more important, if future aircraft tailplane area will be reduced for improved aircraft performance. Meaningful predictions of the complex viscous flow over the tail and the elevator efficiency at low speed manoeuvre conditions are not possible by using aerodynamic analysis tools based on the potential theory, because of the nonlinear characteristics of stabiliser and elevator efficiency for increasing aerodynamic load. High-order numerical simulation tools could cope with this, but cannot simply be integrated into the industrial design process, as there is a lack of experimental data for code validation relating to tailplane stall.

Tailplane geometries of commercial aircraft including the airfoils are usually proprietary behaviour is evaluated within wind tunnel tests using wing-body-tail configurations. Some of the industrial investigations for Reynolds numbers of $Re = 3.5\cdot10^6$ indicate a characteristic stall behaviour of the tailp1ane[4]. In these cases leading edge stall and sudden loss of negative lift and elevator effectiveness. Experimental investigations

of the flow around 2D tailplane airfoils could help to analyse this behaviour, but they are not available either.
The objective of the present work is to design and numerically analyse the layout of a generic transport aircraft wind tunnel model that can provide high quality experimental tail flow data close to the limits of tail stall. The validation of numerical results obtained with the DLR Reynolds-Averaged Navier-Stokes (RANS) solver TAU will be possible using the results of future wind tunnel investigations. By merging both experimental and numerical data, fundamental aerodynamic design sensitivities of tail stall may be identified and analysed, to be applied for future aircraft design work.

2 Aerodynamic design of the wind tunnel model

The present design of a generic research model for tail stall is mainly affected by the demand of attaining maximum Reynolds number at the HTP. For a typical HTP geometry (aspect ratio $A_{HTP} = 4.5$ and taper ratio $\lambda_{HTP} = 0.42$) Reynolds numbers about $Re \approx 1.0 \cdot 10^6$ can be realised in the test section of a wind tunnel with a cross section of 1300x1300 mm^2, such as the low-speed tunnel MUB of Braunschweig Technical University. In our concept the span of the HTP is about 68% of the test section width leading to a mean aerodynamic chord of $l_\mu = 200$ mm. The main wing is clipped and serves as the model support. Out of the variety of possible aircraft geometries an Airbus A319 was chosen to form the geometric base of the generic model. The dimensions of the closed test section determine wind tunnel wall interferences [8] and require an aerodynamic redesign of the main wing to keep the level of downwash at the HTP.
As a result of the geometry dimensions (Tab. 1) a wind tunnel blockage of 20% is generated, which can be accepted due to the fact that the wind tunnel is modelled in the numerical calculations as well. With this approach tail plane configurations may be investigated at Reynolds numbers, which would require twice the test section's dimensions using a conventional full span model.

2.1 Main wing

For the design of the main wing the classical vortex lattice method in TORNADO [9] was extended by the method of images [2]. Thereby, the main system of vortices is mirrored at the tunnel walls, so that the kinematic flow condition is fulfilled not only at the collocation point but also at the tunnel walls (Fig. 1). This simplified approach was deemed to satisfy the demands producing representative downwash conditions at the HTP location. Complex systems of vortices, as shed by extended main wing's high-lift devices, will have an additional influence on the stall behaviour of the HTP, which is neglected in this approach.
Starting from the main wing of the A319 in take-off configuration with known distributions of local lift coefficient and angle of incidence along the span, a reference wing with the same planform area was derived, shown in Fig. 2. Its twist distribution was adjusted in order to approximate the given lift distribution (Fig. 3). For this configuration the downwash distribution at the original HTP location behind the main

wing (MW) was then determined at free-flight conditions (Fig. 4).
By variation of sweep and twist, a clipped trapezoid wing of 1300 mm span and 28° sweep was designed, creating an equivalent downwash distribution in the wind tunnel test section at the original HTP location. In fact, the desired strength of the downwash could merely not be obtained, but this will only affect flow separation to occur at a lower incidence angle of the models HTP in comparison to free-flight. The effect of decreased downwash inside the test section can be explained by the following: The mirror images of the lift generating part of the horseshoe vortices above and below the test section decrease the downwash at the HTP position relative to the free-flight conditions. Reducing the distance between the main wing and the HTP from 100% to 40%, as shown in Fig. 4, can not completely solve this problem though the adverse effects of the mirror images decrease, but they do not disappear. But since too short distances between the main wing and the HTP would make a realistic representation of the fuselage rear end flow impossible, an acceptable compromise seems to be the location of the HTP at the 80% position compared to the reference.
To minimise the manufacturing complexity, a rectangular wing with chord length equal to the mean chord of the trapezoid wing was finally designed. Since this geometry produces the desired downwash distribution in spanwise direction, the rectangular wing is selected as the final geometry of the wind tunnel model. Its c_l distribution along the span in Fig. 3 was confirmed by TAU simulations performed at a slightly lower angle of attack.

2.2 Research-airfoil for the horizontal tailplane

The objective of designing a new HTP research airfoil was to produce a stall behaviour at MUB wind tunnel conditions that is comparable to the phenomena observed in large industrial facilities. The aim was to obtain similar values of $c_{l,max}$ and distributions of nose suction peaks compared to industrial references, however at a reduced Reynolds number.
Therefore, an industrial HTP airfoil was at first analysed using the airfoil code XFOIL [6]. Concerning the pressure distribution, this airfoil was compared to two airfoils of the LWK-series [1] showing flow separation in wind tunnel tests via leading-edge stall (LWK 80-120) and trailing-edge stall (LWK 80-150). From the results displayed in Fig. 5 at equal $c_{p,min}$ and α close to $c_{l,max}$ it is obvious that both LWK airfoils show a steeper gradient in the region of pressure increase. Both LWK airfoils have a shorter laminar separation bubble in comparison to the HTP airfoil, denoted by the plateau length of the pressure distribution. This indicates that the industrial HTP airfoil does not separate via leading-edge stall. It was concluded that leading-edge stall previously observed for the industrial HTP airfoil in wind tunnel tests at $\mathrm{Re}=3.5\cdot10^6$ seems to result from three-dimensional effects.
For the design of the research airfoil with XFOIL, the following design requirements were derived from this analysis: To attain an equivalent load at the model's HTP, the maximum lift for $\mathrm{Re}=1.0\cdot10^6$ has to be increased to the value 1.8 by increasing the camber. Simultaneously, the airfoil contour at the leading edge was redesigned in

order to produce a similar pressure distribution as for the original HTP airfoil. The resulting airfoil named HGR-02 is shown in Fig. 7. Its maximum lift coefficient is close to the target and the pressure distribution is quite similar (Fig. 6). So it may be assumed that both airfoils behave fairly equally, exept that the HGR-02 will have a longer laminar separation bubble at the lower Reynolds number.

2.3 Layout of the wind tunnel model

The structural layout in Fig. 8 shows the resulting wind tunnel model [7]. It can be devided into a front part generating the necessary flow conditions at the HTP, including main wing and the fuselage up to the disjunction ahead of the HTP, and the rear part, consisting of tail and tailplanes. This part is connected to the front via a 6-component strain-gauge balance, in order to measure the resulting airloads on the tail. Furthermore, the HTP has 100 pressure tabs in two sections at 55% and 75% semispan with a high resolution close to the leading edge.

3 Results of the numerical flow simulation

The numerical investigations using the DLR TAU code [5] simulate the flow around the model inside the wind tunnel (Fig. 9). By using the Spalart-Allmaras turbulence model and assuming the boundary layer to be fully turbulent, investigations were carried out to analyse the influence of the angle of attack, stabiliser incidence angle and elevator deflection on the flowfield around the HTP and the pressure distribution at the HTP. These simulations are used to assess the model design based on approximate methods as described above and give an impression of the model's behaviour during future wind tunnel tests.

3.1 Grid generation and sensitivity to grid resolution

The generation of the unstructured hybrid grids was done using Centaur [3]. To keep the number of cells low, only those surfaces were meshed with prisms, where viscous boundary layers were supposed to have an influence on the HTP flow. Therefore only the boundary layers of the HTP and the fuselage were resolved with prisms. The flow over the main wing and the wind tunnel walls was assumed to be inviscid. The viscous flow around the HTP was resolved by at least 308 points in circumferential direction and 48 prism layers in normal direction. The resulting surface grid resolution near the leading-edge is shown in Fig. 10. At the fuselage 32 prismatic layers were used. To minimise prismatic grid defects due to chopping in the junction between fuselage and HTP, the rear fuselage was resolved with 40 prism layers. Grid adaptation locally improved the grid quality within the prismatic and tetrahedral regions in particular at the HTP and increased the total number of nodes to about $10 \cdot 10^6$.
The effect of grid resolution on the flow was investigated by using a grid with four times less points. This grid yielded a difference in c_l of the HTP of 1.2%. The differences in the pressure distributions were rather small.

3.2 Variation of the stabiliser setting

Figures 11 (a) and (b) show pressure distributions and wall streamlines on the lower HTP side for stabiliser incidence angles of $i_t = -11°$ and $i_t = -14°$. For $i_t = -11°$ the flow over the whole HTP is attached with a certain cross flow component towards the trailing edge. Also a local separation at the junction with the fuselage is visible. In the region of the Küchemann wing tip the isobars show some effective reduction of the sweep angle. For an increased incidence angle of $i_t = -14°$ boundary layer separation occurs at the trailing edge and covers about 30% chord. However, this cross flow separation does not lead to a significant loss of lift as shown in Fig. 13.

3.3 Variation of the elevator deflection

For an elevator deflection at incidence angles of about $i_t = -11°$, the elevator flow is characterised by large separations, see Fig. 12 (a) and (b). Even for $\delta_e = -10°$, the flow is separated over almost the whole elevator. By increasing the deflection to $\delta_e = -20°$, reverse flow takes place on the complete suction side of the elevator. The elevator efficiency may be assessed by using Fig. 13. It is seen that the increments of local lift coefficients given by elevator deflection decrease for $\delta_e = -20°$. This indicates a loss of elevator efficiency due to flow separation. Fig. 14 displays, that the reduction of the suction peak at the leading edge by the deflected elevator and thereby a displacement of the lift towards the trailing edge leads to an increase of the average lift coefficient, despite of massive separations on the elevator.

3.4 Comparison with the infinite swept wing

For the HTP section at 75% semispan at $i_t = -11°$ and $\delta_e = 0°$ two-dimensional calculations at constant c_l were carried out using the theory of the infinite swept-back wing. As shown in Fig. 15 a comparable pressure distribution was not achieved, because early separations occured at the trailing edge in the 2D result. It appears, that the underlying assumption of a passive spanwise flow component is not valid at moderate Reynolds numbers and close to trailing-edge separation. Further analysis of this problem is neccessary. For this purpose we plan to set up additional 3-D calculations of a swept-back wing of infinite span.

4 Conclusions

Since the manufacture of the wind tunnel model is not yet completed, only the present computational results allow an estimation of tailplane stall. The results indicate that the new HTP configuration in the wind tunnel is aerodynamically well designed and tail stall occurs gradually from flow separations at the trailing edge. It is thought that the configuration represents a good test bed for code validation.

Acknowledgement

The authors gratefully acknowledge the DFG (Deutsche Forschungsgemeinschaft) for the financial support.

References

[1] D. Althaus: "Niedriggeschwindigkeitsprofile". Vieweg, Braunschweig, 1996.
[2] J.B. Barlow, W.H. Rae, Jr. and A. Pope: "Low speed wind tunnel testing". John Wiley & Sons, 3rd ed., New York, USA, 1999.
[3] Centaur^{TM} by Centaursoft: *www.centaursoft.com*
[4] V. Cleemann: Private communications. Airbus Deutschland GmbH, Bremen, 2004.
[5] DLR, Institut für Aerodynamik und Strömungstechnik: "TAU-Code User Guide. Revision: 1.24". Release 2004.1.1, April 5, 2004.
[6] M. Drela: XFOIL: An analysis and design system for low Reynolds number airfoils. T.J. Mueller (Ed.): "Lecture notes in engineering. — Low Reynolds number aerodynamics". Proceedings of the Conference, Notre Dame, Indiana, USA, 1989. Springer (1989), pp. 1–12.
[7] A. Grote, R. Radespiel: "Aerodynamische Untersuchungen an Leitwerken für Transportflugzeuge". Institutsbericht 2003/7, Institut für Stömungsmechanik, TU Braunschweig, 2003.
[8] A. Krynytzky: Conventional wall corrections for closed and open test sections. B. Ewald (Ed.): "Wind tunnel wall corrections." AGARD-AG-336, RTO/NATO, 1998.
[9] T. Melin: "A vortex lattice Matlab implementation for linear aerodynamic wing applications." Master Thesis, Royal Inst. of Tech., Dep. of. Aerodynamics, KTH Sweden, 2000.

Table 1 Geometric dimensions of an A319, its scaled model used as reference and the resulting wind tunnel configuration

	A319 [mm]	ref. config. [mm]	w.t. config. [mm]
b_{HTP}	12450.0	880.0	880.0
$l_{\mu,\mathrm{HTP}}$			200.0
b_{wing}	33913.2	2397.1	1300.0
l_{fuselage}	33839.4	2391.9	2050.0
r_{fuselage}	$\approx$ 1975.0	139.6	140.0

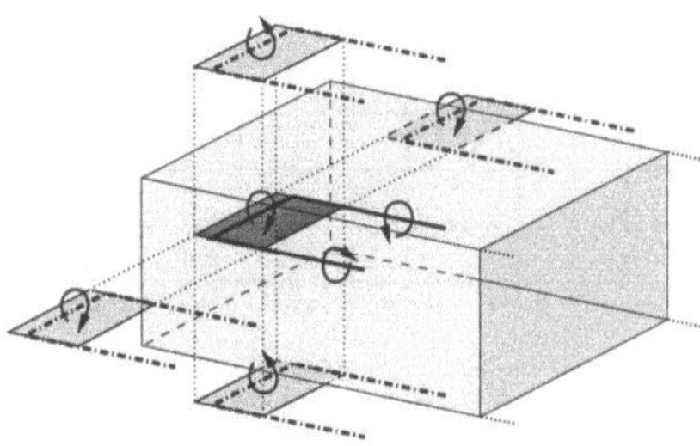

Figure 1 Method of images to account for influences of wind tunnel walls

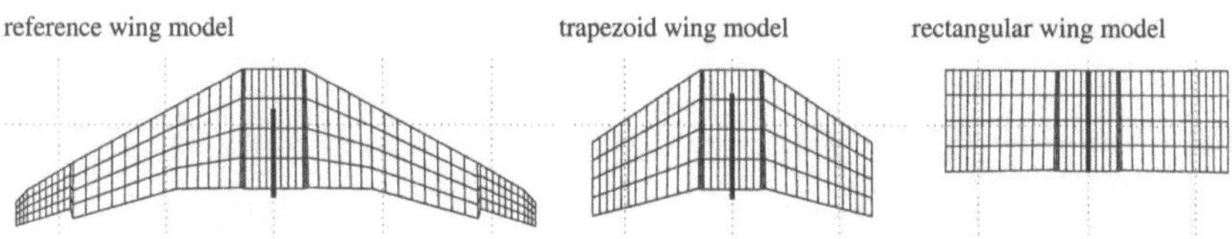

Figure 2 Planform of models used for main wing design

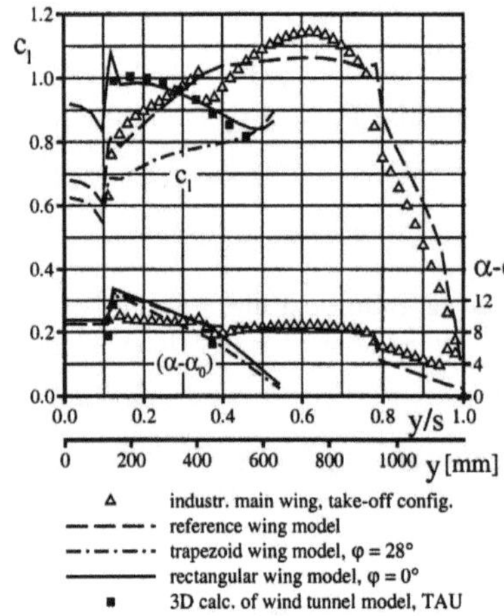

Figure 3 Lift and twist distributions of the wing models supplemented by a TAU solution for the rectangular wing inside the test section

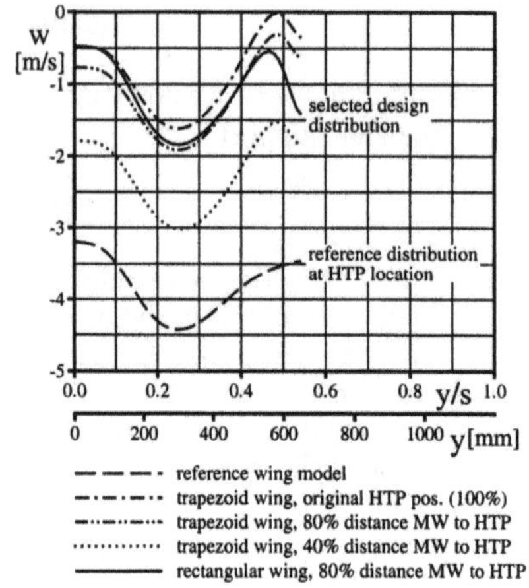

Figure 4 Downwash distribution at the location of the HTP

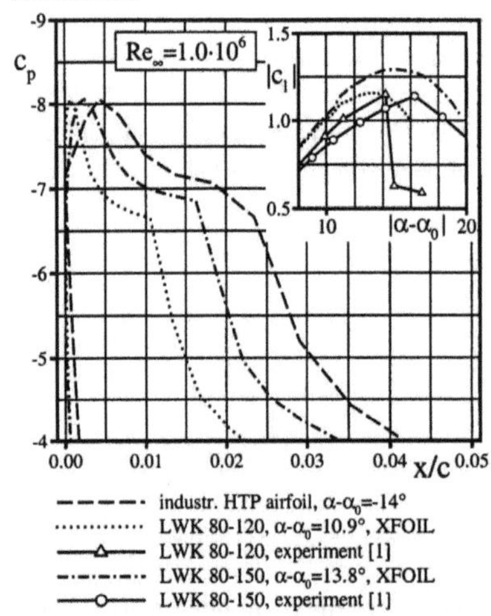

Figure 5 Classification of the industrial airfoil's stall characteristics at low Reynolds numbers

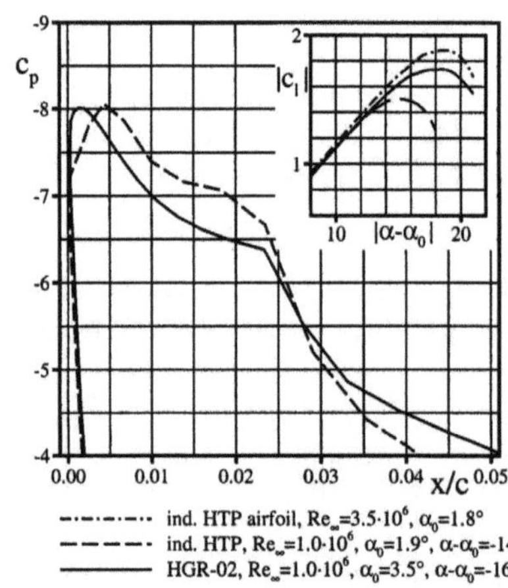

Figure 6 Leading-edge pressure distribution as design criterion for the research airfoil HGR-02

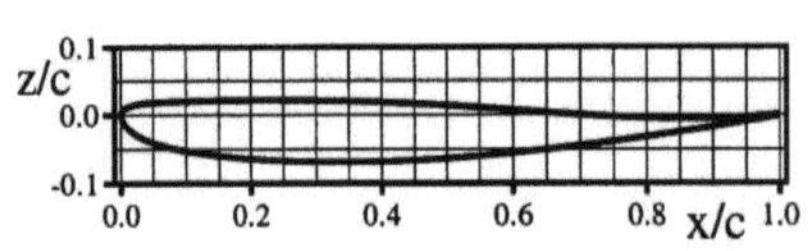

Figure 7 Research airfoil HGR-02

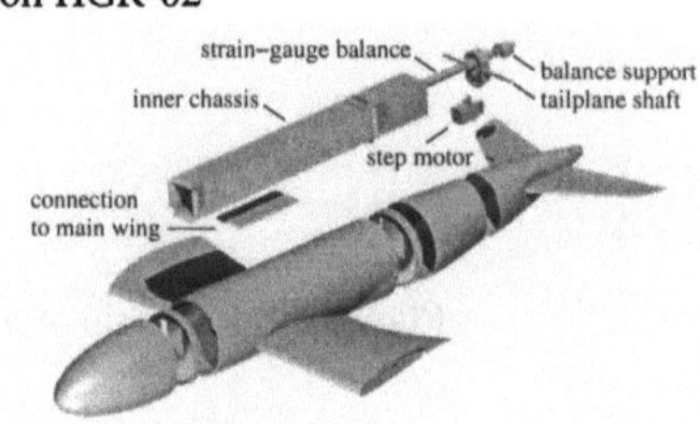

Figure 8 Structural layout of the wind tunnel configuration

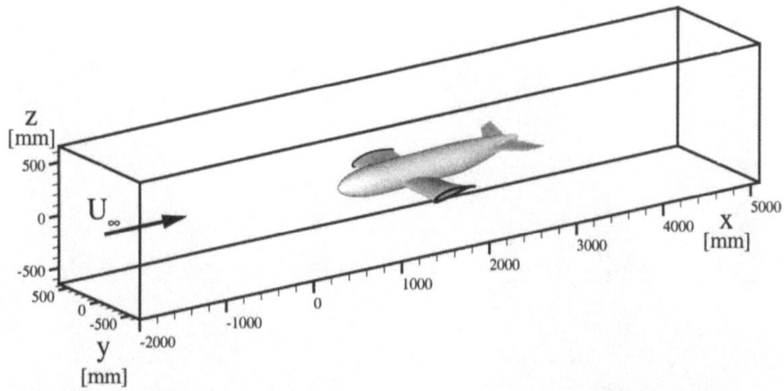

Figure 9 Test section used as outer boundary for the numerical simulation

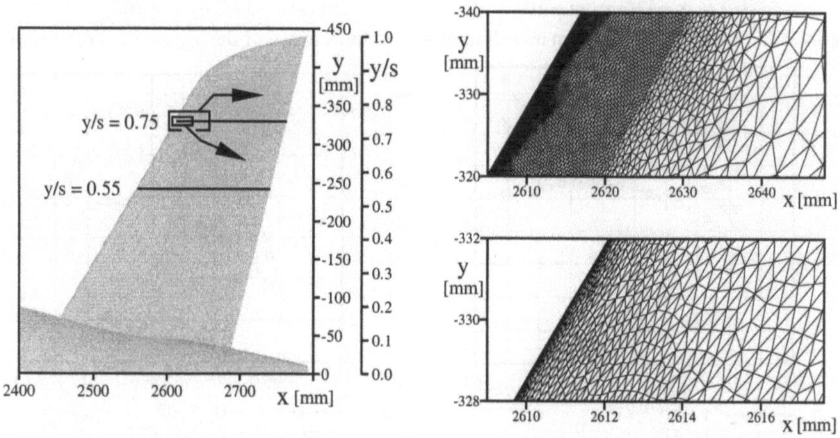

Figure 10 Spanwise position of investigated sections and surface grid resolution in two magnification grades at the lower side of the outer section

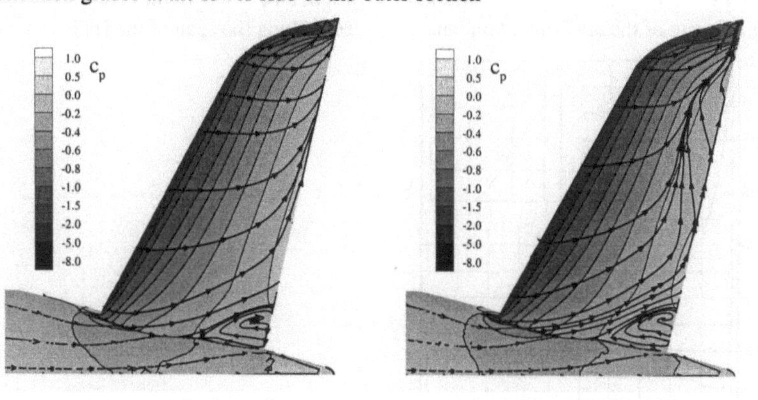

(a) $i_t = -11°, \delta_e = 0°$ (b) $i_t = -14°, \delta_e = 0°$

Figure 11 Pressure distribution and streamlines on the surface of the lower (suction) side of the horizontal tailplane

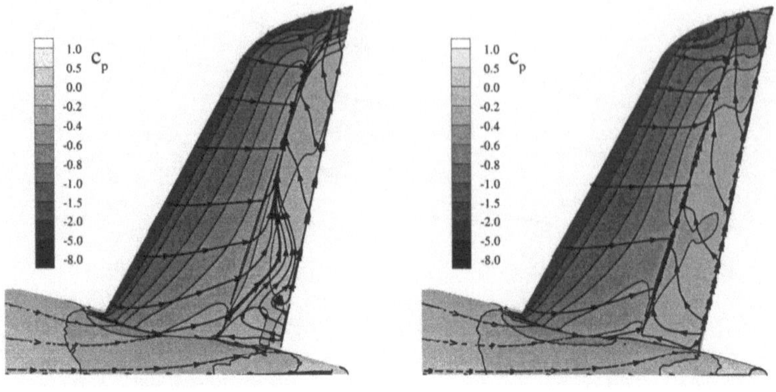

(a) $i_t = -11°, \delta_e = -10°$ (b) $i_t = -11°, \delta_e = -20°$

Figure 12 Pressure distribution and streamlines on the surface of the lower side of the HTP

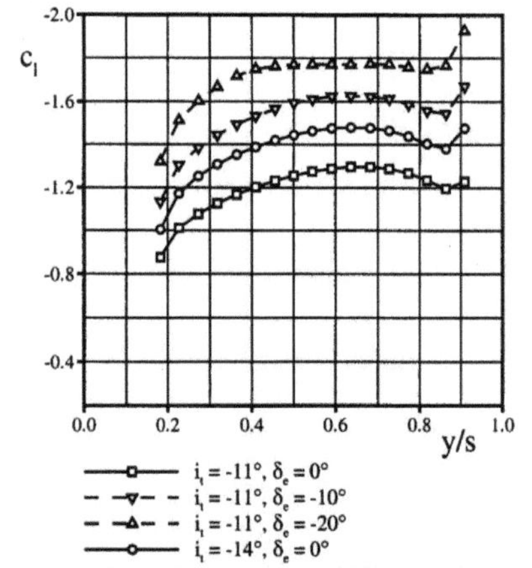

Figure 13 Distribution of lift coefficient along the span of the horizontal tailplane

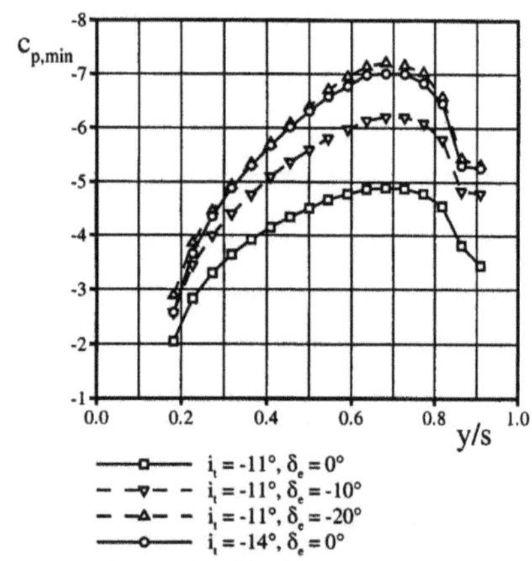

Figure 14 Distribution of minimum pressure along the span of the HTP

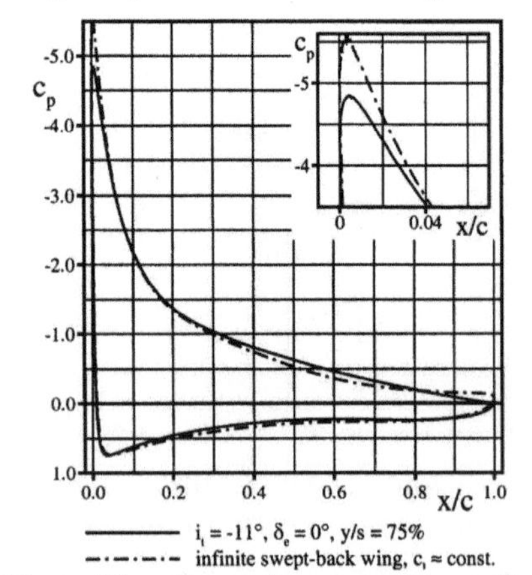

Figure 15 Comparison between results obtained for 3D and 2D simulations

Numerical analysis of transport aircraft using different wing tip devices

TH. STREIT[1], A. RONZHEIMER[1], A. BÜSCHER[2]
[1]Institute of Aerodynamics and Flow Technology, DLR Braunschweig, Lilienthalplatz 7, 38108 Braunschweig, Germany
Th.Streit@dlr.de
[2] Institute of Fluid Mechanics, Braunschweig Technical University, Bienroder Weg 3, 38106 Braunschweig, Germany

Summary

Numerical solutions of the Navier Stokes equations for a modern transport configuration with three different wing tip devices are presented. Solutions are obtained at high speed for the cruise configuration and at low speed for the corresponding take-off configuration. This study is performed within the European Union project M-DAW (Modelling and Design of Advanced Wing Tip Devices). The numerical results are analysed to characterise the performance of the wing tip devices using global force coefficients, local distributions of force and moment, surface contours and skin friction lines. In addition the vortex wake was analysed in the near field tip region for take-off. Numerical results are compared with experimental data obtained in cryogenic wind tunnel tests at high speed in the ETW (European Transonic Wind tunnel) and at low speed in the DNW-KKK (Cryogenic Wind Tunnel Cologne).

1 Introduction

Within the European Union project M-DAW (Modelling and Design of Advanced Wing Tip Devices) DLR and the Braunschweig Technical University obtained RANS CFD solutions for transport configurations with different wing tip devices, provided by Airbus. These are a cruise wing body model for high speed and a corresponding take–off configuration for low speed. Three different wing tip devices were studied within M-DAW. These wing tip devices are: a Küchemann tip (reference), a fence tip and a large winglet, characterising the current wing tip devices used on modern transport aircraft. Within M-DAW, cryogenic wind tunnel tests were performed using both models with the different wing tip devices. High speed tests were performed in the ETW (European Transonic Wind tunnel) and low speed tests were performed in the DNW-KKK (Cryogenic Wind Tunnel Cologne).
The RANS CFD solutions described in this study were obtained on unstructured hybrid meshes at various incidences and at constant lift to evaluate performance

using the DLR-TAU solver. Different adaptation mesh refinements were used to improve the resolution of flow regions of interest. One and two-equation turbulence models were used. The three wing tip devices were compared by analysing results of global force coefficients, local distributions of force and moment, surface contours and skin friction lines. In addition, in case of take-off the vortex wake was analysed in the near-field and compared with wake measurements performed using the PIV technique (Particle Image Velocimetry) [11].
The main objective of this study was to evaluate and characterise the different wing tip devices and compare the results with experimental data. Furthermore, the experimental and numerical knowledge obtained in the M-DAW project from the analysis of the three different wing tip devices will be used in a next step to design and assess an improved wing tip device. Overviews of the results obtained in M-DAW so far are given in [8], [14]. CFD analysis and design results are reported also in [1], [9] and [2].

2 Flow conditions and geometrical configurations

The test model for the cruise configuration is a 1/30th scaled half model wing-body configuration. The wing has a leading edge sweep of 35.92°, an aspect ratio of 8, a taper ratio of 0.21 (tip to root chord) and a span b =1.13167m. The corresponding 1/31.8th scale half model take-off configuration includes engines, slats, flaps and aileron. Figure 1 shows the considered transport aircraft for the take-off configuration with the three different wing tip devices. The winglet has a leading edge sweep of 33.34°, a cant angle of 40.7° and a length 0.155b. The fence is a small device with a length 0.067b. For the CFD analysis a symmetry condition was used, i.e. the numerical solution models the full model. The high speed (HS) solutions were obtained for M=0.85, Re=54.2 mill. at ($-1° \leq \alpha \leq 2°$). For cruise condition c_L is 0.49. The HS wind tunnel model is provided with a jig shape which should deform into the computed flight shape of the reference configuration at the cruise condition at α=1.5°. The low speed (LS) solutions were obtained for M=0.20, Re=7.28 mill. at the high incidence condition of the DNW-KKK PIV measurement ($\alpha \approx 11°$) and for c_L=1.4.

3 Unstructured grid generation and flow solver

The unstructured hybrid grids were generated using the Centaur™ software provided by CentaurSoft. [3]. Due to the large Reynolds number of the high speed flow condition, 36 prism layers were used in order to resolve the boundary layer at this flow condition sufficiently. The initial spacing was chosen in order to obtain a y^+ value lower than 1. The mesh generation of the different wing tip devices was performed in modular form, i.e. a region enclosing the wing tip device was defined. This allows us to consider the different wing tip devices separately while the remaining configuration is treated in the same manner. The wing tip device

surface has been resolved by approx. one third of points the wing surface has been meshed. The meshes for the take off configuration were generated using the same procedure. Initial meshes already included grid refinements in regions with expected large flow gradients (e.g. leading edges of wing components) and in regions of special interest (near tip vortex flow, wing tip device). However, in order to avoid meshes with very large number of points, stretching of cells was used for the wing in spanwise direction and the number of prism layers for the boundary layer was reduced for wing components with small length in flow direction. Mesh generation of the cruise and take-off configuration led to initial mesh sizes of 5 to 7 million points. Starting with these meshes, different adapted grids were obtained by using adaptation in the solution process. Finally, adapted meshes had mesh sizes ranging from 9.5 to 11.5 million points.
Numerical solutions were obtained with the DLR-TAU code, which solves the three dimensional compressible RANS equations on hybrid grids. The TAU-code is described in [6], [12]. In this study, the inviscid fluxes are calculated by employing a central method with scalar dissipation. The viscous fluxes are discretised using central differences. The turbulence model used for the cruise configuration is the one-equation Spalart Allmaras model (SA) [13]. For the LS take-off configuration the one-equation SA model with Edwards modification [5] as well as the two-equation Menter SST [10] model were used. In addition, for regions of vortical flow, the Kok TNT rotation correction was employed [4]. The adaptation module allows local refinement of the hybrid grid based on sensors derived from the flow solution. In the HS study, a refinement of the region containing the shock was performed using the pressure as a sensor. For the LS study the velocity, pressure and total pressure loss were used as sensors.

4 Results

Checking the grid resolution of the near wall prism layer, y^+ values in the order of 1 were found, ensuring that the prismatic layer in the near wall boundary layer has a sufficient resolution of the velocity gradient. For all solutions, time convergence was checked using the density residual. Furthermore, it was checked that force coefficients converged up to their third significant digit. Initial meshes were refined using several adaptation steps.

The obtained results for the pressure distributions at LS, HS and its comparison to the wind tunnel data are not shown here. Instead the reader is addressed to ref. [1], which reports similar results for the cases presented here.

4. 1 High speed results

In order to compare the performance and to describe the flow around the different tip devices, solutions were obtained at constant lift and at constant incidences.
Figure 2 shows the relevant total force and moment coefficients of the wing tip devices. Results are given for the percentual change of drag and wing root bending moment (WRBM) referred to the Küchemann (Kü-tip) case. For comparison with

ETW results, the half model rolling moment was chosen, instead of the (WRBM). For lower lift values (c_L < 0.3 RANS and c_L < 0.2 ETW), the winglet case generates more drag than the reference configuration. This results from additional skin friction produced by the winglet. At higher lift coefficient the large winglet reduces the induced drag in comparison to the Kü-tip, leading to a net drag reduction for c_L > 0.3. To a smaller extent the fence tip device also reduces the drag in comparison to the basic configuration. The opposite is seen for the WRBM. In comparison to the basic configuration, the fence shows a small increase, the winglet shows a larger increase which increases with lift. Compared to the ETW data the percentual change of drag and rolling moment coefficient shows differences for the winglet case. For the drag change this difference is small at the cruise condition, where the deformed wind tunnel model should reach the computed flight shape of the reference configuration. The remaining difference at cruise condition may be attributed to the different deformation of the wing with the Kü-tip and the one with the winglet. At c_L > 0.2 the aeroelastic deformation reduces the drag decrease and the WRBM for the large winglet configuration because the loads decrease at outboard sections. The agreement between RANS and ETW results for the fence tip indicates that the fence performance is not so sensitive on deformations. In the experiment wall interference corrections have been applied, but they are not important for the relative comparison of the wing tip devices. At c_L =0.519 the relative percentual changes for drag, rolling moment and skin friction are given in Table 1. The percentual increase in skin friction is small due to the large Reynolds number.

Skin friction lines at the inner and outer fence surface are shown for α =1.5° in Figure 3. These results show flow separations at the upper inside and at the lower outside of the fence. They indicate that the fence is working like a delta wing. The flow around the wing tip produces nonlinear vortical flow at the inner upper and lower outer surfaces of the fence to increase their thrust components. In the winglet case the flow is attached

Local coefficient distributions are analysed at constant lift condition in Figure 4 as function of the spanwise position $\eta = y/s_{ref}$ (s_{ref} is the wing semi-span of the reference configuration). In comparison to the Kü-tip the local lift force distribution shows that the fence leads to a sharp peak at the tip, while the winglet smoothes the sharp drop of circulation at the reference wing tip, by extending the contribution of lift force to larger span. At the same lift, the winglet case has slightly smaller local lift forces than the two other cases due to its increased wing surface. The reduced local lift leads to smaller shocks and decreases the wave drag. The comparison of local drag contributions shows a region of thrust in the outer wing, for approx. η > 0.70 . This is typical for a swept back wing with circulation distributions as shown in this case [7]. The different wing tip cases show only different local force contributions in the tip region. At the tip the fence leads to two sharp thrust peaks. The first peak (which is larger and placed further inward) corresponds to the upper fence tip; the second peak corresponds to the lower fence tip. The large winglet leads to an increased thrust region extending to the entire additional winglet span. In order to see the influence on WRBM, Figure

4 shows the local distribution of rolling moment for the three tip/wing tip device cases. Note that the fence shows nearly the same distribution as the Kü-tip reference case, with the exception of a small increase in rolling moment for $\eta >$ 0.95 ending with a sharp peak at the fence position. The winglet shows an increase in rolling moment for a span region extending from approx. $\eta = 0.90$ to the entire additional winglet span. However, due to the reduced local lift forces at the inner wing there is also a reduction of rolling moment at the inner sections.

4. 2 Low speed results

In the low speed case the initial mesh was adapted in four steps. The first step was a global adaptation, using the pressure as sensor and increasing the total number of points by 25%. In the following, three successive local adaptations were performed in a box which includes the wing tip devices and the tip vortex wake up to the plane in which PIV measurements were performed. For theses adaptations the total pressure was used as a sensor. In the final mesh the box included 3.6 million points.

Solutions were obtained at the high incidence PIV-measurement flow (α =10.99°/ 10.81°/ 10.98° for Kü-tip / fence / winglet). Solutions are obtained using the SA turbulence model with Edwards's modification. The results for the solution using the 2-equation Menter SST model are similar.

In the following the flow at the wing tip devices is described by showing surface skin friction lines and local force distributions. For the high incidence considered here, the flow separates at the outer wing for sections positioned outboard of the outer slat (see Figure 5). For the winglet the separated region extends to the upper winglet surface. The aileron also shows separated flow. As in the HS case, the fence indicates regions of separated flow. The solution at constant c_L =1.4 shows attached flow in these regions. The local force distributions for lift and drag at the outer wing and tip device region are given in Figure 6. Similar to the HS case, a total thrust force is observed for the outer wing, but the separated tip and winglet region shows drag at high incidence. For lower incidence, with c_L =1.4, thrust is obtained for the tip and winglet. The fence is producing a sharp thrust peak for both high incidence and for c_L =1.4. This peak is of higher level at for c_L =1.4.

The analysis of the RANS solution velocity vector field at high incidence shows following vortical structure: for the Kü- tip a strong tip vortex, for the fence tip an upper inward vortex associated with the upper fence tip and a lower outward vortex associated with the lower fence tip, for the winglet a small vortex localised at the winglet tip and a more extended vortex generated at the junction between the winglet and the wing vortex. These vortices are also obtained in the PIV measurement. This is shown in Figure 7, which compares RANS and DNW-KKK PIV results for the large winglet and Kü-tip cases. Shown are contour plots of the transversal velocity for the PIV-section and transversal velocity vectors. The contour pattern and vortex structure are similar. However, there is a shift in position.

5 Conclusions

RANS solutions have been obtained for a modern transport aircraft with three different tip devices: Küchemann tip (reference), fence tip and large winglet. Solutions were obtained at high speed for the cruise configuration and at low speed for the high lift take-off configuration. These CFD studies were performed within the European Union project M-DAW.
The numerical solutions describe, characterise and evaluate the three different wing tip devices. At cruise condition, comparison to the reference shows that at constant lift the large winglet reduces the induced drag and also the wave drag. The latter, results from the fact that local lift is reduced due to the additional lift surface. The viscous drag increase is small due to the large Reynolds number. The smaller fence device also reduces the drag, but at a lower level. Local force distributions show that the winglet extents the thrust region observed at the outer wing, while the fence produces two sharp thrust peaks. The local force and moment distributions indicate that the changes introduced by the fence are restricted to a small region of the tip, while the winglet influences a larger area. Separated regions are observed at the upper inside and at the lower outside of the fence. These separations are associated with upper and lower fence vortices which may increase nonlinearly the observed thrust peaks. While the large winglet decreases the drag to a larger extent as the fence, the opposite is observed for the wing root bending moment WRBM. Furthermore, the results indicate that the increase in WRBM for the fence is not so sensitive to incidence. Differences with the experimental ETW data arise from the fact that the numerical simulation considers a rigid 1G shape, full aircraft configuration, whereas a half model is used in the experiment which also deforms aeroelastically. CFD studies with deformations will be considered in the second phase of the M-DAW project.
The local force analysis at constant $c_L = 1.4$ shows characteristics for the wing tip devices that are similar to the results explained above in the high speed case. At the large incidence of $\alpha \approx 11°$, the tip region and the winglet show separated flow at the tip and winglet. Instead of thrust, local drag is obtained for this region. The comparison of the numerical velocity vector field with data obtained from PIV-measurements for the high incidence shows the same vortical structure and similar velocity contour plots for the wing tip devices. However, position and scale of vortices show some differences.
The present analysis shows that the RANS solutions obtained using the DLR-TAU code, may be used to characterise the different wing tip devices and to evaluate them with sufficient accuracy. The experience gained through this work will be used in the M-DAW project to design and assess novel wing tip devices.

Acknowledgement

The M-DAW project is partly funded by the European Union and coordinated by

Airbus UK. The authors would like to thank Lars Lekemark for performing the CAD-cleaning work of the geometry models required for the grid generation

References

[1] S. Barakat, E. Elsholz. "CFD Analysis on wing tip devices". DGLR Jahrbuch, Bd. III, pp.1885-1894, 2004, accepted for publication.

[2] A. Büscher, R. Radespiel, Th. Streit. "Two-Point-Design of Nonplanar Lifting Configurations Using a Databased Aerodynamic Prediction Tool". DGLR Jahrbuch, Bd. I, pp. 319-327, 2004.

[3] CentaurSoft™ http://www.centaursoft.de

[4] H.S. Dol, J.C. Kok, B. Oskam. "Turbulence Modelling for Leading-Edge Vortex Flows". AIAA Paper 2002-0843, 2002.

[5] J.R. Edwards, S. Chandra. "Comparison of Eddy Viscosity-Transport Turbulence Models for Three-Dimensional, Shock-Separated Flows", AIAA Journal, Vol. 34, No. 4, pp.756-763, 1996.

[6] T. Gerhold, O. Friedrich, J. Evans, M. Galle. "Calculation of Complex Three-Dimensional Configurations Employing the DLR-τ-Code". AIAA Paper 97-0167, 1997.

[7] D. Küchemann. "The Aerodynamic Design of Aircraft". Pergamon Press, 1978.

[8] A. Mann. "Novel Wingtip Devices – an Overview of the M-DAW Project". KATnet/Garteuer Workshop on High Speed Aerodynamics, 1-4, Univ. of Bath, UK, Sept. 2003.

[9] H. Maseland. "CFD Analysis of the Aerodynamic Characteristics of the M-DAW Datum Wing Tip Devices". KATnet/Garteuer Workshop on High Speed Aerodynamics, 1-4, Univ. of Bath, UK, Sept. 2003.

[10] F.R. Menter, "Two-Equations Eddy-Viscosity Turbulence Models for Engineering Applications", AIAA Journal Vol. 32, No. 8, pp. 1598-1605, 1994

[11] H. Richard, W. Becker, S. Loose, M. Thimm, J. Bosbach, M. Raffel. "Application of Particle Image Velocimetry under Cryogenic Conditions". 20th International Congress on Instrumentation in Aerospace Simulation Facilities, ICIASF 2003, August 2003.

[12] D. Schwamborn, T. Gerhold, T. Hannemann, "On the validation of the DLR TAU Code". Notes on Numerical Fluid Mechanics, Vol. 72, pp. 426-433, Vieweg Verlag, 1999.

[13] P. Spalart, S. Allmaras. "A One-equation Turbulence Model for Aerodynamic Flows". AIAA Paper 92-0439, 1992.

[14] D. Williams, A. Mann, T. Grundy. Large Wingtip Devices – "A Review of Activities and Progress in the NEXUS, M-DAW & AWIATOR Projects" 4th European Congress on Computational Methods in Applied Sciences and Engineering (ECCOMAS), Jyvaskyla, Finland, 24-28 July 2004

Table 1. Drag, skin friction and rolling moment coefficient change relative to reference configuration at M=0.85, Re=54.2 10^6 and c_L=0.519

	RANS (%)		ETW (%)	
	fence	winglet	fence	winglet
$\Delta C_D/C_{D\ ref}$	-1.36	-4.47	-1.93	-3.12
$\Delta C_f/C_{D\ ref}$	0.13	0.80	---	---
$\Delta C_{mx}/C_{mx\ ref}$	0.53	3.38	0.57	1.79

Figures

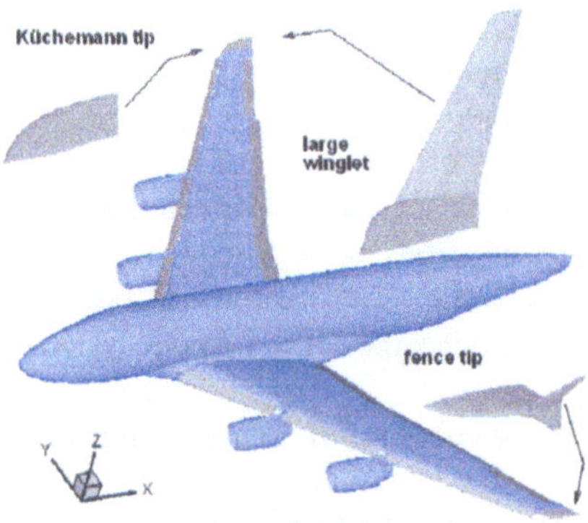

Figure 1 Analysed aircraft in take-off configuration with studied wing tip devices.

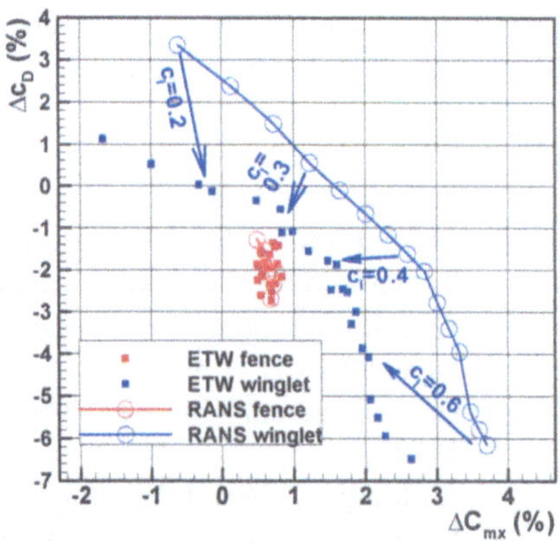

Figure 2 Percentual change of drag versus rolling moment. The percentual differences are based on the Kü-tip and are taken at constant lift.

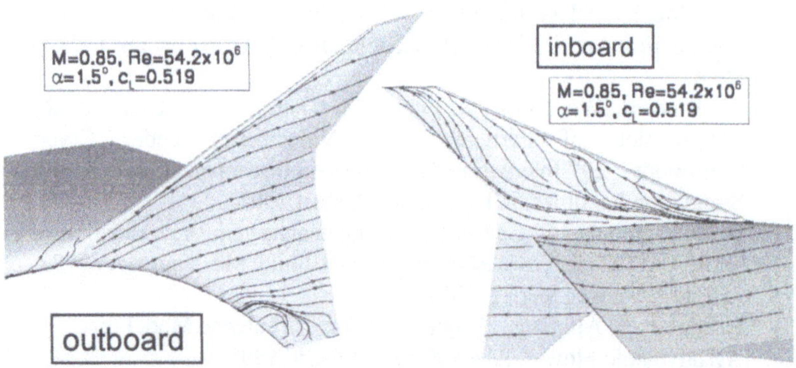

Figure 3 Streamlines for fence tip at cruise condition showing separated regions resulting in upper inboard (upper fence) and lower outboard (lower fence) vortices.

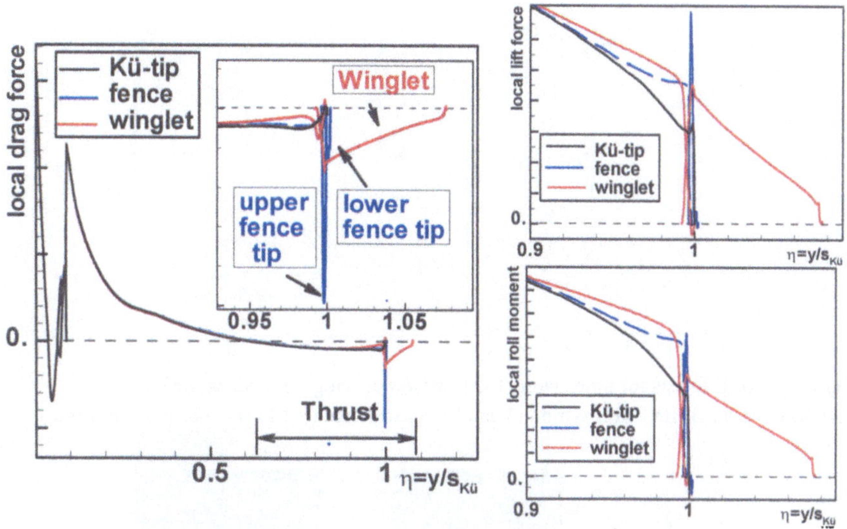

Figure 4 Local coefficient distributions at M=0.85, Re=54.2 10^6 and c_L=0.519.

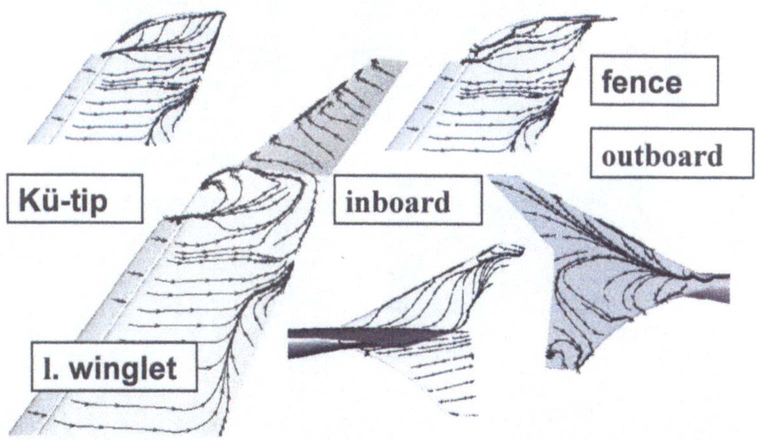

Figure 5 Streamlines for outer wing and tip device region for PIV measurement condition (M = 0.2, Re=7.28 10^6, $\alpha \approx 11°$). The aileron region and region outboard of outer slat show flow separation.

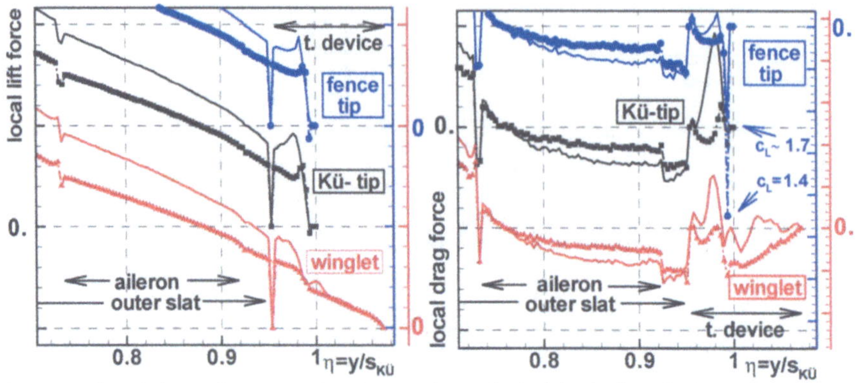

Figure 6 Local forces for outer wing. Comparison of wing tip devices for low speed conditions c_L=1.4 (line with symbols) and high incidence ($\alpha \approx 11°$) condition, (thin line).

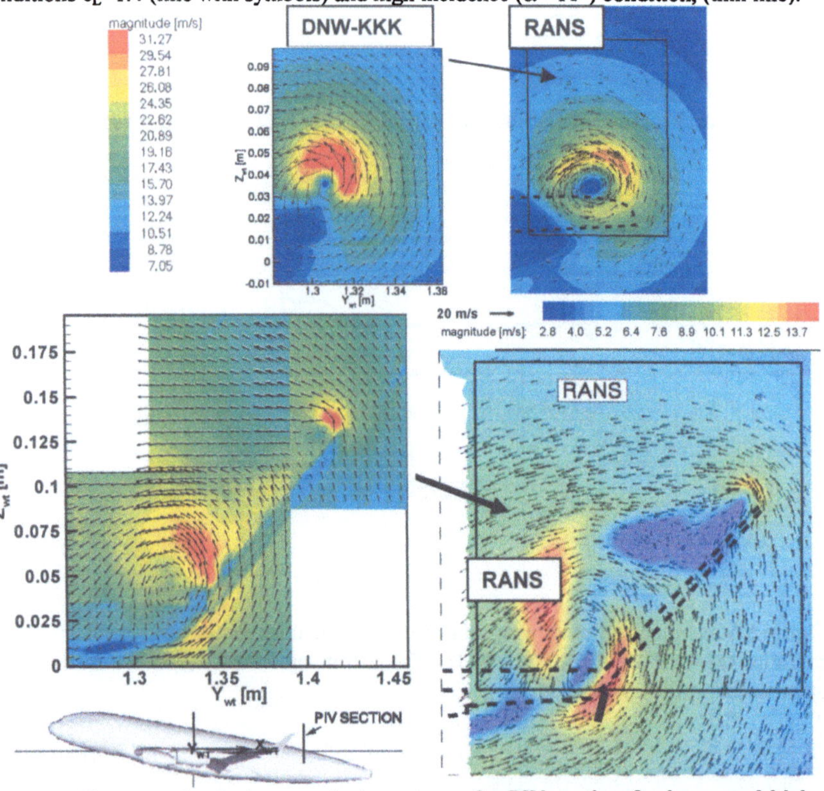

Figure 7 Transversal velocity magnitude contours for PIV section for low speed high incidence ($\alpha \approx 11°$) flow condition. Comparison of DNW-KKK average transversal velocity (left) with RANS result (right) for winglet and Kü-tip case (winglet and Kü-tip projection indicated with dotted line). The DNW-KKK measurement box is indicated with a solid line in the RANS result. The Kü-tip shows the strongest vortex. A small vortex is obtained at the winglet tip and a large vortex at the junction wing/winglet.

Aerodynamic Optimisation of a Flying Wing Transport Aircraft

H. Strüber, M. Hepperle

German Aerospace Center DLR, Institute of Aerodynamics and Flow Technology,
Lilienthalplatz 7, 38108 Braunschweig, Germany, `henning.strueber@dlr.de`

Summary

An approach to optimise the aerodynamic shape of a flying wing transport aircraft is presented. It is split into 2D airfoil design and 3D twist and chord length optimisation. The applied finite volume flow solver is the DLR code FLOWer. The methods are described and results for a flying wing configuration are presented. The paper represents parts of DLR's contribution to the aerodynamics work package of the VELA project, supported by European Commission.

1 Introduction

In the following decades the predicted growth of air traffic requires new ideas and technological solutions with respect to efficiency and environmental impact for the future generation of aircraft. One promising design is the Flying Wing or Blended Wing Body. The wetted surface area of a flying wing can be less than that of a compareable conventional aircraft which ends up in lower drag. With these benefits a significant reduction in fuel consumption could be achieved. A 750 passenger flying wing aircraft is investigated in several disciplines of aircraft design in the European integrated research poject VELA (**V**ery **E**fficient **L**arge **A**ircraft). With AIRBUS as the leading company, 17 European partners from industry and research are working together to extend knowledge about this type of aircraft.

2 Aerodynamic Shape Optimisation

The main task in VELA's aerodynamics work package is to optimise the aerodynamic shape of flying wing configurations. This is done without respect to other disciplines like e.g. structures in order to find the best configuration from the aerodynamic point of view. Each configuration is defined by two leading edge sweep angles (inboard and outboard wing) and a corresponding cabin geometry. The combination of different inner and outer wing leading edge sweep angles defines the design space investigated in VELA.

While the leading edge sweep angles, span and the cabin geometry are prescribed, it is up to the participants to chose and optimise the shape of the airfoils, twist distribution and the planform. Thereby the following constraints have to be considered: the cabin box has to fit into the geometry, a maximum angle of attack should not be exceeded to keep the cabin floor level and a certain stability margin (SM) has to be achieved. The optimisation goal is to maximize L/D for a given lift in transonic cruise flight condition (Ma = 0.85).

The shape optimisation of the flying wing presented in this paper is separated into two steps: 2D airfoil optimisation and 3D twist and chord length optimisation. The different steps are executed separatly and results are fed by the user into the follow on step.

2.1 Flow solver FLOWer

For the 2D and 3D optimisation, the DLR code FLOWer [2] for block structured grids is used for the flow solution. In the present case the FLOWer code solves the three dimensional compressible Euler equations in integral form. The spatial discretization can utilize a central cell-vertex, a cell-centered or an AUSM finite volume formulation. Dissipative terms are explicitly added in order to damp high frequency oscillations and to allow sufficiently sharp resolution of shock waves in the flow field. The time integration is carried out by an explicit hybrid multistage Runge-Kutta scheme. For steady state calculations the integration is accelerated by the techniques of local time stepping, enthalpie damping for inviscid flows and implicit residual smoothing. The solution procedure is embedded into a sophisticated multigrid algorithm which allows standard single grid computations as well as successive grid refinement, with the option of simple or full multigrid, [3]. Although FLOWer is a three-dimensional code it can also be run in a two-dimensional mode. Additionally, quasi-3D modes for infinite swept wings and bodies of revolution are available.

2.2 2D Airfoil Optimisation

The flying wing configuration can be separated in a inner wing housing the payload box and the outer wing, see figure 1. For both regions airfoils with different requirements are needed: thick airfoils for the center body providing enough space for the cabin, operating at low local lift coefficients and airfoils for the outer wing producing high lift coefficients. These airfoils are designed in a first step in a 2D-approach. A multi point design is performed to obtain airfoils which provide low drag for a certain range of lift coefficients which might occur during the following three-dimensional optimisation.

The airfoil geometry is defined by a Bezier polygon consisting of 14 control points. Their z-position is modified by an optimiser to change the airfoil's shape. A polygonal shape is placed inside the airfoil geometry representing e.g. the cross section of a payload box or a spar, figure 2. A geometry check is performed to secure that this shape does not intersect the airfoil geometry and to determine possible overlapping of lower and upper airfoil surfaces. Any overlapping is penalized accordingly.

After building a structured O-grid with 145 points around the airfoil and 97 points in normal direction, the aerodynamic analysis is performed. Swept wing effects are considered by using the FLOWer mode for the infinitely swept wing ($2\frac{1}{2}$D). Due to the taper of the wing, the sweep angle was selected so that the $2\frac{1}{2}$D results match the 3D results closely. A future extension to a bowed airfoil representation is planned. FLOWer is used in Euler mode to solve frictionless, compressible flow fields. To determine friction drag and flow separation, a 2D integral boundary layer method according to Eppler [4] is applied afterwards. Transition was fixed at 5% chord length.

Each geometry generated by the optimiser is analysed at three different lift coefficients (multi point design). The objective function for the optimiser is the sum of the drag coefficients in the design points plus penalties from violated constraints (flow separation, geometric overlapping).

Using four multigrid levels, about 400 time steps are executed for each airfoil shape. Total solution time is about three minutes on a 2.4 MHz PC. Several hundred optimisation steps must be performed to reach a satisfying airfoil shape.

2.3 3D Twist and Chord length optimisation

The application flow consists of parametric grid generation, aerodynamic analysis and a routine to assess the data, figure 3. The optimiser controlling the design variables is the free available Java based *GenOpt*-software [5]. For this work, a Simplex (Nelder-Mead-ONeil) algorithm is applied. Eight design variables are used in the 3D optimisation, see figure 1: for the three outer wing sections, changes of twist and chord length (with constant t/c) are allowed. For the two inner wing sections, only the chord length is a design variable. Thickness and twist are kept constant to provide enough space for the cabin box. The two sections in the fuselage-wing-transition region are coupled by linear laws to the surounding sections.

Grid generation The previously optimised airfoils are used to generate a flying wing geometry. MegaCads, the DLR mesh generator for parametric structured grids, is used to set up the geometry and to generate a finite volume grid [1]. The flying wing consists of seven sections connected by linear surface interpolation. For each section the parameters twist, thickness and chord length can be changed. For each segment between two sections leading edge sweep angle and dihedral angle are free parameters. In batch mode, MegaCads reads the parametric grid generation history from a script file. The user or an optimiser can change the parameters in this file and a structured finite volume grid for inviscid aerodynamic analysis of the new configuration is generated automatically. The grid used for the 3D optimisation is a three level multigrid C-type grid with 892000 points. Grid refinement studies have shown, that a grid at this size delivers sufficiently accurate results. The benefit in accuracy of refined grids does not pay off due to increased CPU-time for the flow solver.

Aerodynamic analysis The structured DLR code FLOWer is used in Euler mode to generate a flow solution. 300 time steps are performed on a three-level multigrid-W-cycle. This takes about 40 minutes on 2.4 GHz PC. A target lift function is applied to analyse the configuration at constant lift. To accelerate convergence the restart capability of FLOWer is used.

Objective function The objective function (OF) which is minimized by the optimiser consists of the following terms:

$$OF = -L/D + \alpha_{penalty} + SM_{penalty} \qquad (1)$$

To derive the objective function a module is used to analyse the aerodynamic data, to start a second FLOWer run with an increased angle of attack to calculate the stability margin and to check the two applied constraints (maximum angle of attack and stability margin).

The drag term in L/D is the sum of the surface pressure drag C_{D_p} delivered by the flow solver and friction drag C_{D_f} which is estimated in the module. Due to the fact that the chord length may change during an optimisation, the wetted surface changes also and it is essential that friction drag is taken into account. A viscious CFD-analysis is not practical, due to the demands of CPU-time. Therefore the local friction drag of a flat plate in turbulent flow, based on the local chord length is approximated using

$$C_{D_f} = 2 \cdot \frac{k_1}{S} \cdot \int_{-\frac{b}{2}}^{\frac{b}{2}} \frac{0.074}{\sqrt[5]{Re(y)}} \cdot c(y)dy. \qquad (2)$$

The factor k_1 is set to 1.25. Of course this a relativly coarse method, but satisfying for the optimisation process where the focus is on the drag increment. Currently, this approach is replaced by a stripwise boundary layer analysis, using an integral method [4].

The stability constraint which must be fullfilled is, that the distance between center of pressure and aerodynamic center should not be larger than ±5% of the mean aerodynamic chord length.

$$SM = \frac{x_{AC} - x_{CP}}{l_\mu} \leq \pm 0.05 \qquad (3)$$

The position of the aerodynamic center is derived by the formula:

$$x_{AC} = x_{ref} - \frac{dC_M}{dC_L} * l_\mu \qquad (4)$$

To estimate the gradient $\frac{dC_M}{dC_L}$ the second run of the flow solver is required, with a slightly increased angle of attack ($\Delta\alpha = 0.25°$). The position of the center of pressure is defined by:

$$x_{CP} = x_{ref} - \frac{C_M}{C_L} * l_\mu \qquad (5)$$

If the maximum angle of attack ($\alpha_{max} = 3°$) or the stability margin is exceeded, the penalty terms are computed with a quadratic function which is for instance for the α-penalty:

$$\alpha_{Penalty} = k_2 * (\alpha - \alpha_{max})^2 \qquad (6)$$

The factor k_2 is set to values between 1.0 and 4.0, depending on the sensitivities of the variables with respect to the objective function.

3 Results

3.1 2D Airfoil optimisation

As a typical result of a multipoint optimisation of an outer wing airfoil figure 2 presents the initial and optimised shape of the airfoil MH 1004. The airfoil shape is changed but the spar box still fits into the optimised section. The aerodynamic performance was improved significantly, as can be seen in figure 4. The maximum L/D is shifted into the region of the design points and a significant reduction in wave drag was achieved. This leads to an airfoil which is well suited for the 3D-optimisation.

3.2 3D twist and chord length optimisation

The optimised configuration presented here is defined by the leading edge sweep angles $\varphi_{inboard} = 55°$, $\varphi_{outboard} = 35°$, a given span and the cabin geometry. An airfoil, which was optimised for the fuselage region is inserted at the wing root at station 1, see figure 1. The outer wing stations six and seven use the MH 1004 airfoil. The remaining sections were taken from a flying wing configuration designed by Airbus. The initial planform from where the 3D optimisation was started is also similar to this aircraft.

The optimisation converges to a maximum in L/D after 220 steps. After 41 simulations the constraint on maximum angle of attack is violated. Therefore, the optimiser changes the twist distribution on the outer wing during the next optimisations. This stops the initial steep increase in L/D, which is continued with a reduced slope after some steps of reorientation . This shows, that the violation of a constraint is applied correctly to the objective function and that it is recognized by the optimiser, which succesfully applies counter measures.

The reduction in drag (remind that lift is constant) is about $\Delta C_{D_{total}} = -16\%$, and the improvement is dominated by the reduction of wave drag ($\Delta C_{D_{wave}} = -63\%$). This was achieved by increasing the chord lengths on the outer wing and by adjusting the corresponding twist angles. To reduce the total drag, the optimiser balances induced, wave and friction drag against each other. The shape of the circulation distribution approximates an elliptic shape which indicates, that wave drag is relatively small. Due to the increased wing area, friction drag was slightly increased ($\Delta C_{D_{friction}} = +4.5\%$).

In figure 5 iso Mach lines are shown on the initial and the optimised planform. The distribution of shocks on the outer wing has developed from a single trailing edge shock to a weaker double shock system, see also figure 6. On the inner wing the change of shock intensity and position is not that significant and caused by the increased angle of attack, the changed chord length and the influence of the outer wing shock to the inner wing.

Figure 6 compares chordwise Mach number distributions at six spanwise stations of the optimised configuration with the initial configuration. The dominating changes take place on the outer wing ($\eta \geq 0.5$). The transformation from a strong single shock close to the trailling edge to a weaker double shock system can be observed. This conributes most to the reduction in wave drag. On the lower side of the outer wing a small suction peak has developed at the leading edge due to changes in twist. This resulting shock does not contribute significantly to the wave drag.

4 Conclusion

An approach to optimise the aerodynamic shape of flying wing transport aircrafts is presented. The procedure splits in 2D airfoil design and 3D twist and chord length optimisation, applying the DLR flow solver FLOWer in Euler mode. It has been shown how constraints like e.g. a cabin box or a stability margin are included in the methods and that the optimiser is able to apply counter measures if the constraints are violated. The results show, that significant improvements in aircraft peformance are achieved by applying the 2D and 3D optimisation to a flying wing configuration. For future work it would be desirable to combine airfoil optimisation and 3D twist and chord length optimisation in a single loop to better account for 3D features of the flow. Improvements in friction drag estimation are currently developed.

References

[1] Brodersen, O., Hepperle, H., Ronzheimer, A., Rossow, C.-C. and Schöning, B.: "*The Parametric Grid Generation System Mega Cads*", in "5th International Conference on Numerical Grid Generation in Computational Field Simulation, National Science Foundation", (1996)

[2] Kroll, N., Radespiel, R., Rossow, C.-C.: "*Structured Grid Solvers / Accurate and Efficient Flow Solvers for 3D Applications on structured meshes*". AGARD Report 807, Special Course on Parallel Computing in CFD (1995)

[3] DLR (ED.): "*FLOWer Installation and User Handbook, Release 115*". Institut für Entwurfsaerodynamik, Braunschweig, (February 1997)

[4] Eppler, R.: "*Airfoil Design and Data*". Springer Verlag (1990)

[5] Wetter, M.: "*GenOpt, Generic Optimization Program*". Lawrence Berkeley National Laboratory (2004), http://SimulationResearch.lbl.gov

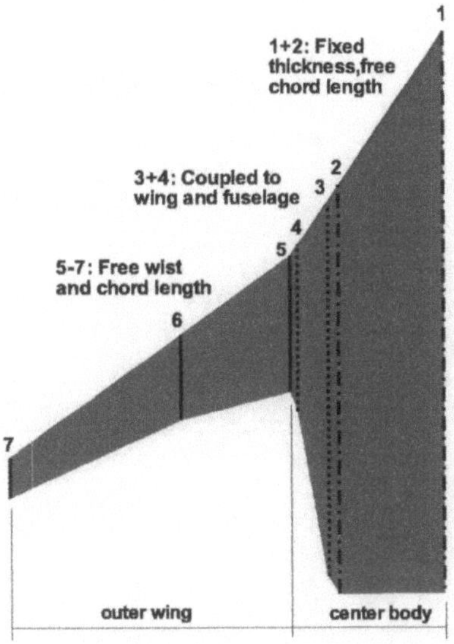

Figure 1 Position of the sections describing the flying wing geometry.

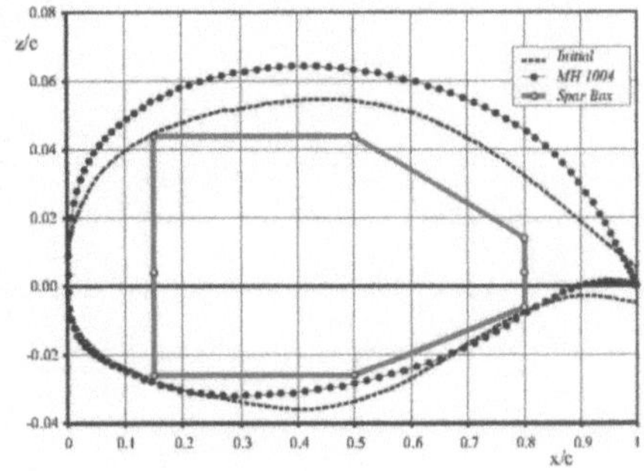

Figure 2 Initial and optimised section with spar box of the outer wing airfoil MH 1004

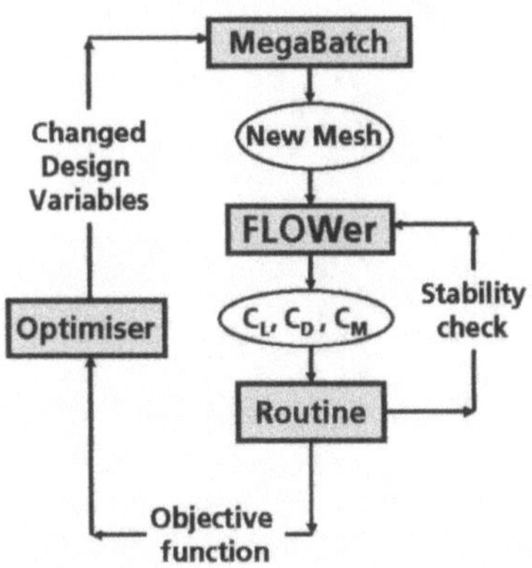

Figure 3 Flow chart of the 3D optimisation.

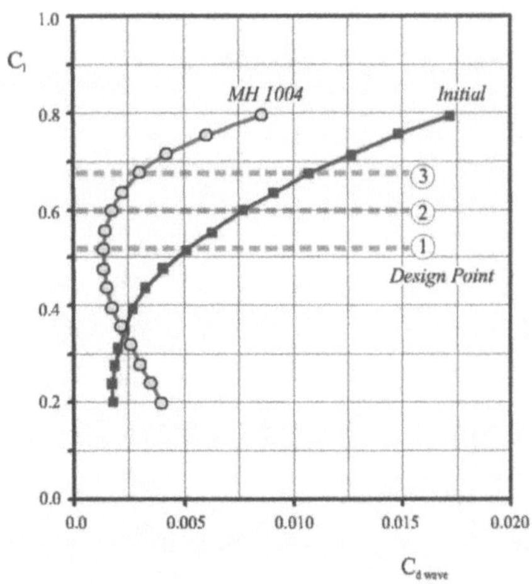

Figure 4 Polar of initial and optimised airfoil MH 1004

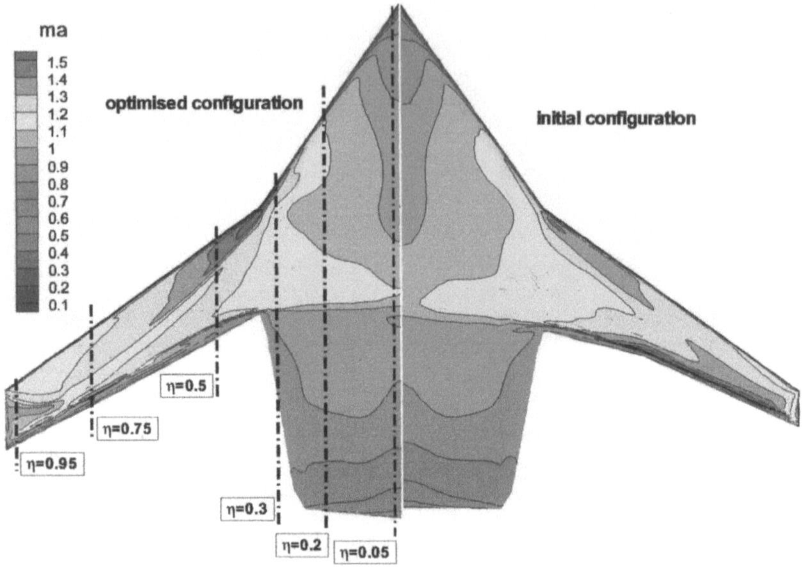

Figure 5 Optimised and initial configuration of the 3D twist and chord length optimisation at Ma=0.85 and constant C_L

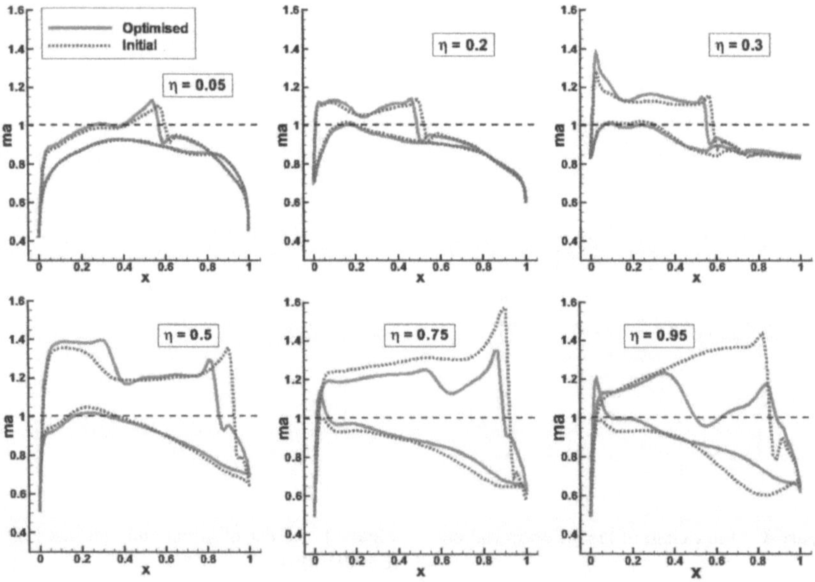

Figure 6 Mach distributions at six spanwise stations at Ma = 0.85.

Delta Wing Steady Pressure Investigations for Sharp and Rounded Leading Edges

Andrej Furman, Christian Breitsamter

Technische Universität München, Lehrstuhl für Fluidmechanik, Abteilung Aerodynamik
Boltzmannstrasse 15, 85748 Garching, Germany
andrej@flm.mw.tum.de

Summary

The design, construction, manufacturing and geometric properties of a generic 65° Delta Wing for the International Vortex Flow Experiment 2 are described. Briefly, the measurement equipment implemented in the model and the external sensors are specified. A wind tunnel investigation has been performed and the results for two different leading edge geometries (sharp and rounded) at three angles of attack are discussed in detail. Further, the results are compared with the ones obtained by NASA and also with computational results. Partly developed ($\alpha = 13^o$) and fully developed ($\alpha = 18^o$) leading edge vortices as well as vortex breakdown ($\alpha = 23^o$) are considered.

1 Introduction

The International Vortex Flow Experiment 1 (VFE–1) led to a broad data base in order to validate inviscid Euler solvers. For delta wings viscous effects are very important as they lead to secondary separation. The increasing computer power over the last decade enables the simulation of viscous flows in acceptable time. For the validation of Navier–Stokes codes many experimental details of the flowfield are needed in order to judge the variety of turbulence models for delta wing application. The experimental knowledge needed is far beyond the scope of the measurements of the VFE–1 configuration. Therefore a new 65° delta wing configuration (VFE–2) was selected [2], [4]. As part of the scientific network the Institute for Fluid Mechanics, Aerodynamics Division, experimentally investigates the flowfield around this configuration.

2 Geometry and Design of delta wing model

A generic delta wing model was designed, to study leading–edge vortex flow features comparing sharp and rounded leading edges. The present model has a root chord length of $c_r = 0.980\ m$, a wing span of $b = 0.914\ m$, a leading edge sweep of $\varphi_{LE} = 65^o$, a wing area of $S = 0.448\ m^2$ and an aspect ratio of $\Lambda = 1.865$ (Fig. 1). The wing was designed using a CAD tool and then posted to the manufacturer as digital three–dimensional model. The delta wing consists of an upper and a lower base plate, the trailing edge with a depth of $x_{TE}/c_r = 10\%$ and the pressure orifices being part of these plates. On the inside of these plates cut–outs are milled to house the tubes and wires from the pressure orifices and unsteady pressure transducers. The thickness is $t = 0.033\ m$, which is constant over the base plate. A sharp and a rounded leading edge ($r_{LE,rounded}/l_{\mu} = 0.0015$) are available, r_{LE} being the leading edge radius and l_{μ} the mean aerodynamic chord. The leading edges are fitted on the left and right hand side of the lower base plate and have a depth of $x_{LE}/c_r = 15\%$. On each of the leading edge elements, five pockets for the pressure sensors have been milled, which are closed with seperate lids. On the mounted wing a model sting is installed, which is attached to the three–axis model support via a model adapter (Fig. 2). All parts are manufactured from aluminium "Certal" and, therefore, enabling measurements in cryogenic wind tunnels also. There are 177 pressure orifices with a diameter of $d = 0.3\ mm$ situated on the entire wing, 44 are equipped with unsteady pressure sensors. The pressure orifices are positioned in five chordwise positions ($x/c_r = 0.2, 0.4, 0.6, 0.8$ and 0.95).

3 Measurements

The measurements have been performed in the large low–speed wind tunnel A of the Institute for Fluid Mechanis, Aerodynamics Division, of the Technische Universität München at a Mach number of $M = 0.12$, a Reynolds number based on the mean aerodynamic chord of $Re_{l\mu} = 1.67 \cdot 10^6$ and angles of attack varying between $\alpha = 0^o$ and $\alpha = 30^o$. The wind tunnel is of closed–return type with an open test section. The test section is 2.4 m in width, 1.8 m in height and 4.8 m long. The freestream turbulence intensity is less than 0.4%. The uncertainty in the temporal and spatial mean velocity distribution is less than 0.6%. The uncertainty in freestream direction is below 0.2^o and static pressure variations are below 0.4%. Steady measurements were performed using only the 133 steady pressure sensors. The affected pressure orifices are connected via flexible pressure tubing with a Scanivalve system from where the data are processed online.

4 Results

Three steady measurement results are shown for $\alpha = 13^o$ (Fig. 3), $\alpha = 18^o$ (Fig. 4) and $\alpha = 23^o$ (Fig. 5). Each figure shows the development of the pressure distribution over the chord for the sharp and the rounded leading edge. A comparison of the pressure distributions for two wind tunnel results, the one obtained by Luckring [6] and the other by Technische Universität München (TUM), and computational results [3] are presented for the two leading edges at the chord station $x/c_r = 0.8$ for $\alpha = 13^o$. For the two higher angles of attack static pressure distributions as well as corresponding iso–lines of the pressure for the overall wing for each leading edge geometry are shown.

The three angles of attack were chosen, as the vortex already exists at $\alpha = 13^o$, the

leading edge vortex is fully developed at $\alpha = 18^o$ and the vortex breakdown is visible at $\alpha = 23^o$. For wings with sharp leading edges the primary separation is fixed to the leading edges. The flow separates there in any case, and there is no significant Reynolds number effect on the primary separation [5].
Leading edge separation occuring on a blunt or rounded leading edge contour is a more complex phenomenon as the primary separation is no longer fixed to a geometric discontinuity as given by a sharp edge. Therefore, the onset of leading edge separation is determined by flow conditions and the particular wing geometry. For low to moderate angels of attack fully attached flow may be present. Leading edge separation will first occur near the wing tip progressing then in direction to the apex with further increase in angle of attack. Consequently, the wing will exhibit partially developed leading edge separation with attached flow on the upstream portion of the wing and leading edge vortex formation on the downstream portion. The separation line of the primary vortex is free to move inboard or outboard depending on Reynolds number and Mach number conditions. This complexity in the flow physics can have considerable impact on the aerodynamic properties and maneuver performance of slender wing geometries.
At $\alpha = 13^o$ (Fig. 3) significant suction peaks are visible, marking the axis of the primary vortex. The suction peaks are reduced with increasing chord station, except for the second station ($x/c_r = 0.4$), where the peak is slightly increased. This is due to the vortex not being fully developed at this angle of attack. The constant suction level over the inner wing area marks the region of reattachment. The secondary vortex is visible in all chord stations in form of a locally higher suction level outboard of the primary vortex, except for $x/c_r = 0.2$ for the rounded leading edge, where the secondary vortex is just forming. The suction peaks for the primary as well as for the secondary vortex also move inboard with increasing chord station. For the sharp leading edge at $x/c_r = 0.8$ all presented results show good conformity, even though the wind tunnel results obtained by Luckring [6] and the computaional results obtained by Ghafarri [3] correspond to a higher Reynolds number. For the rounded leading edge, the computaional results show good conformity with the wind tunnel results obtained by Luckring [6]. In contrast, the wind tunnel results obtained at the Technische Universität München show the primary vortex axis further inboard. This is due to the lower Reynolds number in this investigation.
With increasing angle of attack (Fig. 4) the suction peaks increase in all chord stations. The vortex diameter also increases, being evident in the suction peak broadening. The vortex axis moves inboard with increasing angle of attack and the reattachment lines are moved towards the wing center line. Fig. 4 also shows the overall pressure distribution. The trace of the primary and secondary vortices have been marked.
Further increasing the angle of attack to $\alpha = 23^o$ (Fig. 5), the suction peaks also increase in the chord stations $x/c_r = 0.2, 0.4$ and 0.6. Chord stations further downstream show decreased suction peaks. At $\alpha = 18^o$ the suction peaks at $x/c_r = 0.8$ are $c_{p,sharp,18^o} = -1.65$ and $c_{p,rounded,18^o} = -1.53$. Increasing the angle of attack to $\alpha = 23^o$ reduces the suction peaks to $c_{p,sharp,23^o} \approx c_{p,rounded,23^o} = -1.48$.

This is due to the vortex breakdown in this area [1]. Again, the primary vortex axes have been marked in Fig. 5, also showing the position of vortex breakdown, which is significantly further upstream for the sharp leading edge. The position of primary vortex breakdown for the rounded leading edge was determined at $x/c_r \approx 0.85$ and for the sharp leading edge at $x/c_r \approx 0.68$.
The difference in the pressure distribution between the rounded and the sharp leading edge is most evident at $x/c_r = 0.2$ for all angles of attack illustrated here. This difference decreases with increasing chord station, disappearing completely at the aft station. The primary vortex for the rounded leading edge is located slightly further outboard than for the sharp leading edge. For an angle of attack of $\alpha = 23^o$ the suction peak at $x/c_r = 0.2$ for the rounded leading edge is significantly higher than for the sharp leading edge.
Increasing the angle of attack causes the pressure on the upper surface in the inboard area and also the pressure on the entire lower surface of the wing to slightly decrease. On the lower surface no significant difference between the sharp and the rounded leading edge is evident.

5 Conclusions and outlook

The steady pressure distribution on a generic 65° swept delta wing has been studied experimentally comparing sharp and rounded leading edges. Stages of partly developed and fully developed vortices as well as vortex breakdown are considered. A downstream shift in the onset of leading edge vortex formation is shown for the blunt leading edge resulting also in a delay in vortex breakdown. For the considered Reynolds number, differences in the steady pressures between sharp and rounded leading edge occur mainly at upstream stations ($x/c_r < 0.6$). A further increase in Reynolds number promotes attached flow and hamper the onset and progression of leading edge vortex separation. In the next phase unsteady pressure measurements are to be performed, followed by unsteady three–dimensional flow field measurements and boundary layer investigations.

Acknowledgements

The authors would like to thank the German Research Association (DFG) for supporting the project. Furthermore we would like to thank the VFE–2 network for the good scientific co–operation. The authors would also like to thank the manufacturing company VAU Werkzeug– und Gerätebau KG for their assistance and technically perfect production.

References

[1] C. Breitsamter: "Turbulente Strömungsstrukturen an Flugzeugkonfigurationen mit Vorderkantenwirbeln". Dissertation, Technische Universität München, Herbert Utz Verlag, ISBN 3–89675–201–4, 1997.

[2] J. Chu, J. M. Luckring: "Experimental Surface Pressure Data Obtained on 65° Delta Wing Across Reynolds Nummber and Mach Nummber Ranges". NASA–TM–4645, Volume 1– Sharp Leading Edge, 1996.

[3] F. Ghafarri: RTO–MP–AVT–113, Personal communication, 2004.

[4] D. Hummel, G. Redeker: "A new vortex flow experiment for computer code validation". Paper 8, RTO Symposium on Advanced Flow Management, Part A–Vortex flow and high angle of attack, Leno, Norway, 7–11 May 2001.

[5] D. Hummel: "Effects of boundary layer formation on the vortical flow above slender delta wing". Paper 30, RTO Symposium on Enhancement of NATO Military Flight Vehicle Performance by Management of Interacting Boundary Layer Transition and Separation, Prague, Czech Republic, 4–7 Oct. 2004.

[6] J. M. Luckring: "Reynolds Number, Compressibility, and Leading-Edge Bluntness Effects on Delta–Wing Aerodynamics". ICAS–2004–4.1.4, 24th Congress of the International Council of the Aeronautical Sciences, Yokohama, Japan, 29 Aug.–3. Sept. 2004.

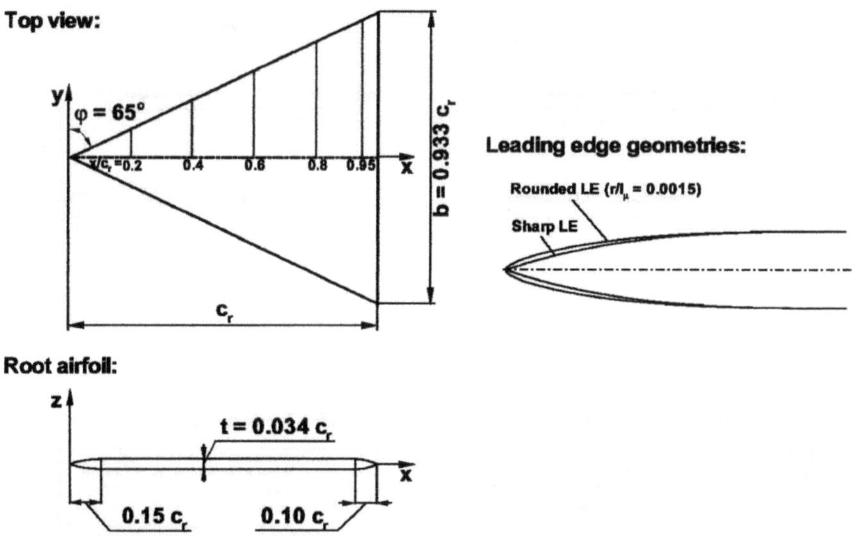

Figure 1 Geometry of delta wing model and leading edge comparison.

Figure 2 Delta Wing in wind tunnel A of the Institute for Fluid Mechanics, Aerodynamics Division, Technische Universität München.

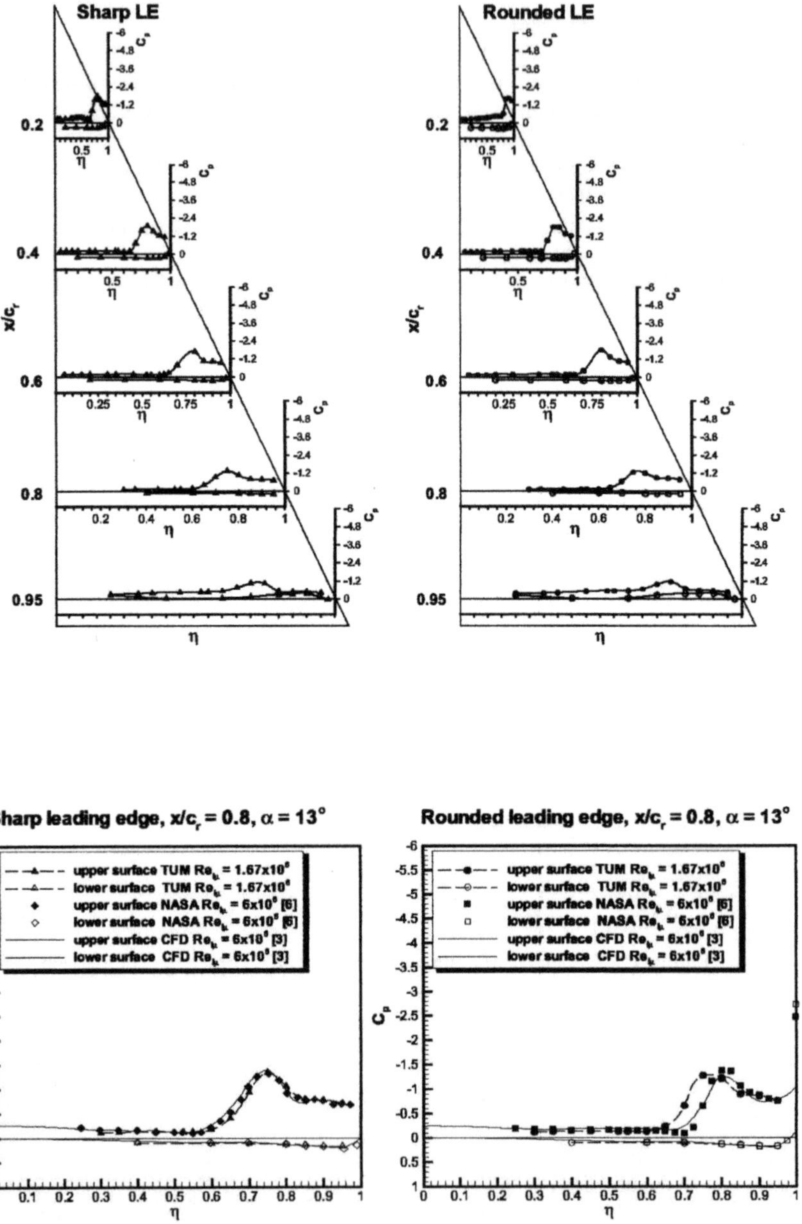

Figure 3 Pressure distribution on 65° delta wing with sharp and rounded leading edge at $M = 0.12$, $Re_{l\mu} = 1.67 \cdot 10^6$ (TUM) and $M = 0.40$, $Re_{l\mu} = 6 \cdot 10^6$ (NASA) and $\alpha = 13°$.

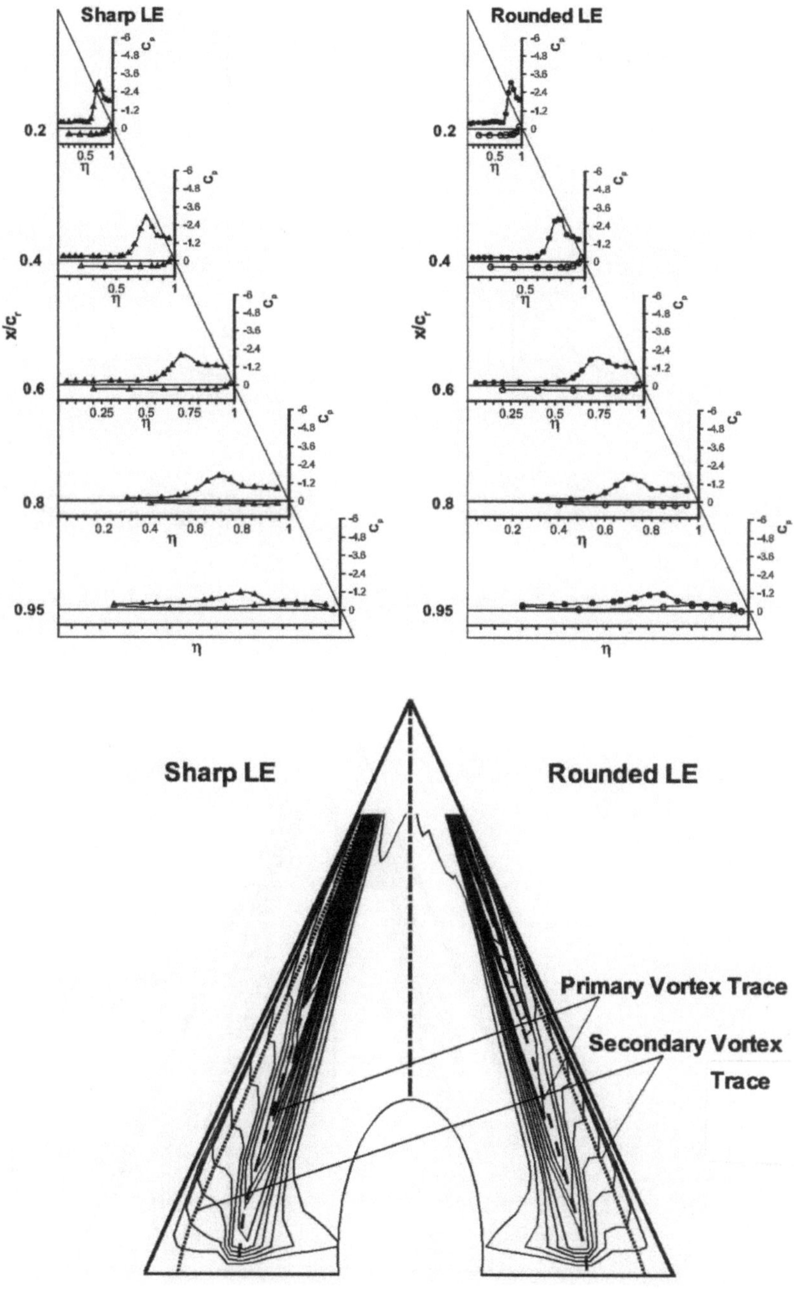

Figure 4 Pressure distribution and pressure iso–lines on 65° delta wing with sharp and rounded leading edge at $M = 0.12$, $Re_{l\mu} = 1.67 \cdot 10^6$ and $\alpha = 18°$.

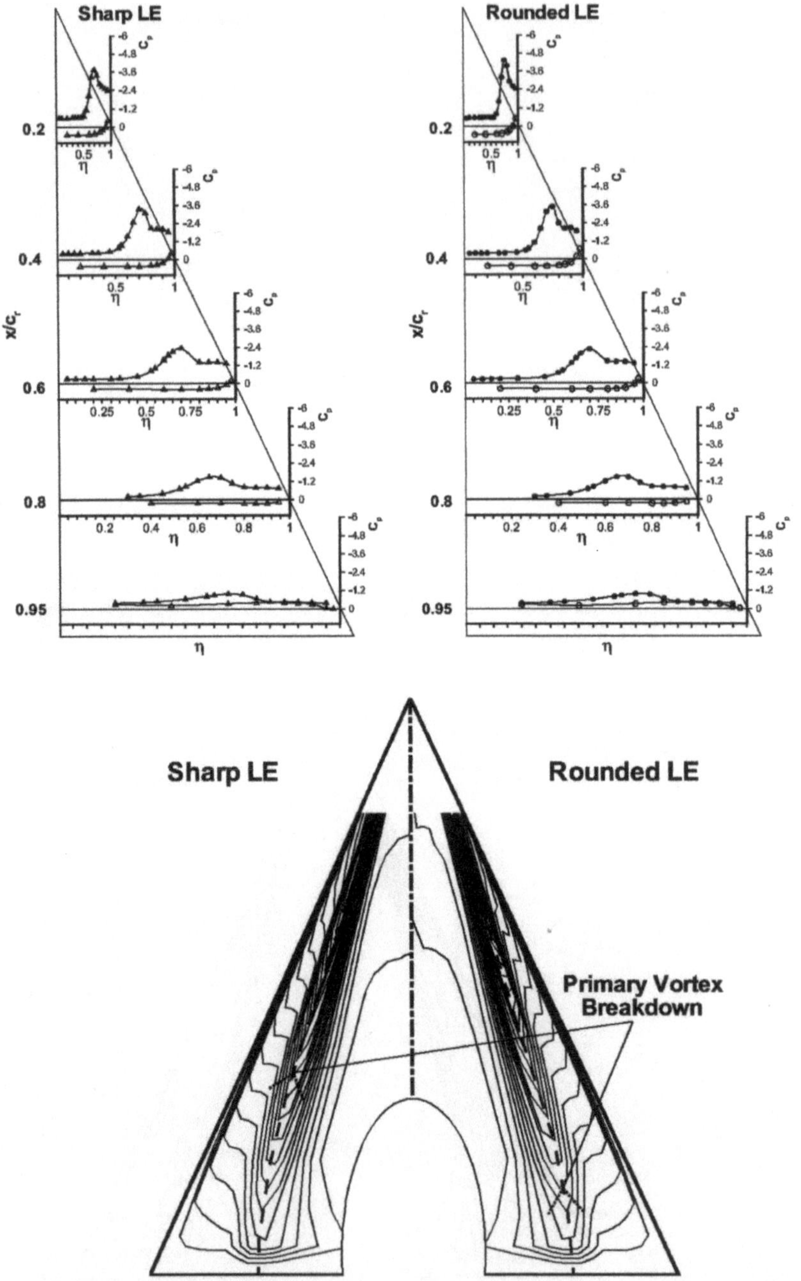

Figure 5 Pressure distribution and pressure iso–lines on 65° delta wing with sharp and rounded leading edge at $M = 0.12$, $Re_{l\mu} = 1.67 \cdot 10^6$ and $\alpha = 23°$.

Computation of Delta Wing Flap Oscillations with a Reynolds–averaged Navier–Stokes Solver

Alexander Allen, Michail Iatrou, Alexander Pechloff, Boris Laschka

Technische Universität München, Lehrstuhl für Fluidmechanik, Abteilung Aerodynamik
Boltzmannstrasse 15, 85748 Garching, Germany
allen@flm.mw.tum.de

Summary

A selection of steady and unsteady validation results for the Reynolds–averaged Navier–Stokes solver FLM–NS with respect to an experimental delta wing test case is presented and compared to other numerical results. Having put the numerical method's validity into evidence, the unsteady flow induced by an oscillating flap is investigated for a Fighter Type Delta Wing.

1 Introduction

Over the past decade considerable efforts have been undertaken at the Institute for Fluid Mechanics to realize computational methods for accurately predicting three–dimensional flow about aircraft configurations. The developed and to this date well established inviscid method FLM–Eu allows for steady as well as for unsteady solution of the Euler equations dependent on the flow problem, the latter case being performed time accurately. Even though this inviscid method has been employed very successfully, extension to the viscous flow regime is highly desirable. In this regard FLM–Eu is evolved to FLM–NS on basis of the Reynolds–averaged Navier–Stokes (RANS) equations, by introducing the viscous fluxes and closure relationships into the code.

2 Theory

The nondimensionalized conservation form of the RANS equations for three dimensions as written in curvilinear coordinates is

$$\frac{\partial \mathbf{Q}}{\partial \tau}+\frac{\partial \mathbf{F}}{\partial \xi}+\frac{\partial \mathbf{G}}{\partial \eta}+\frac{\partial \mathbf{H}}{\partial \zeta}=\frac{\partial \mathbf{F_v}}{\partial \xi}+\frac{\partial \mathbf{G_v}}{\partial \eta}+\frac{\partial \mathbf{H_v}}{\partial \zeta}, \tag{1}$$

with ξ, η and ζ being the curvilinear coordinate directions and τ the nondimensional time. The conservative state vector

$$\mathbf{Q}=J(\rho,\rho u,\rho v,\rho w,\rho e)^T, \tag{2}$$

is comprised of the primitive variables: density ρ, velocity components u, v and w with respect to the cartesian coordinate system and specific total energy e. The determinant of the coordinate transformation's Jacobian is

$$J = det\frac{\partial(x,y,z,t)}{\partial(\xi,\eta,\zeta,\tau)}. \quad (3)$$

The convective flux vectors $\mathbf{F}$, $\mathbf{G}$, $\mathbf{H}$ and viscous flux vectors $\mathbf{F_v}$, $\mathbf{G_v}$, $\mathbf{H_v}$ are expressed in generalized form:

$$\mathbf{E}_\psi = J\begin{pmatrix} \rho\Theta_\psi \\ \rho u\Theta_\psi + \psi_x p \\ \rho v\Theta_\psi + \psi_y p \\ \rho w\Theta_\psi + \psi_z p \\ (\rho e + p)\Theta_\psi - \psi_t p \end{pmatrix}, \quad \mathbf{E_{v\psi}} = J\begin{pmatrix} 0 \\ \psi_x\tau_{xx} + \psi_y\tau_{yx} + \psi_z\tau_{zx} \\ \psi_x\tau_{xy} + \psi_y\tau_{yy} + \psi_z\tau_{zy} \\ \psi_x\tau_{xz} + \psi_y\tau_{yz} + \psi_z\tau_{zz} \\ \psi_x\Pi_x + \psi_y\Pi_y + \psi_z\Pi_z \end{pmatrix}. \quad (4)$$

By substituting ψ with ξ, η or ζ the fluxes in the individual directions are obtained. In equation (4) p represents the static pressure. Θ_ψ stands for the generalized contravariant velocity, τ_{ij} for the components of the Cartesian shear stress tensor and Π_i for the energy flux resulting from shear stress work and heat transfer.
Taking into account the thermal equation of state in nondimensional notation $p = \rho T$, the caloric equation of state $e_i = c_v T$ and the definition for the specific total energy $e = e_i + \frac{1}{2}(u^2 + v^2 + w^2)$ an additional relationship between p and the conservative variables is derived:

$$p = (\kappa - 1)\left[\rho e - \frac{1}{2\rho}\left((\rho u)^2 + (\rho v)^2 + (\rho w)^2\right)\right]. \quad (5)$$

Further closure of the equation system is realized through Sutherland's law for the molecular viscosity.

$$\mu_l = \mu_\infty T^{\frac{3}{2}}\frac{1+S}{T+S}, \qquad \text{with} \qquad \mu_\infty = \sqrt{\kappa}\frac{Ma_\infty}{Re_\infty}l_{Re_\infty}. \quad (6)$$

The Sutherland constant S is defined as $110.4[K]/\check{T}_\infty$, with $\check{T}_\infty$ being the dimensional static freestream temperature. In equation (6) κ represents the ratio of the specific heats, Ma_∞ the freestream Mach number, Re_∞ the freestream Reynolds number and l_{Re_∞} the characteristic length used in the formulation of Re_∞. The eddy viscosity is computed through an appropriate turbulence model, selected at user discretion.

3 Numerical Method

3.1 Properties of FLM–NS

The RANS equations are solved within a finite volume framework. Discrete evaluation of convective fluxes is conducted with Roe's upwind flux difference splitting scheme, in conjunction with MUSCL extrapolation of the conservative values. Calculation of the velocity and temperature gradients as appearing throughout the viscous fluxes is realized with the integral theorem of Gauss as proposed by Chakravarthy for high resolution schemes [4]. An implicit lower–upper symmetric successive overrelaxation (LU–SSOR) technique is chosen for the time integration [3]. Steady state solutions are obtained in pseudo–time, while unsteady solutions are calculated time accurately through dual time–stepping. Spatial as well as temporal accuracy is of second order. The algebraic Baldwin–Lomax [1] and the single equation Spalart–Allmaras [8] turbulence models are implemented in FLM–NS in order to gain the eddy viscosity.

For the unsteady computations in respect to harmonic motions of a body or body parts with FLM–NS two grids are required: one for the reference and one for the extremum position. Mesh movement is realized by interpolating intermediary grids sinusoidally between these positions. Three oscillations are computed in order to eliminate transient phenomena, each discretized with 80 physical time intervalls. For the individual physical time step the solution process towards a pseudo–steady state is terminated after having reached a density residual of 10^{-5}.

Extensive experimental as well as numerical investigations on oscillating control surfaces have been conducted in the past years, e.g. by Bennett and Walker [2] and Obayashi and Guruswamy [7]. Unsteady computations of flap oscillations are presented here in order to extend the available data.

3.2 Properties of FLOWer

For validation purposes an alternative RANS solver has been applied, namely FLOWer [5]. Also constructed as a finite volume method, both convective and viscous fluxes are discretized through second order accurate central–differences with respect to a cell vertex control volume. Both pseudo–time as well as time–accurate integration are conducted explicitly. Either one of two turbulence models is selected for calculation of the eddy viscosity: the algebraic turbulence model by Baldwin–Lomax [1] or the two equation k–ω turbulence model by Wilcox [9].

4 NASA Clipped Delta Wing

Steady calculations of the rigid body and unsteady calculations for the harmonically oscillating flap are performed with the solver FLM–NS. These are compared with experimental data [2] and with results obtained with the solver FLOWer.

4.1 Grid Properties

The NASA Clipped Delta Wing (NCDW) has a sharp leading edge with a sweep angle $\Phi_{LE} = 50.4^o$ and an unswept trailing edge. The wing semi–span $s = 0.7094c_r$ with c_r being the root chord length. The airfoil is a symmetrical circular arc section with a thickness of 6% of the local chord length c. The wing has a flap between the span stations $y/s = 0.566$ and $y/s = 0.829$ with a hinge line at 80%c. Further geometric details are to be found in Ref. [2].

A structured multiblock grid consisting of an upper and a lower block has been generated using ICEM–CFD [10]. The initial volume grid undergoes elliptic smoothing with GRID-FLM, an in–house developed tool. The body surface mesh remains unsmoothed. The mesh is constructed as a CH–Topology with 96 cells in chord direction per block, 48 cells in spanwise direction and 40 cells normal to the wing surface. The overall mesh is comprised of 368640 cells. The distance of the first offbody grid line is fixed to $1 \cdot 10^{-5}s$ for the cell–centered FLM–NS scheme, whereas an offbody distance of $5 \cdot 10^{-6}s$ is used in connection with the FLOWer calculations, due to it being a vertex centered scheme. The wing surface is discretized with 72 cells for each the upper and lower surface in chord direction and 32 cells in spanwise direction. The flap consists of 20 cells in chord direction and 9 cells in spanwise direction. In the area of the flap hinge line and the flap side edges the grid points are concentrated in order to achieve an optimum resolution of the flow gradients, while modelling of the flap gaps is abstained from. As a consequence the discretized wing surface retains its character as a continous surface (Fig. 1). In this manner the cells modelling the flap gaps are stretched and skewed during deflection.

4.2 Results

The ratio of the specific heats and the Prandtl number are set in accordance to the experimental conditions (heavy gas), $\kappa = 1.132$ and $Pr = 0.775$ respectively. A steady calculation was performed at $Ma_\infty = 0.899$, $Re_\infty = 9.77 \cdot 10^6$ and an angle of attack $\alpha = 0.05^o$. Fig. 2 shows the chordwise pressure distributions at selected span stations for the upper surface (Fig. 1). In all presented span stations both solvers show very good conformity with the experimental results. No differences between FLOWer k-ω and FLOWer Baldwin–Lomax (B/L) are visible, yet surprisingly deviations between the Baldwin–Lomax models implemented in FLM–NS and FLOWer are evident. Another transonic computational result is shown in Fig. 3 ($Ma_\infty = 0.965$, $Re_\infty = 9.81 \cdot 10^6$, $\alpha = 0.00^o$). Again very good conformity and hardly any differences between the two turbulence models used by FLOWer are visible. The shock lies within a position margin of 4%c for both solvers. Further detailled investigations can be found in Ref. [6].
Two unsteady calculations with an oscillating flap are depicted in Fig. 5 ($Ma_\infty = 0.924$, $Re_\infty = 10.25 \cdot 10^6$, $\alpha = 0.05^o$). Fig. 4 illustrates the corresponding steady measurement, which was performed at the same angle of attack, but at slightly different Mach and Reynolds numbers corresponding to the windtunnel test. Differences between FLOWer B/L in respect to FLM–NS and the experimental data points are visible. The flap oscillates harmonically around its neutral position by $\Delta\eta = \pm 3.89^o$ with a dimensional frequency of $f = 22[Hz]$, yielding a reduced frequency of $k_{red} = 0.918$. Results for FLM–NS and FLOWer B/L and again restricted to the upper surface are shown, as no experimental data for the lower surface are provided. The results match the experimental data very well in all span stations. For the real part of the first harmonic pressure coefficient a peak in the area of the hinge line is evident, caused for one by the discontinuity at the flap edge and second by the movement of the shock, due to the flap oscillation. The imaginary part of the first harmonic

pressure coefficient shows a change of sign in the area of the flap hinge line due to the flap oscillation. In the imaginary part the peak located at the flap hinge line is due only to the motion of the shock. The prominent pressure peak evident in the imaginary part indicates that the shock motion is out of phase with the flap oscillation. FLOWer B/L predicts the contribution of the shock to the imaginary part further upstream and at a reduced value. The influence of the oscillating flap on the upstream flow is limited, as a result of the freestream Mach number in conjunction with the leading edge sweep.

5 Fighter Type Delta Wing

5.1 Grid Properties

The Fighter Type Delta Wing (FTDW) has a round leading edge with a sweep angle of $\Phi_{LE} = 53^o$, a sharp trailing edge with a sweep angle of $\Phi_{TE} = -3^o$, a wing semi–span of $s = 0.621c_r$ and a thickness of 7.2% at the root chord. The airfoil and thus the local wing thickness varies in spanwise direction. An inboard and an outboard flap are present in the trailing edge area. The inboard flap begins at $y/s = 0.239$ and ends at $y/s = 0.617$. The outboard flap begins at this position and stretches on to the wing tip.
The FTDW is embedded in the same grid topology as the NCDW, retaining all characteristics, such as block structure, surface resolution and offbody distance, as mentioned previously. Both flaps are comprised of 20 cells in chord direction. In spanwise direction the inboard flap is discretized with 11 cells and the outboard flap with 13 cells. Again flap gaps are not explicitly considered in the modelling (Fig. 1).

5.2 Results

The investigation is performed with $\kappa = 1.4$, $Pr = 0.72$, $Ma_\infty = 0.8$, $Re_\infty = 10.0 \cdot 10^6$ and $\alpha = 5.0^o$ in order to also use the solver on a industry relevant wing configuration. The steady state pressure distribution for the spanstations given in Fig. 1 are composited in Fig. 6.
With these steady results unsteady calculations were initiated for an oscillating inboard flap with FLM–NS using both turbulence models with good agreement (Fig. 7). The flap oscillates harmonically around its neutral position by $\Delta\eta = \pm 0.4^o$ at a dimensional frequency of $f = 6[Hz]$, yielding a reduced frequency of $k_{red} = 0.7283$. For the span station through the oscillating inboard flap ($y/s = 0.35$) the characteristic peak in the real part of the first harmonic pressure coefficient can be seen in the area of the flap hinge line. Likewise, the imaginary part of the first harmonic pressure coefficient shows a change of sign in this vicinity. In comparison to the NCDW case the lower freestream Mach number limits the flow on the entire wing to the subsonic region. The oscillating flap influences the unsteady flow over the entire wing, as visible in span station $y/s = 0.75$. The imaginary part shows a change of sign in the area of the flap hinge line, as was evident in the NCDW case. However, with the absence of a shock the influence of the oscillating flap is also seen upstream.

6 Conclusion

The validity of the Reynolds–averaged Navier–Stokes solver FLM–NS developed at the Institute for Fluid Mechanics has been put into evidence. Computational results are in good agreement with experimental data and results obtained with the DLR code FLOWer. The two implemented turbulence models allow for good shock prediction in both steady and unsteady cases. In regard to the Fighter Type Delta Wing more unsteady computations especially at higher angles of attack and also regarding outboard flap oscillations are underway.

Acknowledgements

The authors would like to thank Dr.–Ing. Jürgen Becker and Dr.–Ing. Caroline Weishäupl (both EADS Ottobrunn) for their support and good cooperation over the years. Furthermore the authors would like to thank Dipl.–Ing. Johannes Markmiller (Technische Universität München) for his significant contribution towards the validation of the FLM–NS code with the NCDW.

References

[1] Baldwin B. S., Lomax H.: Thin Layer Approximation and Algebraic Model for Separated Turbulent Flows. AIAA Paper 78-257, NASA Ames Research Center, Moffett Field, CA, USA, 1978.

[2] Bennett R. M., Walker C. E.: Computational Test Cases for a Clipped Delta Wing With Pitching and Trailing–Edge Control Surface Oscillations. NASA-TM-1999-209104, NASA Langley Research Center, Hampton, Virginia, 1999.

[3] Blazek J.: Investigations of the Implicit LU–SSOR Scheme. DLR-FB 93-51, DLR, 1993.

[4] Chakravarthy S. R.: High Resolution Upwind Formulations for the Navier–Stokes Equations. Rockwell Int. Science Center, USA, 1988.

[5] Kroll N.: FLOWer Installation and User Handbook, Release 116. Institut für Entwurfsaerodynamik, DLR, Braunschweig, 2001.

[6] Markmiller J.: Validierung eines zeitechten und eines small disturbance Navier–Stokes Verfahrens an einem schwingenden Deltaflügel. Diploma thesis, Technische Universität München, TUM-FLM-2003/28, 2003.

[7] Obayashi S., Guruswamy G. P.: Navier–Stokes Computations for Oscillating Control Surfaces. Journal of Aircraft, Vol. 31, pp. 631-636, 1994.

[8] Spalart P. R., Allmaras S. R.: A one–equation turbulence model for aerodynamic flows. AIAA Paper 92-0439, 30th Aerospace Sciences Meeting & Exhibit, Reno, NV, USA, 6.-9. January, 1992.

[9] Wilcox D. C.: Reassessment of the Scale–Determining Equation for Advanced Turbulence Models. AIAA–Journal, Vol. 26, pp. 1299-1310, 1988.

[10] ICEM–CFD Hexa, Operating Manual. ICEM CFD & Structural Engineering GmbH, 1998.

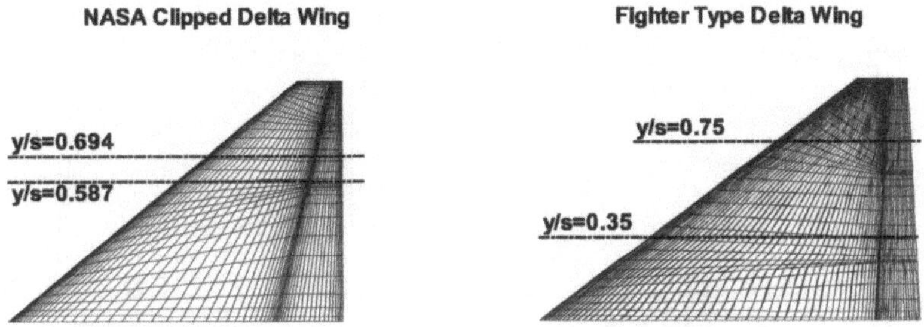

Figure 1 Planform view and span stations

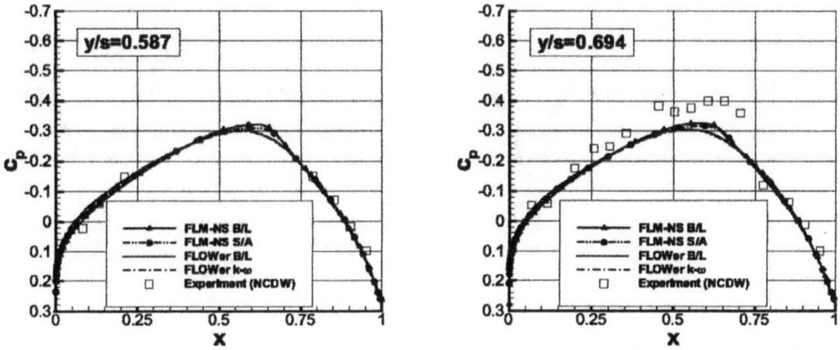

Figure 2 Pressure coefficient at $Ma_\infty = 0.899$, $Re_\infty = 9.77 \cdot 10^6$ and $\alpha = 0.05^o$

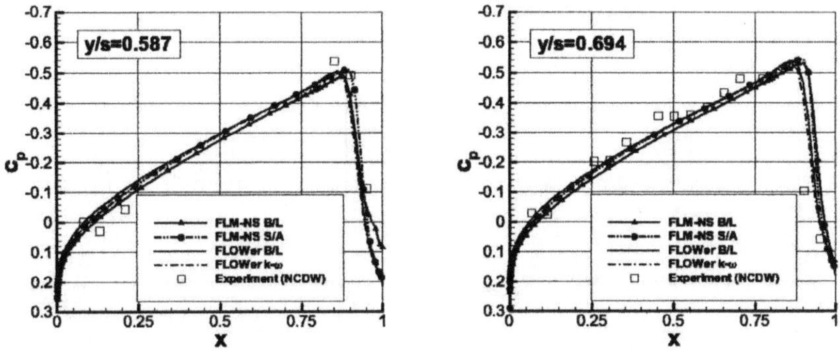

Figure 3 Pressure coefficient at $Ma_\infty = 0.965$, $Re_\infty = 9.81 \cdot 10^6$ and $\alpha = 0.00^o$

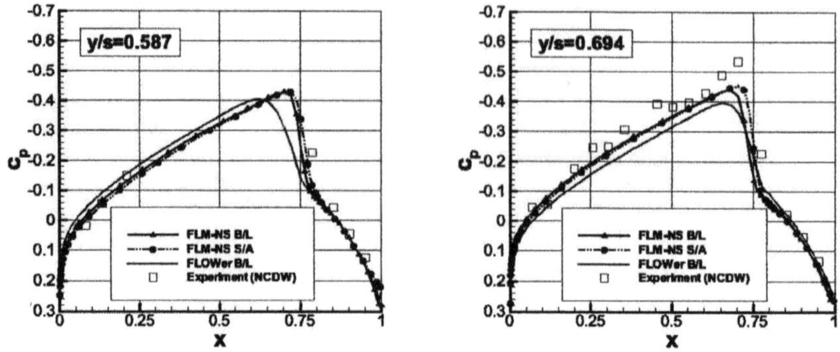

Figure 4 Pressure coefficient at $Ma_\infty = 0.921$, $Re_\infty = 10.35 \cdot 10^6$ and $\alpha = 0.05^o$

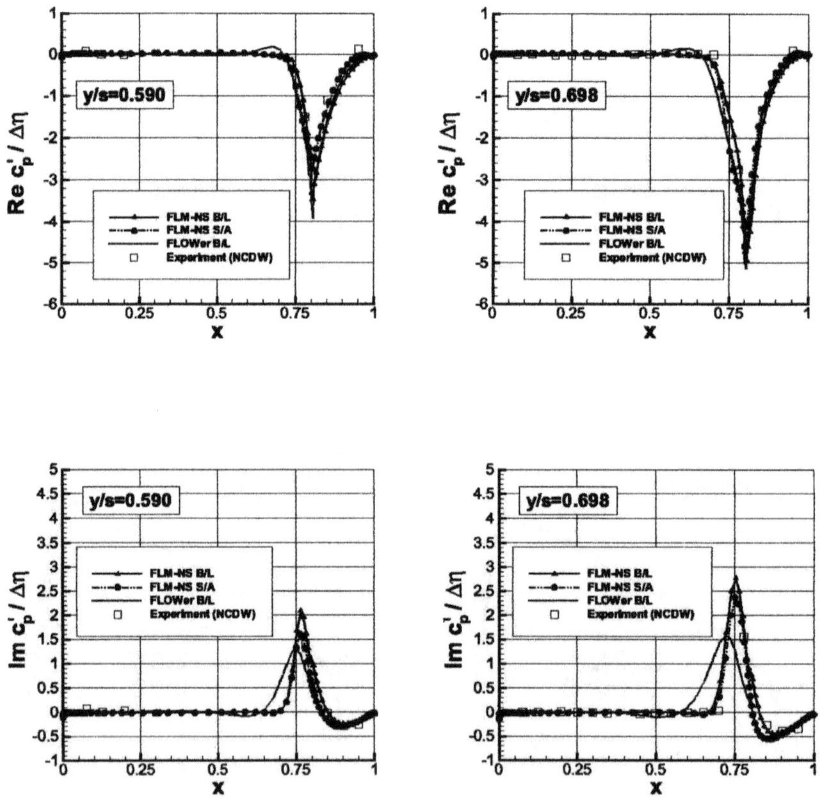

Figure 5 First harmonic pressure coefficient at $Ma_\infty = 0.924$, $Re_\infty = 10.25 \cdot 10^6$ and $\alpha = 0.05^o$ for oscillating flap ($\Delta\eta = \pm 3.89^o$, $f = 22[Hz]$, $k_{red} = 0.918$)

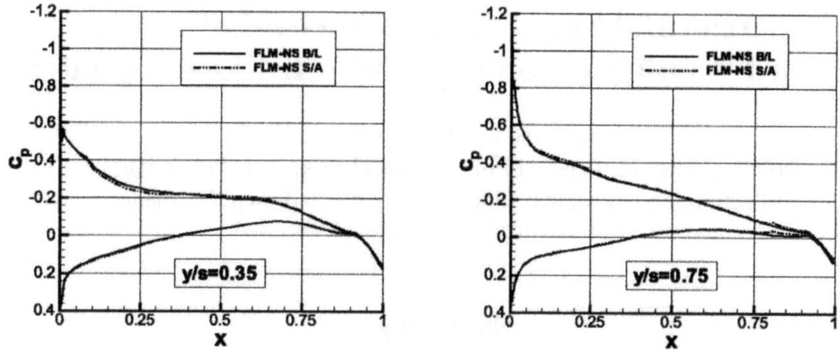

Figure 6 Pressure coefficient at $Ma_\infty = 0.8$, $Re_\infty = 10.0 \cdot 10^6$ and $\alpha = 5.0^o$ (FTDW)

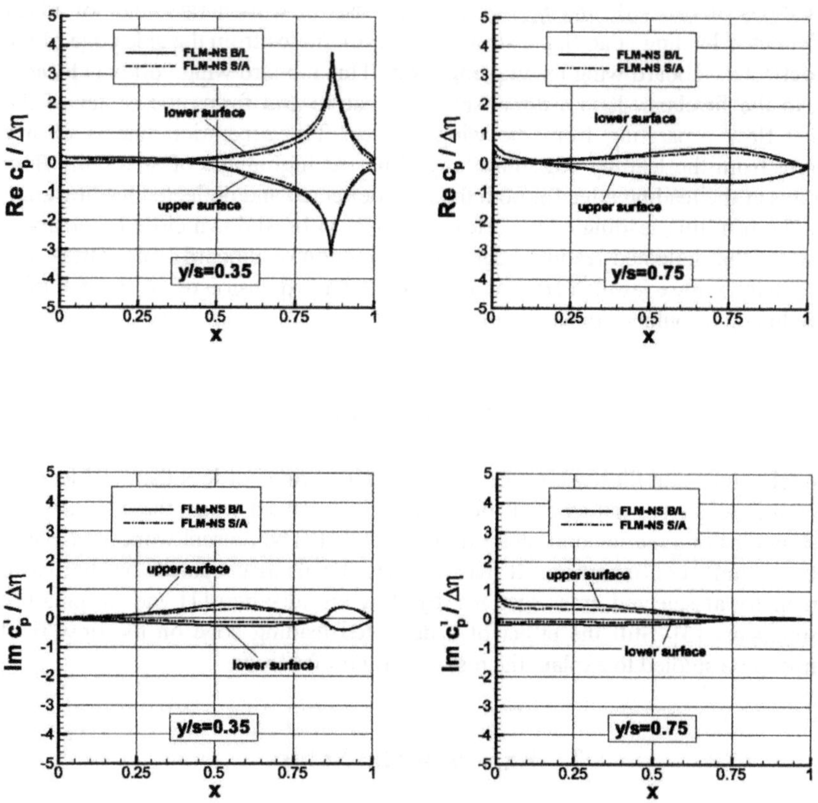

Figure 7 First harmonic pressure coefficient at $Ma_\infty = 0.8$, $Re_\infty = 10.0 \cdot 10^6$ and $\alpha = 5.0^o$ for oscillating inboard flap ($\Delta\eta = \pm 0.4^o$, $f = 6[Hz]$, $k_{red} = 0.7283$) (FTDW)

Experimental Study on the Flowfield of a Delta–Canard–Configuration with Deflected Leading Edge

A. SCHMID, C. BREITSAMTER
Boltzmannstr. 15, Lehrstuhl für Fluidmechanik, Abt. Aerodynamik,
Technische Universität München, 85747 Garching, Germany
Arne.Schmid@aer.mw.tum.de

Summary

The turbulent flowfield above the wing of a delta–canard–configuration at moderate ($\alpha = 15°$) and high ($\alpha = 24°$) angle of attack was measured at a Re–number of $0.97 \cdot 10^6$ in a wind tunnel by hotwire anemometry. Leading edge flap settings of $\eta_{l.e.} = 0°$ and $\eta_{l.e.} = -20°$ were used . At moderate angle of attack and deflected leading edge flap a strong vortex originates from the side edge of the non–deflected inboard wing leading edge part. This inboard wing vortex is located close to the fuselage. It is a dominant flow feature and forms the center of the vortical flow separating from the wing surface. The separation line is clearly different from the leading edge flap hinge line. At high angle of attack the flow separates at the leading edge for both the non–deflected and deflected leading edge case. The resulting leading edge vortex is subject to breakdown close to the apex. In case of the deflected leading edge, the interaction of inboard wing vortex and leading edge vortex results in decreased downstream expansion of the burst vortex, also reducing turbulence intensity levels.

1 Introduction

Although studies on the vortex system above delta wings of less than 60° leading edge sweep angle are not as numerous as for highly swept wings there are several investigations of generic configurations, e.g. [2]–[4]. The present work uses a high agility aircraft configuration. It has already been investigated by hot wire anemometry at selected cross sections, by flow visualisation [1] and fin pressure measurements [5]. Still the effect of a deflected leading edge on the flow field remains to be studied to explain the resulting fin buffet loads.

2 Experimental Setup

The turbulent flow field was measured above the 50° swept delta wing of the steel model of a modern fighter type aircraft as sketched in Fig. 1. Two different leading edge settings of $\eta_{l.e.} = 0°$ and $\eta_{l.e.} = -20°$ were studied by changing the leading edge parts of the model. The investigated leading edge deflection is a rotation around a hinge line without streamwise movement. The deflected leading edge forms a kink at the junction to the main wing without any gap between leading edge part and wing. Therefore, there is no through–flow between wing lower and upper side in the area of the hinge line. The local chord of the leading edge flap at the innermost section is a tenth of the local wing chord. The leading edge flap chord decreases with the wing chord to the wing tip. The wing span is $2s = 0.74$ m and the wing mean aerodynamic chord is $l_\mu = 0.36$ m.

The turbulent flow field was measured by constant temperature anemometry. A crosswire probe was used to measure two unsteady velocity components and rotated 90° to measure the component perpendicular to the initial plane. The two perpendicularly arranged platinum plated tungsten wires of 5 µm diameter cover a measurement area of approximately 1 mm^2. The probe was positioned by means of a computer controlled traversing system. A regular measurement pattern of $\Delta Y/s$ and $\Delta Z/s = 0.027$ was adapted to the shape of the wing using the CAD data of the wing allowing a minimal distance to the model below 10 mm. Since this data were not available when the moderate angle of attack reference case was investigated, the corresponding flow field survey misses most of the measurement points very close to the model surface. An additional thermal probe of Pt100 type was used to measure flow temperature close to the hotwire probe for temperature correction of hot wire signals.

The experiments were conducted at a freestream reference velocity of 40 m/s at the Göttingen–type low speed wind tunnel B and the wind tunnel 1 (moderate angle of attack reference case) of the Lehrstuhl für Fluidmechanik. Both facilities reach a freestream turbulence intensity below 0.5 %. The Re–number based on the wing mean aerodynamic chord is $0.97 \cdot 10^6$.

Two angle of attack conditions are investigated for both leading edge settings. A moderate angle of attack of $\alpha = 15°$ and a high angle of attack of $\alpha = 24°$ are selected representing partially burst leading edge vortex flow and mostly burst leading edge vortex flow within the range of nonlinear lift increase but well below stall conditions, Fig. 2.

After passing a 1000 Hz low–pass filter, the signals were sampled at a rate of 3000 Hz and 12 bit resolution. Sampling time was approximately 7 seconds. Since no dominant flow phenomena were recorded at frequencies higher then 800 Hz, the filter frequency to sampling rate ratio was sufficient. The velocity components were calculated by applying a look–up table from prior calibration [1].

3 Results

In Fig. 3 an overview over the turbulent flowfield is given for all four cases by plotting the streamwise turbulence intensity at all cross sections. The following discussion of the flowfield is based on the total velocity distribution with crossflow velocity vectors in Figs. 4, 5 and the distribution of streamwise turbulence intensity in Fig. 6. The values are normalized by freestream velocity and squared freestream velocity, respectively. The velocities are based on the model coordinate system allowing better estimation of the flow conditions close to the surface then wind tunnel system based velocities. The selected cross section $X/c_r = 0.3$ is positioned near the inboard hinge line edge, $X/c_r = 0.6$ at the midwing and $X/c_r = 1.0$ slightly downstream of the trailing edge, Fig 1. Positions of specific flow features will be given as (Y/s, Z/s).

3.1 Reference Case at Moderate Angle of Attack

Since the measurement pattern for this case becomes coarser at greater distance from the wing, the wake of the canard vortex system is not resolved, Fig. 4.

At $X/c_r = 0.3$, the leading edge vortex is within the measurement area and the maximum total velocity of $W = 1.6$ is located above and outboard of the leading edge vortex (0.33, 0.2). The maximum streamwise turbulence intensity occurs below that point close to the wing surface ($Y/s = 0.33$), Fig. 6a.

The leading edge vortex of the clean wing bursts around $X/c_r = 0.4$. At $X/c_r = 0.6$, the velocity distribution exhibits the typical pattern of a helically burst leading edge vortex. Inboard close to the wing surface at $Y/s = 0.5$, the total velocity becomes a maximum. Futher outboard and above, there is a minimum in total velocity (0.56, 0.18) followed by a maximum at the upper outer part of the vortical flow (0.6, 0.26). Close to the wing surface near the leading edge at $Y/s = 0.7$, another area of low total velocity in the wake of the boundary layer and the assumed secondary vortex can be found.

At $X/c_r = 1.0$, the area of low total velocity at the center of the vortical structure has merged with that near the leading edge and is increased in lateral direction. The center of the vortical crossflow is located at (0.65, 0.25). The area of accelerated flow is relatively small while most of the vortical flow is of freestream velocity and decelerates in the upwind region, respectively. The same region exhibits the lowest streamwise turbulence intensity with areas of high turbulence above and below that region, Fig. 6c. Velocity and turbulence intensity distribution show a concave shape at the wing tip, caused by the complex vortex structures around the tip pod.

3.2 Deflected Leading Edge at Moderate Angle of Attack

The cross section above the non–deflected inboard and deflected outboard part of the leading edge at $X/c_r = 0.3$ shows a combined vortical structure. It is formed by a vortex shed at the non–deflected part of the leading edge and the corresponding side edge vortex. The center of this inboard wing vortex is marked by local deceleration, Fig. 4, while streamwise turbulence intensity reaches a level of 0.15, Fig. 6b. The flow above the leading edge flap ($Y/s = 0.3 \dots 0.4$) is deflected upwards and accelerated without formation of a leading edge vortex.

At $X/c_r = 0.6$, the center of the inboard wing vortex has moved to (0.28, 0.15) with a low velocity region extending from the surface to the vortex center in the upwash region. Close to the wing surface, a flow pattern extends from this vortex in direction to the wing tip. Above this layer there is a region of opposite flow direction. The associated separation takes place on the wing between $Y/s = 0.4$ and the hinge line. Further downstream, this flow pattern extends laterally and in vertical direction.

At $X/c_r = 1.0$, the inboard wing vortex is approximately located at the same position as further upstream. The layer of outward and inward flow extending from there is limited by a region of decreased velocity close to the surface at $Y/s = 0.6$. This flow structure can be found at all cross sections of $X/c_r = 0.7 \ldots 1.0$ indicating separation on the wing surface at this position instead at the hinge line. Vortex area and lateral extension correspond to an increased streamwise turbulence intensity with a maximum of 0.1 at the separation position at $Y/s = 0.6$. The region of low velocity and increased streamwise turbulence intensity ($Y/s = 0.8 \ldots 1.2$) indicates the wake of the outboard flap vortex and tip pod vortex system.

Instead of a vortex originating from the leading edge with high velocity and a large area of straight flow close to the body, the flow caused by the deflected leading edge flap is completely different. Here, an inboard wing vortex arising from flow separation at the non–deflected wing leading edge part and a strong vortex shed at the exposed side edge of this wing part constitutes a dominant flow field structure. The separation on the wing does not result in the formation of a separate wing vortex. It forms an extended vortical structure with the inboard wing vortex as the center. The peak velocities and streamwise turbulence intensities are generally lower than those of the reference case.

3.3 Reference Case at High Angle of Attack

At $X/c_r = 0.3$, the center of the burst leading edge vortex moves closer to the fuselage and wing surface compared to the moderate angle of attack case. Peak velocity and streamwise turbulence intensity is higher, Fig. 6e.

Further downstream the leading edge vortex increases in size but keeps its appearance. The vortex center moves outboard and upward until $X/c_r = 0.6$. From there the center moves upward without further lateral movement.

3.4 Deflected Leading Edge Flap at High Angle of Attack

Above the station between non–deflected and deflected leading edge at $X/c_r = 0.3$, the center of the combined inboard wing vortex and leading edge vortex is located closer to the fuselage and exhibits higher streamwise turbulence intensities compared to the moderate angle of attack case, Fig. 6f.

Above the wing at $X/c_r = 0.6$, the flow pattern indicates a separate vortical structure although no flow downwards to the wing surface is measured between leading edge and inboard wing vortex.

Downstream at $X/c_r = 1.0$, the influence area of the vortical structure shed at the deflected leading edge is increased in lateral and vertical direction, however its extent is limited by the inboard wing vortex.

4 Conclusions

The flowfield over a delta–canard fighter aircraft wing has been investigated experimentally comparing cases of clean wing and deflected wing leading edge.

At moderate angle of attack and deflected leading edge, the vortex originating from the non–deflected inboard leading edge part and, in particular, from the corresponding side edge dominates the wing flow field. There is no strong separation at the hinge line, but separation takes place further downstream on the wing surface. No separate wing vortex is formed but a flat extension of vortical flow develops around the wake of the inboard wing vortex.

At high angle of attack the deflected leading edge flap results again in formation of the inboard wing vortex which remains of constant strength above the wing. Instead of the weak wing vortical structure of the moderate angle of attack case a strong leading edge vortex shed at the deflected leading edge develops. The latter becomes then dominant in the rear part of the wing.

Acknowledgements

The authors would like to thank EADS Deutschland GmbH for supporting this investigation.

References

[1] C. Breitsamter: ”Turbulente Strömungsstrukturen an Flugzeug–konfigurationen mit Vorderkantenwirbeln“. Dissertation, Herbert Utz Verlag 1997.

[2] P. B. Earnshaw, J. A. Lawford: “Low–Speed Wind–Tunnel Experiments on a Series of Sharp–Edged Delta Wings”. RAE Reports and Memoranda No. 3424, Her Majesty's Stationary Office 1966

[3] J. J. Miau et al.: “Flow Developments Above 50–Deg Sweep Delta Wings with Different Leading–Edge Profiles”. J. Aircraft Vol. 32, No. 4, 1995, pp. 787–794.

[4] M. V. Ol: “An Experimental Investigation of Leading Edge Vortices and Passage to Stall of Nonslender Wings”. RTO–MP–069(I) (SYA) 2, 2003.

[5] A. Schmid, C. Breitsamter: “High Incidence Buffet Flow over Fighter Type Aircraft”. In Notes on Numerical Fluid Mechanics and Multidisciplinary Design Vol. 87, 2004, pp. 107–115.

Figures

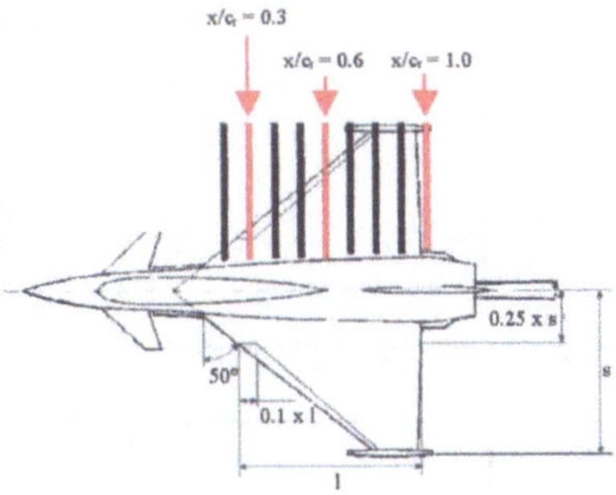

Figure 1 Model geometry and measurement planes.

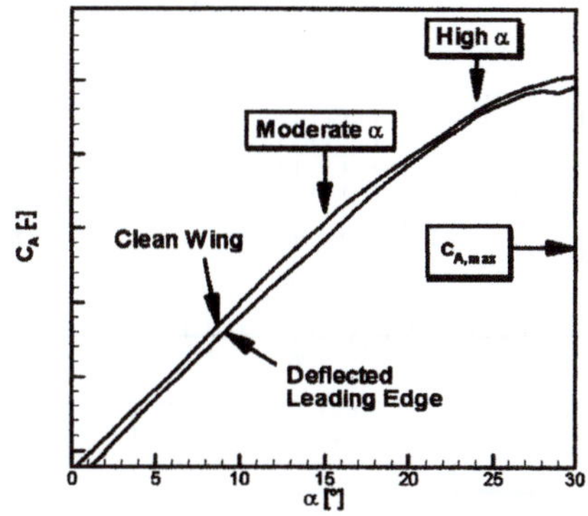

Figure 2 Angle of attacks selected for flowfield measurements.

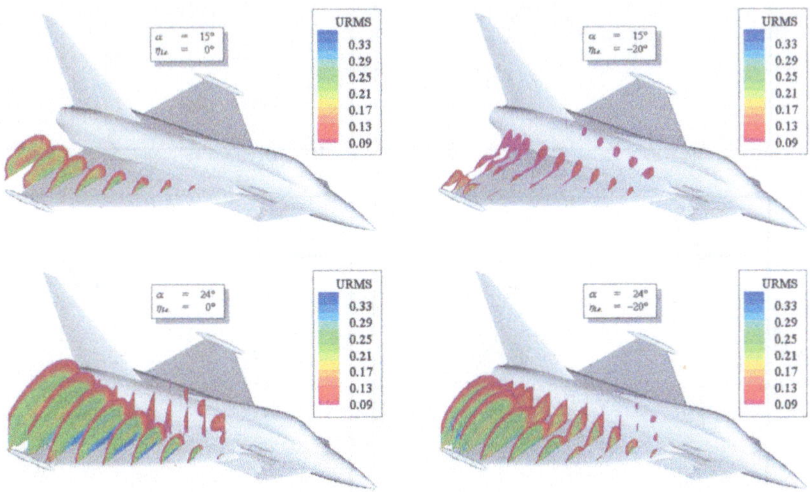

Figure 3 Streamwise turbulence intensity distribution at different cross sections above the wing at $\alpha = 15°$ (top) and at $\alpha = 24°$ (bottom), no flap deflected (left) compared to leading edge flap setting $\eta_{l.e.} = -20°$ (right).

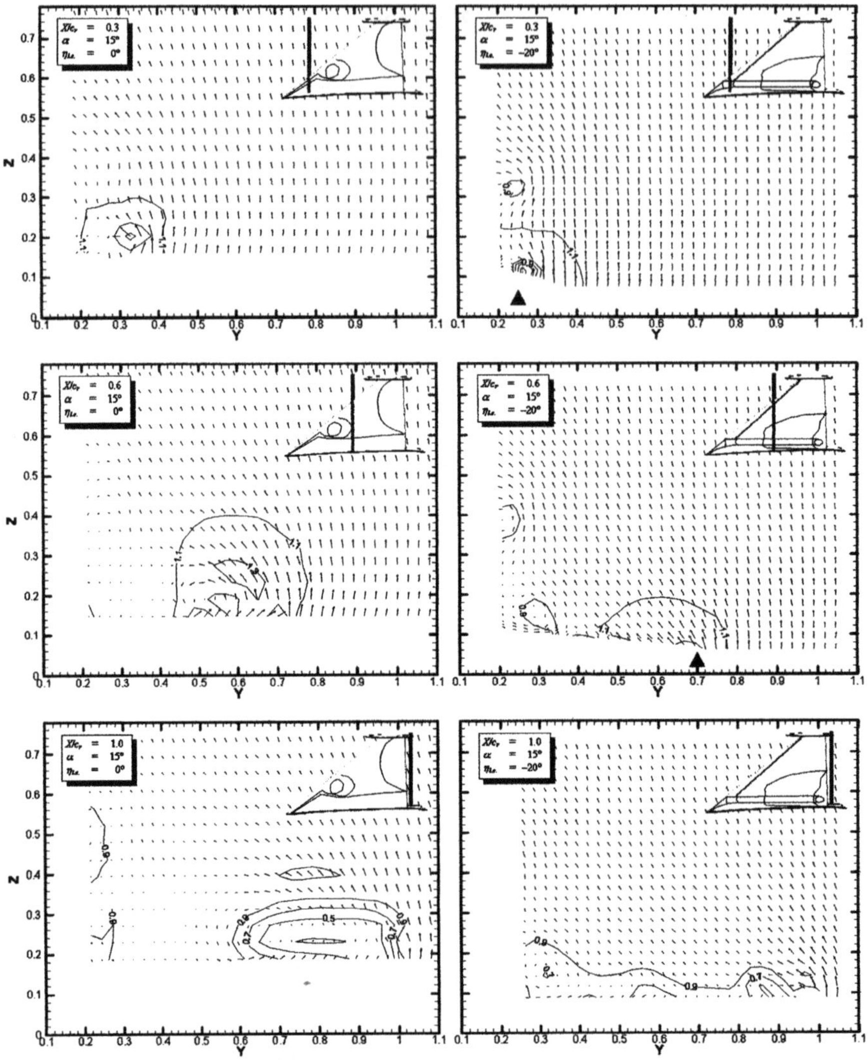

Figure 4 Total velocity distribution and crossflow velocity vectors (body fixed coordinate system) at different cross sections above the wing and $\alpha = 15°$, clean wing (left) and deflected leading edge flap $\eta_{l.e.} = -20°$ (right). The symbol "▲" indicates lateral position of the side edge of the non–deflected leading edge part at cross section $X/c_r = 0.3$ and hinge line intersection at $X/c_r = 0.6$, respectively.

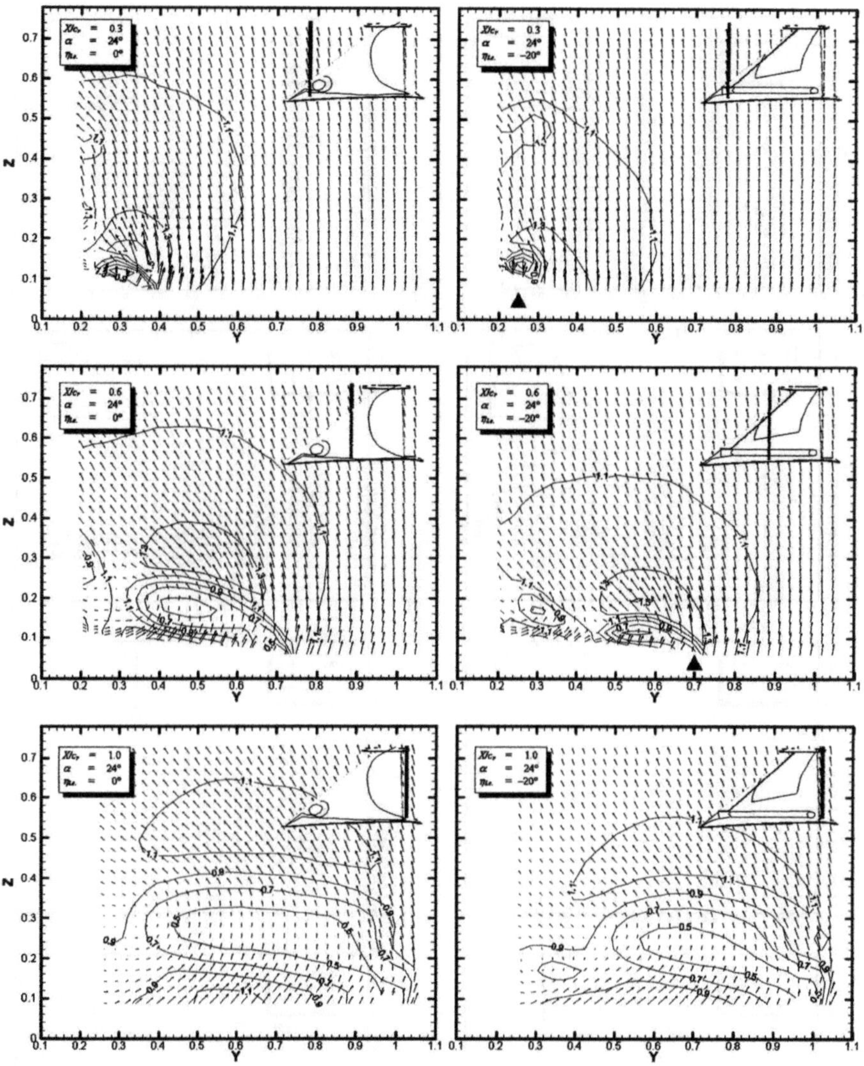

Figure 5 Total velocity distribution and crossflow velocity vectors (body fixed coordinate system) at different cross sections above the wing and $\alpha = 24°$, clean wing (left) and deflected leading edge flap $\eta_{l.e.} = -20°$ (right). The symbol "▲" indicates lateral position of the side edge of the non–deflected leading edge part at cross section $X/c_r = 0.3$ and hinge line intersection at $X/c_r = 0.6$, respectively.

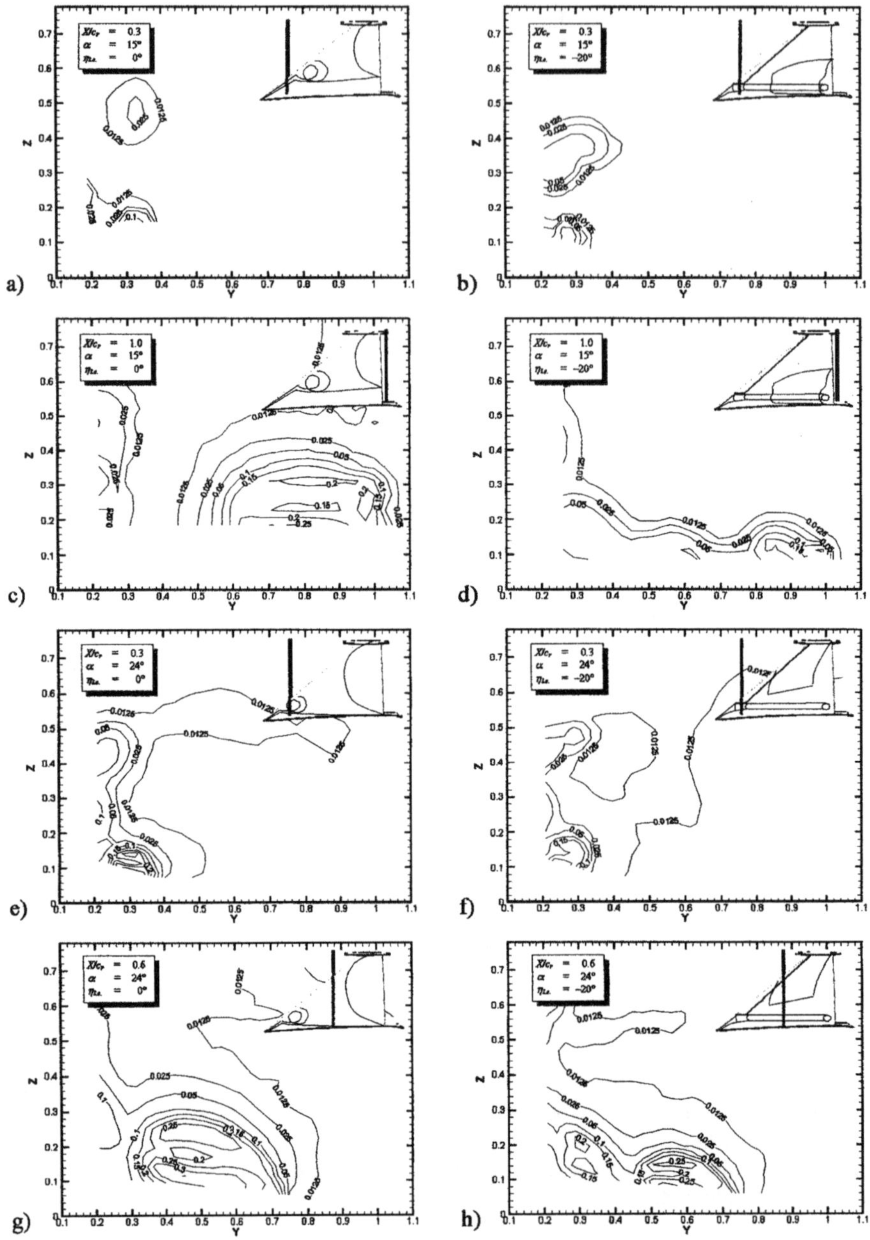

Figure 6 Streamwise turbulence intensity distribution at different cross sections X/c_r above the wing for different leading edge flap settings $\eta_{l.e.}$ at $\alpha = 15°$ (figures a–d) and at $\alpha = 24°$ (figures e–h).

Numerical simulation of maneuvering combat aircraft

Andreas Schütte[1], andreas.schuette@dlr.de
Gunnar Einarsson[1], Britta Schöning[1], Axel Raichle[1], Thomas Alrutz[3],
Wulf Mönnich[2], Jens Neumann[4], Jörg Heinecke[5]

[1] DLR, Institute of Aerodynamics and Flow Technology, Braunschweig
[2] DLR, Institute of Flight Systems, Braunschweig
Lilienthalplatz 7, 38108 Braunschweig
[3] DLR, Institute of Aerodynamics and Flow Technology, Göttingen
[4] DLR, Institute of Aeroelasticity, Göttingen
Bunsenstr. 10, 37073 Göttingen
[5] DLR, Simulation and Software Technology, Köln
Linder Höhe, 51147 Köln

Summary

An overview about recent results of the DLR-Project SikMa-"Simulation of Complex Maneuvers" is presented. The objective of the SikMa-Project is to develop a numerical tool to simulate the unsteady aerodynamics of a free flying aeroelastic combat aircraft, by use of coupled aerodynamic, flight-mechanic and aeroelastic computations. To achieve this objective, the unstructured, time accurate flow-solver TAU is coupled with a computational module solving the flight-mechanic equations of motion and a structural mechanics code determining the structural deformations. By use of an overlapping grid technique (chimera), simulations of a complex configuration with movable control-surfaces are possible.

Nomenclature

Θ	Incidence angle, pitch angle at $\Phi = 0°$	Ma	Mach number
α	Angle of attack	$Re = \frac{V_\infty \cdot l_i}{\nu}$	Reynolds number
Φ	Roll angle	$q_\infty = \frac{\varrho_\infty}{2} V_\infty^2$	Dynamic pressure
Φ_0	Initial roll angle	$\omega^* = 2\pi f \cdot l_i / V_\infty$	Reduced Frequency
$\Delta\alpha$	Angle-of-attack amplitude	$c_M = \frac{M}{q_\infty F l_i}$	Pitching moment coefficient
η	Flap deflection angle	$c_L = \frac{L}{q_\infty F}$	Lift coefficient
t	Time	$c_p = \frac{p - p_\infty}{q_\infty}$	Pressure coefficient
F	Reference area	$c_l = \frac{l}{q_\infty F l_i}$	Rolling moment coefficient
l_i	Chord length of the model		

1 Introduction

The improvement of maneuverability and agility is a substantial requirement of modern fighter aircraft. Currently, roll-rates of $200°/s$ and more can be achieved, especially if the design of the aircraft is inherently unstable. Most of today's and probably future manned or unmanned fighter aircraft will be delta wing configurations. Already at medium angles of attack the flow field of such configurations is dominated by vortices developed by flow separation at the wings and the fuselage. The delay in time of vortex position and condition to the on-flow conditions of the maneuvering aircraft can lead to significant phase shifts in the distribution of loads. In such a case, reliable results for the analysis of the flight properties can only be achieved by a combined non-linear integration of the unsteady aerodynamics, the actual flight motion, and the elastic deformation of the aircraft structure.

Today, these types of data can only be obtained by flight tests, and not during the design period. Flight tests, as well as modifications after the design phase, lead normally to an increase in costs. In order to decrease the costs incurred by extensive flight-tests and the post-design phase modifications, it would be helpful to have a tool which enables aircraft designers to analyze and evaluate the dynamic behavior during the design phase.

The main objective of this paper is to focus on the necessity for developing an interactive, multidisciplinary engineering tool for predicting the unsteady critical states of complex maneuvering aircraft. Such a simulation environment has to bring together aerodynamics, aeroelasticity and flight mechanics in a time accurate simulation tool. In order to deliver such a tool in the near future, the DLR Project SikMa-"Simulation of Complex Maneuvers" has been initiated to combine these three disciplines into one simulation environment.

For validating the numerical simulations several wind tunnel experiments in both the low speed and transonic regime will be done within the SikMa project.

2 Numerical Approach

2.1 CFD Solver TAU

The behavior of the fluid-flow affecting the object of interest is simulated with the TAU-Code, a CFD tool developed by the DLR Institute of Aerodynamics and Flow Technology [3][4]. The TAU-Code solves the compressible, three-dimensional, time-accurate Reynolds-Averaged Navier-Stokes equations using a finite volume formulation. The TAU-Code is based on an hybrid unstructured-grid approach, which makes use of the advantages that prismatic grids offer in the resolution of viscous shear layers near walls, and the flexibility in grid generation offered by unstructured grids. The grids used for simulations in this paper were created with the hybrid grid generator Centaur, developed by CentaurSoft [1]. A dual-grid approach is used in order to make the flow solver independent from the cell types used in the initial grid. The unstructured grid approach is chosen due to its flexibility in creating grids for complex configurations, e.g. a full-configured fighter aircraft with control surfaces and armament, the capability of grid adaptation and straightforward parallelization of all the main TAU modules.

The TAU-Code consists of several different modules, among which are:

- The Preprocessor module, which uses the information from the initial grid to create a dual-grid and the coarser grids for multigrid.
- The Solver module, which performs the flow calculations on the dual-grid.
- The Adaptation module, which refines and derefines the grid in order to capture flow phenomena like vortex structures and shear layers near viscous boundaries, among others.
- The Deformation module, which propagates the deformation of surface coordinates to the surrounding grid.
- The Post-processing module, which is used to convert TAU-Code result files to formats usable by popular visualization tools.

In the Solver module, several upwind schemes, as well as a central scheme with artificial dissipation, are available for the spatial discretization. Both Spalart-Allmaras and $k\omega$ turbulence models are implemented. For steady calculations an implicit LU-SSOR multistage Runge-Kutta time stepping scheme is used [2]. For time accurate computations, an implicit dual-time stepping approach is used. The TAU-Code is parallelized using grid partitioning, and a multigrid approach is used in order to increase the performance.

The TAU-Code can handle simulations containing multiple bodies in relative motion with one another, e.g motion of control surfaces with respect to the aircraft, by use of a hierarchical motion-node structure. The motion of each body can either be calculated internally by the TAU-Code, or supplied by an external program through a Python implemented external interface.

2.2 TAU-Code Extension: Chimera Technique

The chimera technique provides the capability to perform calculations with systems of overset grids. By allowing large relative body movement without the need for local remeshing or grid deformation, the technique is invaluable for the simulation of maneuvering combat aircraft, where large-amplitude control surface deflections and/or store release are a standard part of the simulation. The current implementation can handle multi-body simulations where the overset-grid boundaries have been predefined; a version that allows 'automatic-hole-cutting' is currently under development. The chimera search algorithm, which is based on a state-of-the-art alternating digital tree (ADT), is available for both sequential and massively parallel architectures (Linux Clusters). A more detailed description of the chimera approach is given in [5].

2.3 Flight Mechanics

For the numerical simulation of the flight mechanics, the simulation environment SIMULA developed at the DLR Institute of Flight Systems is used [6]. SIMULA provides the three basic functionalities necessary for flight simulation and flight control purposes: trimming, i.e. the determination of the initial state and control values, linearization and stability analysis, and simulation, i.e. the numerical integration of the equations of motion.

Single and multi-body flight-mechanic models, ranging from 1 to 6 degrees of freedom, are made available to the simulation by SIMULA. The amount of data that is exchanged between SIMULA and TAU is of a scale that can be easily transported directly through a TCP/IP socket connection, which is offered by the TENT simulation environment.

2.4 CSM-Code

For the coupling of the aerodynamic and structural dynamic simulations in the time domain, a loose coupling scheme has been implemented. The coupling scheme is conservative with regards to the forces, moments and the work performed on both the aerodynamic and structure dynamic side. Furthermore, it is verified that no dissipation or accumulation of net energy occurs.

The main characteristics of the aeroelastic fluid structure interaction in the time domain are as follows:

- loose coupling of computational fluid dynamics (CFD) and computational structure dynamics (CSD) through file input/output,
- use of an implicit or explicit Newmark algorithm for the time integration of the CSD equations of motion,
- use of different scattered data interpolation methods with and without compact support radius for coupling in space domain,
- data exchange based on adjusted conventional serial staggered (CSS) algorithm modified with a predictor-corrector scheme,
- structural behavior is described by the complete FE-Model of the delta-wing and the support.

2.5 Integration Framework

The integration framework TENT [8] provides a graphical user interface for controlling and monitoring coupled simulation workflows. The various codes used in the SikMa simulations will be made available in the TENT system, where a simulation workflow can be built by connecting icons representing each code using a graphical workflow editor. Java wrappers containing the basic control functionality for the TAU and SIMULA applications are already integrated in the TENT environment. The wrapper for the CSM-Code as well as the extension of the functionality to handle the coupling between all three disciplines is under development.

While TENT is providing the data transfer and the communication between the applications, the communication logic for the simulation workflow is contained within a coupling manager script. The coupling manager is a user-extensible script based on a Python and Java interface, where functionality to control the flow of the simulation has been implemented.

3 Experimental Data

For the validation of the numerical simulation software, various wind-tunnel experiments, designed specifically for the SikMa project, are performed. Experimental data, both steady and unsteady, are available for a 65°-swept delta-wing-fuselage-model-configuration which has been tested in the DNW Transsonic-Wind-Tunnel Göttingen (DNW-TWG). The model has movable trailing-edge flaps an can be used for both guided and free-to-roll maneuver simulations around its longitudinal axis. The model has a chord length of 482mm and a span of 382mm. For the verification of the aerodynamic-structure coupling a steady and dynamic system identification of the delta-wing and the support within the wind tunnel is done. The system parameters are used to setup the FE-model for the coupled simulation.

The main experiments are done in the DNW Low-Speed-Wind-Tunnel Braunschweig (DNW-NWB). In order to perform these experiments, a wind-tunnel model has been designed and built for the SikMa project. The model, shown in **Fig. 1**, is based on the X-31 experimental high angle-of-attack aircraft configuration. The X-31 model is a 1:7 scaled model with a span of 1000mm and an overall length of 1800mm. The model is equipped with remote controlled moveable control devices driven by internal servo-engines. Measurement equipment is installed to determine the aerodynamic forces and moments on the model, as well as span-wise pressure distributions at locations of 60% and 70% chord length. The experiments include steady-state measurements using PSP-"Pressure Sensitive Paint", which provide detailed information on the surface pressure distribution for the whole wing. The experiments will culminate with maneuver simulations, where the movement of the aircraft and the control devices will be synchronized. For the maneuver experiments the model will be mounted on the MPM-"Model Positioning Mechanism" of the DNW-NWB.

4 Results

For the verification and validation of the simulation environment the results of the numerical simulations are compared against data collected from various experimental simulations. To show the capability of the TAU-Code to predict the unsteady aerodynamic behavior of configurations with vortex dominated flow fields the delta-wing-configuration described in section 3 is used. **Fig. 2** shows a result of a delta-wing in rigid body pitching motion. In this calculation the $k\omega$ turbulence model is used. For all simulations presented a central spatial dicretization scheme is used. Depicted is the hysteresis loop of the lift coefficient over the angle-of-attack α. The wing is pitching around $\alpha = 9°$ with an amplitude of $\Delta\alpha = \pm 6°$. The wing is oscillating with a reduced frequency of $\omega^* = 0.56$. It is seen that the calculation compares well with the experimental data. For higher angles-of-attack the calculation tends to predict higher lift, because with the common turbulence models a different load distribution over the wing is predicted compared to the experiment. At higher angles of attack the vortex structure and the location of vortex breakdown will not be correctly predicted. The same behavior has been found to occur in simulations using the structured DLR FLOWer-Code.

In **Fig. 3** the result of a coupled simulation between CFD and flight-mechanics is shown using the delta-wing with trailing-edge flaps. To simulate the motion of the control surfaces (trailing-edge flaps) the chimera approach has been used. The maneuver shown in Fig. 3 is a 1 DoF rotation around the longitudinal axis of the delta-wing induced by an asymmetric deflection of the flaps by $\eta = \pm 3°$. The initial aircraft attitude is at $\alpha = 9°$ and $\Phi_0 = 0°$. It is seen that the wing enters a periodic roll motion in both the numerical and experimental simulations. One of the sources for the difference seen between the numerical and experimental results is the mechanical friction in the experimental setup, which is not taken into account in the numerical simulation, thus leading to a higher frequency of rotation. Currently only a one equation turbulence model is used in the coupled simulations. It is known that for sharp leading-edge delta-wings the $k\omega$ turbulence model delivers better results and will be used in further simulations.

The capability to predict the elastic deformations of the delta-wing configuration during a guided roll maneuver is shown in **Fig. 4**. In the coupled simulation between the TAU-Code and the structural mechanics tool developed within SikMa the delta-wing and the rear-sting support are considered to be elastically deformable. For the coupled simulation the finite element model [7] takes into account both the delta wing configuration as well as the the flexible support. The FE-Modell is validated based on results of both ground vibration and static deformation tests.

In **Fig. 5** the corresponding history of the model deflection due to the elasticity is seen. Depicted is the displacement of the delta-wing nose-tip and sting relative to the rigid-body motion case. It is seen that the wing tip is describing an elliptic motion during the rotation. The green loop in Fig. 5 shows the tip movement from the system identification in the experiment due to the integration of the acceleration. Because of the integration the shift in z-direction can not be captured. However, it is shown that the numerics predict the characteristic movement of the wing tip accurately. Due to this deformation the effective angle-of-attack at the same roll angle is higher in the elastic case. This leads to a higher amplitude of the rolling moment, as is seen in Fig. 4.

For the X-31 configuration, results from steady-state numerical simulations have been obtained. These simulations show the capability of the TAU-Code to simulate complex delta-wing configurations with rounded leading-edges. **Fig. 6** shows the numerically simulated 3D flow field over the X-31 configuration, which is a good indication of the complexity of the vortex flow topology over the wing and fuselage. Comparisons with experimental data show good agreement regarding the vortex topology. In **Fig. 7** an oil flow picture of the X-31 clean-wing from low speed experiments is shown. The angle-of-attack is $\alpha = 18°$ at a Reynolds number of 1.0Mio. The separation line of the strake vortex and the main wing vortex as well as the attachment line of the main wing vortex near the leading-edge is emphasized. In **Fig. 8** the corresponding CFD calculation is depicted. It is seen that the flow topology from the calculation fits quite well with the experiment. Further experimental results delivering steady pressure distributions upon the wing were done within another X-31 test campaign. The PSP result at $\alpha = 18°$ at a Reynolds number of 2.07Mio is shown in **Fig. 9**. Comparing the pressure distribution in Fig. 9 with the CFD calculation in **Fig. 10** shows that the main footprints of the vortices are captured by the numerics, but the suction-strength and the vortex location are not. However, in this case the experiment is done with mounted leading-edge flaps. The gaps between the leading-edge flaps probably influence the vortex formation considerably. Additional calculations will follow taking all geometric features into account.

5 Conclusions

In this paper the activities and recent results of the DLR-Project SikMa were presented. In SikMa a simulation tool will be developed that is capable of simulating a maneuvering elastic aircraft with all its moveable control devices. The simulation tool combines time-accurate aerodynamic, aeroelastic and flight-mechanic calculations to achieve this objective. Preliminary verification of the functionality of the simulation tool has been shown by simulating a sharp leading-edge delta-wing during a guided motion pitching maneuver and free-roll maneuvers due to flap deflections. Furthermore, first perspectives were presented regarding the time accurate coupling between the TAU-Code and the numerical Structure-Mechanical Tool. Initial results of the steady flow field around the X-31 configuration were presented.

Acknowledgements

The authors would like to thank Dr. Burkhard Gölling for providing the wind-tunnel data and for his engagement managing the experimental simulation tasks at DLR Göttingen within the project SikMa.

References

[1] *Centaur Soft:* http://www.Centaursoft.com

[2] Dwight, R.: Time-Accurate Navier-Stokes Calculations with Approximately Factored Implicit Schemes. Proceedings of the ICCFD3 Conference Toronto, Springer, (2004).

[3] *Galle, M.; Gerhold, T.; Evans, J.:* Technical Documentation of the DLR TAU-Code DLR-IB 233-97/A43 1997

[4] *Gerhold, T.; Galle, M.; Friedrich, O.; Evans, J.:* Calculation of Complex Three-Dimensional Configurations employing the DLR TAU-Code AIAA-97-0167 1997

[5] *Madrane, A.; Raichle, A.; Stürmer, A.;:* Parallel implementation of a dynamic overset unstructured grid approach. ECCOMAS Conference Jyväskylä Finland, 24.-28. July 2004.

[6] *Mönnich, W.; Buchholz, J. J.:* SIMULA - Ein Programmpaket fuer die Simulation dynamischer Systeme - Dokumentation und Benutzeranleitung - Version 2 DFVLR Institutsbericht IB 111-91/28, 1991

[7] *Neumann, J.:* Strukturmechanische und strukturdynamische Finite Element Modelle des Windkanalmodells "AeroSUM" mit Halterung. DLR-IB, IB 232-2003-J01, (2003).

[8] *Schreiber, A.:* The Integrated Simulation Environment TENT. Concurrency and Computation: Practice and Experience, Volume 14, Issue 13-15, S.1553-1568, 2002.

[9] *Schütte, A.; Einarsson, G.; Schöning, B.; Madrane, A.; Mönnich, W., Krüger, W.:* Numerical simulation of manoeuvring aircraft by aerodynamic and flight mechanic coupling. RTO AVT-Symposium Paris, 22.-25. April 2002.

Figures

Figure 1 X-31 Remote control model in the DNW-NWB Braunschweig.

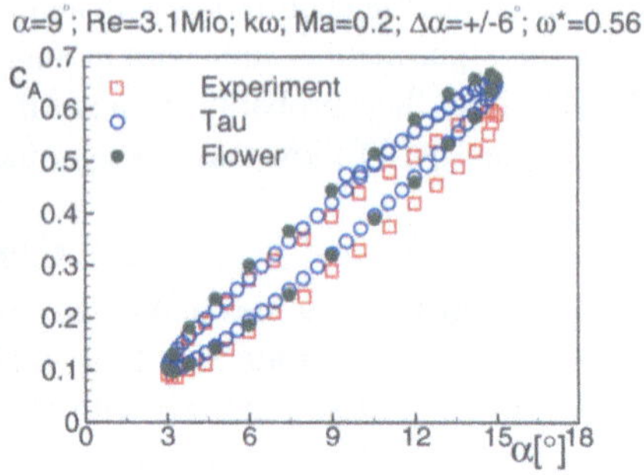

Figure 2 Lifting coefficient vs. angle of attack in pitching motion of a 65°-swept delta-wing.

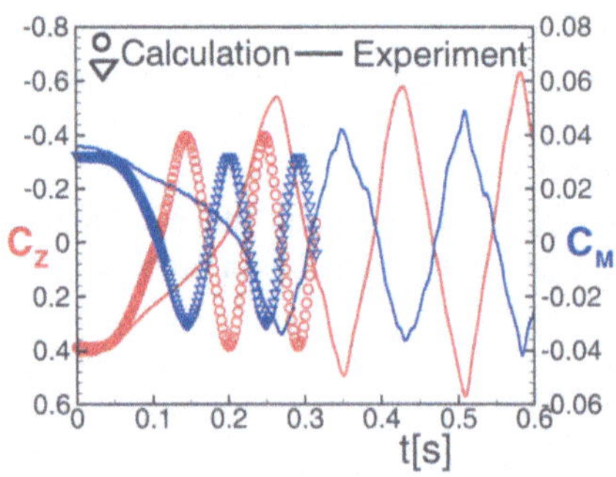

Figure 3 1 DoF Free-to-Roll maneuver of delta-wing-flap conf. through trailing-edge flap deflection of $\eta = \pm 3°$.

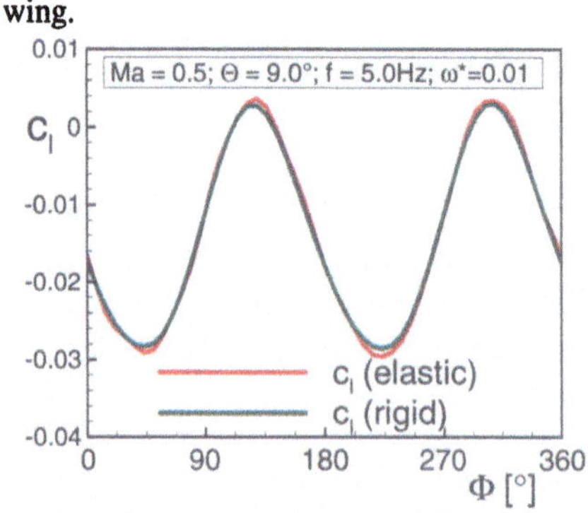

Figure 4 Comparison of rolling moment between rigid- and elastic-body motion of delta-wing during constant rotation. Time accurate coupled CFD(Euler)-CSM simulation.

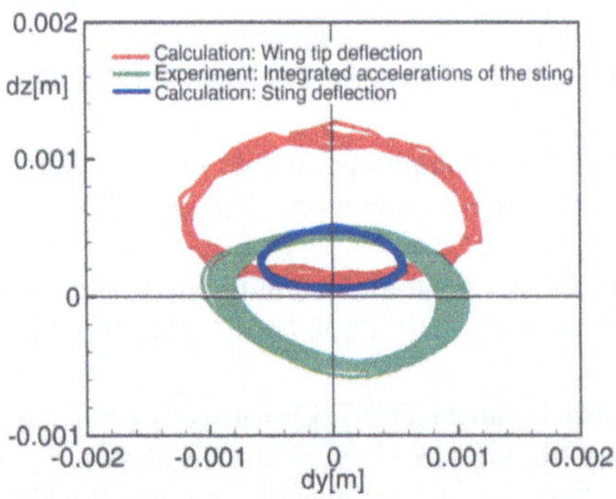

Figure 5 History of the delta-wing nose and sting deflection during elastic-body motion comparison with experiment. Time accurate coupled CFD(Euler)-CSM simulation.

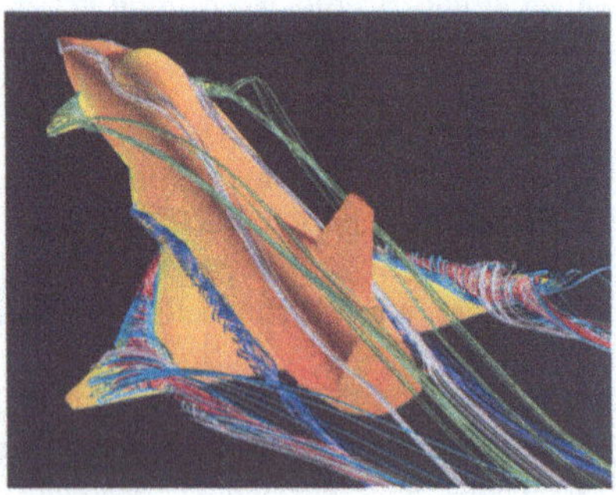

Figure 6 3D flow field over the X-31-Configuration at 18° angle-of-attack.

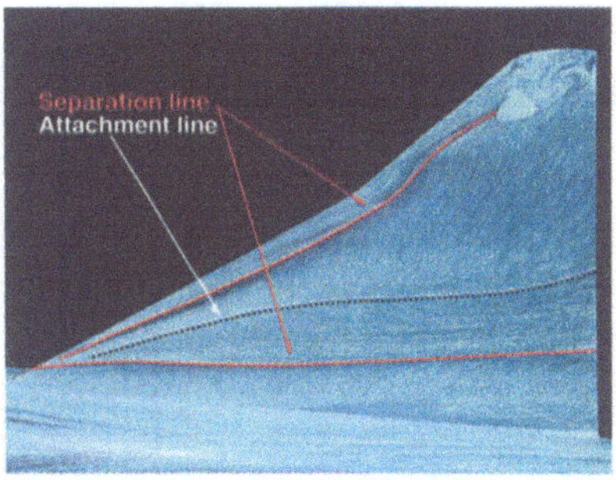

Figure 7 Oil flow visualization of the X-31 clean wing at $\alpha = 18°$, Re=1.0Mio.

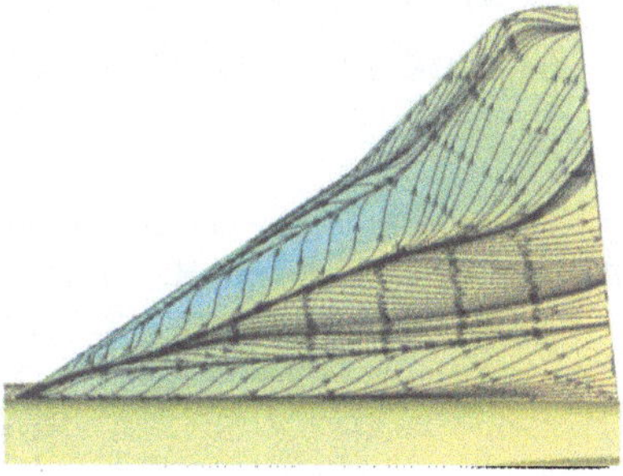

Figure 8 TAU calculation: Visualization of surface streamlines at $\alpha = 18°$, Re=1.0Mio.

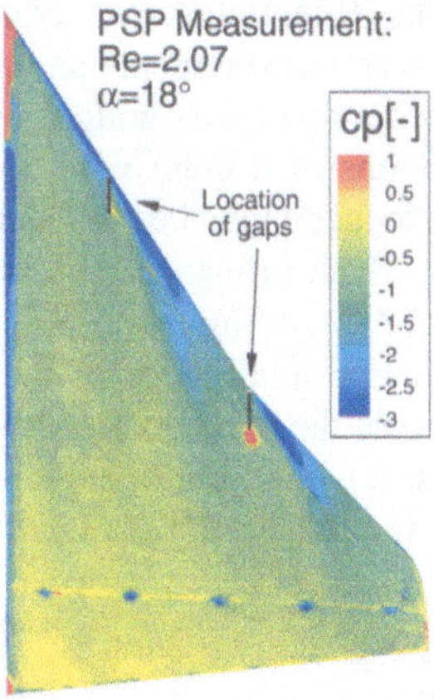

Figure 9 Steady PSP measurement of the pressure distribution over the X-31 wing.

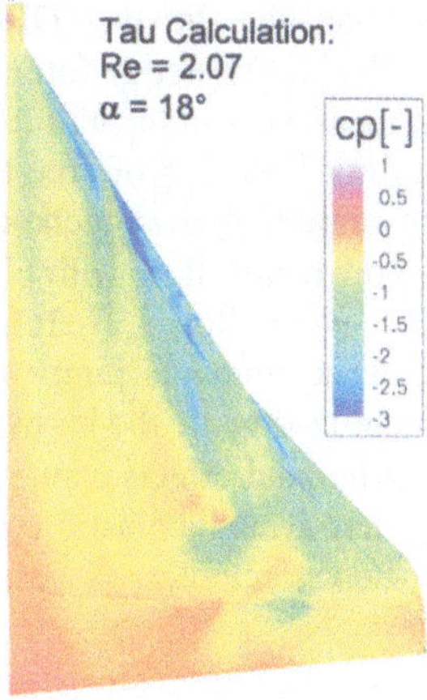

Figure 10 Steady TAU RANS calculation of the pressure distribution over the X-31 wing (clean wing).

Experimental investigation on periodic rolling of a delta wing flow at transonic Mach numbers

B. GÖLLING[1] and O. ERNE[2]
[1] Airbus Deutschland GmbH, Hünefeldstr. 1-5, 28199 Bremen; info@goelling.de;
[2] GOM mbH, Mittelweg 7-8, D-38106 Braunschweig; o.erne@gom.com.

Summary

Experimental simulation was conducted in the Transonic Wind-Tunnel Göttingen (TWG) of DNW with the 65 degrees cropped delta wing **AEROSUM** model, which was managed in the DLR project **SikMa** (Simulation of complex maneuvers). Besides unsteady force and surface pressure measurements an optical deformation measurement technique was applied. In order to a better understanding of the instability behavior of free-to-roll rolling motions of a delta wing flow a high-time resolution optical technique was applied to get hints i.e. to coupling effects of aerodynamic flow with elastic wing model which influence the stability of the delta wing flow. Therefore defined periodic rolling motions were investigated. Discontinuous jumps of normal deformation amplitudes of the rear flaps are recognized while the harmonic rolling motion of the wing body. Although technical or frictional effects cannot be excluded, it seems possible to explain this behavior by aerodynamic-elastic coupling effects of the delta wing flow with the elastic wing-flap structure.

1 Introduction

In a series of wind tunnel experiments at transonic Mach numbers up to Ma=0.85 various complex maneuver-like delta wing movements were simulated experimentally supported by a roll-rig. The complete experimental set-up – AEROSUM model and roll rig with its opportunities – is well described in **[1]**. However, new experimental results in analysis of rolling instability of a delta-wing flow in comparison with numerical computations can be given by further studying the coupling interaction of vortex flow instabilities with the structure of the elastic delta-wing body **[2]**. Especially, the rear wing flaps initializing rolling movements and changing vortex development of the delta wing flow are elastic structure elements. The defined flap angle will not be aimed exactly, caused by aerodynamic loads, which lead to another, heavily predictable rolling movements. Furthermore, it has to be separated into stationary and unsteady flow-structure interaction due to the response of the flow-structure system. Therefore, unsteady force and surface pressure measurements were combined with an optical technique, which measured normal surface deformation of the wings and the rear flaps during the maneuver movements of the model. Deformation results, especially of the rear flaps, will be discussed with the help of unsteady aerodynamic data as of rolling angle and moment coefficients.

2 Deformation measurement and experimental set-up

The measurement of 3D displacement and deformation of parts under load is a complex task, which is solved with many kinds of different sensors like Linear Variable Displacement Transducer (LVDT) and draw wire sensors. In case of dynamic applications accelerometers are often used. However, in case of aerodynamically loaded parts non-contact measuring techniques are essential, which cope for the complex task and do not influence the aerodynamic performance. While two-dimensional estimations are covered using a simple video camera, precise measurements are in need of non-contact 3D measurement tools.

2.1 Principle of optical deformation measurement technique

In this application, PONTOS **[3]**, an optical measuring system, is used to determine the 3D deformation of the model under specific wind loads. The system is based on photogrammetry using a calibrated stereo-camera set-up. In combination with an embedded digital image processing, PONTOS is able to determine the 3D coordinates of optical markers, which are applied on the object surface. Measuring the 3D coordinates at specific states of loading, i.e. before and after loading, 3D displacements of the optical markers and therefore object displacements and deformations can be calculated.

The photogrammetric arrangement of the PONTOS system consists of two area cameras, mounted on a frame. The position of each marker at the 2D image is determined with sub-pixel accuracy utilising the embedded digital image processing. Both cameras and the object build up a triangle, using both cameras as a base (**Fig. 1**). Knowing the position of the image point P1(x1,y1) and P2(x2,y2) on the camera chip and the geometrical and optical properties of the set-up, the position of the object point P(X,Y,Z) can be determined. The geometrical and optical properties of the set-up are determined by a calibration. In this calibration a well-known object is imaged in different positions in front of the camera set-up. Using the images, a geometrical model of the camera set-up can be described, including the optical properties of the imaging set-up. This model and out of it the transformation matrix is used to determine the co-ordinates of point P(X,Y,Z).

2.2 Experimental set-up of optical system

In our experiment, the PONTOS system was directly mounted at the outside of the wind tunnel, viewing through a 15 mm thick glass window (**Fig. 2**). The stereo-camera set-up consists of two high-speed CMOS cameras, which are able to acquire images at a resolution of 1.3 million pixel with up to 480 fps and a triggered LED light system. Both camera modules were aligned to image the airplane model inside the wind tunnel. After alignment, the calibration procedure of the set-up was performed using a flat calibration artefact. This procedure is semi-automatically done using the PONTOS software. The artefact was positioned inside the wind tunnel to cope for the glass window and take its optical properties into account.

After calibration, the model was prepared for measure-ment. A number of 64 optical markers were applied on the main wing body and, especially, on the rear edge flaps. These markers are 0.1mm thin, lightweight (0.1g) and self-adhesive retro-reflective targets. Its optical properties in combination with the triggered LED light system are perfectly suited to cope for the very short illumination times of about 100μs at the fast frame rates.

Measurements were taken under specific loads and movements of the model. For each specific load scenario, a series of up to 200 image pairs was taken. At each image pair the position of the optical markers was automatically determined and therefore the 3D coordinates of each marker at each stage automatically calculated. Then, the evaluation was split into a determination of displacement and a deformation analysis. For the displacement analysis of the markers on the model the displacements will be computed relative to sensor position. Consequently the deformation analysis was performed utilizing global reference points, which were defined at the stiff section at the centre of the model. An automatic transformation of each stage relative to these global reference points was calculated. Thus, the rigid body motion of the model due to vibration or forced model displacement is compensated and the pure model deformations are displayed. This deformation can be mapped onto the 2D-image series and for each stage.

3 Results

In case of free-to-roll motions (initiated by flap movements) we obtain asymmetric behavior of the rolling moment coefficient (here not shown), e.g. that the asymmetric vortex bursting occurs on the luv side at different rolling angle dependently of the direction of the movement. In order to explain this asymmetric behavior there were used defined experimental conditions for periodic rolling motions (controlled by roll-rig motions) in order to investigate the deformation of wings and flaps of the model. Mechanical or friction problems could be addressed for these differences due the mechanical bearing of the flaps on the main body of the delta wing and due to the mechanical construction of momentum transfer from electric motor to the flaps, but more probably seem flow-structure interactions.

For defined periodic rolling motions it is shown in **Fig. 3** that at symmetric flow conditions, e.g. at Ma=0.50, AOA of 0 deg. and starting rolling angle at 90 deg. the normal deformation of both flaps, presented by the deformation of the outer side edges of the flaps, have the same amplitude, roughly, and are working in an opposite manner, which is expected due to the periodic motion of the delta wing around 90 degrees with an amplitude of roughly plus/minus 30 degrees. By applying an AOA of 17 degrees the normal deformation of the left rear flap becomes stronger (luv side) and the right rear flap (lee side) are smaller which is reasonable due to the corresponding strength of the mean aerodynamic loads. In comparison of the delta wing flow at Ma=0.50 (**Fig. 4**) and Ma=0.80 (**Fig. 5**) the deformation amplitude of the luv side flap increases with increasing Mach number. In the last case the corresponding rolling angle and rolling moment coefficient is given in **Fig. 6**, which underlines the periodic motion.

However, the deformation of the flaps (please take a zoom view on **Fig. 5** and **Fig. 6**) is characterized as a periodic function in time or rolling angle of a first order, only. Exactly, there are some discontinuities in normal deformation amplitude and rolling moment coefficient. While the rolling angle amplitude is increasing continuously until its maximum is reached, the deformation of the luv side flap acts in steps. By a small rising rolling angle the deformation of this flap is changed discontinuously, and this happens several times until the maximum rolling angle is reached. The series of discontinuous deformation D_Z are displayed in **Fig. 4** and **Fig. 5**. Especially for high Mach numbers like Ma=0,80 (**Fig. 5**) it was observed that the changes of the rolling angle for one series of discontinuous deformations is about only 5 degrees, with respect of continuous deformation about 10 degrees, roughly. There must be an additional energy effect, which cause these discontinuities. In **Fig. 6** it is shown that these discontinuities are observed for two cases: (1) increasing rolling angle with increasing positive rolling moment, and (2) decreasing rolling angle with decreasing and negative rolling moment.

One assumption could be that there are some mechanical or friction effects which act on the flap control, additionally, to the aerodynamic loads, which lead to this discontinuous steps in normal deformation amplitudes. Otherwise this should also happen at lower Mach numbers, which was not observed. Another point of view could be the response of the elastic structure of the wing-flap model system due to the unsteady flow, which were induced by unsteady aerodynamic loads during periodic rolling movements. Typical *eigenmodes* of the delta wing starts at 25 Hz. A detailed report is given in **[2]**. An elastic resonant mode could trigger the evolution of vortex flow for this short time period of response of the flow-structure-interaction (see also **[4]**), so that the resulting discontinuous jumps of normal deformation amplitudes could be formed by one or more elastic modes of the wing-flap model system.

Conclusions

A new phenomena in dynamic rolling motions of a delta wing was explored by the help of a modern optical deformation measurement technique with a rapid data management (PONTOS). The most acceptable explanation was given with respect to coupling mechanisms of aerodynamics and aeroelastics, so that the high resolution of this deformation measurement technique in connection with unsteady force and pressure measurements allows the direct validation of aeroelastic numerical codes. Therefore this kind of technique is a sophisticated tool for further investigations and deeper physical understanding.

Acknowledgements

The authors thank Director Professor Dillmann of Institute of Aerodynamics and Flow Technology and section High-speed Configuration of Dr. Rosemann for grants and for co-working. Further, we thank the DNW teams of Dr. Jacobs and Dr. Bruse. Also, we thank EADS-M for cooperation.

References

[1] H. Psolla-Bress, H. Haselmeyer, A. Heddergott, G. Höhler, H. Holst; High-roll rate experiments on a delta wing in transonic flow. ICIASF, 2002.

[2] D. Hoffmann, J. Neumann, M. Sinapius; Strukturmechanische Identifikation von Halterung und Windkanalmodell „AeroSUM", IN-232-2002 C06; DLR, Institut für Aeroelastik, Göttingen, 2002, 40 pp.

[3] R.C. Nelson, A. Pelletier; The unsteady aerodynamics of slender wings and aircraft undergoing large amplitude maneuvers. Progress in Aerospace Sciences, vol. 39, 2003, pp. 185-248.

[4] O. Erne, PONTOS – a modern optical deformation measurement technique for application in research and development. Presentation, Workshop, 2004. GOM, Brunswick, 2004.

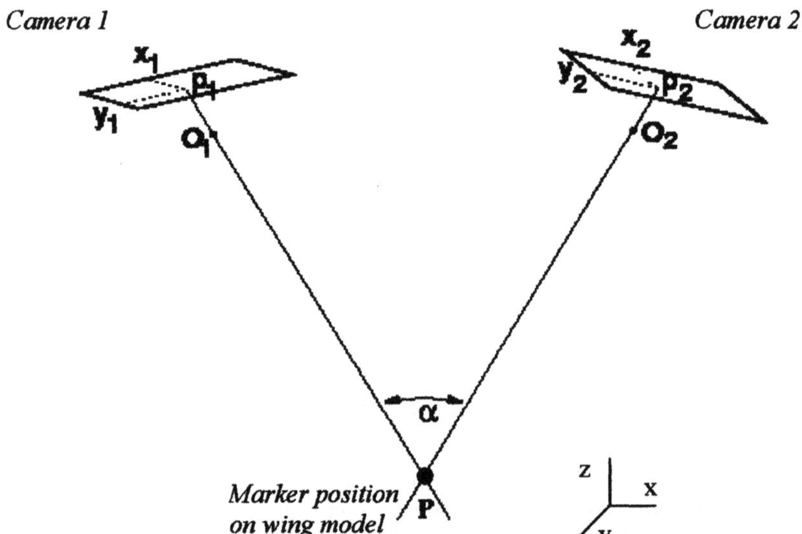

Figure 1 Measurement principle to determine the object coordinates using two known image points.

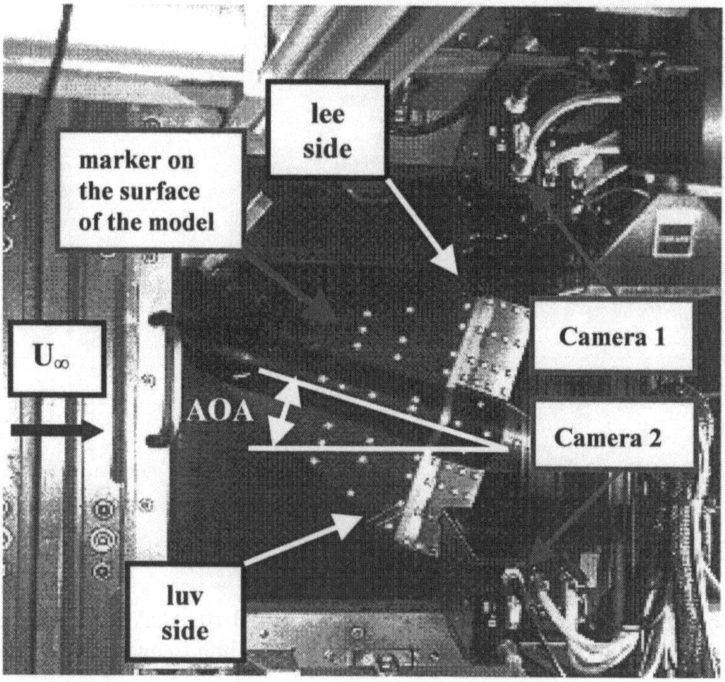

Figure 2 PONTOS camera system with marker on delta wing model in TWG in the adaptive test section with $1m^2$. Model sizes: chord=483mm, span=383mm.

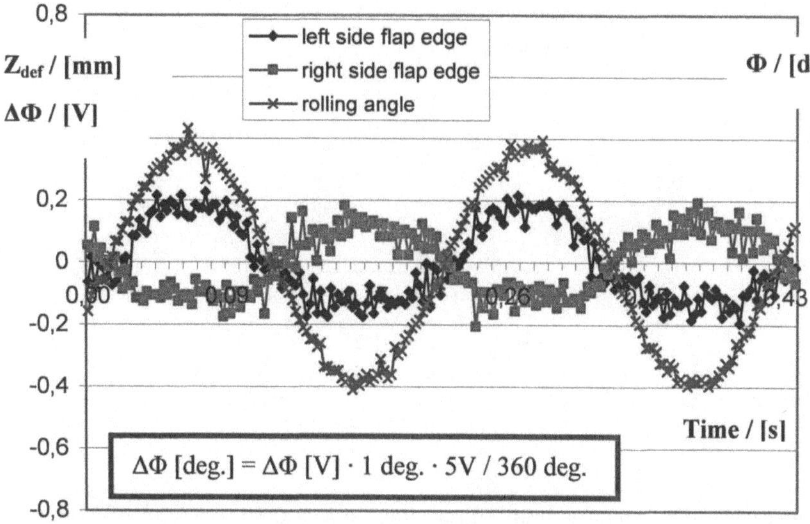

Figure 3 Z-Deformation [mm] of rear flaps of a delta wing at Ma=0.50, no AOA, rolling ampl. ±30 deg. (**starts at 90 deg.**), rolling frequency 4.77 Hz.

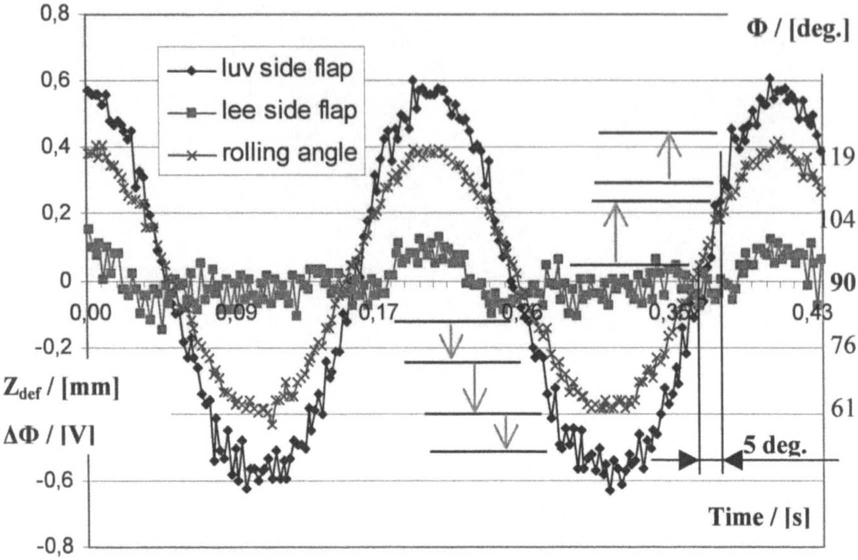

Figure 4 Z-Deformation [mm] of rear flaps at Ma=0.50, AOA of 17 deg., rolling angle amplitude ±30 deg. (***starts at 90 deg.***), rolling frequency 4.77 Hz.

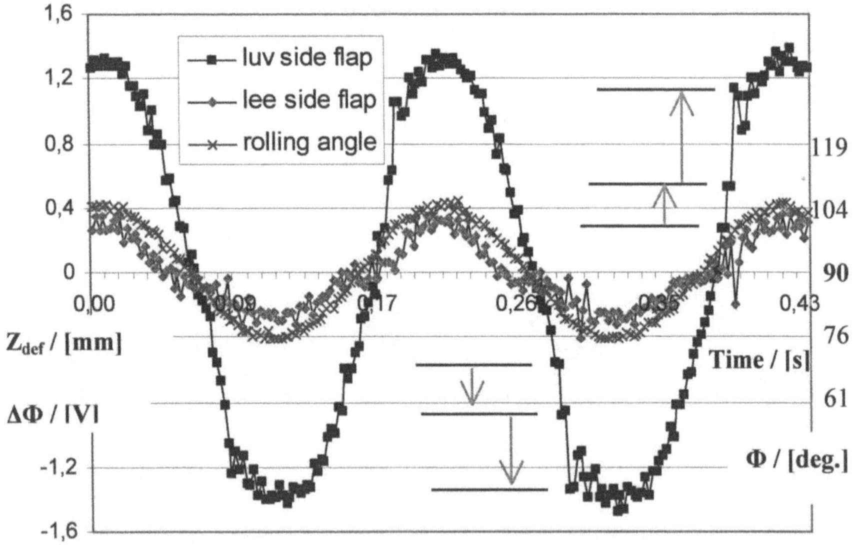

Figure 5 Z-Deformation [mm] of rear flaps at Ma=0.80 and AOA of 17 deg., rolling angle ampl. ±30 deg. (***starts at 90 deg.***), rolling frequency 4.77 Hz.

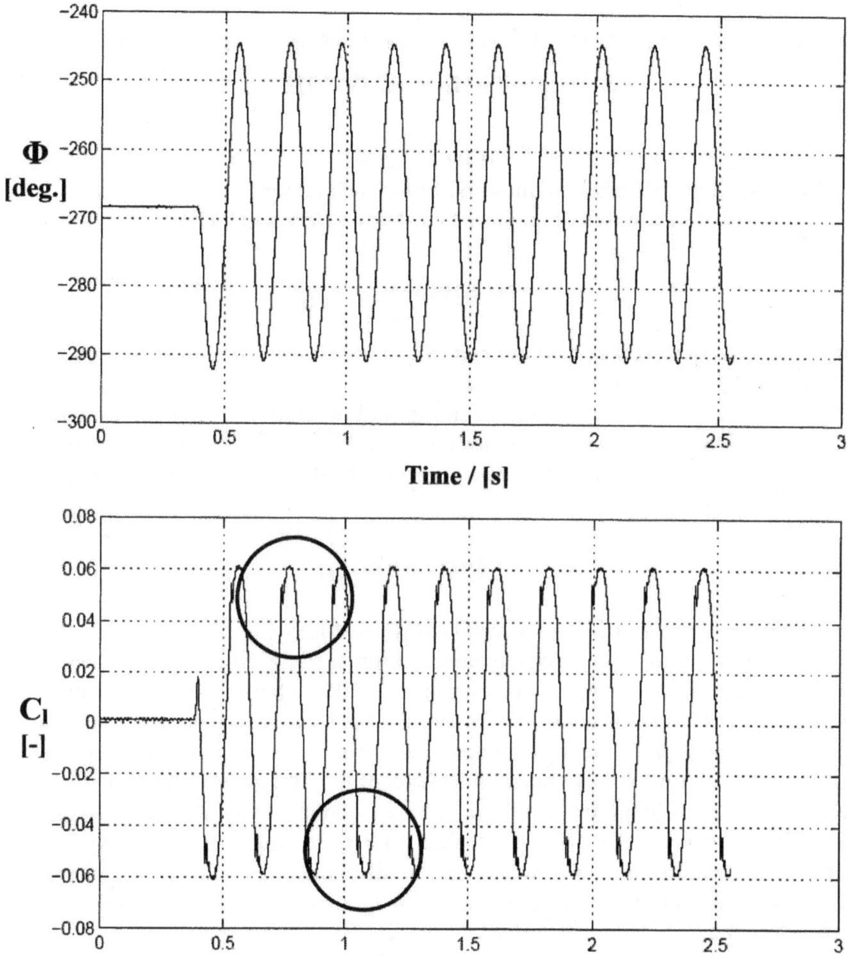

Figure 6 Rolling angle and rolling moment coefficient of the delta wing at Ma=0.80 and AOA of 17 deg., rolling amplitude ±30 deg. (***starts at 90 deg.***), rolling frequency 4.77 Hz.

Numerical Design of a Low-Noise High-Lift System for a Supersonic Aircraft

U. HERRMANN
DLR, Institute of Aerodynamics and Flow Technology, Lilienthalplatz 7, D-38108 Braunschweig, Germany, Ulrich.Herrmann@DLR.de

Summary

Supersonic commercial transport aircraft (SCT) are known for their short travel times and also for their low-speed noise.
This paper focuses on the aerodynamic design of low-drag SCT high-lift systems contributing significant low-speed noise reductions by enabling steeper climbing. The high-lift system design - relying on separated flow- and validation work at DLR will be reported.

1 Introduction

Supersonic commercial transport aircraft (SCT) enable economic supersonic flight with low aspect ratio, thin wings featuring high leading edge sweep. Even for moderate angle of attack (AoA) at low-speed the flow starts to separate from the leading edge (LE) and forms stable vortices. These vortices generate favourable additional lift with non-linear characteristics but add significant drag. They are therefore a reason for the low-speed noise level of the first generation SCT's.
Several disciplines (engines, structure, ..) can contribute to reduce low-speed noise. The EC-Project EPISTLE [1] aims to reduce noise using aerodynamic means only. Because thin SCT wings can hardly house retractable, slot opening high-lift systems -similar to subsonic aircraft- modern high-lift systems for SCT are therefore mechanically of a different kind.
Droop nose based high-lift systems will most likely be applied and are analysed here regarding their drag reducing potential. The obtained drag reductions lead to significant noise reductions by enabling steep climbing near airports. This paper presents the aerodynamic design work at DLR resulting in a low-drag and thus low-noise SCT high-lift system. A validation of the predicted aerodynamic performance gain of the DLR design is added at the end of the paper.

2 SCT High-lift system principle and design objective

Modern SCT high-lift systems will use LE and trailing edge (TE) flaps to adapt the wing camber to the low-speed condition. **Fig. 1** shows a configuration with a segmented LE (five flaps) and TE (three flaps). Because slotted devices as on subsonic aircrafts are mechanically unfeasible on thin SCT wings these flaps will be realized as drooping devices. As no slots open the boundary-layer cannot be energized. This needs to be considered in the design.

The segmented LE-flaps adapt the wings camber and in addition enable drag reductions through high suction forces. The task of the TE-flaps is to increase the lift and to guarantee stability and control during all flight phases. To limit the numerical effort EPISTLE concentrates on the drag reducing aspects of LE-flaps only.

Drag reduction due to deflected LE-flaps is possible because the local force vector is rotated with the deflection angle. This vector rotation transforms some of the local lift into a suction force that reduces the drag. The suction force may be amplified in case a vortex forms and can be kept on the deflected part of the LE-flap. This drag reduction principle causes a small lift degradation compared to a clean wing (no LE-flap deflection).

Applying this drag reduction principle on a low aspect ratio SCT wing means to maximize the suction force in a 3D flow. Designing low-drag high-lift systems therefore requires designing 3D flow topologies using viscous CFD methods. The low-speed design condition is characterised by a high lift coefficient at high AoA producing flows near separation. As vortices may and will appear it is essential that the viscous CFD method is able to predict embedded vortices (topology and drag level) and separation onset.

Within the EPISTLE project viscous CFD codes (RANS methods) are investigated and proven to predict such flows to industrial requested accuracy [2]. The challenging design objective for the new high-lift system is to reduce the drag by 15-20% compared to a datum high-lift system. The datum system has previously been designed using empirical methods by the EU project EUROSUP [3] and already obtained a drag reduction of 18% [1] compared to a clean wing.

Authority regulations define to measure the low-speed noise at a so-called fly-over point. This point is located under the aircrafts climb path 6.5km behind the break release point on the runway. Improved aerodynamic performance of the aircraft allows the obtainment of a greater altitude at the measurement point. This is how aerodynamics reduces noise.

The new high-lift system design is conducted for the lift at the fly-over point. The Mach and Reynolds number is obtainable in the wind tunnel ($M=0.25$; $Re_{max}=22.5*10^6$) selected for validation later on. The constraints for the design are a limitation of the local LE-flap chord and a maximal allowed AoA.

3 Design variables and design sensitivities

The design variables for each of the five flap segments per wing half are depicted in **Fig. 2**. In the project the full design space is explored in a co-ordinated action aiming at extracting simple design rules. Partners CFD work revealed the variables flap deflection angle, flap chord and vertical hinge line position (flap knuckle geometry) as most sensitive. The remaining two design variables are evaluated as less sensitive in EPISTLE.

Extracting quantifiable design rules from the numerical data provided by partners failed due to strong numerical influences (separated flow). Results are therefore thoroughly analyzed and flow physics based design guidelines are formulated instead.

All CFD work at DLR is carried out on a purpose-adapted mesh having 1.8 million cells (48 cells in the boundary-layer) to resolve all drag relevant flow details. A manually design approach is applied as the compute time for each solution (NEC SX-5) prevents an automated procedure. Quantification of some design sensitivities is performed before the design work is started at DLR. The DLR flow solver FLOWer [4] is used on structured meshes generated by the DLR mesh generator MegaCads [5]. The Baldwin-Lomax [6] turbulence model (TM) with the Degani-Schiff extension [7] for vortical flows is used for all fully turbulent computations. This decision is based on the models low dissipation behaviour [2]. An important feature of DLR's computations is the use of pre-conditioning [8] in order to increase the accuracy for low Mach number computations. In **Fig. 3**, two viscous flow solutions (M=0.25, α=11.4°, Re=22.5*10^6) are obtained without (left) and with (right) low Mach number pre-conditioning (LMPC). A smeared vortex structure with increased cross flow is present without LMPC.

The first sensitivity analyzed is the knuckle geometry. Three different knuckle geometries are generated for the datum high-lift system (at a slightly higher flap deflection). The default knuckle shape is called K1. Two more rounded knuckle geometries are generated simulating a knuckle hinge located below the wing. The variation of the local geometry is small; see the wing cross-section in **Fig. 4**.

The aerodynamic performance for these geometries is given in **Fig. 5**. The design lift coefficient is indicated. Increasing the knuckle radius (K1→K2) reduces the drag because the associated lower pressure peak prevents flow separations from the flap knuckle. Using the moderate knuckle rounding (K3) still improves the performance by half a $\Delta C_L/C_D$ point. This corresponds to about 6% drag reduction, being significant.

Next the LE-flap deflection is varied using the maximal allowed LE-flap chord. Instead of varying all five LE-flaps individually two flap deflection distributions are compared. The first deflection distribution is the datum one. The flaps (from inboard to outboard) are deflected for: 2x14.3°, 15.1° and 2x19.8° (measured in streamwise direction). The other variant increases the flap deflection significantly to: 2x23.4°, 19.7° and 2x28° and is called high LE deflection.

The impact of both deflection distributions (low and high) on the flow is given as spanwise lift and drag distribution. **Fig. 6** shows the local lift times the local wing chord (C_L*CCM) over the wingspan η. Drag times the local wing chord (C_D*CCM) is plotted over the span in **Fig. 7**. Results are taken from 12° AoA solutions and are summarized in addition below:

Configuration	**AoA**	$C_L/C_{L\,Target}$	**L/D**
Datum: min. chord, low LE deflection	12°	1.193	6.73
max. chord, low LE deflection	12°	1.052	8.56
max. chord, high LE deflection	12°	1.055	7.64

The high LE-flap deflection solution indicates vortices on the inner and on the outer wing by local maxima in the lift distribution. Especially on the outer wing a

significant drag increase compared to the low LE-flap deflection solution is observed. Both configurations outperform the datum wing.
Results for two different LE-flap chords using the datum (low) flap deflection settings are presented next. The wing featuring the minimal flap chord produces at this AoA more (vortical) lift, especially on the outer wing; see **Fig. 8** and the table above. This lift benefit unfortunately results in significant higher drag; compare **Fig. 9**. The configuration with increased flap chord achieves higher aerodynamic performance.
The presented variable sensitivities show that their influence is highly linked because of the complex 3-D flow near separation. Furthermore a flap chord increase at moderate deflection angles seems preferable.

4 Low-drag high-lift system design

The high-lift system design in EPISTLE is organized as a competition between the project partners. This maximizes the probability to obtain a high-lift system with minimal drag reaching the additional 20% performance gain objective.
DLR followed a manually design procedure as sketched in **Fig. 10**. The three most sensitive design variables for each of the five flaps form a design problem with 15 variables. Over night computations followed by flow analysis (and discussions), design decisions and mesh generation close the sketched manual loop. Typical turnaround time is just one day for a new high-lift system variant.
About thirty different high-lift system configurations are designed and analysed (each at three AoA enabling interpolation of the target lift) before the requested 20% performance improvement compared to the datum system is obtained.
The aerodynamic performances of all geometries (including datum and final design) fill the performance map in **Fig. 11**. The C_L/C_D values (at the target lift) are given over the AoA needed to obtain this lift. The value of the AoA constraint coincides with the y-axis of this figure.
As can be seen in Fig. 11 the performance of high-lift systems configurations that fulfil the AoA constraint is degraded compared to the performance of the datum configuration (red symbol). Further effort concentrated on aerodynamic performance improvements of the high-lift system while accepting higher AoAs. The horizontal line in this graph indicates the design target. The green bullet just above this line gives the performance of the final DLR design.
Many solutions cluster just below the 20% improvement line. The magnitude of solutions in Fig. 11 reflects the difficulty to develop a flow topology producing such high aerodynamic low-speed performance.
Next the changes in the flow topologies from the datum to the designed low-drag high-lift systems are highlighted in **Fig. 12** and Fig. 13. The presented solutions are obtained at 10° AoA. The streamlines and pressures (colour) at the surface are given. Besides this total pressure loss contours (at the wing trailing edge) are plotted. The main geometric differences of the DLR design to the datum high-lift system is the flap depth, which increases with span.
The flow on the datum configuration is characterized by strong cross flow and a large area of high total pressure loss toward the wing tip. The streamlines reveal a

small vortex of the inner LE-flap starting about midspan. Another small vortex is depicted running across the outer wing part. All other wing parts are covered by a large merging vortex structure starting at the wing apex. The knuckle is sharp with partial flow separations.
The DLR low-drag configuration, **Fig. 13**, shows partially separated flow too. Nevertheless the amount of cross flow is significantly reduced. This prevents the wing apex vortex to interact with the vortices developing on the deflected LE-flaps. Therefore the total pressure loss plot clearly shows a tip vortex and the footprint of the wing apex vortex at about midspan. The flow topology depicted here is obviously vortical but the vortices do not merge.
The appropriate spanwise LE-flap depth and deflection distribution enables to generate suction force increasing vortices on the deflected LE-flaps and to reduce the cross flow on the wing. This prevents vortex merging and helps in essence to reduce drag. The applied knuckle rounding prevents flow separations from the knuckle. These are the main ingredients to obtain a low-drag high-lift flow for this SCT wing. Using an alternate two-equation TM on the same mesh (not shown here) confirmed the numerically designed flow topology.

5 Experimental design validation

Two configurations, a double knuckle and the DLR designed system, are qualified for experimental validation on a large-scale modular wind tunnel model. Care is taken to accurately manufacture the small radii of curvature of the different LE-shapes. The high-lift systems are tested in a pressurized low-speed wind tunnel in 2002.
A comparison of the predicted drag reductions and the measured drag on the DLR configuration is given in **Fig. 14**. The measured polar of the datum configuration is given as the red broken line. The numerically estimated drag reduction at the design condition is plotted as the blue horizontal bar (broken line, upper right). This prediction is fully confirmed by the measured drag polar of the DLR configuration, given as green solid line in Fig. 14. The requested performance increase of 20% is obtained and experimentally confirmed. Please note that the drag reduction difference between prediction and experiment is only three drag counts. Such excellent agreement of the aerodynamic coefficients is obtained only if the designed physics is accurately met. The designed flow topology in Fig. 13 is confirmed by another experiment conducted in the DNW-NWB, **Fig. 15**. The oil-flow picture shows the apex vortex and vortices located on the deflected LE flaps. Small differences regarding local details do exist but the overall agreement is excellent.
Further work in EPISTLE estimated the obtainable noise reduction due to the validated aerodynamic performance gains. The different climb trajectories for the datum and the low-drag high-lift system are computed. The additional altitude for the DLR low-drag design at the fly-over noise measurement point results in accumulated noise reduction of about 4 EPNdB (effective perceived noise decibel).

6 Conclusion

All designed physical effects in the flow topology of the DLR high-lift system are successfully obtained. This holds for the design of the LE-flap vortices, the applied knuckle rounding and the selection of flap depth and deflection distribution keeping the vortices separated. It is clear that the simple suction force principle is only part of the success story. It requires numerical tools of known high accuracy characteristics on sufficient fine meshes to design such a complex and separated flow topology. The design's performance is validated by the experiment. It is estimated to enable significant low-speed noise reductions.

Acknowledgements

The author would like to thank his colleagues S. Heinrich and A. Traore for their valuable contribution to the successful design presented here.

References

[1] U. Herrmann. "Low-Speed High-Lift Performance Improvements obtained and validated by the EC-project EPISTLE". ICAS 2004-4.1.1. 24th ICAS congress, Yokohama, Japan. 29.8.-3.9. 2004.

[2] U. Herrmann et. al. "Validation of European CFD Codes for SCT low-speed high-lift Computations". AIAA 2001-2405. 2001.

[3] D. A. Lovell. "European Research of Wave and Lift dependent Drag for Supersonic Transport Aircraft. AIAA 99-3100. 1999.

[4] N. Kroll, C.-C. Rossow, K. Becker and F. Thiele: "MEGAFLOW – A Numerical Flow Simulation System" Aerospace Science Technology, Jahrg. 4 (2000), pp. 223-237.

[5] O. Brodersen, M. Hepperle, A. Ronzheimer, C.-C. Rossow and B. Schöning. "The Parametric Grid Generation System MegaCads". 5th International Conference on Numerical Grid Generation in Computational Field Simulation". National Science Foundation (NSF) pp. 335-362. 1996.

[6] B.S. Baldwin and H. Lomax. "Thin Layer Approximation and Algebraic Model for separated Turbulent Flows". AIAA Paper No. 78-257. 1978.

[7] D. Degani and L. Schiff. "Computations of Turbulent Supersonic Flows around pointed Bodies having Crossflow Separation". Journal of Computational Physics 66, pp. 173-196. 1986.

[8] R. Radespiel and N. Kroll, "Assessment of Preconditioning Methods". DLR-FB 95-29. 1995.

Figures

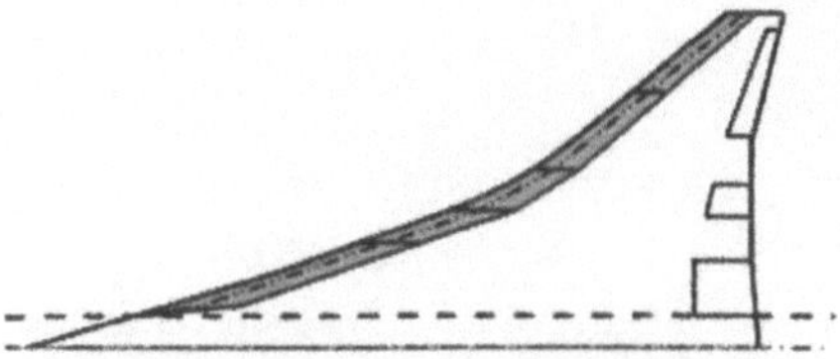

Figure 1 Leading edge flap based high-lift system, five flaps in spanwise direction.

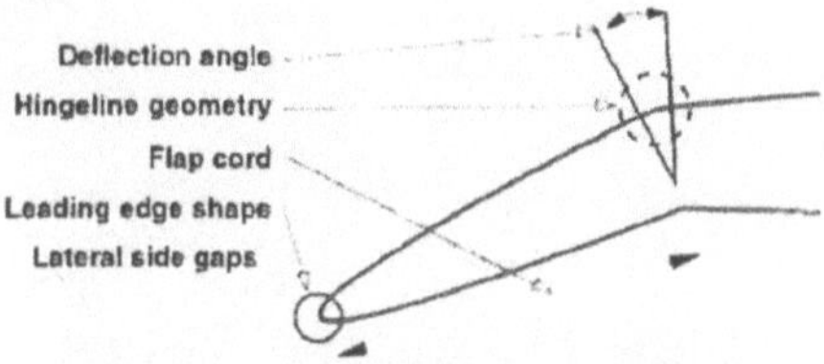

Figure 2 Design variables of the leading edge flaps

Figure 3 Computed flow topology at M=0.25, α=11.4° without (left) and with (right) low Mach number pre-conditioning

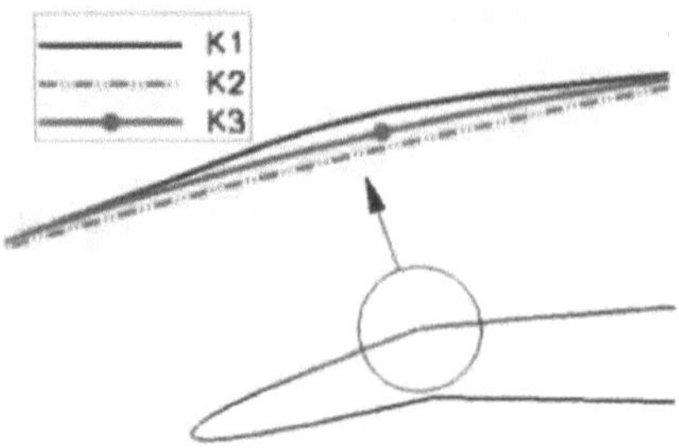

Figure 4 Variation of the flap hinge geometry

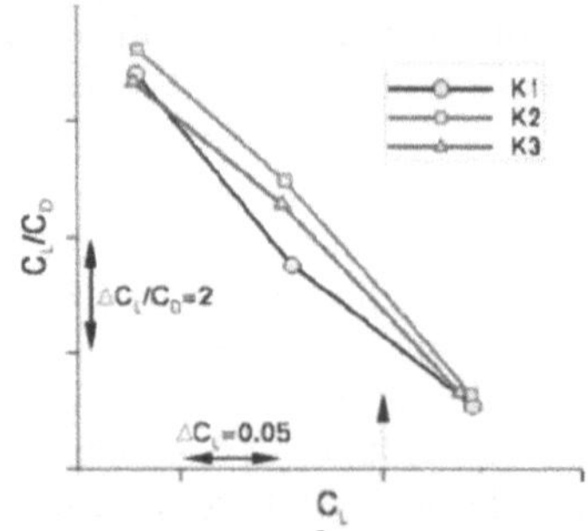

Figure 5 Aerodynamic performance improvements due to increasing flap knuckle radius

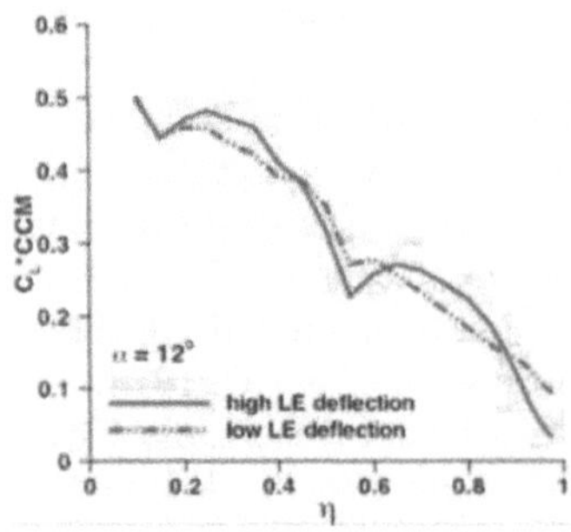

Figure 6 Local lift distribution over span; two leading edge flap settings

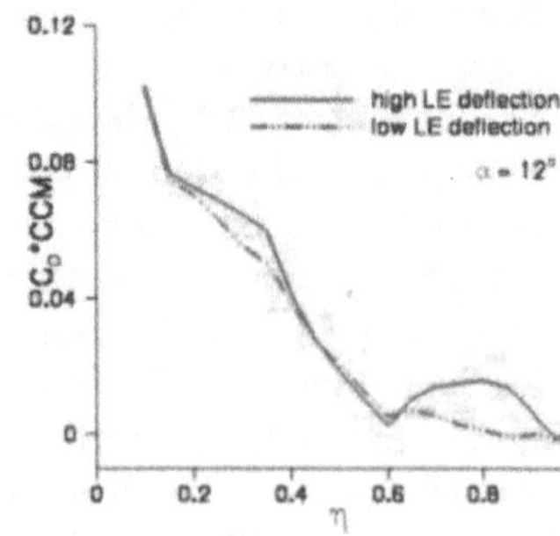

Figure 7 Local drag distribution over span; two leading edge flap settings

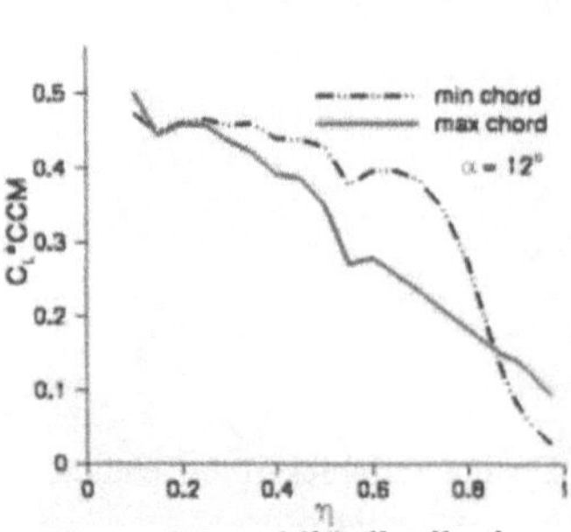

Figure 8 Local lift distribution over wingspan; two leading edge flap chords

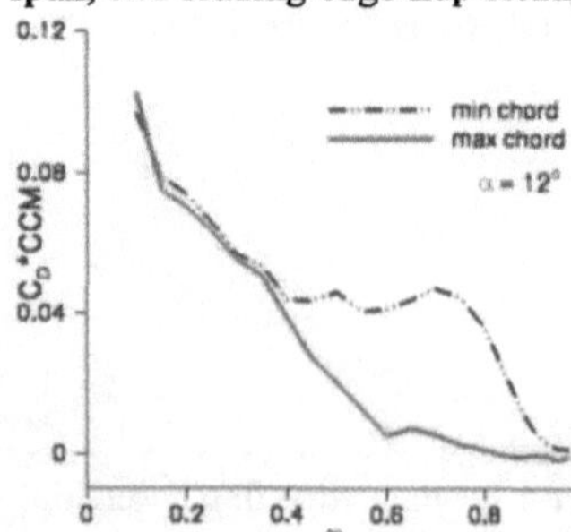

Figure 9 Local drag distribution over wingspan; two leading edge flap chords

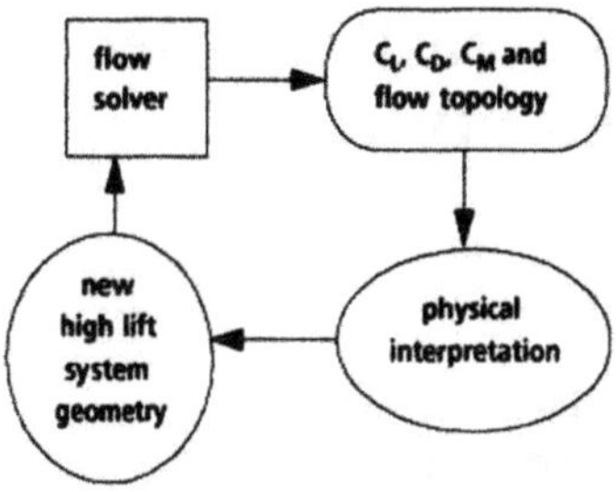

Figure 10 Manual SCT high-lift design scheme

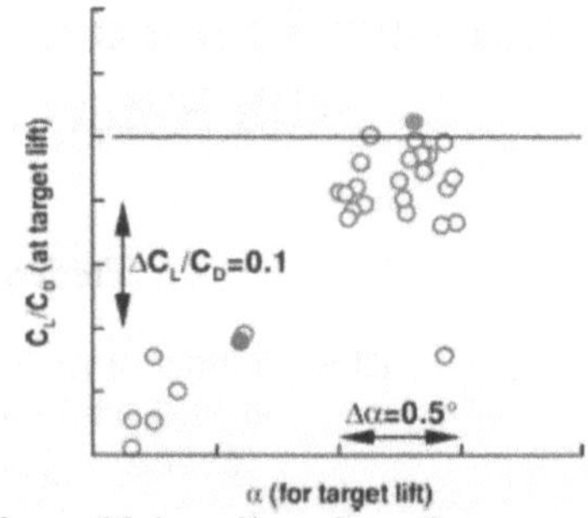

Figure 11 Aerodynamic performance map of all SCT high-lift system variants at target lift

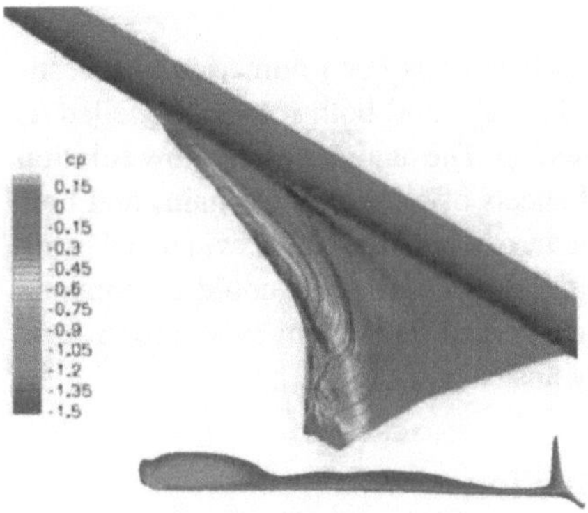

Figure 12 Surface pressure, streamlines and total pressure loss contours at the wing trailing edge for the datum high-lift system

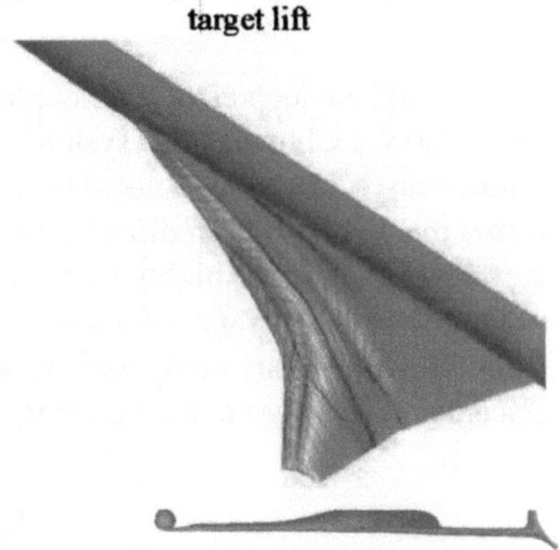

Figure 13 Surface pressure, streamlines and total pressure loss contours at the wing trailing edge for the DLR designed high-lift system

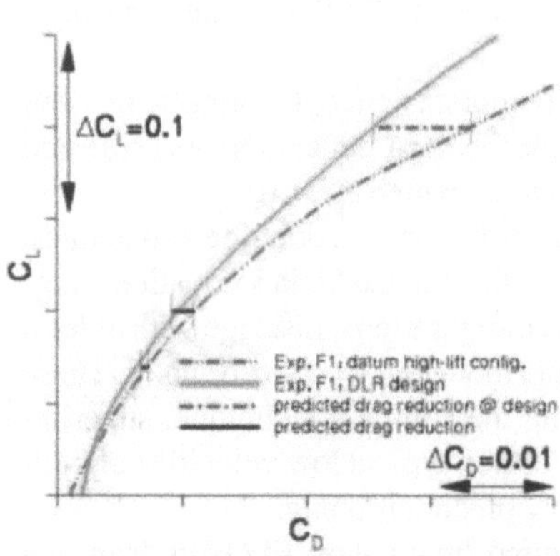

Figure 14 Experimental results of the designed and datum high-lift system in caparison to the predicted drag reduction

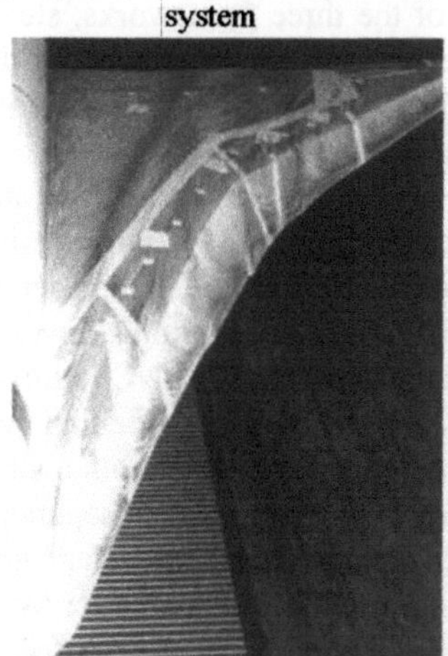

Figure 15 Oilflow pattern on the DLR designed high-lift system at $\alpha=12°$ in the DNW-NWB wind tunnel

Chimera Simulation of a Complete Helicopter with Rotors as Actuator Discs

Frédéric Le Chuiton

DLR – Institute of Aerodynamics and Flow Technology,
Pf 3267, 38022 Braunschweig, Germany,
frederic.lechuiton@dlr.de

Summary

The EC-145 helicopter of Eurocopter Deutschland has been numerically investigated. How a Chimera grid system for the fuselage and both rotors, modelled as actuator discs, can be assembled is commented on. The analysis of the flow solution shows the wakes of both discs by means of sheets of total pressure gains and how they develop. Also the highly vortical flow pattern behind the cell is evidenced using stream ribbons. Average values for the lift and drag coefficients could be obtained from the steady-state computation, which converged only down to a certain level since the flow proves to be inherently highly unsteady.

1 Introduction

The French-German helicopter project Complete Helicopter AdvaNced Computational Environment (CHANCE) [8] is now, after six years runtime, being completed. Out of the three frameworks: steady (isolated rotors and fuselages), quasi-steady (fuselage with rotors as actuator discs) and unsteady (rotors in forward flight and complete configurations), the present paper reports on the DLR activities related to the second approach.

State-of-the-art quasi-steady and unsteady complete helicopter simulations can be seen in [5, 3] and [4, 6] respectively, where the ONERA developments presented in [6] are also CHANCE-representative of similar activities at DLR.

The quasi-steady simulation of fuselage/rotor wakes interactional aerodynamics, as planned within CHANCE, aims at integrating the Navier-Stokes equations augmented by turbulence modelling over complete configurations: fuselage with at least rear control surfaces and both main and tail rotors modelled by actuator discs. These actuator discs are meshed separately and, using the Chimera technique, superimposed on a pre-existing fuselage grid system. Further, typical low velocities of helicopters are to be accounted for by a low-velocity preconditioning.

The numerical platform used is the structured flow solver FLOWer from the MEGAFLOW project and is presented at length in [2]. Roughly: a cell-centred finite volume formulation of the Navier-Stokes equations has been used together with the 2-transport equation LEA turbulence model; the main equations are second order integrated in the frame of a multigrid algorithm using the scheme by Jameson and a

Runge-Kutta time-stepping method, while use is made for the turbulence equations of a first order Roe scheme and an implicit DDADI-method. Chimera and low-velocity preconditioning along with the actuator disc feature developed in FLOWer [3] have also been used.

The present configuration, the Eurocopter EC-145 helicopter, has already been considered in [1] where the fuselage with and without an actuator disc for the main rotor only was studied. This paper deals with the extension to the presence of an actuator disc for the tail rotor too. First, the technique for meshing such a geometry is explained and commented on. Second, the flow solution is presented and analysed. Third, the course of the computation run and the convergence level are detailed. Finally, concluding remarks close the paper.

2 Chimera Grid System

Setting up the global grid system for such a configuration relies on three steps: generation of the fuselage grid, generation of grids for both actuator discs and a certain expertise for binding them altogether in a Chimera setting.

In the Chimera framework retained at DLR, the global grid consists of several subgrids, which communicate with each other solely via interpolations. That is, each subgrid is a multi-block structured grid in itself, the boundary points of which are set in interpolating their flow values from other overlapping subgrids. Such points are termed Chimera boundary points. A subgrid generally houses a solid body that is solved for in its own subgrid but must be hidden in all other subgrids in order to avoid multiple definitions of the body. To this end use is made of so-called masks. A mask is an auxiliary volume enclosing the body; points of other subgrids overlapped by the mask are removed from the solution process, thus creating so-called holes in the respective subgrids. A layer of points wrapping up each hole is defined, the flow values of which are also interpolated from other subgrids. Such points are termed Chimera fringe points. Chimera points must be interpolated from so-called donor cells, which are searched for in the global grid using an alternating digital tree (ADT) for greatest search speed. In case more than one donor cell for a given Chimera point (boundary or fringe) are available, the one with the smaller volume is retained for greater accuracy. Next this donor cell is subdivided in six tetrahedra, one of which encloses the Chimera point and, with its four vertices, serves as basis to the interpolation. If no donor cell can be found, e.g. because of poor mesh topology and/or quality, the Chimera point is said to be an orphan point and is assigned the flow values of the nearest grid point. The Chimera feature implemented in FLOWer does not require any hierarchical structure of the various subgrids, any combination can be handled. The interested reader is refered to the work described in [7].

The grid around the isolated fuselage has been generated by Eurocopter Deutschland with the ICEM-Hexa mesh generator, see [1]. It consists of 64 blocks and approximately 5.3 million points and is arranged in a C-O topology in order to account for boundary layers and to spare points in the fore part of the mesh. Concerning the capturing of thin boundary layers, it proved to be well adapted since results showed values of y^+ of at most unity almost everywhere on the surface.

Both grids for the main and tail actuator discs have been generated using an ad hoc generator. Both consists of 4 blocks and have about 1.5 and 0.5 million points respectively. The block topology is arranged as follows: two blocks lie on the top side of the disc (where the thrust vector points to) and the other two below. On each side, the major part is meshed by a cylindrical block, hence accouting for grid quality, and an additional block is inserted in the middle so as to avoid a grid singularity.

Next, using the Chimera technique, actuator disc grid systems are superimposed on the fuselage grid, which consequently also plays the role of a background grid. Positioning both actuator discs has been done using the translation/rotation capability of the mesh generator. As a matter of fact, very few space was left between the various components: fuselage, main and tail disc. This led to a series of difficulties in setting up the Chimera feature, since it has always been striven for no orphan points at all. First, the definition of masks had to be paid sufficient care. Masks for the main and tail discs have been defined by cylindrical boxes in such a way that they do not overlap and allow for a handful of computed points inbetween in their own subgrids. As for the fuselage more than one mask has been necessary. Most of them are simply very coarse auxiliary meshes consisting of a few cells. On the other hand, the top part of the engine casing required a very fine tuning: a thin layer of sixteen cells has been extracted out of the original fuselage grid in order to complete the fuselage masking. Doing so, masks for the fuselage and the main disc did not overlap and allow for enough computed points inbetween. No special treatment was necessary between the fuselage and the tail disc. The final setting of masks is displayed in figure 1. The second difficulty dealt with the remaining orphan points which could be completely removed in adding two grid blocks. It is here to be mentioned that an automatic hole definition procedure would have been of course of great help.

A median slice of the final grid system can be seen on picture 2 where only the third grid level has been displayed for the sake of clarity. In total there are 74 blocks, partitioned in 5 grid subsystems, with approximately 8.3 million points.

3 Computation Results

The global configuration is an EC-145 helicopter from Eurocopter Deutschland in high-speed forward flight with the following flight parameters: Mach-number $M_\infty = 0.21$, no incidence angle $\alpha = 0^o$ and Reynolds-number $\mathrm{Re}_\infty = 4.3 \; 10^6$ (m^{-1}). Here engines have not been modelled and are closed with non-permeable walls.

Rotors have been simulated by actuator discs, on the surface of which source terms (a distribution of force density over the disc surface) are introduced in the impulse and energy equations. For numerical details, the interested reader is refered to [3]. Forces for the main disc stem from a previous FLOWer computation of the ATR-A rotor (developed by Eurocopter Deutschland) in high speed forward flight, where the retreating blade side is on the left. Such a pre-computation requires CPU-time of the same order as that of the present simulation or may be even more expensive, which has been done only because the corresponding results were already available. Instead, a more simple computation with a so-called rotor simulation code (e.g. lifting line theory with 2D-aerodynamics and balanced rotor wake, see [3] for a brief review of their use in an actuator disc context), with accordingly negligble computational cost, can be used for initialisation of the actuator disc. More simply, forces for the tail disc correspond to a uniform pressure jump over the disc surface. Both force distributions can be seen on figure 3 and global run parameters are: thrust coefficient $C_T = 0.007673$, tip Mach-number $M_{\text{tip}} = 0.637$, radius $R = 5.5$ (m), advance ratio $\mu = 0.3265$ and shaft angle $\alpha_q = -5^o$ for the main rotor; thrust coefficient $C_T = 0.008927$, tip Mach-number $M_{\text{tip}} = 0.636$ and radius $R = 0.97$ (m) for the tail rotor. As this approach really represents the time-averaged rotor loading on the disc, it is claimed in the literature the flow solution be the time-averaged representation of the underlying unsteady flow, a proof of which the author is not aware of. Nevertheless, as illustrated in [5] by means of vorticity experimental data, a quasi-steady flow solution reproduces meaningfully the actual physics.

The solution on the fuselage surface is to be seen on picture 4 where the distribution of the pressure coefficient and friction lines are plotted. Particularly conspicuous on the boot is the pressure recovery area with its associated large separation zone. Notice also the longitudinal high and low pressure spots on the tail boom that may cause the boom to shake, if in reality highly unsteady. Further it is of interest to analyse the three-dimensional flow-field, since the interaction of the wake of both rotors on the rear control surfaces represents the major motivation for such simulations. A quantity very well suited for tracing back the influence of rotors proves to be total pressure gains $p_o/p_{o_\infty} - 1$ (mechanical energy supply). Sheets of total pressure gains, for a value of 10^{-3}, have been generated separately in each actuator disc grid subsystem: main and tail, and are displayed in figure 5. Considering the main disc, it is to be mentioned that, as a consequence of the convection, the lower part of the sheet has been emitted from the front half of the disc and the upper part from the rear half. The rolling up of the marginal vortices is especially distinct on the retreating side where the circulation is maximum. As for the tail disc, the interaction of the wake of the main actuator disc with that of the tail actuator disc provides a somewhat more distorted picture. Still, anything that is approximately horizontally oriented can be attributed to the main actuator disc and, conversely, anything that is vertically directed can be assigned to the wake of the tail actuator disc. Finally, stream ribbons are plotted in figure 6 where the ribbon colouring variable is the norm of the vorticity vector. The vortex system is here easily identifiable: two trailing vortices, left and right of the boom, are generated on the boot surface by the flow separation. In this case, they are also artificially fed by the vortices issued from the separation areas behind the engine outlet surfaces. In this context, the deformation of the lower part of the total pressure sheet in figure 5 is no surprise.

4 The Convergence

The computation has been run on one processor of a NEC SX5-16Be machine and took circa 2 days 4 hours of CPU-time for 3100 iterations. The level of convergence reached 2.5 orders of magnitude and may appear moderate but considering the complexity of the flow configuration and the tolerance achieved on the lift and drag coefficients, the present results are very much of engineering interest. Indeed, lift and drag coefficients are $C_L = -1.123 \pm 0.029$ and $C_D = 0.5349 \pm 0.0075$, which represent 2.6% and 1.4% RMS-tolerance respectively.

It has been tried, unfortunately in vain, to improve the level of convergence in switching off either the implicit residual smoothing or the multigrid algorithm or even both, since these are known techniques that leave perturbations in the flow-field as long as enough convergence has not been reached. But on the other hand, upon inspection of the locations where the highest density residuals occur it appears clearly that this computation cannot converge since the flow-field is very likely to be inherently unsteady. As a matter of fact, as shown in figure 7, they arise at the engine outlet surfaces and on the top part of the fin. Consequently one is allowed to speculate on the easing property of the simultaneous simulation of the engine exhaust gases.

However, this kind of quasi-steady computation represents in an industrial context the only possibility to investigate such flow-fields since corresponding unsteady Navier-Stokes simulations still remain prohibitively expensive.

5 Conclusion

The complete configuration, apart from the engines, of the EC-145 helicopter has been simulated in modelling both main and tail rotors as actuator discs. The grid system used consists of five grid subsystems for the fuselage, the actuator discs and two others. The fuselage grid served also as background grid to the others that were coupled to it with Chimera interpolations. In doing so it has been possible to remove all orphan points.

Source terms for the actuator discs have been initialised with a previous FLOWer computation for the ATR-A rotor in high speed forward flight and with a constant pressure jump for the main and tail actuator discs respectively.

The flow solution shows high physical relevance and it has been possible to trace back the wakes of both actuator discs using sheets of total pressure gains. Furthermore stream ribbons aft the cell visualise the strong separated and vortical flow pattern experienced by rear control surfaces.

Only a certain level of convergence could be reached but it has been shown that the flow field is inherently highly unsteady, so that only average values for the lift and drag can be determined.

Acknowledgements

The author acknowledges gratefully the contribution of Alessandro d'Alascio of Eurocopter Deutschland who provided the fuselage grid system of the EC-145 configuration.

References

[1] D'Alascio, A., Berthe, A., Le Chuiton, F., *Application of CFD to the fuselage aerodynamics of the EC145 helicopter*, Proceedings of the 29th European Rotorcraft Forum, Friedrichshafen, Germany, 2003

[2] Kroll, N., Rossow, C., Schwamborn, D., Becker, K., Heller, G., *MEGAFLOW – a numerical flow simulation tool for transport aircraft design*, ICAS 2002 Congress

[3] Le Chuiton, F., *Actuator disc modelling for helicopter rotors*, Aerospace Science and Technology, Volume 8, Issue 4, 2004, pp. 285-297

[4] Meakin, R. L., Wissink, A. M., *Unsteady aerodynamic simulation of static and moving bodies using scalable computers*, AIAA 99-3302-CP, 14th AIAA Computational Fluid Dynamics Conference, Norfolk, VA, June-July 1999

[5] Renaud, T., O'Brien, D., Smith, M., Potsdam, M., *Evaluation of isolated fuselage and rotor-fuselage interaction using CFD*, Proceedings of the 60th American Helicopter Society Forum, Baltimore, MD, June 2004

[6] Rodriguez, B., Benoit, C., Gardarein, P., *Unsteady computation of the flowfield around a helicopter rotor with model support*, 43rd AIAA Aerospace Sciences Meeting and Exhibit, Reno, January 2005

[7] Schwarz, T., *Development of a wall treatment for Navier-Stokes computations using the overset-grid technique*, Proceedings of the 26th European Rotorcraft Forum, The Hague, The Netherlands, 2000

[8] Sidès, J., Pahlke, K., *Progress towards the CFD computation of the complete helicopter: recent results obtained by research centers in the framework of the french-german CHANCE project*, in: CEAS Aerospace Aerodynamics Research Conference, Cambridge, UK, 2002

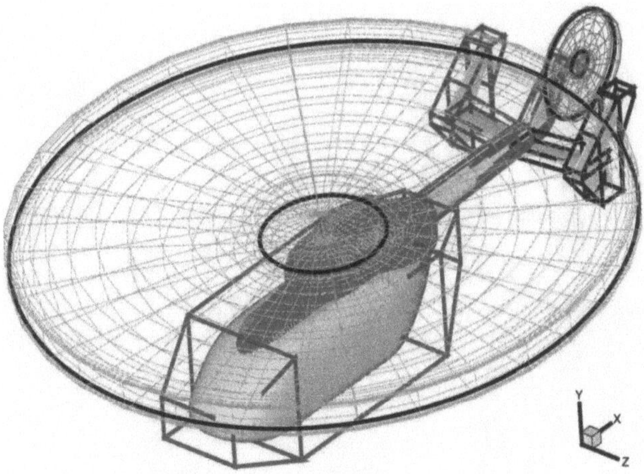

Figure 1 System of masks for the fuselage and actuator discs.

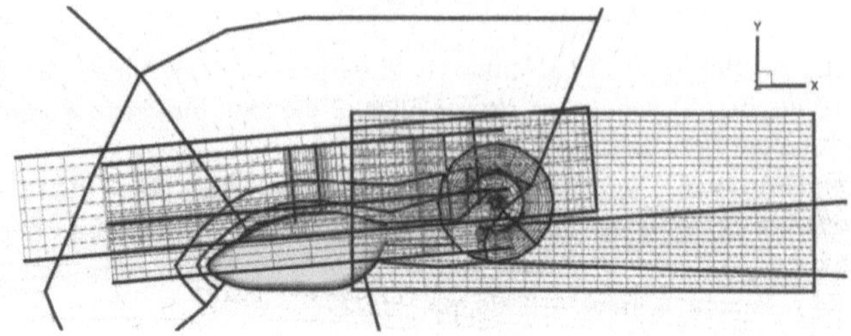

Figure 2 Global view of the composite grid system, third grid level.

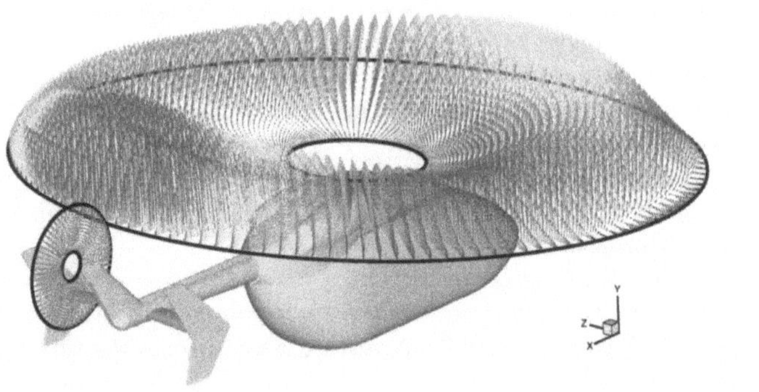

Figure 3 Actuator disc forces applied as source terms in the equations.

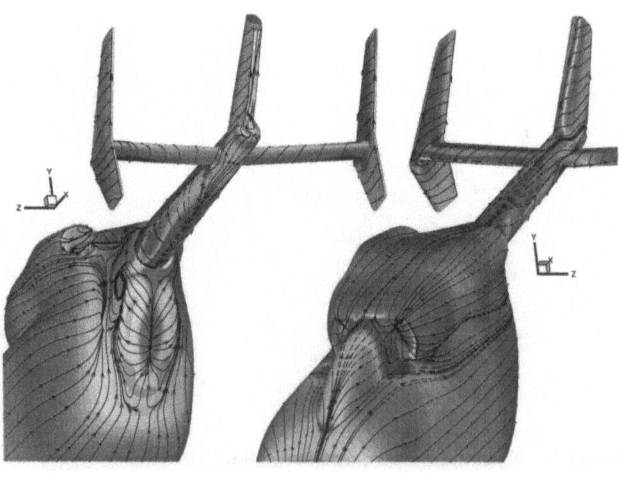

Figure 4 Surface distribution of c_p and surface friction lines.

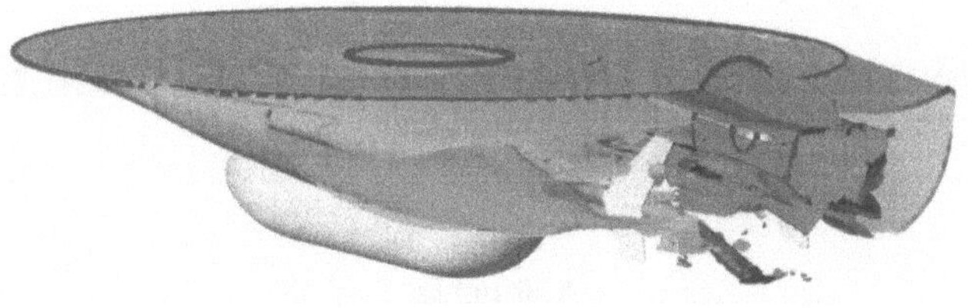

Figure 5 Total pressure gains; only grid subsystems of the main and tail actuator discs.

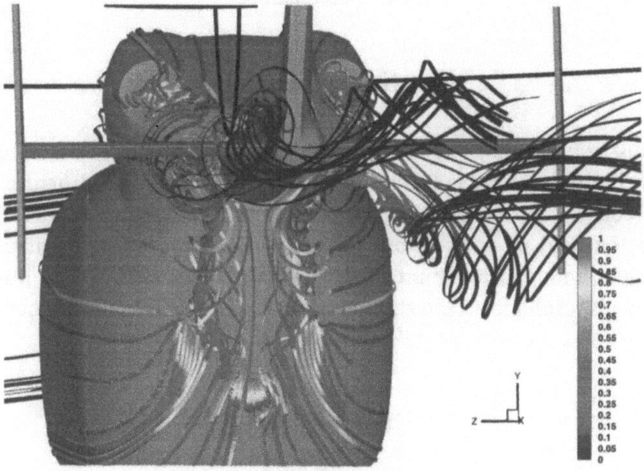

Figure 6 Stream ribbons aft the cell

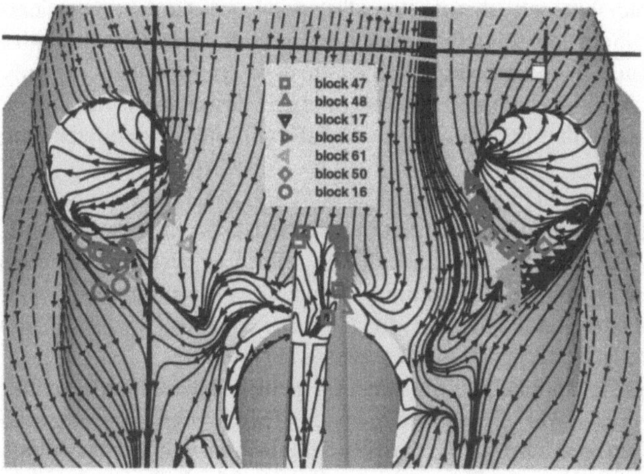

Figure 7 Location of some of the largest density residuals.

Extension of the Unstructured TAU-Code for Rotating Flows

A. Raichle

DLR Institute of Aerodynamics and Flow Technology
Lilienthalplatz 7, 38108 Braunschweig, Germany
Axel.Raichle@dlr.de

Summary

A method has been developed how to integrate numerical fluxes of velocity fields of rigid body motion exactly in the context of an edge based data structure to ensure free stream consistency for rotating flows. For the unstructured DLR TAU code this extends the field of applications to rotor and helicopter aerodynamics. This is demonstrated with a calculation of the 4-bladed rotor HELI7A in hover and a comparison with results from the structured DLR FLOWER code and experimental data.

1 Introduction

The calculation of flows around helicopters is a challenging task. There are special requirements to the precision of the numerical scheme and to the grid generation. The linear radial velocity distribution of a rotor causes strong tip vortices which influence the whole flow field. For rotary-wing aircraft the relative strength of vortices is large because incident flows are either small or zero and there are only small induced velocities. Because of their more complex design the vortex interaction with the blades and with other components like the tail rotor or stabilizer is of great importance.

With unstructured methods the process of grid generation can be automated and considerably be simplified. Unstructured methods offer the possibility of local grid adaptation. The grid can automatically be refined or derefined depending on the details of the flow field. This optimizes the number of grid points. For rotary-wing applications unstructured methods for the Euler equations are presented in [4] and [7] with local grid refinement for tetrahedral meshes.

For industrial applications unstructured methods have a high potential to reduce turnaround time and to increase the accuracy of predictions. For rotor aerodynamics they have been used hardly. One reason is still the high demand of computation time in particular for unsteady flows. The resolution of the boundary layer for unstructured hybrid grids for the Navier-Stokes equations needs much more points compared to structured methods. To reduce the complexity in [2] the boundary layer is resolved with a structured grid embedded into an unstructured background mesh.

The boundary points are interpolated with the chimera technique. Instead of resolving the boundary layer wall functions are used in [5] and the code has been parallelized. Massive parallelization is a good means to improve the performance in future.

At DLR the flow solver TAU [3] has been developed for unstructured hybrid grids. The code is parallelized, has multigrid to accelerate the convergence, uses the chimera technique for relative grid motion and offers the possibility of local grid adaptation. The code is already in industrial use for fixed-wing applications. The objective was to apply the TAU code for rotary-wing applications.

From experiences with structured methods it is known that a key requirement is free stream consistency. For rigid body motion the code must reproduce constant solutions exactly. For rotating flows there are otherwise disturbances in the flow field which grow proportionally with the distance to the axis of rotation and the angular velocity. Errors are of the same order of magnitude as the wake of a rotor. Tests showed that the TAU code was not free stream consistent for rotating flows and that the error was caused by an algorithmic problem. An extension had to be found compatible with the existing data structure which was not disturbing the existing features.

2 Free Stream Consistency

To understand the problem and how to solve it the focus has to be set on the discretization and the integration of the numerical fluxes. The system of governing equations are the Euler/Navier-Stokes equations. The equations conserve mass, momentum and energy for a control volume. They set into relation the change of the state inside the control volume with the fluxes through the surface of the control volume. For the discretization the domain is decomposed into a set of control volumes and the equations are balanced for each control volume. To construct the control volumes the domain is subdivided into simple polygons of the same type. The polygons are the elements of the primary grid. If more than one elment type exists the grid is hybrid. Standard types are tetrahedrons, prisms, hexahedrons and pyramids. Cell centered based metrics operate with the elements of the primary grid but the TAU code uses a dual metric. With this metric the control volumes are constructed from the elements of the primary grid. The set of control volumes is denominated the dual grid. The dual metric has the advantage that there is only one type of control volumes and the number of control volumes scales with the number of grid points which is smaller than the number of elements of the primary grid. The control volumes of the dual grid are constructed by connecting the center of an element with the center of its sides and the centers of the sides with the centers of its edges. At the boundary the centers of the edges are connected with their corner points. In 3D the surfaces are quadrilateral patches which are subdivided into two triangular facets. In 2D the surfaces are line segments. Figure 1 shows the elements of the primary grid and the control volumes of the dual grid in 2D.

For moving grids a vector function $\mathbf{q}$ has to be prescribed describing the grid motion. For rigid body motion the function is linear in space and has the following form

$$\mathbf{q}(t,\mathbf{r}) = \omega(t) \times (\mathbf{r} - \mathbf{r_0}) + \mathbf{v_t}(t). \quad (1)$$

The vector ω is the rotation vector, $\mathbf{v_t}$ the translation vector, $\mathbf{r}$ the displacement vector and $\mathbf{r_0}$ the center of rotation. The basic form of the Euler/Navier-Stokes equations for moving grids is

$$\int_V \frac{\partial \mathbf{W}}{\partial t} + \omega \times \mathbf{v}\, dV + \int_{\partial V} \mathbf{F}\mathbf{n_0} - \mathbf{W}(\mathbf{q}\mathbf{n_0})\, dS = 0. \quad (2)$$

The derivation is described in detail in [6]. $\mathbf{W}$ is the vector of conservative variables, $\mathbf{v}$ the velocity vector and $\mathbf{n_0}$ the normal unit vector of the surface of the control volume. The matrix $\mathbf{F}$ is denominated as flux density tensor. The components are composed of the components of $\mathbf{W}$. From the mathematical point of view only the submatrix of the momentum equations represents a tensor but the expression is very popular. For constant solutions $\mathbf{W}$ in space and time the velocity vector $\mathbf{v}$ must either be parallel to the rotation vector ω or zero. The volume integral vanishes and the system can be written in the following form

$$\mathbf{F} \underbrace{\oint_{\partial V} \mathbf{n_0}\, dS}_{=0} - \mathbf{W} \underbrace{\oint_{\partial V} \mathbf{q}\mathbf{n_0}\, dS}_{=0} = 0. \quad (3)$$

Both $\mathbf{W}$ and $\mathbf{F}$ can be put before the integral because $\mathbf{W}$ is constant and $\mathbf{F}$ consists of components of $\mathbf{W}$. The remaining integrals contain only geometrical data and are identical zero. For the first integral this is obvious. The second integral is according to the theorem of Gauss identical to the volume integral of div $\mathbf{q}$ and div $\mathbf{q} \equiv 0$ for rigid body motion in the form of (1). This integral defines the whirl flux

$$\int_{\partial V} \mathbf{q}\mathbf{n_0}\, dS. \quad (4)$$

Free stream consistency is given if these identities are fulfilled by the discretization. The first integral is a piecewise constant function of the normal unit vector $\mathbf{n_0}$ over the facets f_i. The integral can be split into a sum of normal vectors $\mathbf{n_i}$ scaled with the area of the facets S_{f_i}

$$\int_{\partial V} \mathbf{n_0}\, dS = \sum_{f_i \in \partial V} \mathbf{n_i}. \quad (5)$$

The second integral is a piecewise linear function over the facets f_i. The integral can be split into a sum over the facets f_i and the linear function can be integrated exactly with the midpoint rule of second order accuracy. The integration point is the barycenter $\mathbf{r_i}$ of each facet f_i. If $\mathbf{q_i}$ is the function $\mathbf{q}$ evaluated at the barycenter r_i the integral can be written

$$\int_{\partial V} \mathbf{q}\mathbf{n_0}\, dS = \sum_{f_i \in \partial V} \mathbf{q_i}\mathbf{n_i}. \tag{6}$$

Because both integrals are evaluated exactly each sum is zero if the surfaces of the control volumes are closed. This is given by the construction of the control volumes. Unfortunately the edge based data structure of the TAU code does not store any of these integration points. Instead a reduced integration scheme is used. Figure 2 shows that the normals of all facets f_i touching one edge are added up to the normal of the face F_i with the edge midpoint as common integration point. The integration of the fluxes is formally not of second order accuracy. It can achieve second order accuracy depending on the regularity of the grid. For tetrahedrons the integration over a closed surface is of second order accuaracy but for general hybrid grids this is not the case. The amount of data to store reduces about one order of magnitude with this scheme.

The workflow which is shown in figure 3 is subdivided into a preprocessing step and the work of the flow solver. In the preprocessing step the edge based data structure of the dual grid is calculated from the element-point connectivity of the primary grid which is then discarded. The whirl flux cannot be calculated in the preprocessing step on its own because the rotation vector $\omega(t)$ and the translation vector $\mathbf{v_t}(t)$ are time dependent and the whirl flux has to be recalculated for each physical time step. The operations have to be split so that the geometrical data are calculated in the preprocessing step and the exact information will be maintained in the flow solver. This is possible by taking into account the special form of the function $\mathbf{q}$

$$\sum_{f_i \in F_i} \mathbf{q_i}\mathbf{n_i} = \sum_{f_i \in F_i} (\omega \times \mathbf{r_i})\mathbf{n_i} + (\mathbf{v_t} - \omega \times \mathbf{r_0}) \sum_{f_i \in F_i} \mathbf{n_i}. \tag{7}$$

The sum can be split into a sum of the spatial product $(\omega, \mathbf{r_i}, \mathbf{n_i})$ and the sum of a scalar product. With a cyclic permutation the first sum can be factorized with ω so that the remaining sum contains only geometrical data which can be calculated in the preprocessing step. The additional vector is called **pre whirl flux**. Factorization and calculation of the second sum gives the normal of the face F_i.

$$\sum_{f_i \in F_i} \mathbf{q_i}\mathbf{n_i} = \omega \underbrace{\sum_{f_i \in F_i} (\mathbf{r_i} \times \mathbf{n_i})}_{\textbf{Pre Whirl Flux}} + (\mathbf{v_t} - \omega \times \mathbf{r_0}) \underbrace{\sum_{f_i \in F_i} \mathbf{n_i}}_{\textbf{Face Normal}} . \tag{8}$$

The data structure of the original TAU code has to be extended with the new vector "pre whirl flux". The vector is built in the preprocessing step similar like the face normal vector. With multigrid the whirl flux is exact for each coarser grid level.

Coarser grid levels are constructed by agglomeration and adding up faces. If the components of the vector "pre whirl flux" are added up the exact surface integral (8) does not degrade.

3 Numerical Results

The new integration of the whirl fluxes is demonstrated with a parallel Navier-Stokes calculation with the TAU code of a 4-bladed isolated rotor in hover. The calculation has been done with 6 prozessors on a Xeon Myrinet Cluster with 32 nodes. Each node has 2 CPUs with 3.06 GHz and 2GB RAM. The unstructured hybrid grid has 3.4×10^6 points. The calculation has been done with periodic boundary conditions and a Spalart Almaras turbulence model. To accelerate the convergence a 3-V multigrid cycle has been used. Parameters are the tip mach number $M_{Tip} = 0.662$, the Reynolds number $Re = 2.04 \times 10^6$, the temperature $T = 297.4K$ and the pitch angle $\theta = 7.5°$. The wall clock time for one iteration was 38.42 seconds. The results are compared with a calculation of the Flower code on a structured grid with 1.4×10^6 points. The calculation has been done on a PC with an Intel Pentium 4 processor with 2.6 GHz and 2 GB RAM. The wall clock time for one iteration was 30.65 seconds. Figure 5 shows the pressure distribution of the calculations and the experiment at four blade sections. The pressure coefficient is defined $Cp = (p - p_\infty)/(\frac{1}{2}\rho_\infty(\omega R)^2)$ with the pressure p_∞ and density ρ_∞ of the farfield state, the rotational velocity ω and the radius of the rotor R. The suction peak is not fully resolved and refinement is needed in this area. Figure 4 shows the residual of ρ and the thrust coefficient $Z_b = 100T/(\frac{1}{2}\rho_\infty(\omega R)^2 S\sigma)$ with the thrust of the rotor T, the rotor disk area S and the solidity of the rotor $\sigma = b\,C_{ref}R/S$ where b is the number of blades and C_{ref} the reference chord length. The residual and the thrust coefficient in figure 4 do not completely converge to steady state values, but oscillate in a low-frequency limit cycle. This behaviour is characteristic for rotors in hover and also has been shown in [1]. It vanishes with high dissipation where the tip vortices are destroyed. Figure 7 shows the grid and the vorticity in the axial cut plane of the blade. In the refined area the tip vortices and the downwash of the wakes of the previous blades are in evidence. The streamlines of Figure 6 show a vortex at the bottom. The vortex is weak as can be seen from the distribution of the velocity magnitude. This may come from difficulties with the farfield boundary condition to detect inflow and outflow with a free stream velocity of zero.

4 Conclusion

A procedure to ensure free stream consistency for rigid body motion with hybrid grids has been derived and successfully implemented in the DLR TAU code. The extension is consistent with the edge based data structure of the code and also valid on each coarse multigrid level. All features of the code are maintained as well as the scalability of the extended data structure for multigrid and parallelization. The expensive geometrical part of the calculation could be split off and shifted to the preprocessing step without loss of accuracy. A parallel Navier-Stokes test case of a 4-bladed isolated rotor in hover has been calculated and the results are in good agreement with the results of the structured Flower code and with experimental data.

References

[1] Wake, B.E. and Baeder, J.D.: "Evaluation of the TURNS Analysis for Hover Performance Prediction", American Helicopter Society Aeromechanics Specialists Conference, San Francisco, Jan. 1994.

[2] Duque, N.: "A Structured / Unstructured Embedded Grid Solver for Helicopter Rotor Flows" 50th AHS Forum, Washington, DC, USA, 1994.

[3] Gerhold, T., Friedrich, O., Evans, J., Galle, M.: "Calculation of Complex Three–Dimensional Configurations Employing the DLR–TAU–Code" AIAA-Paper 97–0167, 1997.

[4] Hee Jung Kang, Oh Joon Kwon: "Effect of Wake Adaptation on Rotor Hover Simulations Using Unstructured Meshes" Journal of Aircraft Vol.38, No.5, September - October 2001.

[5] Hee Jung Kang, Oh Joon Kwon: "Viscous Flow Simulation of a Lifting Rotor in Hover Using Unstructured Adaptive Meshes" 57th AHS Forum, Washington, DC, USA, 2001.

[6] Pahlke, K.: "Berechnung von Strömungsfeldern um Hubschrauberrotoren im Vorwärtsflug durch die Lösung der Eulergleichungen" DLR - FB 1999-22, 1999.

[7] Strawn, Roger C.: "A Finite-Volume Euler Solver for Computing Rotary-Wing Aerodynamics on Unstructured Meshes" 48th AHS Forum, Washington, DC, USA, 1992.

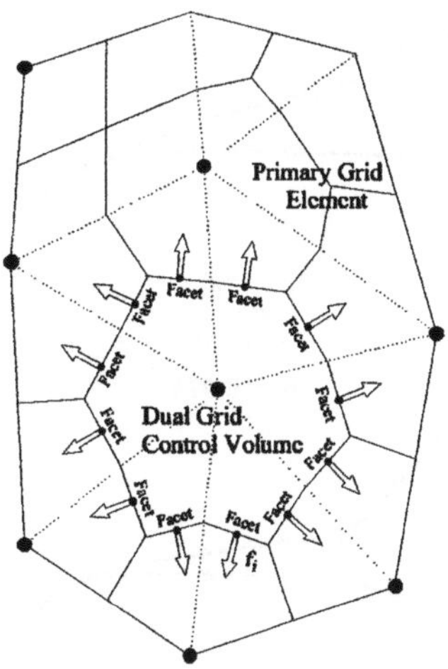

Figure 1 Integration of fluxes with the midpoint rule.

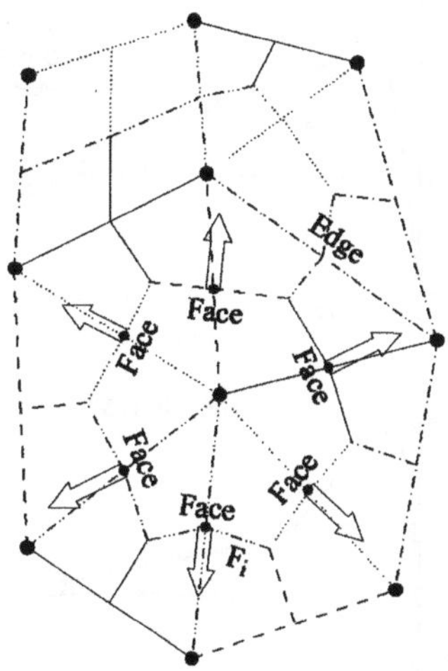

Figure 2 Reduced integration scheme with edge based data structure.

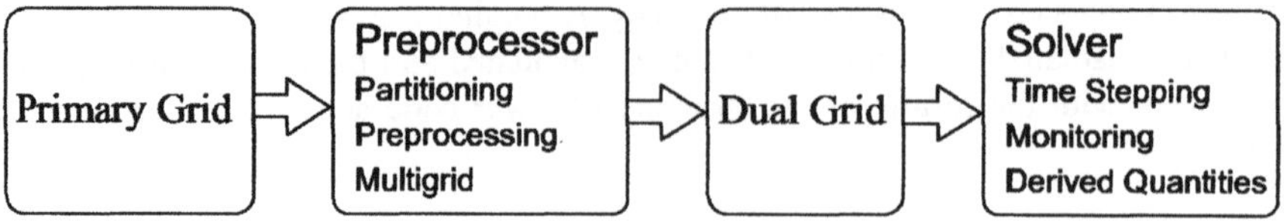

Figure 3 Workflow of Tau-Code.

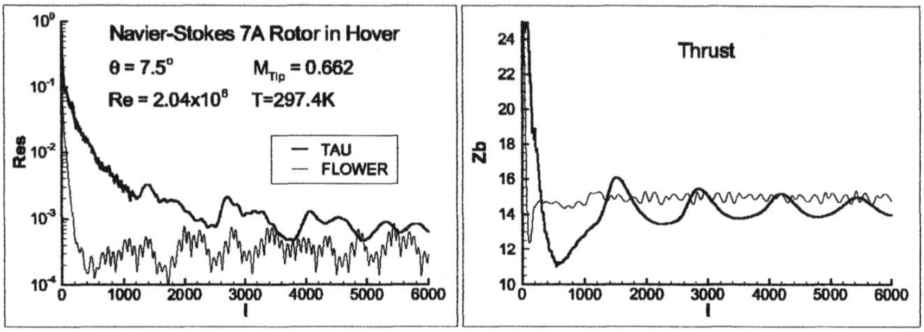

Figure 4 Residual and thrust coeffi cient of 4-bladed rotor HELI7A in hover.

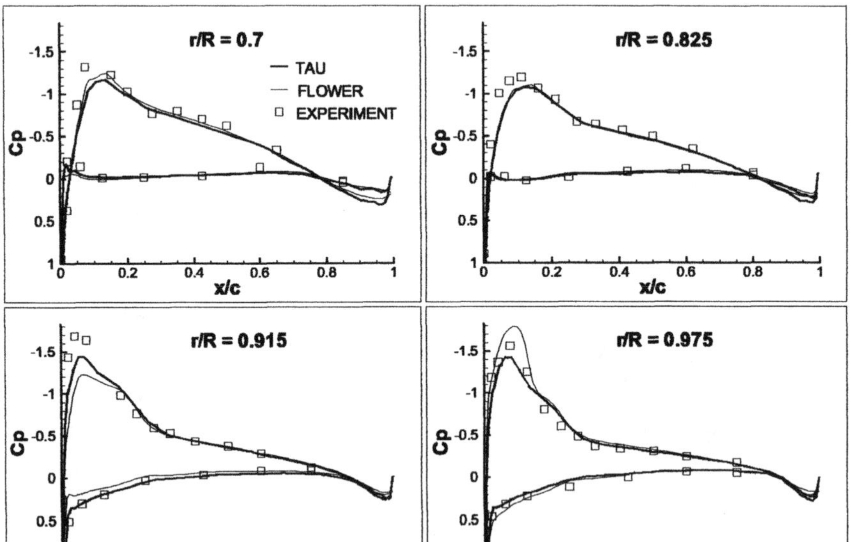

Figure 5 Pressure distribution along blade of 4-bladed rotor HELI7A in hover.

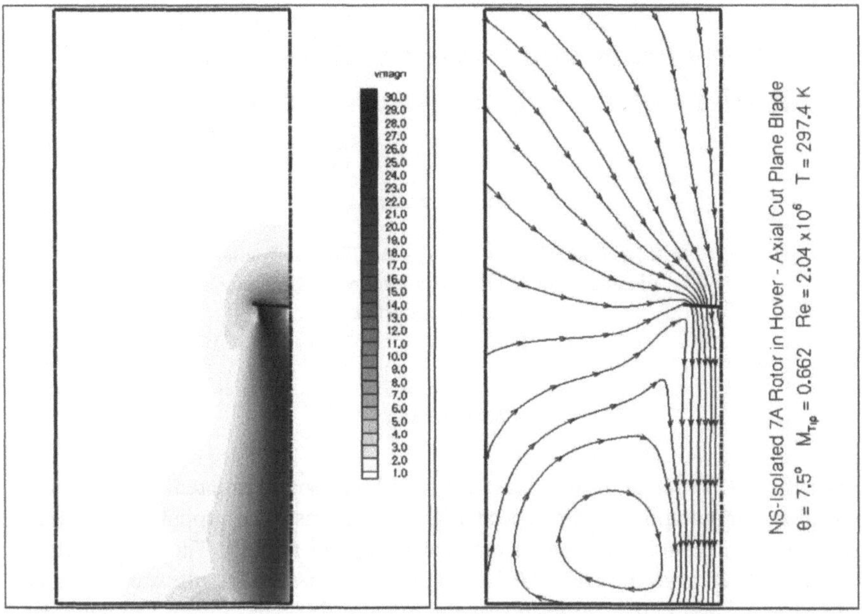

Figure 6 Streamlines and velocity magnitude of 4-bladed rotor HELI7A in hover.

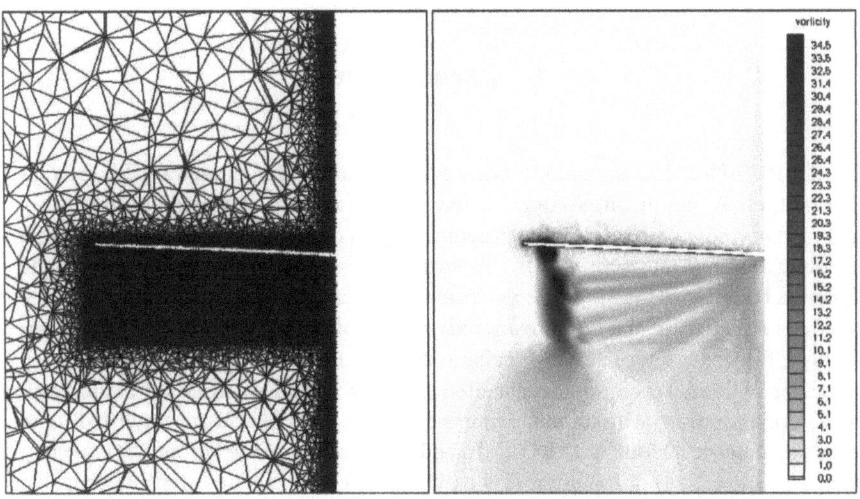

Figure 7 Grid and vorticity of 4-bladed rotor HELI7A in hover.

Unsteady Euler and Navier-Stokes Simulations of Propellers with the Unstructured DLR TAU-Code

Arne Stuermer

DLR Institute of Aerodynamics and Flow Technology
Lilienthalplatz 7, 38108 Braunschweig, Germany
arne.stuermer@dlr.de

Summary

Unsteady CFD simulations have been conducted for generic isolated- and installed-propeller configurations at low-speed flight conditions. The propeller is a four-bladed design typical for modern regional turboprop aircraft. The computations were performed with the DLR TAU-code and the numerical results are compared with experimental data. The results of the Euler and Navier-Stokes computations agree well with the wind tunnel wake data and surface pressure distributions. Additionally, an analysis of the forces acting on the wing and the propeller is performed.

1 Introduction

The DLR TAU-code has recently been extended to handle Chimera grids in relative motion. A significant amount of work has gone into retaining the flexibility and memory advantages of the unstructured codes edge-based data structure while enabling fast and efficient donor cell search and interpolation techniques at the Chimera boundaries [1, 2]. These extensions have enabled the code to be employed for the simulation of the complex aerodynamics of propeller flows [5]. Here, unsteady CFD computations of two generic configurations are presented: an isolated propeller with an axisymmetric nacelle (AGARD 1) and an installed tractor propeller configuration, with an axisymmetric nacelle mounted on an untwisted wing with a symmetric airfoil (AGARD 2). In both cases a four-bladed propeller typical of modern regional turboprop aircraft is used. These configurations were investigated in wind tunnel tests conducted in the 1980s by FFA, the Aeronautical Research Institute of Sweden, now a part of the Swedish Defense Research Agency FOI [3, 4], and are also included in an AGARD report on CFD test cases. A typical low-speed flight condition is investigated with a Mach number of $Ma = 0.15$, an angle of attack of $\alpha = 0°$, and an advance ratio of $J = 0.705$. The Reynolds number of the viscous computation is $Re = 3.7 \cdot 10^6$.

2 Computational Strategy

The unstructured grids, shown in fig. 1, are created using the CentaurSoft Centaur grid generation software. The two-block grids require taking special care during the creation of the CAD geometry for use with the Chimera grid approach, which necessitates leaving a cylindrical hole in the nacelle grid block at the position of the propeller. In addition, an adequate overlap of the two grids needs to be assured. The first grid block was created for the nacelle part of each configuration. As the nacelle or the nacelle-wing geometry is symmetric to the cartesian xz-plane, the mesh was generated for half of the domain and subsequently joined with its mirror image. To ensure an adequate resolution of the propeller wake and the blade tip vortices, very small grid elements were generated in a cylindrical area downstream of the propeller disc. For the second block a grid was generated for the rotating parts of the geometry, i.e. the propeller. Again, use was made of symmetry and a 90° slice was meshed and subsequently copied and rotated to complete the cylindrical propeller grid block. This strategy guarantees an identical spatial discretization for all four blade passages, greatly enhancing the ability of the CFD solver to resolve the expected periodic fluctuations of the flow. The Chimera grid of the isolated propeller in fig. 1(a) has 3.807.954 nodes for the inviscid and 8.373.914 nodes for the viscous computations. The complete Euler grid for the configuration with the wing, shown in fig. 1(b), has a total of 5.632.813 nodes. Unsteady Euler computations have been performed for both configurations and an unsteady Navier-Stokes simulation using the Spalart-Allmaras turbulence model has been completed for the isolated nacelle geometry. The computations used the dual time method and a central spatial discretization with scalar dissipation. One propeller revolution was resolved with 360 dual time steps with 100 inner iterations each. To achieve convergence and periodicity in the propeller forces and to ensure a proper slipstream development, five full propeller revolutions were computed. The simulations were run on 32 processors of a Linux cluster and took between $105h$ and $245h$. These runtimes are quite acceptable for unsteady computations and due to the good parallel scaling of the TAU-code and the rapid development of faster computer hardware these propeller CFD simulations are becoming increasingly more viable for industrial application.

3 Results and Validation

3.1 Propeller Wake Development

To validate the numerical results, the computed propeller wake development is compared with experimental data. Fig. 2 and 3 show the normalized axial and tangential velocities for the two configurations at a position of $x/R = 1.73$ behind the propeller and along a ray extending out above the nacelle. In the experiments the velocities were measured with a 5-hole probe and are in effect time averaged data. The numerical results allow an analysis of the unsteady fluctuations of the slipstream, which are periodic for every 90°-rotation of the propeller due to the blade count of

4. The CFD results are plotted as a time averaged result as solid lines and at two selected times during a propeller rotation of 90° as the dashed lines. The time accurate profiles show the extrema of the unsteady fluctuations in the propeller slipstream. The time averaged axial velocity profiles from the inviscid computations, shown in the top half of fig. 2(a) and in fig. 3(a), are in excellent agreement with the wind tunnel data with both profile shape and magnitudes of the velocity being correctly predicted. The most notable deviations are near the surfaces of the nacelle, where the neglect of viscous effects lead to higher velocities than found in the experiment. The presence of the wing leads to a slight increase in axial velocity magnitude. The time accurate axial velocity profiles at blade positions of $\psi_{bld} = 30°$ and $\psi_{bld} = 80°$ show strong unsteady variations in velocity magnitude. These fluctuations are due to the periodic passage of the blade wake while the profile in the boundary region of the slipstream at the blade position of $\psi_{bld} = 80°$ shows the passage of the blade tip vortex. The time averaged tangential velocity profiles from the inviscid simulations in fig. 2(b) and 3(b) also match the measurements very well. Again, the only notable discrepancy near the nacelle surface is due to the inviscid approach of the CFD simulations. The wing has a pronounced effect on the tangential velocities, which are reduced in peak magnitude by around 18%. This is due to the straightening effect of the wing on the propeller slipstream. The time accurate profiles for blade positions of $\psi_{bld} = 40°$ and $\psi_{bld} = 80°$ show the significant unsteadyness of the flow. Along with the fluctuations in velocity magnitude caused by the blade wakes, the passage of the blade tip vortex can be seen as a kink in the velocity profile near the slipstream boundary for the blade position of $\psi_{bld} = 80°$. The bottom half of fig. 2(a) and 2(b) shows the axial and tangential velocity profiles from the Navier-Stokes simulation of the isolated propeller. Comparing the axial velocities with those from the inviscid computations shows that while agreement in the boundary layer region is improved, the maximum velocity magnitude is slightly lower than found in both the experiments and the Euler simulations. The exact causes of this discrepancy will be investigated in the future, but the likely origins are thought to be the impact of grid density and numerical dissipation as well as the turbulence model. The time averaged tangential velocity profile from the Navier-Stokes computation shown in fig. 2(b) is in good agreement with both the experimental data and the Euler results. Velocity magnitude is only marginally lower than in the inviscid computations but the agreement with the experimental data close to the nacelle is very good.

3.2 Propeller Forces

Fig. 4 shows the development of the thrust force acting on a blade during one propeller revolution. For the isolated propeller configuration the blade produces a constant thrust force as the relative flow for each blade section, the vector sum of the rotational and the onflow velocity component, is constant for this axial flow case. The Navier Stokes computations show a reduction in the blade thrust of around 6% versus the Euler results, which is due to the viscous drag acting on the blades. The blade thrust of the installed configuration exhibits a sinusoidal oscillation with two force cycles per revolution. This fluctuation can be attributed to the interaction of the wing and the propeller, which leads to the blade being affected by an altered flowfield as it passes in front of the wing during its rotation. This causes a small increase in the overall blade thrust force for this configuration versus the isolated propeller case. Table 1 lists the thrust coefficients as a comparison of the experimental data with the CFD results. The ranges of the experimental values are the minimum and maximum values measured at the advance ratio of $J = 0.705$, while the value in parentheses is the median of these values. The inviscid computation of the isolated propeller configuration overpredicts the experimental thrust coefficient by 7.3%. The thrust coefficient taken from the Navier-Stokes computations falls within the range of values found in the wind tunnel tests and is 6.5% lower than the value from the Euler computations. Both the experimentally and the numerically determined thrust coefficients indicate that the presence of the wing leads to a small increase in thrust. The inviscid computations overpredict the experimental value by 6%, which is in line with the tendency seen for the isolated propeller configuration.

3.3 Propeller-Wing Interaction

The swirl of the propeller slipstream affects the local angle of attack for wing sections of the installed propeller configuration lying within the propwash. On the side of the nacelle with an upward rotation of the blades a locally positive angle of attack results, which leads to a positive wing lift, while on the other side of the nacelle the situation is reversed. The unsteady propeller slipstream leads to periodic fluctuations in the pressure and lift force distributions, shown in fig. 5. The amplitude of the fluctuations are shown for two wing sections as dashed and dotted lines representing two time accurate pressure distributions from the CFD results, while a time averaged pressure distribution is shown as a solid line for comparison with the experiments. To illustrate the impact of the slipstream on the wing fig. 5(a) shows pressure distributions for the same sections from a computation of the configuration without propeller. The symmetric airfoil and the straight and untwisted wing geometry lead to symmetrical distributions which are identical in all sections for this case. The section at $y/b = 0.0975$ in fig. 5(b), which lies at a propeller radial position of $r/R = 0.61$, exhibits a slight variation in the pressure levels in the suction peak caused by the fluctuations in the velocity field of the slipstream. This correlates well with the small fluctuations found in the axial velocities at this radial position in fig. 3(a). The high velocity of the propwash here results in strong suction peaks being produced. A section further outboard at a spanwise position of $y/b = 0.13$, which lies within the propwash at a radial position of $r/R = 0.8125$, is shown in fig. 5(c). The impact of the periodic passage of the blade tip vortex leads to a pronounced fluctuation in the suction peak pressures as well as an oscillation of the pressures in chordwise direction. Due to the lower slipstream velocities at this station (see fig. 3(a)) the pressure levels are lower than those seen for the previous section. In all sections agreement between the time averaged CFD results and the experimental data is good with differences limited to the suction peak levels which tend to be overpredicted slightly in the computations. Fig. 5(d) shows the spanwise distributions of the lift as a time averaged value and two time accurate results from the CFD simulation.

The slipstream induced local angles of attack for the wing within the propwash lead to large local lift forces being produced. As the wing consists of a symmetric airfoil with no twist, lift is produced on the side of the nacelle of upward propeller rotation while on the opposite side a downforce of equal magnitude is produced. In addition to this local impact, a spanwise influence is visible with the lift distribution gradually dropping off towards the tips. The unsteady nature of the propeller slipstream leads to an oscillation of the lift force distributions both in the propwash and along the entire span of the wing.

4 Conclusion

The DLR TAU-code has been applied to the unsteady simulation of isolated and installed propeller configurations. A comparison with propeller slipstream data obtained in wind tunnel tests shows good agreement between the numerical results and the experimental data. Except for small deviations close to the surfaces of the nacelle, the Euler results match the experimental data extremely well, while the viscous simulation gives a slight underprediction of axial velocity but improves the results in the boundary layer regions. The effect of grid density, numerical dissipation and the turbulence model is thought to be a cause of the reduced axial velocity magnitude and will be investigated in the future. An analysis of the complex aerodynamic interactions for a wing mounted propeller configuration shows both an unsteady fluctuation in blade thrust due to the presence of the wing as well as an effect of the propeller on the wings lift. A small increase in propeller thrust has been shown to result form these interactions. The inviscid computations overpredict the propeller thrust by around 7%, but the inclusion of viscous effects in the simulation brings the numerical values within the scatter range of the experimental data. Future work will continue with Navier-Stokes simulations to determine the impact of such numerical parameters as grid density, numerical dissipation and the turbulence model on the results as well as to enhance understanding of the aerodynamic interactions at different advance ratios, angles of attack, sideslip and Reynolds numbers.

References

[1] Madrane, A. and Heinrich, R. and Gerhold, T.: "Implementation of the Chimera Method in the Unstructured Hybrid DLR Finite Volume TAU-Code", 6th Overset Composite Grid and Solution Technology Symposium, Ft. Walton Beach, FL, USA, 2002.

[2] Madrane, A. and Raichle, A. and Stuermer, A.: "Parallel Implementation of a Dynamic Unstructured Chimera Method in the DLR Finite Volume TAU-Code", 12th Annual Conference of the CFD Society of Canada, Ottawa, Ontario, Canada, 2004, pp. 524–534.

[3] Samuelsson, I.: "Low Speed Wind Tunnel Investigation of Propeller Slipstream Aerodynamic Effects on Different Nacelle/Wing Combinations Part 1", FFA TN 1987-22, 1987.

[4] Samuelsson, I.: "Low Speed Wind Tunnel Investigation of Propeller Slipstream Aerodynamic Effects on Different Nacelle/Wing Combinations Part 2", FFA TN 1990-24, 1990.

[5] Stuermer, A.: "Validation of an Unstructured Chimera Grid Approach for the Simulation of Propeller Flows", AIAA 2004-5289, 2004.

Table 1 Propeller thrust coefficients C_T

	Exp	TAU-Euler	TAU-NS
AGARD 1	0.236 - 0.248 (0.242)	0.2598	0.2429
AGARD 2	0.233 - 0.260 (0.2465)	0.2611	-

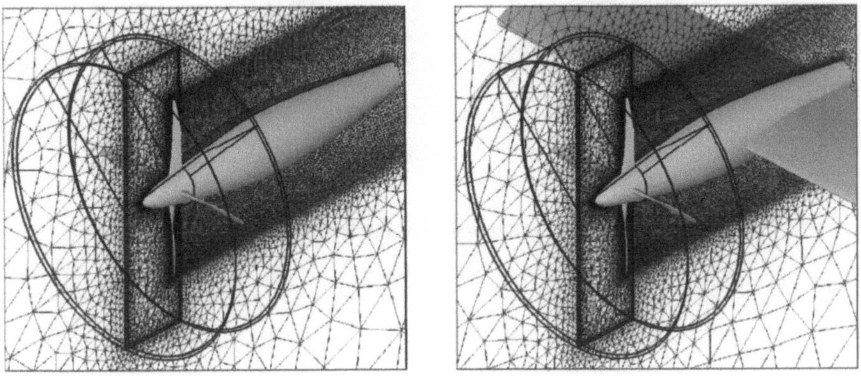

(a) Isolated propeller Chimera grid (b) Installed propeller Chimera grid

Figure 1 Two-block unstructured Chimera grids

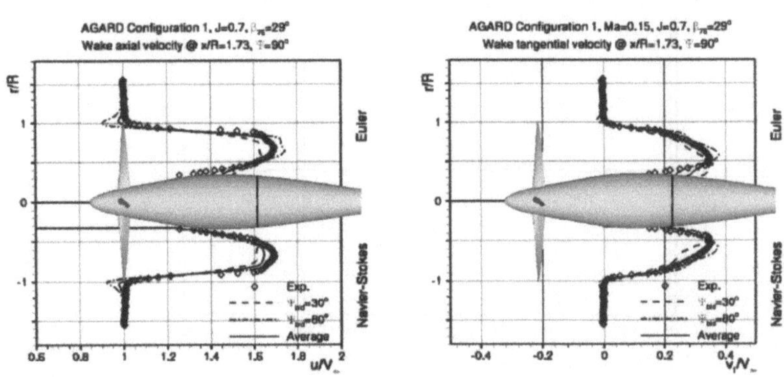

(a) Axial velocity at $x/R = 1.73$ (b) Tangential velocity at $x/R = 1.73$

Figure 2 Isolated propeller wake velocity profiles

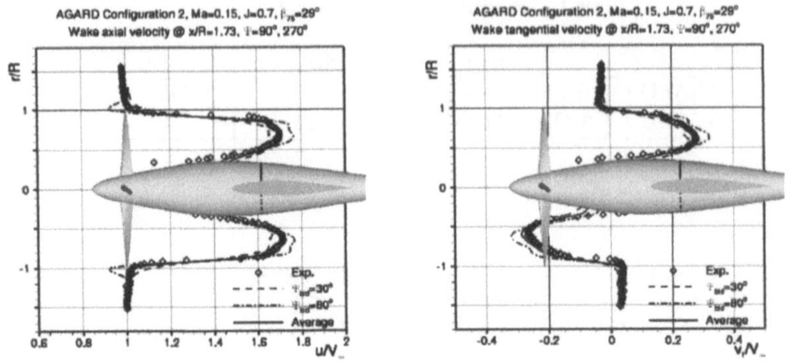

(a) Axial velocity at $x/R = 1.73$ (b) Tangential velocity at $x/R = 1.73$

Figure 3 Installed propeller wake velocity profiles, TAU Euler

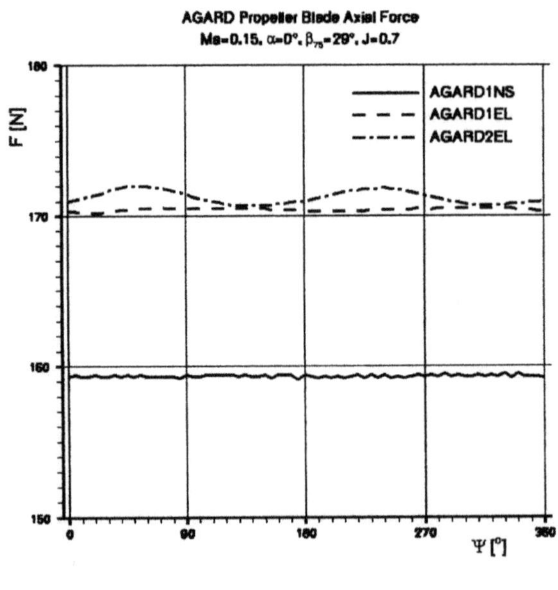

Figure 4 Development of blade thrust during the propeller revolution

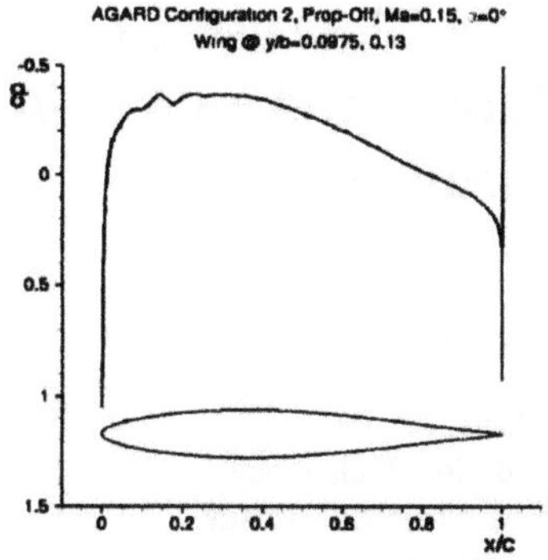

(a) Prop-off pressure distributions

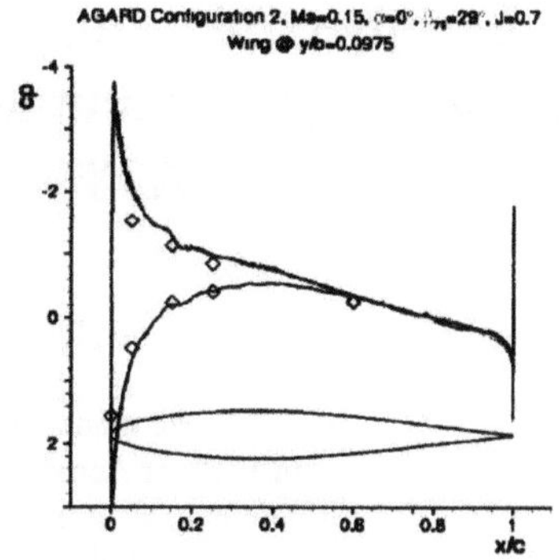

(b) Pressures at y/b=0.0975, r/R=0.61

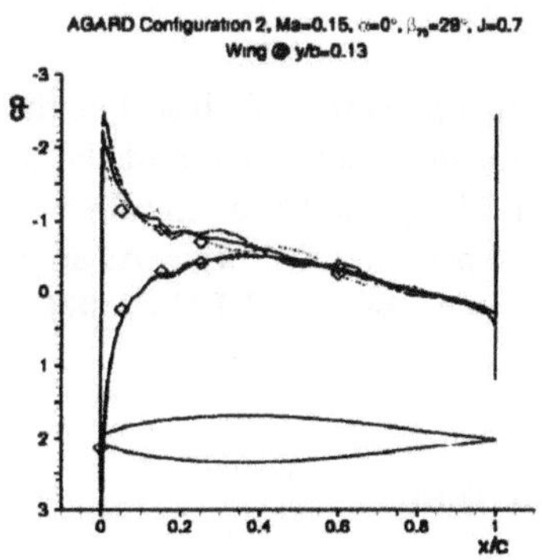

(c) Pressures at y/b=0.13, r/R=0.8125

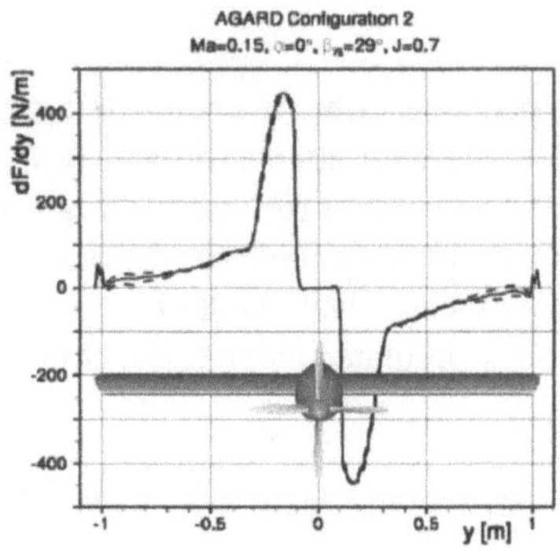

(d) Wing lift distributions

Figure 5 Propeller slipstream effects on wing

Aerodynamic Analysis of Jet-Blast using CFD considering as example a Hangar and an AIRBUS A380 configuration

S. Melber-Wilkending

German Aerospace Center (DLR)
Institute of Aerodynamics and Flow Technology
Lilienthalplatz 7, D-38108 Braunschweig, Germany
EMail: Stefan.Melber@DLR.de

Summary

The paper describes the setup and validation of a process-chain based on the DLR-TAU code with adaptation and flow solver for the determination of the jet-wake behind jet engines for viscous flows is shown. This jet-wake (often called jet-blast) plays an important role for the design and the operation of airports. As an example of the process-chain the jet-blast impact on a hangar by an AIRBUS A380 configuration is simulated using this technique.

1 Introduction

For the design and operation of an airport the volume of traffic, available amount of space and legal requirements can play an important role. Also aerodynamic questions may have significant impact for ground operations. Considering as an example the introduction of the new AIRBUS A380 in the following methods to get answers on these questions using numerical flow simulation are shown.

For the operation of the apron area and runways the jet-blast is one of the limiting factors. The jet-blast is caused by highly accelerated air flow and exhaust gas behind jet engines. To prevent the hazard of persons, ground vehicles and other aircraft by jet-blast, security areas and the safe operations of taxiways have to be determined. Furthermore, the aerodynamic loads on buildings (e.g. blast-fences) caused by jet-blast and also possible additional wind-load is a priori unknown.

The strength of jet-blast behind aircrafts with jet-engines is a function of the maximum thrust of an airplane. The civil aircraft with the strongest thrust to date is the Boeing B747-400 with a maximum thrust of 4 x 282 kN which is far of the maximum thrust of the future AIRBUS A380 with 4 x 340 kN thrust. In the sector of twin-engines more thrust per engine is realized, e.g. 2 x 436 kN of a Boeing B777-300. This increasing thrust of modern jet-engines affects the design of buildings or the determination of security areas and leads to the inspection or modification of existing facilities at airports.

Currently only a few methods for the solution of jet-blast problems are available. The methods for the prediction of jet flows used by aircraft and engine manufacturers can be used for the determination of the propagation of undisturbed jets behind engines, but interactions of the jets with objects (e.g. blast-fences) are not covered by them. On the other side, experiments with existing aircraft are too cost-intensive or, in case of aircraft in development like the AIRBUS A380, are not possible. Also model experiments with reduced size often suffer from considerable costs and scaling problems.

In the following the numerical flow solution (CFD) is shown as a possible alternative to address the problem of jet-blast. For this purpose the layout and validation of a process-chain based in the DLR-TAU code is described. This process-chain has been successfully used for the inspection of blast security facilities of an international airport. Due to confidentiality restrictions the results cannot be shown here. For the demonstration of the capabilities of this technique an example of a hangar located on the apron area is shown encountered by jet-blast of an AIRBUS A380 in take-off. Further more cross-wind with 20 kts is taken into account for the simulation. The numerical investigation includes the complete aircraft configuration, the fictive hangar and apron area. Physical features like e.g. the ground effect are incorporated in the simulation.

2 Geometry

In the considered take-off case usually the complete high-lift devices (e.g. slats, flaps) of a transport aircraft are deployed. The interaction of the engine-jets with these elements or the empennage cannot be disregarded a priori. For this reason the complete high-lift configuration of the AIRBUS A380 is considered. The landing gears are included, because in the considered ground case they are not affected by the jet and thus they have no effect on the jet flow. The engines used examplarily below are of the Trent 900 engine family with a "maximum take-off thrust" of 4 x 340 kN for the AIRBUS A380. The configuration of the AIRBUS A380 is provided by AIRBUS Germany.

The obstacle in the engine-jet of the AIRBUS A380 is a fictive hangar, consisting of the square main part with a dimension of 150 m x 85 m x 37 m. On the forefront a balustrade with a height of 10 m is added, see figure 1. The positioning of the aircraft in relation to the hanger is shown likewise in this figure.

3 Flow Solution Method

The solution of the Reynolds-averaged Navier-Stokes equation (RANS) is done using the hybrid unstructured DLR TAU-code [1]. For the solution of the viscous flow the $k-\omega$-SST turbulence model of Menter [2] was used, which combines rubustness with suitability for jet flows. The grids necessary for the discretisation of the volume around the configuration are based on the geometry discussed in the previous section. They are built using the grid generator CENTAUR from CentaurSoft [3].

In order to reduce the number of grid points only such boundaries are simulated with a viscous boundary layer resolution for which an estimated influence on the flow field is anticipated. These are the engines including the pylons, because the jets arise from them and initial mixing processes of the fan- and core-jets can be found there. All other boundaries like wing, high-lift systems, fuselage or ground are simulated by inviscid surfaces. The aircraft except for the flaps in the engine area does not face a flow and the boundary-layer influence of parts in recumbent flow can be disregarded. At the ground and the hangar the influence of the viscosity is also small, because the interaction with the jet is impulse-dominated.

The initial hybrid unstructured grids are comparatively coarse in the area of the engine-jet. This leads to an increased numerical dissipation in these areas and an unphysical dissipation of the jet and smearing out of the gradients in the flow field. An general refinement is not reasonable because of the big volume influenced by the jet and the resulting high number of grid points. Particularly the interaction of the jet with obstacles cannot be captured with suitable discretisation. Hence automatic grid adaptation is used, based on the flow solution of the initial grid using gradients of flow variables for a local grid refinement.

The disadvantage of adaptive grid refinement used at long-range flow phenomena like engine jets is the slow propagation of the increased grid resolution in the area of the coarse grid. The reason is that the gradient sensors respond only weakly in the coarse grid because of the numerical dissipation and the smeared out flow solution. This behavior can be improved by using a multiple adaptation - a so called adaptation-chain.

The gradient sensor used for the jet adaptation is the total pressure loss, because inside the jet a pressure gain can be found whereas in the stagnating flow around the jet a total pressure loss takes place. In the area between both significant gradients can be found which lead to an increased grid resolution there by grid adaptation. To resolve the interaction of the jet with obstacles by grid adaptation the total pressure loss is not well suited because of its small value there. Using an speed-sensor based on the absolute value of the flow velocity in combination with the total pressure loss sensor improves this situation.

4 Results

4.1 Comparison of numerical results with reference methods

For the comparison of the numerical simulation of jet flows behind jet-engines with other methods reference-data of AIRBUS [4] as well as analytical results of an axisymmetric free jet in turbulent flow [5] are used. The data for comparison were analyzed concerning uncertainties and model assumptions and compared to CFD-simulations of an isolated jet-engine to demonstrate the applicability of the DLR TAU-code for the task "ground simulation".

As one example in figure 2 the lines of constant velocity of the analytical data and the numerical simulation of an isolated jet-engine for a thrust of 340 kN and 391 kN are compared. Whereas in figure 3 the reference-data used by AIRBUS for a thrust of 340 kN per engine is shown. Comparing both figures the increased jet range for the lines of constant velocity $|V| = 46\ m/s$ (17 %) and $|V| = 30\ m/s$ (30 %) of the reference-data can be found, whereas the lines of constant velocity $|V| = 15\ m/s$ show more jet range in the numerical simulation. The differences can be attributed on the different number of engines: one in the numerical simulation and two in the reference-data of AIRBUS.

In figure in addition lines of constant velocity of 340 kN thrust also the lines of constant velocity of 391 kN thrust are shown. The thrust increased by 15% can be used to determine the sensitivity of the flow field to the thrust. Compared to the results of the numerical simulation at 340 kN thrust an increased jet range can be found as expected. In figure 2 further more the analytical results are depicted. There a slightly reduced extension is found for all lines of constant velocity. The reason is the assumption of a constant viscosity in the complete flow field in the derivation of the analytical model.

Altogether the comparisons of the numerical simulation to other methods shows the applicability of the choosen method DLR TAU-code for the task "jet-blast". Limiting assumptions necessary in the reference-data or the analytical method can be omitted in the RANS simulations.

4.2 Configuration AIRBUS 380 / Hangar

In this section the numerical results of the simulation of an AIRBUS A380 in take-off configuration including a hangar and the ground presence are shown and discussed. A thrust of 4 x 340 kN and a wind with 20 kts and a direction of 20° to the aircraft center-line are used for the simulation.

To show the complete flow field in a three-dimensional view in figure 4 (left) the area of constant velocity of $|V| = 15\ m/s$ is depicted. In this figure the process of jet-flow mixing of the engines of the left / right wing can be seen: In the area of the elevator the jets are already mixed. At the impinging on the hangar the jets of the right wing evade laterally and turn upwards. Behind the hangar the jets flow off in one front. A significant influence of the cross wind cannot be found.

Important for the operation of an aircraft is the avoidance of recirculation, so that exhaust gas can re-enter in the inlet of an jet engine. Based on the shown area of constant velocity this effect cannot be found in the presented configuration.

To determine security areas, both around the hangar and the aircraft, the velocity distribution directly above the ground is accounted for. In figure 4(right) flow velocities of $|V| \geq 15m/s$ are depicted. Beside the trace of the jet behind the aircraft to the hangar the raise and lateral deflection of the jet starting from the left wing can be clearly seen. Furthermore a part of the flow is deflected on the front of the hangar. For the definition of security areas against jet-blast on the apron limiting flow velocity of 15 m/s is selected. As shown in figure 4 on the ground an area behind the aircraft and a part in front of the hangar falls in this definition. Considering the three-dimensional area of constant velocity, the critical velocities $V \geq 15\ m/s$ in the environment of the hangar cannot only be found on the ground, but also laterally on the hangar. This effect must be keept in mind when determining security areas.

The flow topology of the flow field close above the ground is shown in figure 5. The streamlines in front of the configuration reflect the wind direction. Further more, the streamlines converge in front of the aircraft as a result of the suction of the engines. Behind the aircraft the streamlines are absorbt and accelerated respectively due to the entrainment of the jet. The entrainment effect is caused by the lower static pressure in the jet compared to the ambient pressure and thus the flow enter this low pressure area.

Behind the hangar across from the aircraft side two strong vortices can be found. They are caused by the nearly normally directed wind on the front of the hangar and wake behind it. In contrast on the aircraft side of the hangar both vortices are nearly completely suppressed by the combination of jet- and incoming flow.

Beside the calculation of the flow field the static load on the hangar due to the jet-flow and the wind can be determined from the numerical simulations. From the simulation a maximal static pressure of 104005 Pa can be found on the hangar based on an ambient pressure of 101325 Pa. The hangar-doors should be dimensioned for example with this pressure load. It must be keept in mind, that the pressure inside the hangar may be different from the ambient pressure because of the dynamic pressure in front of the hangar and with it, the load on the hangar-doors can vary.

5 Conclusion

In the present paper aerodynamic questions for the design and the operation of airports are discussed. The problem of jet-blast of aircraft at take-off is shown in detail. This problem can be separated in determining security areas for persons, vehicles and aircrafts and the determination of jet- and wind-loads on buildings and facilities. These aspects have to be considered in the design of buildings close to run or taxiways but likewise by a inspection or modification when new aircraft type (e.g. AIRBUS A380) is introduced.

As an example of the aerodynamic task "jet-blast" the interaction of the engine-jets with a building was shown: As typical obstacle a hangar on the apron area was choosen with jet-blast by a taking-off AIRBUS A380. Further on, cross wind with 20 kts was simulated to show the effect of combinated blast- and wind-loads. Among the complete aircraft configuration and in front of a hangar the apron area and coupled with it the ground effect was simulated.

The task jet-blast was addressed by the solution of the Reynolds averaged Navier-Stokes equations using the flow solver DLR TAU. The results for the configuration AIRBUS A380 in front of a hangar the numerical results are discussed and used to determine security areas against jet-blast or the aerodynamic loads on a building.

The application of CFD-methods on aerodynamic questions for the design and the operation of airports is not limited on the application range shown here. In fact, it can be used for other applications, which cannot be solved with approximative- or handbook methods due to the underlying simplifications.

References

[1] Kroll, N.; Rossow, C.-C.; Becker, K.; Thiele, F.: *MEGAFLOW - A Numerical Flow Simulation System.* 21st ICAS congress, 1998, Melbourne, 13.09-18.09.1998, ICAS-98-2.7.4, 1998.

[2] Menter, F.R.: *Zonal Two Equation k-ω Turbulence Models for Aerodynamic Flows.* AIAA-Paper 93-2906, 1993.

[3] Kallinderis, Y.: *Hybrid Grids and Their Applications.* Handbook of Grid Generation, CRC Press, Boca Raton / London / New York / Washington, D.C., pp. 25-1 - 25-18, 1999.

[4] AIRBUS S.A.S: *A380 - Airplane Characteristics For Airport Planning AC (Preliminary Issue).* AIRBUS S.A.S, Custumer Service, Technical Data Support and Services, 31707 Blagnac Cedex, France, September 2003.

[5] Schlichting, H.; Gersten, K.: *Grenzschichttheorie.* Springer Verlag Berlin, Heidelberg, Auflage 9, 1997.

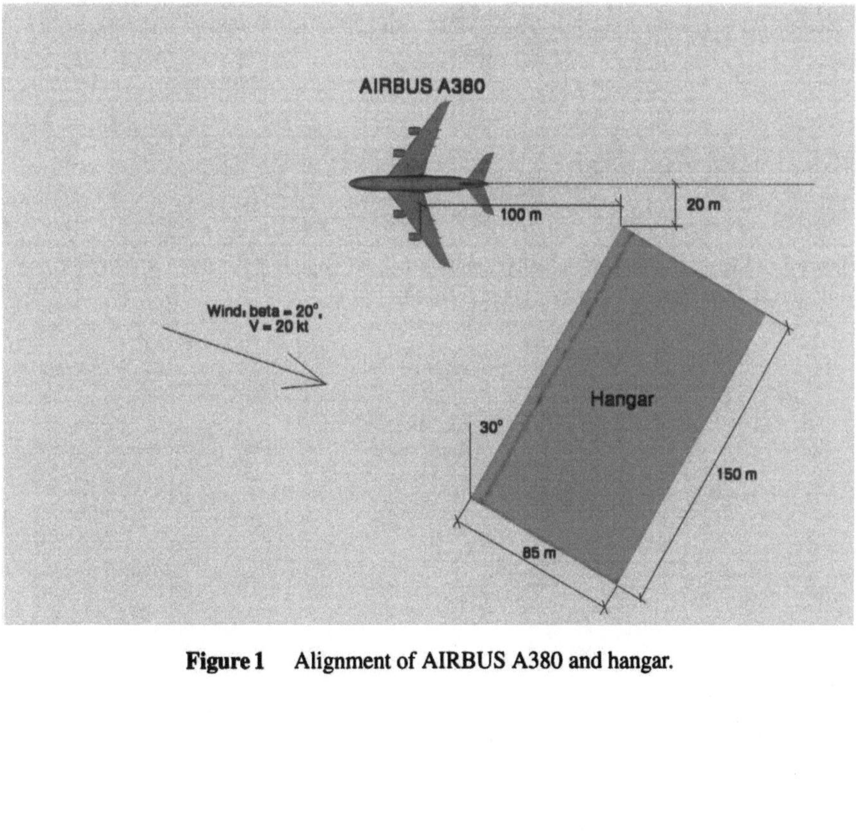

Figure 1 Alignment of AIRBUS A380 and hangar.

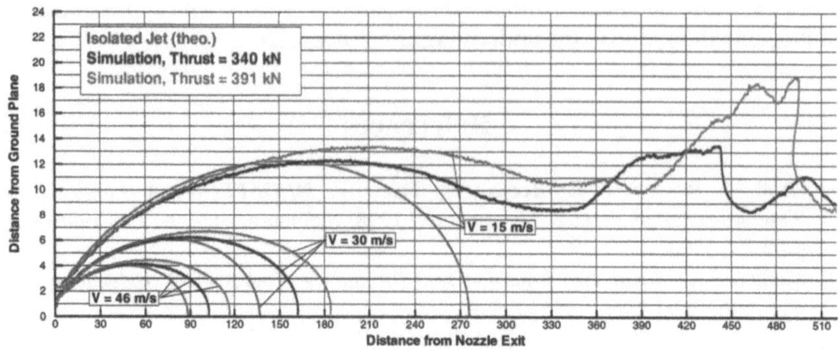

Figure 2 Lines of constant velocity diagram, engine Trent 900, engine-axis on aircraft center-line, comparison of analytical / numerical solution.

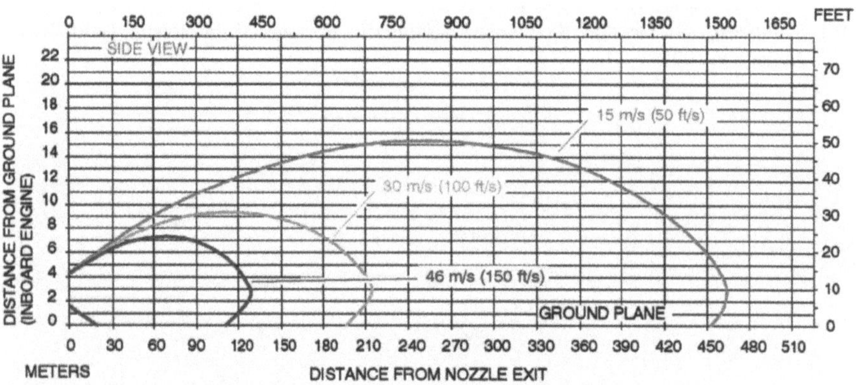

Figure 3 Lines of constant velocity diagram behind engine Trent 977, AIRBUS A380-843F, thrust "max. take-off power", taken from [4].

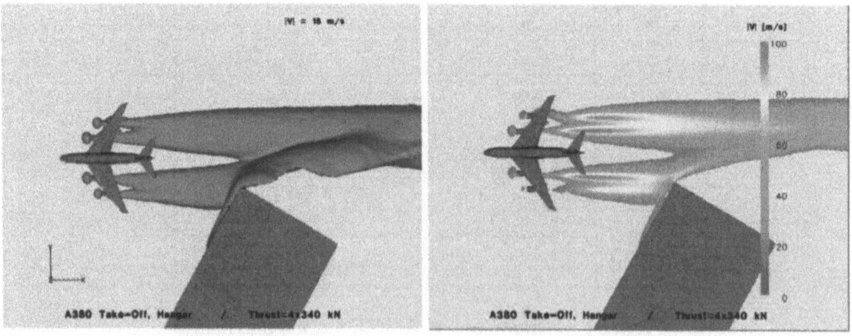

Figure 4 AIRBUS A380 take-off configuration with hangar, thrust 4 x 340 kN, top view, left: area of constant velocity $V = 15\ m/s$, right: velocity.

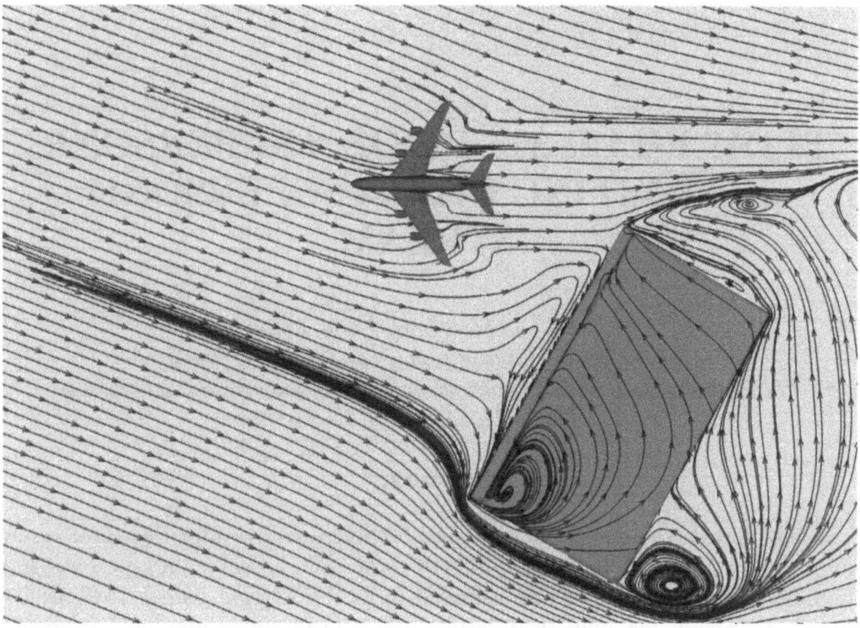

Figure 5 AIRBUS A380 take-off configuration with hangar, thrust 4 x 340 kN, top view, stream lines.

FLOW FIELD STUDY OF A SUPERSONIC JET EXITING INTO A SUPERSONIC STREAM

R. Adeli(*)*, J.M.A. Longo*(*)*, H. Emunds*(**)
DLR, Institute of Aerodynamics and Flow Technology,
(*)*Lilienthalplatz 7, 38108 Braunschweig, Germany ; Reza.Adeli@dlr.de*
(**)*Linder Höhe, 5114 Köln, Germany*

Summary

The paper present a numerical / experimental investigation of the flow field resulting around a generic vehicle controlled by means of a lateral jet in supersonic motion. The experimental data are acquired in the DLR wind tunnel TMK in Cologne. To obtain the numerical results, the 3-D viscous, turbulent, Reynolds-averaged hybrid Navier-Stokes Code TAU of DLR is used. Calculations are made for free stream Mach number of M_∞=2.8 and various angles of attack. Several grids of varied density and structure and different turbulence models are investigated. The CFD results in terms of surface pressure, normal force coefficients, and jet interaction are in good agreement with the results from wind-tunnel tests. On hand the numerical results it is shown important features of the interaction mechanisms between the lateral jet and the flow field originated by the motion of the vehicle.

1. Introduction

During the homing phase of an interceptor missile, a short response time of controls is mandatory. Classical aerodynamic control surfaces are not efficient at low dynamic pressures which occur at low speeds (launch phase) or at low density (high altitudes). Improved agility and maneuverability can be achieved with quick-response control jets, which are an attractive alternative to conventional aerodynamic control surfaces. However, using reaction control jets on a flight vehicle in the atmosphere causes interactions between the jet and the external flow. Indeed, while for some specific orientations and flow conditions, the interaction effects have been found to amplify the lateral force of the jet, under certain other flow conditions and for other orientations the control jets may cause adverse effects like a de-amplification of the thrust. To gain a better understanding of the phenomena that occur with missiles or re-entry vehicles using reaction control systems, a numerical / experimental investigation of the interaction between a non-reactive, perfect gas jet control system and an external ideal onset gas flow is conducted for a generic vehicle.

2 Numerical Method

2.1 DLR TAU-Code

The DLR-developed TAU-Code [1] is used to solve the Reynolds-averaged Navier-Stokes equations. The discretization of space and time are done independently. The space discretization was an unstructured, finite volume approximation while the time discretization is achieved using an explicit Runge-Kutta method. The grid can be made up of tetrahedrons, prisms, pyramids and hexagons. The flow is calculated using a dual grid generated by preprocessing modules. The discretization of the convective terms (Euler terms) is done using centralized- or upwind methods while viscous flow is discretized using a centralized method. For more detailed information see Refs. [1, 2].

2.1.1 Grid Generation

In order to perform the numerical calculations, a geometrical discretization of the region is required. A grid is generated using the CENTAUR-program [3]. It uses CAD-data in form of IGS / IGES files. With this program the boundary conditions, for differential equations, can be defined for each panel. CENTAUR provides structured as well as unstructured 3D-hybrid grids for numerical analysis consisting of cells of varying shapes.

3 Experimental Testing

3.1 Wind-Tunnel Model

The experiments are conducted at the Trisonic wind tunnel (TMK) in Köln-Porz [4]. For the investigation of the interaction characteristics, a vehicle ***40mm*** diameter model of generic shape is selected. It consists of a cone shaped nose, a cylindrical body and an adjacent conical transition with an increasing diameter, which in turn connects to a cylindrical aft extension. **Fig. 1** displays the wind tunnel model with holes for pressure measurements. The holes are shown in cross sections 1-4 and in the longitudinal sections 5, 6 and 7. The orifice of the supersonic jet is perpendicular to the surface of the cylindrical body and is positioned at an azimuth of $\varphi = 180°$.

3.1.1 Test Condition

Mach number:	$M_\infty = 2.8$
Reynolds number associated with the diameter *D*:	$\mathrm{Re} = 1.9 \cdot 10^6$
Angle of attack:	$-10° \le \alpha \le 15°$
jet pressure ratio:	$1 \le P_{0D}/P_\infty \le 300$
jet medium:	cold air
jet stagnation temperature:	ambient temperature $T_0 \approx 280K$

stagnation pressure: $P_0 = 5.6bar$

Where P_{0D} is the total pressure of the lateral jet.

4 Numerical Simulation of Jet Interactions with a Super sonic Free Stream

4.1 Assessment of the numerical solutions

Since the flight vehicle has a body of revolution, the geometry is reduced to a circular-arc-cylinder. Using CENTAUR, several grids are generated around this geometry. Geometric sources in the shape of the hollow cylinders are defined at the transitions from nose cone to cylindrical body and from the cylindrical body to the flare. The forward parts of these cylinders are refined, thus creating a hybrid grid with a prismatic surface. This can be seen in **Fig.2**. For the field discretization, the height of each successive layer equals the height of the previous layer multiplied by a scaling factor of *1.25*. In this manner, an expansion ratio is attained. For a supersonic flow of $M_\infty = 2.8$ the boundary layer fits nicely into 16 prismatic cells. This assertion is based on calculations executed with another grid containing approximately the same number of nodes, but twice as many prismatic cells, which delivered the same results for the forces, pressure coefficients and amplification factors. Indeed, several grids and various turbulence models are investigated to assess the numerical solutions with respect numerical errors and physical modeling. About 24% of the cells of the coarse grid, coarse grid Nr.1, (**Fig. 3 - center**) is refined with the TAU-code to enhance the resolution of the shocks. After calculating the flow with a relatively coarse grid, course grid Nr. 1, featuring only local grid refinements, a new grid is generated with CENTAUR, which in field close to the body results about 24% finer than the coarse grid. The flow is then recalculated using this refined grid. A geometric source in the shape of a hollow cylinder is added at the jet exit position with CENTAUR (**Fig. 3 - left**). Using the TAU-code, further numerical refinements are defined in regions where large flow gradients are present. Local adaptations are added to this finer grid resulting in additional 6% in grid nodes in certain areas. Finally, a super fine grid with slightly more than *2.5mio* nodes is generated (see **Fig. 3 - right**). **Table 1** presents an overview of the grids and turbulence models used for the present study. The distribution of the pressure coefficient C_p in a jet-off environment for a variety of grids is displayed in **Fig. 4**, while the effect of the turbulence models is shown in **Fig. 5**. It turns out that under jet-off conditions grid convergence is demonstrated and also, all turbulence models provide a good correlation with the experimental data. In **Fig. 6**, the distribution of the pressure coefficient C_p is shown using different grids for the jet-on case. While for the attached flow parts the results converge regardless of the used grid, slight differences can be observed on the separated regions. A good correlation of the numerical results with experimental data is obtained using a fine grid with local adaptations of about 760000 nodes (not shown here). Finally, **Fig. 7** displays distributions of pressure coefficients using various turbulence models for the jet-on case. The calculations are executed for a total pressure ratio of $P_{OD}/P_{\infty} = 50$. It turns out that the $k-\varepsilon$ model leads to an additional local boundary layer separation upstream of the jet not obtained with the $k-\omega$ and the Spalart-Almaras model and also not shown by the experimental data. Finally, **Figs. 8 and 9** display distributions of surface pressure coefficient for jet-off/on – conditions for various angles of attack obtained numerically and compared with wind tunnel experiments. It can be seen, that the results from the CFD-calculations are in good agreement with the experimental data.

4.2 Flow Interaction

The analysis of the flow interaction is based mainly on results obtained from CFD. **Fig. 10** displays the flow field, the jet exhaust and its interaction with the free stream in the upper plot; the pressure coefficients along a critical profile at an azimuth of $\varphi = 180°$ are presented in the lower diagram. A circle in Fig. 10 highlights the region, in which some important features of the interaction between the jet and the free stream occur. This region is enlarged in the fig.. The jet plume presents an obstacle to the free stream. In a supersonic free stream flow, this results in a bow shock, which in turn leads to a boundary layer separation upstream of the jet. As a result of the obstacle, which the jet presents to the external flow, a region of increased pressure is created upstream of the jet and a low-pressure region occurs downstream of the jet. Where the boundary layer of the external flow meets the sheer layer of the jet, they create a mixing layer, which separates the jet plume from the external flow. Due to the high pressure directly downstream of the bow shock, the jet is deflected so that it flows parallel to the axial free stream. Inside the jet flow, close to the jet exit, another structure is noted to abruptly increase the pressure. This structure is known as the mach disk. Since the jet exits normal to the supersonic free stream from the surface of the flight vehicle, the mach disk too is deflected downstream. This flow topology obtained from the CFD is in good correlation with the sketch shown in **Fig. 11** based on experimental finding [5-6].

Fig. 12 presents a 3D picture of the flow in the surrounding of the jet exit resulting from a composition of several cross sections plane perpendicular to the main stream. While only one side of the symmetry plane is displayed in the Fig. the topology is symmetric. It turns out, that the interactions in a supersonic flow reach far downstream of the jet plume. Originating at a relatively short deflection section of the jet, a field vortex can be identified, which is retained until very far downstream. A pair of wake-vortices dominates the flow field downstream of the jet exit. Also, the bow shock introduced by the jet interacts with the vehicle boundary layer ahead of the jet exit. The boundary layer separates and due to the supersonic character of the flow a Lambda shock takes place, indicated with the marker B on Figs. 10 and 12 (see also sketch on Fig. 11). Additional calculations of various flow conditions show an increase in the distance between the bow shock and the jet exit with increasing jet pressure and therefore an increased size of the boundary layer separation region. A second separated region downstream of the jet, Fig. 10, shows the same characteristics as a separation caused by a backward-facing step. Due to the supersonic character of the flow a recompression shock in this region is necessary to changes the direction of the flow in relation to the onset flow. Fig. 12 shows also that the jet creates an unbound ring vortex. Downstream of the jet exit and close to the vehicle surface a horseshoe vortex appears induced by the wake vortex. This horseshoe vortex propagates along the plane of symmetry until finally it dissolves into the main flow.

Conclusion

The numerical solution here obtained shows good correlation with the pressure distribution data form wind tunnel experiments. The main features of the interaction between the jet-flow and main flow are well capture by the numerical solution. Grid convergence solutions have been achieved for jet-off conditions, while for jet-on almost grid convergence has been shown with the exception on areas of flow separation. For jet-off conditions, the present investigation shows that the solutions are almost independent of the turbulence model, but are highly sensitive to the turbulence model employed for jet-on conditions. In general main differences with the experimental data are observed only at short distance ahead of the jet exit and mainly due to the unsteady character of the flow in that region. Furthermore, essentials details about the dominant features of the interaction process like shock induced boundary layer separation, mach disk, recompression shock and wake and horseshoe vortices are well captured.

References

[1] Technical Report Technical Documentation of the DLR Tau -Code, version: 2002.1.0,

[2] Andreas Mack, Volker Hannemann; Validation of the Unstructured DLR-TAU-Code for Hypersonic Flow; AIAA 2002-31111, 2002

[3] http:// www.centaursoft.com; Centaur V5.0 Online Documentation, 2004

[4] H. Esch: Druckverteilungsmessungen an einem Flugkörperrumpf im Überschall Messdaten; DLR, IB 39113 97 C 0 3

[5] Louis A. Cassel; Applying Jet Ineraction Technolog; Journal of Spacecraftand Rockets, Vol.40, No 4, Juli-August 2003

[6] P. Champigny; R. Lacau; Lateral Jet control for tactical Missiles; Proceedings of the AGARD Special; AGARD-R-74, pp. 3-1 to 3-57

Table 1: Chosen turbulence model, grid and test condition

Angle-of-Attack α	**-5, 0, 10, 15**	$\frac{P_{0D}}{P_\infty}$	**Number of prismatic layers**
Turbulence model for jet-on/jet-off condition	$k-\omega$ model, $k-\varepsilon$ model, Spalart-Almara		
grid	coarse grid; coarse grid, adapted (Nr. 1, 2) fine grid fine grid, adapted (Nr. 1, 2) super-fine grid	50	16 16 16 16

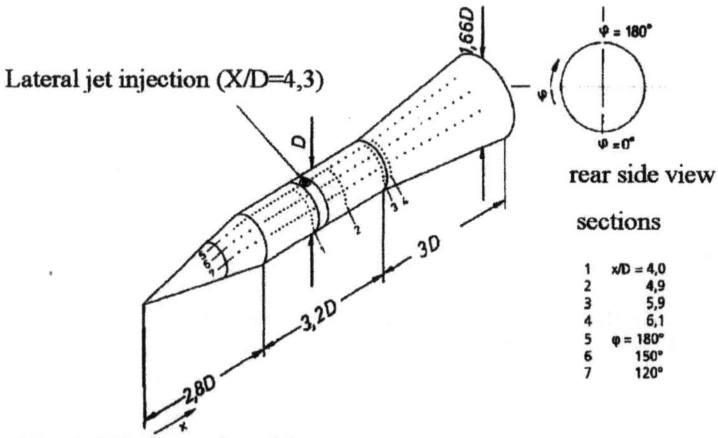

Fig. 1: Wind tunnel model

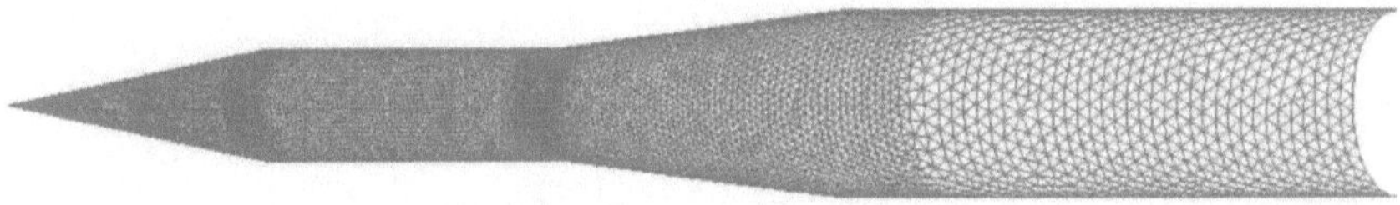

Fig. 2: Surface grid generated with CENTAUR

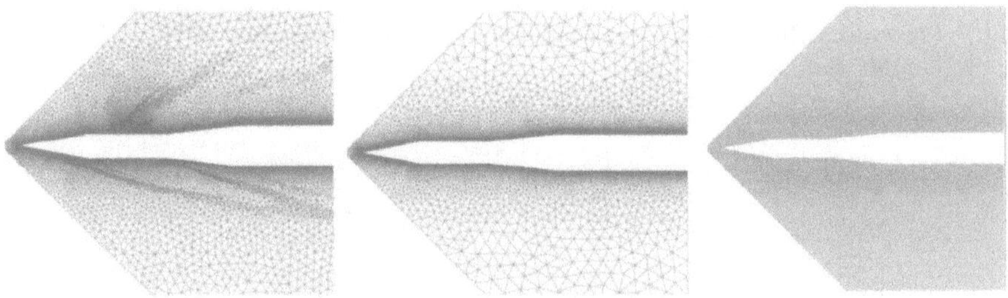

Fig. 3: Left: fine grid Nr.2 with 909502 nodes; middle: course grid Nr.1with 543062 nodes, right super fine grid with2589107 node

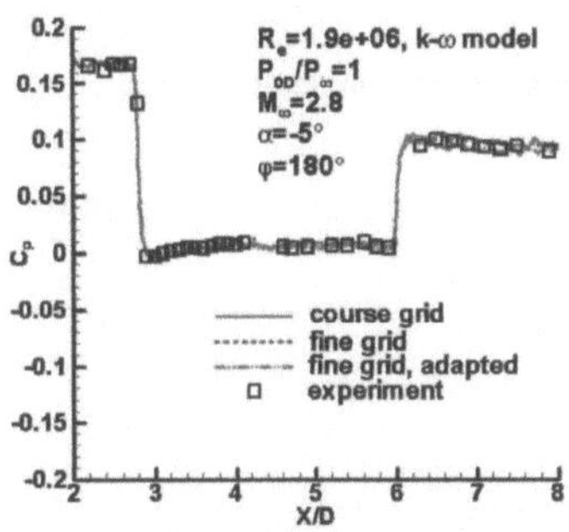

Fig. 4: Pressure coefficient for various grid in symmetry plane

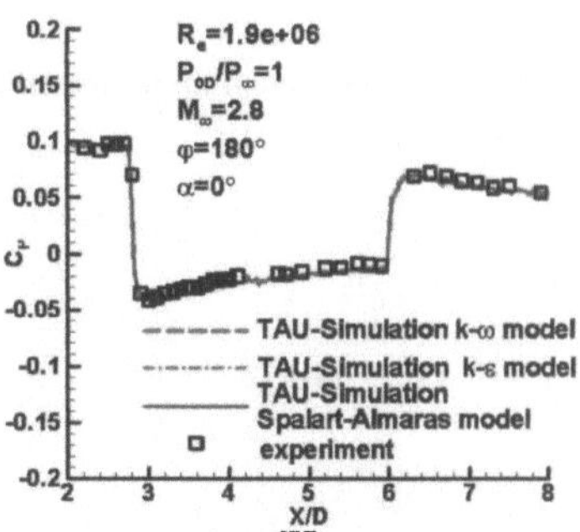

Fig. 5: Pressure coefficient for various grid in symmetry plane

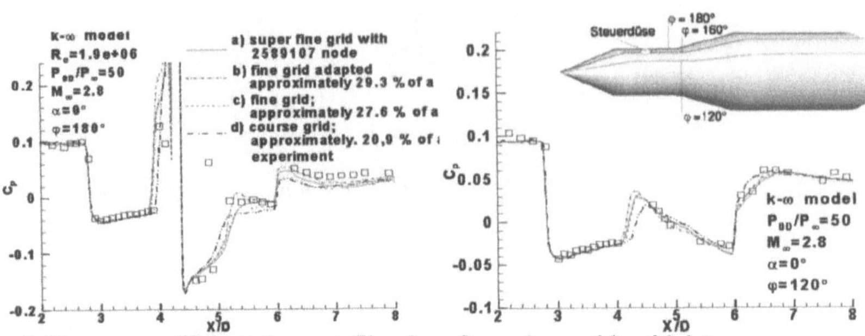

Fig. 6: Pressure coefficient along profile edges for various grids with jet-on condition φ=180°(left), φ=120° (right)

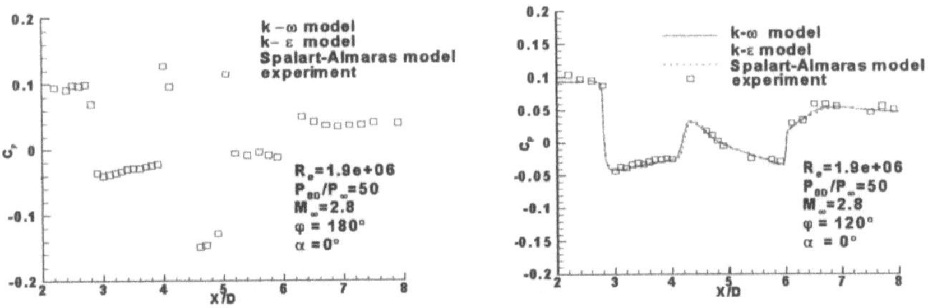

Fig. 7: Pressure coefficient with jet-on condition for various turbulence models φ =180° (left), φ =120° (left),

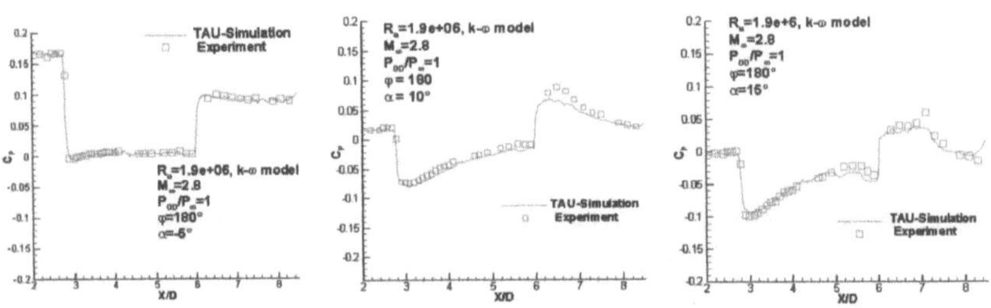

Fig.8: Pressure coefficient under jet-off condition at various angles-of-attack (α = -5°, 10°, 15°)

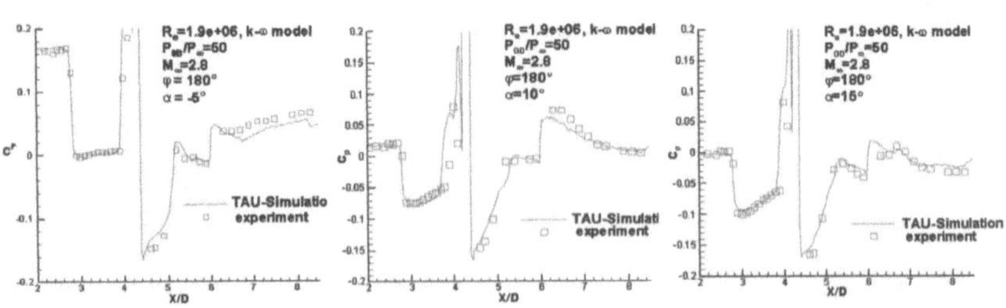

Fig. 9: Pressure coefficient under jet-on condition at various a angles-of-attack (α = -5°, 10°, 15°)

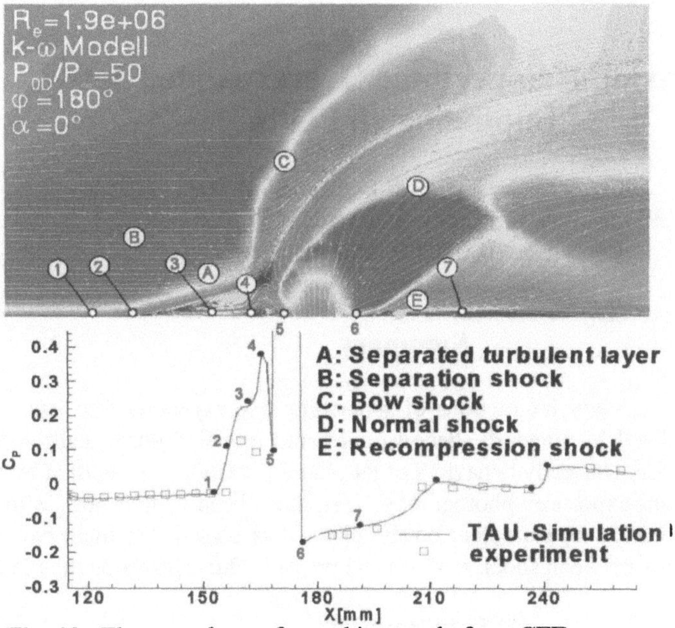

Fig. 10: Flow topology of round jet nozzle from CFD

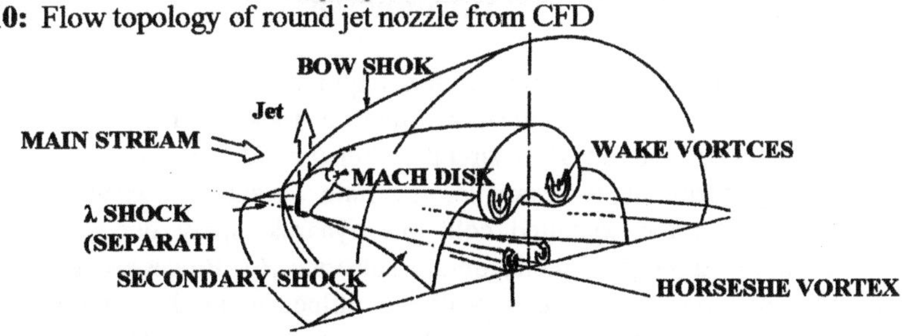

Fig. 11: Sketch of flow field topology in the vicinity of the jet from wind tunel

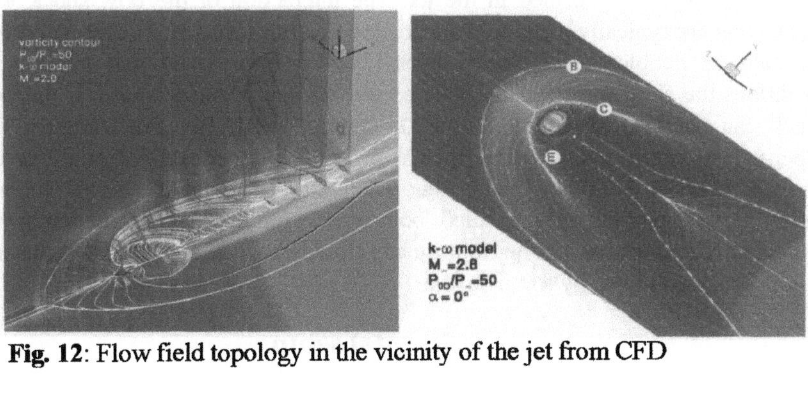

Fig. 12: Flow field topology in the vicinity of the jet from CFD

Behavior of Supersonic Overexpanded Jets Impinging on Plates

K.V. Klinkov [1], A. Erdi-Betchi and M. Rein
DLR – Institute of Aerodynamics and Flow Technology
Bunsenstr. 10, 37073 Göttingen, Germany
[1] *e-mail: konstantin.klinkov@dlr.de*

Summary

Steady and unsteady interactions of supersonic overexpanded free jets with plates are visualized by standard shadowgraphy and modeled numerically using the DLR Tau Code. Unsteady behaviors of jet-plate interactions are studied by the method of multi-exposure photography that has been combined with a synchronized pressure measurement on the plate. It is shown that there exists a distinct correlation between shock motions and pressure fluctuations on the plate.

Introduction

The behavior of supersonic jets that impinge on a solid surface is of a great importance in aerospace applications such as rocket launching systems, lunar landing modules and STOVL aircrafts. Supersonic jets are also employed in process engineering as, for example, in certain variants of thermal spray coating. In particular, in the so-called cold gas dynamic spray deposition method a supersonic jet impinges directly onto a substrate [6]. Processes occurring in the impingement zone can have a profound influence on the quality of coatings produced. In this context we have studied phenomena characteristic of the normal impingement of a supersonic jet onto a plate on a laboratory scale. Ahead of the plate a bow shock is formed in the jet. The interaction of the bow shock with shocks that are typically present in supersonic jets can result in the formation of a recirculation bubble in the shock layer [2]. We have found that under such conditions the flow ahead of the plate as well as shape and position of the bow shock can oscillate strongly in time. In the following we will introduce the experimental setup and an experimental procedure developed for studying the unsteady flows in the impingement zone. Thereafter, first characteristic features of steady flows that we have studied both experimentally and numerically are discussed. Then, results relating to unsteady flow behaviors are presented in some detail. This is followed by a conclusion.

Experimental setup

Air is expanded by a slender axisymmetric Laval nozzle. The diverging part of the nozzle is conically. The length L of the diverging part equals about $L \approx 30\ r_e$

where the radius r_e of the nozzle exit is given by $r_e = 3.3$ mm. At the nozzle exit the flow Mach number is in the range of M = 2.6-2.8 (small variations in the Mach number are caused by a displacement effect due to the boundary layer formed in the slender nozzle). The reservoir temperature T_0 can be varied between $T_0 = 20$-500 C° and the reservoir pressure p_0 can be adjusted up to $p_0 = 50$ bar. In the experiments the degree of off-design n that is defined as the ratio of the static pressure p_e at the nozzle exit and the ambient pressure p_a outside of the nozzle is $n = p_e/p_a = 0.3$-1.2. In the past usually the impingement of underexpanded jets was studied, e.g., [2, 3, 5]. In the following we consider overexpanded jets, i.e., $n < 1$. Those are typically applied in cold gas dynamics sprays.

The flow field is visualized by shadowgraph imaging. A semiconductor laser is used as the source of light. Images are received and digitized by a CCD-camera. Control signals for laser, camera and other devices are formed by a programmable signal sequencer. The whole system, including storage of images, is controlled by a personal computer. In order to study unsteady behaviors of jet-plate interactions a method of multi-exposure visualization has been developed (cf. [4]). The diagnostic system enables one to produce a sequence of light pulses of a wide variability of pulse shapes. The duration of single pulses is as low as 20 ns and the interval between different pulses is greater than 20 ns. The exposure time of the CCD-camera is as small as 10 μs. In addition to multi-exposure photography, pressure variations are measured on the plate on the axis. The pressure signal and the light pulse expositions are synchronized by the signal sequencer. A high-frequency pressure transducer having a bandwidth of 0 - 1 MHz is used. The sensitive element of the transducer has a diameter of $d_{tr} = 2.3$ mm. This is sufficient for determining frequencies. In addition, on the axis the pressure equals the one measured with smaller devices ($d_{tr} = 0.5$ mm).

Steady flows

In Figure 1 typical shadowgraph pictures of an overexpanded supersonic jet impinging normally on a plate are shown for different nozzle-to-plate distances x. The degree of off-design is $n = 0.76$ in all cases. As can be seen the geometry of the bow shock and its stand-off distance h depend strongly on the nozzle-to-plate distance. For example, the shock is flat at a distance of $x = 2.7r_e$ and convex for $x = 2.0r_e$ (cf. Figure 1). For an intermediate nozzle-to-plate distance of $x = 2.3r_e$ the bow shock shows a highly unsteady behavior and changes from a flat to a convex shape periodically. This cannot be seen on the shadowgraph of Figure 1 (here, the exposure time is too large) but will be studied in the next section.

The convex shape of the bow shock is usually attributed to a flow field with a recirculation bubble that is formed in the shock layer between the bow shock and the plate. A convex shape of the bow shock corresponds locally with a positive radial pressure gradient on the plate. This pressure gradient enables the formation of a recirculation bubble [2, 3]. The positive pressure gradient is caused by different losses in the stagnation pressure along streamlines passing either through the normal bow shock (greater losses) near the axis or through the system of two oblique shocks (smaller losses) at radii that are greater than the radius of the location of the triple point formed by the interaction of the jet and the bow shock.

In Figure 2 the stand-off distance h of the bow shock is presented as function of the nozzle-to-plate distance x. Again, the degree of off-design of the jet equals $n = 0.76$. Note that in the case of convex bow shocks the maximum distance (which always occurs on the jet axis, cf. Figure 1) is taken. For a wide range of x the stand-off distance h is nearly constant. This corresponds with a flat shape of the bow shock. Regions with increased values of h coincide with convex bow shocks. We note that an unsteady behavior of the bow shock is typical of x corresponding with the transition from large and to small h.

We have modeled the impingement of supersonic jets numerically using the DLR Tau code. The flow is assumed to be axisymmetric and steady. In order to correctly model the development of turbulent boundary layers in the slender nozzle the flow field considered includes the nozzle flow. The Spalart-Allmaras turbulent model has been used. Outside of the nozzle the computational flow field extends to large radii where an outflow boundary condition is applied. A detailed presentation of the boundary conditions can be found elsewhere [1]. In Figure 3 numerical results obtained for two different nozzle-to-plate distances are shown. As in the shadowgraphs of Figure 1 the bow shock is flat for the larger nozzle-to-plate distance x (cf. also Figure 2) and convex in the case of $x = 2.0r_e$. Furthermore, the numerical calculation clearly shows that a recirculation bubble is formed when the shock is convex while a simple stagnation point flow is present in the other case.

Unsteady jet-plate interactions

In order to study unsteady flows caused by the interaction of a supersonic jet and a plate we have taken multi-exposure shadowgraphs of the flow field in the impingement zone. This happened simultaneously with synchronized high frequency pressure measurements. The nozzle-to-plate distance was varied while the degree of off-design of the jet was kept constant at $n = 0.76$.

The motion of the shock wave has been traced by combining several light pulses with different time intervals between the pulses. This allows for determining the direction of the shock motion. A typical multi-exposure shadowgraph image of an unsteady bow shock is shown in Figure 4. The five positions of the bow shock are enhanced by dashed lines. The sequence of five light pulses applied is also sketched in Figure 4. In Figure 4 a typical case with strong oscillations of the shock is shown. This case corresponds with the formation of a recirculation zone as, for example, at a nozzle-to-plate distance of $x \sim 6r_e$ (cf. Figure 2).

The variation with time of the pressure on the plate is determined on the axis. In Figure 5 a section of the pressure oscillogram that was obtained at the same time as the shadowgraphs of Figure 4 is shown. The exact times corresponding with the five exposures of the shadowgraph image are marked by vertical lines. In this manner several hundreds of synchronized shadowgraph images and pressure oscillograms have been obtained for different nozzle-to-plate distances. From the shadowgraphs the stand-off distances h_{ij} of the bow shocks were determined. Here, the index i denotes the shadowgraphs and j, j=1,...,5, the different shock locations on shadowgraph number i.

The synchronized measurement of pressure and shock locations enables us to correlate these two quantities for every nozzle-to-plate distance. The correlation function $r(t)$ relating to a particular nozzle-to-plate distance is defined as follows:

$$r(t) = \frac{1}{\sigma_p \cdot \sigma_h} \sum_{i,j} \left(p_i(t - \tau_j) - \bar{p}\right)\left(h_{i,j} - \bar{h}\right)$$

where t is the time, τ_j is the time of each exposition of the shock on shadowgraph and $p_i(t)$ is the pressure oscillogram. Statistical means $\bar{p}, \bar{h}$ and standard deviations σ_p, σ_h are calculated from the same sample of shadowgraph-oscillogram pairs.

In Figure 6 pressure oscillograms and correlation functions are plotted for two nozzle-to-plate distances, $x = 6.3r_e$ (1) and $x = 6.5r_e$ (2), respectively. In both cases the pressure oscillates about sinusoidally. However, in the case of the greater nozzle-to-plate distance also pressure bursts occurring non-regularly are present. Nonetheless, the correlation functions both have physically significant extremums at the same time: $t_{corr} \approx 7$ µs. This is also true for other nozzle-to-plate distances not considered in Figure 6. At the time $t_{corr} \approx 7$ µs values of the correlation functions are always close to minus one ($r(t_{corr}) \approx -1.0$). This shows that there exists a strong correlation between the oscillation of the shock and the variation of the pressure. Furthermore, the pressure change is in anti-phase to the shock oscillation and has a time lag of about $t_{corr} \approx 7$ µs. This is sketched in Figure 7.

Conclusion

The time lag of the pressure ($t_{corr} \approx 7$ µs) is of the same order of magnitude as the time required for a pressure pulse to traverse the shock layer, i.e., the stand-off distance h. Therefore, a probable cause of the response of the pressure on the plate to the shock motion is the periodic variation of the shock strength. When the shock moves upstream the relative velocity of the flow with respect to the shock and, thus, losses in the stagnation pressure increase, and *vice versa*. An estimate of the change in shock strength due to the shock motion shows that the amplitude of the pressure oscillations ($\Delta p \approx 1 - 2$ bar) is of the correct order of magnitude. Hence, pressure oscillations on the plate are caused by shock oscillations.

It remains the question: what causes shock oscillations? In supersonic jet flows (both, overexpanded and underexpanded jets) typical mechanisms are due to instabilities of the shear layer between the jet and the ambient air. These are typically excited at the nozzle exit by acoustic waves that originate in the impingement zone and propagate upstream in the ambient air [5]. In order to obtain a more exact answer to our question both processes in the shear layer and acoustic wave propagation outside of the jet need to be studied. This should be done both, experimentally and numerically. In order to correctly consider feedback mechanisms that are based on acoustic wave propagation suitable numerical procedure need to be applied.

Acknowledgement

Support of this work by the Graduate College "Flow Instabilities and Turbulence" of the German Science Foundation (DFG) is gratefully acknowledged.

References

[1] Erdi-Betchi, A., Beschichtungsverfahren Kaltgasspritzen: Experimentelle und numerische Untersuchungen der Dynamik der partikelbeladenen Überschallströmung,.DLR FB 2004-20, Köln (2004).
[2] Ginzburg, I. P., Semiletenko, B. G.; Terpigor'ev, V. S., and Uskov, V. N., Journal of Engineering and Physics **19**, 1081-1084 (1973).
[3] Kalghatgi, G. T. and Hunt, B. L., Aeronautical Quarterly **27**, 169-186 (1976).
[4] Klinkov, K. V. and Rein, M., Proc. ICMAR XII, Part **2**, 101-106, Novosibirsk (2004).
[5] Krothapalli, A., Rajkuperan, E., Alvi, F. S., and Lourenco, L., Journal of Fluid Mechanics **392**, 155-181 (1999).
[6] Rein, M., Erdi-Betchi, A., and Klinkov, K. V. In: Sobieczky, H. (ed.), IUTAM Symposium Transsonicum IV, 177-182, Kluwer, Dordrecht (2003).

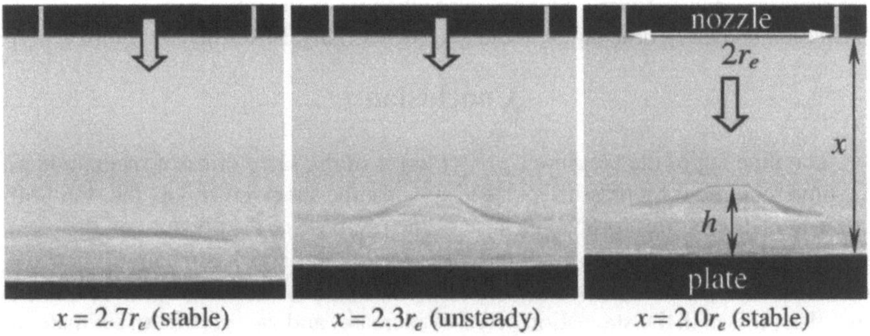

Figure 1 Shadowgraphs of bow shock ahead of a plate at different nozzle-to-plate distances x.

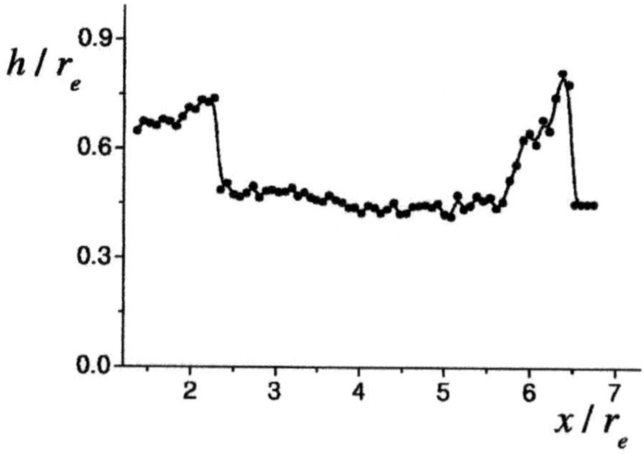

Figure 2 Stand-off distance h of the bow shock as a function of the nozzle-to-plate distance x.

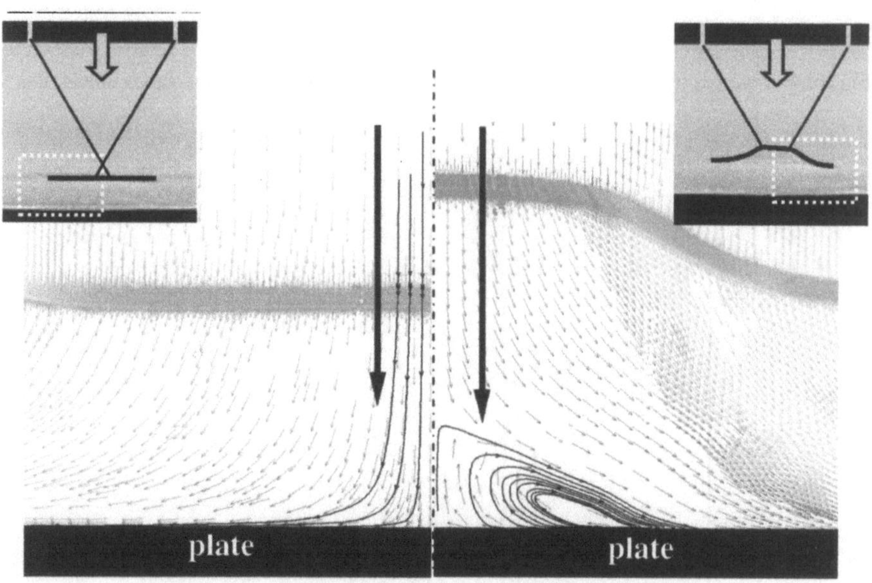

Figure 3 Results of a computation of the flow field in the impingement zone for two nozzle-to-plate distances: $x = 4r_e$ (left) and $x = 2r_e$ (right).

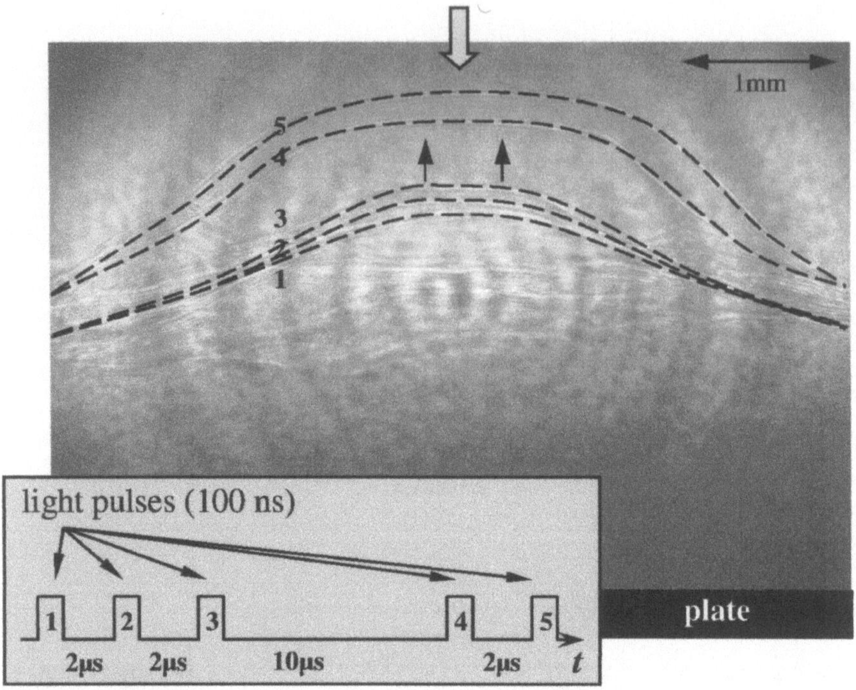

Figure 4 Five-exposure shadowgraph image of a strongly unsteady bow shock ahead of a plate.

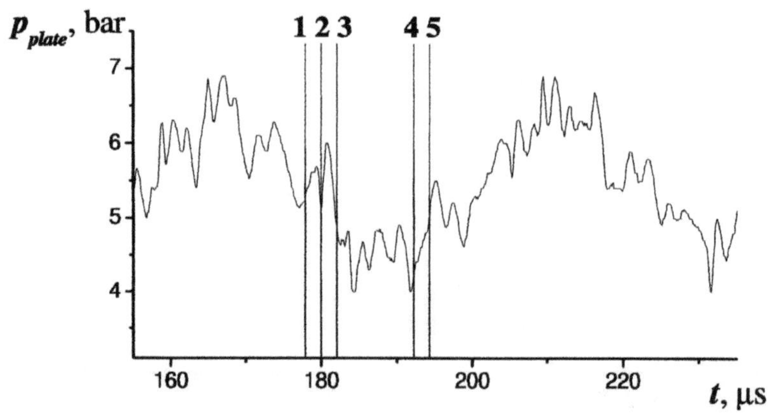

Figure 5 Pressure on the plate (at the axis of the jet) as a function of time. Vertical lines indicate times of exposure in the corresponding shadowgraph (cf. Figure 4).

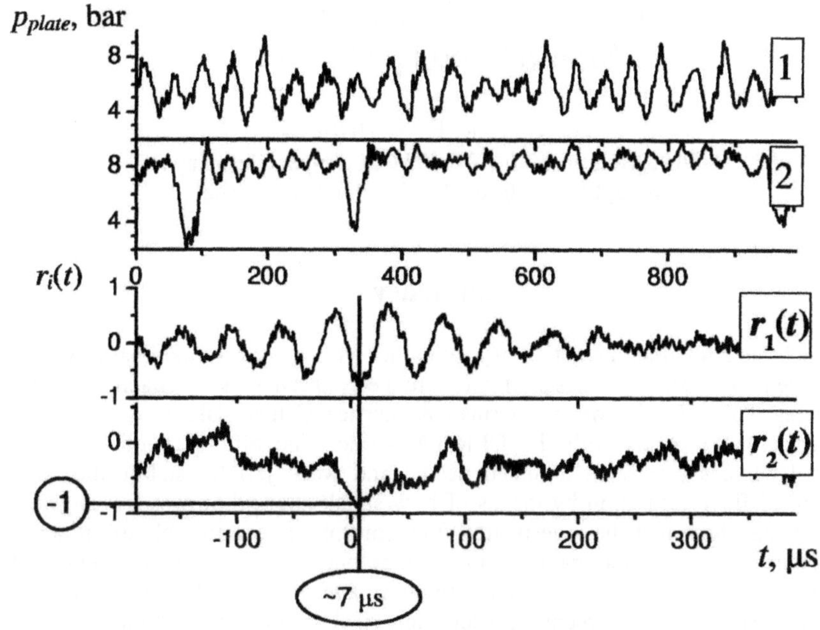

Figure 6 Oscillograms of the plate-pressure for different nozzle-to-plate distances $x = 6.3r_e$ (1) and $x = 6.5r_e$ (2) as well as correlation functions $r_1(t)$ and $r_2(t)$, corresponding to the pressure oscillograms. Significant extremums of the correlation functions are marked.

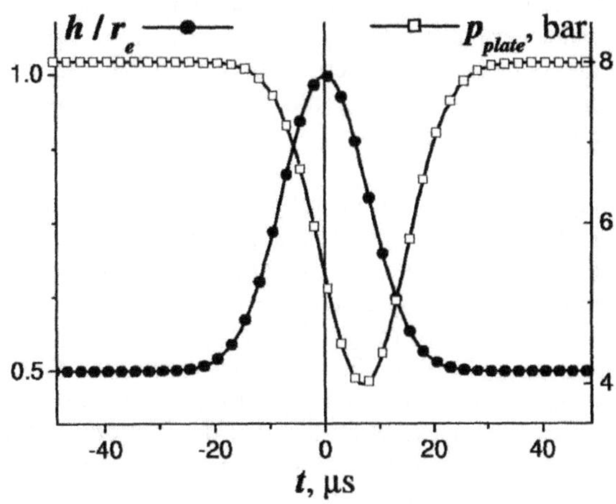

Figure 7 Sketch of the response of the pressure on the plate to shock motions.

Study of Supersonic Flow Separation Induced by a Side Jet and its Control

A. KOVAR, E. SCHÜLEIN
DLR, Bunsenstr. 10, D-37073 Göttingen, Germany
Anke.Kovar@dlr.de, Erich.Schuelein@dlr.de

Summary

In this basic study the interaction of a lateral jet with supersonic external flow was examined. The study consists of wind tunnel experiments at Mach 5 in the Ludwieg Tube Facility of the German Aerospace Center (DLR) as well as of numerical simulations with the DLR-TAU-code. The efficiency of two flow control devices (upstream mounted needle and weak jet) for stabilization and control of flow separation by means of bow shock attenuation was investigated. The results show that the investigated flow control devices are able to influence the shape and size of the main separation zone but are not capable to suppress it completely. A comparison of wall pressures and limiting streamline patterns showed reasonable agreement between numerical simulations and experimental data.

1 Introduction

Fast-reacting side jet thrusters are an attractive tool for flight control of highly agile missiles. Under certain conditions they succeed over conventional control surfaces, which create higher drag at cruise flight and become ineffective at low stagnation pressures typical for high altitudes or at low speeds of the missile e.g. during its starting phase.

In the external supersonic flow the lateral jet acts as an obstacle. A bow shock develops in front of the blowing jet. On the missile body upstream of the jet the boundary layer experiences a pressure rise and separates, forming the so called "horseshoe" vortex. This vortex around the jet marks the three-dimensional separation zone. Downstream of the jet the wall pressure drops as the jet blocks the wake area from the outer flow. Further downstream, the flow reattaches and the wall pressure increases approximately up to the undisturbed incoming value due to the recompression shock. More detailed descriptions of this flow pattern can be found amongst others in [4, 6]. A good historical review about side jet applications and their studies is given also in [5].

The interaction of the transverse jet with the outer flow is strongly dependent on a number of parameters, such as Mach and Reynolds number as well as the ratio of the total jet pressure over the static main flow pressure. The dependence on those parameters can be seen in a strong sensitivity concerning the size and the shape of the separation zone, also making the prediction of the expected effects very difficult. This sensitivity led to the idea of influencing or probably even stabilizing the separation zone by means of flow control.

In general, there are two ways to suppress the separation induced by jet-boundary layer interaction. Firstly, flow control devices like vortex generators can be used to energize the near-wall flow through a momentum transfer from the outer flow into the wall region. Thus, larger adverse pressure gradients or stronger shock waves can be overcome without causing the boundary layer to separate. A second possibility is to weaken the shock wave itself. This effect was demonstrated earlier in [3], where a three-dimensional separation zone, induced by a side jet, was successfully suppressed by positioning a thin liquid jet or needle upstream of the side jet. The interaction of the obstacle's wake flow with the bow shock, which induces separation, leads to the attenuation of the bow shock and the main separation zone shrinks.

The aim of the present study was to examine the efficiency of the means proposed in [3] in terms of stabilization and control of the side jet induced separation.

2 Test conditions and applied tools

The experimental investigations were conducted in the Ludwieg Tube Facility (DLR Göttingen) at Mach 5. The basic test model (Figure 1) was a flat plate of 400 mm in width and 660 mm in length with a circular sonic nozzle of Ø6 mm located on the plate's centerline at a distance of 354 mm from its leading edge. The axis of the nozzle was oriented perpendicularly to the plate's surface. For separation control tests this basic model was additionally equipped with a second small nozzle and a 15 mm long needle respectively, both of Ø0.6 mm in diameter. They were mounted 25 mm upstream of the main jet nozzle axis, perpendicularly to the model's surface. The tests comprised steady surface pressure measurements of 160 taps distributed over the area of interest, flow-field pitot pressure measurements at 280 mm downstream of the jet at the end of the plate, and oil flow visualizations. Heat flux measurements were taken by infrared camera but are not discussed in this paper. The ratio of the total jet pressure over the static free stream pressure was adjusted to $p_{oj}/p_\infty = 110$ and kept constant for the discussed cases. The unit Reynolds number was $Re_{1\infty} = 38\times10^6\, m^{-1}$ for all presented cases, resulting in a turbulent boundary layer on the flat plate at the interaction area.

The numerical calculations have been performed with the DLR-TAU-code. It is a finite-volume Navier-Stokes solver working with hybrid, unstructured or structured grids. The structured grids are used to resolve the boundary layer and an unstructured grid for all other areas. The initial grids were produced with the commercial grid generator Centaur® and consisted of about 1.7 million grid cells.

The calculations were performed with the 1-equation turbulence model of Spalart-Allmaras with Edwards' modification. Further information as well as some examples for applications of the code can be found in [1, 2]. The applied grids were y^+-adapted as well as cell refined to gain the optimal resolution of the boundary layer as well as areas of large gradients, e.g. shock waves.

3 Discussion of the results

In the following, experimental and numerical results are discussed and compared for the general surface flow pattern and pressure distributions in axial and lateral direction. The reference case with the single jet-flow-interaction is presented in Figure 2. The top half of the figure shows wall streamlines derived from oil flow visualizations, the flow pattern in the bottom half was integrated from the calculated skin friction. A separation line S_1 limits the horseshoe vortex and devides the recirculation area from the incoming flow. The wall streamline R indicates where the main flow is reattaching in front of the jet as it is acting as an obstacle. Along the centreline in the wake of the jet the wall streamlines are diverging. Where the streamlines become parallel again, the jet reattaches. The wall streamlines are directly related to the surface pressure distributions shown in Figure 4a (axial) and b (lateral direction). In axial direction the recirculation zone shows up in the area of increased pressure upstream of the jet position. The reattachment line corresponds to the peak of the surface pressure distribution. Where the streamlines in the wake are diverging, low pressure is measured, as the jet is blocking the wake from the main flow. The surface pressure increases to its approximately undisturbed upstream value where the flow is reattaching with a recompression shock. Also in lateral direction, the reattachment line is related to the surface pressure peak (Figure 4b) and the wide separation zone is visible in the pressure plateau.

Differences between numerical calculation and experiment are seen in both surface pressure and the related wall streamlines. The size of the separation zone in the numerical simulation is bigger than in the experiment. The authors assume that this is caused by the fully turbulent computation, whereas in the experiment a new laminar boundary layer develops from the reattachment line later turning into a turbulent one. Hence, the flow close to the wall underneath the horseshoe vortex seems to be more energetic in the computation than it is actually the case in the experiment. This explanation is supported by the representation of the streamlines. Where the streamlines are pointing against the flow direction towards the primary separation line S_1, the angle in which they are bent into the primary separation line is bigger for the simulation than for the experiment. The presentation of the limiting streamline pattern shows that more streamlines from an inner area of the separation zone of the simulation are still leading into the primary separation line, than in the experiment. This is also a hint for a more energetic flow. The existence of the secondary convergence line S_2, that is not predicted in the numerical calculation is also most likely caused by this reason.

The first flow control case under investigation is the weak jet interaction. The main features of the wall streamlines in Figure 3a remain unchangend compared to the reference case, but a difference can be seen in the shape of the separation zone that shrinks in lateral direction. Around the weak jet, a second separation zone develops as well as a reattachment line upstream of the weak jet. The wall streamlines in the wake of the main jet stay very similar to the single jet case. The surface pressure distributions in Figure 5a and b show a slightly smaller separation zone than for the reference case. The surface pressure distribution also shows the peak from the reattachment line due to the weak jet at roughly x= -25 mm. In this case both horseshoe vortices are forming one big separation zone, where the wake of the weak jet is interacting with the recirculation flow ahead of the main jet. The distribution of the pressure in the wake remains unchanged as obviously the weak jet is not strong enough to influence the pressure here. Apart from the additional horseshoe vortex and the change of the size of the separation, the wall streamlines as well as the wall pressure show the same behaviour as for the single jet interaction.

In Figure 3b the wall streamlines for the second flow control case using the needle are displayed. The interaction with the needle also causes a laterally reduced main separation and a second separation zone upstream of the needle. However, the second separation zone caused by the needle, is detached from the jet's separation zone. The flow behaviour in the wake of the main jet of the needle configuration is totally different from the wake of the reference case as it takes much longer for the streamlines to get parallel again. Differences can also be seen in the pressure distribution in Figure 6a and b. The pressure increase and herewith the separation upstream of the needle is totally decoupled from the low pressure area and the following separation of the main jet. Also the different behaviour of the wake flow can be seen in the surface pressure measurement, where the pressure recovery downstream of the jet takes place over a much longer distance.

Figure 7 represents the normalised pitot pressure as it was taken 280 mm behind the jet position. The difference between the case of the weak jet separation control and the reference case is small for the experimental as well as for the numerical results. With increasing height the differences nearly vanish. This indicates that the weak jet has an effect only close to the wall, which can also be seen on a slightly smaller separation zone on Figure 3a. The case of the needle separation control exhibits a higher pressure close to the wall, whereas in heights between 30 mm $< z <$ 60 mm the pressure is lower than for the other two cases. This comparison shows that the needle separation control is the case with the greatest impact throughout the whole profile.

The comparison between experimental and numerical results in Figure 7 does not look good at the first view, as close to the wall they deviate by a factor of two. But despite the differences in absolute values, the overall shapes of the curves are very similar and show a similar behaviour for all three cases. The agreement between experiment and numerical simulation becomes better with increasing height. This is most pronounced for the case of the needle separation control. The general agreement of the shapes of the curves is still remarkable, as this comparison was performed at the most downstream position at the end of the plate, which means that any error that may have occurred in the prediction of the wake of the jets has been "integrated" at that position. The similar behaviour of the curves indicates that the calculation still represents the main physics of the flow structures.

The results discussed show that the devices used were not able to suppress the separation zone as was suggested earlier in [3]. It can be seen from the streamlines that the weak jet and the needle were chosen to be on a location where they were keeping the main separation zone at that spot rather than reducing the size of it or suppressing it entirely. Here the question arises whether this behaviour depends on the distance between obstacle and jet.

4 Conclusions

A study was conducted with the target to influence or stabilize the shape and size of the separation zone induced by the interaction of a lateral jet in a supersonic flow. According to a former study, this was tried with a weak jet and a needle mounted upstream of the main jet. The height as well as the ratio of the diameter of the jet and the obstacles were within the limits of this former study whereas the distance between the obstacle and the main jet was fixed at one position. The investigation showed that a weak jet and a needle have an influence on the shape and the lateral size of the separation zone. This tendency was also predicted in the numerical simulations, but the absolute size was overpredicted. Problems occured also in the calculation due to a fully turbulent computation, resulting in missing secondary convergence lines which were present in the experiment.

A total suppression of the separation zone, as stated in the former investigation, was not observed, neither in the experiments nor in the numerical calculation. The influence of the weak jet on the pitot pressure distribution was only weak in comparison to the reference case, whereas the needle showed a strong impact. The numerical pitot pressure showed larger deviations in absolute values but exhibited similarity in the curve shape. This was interpretated in a way that the main flow features were represented correctly in the calculation.

However, for the chosen test parameters and geometries it was not possible to reduce the separation zone to an extent found in former literature. Hence, with the limited set of parameters investigated, this study does not yet allow conclusive statements. In future, additional measurements of the wall shear stress are planned to gain a better understanding of the flow structures in the questionable areas. A further study would be necessary, in which a broader range of the decisive parameters e.g. a variation of the pressure ratio, size of obstacles and distance to it in relation to the boundary layer thickness should be investigated.

References

[1] T. Gerhold and J. Evans: "Efficient Computation of 3D-Flows for Complex Configurations with the DLR TAU-Code Using Automatic Adaptation", in Notes on Numerical Fluid Mechanics, Vol.72, Vieweg, 1999, pp. 178-185

[2] T. Gerhold, O. Friedrichs, J. Evans and M. Galle: "Calculation of Complex Three-Dimensional Configurations Employing the DLR-τ-Code", in AIAA 97-0167, 35th Aerospace Sciences Meeting & Exhibit, Reno, Nevada, January 6-10, 1997

[3] G.F. Glotov and Y.F. Korontsvit: "Investigation of the Method for Three-Dimensional Separation Zone Control", in Uch. Zap. TSAGI, Vol. 14, No. 3, 1983, pp. 126-131 (in Russian)

[4] A. Heyser and F. Maurer: "Experimentelle Untersuchungen an festen Spoilern und Strahlspoilern bei Machschen Zahlen von 0.6 bis 2.8", in Z. Flugwiss. 10, Heft 4/5, 1962, pp. 110-130

[5] R.J. Margason: "Fifty years of Jet in Cross Flow Research", in AGARD-CP-534, Computational and Experimental Assessment of Jets in Cross Flow, Symp. in Winchester, UK, 19th-22nd April, 1993

[6] F.W. Spaid and E.E. Zukoski: "A Study of the Interaction of Gaseous Jets from Transverse Slots with Supersonic External Flows", in AIAA Journal Vol. 6, No. 2, 1968, pp. 205-212

Figures

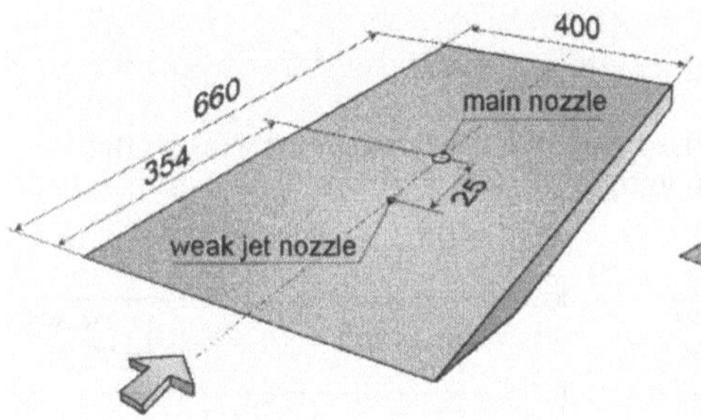

a.

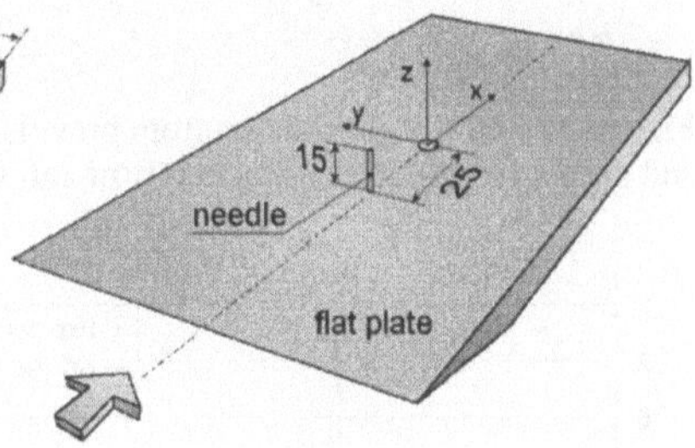

b.

Figure 1 Sketches of the test model configurations for investigations of flow control by weak jet (a) and needle (b) mounted upstream of the side jet nozzle.

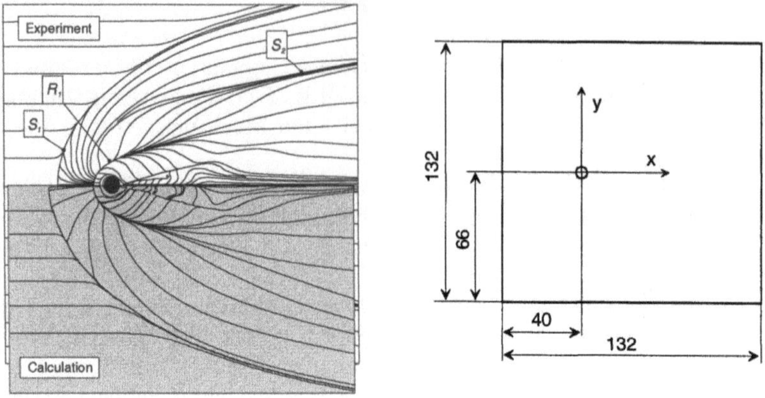

Figure 2 Top view of comparison between experimental (top half) and numerical (bottom half) wall streamlines for single jet. Right sketch shows scaling of streamline pattern.

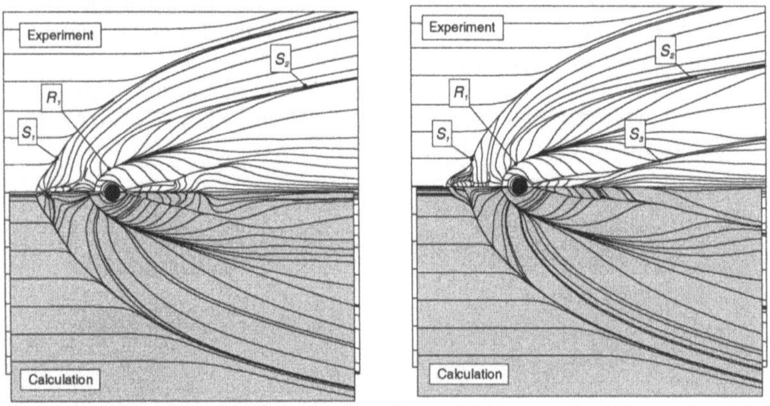

Figure 3a and b Continuation from previous figure. Wall streamlines for weak jet (left) and needle (right). Flow direction from left to right.

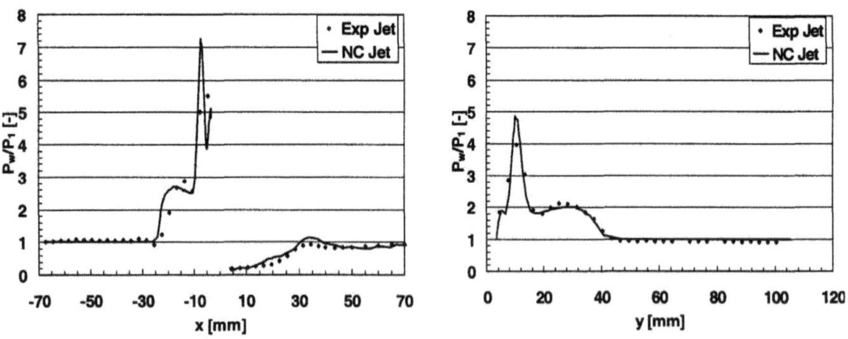

Figure 4a and b Wall pressure distribution in axial (left, jet at x=0mm) and lateral (right, jet at y=0mm) cross sections for jet alone (without flow control devices).

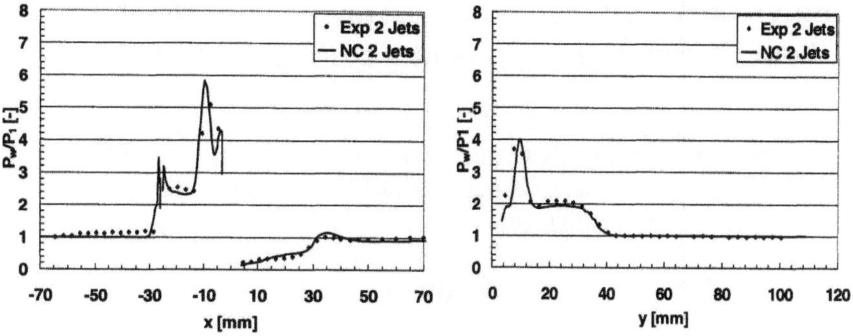

Figure 5a and b Wall pressure distribution in axial (left, weak jet at x=-25mm) and lateral (right, weak jet at y=0mm) cross sections for the weak jet flow control case.

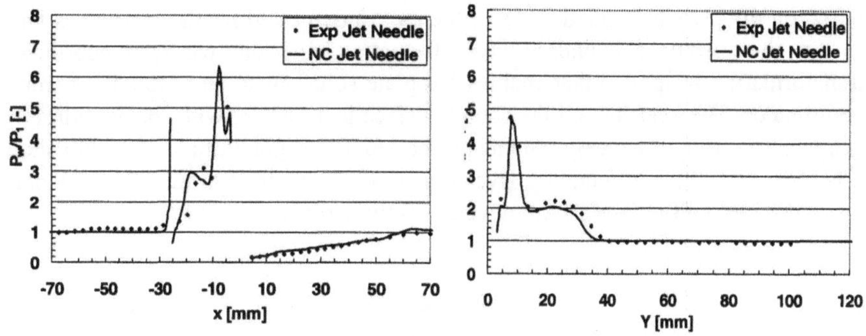

Figure 6a and b Wall pressure distribution in axial (left, needle at x=-25mm) and lateral (right, needle at y=0mm) cross sections for the needle flow control case.

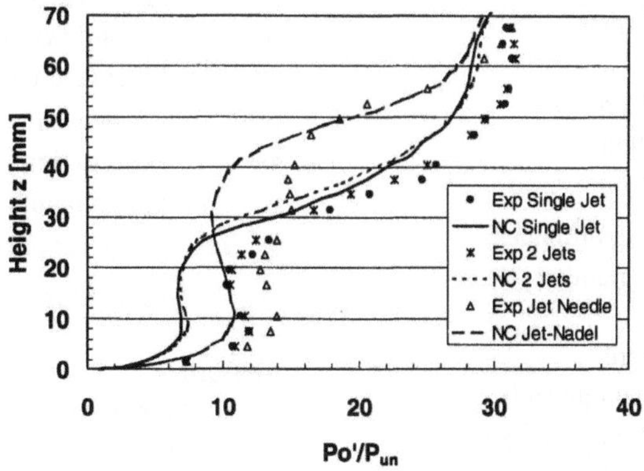

Figure 7 Vertical profile of pitot pressure at x = 280 mm. Comparison of all three cases.

Analytically obtained data compared with shock tunnel heat flux measurements at a conical body at M = 6

J. SRULIJES, F. SEILER
French-German Research Institute of Saint-Louis (ISL)
5, Rue du Général Cassagnou – 68301 Saint-Louis, France
srulijes@isl.tm.fr, seiler@isl.tm.fr

Summary

This paper presents analytical methods for calculating the heating of a hypersonic projectile equipped with a conical forebody. Calculated heat fluxes are compared with data obtained from ISL shock tube experiments. Two theoretical approaches based on the classical boundary layer theory are described for the laminar as well as for the turbulent boundary layer formation. In both cases, a coordinate transformation is applied that enables flat plate solutions to be adapted to conical coordinates. The heat flux on the cone's surface is measured with special thin film gauges. The heat flux measurements performed at a flight altitude of 15 km show a laminar to turbulent boundary layer transition. In contrast to this, the boundary layer at a flight altitude of 21 km develops fully laminar.

Nomenclature

Symbols

a_f	transfer factor
B	function of n
f	function
M	Mach number
n	factor
p	static gas pressure
Pr	Prandtl number
$\dot{q}$	heat flux
r, ϑ, φ	spherical coordinates
Re	Reynolds number
t	time
T	temperature
u, v	velocity components
$\tilde{v}$	transformed velocity comp.
w	transformed velocity comp.
x, y, z	Cartesian coordinates
α	temperature coefficient of the thin film
δ	boundary layer thickness
δ^{**}	momentum thickness
η	similarity parameter
λ	heat conductivity
μ	dynamic viscosity
ν	kinematic viscosity
ρ	gas density
τ	skin friction
φ	gas compressibility factor
χ, $\tilde{\chi}$	norm. space coordinate
ψ	stream function

Subscripts

e	outer edge of boundary layer
g	gas
p	projectile
r	recovery state
w	cone (plate) surface
∞	flow upstream of bow shock

1 Introduction

A projectile with a conical nose flying in the atmosphere at a high Mach number $M_\infty >> 1$ develops a narrow tapered bow wave around its nose. The incoming airflow turns parallel to the projectile's surface in the reference frame of the projectile, as depicted in Figure 1. Through the bow wave the quiescent air conditions change from upstream p_∞, T_∞ and u_∞ (u_∞ equals the projectile flight velocity u_p) to the downstream parameters p_e, T_e and u_e. As a result, a heat flux develops along the surface of the projectile and directs it into the projectile's material. Especially the nose of the projectile can be extremely heated.
The aim of the work is to model theoretically and measure experimentally the heating of a projectile's nose during its flight in the earth's atmosphere. Surface temperature measurements in flight are very difficult to carry out. Therefore, the shock tube STA providing flow durations of about 2 ms and operated in the shock reflection mode was used. The measurements were carried out at the flight conditions present in 15 and 21 km of altitude. In this test facility the sharp-nosed-body is fixed inside a measuring chamber. A boundary layer develops either laminar or turbulent at the cone's surface. Depending on the Reynolds number a transition takes place at a certain distance from the nose tip. The boundary layer can be modeled as laminar or turbulent. In this study a fast engineering approach was developed to find solutions for the boundary layer formation at the surface of the nose of a projectile in flight. This is done, both for the laminar and for the turbulent case. We used the boundary layer approximation of Prandtl (see Schlichting [1]) to obtain solutions in terms of analytical relations. These results are compared with the shock tube heat transfer measurements.

2 Theory

2.1 Turbulent boundary layer formation

For a theoretical description of the formation of the turbulent boundary layer, the conservation equations, i.e. the mass, momentum and energy equations have to be applied. The Reynolds number ranges from zero at the nose tip to more than $Re \approx 6 \cdot 10^6$ at the end of the cone's surface. Therefore, transition from laminar to turbulent boundary layer behavior occurs somewhere downstream the tip of the projectile. The boundary layer equations are applied in spherical coordinates having its origin located at the tip of the nose, see Figure 2. The coordinate r is directed along the cone's surface having originated at the projectile's tip.
For finding a solution to the appropriate boundary layer equations of Prandtl as given in [1], these equations are transformed as explained in detail by Hantzsche and Wendt [2]. As the boundary layer here is considered to be turbulent the corresponding parameters in the conservation equations are written time averaged with an upper line.

$$\text{Mass:}\ -\frac{\tilde{\chi}}{2}\frac{d}{d\tilde{\chi}}(\bar{\rho}\,\bar{u})+\frac{d}{d\tilde{\chi}}(\bar{\rho}\,\bar{w})=0\ , \tag{1}$$

$$\text{Momentum:}\ -\frac{\bar{\rho}\,\bar{u}\,\tilde{\chi}}{2}\frac{\partial\bar{u}}{\partial\tilde{\chi}}+\bar{\rho}\,\bar{w}\frac{\partial\bar{u}}{\partial\tilde{\chi}}=\frac{\partial}{\partial\tilde{\chi}}\left(\mu\frac{\partial\bar{u}}{\partial\tilde{\chi}}\right), \tag{2}$$

$$\text{Energy:}\ -\frac{\bar{\rho}\,\bar{u}\,\tilde{\chi}}{2}\frac{\partial}{\partial\tilde{\chi}}\left(c_p\,\bar{T}\right)+\bar{\rho}\,\bar{w}\frac{\partial}{\partial\tilde{\chi}}\left(c_p\,\bar{T}\right)=\frac{\partial}{\partial\tilde{\chi}}\left(\lambda\frac{\partial\bar{T}}{\partial\tilde{\chi}}\right)+\mu\left(\frac{\partial\bar{u}}{\partial\tilde{\chi}}\right)^2. \tag{3}$$

The differential Eqs. (1) to (3) have to be solved using the boundary conditions for the temperature $\bar{T}$ and the velocity components $\bar{u}$ and $\bar{v}$.
From a formal point of view Eqs. (1) to (3) are similar to the Eqs. valid for the flat plate boundary layer. In this case the simplified conservation equations, see Schlichting [1] as well, are given in Cartesian coordinates, see Figure 3.
A set of boundary layer equations describing the boundary layer formation on a flat plate can be obtained by transforming the plate boundary layer equations to the following form:

$$\text{Mass:}\ -\frac{\chi}{2}\frac{d}{d\chi}(\bar{\rho}\,\bar{u})+\frac{d}{d\chi}(\bar{\rho}\,\bar{\bar{v}})=0\ , \tag{4}$$

$$\text{Momentum:}\ -\frac{\bar{\rho}\,\bar{u}\,\chi}{2}\frac{\partial\bar{u}}{\partial\chi}+\bar{\rho}\,\bar{\bar{v}}\frac{\partial\bar{u}}{\partial\chi}=\frac{\partial}{\partial\chi}\left(\mu\frac{\partial\bar{u}}{\partial\chi}\right), \tag{5}$$

$$\text{Energy:}\ -\frac{\bar{\rho}\,\bar{u}\,\chi}{2}\frac{\partial}{\partial\chi}\left(c_p\,\bar{T}\right)+\bar{\rho}\,\bar{\bar{v}}\frac{\partial}{\partial\chi}\left(c_p\,\bar{T}\right)=\frac{\partial}{\partial\chi}\left(\lambda\frac{\partial\bar{T}}{\partial\chi}\right)+\mu\left(\frac{\partial\bar{u}}{\partial\chi}\right)^2. \tag{6}$$

The transformed Eqs. (1) to (3) for the cone boundary layer and (4) to (6) for the plate boundary layer show that the two systems appear to be equivalent. The difference between the two Eqs. systems can be found in the different meaning of the coordinate $\chi(x,y)$ for the plate and $\tilde{\chi}(r,\vartheta)$ for the cone. There are also differences in the velocity component definitions of $\bar{\bar{v}}(x,\bar{v})$ (plate) and $\bar{w}(r,\bar{u},\bar{v})$ (cone). The meaning of these parameters is explained by Seiler et al. [3] The similarity of the transformed cone boundary layer equations to those for the plate boundary layer shows that a solution for the formation of the plate boundary layer and appropriate coordinate transformations, permit to relate the cone and the plate boundary layers. Hantzsche and Wendt [2] gave some data on these relations concerning laminar boundary layer development. For a turbulent flow the transfer factor is determined with data given by Chien [4]. The similarity considerations discussed above justify solving the plate boundary layer equations (4) to (6) instead of the more complex cone boundary layer equations.

A solution for the plate boundary layer Eqs. (4) to (6) using an integral approach can be found in Schlichting [1]. Integration of the momentum equation for the plate boundary layer from y = 0 (wall) up to y = δ (outer edge of the boundary layer) gives:

$$\int_{y=0}^{\delta}\left(\overline{\rho}\,\overline{u}\frac{\partial\overline{u}}{\partial x}+\overline{\rho}\,\overline{v}\frac{\partial\overline{u}}{\partial y}\right)dy=-\tau_w\,. \tag{7}$$

Replacing in equation (7) the velocity component $\overline{v}$ by the conservation of mass equation the following integral relation is obtained; see Oertel [5]:

$$\tau_w=\frac{d}{dx}\int_0^{\delta}\rho\overline{u}(\overline{u}_e-\overline{u})\,dy+\frac{d\overline{u}_e}{dx}\int_0^{\delta}(\rho_e\overline{u}_e-\rho\overline{u})\,dy. \tag{8}$$

In turbulent boundary layers velocity profiles are described approximately by the power-law equation as discussed for example by Schlichting [1]:

$$\frac{\overline{u}}{\overline{u}_e}=\left(\frac{y}{\delta}\right)^{\frac{1}{n}},\quad 5\le n\le 10. \tag{9}$$

Introducing equation (9) into (8) a relation for the boundary layer thickness δ as a function of x along the plate is obtained:

$$\frac{\tau_w}{\rho_e\,\overline{u}_e^{\,2}}=\frac{\delta^{**}}{\delta}\frac{d\delta}{dx}\,. \tag{10a}$$

Using the relation given by Schlichting [1] for turbulent flows, as

$$\frac{\tau_w}{\rho_e\,\overline{u}_e^{\,2}}=B(n)\varphi\left(\frac{\nu_e}{\overline{u}_e\,\delta}\right)^{\frac{2}{n+1}}, \tag{10b}$$

the following differential equation can be found:

$$B(n)\varphi\left(\frac{\nu_e}{\overline{u}_e\,\delta}\right)^{\frac{2}{n+1}}-\frac{\delta^{**}}{\delta}\frac{d\delta}{dx}=0\,. \tag{10c}$$

Equation (10c) is solved with the outer core flow conditions, i.e. flow speed u_e, pressure p_e and temperature T_e. Using the Reynolds analogy,

$$\dot{q}_g=\frac{c_p\left(\overline{T}_r-\overline{T}_w\right)Pr^{-\frac{2}{3}}}{\overline{u}_e}\tau_w\,, \tag{11}$$

an analytical solution for predicting the heat flux $\dot{q}_g$ from gas to surface as a function of the coordinate x along the surface of the flat plate is given as:

$$\dot{q}_g(x)=a_f\left(\frac{n+1}{n+3}\right)^{\frac{2}{n+3}}\left(B(n)\varphi\right)^{\frac{n+1}{n+3}}c_p\left(\overline{T}_r-\overline{T}_w\right)\cdot Pr^{-\frac{2}{3}}\,\overline{\rho}_e\left(\frac{\delta^{**}}{\delta}\right)^{\frac{2}{n+3}}\overline{u}_e^{\,\frac{n+1}{n+3}}\left(\frac{\nu_e}{x}\right)^{\frac{2}{n+3}} \tag{12}$$

All the parameters used in equation (12) are explained in the Nomenclature and more in detail by Seiler [3]. Equation (12) works for a perfect gas. Real gas effects as e.g. vibration and dissociation of the molecules are not relevant in the study considered herein. For taking into account the heat flux $\dot{q}_g$ on the sharp-cone geometry of the projectile's nose, the calculated flat-plate heat flux of equation (12) is transformed to cone geometry using the transfer factor a_f, with $a_f > 1$ and the plate coordinate x is transformed into the cone coordinate r. A theoretical estimation of this factor is not available for turbulent flows. For determining a_f, the heat flux $\dot{q}_g$ was fitted to the experimental heat transfer results of Chien [4], which were obtained at hypersonic flow conditions similar to those present in this study. The best agreement with Eq. (12) is found for $a_f = 1.07$ using $n = 9$.

2.2 Laminar boundary layer formation

For calculating the formation of a laminar boundary layer on a flat plate, we start at $x = 0$ (Figure 3). The flow as well as the boundary layer is considered to be two-dimensional in the x- and y-coordinates, therefore the z-coordinate is not considered. As described by Schlichting [1] Prandtl's boundary layer equations for mass, momentum and energy are transformed by applying a similarity parameter η and a stream function ψ as:

$$\eta = \sqrt{\frac{u_e}{2 x \nu_w}} \int_0^y \frac{\rho}{\rho_w} dy, \tag{13}$$

and

$$\psi = \sqrt{2 x \nu_w u_e}\, f(\eta) . \tag{14}$$

As a result, see Schlichting [1], the following differential equation is obtained for the function $f(\eta)$

$$f''' + f f'' = 0 . \tag{15}$$

From the energy equation a second differential equation becomes available which describes the behavior of the temperature T(y) inside of the boundary layer with the variable $g(\eta) = T/T_w$, as:

$$g'' + Pr f g' = -\frac{Pr\, u_e^2}{c_p T_w} (f'')^2 . \tag{16}$$

Both differential equations have to be solved with the given boundary conditions for the functions f, f' and g. The functions f and g cannot be solved analytically. They have been determined numerically for calculating the heat flux transferred from the laminar boundary layer gas flow to the plate surface with the relation $\dot{q}_g = -\lambda_w \partial T/\partial y$. Replacing $\partial T/\partial y$ by

$$T_w \frac{\partial (T/T_w)}{\partial y} , \tag{17}$$

the heat flux can be expressed as follows:

$$\dot{q}_g = -\lambda_w T_w \frac{g'(0)}{g(0)} \sqrt{\frac{u_e}{2 x \nu_w}} . \tag{18}$$

For applying the solution (18), valid for a flat plate boundary layer, to the sharp cone geometry, the Mangler transformation [1] is used. Mangler's theory predicts that the heat flux on a sharp cone is given by the flat plate heat flux multiplied by the factor $\sqrt{3}$.

3 Experiments

3.1 Shock tunnel facility STA

Two high energy shock tubes, STA and STB, having 100 mm inner diameter, are located in the shock tube laboratory of ISL. For these experiments the flow is accelerated inside of a contoured nozzle of 218.7 mm exit diameter, designed for $M = 6$. Two flight altitudes have been generated: a) 15 km and b) 21 km, respectively. A schematic diagram of the shock tunnel is given in Figure 4. The average measured flow conditions at the nozzle exit are:

a) $p = 12.2$ kPa, $\rho = 0.186$ kg/m^3, $T = 222$ K, $u = 1800$ m/s, $M = 5.93$.
b) $p = 4.7$ kPa, $\rho = 0.074$ kg/m^3, $T = 215$ K, $u = 1778$ m/s, $M = 5.95$.

The free stream unit Reynolds number has been calculated for condition a) $23.5 \cdot 10^6\,\text{m}^{-1}$ and for condition b) $9.6 \cdot 10^6\,\text{m}^{-1}$. A conical body of 7,125° half angle and length L of 210 mm, measured along the cone's surface, was used for the experiments. The unit Reynolds number at the cone surface has been calculated to be $33.9 \cdot 10^6\,\text{m}^{-1}$ for condition a) and $13.0 \cdot 10^6\,\text{m}^{-1}$ for condition b).

3.2 Surface heat flux measurements

Special temperature sensors, bought from Stoßwellenlabor RWTH Aachen in Germany, were used to measure the surface heat flux. The sensor surface is flush mounted with the outer contour of the model. When heat flux is transferred from the gas to the gauge, the gauge body is heated and the thin film layer temperature rises by the same amount. Therefore, by measuring the film layer temperature increase ΔT, the temperature rise of the gauge is measured too. The variation ΔR of the resistance R_0 of the film layer is related to the temperature change as $\Delta R = \alpha R_0 \Delta T$. Applying a one-dimensional solution of the heat conduction equation, see Oertel [6], the heat flux $\dot{q}$ at $y = 0$ can be calculated as:

$$\dot{q}(0,t) = \sqrt{\frac{\rho c \lambda}{\pi}} \frac{d}{dt} \int_0^t \frac{\Delta T(0,\tau)}{\sqrt{t-\tau}} d\tau .$$

With the calibration factor $E = \sqrt{\pi \rho c \lambda} / 2\alpha$ for each gauge the heat flux $\dot{q}$ is calculated from the recorded temperature change ΔT. The heat flux determined for

15 km flight altitude is shown in Figure 5 along the surface of the cone. It can be seen that for x < 90 mm, i.e. for x/L = 0.43, the measured heat flux matches quite well with the laminar boundary layer solution. For x > 90 mm transition to turbulent boundary layer starts, finally reaching a heat flux about three times the amount of the laminar one.

The transition Reynolds number range at the cone's surface is $3.05 \cdot 10^6 \leq Re_{TR} \leq 3.73 \cdot 10^6$. The agreement of the measured values with the turbulent boundary layer solution is quite good within the given error bars of about ±10%. A temperature overshoot at transition near the edge of the boundary layer was not found, probably due to an insufficient resolution of the experiments in x direction, see Schneider [7].

For the 21 km altitude flight condition the experimental and theoretical heat flux results are given in Figure 6. The calculation is just done for a laminar boundary layer formation. Looking at the measured data points, there are only small deviations between the two. The boundary layer is fully laminar along the projectile surface. No transition occurs.

4 Conclusions

The heat flux at the surface of a 7,125° cone of a projectile flying at Mach number 6 in earth atmosphere was simulated with the high energy shock tube STA of ISL. For comparison, theoretical data for the surface heat flux on a sharp cone model of the nose of a projectile have been calculated using relations given by solutions for the laminar and turbulent boundary layer formations. The 15 km altitude flight duplication shows boundary layer transition from laminar to turbulent for a Reynolds number greater than $3.05 \cdot 10^6$ at about a distance of 90 mm from the cone tip. At 21 km altitude the boundary layer develops fully laminar. The theoretical determined heat fluxes and the experimental results are, within the given error bars, in good agreement.

References

[1] H. Schlichting. "Boundary-Layer Theory", McGraw-Hill, New- York, 1960

[2] W. Hantzsche and H. Wendt. "Die laminare Grenzschicht bei einem mit Überschall-geschwindigkeit angeströmten nicht angestellten Kreiskegel". In Jahrbuch der deutschen Luftfahrtforschung I, 1941

[3] F. Seiler, U. Werner and G. Patz. "Theoretical and experimental modeling of real Projectile Flight Heating". J. of Sp. and Rockets, Vol. 38, No. 4, pp. 497-503, 2001

[4] K.-Y. Chien. "Hypersonic, Turbulent Skin-Friction and Heat-Transfer Measurements on a Sharp-Cone". AIAA Journal, Vol. 12, No. 11, 1974

[5] H. Oertel. "Stoßrohre", Springer Verlag, Wien-New York, 1966

[6] H. Oertel. "Wärmeübergangsmessungen". In Kurzzeitphysik, *Springer Verlag*, Wien-New York, 1967

[7] S. P. Schneider. "Hypersonic Laminar-Turbulent Transition on Circular Cones and Scramjet Forebodies". Aerospace Sciences, Vol. 40, Issue 1-2, pp. 1-50, 2004

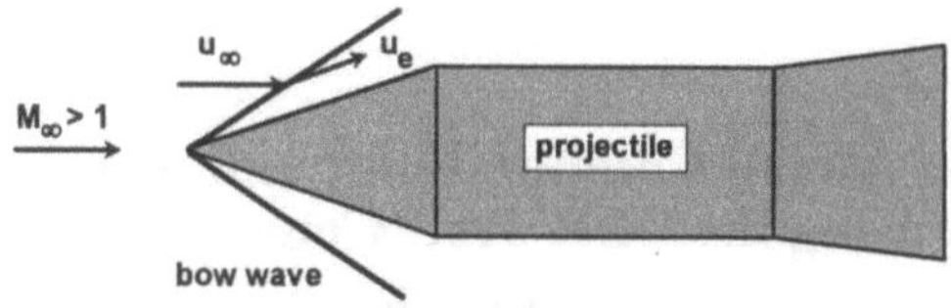

Figure 1 Bow wave formation at the nose in supersonic flight

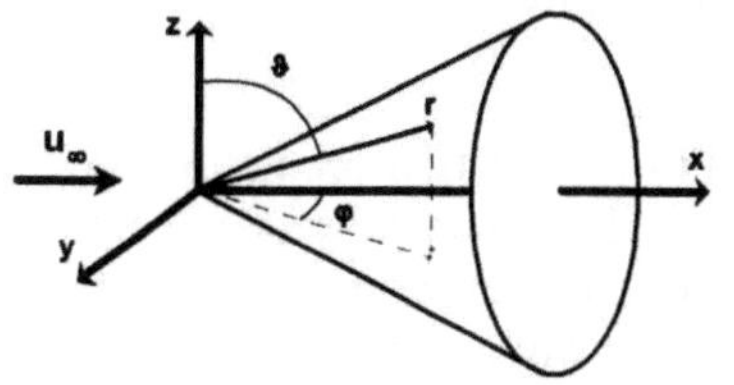

Figure 2 Cone in spherical coordinates

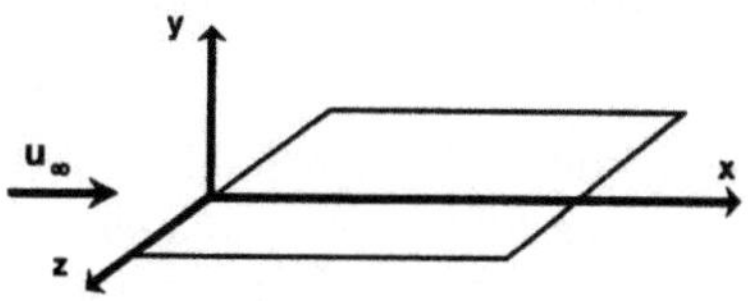

Figure 3 Flat plate in Cartesian coordinates

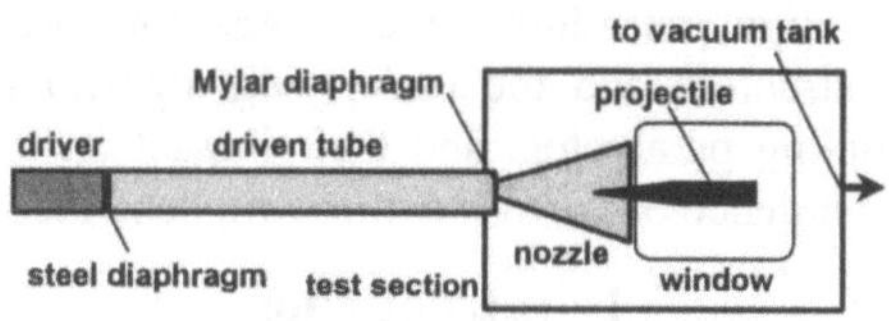

Figure 4 Schematic of shock tunnel STB

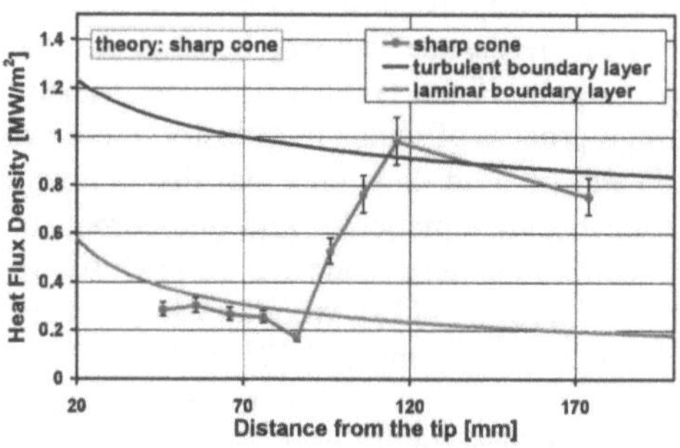

Figure 5 Heat flux density in 15 km flight altitude for the 7.125° cone at Mach 6

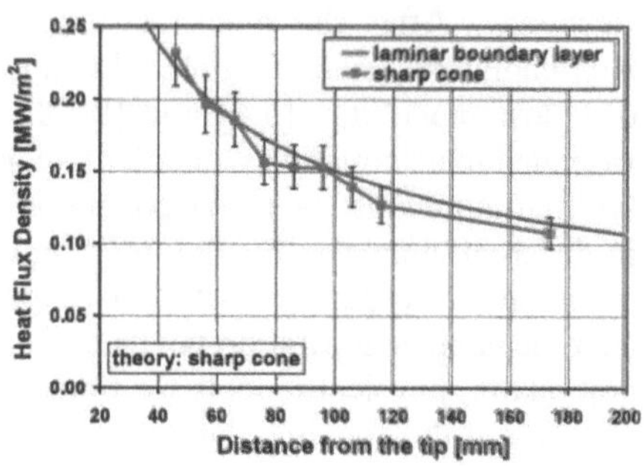

Figure 6 Heat flux density in 21 km flight altitude for the 7.125° cone at Mach 6

Navier-Stokes Airfoil Computations with Automatic Transition Prediction using the DLR TAU Code – A Sensitivity Study

A. KRUMBEIN
DLR, Institute of Aerodynamics and Flow Technology
Lilienthalplatz 7, 38108 Braunschweig, Germany, andreas.krumbein@dlr.de

Summary

The hybrid DLR RANS solver TAU coupled to a transition prediction module was successfully applied to a single-element airfoil automatically taking into account the locations of laminar-turbulent transition. The experimentally measured transition locations could be reproduced with very high accuracy. A sensitivity study of the parameters of the coupling procedure was performed in order to investigate the behaviour of the coupled system with respect to the accuracy and robustness of the iteration procedure for the transition locations. The transition prediction coupling structure and the underlying algorithm are described. The functions of the coupling parameters and their impact on the transition location iteration and the convergence of the simulations are described and documented.

1 Introduction

The modeling of laminar-turbulent transition in Reynolds-averaged Navier-Stokes (RANS) solvers is a necessary requirement for the computation of flows over airfoils and wings in the aerospace industry because it is not possible to obtain quantitatively correct results if the laminar-turbulent transition is not taken into account. For the design process of wings in industry, there exists the demand for a RANS-based computational fluid dynamics tool that is able to handle flows with laminar-turbulent transition automatically and autonomously. The first steps towards the setup of such a tool were made e.g. in [11], where a structured RANS solver and an e^N-method, [12] and [18], were applied, and in [15], where a RANS solver, a laminar boundary-layer method [2] and an e^N-method were coupled. There the boundary-layer method was used to produce highly accurate laminar, viscous layer data to be analysed by a linear stability code. The use of an e^N-database method [14] results in a coupled program system that is able to handle transition prediction automatically. After the block-structured DLR RANS solver FLOWer [1] is in a well-engineered state with regard to automatic transition prediction and transitional flow modelling, [5-6] and [8], the DLR hybrid RANS solver TAU [3] was extended in a similar way in order to combine the benefit of a hybrid RANS code with an automatic transition prediction capability, [9-10]. In contrast to [9], where different strategies for the iteration of the transition locations were applied, in the present paper the established algorithm from [5-6] and [8] – taking into account the laminar separation points in the RANS computational grid – is applied. Additionally, physical models for transitional flow regions are introduced. Finally, a setting of the coupling parameters is specified which allows for a fast and accurate iteration of the transition locations.

2 Coupling of the TAU Code and the Transition Prediction Module

The transition prediction module consists of a laminar boundary-layer code for swept, tapered wings [2], and an e^N-database method for Tollmien-Schlichting (TS) instabilities [14]. The coupled system can be run in two different modes. Either the TAU code communicates the surface pressure distribution of the configuration to the laminar boundary-layer method, the laminar boundary-layer method computes all of the boundary-layer parameters that are needed for the e^N-database method and the e^N-database method determines new transition locations that are given back to the RANS solver (*1st mode*). Or the TAU code computes the boundary-layer parameters δ, H_i and Re_{δ^*} internally and communicates them directly to the e^N-database method (*2nd mode*). In [9-10] the influence of the cell number in wall normal direction and a comparison with results from a boundary-layer code are shown.

This coupled structure results in an iteration procedure for the transition locations within the iterations of the RANS equations. During the solution process of the RANS equations, the transition prediction module is called after a certain number of iterations, k_{cyc}, of the RANS iteration process. With the call of the module the solution process is interrupted and the module analyses the laminar boundary layers of both sides of an airfoil configuration. The determined transition locations, $x_j^T(\text{cycle} = k_{cyc})$ with $j = 1, ..., n_{loc}$, where n_{loc} is the number of transition points, are communicated to the RANS solver, which continues the solution process of the RANS equations. In so doing, the determination of the transition locations becomes an iteration process itself. The structure of the approach is outlined graphically in Fig. 1. At every call of the module the surface pressure, $c_p(\text{cycle} = k_{cyc})$, or the internally computed boundary-layer parameters along the surface of an airfoil are used as input for the transition prediction module. The viscous data is then subsequently analysed by the e^N-database method. The algorithm for the transition prediction iteration works as follows:

a) The RANS solver is started as if a computation with prescribed transition locations should be performed. At this moment, the transition locations are set very far downstream on the upper and lower sides of the airfoil, e.g., at the trailing edge. The RANS solver now computes a fully laminar flow over the airfoil.

b) During the solution process of the RANS equations the laminar flow is checked for laminar separation points by the RANS solver. In the case that laminar separation is detected, the separation point is used as an approximation of the transition location and the computation is continued.

c) The RANS equations are iterated until the lift coefficient $c_l = c_l(\text{cycles})$ has become constant with respect to the iteration cycles.

d) The transition prediction module is called. The e^N-database method determines the transition locations on upper and lower sides of the airfoil. The procedure acts differently in the *1st mode* and in the *2nd mode*. In the case that the e^N-database method does not detect a transition location due to TS waves, two possibilities are implemented: Either the laminar separation point from the boundary-layer code is used as an approximation of the transition point (*1st mode*) or the previous transition locations is kept unchanged (*2nd mode*).

e) The current coordinate $x_j^T(\text{cycle} = k_{cyc})$, which is used as transition location, is underrelaxed. That is, as new transition locations the coordinates $x_j^{*T}(\text{cycle} = k_{cyc})$, which are located downstream of the coordinates x_j^T, are used according to,

$$x_j^{*T}(k_{cyc}) = C_j^T(k_{cyc})\, x_j^T(k_{cyc}) \quad \text{with} \quad j = 1, ..., n_{loc}, \tag{1}$$

with $C_j^T(k_{cyc}) > 1$. The underrelaxation of the determined transition locations prevents the case that at an unconverged stage during the transition location iteration transition coordinates which are determined too far upstream might not be shifted downstream again.

f) As convergence criterion $\Delta x_j^{*T,l} < \varepsilon \approx 1\%$ with $\Delta x_j^{*T,l} = | x_j^{*T}(k^l_{cyc}) - x_j^{*T}(k^{l-1}_{cyc}) |$ is applied, where l is the current iteration step. In the case that the criterion is satisfied the iteration for x_j^T is finished, else the algorithm loops back to station b).

3 Generation of Transitional Flow Regions

In the case that a new transition location has been determined, the laminar, transitional and turbulent flow regions must be generated within the computational grid. The generation of the different regions is done by setting a real value flag FLG_{lt} at each point in the computational grid that is applied to the value of the source term S_{tp} of the turbulence production, which is computed for every point in the flow field. FLG_{lt} is applied in the following way to all of the points of the computational grid,

$$S_{tp}^{code}(P_F) = S_{tp}(P_F) * \text{MIN}\,[\, FLG_{lt}(P_F), 1\,], \tag{2}$$

with $FLG_{lt}(P_F) = 0.0$ for a laminar surface point, $FLG_{lt}(P_F) = 1.0$ for a turbulent surface point and $FLG_{lt}(P_F) = \gamma(P_F)$ for a transitional surface point, where $\gamma(P_F)$ is the value of the intermittency function γ at the field point P_F, which takes the value of the intermittency function at the nearest surface point $P_S^{nst}(P_F)$ within a limiting wall normal distance. According to [6] and [8] the intermittency function can be expressed as

$$\gamma(s_q) = 1 - \exp\left[-0.412\left(3.36\,\frac{\text{MAX}\,(s_q - s_{q,tr}^{beg}, 0)}{s_{q,tr}^{end} - s_{q,tr}^{beg}}\right)^2\right], \tag{3}$$

where s_q is the arc length on the airfoil side q starting at the stagnation point and $s_{q,tr}^{beg}$ is the location of transition onset. For the determination of the extent of the transitional region, the transition length $l_{tr} = s_{tr}^{end} - s_{tr}^{beg}$, the formula from [19],

$$Re_{l_{tr}} = a\left(Re_{\delta_{tr}^{*beg}}\right)^{\frac{3}{2}}, \tag{4}$$

with a = 4.6 for flows with pressure gradient, is applied. For testing purposes, equation (4) with a = 2.3 is applied, as is recommended in [16] for flows in which transition does not occur before laminar separation. The thickness of the laminar boundary layer δ is evaluated directly from the Navier-Stokes grid using the 99%-criterion,

$$\delta(x) = y\Big|_{U=0.99U_e}, \tag{5}$$

and the compressible Bernoulli equation is used to evaluate the edge velocity U_e,

$$U_e = \sqrt{U_\infty^{\,2} - \frac{2\kappa}{\kappa-1}\frac{p_\infty}{\rho_\infty}\left[\left(p_{wall}/p_\infty\right)^{\frac{\kappa-1}{\kappa}} - 1\right]} \ . \tag{6}$$

Thus, the displacement thickness δ^* and ρ_e and U_e, the values of the density and the tangential flow velocity at the boundary-layer edge, can be determined. In the case that the value of U_e is not reached, the y-location of the maximum value of U(y) is taken as the value of the boundary-layer thickness.

4 Results

The coupled system was applied to the NLF(1)-0416 laminar airfoil [13] with $M = 0.3$, $Re = 4\times10^6$, $\alpha = 2.03°$ and $N_T = 11$ [17], the limiting N-factor for the TS-database method. In [16] the transition locations $x^T_{upper}/c = 0.35$ on the upper side and $x^T_{lower}/c = 0.6$ on the lower side are reported for $M = 0.1$, a Mach number which could not be set in the computation because the TAU code version used does not yet provide a preconditioning capability. All computations were started with initial transition locations set at 75% of chord on upper and lower sides, and the Spalart-Allmaras turbulence model with Edwards modification (SAE) was used for all tests. The transition prediction procedure was run in the 2nd mode.

To ensure the high quality of the laminar boundary-layer data necessary for accurate transition predictions the boundary layer was resolved using 64 cells in the wall normal direction. Within a constant wall distance in the prismatic layer part of the hybrid grid 80% of the maximum turbulent boundary-layer thickness at the upper side trailing edge are embedded, Fig. 2. Additionally, it seems to be sufficient to use a grid spacing with near wall clustering which is optimised for the resolution of a fully turbulent boundary layer also in the laminar part of the flow. Thus, the very expensive grid adaptation procedure [15], which is normally necessary after each iteration step of the transition locations iteration, could be avoided.

The basic parameter settings for the coupling procedure were: a) the overall number of iteration cycles for the RANS computation, $cycle_{max} = 30{,}000$; b) the cycle interval for the iteration of the transition locations, $\Delta k_{cyc} = 3{,}000$; and c) the relaxation factor f^{relax} for the underrelaxation of a transition location x^T, $f^{relax} = 0.7$, while the underrelaxation formula reads

$$x^{*T}(k^l_{cyc}) = x^{*T}(k^{l-1}_{cyc}) - f^{relax}\left[x^{*T}(k^{l-1}_{cyc}) - x^T(k^l_{cyc})\right]. \tag{7}$$

The sensitivity aspects considered are: 1) the surface grid point where the turbulence model is activated first, which either can be the nearest point upstream or the nearest point downstream of x^{*T}; 2) the possible downstream movement of the transition points; 3) the consideration of the RANS grid laminar separation points as transition points; 4) the application of point transition vs. the application of transition lengths.

Three cases with different combinations of the sensitivity parameters are presented here. The results of the *1st case*, applying point transition, no laminar separation and downstream movement are depicted in Fig. 3. On the left hand

side, for the nearest point upstream and on the right hand side, for the nearest point downstream, the convergence histories of the transition locations vs. the RANS iteration cycles are shown. In the figures, the transition locations coming directly from the database and the underrelaxed values are depicted. *Left:* the upper side transition point has overrun the experimental transition point and still has a clear tendency to further move upstream. The lower side transition point has converged to a value which is located 8% downstream of the experimental value based on the following specification of $b\% = [(x^T_{comp}/x^T_{exp}) - 1] \times 100$. *Right:* The upper side transition location converges directly to a value which approximates the experimental transition point with an error of 2.86%. The lower side transition location, however shows perturbations which seem to be a more or less periodic back-and-forth oscillation of the transition point from the database.
The results of the *2nd case*, applying point transition at the nearest surface grid point downstream of x^{*T} with suppressed downstream movement are depicted in Fig. 4, without laminar separations on the left hand side and with laminar separation on the right hand side. Additionally, the points of laminar separation as they occurred during the computation in RANS grid and the grid points that represent the transition location in the RANS computational grid are depicted in the plots. *Left:* The back-and-forth oscillation of the transition location on the lower side does not occur, and the transition point converges to a value which is located 7.5% too far downstream compared to the experimental value. On the upper side, the value of the transition location is the same as in the computations which permitted downstream movement. *Right:* The lower side transition location is hit almost exactly by a laminar separation point. The convergence of the transition location iteration is accelerated from 3 calls of the transition prediction module to 2 calls.
The results of the *3rd case*, applying the intermittency function and the transition length model from equation (4) at the nearest surface grid point downstream of x^{*T} with suppressed downstream movement and the use of the RANS grid laminar separation points with $a = 4.6$ on the left hand side and with $a = 2.3$ on the right hand side are depicted in Fig. 5. *Left:* The results are characterised by strong oscillations of the lower side transition location given by the database. The final lower side transition location, which is based on a laminar separation point, overruns the experimental value slightly and converges to a value which is located about 3.5% too far upstream. An investigation of the flow field solution [7] reveals a large separation bubble on the upper side starting slightly downstream of the transition point. This was due to a transition length of 13% of the chord length which is too long – expected is an extent of about 5% of the chord length – such that the influence of the turbulence model, which should result in an increase of the skin friction, is delayed. *Right:* The final lower side transition location, which is again based on a laminar separation point, overruns the experimental value slightly and converges to a value which is located about 1.7% too far upstream. The lower side transition points which are determined by the database after the last laminar separation point was set as a transition point, are located about 7% downstream of the experimental value. The separation bubble on the upper side

does not occur [7]. Instead, on the lower side another separation bubble has developed, because now the value of the transition length on the lower side has become significantly larger.

Finally, the effects of the basic coupling parameters Δk_{cyc} and f^{relax} were investigated. Δk_{cyc} was changed to $\Delta k_{cyc} = 5{,}000$ to see if the accuracy of the transition locations might be improved by a longer convergence of the RANS solution between two calls of the transition prediction module. f^{relax} was changed to $f^{relax} = 0.5$ to see if a slower convergence of the transition location iteration has an influence on the final values of the transition locations. The tests were carried out only for the settings without transition lengths and without taking into account the laminar separation points. Each of these changes did not influence the final values of the transition locations on upper and lower sides of the airfoil [7].

With respect to the accuracy of the predicted transition locations it seems to be justified to compare the results from a computation at M = 0.3 and the experimental values at M = 0.1, especially because an incompressible stability analysis using the database method was performed such that the only compressible influences possible may come from the RANS solution. On the other hand, the predicted transition locations are of the same - very good - order of accuracy that were obtained using the DLR block-structured FLOWer code and the transition prediction module applying them to the same test case at M = 0.1 [4-5].

5 Conclusions

The hybrid TAU code coupled to a transition prediction module was successfully applied to a subsonic airfoil test case automatically taking into account the transition locations predicted by an e^N-database method or which were based on laminar separation points determined by the TAU code. The experimentally measured transition locations could be determined with very high accuracy when the nearest surface grid point downstream of the predicted transition location is used as transition point, when a possible downstream movement of the transition points is suppressed and when point transition is applied instead of transitional flow models. A sensitivity study of the parameters of the coupling procedure showed that the accuracy of the converged transition locations can be highly improved and the convergence of transition location iteration itself can be significantly accelerated when the laminar separation points which occur in the RANS computational grid during the transient phase of the computation are immediately used as transition locations. In contrast to these encouraging results, the application of the intermittency function based on algebraic transition length models for the modelling of transitional flow regions led to large separated regions directly downstream of the transition point. These separated flow regions are contrary to experimental findings on the one hand, and lead to strong disturbances in the convergence of the RANS computation on the other. Therefore, at present the application of these models in the TAU code together with the SAE turbulence model can not be recommended. Further investigations are necessary here to clarify why the transitional flow models showed such an unexpected behaviour.

Acknowledgements

This work has been carried out within the HiAer Project (High Level Modelling of High Lift Aerodynamics) and documented in detail in [7]. The HiAer project was a collaboration between DLR, ONERA, KTH, HUT, TUB, Alenia, Airbus Deutschland, QinetiQ and FOI. The project was managed by FOI and was partly funded by the European Union (Project Ref: G4RD-CT-2001-00448).

References

[1] FLOWer - Installation and User Handbook, Release 116, Doc.Nr. MEGAFLOW-1001, Institut für Entwurfsaerodynamik, Deutsches Zentrum für Luft- und Raumfahrt e.V., 2000

[2] Horton, H. P., Stock, H. W., "Computation of Compressible, Laminar Boundary Layers on Swept, Tapered Wings", *Journal of Aircraft*, Vol.32, No. 6, 1995, pp.1402-1405

[3] Kroll, N., Rossow, C.-C., Schwamborn, D., Becker, K., Heller, G., "MEGAFLOW - A Numerical Flow Simulation Tool For Transport Aircraft Design", ICAS Congress 2002, Toronto (can), 09.-13.09.2002, ICAS, CD-Rom, pp. 1.105.1-1.105.20, 2002

[4] Krumbein, A., Stock, H. W., "Laminar-turbulent Transition Modeling in Navier-Stokes Solvers using Engineering Methods", Barcelona, *ECCOMAS 2000-CD-Rom Proceedings*

[5] Krumbein, A., "Coupling of the DLR Navier-Stokes Solver FLOWer with an e^N-Database Method for laminar-turbulent Transition Prediction on Airfoils", *Notes on Numerical Fluid Mechanics · Volume77*, pp. 92-99, Germany 2000, Springer Verlag, 2002

[6] Krumbein, A., "Transitional Flow Modeling and Application to High-Lift Multi-Element Airfoil Configurations", *Journal of Aircraft*, Vol. 40, 2003, pp. 786-794

[7] Krumbein, A. et al., HiAer Deliverable D3.1-2: "Implementation of transition/turbulence models", Technical Report, April 2004

[8] Krumbein, A., "Automatic Transition Prediction and Application to High-Lift Multi-Element Airfoil Configurations", AIAA-2004-2543, July 2004 (sub. to *Journal of Aircraft*)

[9] Nebel, C., "Transitionsberechnung an einem 3D Rumpfkörper", Institutsbericht 2003/1, Institut für Strömungsmechanik, Technische Universität Braunschweig, 2003

[10] Nebel, C., Radespiel, R., Wolf, T., "Transition Prediction for 3D Flows Using a Reynolds-Averaged Navier-Stokes Code and N-Factor Methods", AIAA-2003-3593

[11] Radespiel, R., Graage, K., Brodersen, O., "Transition Predictions Using Reynolds-Averaged Navier-Stokes and Linear Stability Analysis Methods", AIAA Paper 91-1641, 1991

[12] Smith, A. M. O., Gamberoni, N., "Transition, Pressure Gradient and Stability Theory", Douglas Aircraft Company, Long Beach , Calif. Rep. ES 26388, 1956

[13] Somers, D. A., "Design and Experimental Results for a Natural-Laminar Flow Airfoil for General Aviation Applications", NASA Technical Paper 1861, Scientific and Technical Information Branch, 1981

[14] Stock, H. W., Degenhardt, E., "A simplified e^N method for transition prediction in two-dimensional, incompressible boundary layers", *Zeitung für Flugwissenschaft und Weltraumforschung*, Vol.13, 1989, pp.16-30

[15] Stock, H. W., Haase, W., "A Feasibility Study of e^N Transition Prediction in Navier-Stokes Methods for Airfoils", *AIAA Journal*, Vol.37, no. 10, 1999, pp. 1187-1196

[16] Stock, H. W., Haase, W., "Navier-Stokes Airfoil Computations with e^N Transition Prediction Including Transitional Flow Regions", *AIAA Jour.*, Vol.38, no.11, 2000, pp.2059-2066

[17] Stock, H. W., "Airfoil Validation Using Coupled Navier-Stokes and e^N Transition Prediction Methods", *Journal of Aircraft*, Vol.39, No. 1, 2002, pp.51-58

[18] van Ingen, J. L., "A suggested Semi-Empirical Method for the Calculation of the Boundary Layer Transition Region", University of Delft, Dept. of Aerospace Engineering, Delft, The Netherlands, Rep. VTH-74, 1956

[19] Walker, G. J., "Transitional Flow on Axial Turbomachine Blading", *AIAA Journal*, Vol.27, No. 5, 1989, pp. 595-602

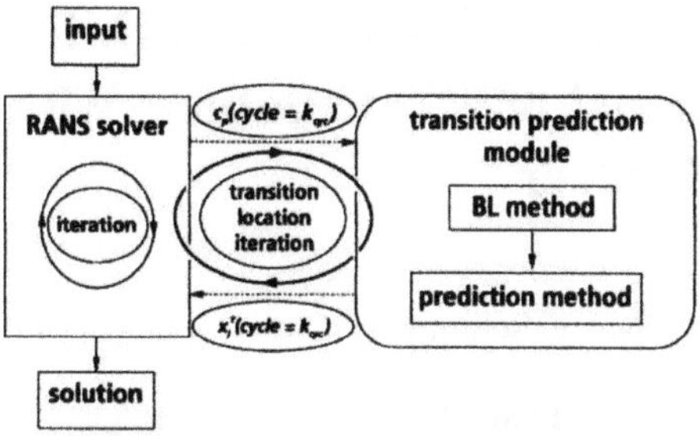

Fig. 1 Transition prediction coupling structure

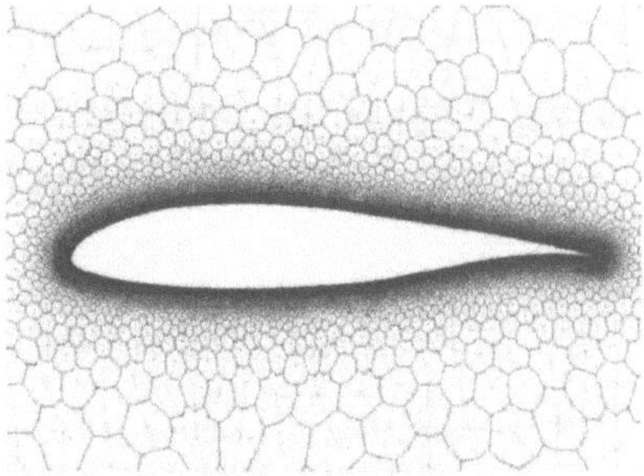

Fig. 2 Dual grid and hybrid primary grid

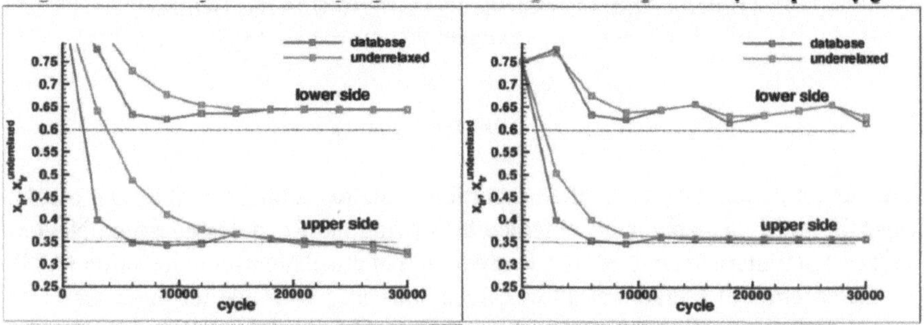

Fig. 3 Convergence of the transition locations, no laminar separation, no transition lengths, downstream movement, left: nearest point upstream, right: nearest point downstream

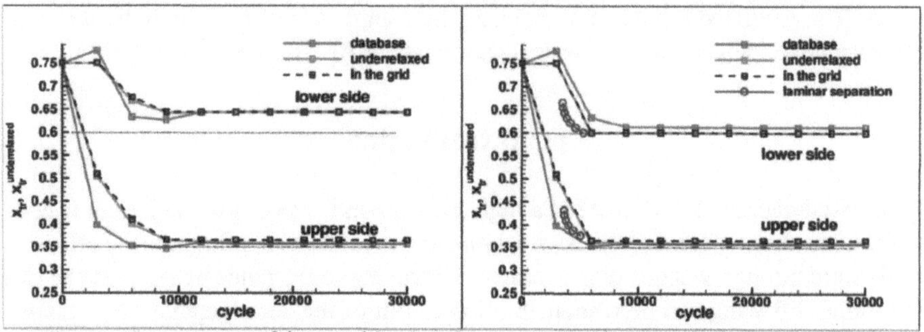

Fig. 4 Convergence of the transition locations, nearest point downstream, no transition lengths, no downstream movement, left: no laminar separation, right: with laminar separation

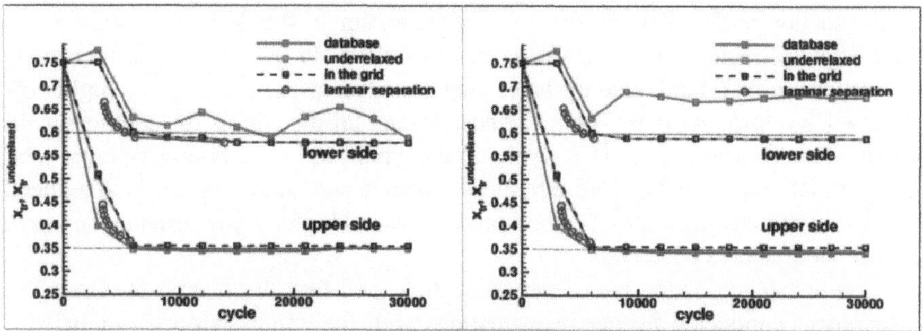

Fig. 5 Convergence of the transition locations, nearest point downstream, with laminar separation, no downstream movement, transition lengths, left: a = 4.6, right: a = 2.3

Prediction of attachment line transition for a high-lift wing based on two-dimensional flow calculations with RANS-solver

Jochen Wild[1] and Oliver T. Schmidt[2]

[1] DLR Braunschweig, Institute of Aerodynamics and Flow Technology, Lilienthalplatz 7, D-38108 Braunschweig, Germany, jochen.wild@dlr.de
[2] TU Berlin, ILR, Marchstraße 12-14, D-10587 Berlin, Germany, Oliver.T.Schmidt@TU-Berlin.de

Summary

This investigation shows that the properties of the two-dimensional flow around a high-lift multi-element airfoil obtained by solving the Reynolds-averaged Navier-Stokes equations can be used for the prediction of attachment-line transition (ALT) by the criterion of Pfenninger. Flow calculations are performed for three spanwise sections of a three-dimensional swept and tapered high-lift wing for which the occurrence of ALT is assumed at higher Reynolds numbers. It is shown that the onset of ALT is predicted reliably for changes of the angle of attack. The method is also applicable to indicate the spanwise location where ALT occurs first.

1 Introduction

The aerodynamic performance of a high-lift wing in terms of lift coefficient is coupled to the development of the boundary layer on the slat as it influences the effective curvature through viscous displacement. Hereby the occurrence of ALT on the slat of a high-lift wing can be responsible for a limit of the aerodynamic performance with increasing Reynolds number.

Without ALT the boundary layer starting from the attachment line downstream to the suction peak is laminar due to the acceleration of the flow and transition occurs slightly behind the suction peak at the earliest. In this case, with increasing Reynolds number, the boundary layer thickness decreases resulting in a higher effective curvature and therefore more suction and an increasing lift coefficient. In contrast the occurrence of ALT results in an overall turbulent boundary layer with increased thickness and therefore reduced effective curvature. This leads to reduced suction at the leading edge of the slat and a reduced high-lift performance in terms of lift coefficient is observed.

The existence of the phenomenon of ALT was first discovered by Gray [5]. Pfenninger conducted further investigations with the emphasis on avoiding ALT [9, 10, 11].These experiments led to a criterion, described below, that was assumed to characterize the occurrence of ALT. Experimental investigations by Poll [12, 13] and Arnal and Juillen [1, 2] validated this criterion for the assumption of infinite swept wings. Newer experiments of Seraudie et. al. [14] also verified the criterion for swept finite high-lift wings at high angles of attack. In addition to the wind tunnel experiments mentioned above flight tests also show the relevance of the criterion [9, 4].

Starting point of this investigation are measurements of the ProHMS high-lift wing-body half-model of a transport aircraft in the cryogenic wind tunnel Cologne DNW-KKK for a Reynolds number range of $Re_\infty = 1.4 \times 10^6 \ldots 6.2 \times 10^6$ [15]. With increasing Reynolds number the maximum lift coefficient of the high-lift configuration decreased after having reached a maximum at a medium Reynolds number of $Re_\infty = 3.0 \times 10^6$. This led to the assumption that ALT occurs above $Re_\infty = 3.0 \times 10^6$. Since transition measurements were not performed, the attempt is made to predict the occurrence of ALT by numerical methods.

Instead of computing the whole three dimensional flow field, here the attempt is made to use only two-dimensional flow calculations of selected wing sections. There are at least two reasons for this approach: a) the two-dimensional flow calculations are cheaper in terms of computational resources and easier to set up; b) due to the fact that high-lift design is still mainly performed based on two-dimensional computations a validated prediction method based on this data can be easily implemented into the design process, giving hints to avoid ALT.

2 Attachment-Line Transition Prediction

The prediction criterion for the onset of ALT of Pfenninger [9] used within this work distinguishes between the flow normal to the leading edge and the crossflow. It correlates the crossflow velocity outside the boundary layer w with the acceleration of the flow out of the stagnation line $\frac{\partial u_e}{\partial s}$. Pfenninger formulates an attachment line boundary layer Reynolds number

$$Re_{\theta_{a.l.}} = 0.405 \frac{w}{\sqrt{\nu \frac{\partial u_e}{\partial s}}} \tag{1}$$

with the kinematic viscosity ν, the local arc length s in the coordinate system normal to the leading edge and the velocity component at the edge of the boundary layer normal to the leading edge u_e.

A number of experiments on swept wings and cylinders (e.g. of Pfenninger [11] and Poll [13]) verified a lower limit of $Re_{\theta_{a.l._{krit.}}} = 100$ below which no ALT occurs. Experiments of Arnal [1] with a swept wing at incidence showed, that due to the limited extent of the wing an upper critical value of $Re_{\theta_{a.l._{krit.}}} = 133$ exists. Between both values existing turbulence is only propagated, above the value of Arnal the turbulence is amplified. All investigations showed that the main mechanism for ALT in real applications is the propagation of turbulence from upstream along the attachment line, the so called bypass transition or leading edge contamination. Transition due to instabilities, which may occur at higher values of $Re_{\theta_{a.l.}}$ as has been shown experimentally by Poll [13] and has also been computed by Joslin [6, 7] using direct numerical simulation, has never been observed for real aircraft.

As can be seen from (1) the only term that takes into account three-dimensional flow is the crossflow velocity w outside the viscous flow, which for an infinite swept and untapered wing is constant

$$w = U_\infty sin\phi \tag{2}$$

with the onflow velocity U_∞ and the sweep angle ϕ. All other terms of eq. 1 correspond to the flow components normal to the leading edge. By assuming that the effects of tapering and the limited span of the wing only play a minor role for the most part of a high aspect ratio wing, this criterion for ALT can be based on the evaluation of calculations of the flow normal to the leading edge.

3 Flow Calculation

For the calculation of the flow the structured DLR RANS-solver FLOWer [8] is used. It solves the unsteady compressible Reynolds-averaged Navier-Stokes equations in applying an explicit 5-stage Runge-Kutta time-stepping method. Turbulence modeling is done using the Spalart-Allmaras model with Edwards-modification. The turbulence equations are solved using a fully implicit scheme that allows for high CFL numbers.

In order to minimize grid dependencies of the flow solution, separate grids are generated using the DLR grid generator MegaCads [3] for each wing section at each Reynolds number. In particular the boundary layer resolution is adjusted in order to have approximately the same resolution in terms of the number of cells in the boundary layer and to obtain a value of the dimensionless wall distance y^+ on the order of 1.

For this investigation three spanwise sections of the wing of the ProHMS high-lift model are used, denoted inboard, midboard and outboard, corresponding to the pressure tab rows of the model. They are each located in the middle of the slat elements, far enough from any model tracks, so that 2D flow assumptions are most likely to apply.

In order to perform 2D calculations comparable with the 3D flow the assumption of an infinite swept wing is used. This leads to a scaling of the wing section into a coordinate system normal to the leading edge

$$y_{2D} = z_{3D}/cos\phi\,.^1 \tag{3}$$

Now only the onflow components normal to the leading edge are of interest

$$\begin{gathered} Ma_{\infty 2D} = Ma_{\infty 3D} cos\phi,\ Re_{\infty 2D} = Re_{\infty 3D} cos^2\phi, \\ \alpha_{2D} = atan\left(\frac{tan\alpha_{3D}}{cos\phi}\right). \end{gathered} \tag{4}$$

[1] in 3D usually the z-coordinate points in the vertical direction while in 2D the y-coordinate is used and $z_{2D} = 0$.

To compare coefficients of 2D calculation and 3D experiment an additional scaling applies due to the scaled onflow

$$c_{p_{3D}} = c_{p_{2D}} cos^2\phi,\ C_{L_{3D}} = C_{L_{2D}} cos^2\phi. \tag{5}$$

It is a priori unknown if the flow around the slat leading edge is turbulent or laminar. Only for conditions where ALT has already occured at a lower Reynolds number fully turbulent flow can be assumed. For all other conditions two calculations are performed, the first fully turbulent, the second with a prescribed laminar region on the slat lower side fixing transition at the slat leading edge. All other wing elements are calculated fully turbulent in all cases.

4 Correlation method

The main issue when comparing two-dimensional data of a wing section with data from the three-dimensional wing is that a priori the local angle of attack for the wing section is unknown as long as no spanwise lift distributions are available. To be able to compare the data it is necessary to find the right 2D data for the corresponding 3D data. Throughout this work a comparison is done based on the measured and computed pressure distributions. Calculations are performed for distinct angles of attacks, and afterwards the experimental data is screened to find a corresponding angle of attack of the 3D high-lift wing where the pressure distributions match. Since this is not exact for the whole wing, primarily the pressure distributions on the slat are considered for comparison.

Fig. 1 shows such a correlation for each of the three sections, where the calculated 2D pressure distribution is plotted against the pressure distribution of the 3D experiment for $Re_{\infty_{3D}} = 6.0 \times 10^6$. It can be seen that the pressure distributions match the experiment especially on the slat. The biggest deviations can be seen for the midboard section for the rear part of the main element and the flap. The pressure distributions of the experiment indicate attached flap flow, while the flap flow is separated in the calculations. This might be a result of the fully turbulent calculation of the main element and the flap, since the transition locations are not known and the effort of transition prediction is not undertaken. The increased boundary layer thickness around the flap tends to slightly close the effective gap, which leads together with the increased friction losses to a separation of the flap. Another reason for the differences may be that the local sweep angle in the rear part of the high-lift wing is different from the angle assumed for the normalization due to the taper of the wing. This is observed especially on the flap in the different levels of stagnation pressures. Nevertheless the slat pressure distribution is matched very well, and since the slat flow is the primary interest in this investigation, all cases in the following are treated using the described method.

5 Results

A first point of interest is if the predicted occurrence of ALT correlates with the dependency of the lift coefficient on the onflow Reynolds number. Since it is in all likelihood the occurrence of ALT on the slat influencing the lift coefficient behavior, in the following only the slat is regarded. Fig. 2 shows the lift coefficient of the 3D measurements over the Reynolds number for a fixed angle of attack of $\alpha_{3D} = 15°$ together with the corresponding calculated values of $R_{\theta_{a.l.}}$ for the three sections. It is observed that with increasing Reynolds number the lift coefficient decreases. There are two major changes in the slope of $C_{L_{3D}}$ vs. $Re_{\infty_{3D}}$ that correspond to $Re_{\theta_{a.l.}}$ to the first and last wing section exceeding the upper limit of $Re_{\theta_{a.l._{krit.}}} = 133$. Thus it can be seen that the predicted occurrence of ALT is in accordance with the measured effect on the lift coefficient. Summarizing it is proposed that the prediction of ALT by using 2D calculations can be applied.

The ALT criterion for the slat is expected to be highly sensitive to the angle of attack due to the variation of the attachment line position. For low incidences the attachment line is known to be at the leading edge while at high angles of attack the attachment line moves upstream towards the lower trailing edge of the slat. On the other hand the region of maximum suction is located at the leading edge where the curvature of the slat is the highest. So the distance between attachment line and suction peak varies as does the suction itself. The development of $Re_{\theta_{a.l.}}$ with respect to the local 2D angle of attack is shown for the inboard section in fig. 3, for the midboard section in fig. 4 and for the outboard section in fig. 5. The general trend is an increasing value of $Re_{\theta_{a.l.}}$ with increasing angle of attack. The only case where $Re_{\theta_{a.l.}}$ is decreasing is for high α_{2D} for the inboard and midboard section, where the flow indicates being beyond $C_{L_{max}}$.

For the evaluation of ALT occurrence it is worthwhile to investigate at which spanwise section ALT begins. Fig. 6 shows the values of $Re_{\theta_{a.l.}}$ over the relative wing span $\eta = y/b$ for a specific angle of attack $\alpha_{3D} = 15°$ at different onflow Reynolds numbers. Fig. 7 shows the same for a specific Reynolds number $Re_{\infty_{3D}} = 3 \times 10^6$ at different angles of attack. It is observed that ALT starts at the outboard section propagating towards the inboard region with increasing Reynolds number and angle of attack.

6 Conclusion

The present investigation is an attempt to predict the 3D flow phenomenon of attachment line transition on the slat of a transport aircraft in high-lift configuration through two-dimensional flow calculations of distinct wing sections. This was expected to be possible as the unknown values of the used ALT criterion of Pfenninger correspond to the flow normal to the leading edge. It has been shown that the predicted occurrence of ALT correlates to the measured behavior of the lift coefficient for the ProHMS 3D half model. The method reliably predicts the onset of ALT when changing the angle of attack and gives an indication of the spanwise section at which ALT occurs first.

Acknowledgments

The authors would like to thank Airbus Germany GmbH for the supply oft he 3D high-lift model and data. embedded in the joint projects ProHMS and ProHMS+. They would also like to thank H. W. Stock for sharing his experience in transition prediction.

References

[1] Arnal., D., Juillen, J.: Leading-edge contamination and relaminarization on a swept wing at incidence. In Cebeci, T., ed.: Numerical and Physical Aspects of Aerodynamic Flows. Springer-Verlag Berlin Heidelberg (1990) 391–402

[2] Arnal, D., Juillen, J., Reneaux, J., Gasparian, G.: Effect of wall scution on leading edge contamination. Aerospace Science and Technology **1** (1997) 505–517

[3] Brodersen, O., Hepperle, M., Ronzheimer, A., Rossow, C.C., Schöning, B.: The parametric grid generation system megacads. In Soni, B., Thompson, J., Häuser, J., Eiseman. P., eds.: 5th International Conference on Numerical Grid Generation in Computational Field Simulation, National Science Foundation (NSF) (1996) 353–362

[4] van Dam, C., Los, S., Miley, S., Rooock, V., Yip, L., Bertelrud, A., Vijgen, P.: In-fligh boundary-layer state measurements on a high-lift system: Slat. Journal of Aircraft **34**, (1997) 748–756

[5] Gray, W.: The effect of wing sweep on laminar flow, TM 255, RAE(1953)

[6] Joslin, R.: Simulation of three-dimensional symmetric and asymmetric instabilities in attachment-line boundary layers. AIAA Journal **34** (1996) 2432–2434

[7] Joslin, R.: Simulation of nonlinear inb1abilities in an attachment-line boundary layer. Fluid Dynamics Research **18** (1996) 81–97

[8] Kroll, N., Rossow, C.C., Becker, K., Thiele, F.: Megaflow – a numerical flow simulation system. Proceedings 1998-2.7.4, ICAS(1998)

[9] Pfenninger, W.: Flow phenomena at the leading edge of swept wings. AGARDograph 97, AGARD (1965) Part IV.

[10] Pfenninger, W., Bacon jr., J.W.: Amplified laminar boundary layer oscillations and transition at the front attachment line of a 45° swept flat-nosed wing with and without boundary layer suction. In Wells, C., 00.: Viscous Drag Reduction, Plenum Press New York (1969)

[11] Pfenninger, W.: Laminar flow control – laminarization. Report AGARD-R-654, AGARD (1977)

[12] Poll, D.: Three-dimensional boundary layer transition via the mechanism of "attachment line contamination" and "cross flow instability". In Eppler, R., Fasel, H., eds.: Laminar-Turbulent Transition. Springer-Verlag Berlin Heidelberg New York (1979) 253–262

[13] Poll, D.: Some observations of the transition process on the windward face of a long yawed cylinder. Journal of Fluid Mechanics **150** (1985) 329–356

[14] Séraudie, A., Perraud, J., Moens, F.: Transition measurement and analysis on a swept wing in high lift configuration. Aerospace Science and Technology **7** (2003) 569–576

[15] Wild, J., Puffert-Meissner, W., Sitzmann, M., Lekemark, L.: Messung des FNG Hochauftriebs-Modells bei hohen Reynoldszahlen im Kryogenischen Windkanal Köln. DLR-IB 124-2003/37, DLR (2003)

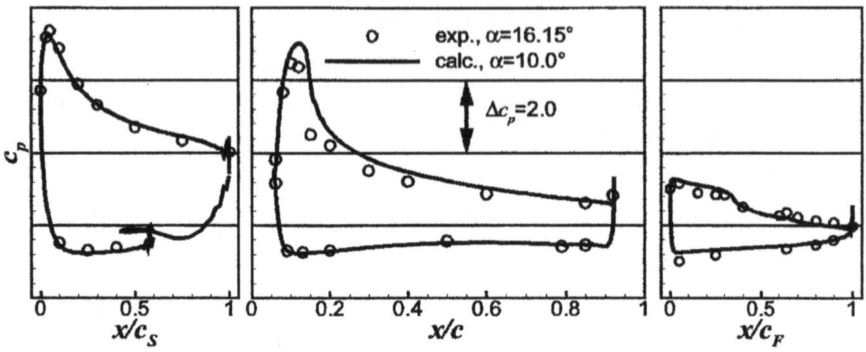

(a) inboard section

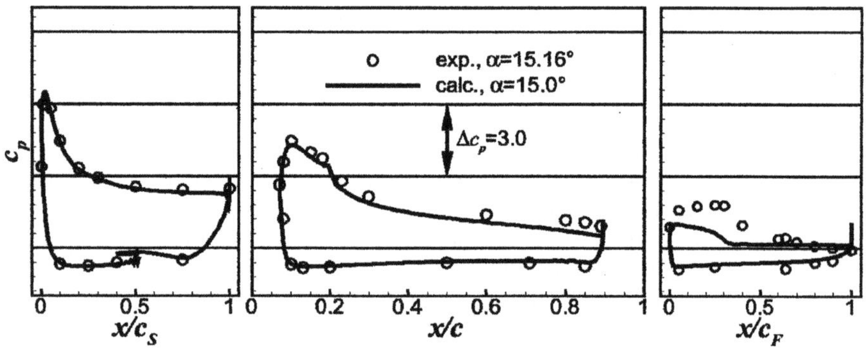

(b) midboard section

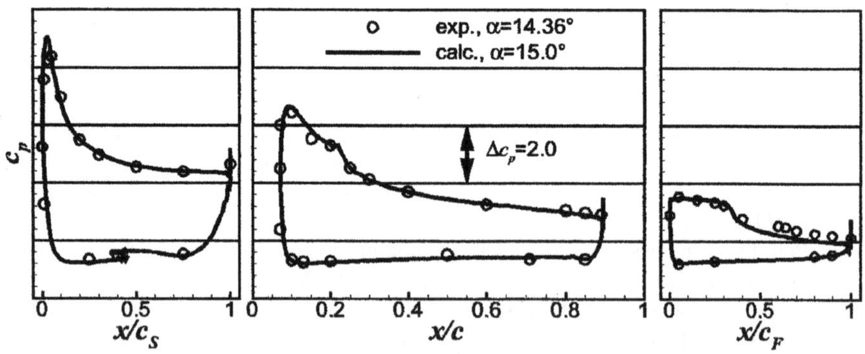

(c) outboard section

Figure 1 Correlation of calculated 2D pressure distributions with experimental data for $Re_{\infty 3D} = 6.0 \times 10^6$

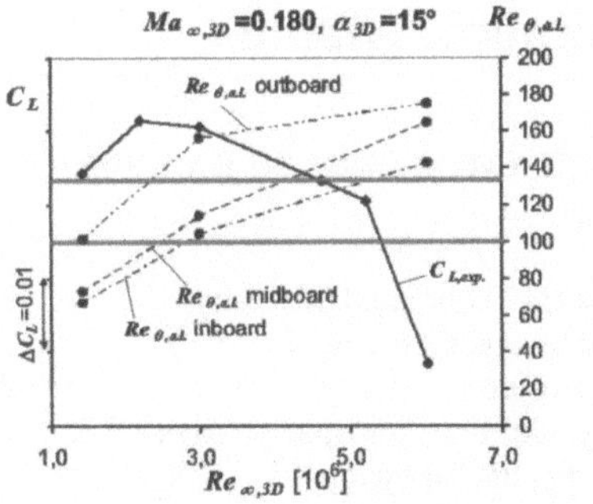

Figure 2 Variation of the lift coefficient and the attachment line Reynolds number with the onflow Reynolds number for $\alpha_{3D} = 15°$

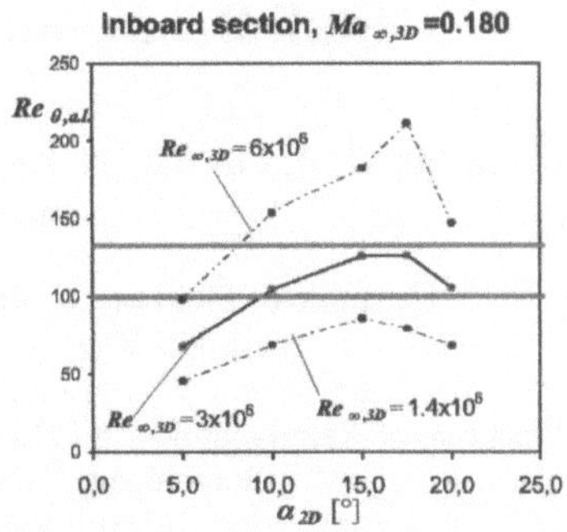

Figure 3 Variation of the attachment line Reynolds number with the local angle of attack for the inboard section

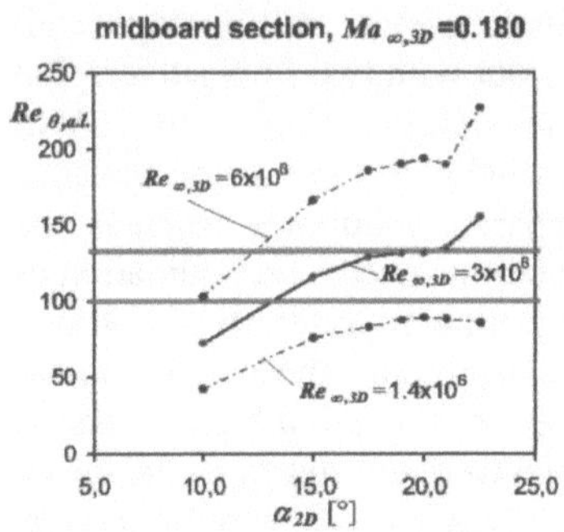

Figure 4 Variation of the attachment line Reynolds number with the local angle of attack for the midboard section

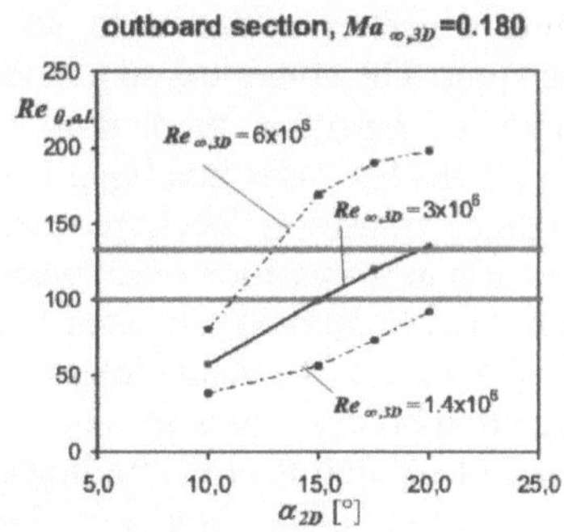

Figure 5 Variation of the attachment line Reynolds number with the local angle of attack for the outboard section

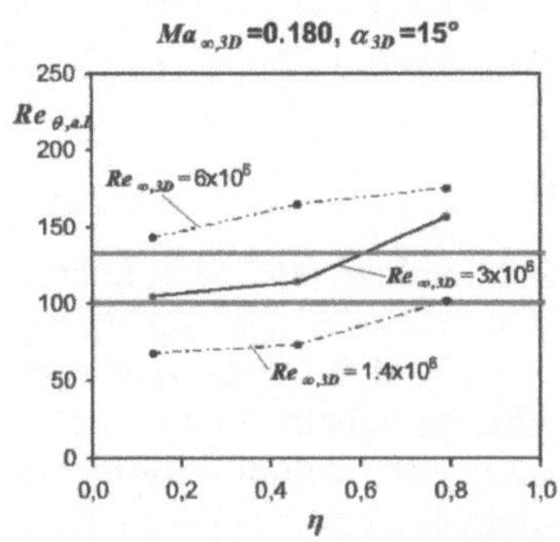

Figure 6 Spanwise variation of $Re_{\theta_{a.l.}}$ with the onflow Reynolds number for $\alpha_{3D} = 15°$

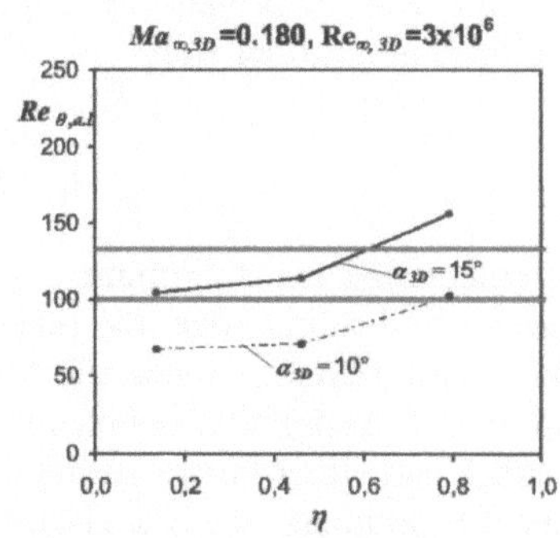

Figure 7 Spanwise variation of $Re_{\theta_{a.l.}}$ with the angle of attack for $Re_{\infty_{3D}} = 3 \times 10^6$

RANS Simulation and Experiments on the Stall Behaviour of a Tailplane Airfoil

R. WOKOECK[1], A. GROTE[1], N. KRIMMELBEIN[1], J. ORTMANNS[1], R. RADESPIEL[1] and A. KRUMBEIN[2]

[1] Institute of Fluid Mechanics, Braunschweig Technical University,
Bienroder Weg 3, 38106 Braunschweig, Germany
e-mail: r.wokoeck@tu-bs.de
[2] Institute of Aerodynamics and Flow Technology, German Aerospace Center DLR,
Lilienthalplatz 7, 38108 Braunschweig, Germany
e-mail: andreas.krumbein@dlr.de

Summary

Measurements and simulations are presented of the flow past a tailplane research airfoil which is designed to show a mixed leading-edge trailing-edge stall behaviour. The numerical simulations were carried out with two flow solvers that introduce transition prediction based on linear stability theory to RANS simulations for cases involving laminar separation bubbles. One of the methods computes transition locations across laminar separation bubbles whereas the other assumes transition onset where laminar separations occur. For validation of the numerical methods an extensive measurement campaign has been carried out. It is shown, that the methodology mentioned first can simulate the size of laminar separation bubbles for angles of attack up to where the separation bubble and the turbulent separation at the trailing edge are well behaved and steady in the mean. With trailing edge separation involved, the success of the new numerical procedure relies on the diligent choice of a turbulence model. Finally, for flows with increased unsteady behaviour of both, separation bubble and turbulent separation, which were observed at higher angles of attack in the experiment between maximum lift and leading-edge stall, steady state prediction methods for transition can no longer be applied and time-accurate methods have to be developed in a further step.

1 Introduction

For calculating the maximum lift of airfoils with a laminar separation bubble close to the leading edge, the precise prediction of transition location along the bubble is important. According to the classification [3] of the stall-types with leading-edge separation bubbles involved, the separation bubble at large angles of attack can either burst causing the so-called leading-edge stall or – for higher Reynolds numbers – will lead to a mixture of leading and trailing edge stall by interacting with the turbulent separation that is moving upstream from the trailing edge before the burst. For the latter case, various associated phenomena have been observed: The trailing-edge separation can be of strong unsteady character [4, 5]. The separated area near the trailing-edge viewed in wind-tunnel experiments usually appears in a spanwise irregular shape [6] and bears inside three-dimensional cells of circulating flow [16, 10].

Widely used numerical procedures consist of utilising a boundary-layer code for calculating the viscous, surface near flowfield in interaction with the inviscid outer flowfield [8, 9]. Determination of the transition position is handled either by databases of solutions of the Orr-Sommerfeld differential equation or by envelope-methods derived from such databases. Also semi-empirical approaches are used to determine the bubble size [17] and burst [2]. The prediction of transition on a laminar separation bubble is new in the context of flow solvers based on the Reynolds-averaged Navier-Stokes equations (RANS). Addressing this subject, the Institute of Fluid Mechanics (ISM) of Braunschweig Technical University has developed a methodology that couples a RANS flow solver to a stability method for the Tollmien-Schlichting instabilities, which occur in the 2D shear layer. For obtaining a solid and comprehensive data base for validation purposes, a dedicated research airfoil was designed, built and tested in a subsonic wind tunnel.

2 Design of the research airfoil

The aerodynamic design of the new research airfoil, named HGR-01, was defined to realize the mixed stall type. This was challenging due to the lack of adequate tools for predicting these stall types (as the acquirement of the latter is the matter of subject of the project). The chosen modus operandi described in [11] was to use the panel-method code XFOIL [9] for step-by-step reproduction of the near-nose pressure distribution of members of the well documented LWK airfoil family [1], which have shown the desired stall behaviour with laminar separation bubbles involved.

3 Experimental set up

The model of the research airfoil was manufactured of carbon fibre reinforced epoxy with a flap size of 30% chord for influencing the pressure distribution at the nose and so affecting the stall behaviour. It features 55 pressure tabs distributed on the surface with higher density at the nose and around the upper leading-edge area for resolving transition and separation of the flow. The experiments took place in the subsonic wind tunnel MUB of ISM with a test section of 1.3m squared and a characteristic turbulence level of about 0.2%. The experiments provided data for three different Reynolds numbers: 0.35, 0.7 and 1.4 million. The angle of attack was varied from negative lift to beyond stall. With the help of 2cm long silk tufts attached in a 2cm spacing a quick overview of the flow behaviour at different angles of attack was achieved. Also the tufts were chosen for measuring the turbulent separation. Then the more inert technique of oil-flow visualization that delivers a far better spatial resolution of the surface friction was applied. With oil flow the transition line for low angles of attack as well as the occurrence, position and size of the laminar separation bubble at higher angles were determined. Oil was also used to visualize the structures within the separated area and to detect the unwanted interaction between the boundary layers of the tunnel side walls and the flow past the model at very high angles of attack. The surface pressure distribution in the centre region of the model was delivered by the pressure tabs. Lift and drag were to be derived according to Jones' approach with the pressure distribution in the wake to be measured by pressure rakes. Due to strong unsteadiness in the wake at high angles of attack the present work utilizes only the pressure-induced normal force coefficient, $C_{n,p}$ for comparison to numerical results instead of the more commonly used lift and drag coefficients. It should be mentioned, that the curve of the pressure-induced normal force is very similar to the lift curve. $C_{n,p}$ is derived by integrating the measured pressure distribution shown exemplarily in fig. 5 around the surface. With the existing pressure tabs resolution obviously being too coarse for resolving the lower side pressure correctly, a correction (adding the tinted areas) to the measured normal force has been applied. This correction was deduced by projecting the measured pressure values on the numerical pressure distribution for several angles of attack so that the true differences between the measured and the calculated pressure forces are preserved. Finally, the non-intrusive method PIV was applied for investigation of characteristics in the area of turbulent trailing edge separation. For each flowfield measurement 1000 pictures illuminated by two Quantech brilliant Nd:YAG lasers with 150mJoul output per pulse at 532nm were taken by a peltier cooled LaVision Flowmaster camera with a resolution of 1280x1024 pixels and an effective 360mm focal length. This recording system was swivel mounted following the angle of attack to assure constant display windows. For processing the velocity fields, the Davis 6.2 software of LaVision was applied with an iterative multigrid scheme of second order and 0% overlap.

4 Numerical methods

4.1 Methodology of transition prediction of ISM

The transition prediction methodology developed by ISM is carried out via a coupled program system shown in fig. 1. This system consists of the RANS solver TAU [7] originated by German Aerospace Center (DLR) that is coupled with a stability method based on the linear stability theory via the transition prediction module. The transition prediction module interrupts the RANS solver TAU after a certain number of iterations (which can be decided, for example, upon a residual) performed with a transition position defined by the user as initial guess. The module extracts all information of the actual boundary-layer flow field necessary for the stability method, hands it over and starts this method. The output returned by the linear stability method consists of the N-factors for the amplified frequencies and is analysed by the transition prediction module that generates an envelope above all N-factors and compares this to a critical value.

This value is determined by free-stream conditions and can be found for example via Mack's formula [14]. The new transition position is assumed to be located where this value is exceeded and is then returned with under-relaxation into the flow field of the solver TAU for restart, closing one work loop. The number of loops can be fixed by the user or decided upon the convergence of the transition positions between work loops. As stability method the fast and proven stability solver COAST3 by G. Schrauf [18] is chosen. The transition length itself is then determined by the growth of the turbulent Reynolds stresses, as given by the turbulence model used in the RANS solver. Previous investigations have shown that, for this setup, a resolution of the laminar boundary-layer normal to the surface with at least 25 grid points is necessary. To fulfil this condition, a set of three grids with different overall resolution for grid refinement study purposes was generated for the computations. The grids (see tab. 1) were specially refined in areas of interest, and the medium grid resolution (fig. 11) was found to be sufficient for resolving the laminar separation bubble and the turbulent separation.

4.2 Methodology of transition prediction of DLR

The methodology developed and applied for the computations of DLR is described in detail in another article of this book [12] and has been published in [13]. Thus, only the differences compared to the methodology of ISM shall be highlighted here: Instead of feeding the flow field of the RANS Solver (here FLOWer) directly into the stability method, the necessary boundary-layer data are obtained by a boundary-layer method, that is fed with the pressure field generated by the RANS solver. This procedure allows for much coarser grid resolution (see fig. 9 and table 1) compared to ISM methodology. As stability method, a database by Degenhart and Stock [19] for the Tollmien-Schlichting instabilities is consulted. Depending on a boundary-layer method, this methodology can only analyze the region up to the laminar separation for stability. So, for laminar separation bubbles, contrary to the ISM methodology, here the point of laminar separation is utilized as transition location and an intermittency function and algebraic transition length for modelling the transitional flow regions is applied.

5 Results and Discussion

As the amount of data acquired in experiment and numerical simulation can not extensively be covered here, along with the general description of the overall performance of the research airfoil, the Reynolds number 0,7 million case is chosen as an example of the observed flow phenomena. Detailed insight and comparisons to numerical results are then given for 12° angle of attack. Flap deflection and its associated flow phenomena are beyond the scope of this article and therefore not covered here. For all three Reynolds numbers measured at relatively low, positive angles of attack (alpha), the flow on the upper surface (suction side) is fully attached with the transition marching upstream with increasing angle. The laminar separation bubble occurs in the nose area at medium angles long before the turbulent separation at the trailing edge starts. Figure 7 gives an example of the flow past the upper surface of the model visualized with oil paint. The extent of the separation bubble is clearly visible as a straight line over the model span as there is paint captured within the bubble through the experiment duration. In general, the separation bubble shrinks and moves towards the nose as the angle of attack is increased (see fig. 3) and it is larger for lower Reynolds numbers (not shown here). With further increased angle, the turbulent separation starts from the trailing edge and moves upstream (see fig. 4). At this point the pressure-induced normal force in fig. 2 deviates from the linear distribution. The growth of the separated area is what limits the maximum normal force (and thereby the maximum lift) in this mixed stall behaviour, that appears at all three Reynolds numbers. For higher Reynolds numbers the maximum normal force coefficient is larger (not shown here). After the location of the turbulent separation has moved to upstream to about 75% of the chord the flow changes for higher angles to a rather unsteady character revealed by the quick tufts and PIV. The turbulent separation then moves back and forth in between 25% and 75% of the chord. The full airfoil stall is then obtained for even higher alphas by the burst of the separation bubble, which takes place earlier (lower angles) for lower Reynolds numbers (not shown here). Here, a significant hysteresis is measured in the normal force. That is, leading-edge stall occurs at larger angles for increasing alpha whereas it occurs at smaller angle for decreasing alpha. Several owl-eyes-called patterns fill the area of turbulent separation depicted in fig. 8. These structures remain uniform and even in number as long as the separated area is small but they change into an uneven low number in the unsteady phase mentioned above and strongly interact with the tunnel side-wall boundary layers. This interaction has been observed in many wind tunnels. It impairs the two-dimensional character of the flow and thereby affects the measurements – a problem, that will be investigated and hopefully reduced by tangentially blowing air into the junction of airfoil and side wall during future experimental work.

The normal forces yielded by the two numerical methods for the 0.7 million Reynolds number case show the same slope up to angles of attack of 10° (fig. 2). For the computations run with ISM methodology and $\alpha \leq 10°$ position and size of the laminar separation bubbles match the experimental data resolved by oil and pressure measurements very well (fig. 3). Pressure distributions of ISM computations for $\alpha \leq 10°$ are also in good agreement with the experiment for the whole surface (not shown here). The ISM computation at $\alpha = 12°$ had to be performed in the time-accurate mode with rather small time steps $\Delta t(\text{chord}/U_\infty) = 0.25\%$ in order to resolve the fluid motion in space and time. However, this way an almost steady pressure distribution was obtained as shown in figure 6. Note that the fine mesh with 82559 nodes gave almost identical results as the medium mesh with 46462 nodes. The computations at 12° with DLR methodology did not yield a separation bubble (fig. 6 and 10). This may be caused by the much coarser grid and by the fact, that transition onset is forced at laminar separations. The experimentally viewed unsteady phase around 12° and the onset of leading-edge stall is not rendered correctly by any of the numerical setups yet (fig. 2). For higher angles both DLR and ISM computations could only be carried out to convergence using the time accurate mode of the RANS solvers. These DLR computations at high angles show irregular oscillations of the normal force [12]. In the computations with the ISM method for $\alpha > 10°$ the chosen Spalart-Allmaras (SA) turbulence model underpredicts the turbulent separation (figs. 13, 14), which leads to a higher suction peak at the nose, and larger overall suction on the upper side (fig. 6). The DLR computation for 12° does not show any turbulent separation either (not shown here). This reveals the necessity to identify a turbulence model that resembles the experimentally observed flow correctly. First improvements could be found with the SALSA turbulence model [15] (fig. 4) but more potential is expected from currently tested, more advanced second-moment closure Reynolds stress models (RSM), which in addition deliver the turbulent shear stresses directly comparable to the particle image measurements. It may appear, that the unsteady behaviour of the laminar separation bubble and the turbulent separation and finally the burst itself can only be captured if the temporal history of the amplification rates is taken into account by a time-accurate formulation of the e^N method. This is an extension of the numerical method that will be attempted in future work.

Acknowledgment

The numerical and experimental investigations were partly founded by the NRC-Helmholtz Collaborative Research Program and by Airbus Deutschland GmbH. This support is thankfully acknowledged by the authors.

References

[1] Althaus, D.: "Niedriggeschwindigkeitsprofile", Vieweg, Braunschweig, 1996.

[2] Baragona, M., Boermanns, L. M. M., van Tooren M. J. L., Bijl H., Beukers A.: "Bubble Bursting and Stall Hysteresis on Single-Slotted Flap High Lift Configuration", AIAA Journal, Vol. 41 (2003), No. 7, pp. 1230-1237.

[3] van den Berg, B.: "Reynolds number and Mach number effects on the maximum lift and stalling characteristics of wings at low speeds", NLR TR 69025 U, Amsterdam, Netherlands 1969.

[4] Bragg, M. B., Heinrich, D. C., Balow, F.A.: "Flow Oscillation over an Airfoil Near Stall", AIAA Journal, Vol. 34 (1994), No. 1, pp. 199-201.

[5] Broeren, A. P., Bragg, M. B.: "Flowfield Measurements over an Airfoil During Natural Low-Frequency Oscillations near Stall", AIAA Journal, Vol. 37 (1999), No. 1, pp. 130-132.

[6] Broeren, A. P., Bragg, M. B.: "Spanwise Variation in the Unsteady Stalling Flowfields of Two-Dimensional Airfoil Models", AIAA Journal, Vol. 39 (2001), No. 9, pp. 1641-1651.

[7] DLR, Institute of Aerodynamics and Flow Technology: "TAU-Code User Guide Revision: 1.24, Release 2004.1.0", April 5, 2004.

[8] Drela, M.: "Two-Dimensional Transonic Aerodynamic Design and Analysis Using the Euler Equations, Ph.D. Dissertation, Dept. of Aeronautics and Astronautics, Massachusetts Inst. Of Technology, Cambridge, MA, 1985.

[9] Drela, M.: "XFOIL: An analysis and design system for low Reynolds number airfoils" in: Mueller, T.J. (Ed.): "Low Reynolds number aerodynamics", Springer (Lecture notes in engineering), 1989, pp. 1-12.

[10] Gleyzes, C., Capbern P.: "Experimental study of two AIRBUS/ONERA airfoils in near stall conditions. Part I: Boundary layers", Aerospace Sience and Technology 7 (2003), pp. 439-449.

[11] Grote, A., Ortmanns J., Radespiel, R.: "Entwurf eines Höhenleitwerkprofils mit Leading-Edge Stall und Konstruktion eines Windkanalmodells mit Ruder für die Messung der Profileigenschaften", Institutsbericht 2003/2, Institut für Strömungsmechanik, TU Braunschweig, 2003.

[12] Krumbein, A.: "Navier-Stokes Airfoil Computations with Automatic Transition Prediction using the DLR TAU code – A Sensitvity Study", 14th Symposium of STAB, Bremen 2004.

[13] Krumbein, A.: "Automatic Transition Prediction and Application to High-Lift Multi-Element Configurations", AIAA-2004-2543, 34th AIAA Fluid Dynamics Conference, Portland, Oregon, USA, June/July 2004 (accepted for publication in Journal of Aircraft).

[14] Mack, L. M.: "Transition and Laminar Instability", Jet Propulsion Laboratory Publication 77-15, Pasadena, CA, 1977.

[15] Rung, T., Bunge, U. ,Schatz, M., Thiele, F.: "Restatement of the Spalart-Allmaras Eddy-Viscosity Model in Strain-Adaptive Formulation", AIAA Journal, Vol. 41 (2003), No. 7, pp. 1396-1399.

[16] Schewe, G.: "Reynolds-number effects in flow around more-or-less bluff bodies", Journal of Wind Engineering and Industrial Aerodynamics 89 (2001), pp. 1267-1289.

[17] Schmidt, G. S., Mueller, T. J.: "Analysis of Low Reynolds Number Separation Bubbles Using Semiempirical Methods", AIAA Journal, Vol. 27 (1989), No. 8, pp. 993-1001.

[18] Schrauf, G.: "COAST3 – A compressible stability code. User's guide and tutorial", Daimler-Benz Aerospace Airbus GmbH, Bremen, Technical Report EF 040/98, 1998.

[19] Stock, H. W., Degenhart, E.: "A simplified e^n method for transition prediction in two-dimensional, incompressible boundary layers", Z. Flugwissenschaft Weltraumforschung 13, 1989.

Table 1 Grids

grid	structured area				hybrid area
	resolution tangential x normal	1st spacing	height	nodes	nodes
ISM-coarse	288 x 48	$2.1x10^{-5}$	0.012	11759	17832
ISM-medium	576 x 96	$1.05x10^{-5}$	0.012	46462	55535
ISM-fine	768 x 128	$7.875x10^{-6}$	0.012	82559	93195
DLR	256 x 73	$2.5x10^{-5}$	up to farfield	32466	structured

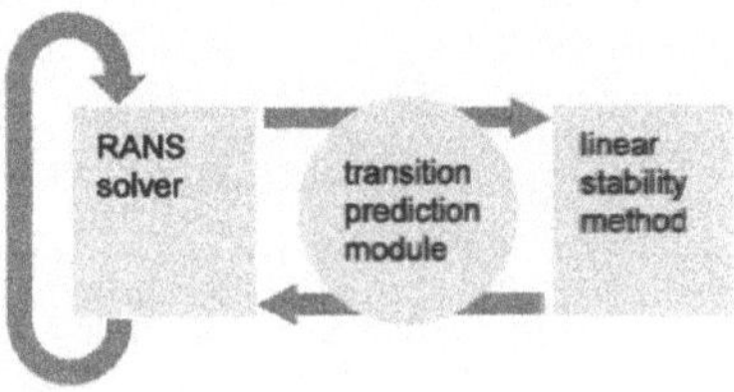

Figure 1 Coupled program system

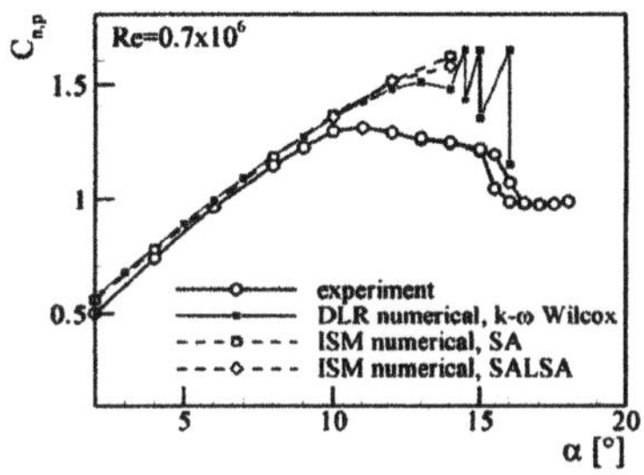

Figure 2 $C_{n,p}$

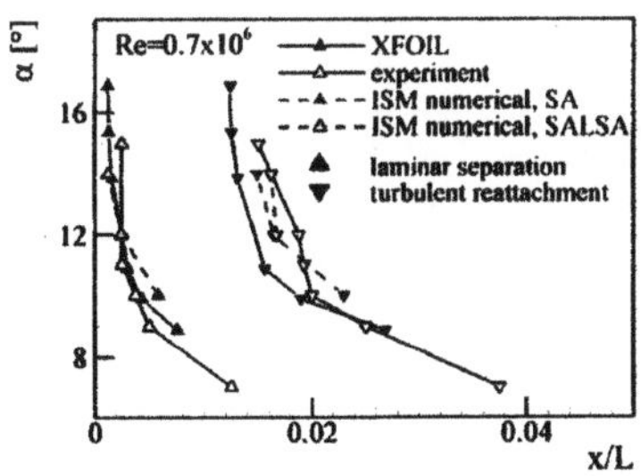

Figure 3 Laminar separation bubble

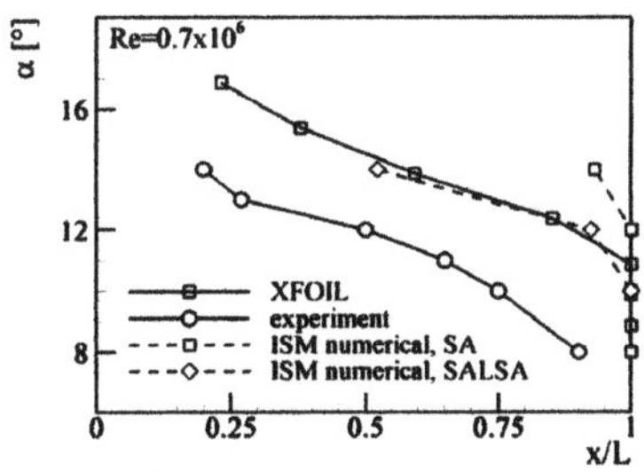

Figure 4 Turbulent separation

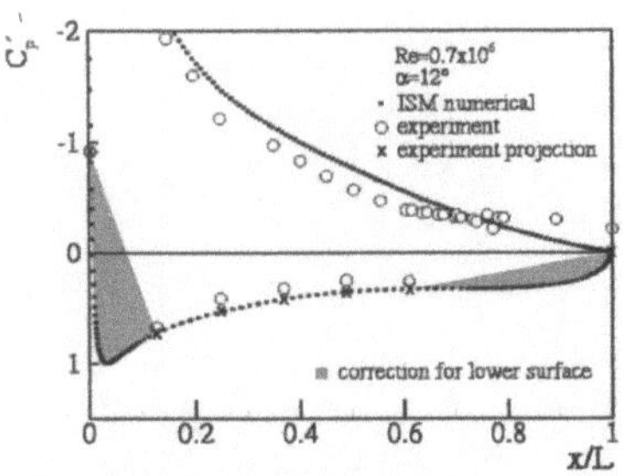

Figure 5 Pressure distribution correction of $C_{n,p}$

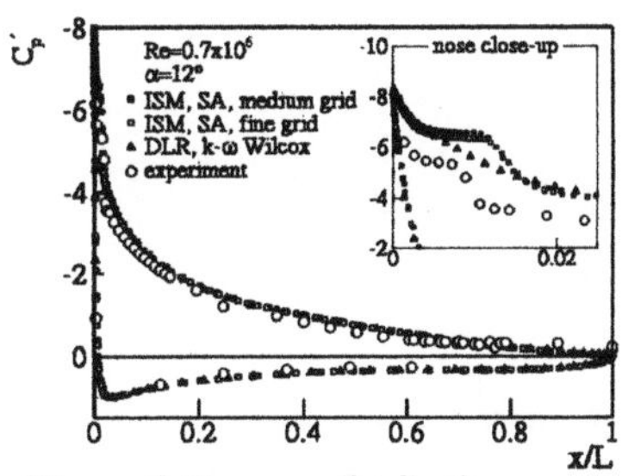

Figure 6 Pressure distribution

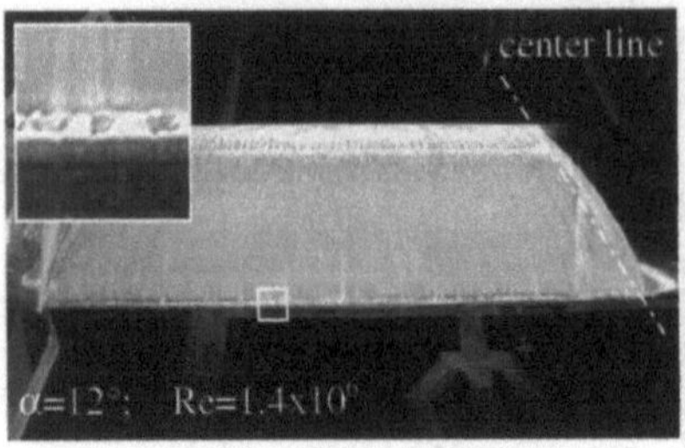

Figure 7 Oil flow visualization front view

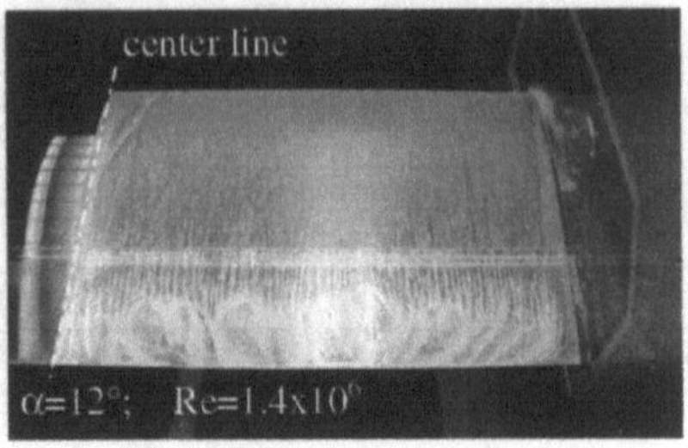

Figure 8 Oil flow visualization rear view

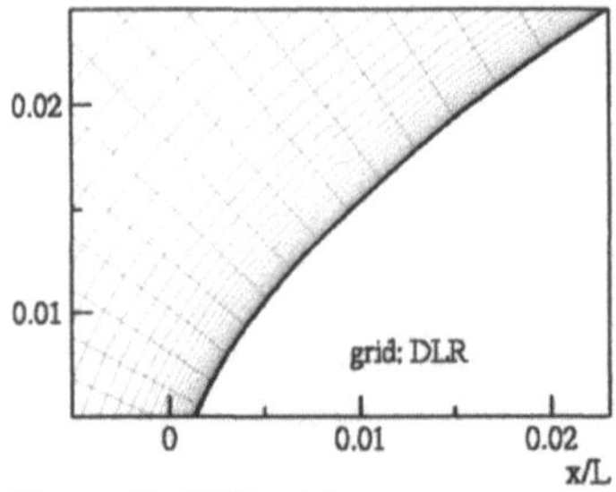

Figure 9 DLR-grid nose region

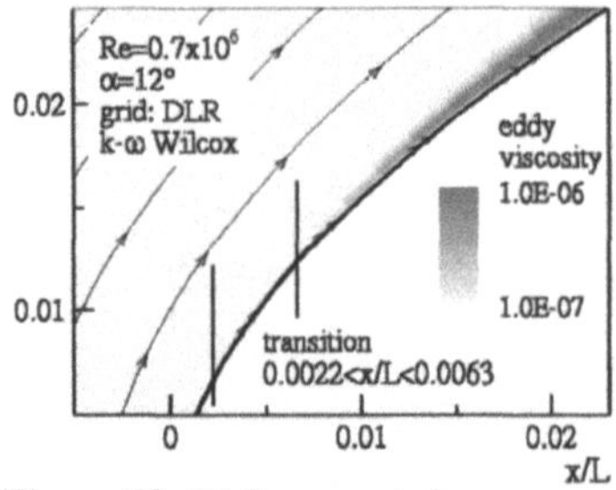

Figure 10 DLR-computation nose region

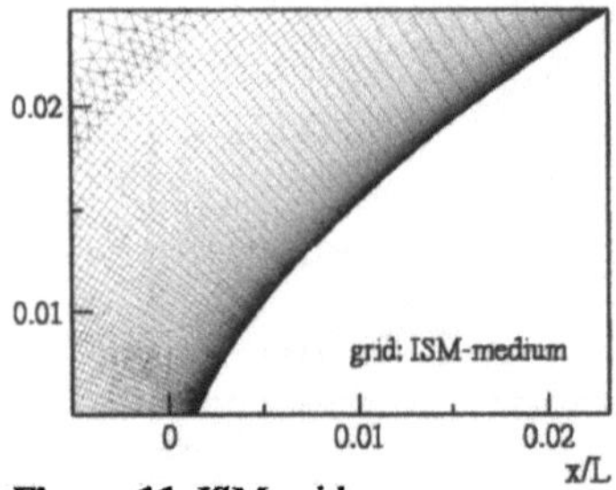

Figure 11 ISM-grid nose region

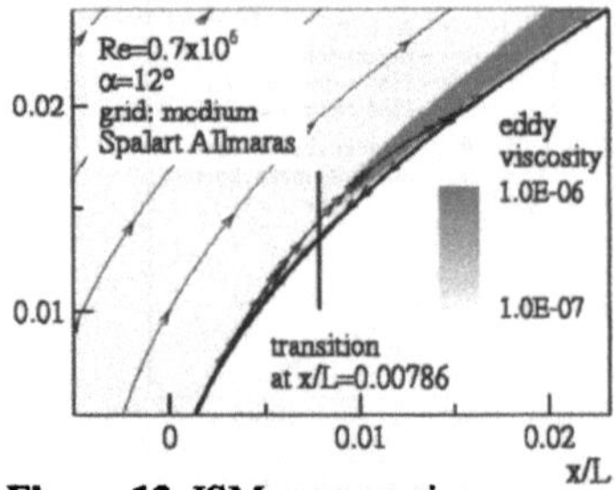

Figure 12 ISM-computation nose region

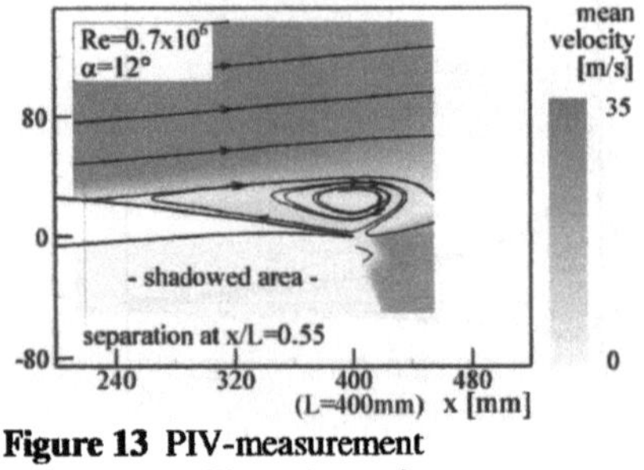

Figure 13 PIV-measurement trailing edge region

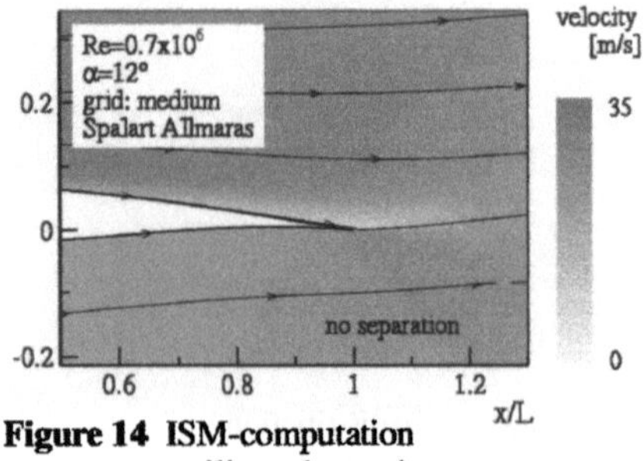

Figure 14 ISM-computation trailing edge region

Experimental and Numerical Investigations of Flow Separation and Transition to Turbulence in an Axisymmetric Diffuser

L. Hoefener[1], W. Nitsche[1], A. Carnarius[2], and F. Thiele[2]

[1] Technical University Berlin, Institute for Aeronautics and Astronautics (ILR)
Marchstr. 12-14, 10587 Berlin, Germany
*Phone: *49-30-314-22954, Fax: *49-30-314-22955*
lars.hoefener@tu-berlin.de,

[2] Technical University Berlin, Hermann-Föttinger-Institute of Fluid Mechanics (HFI)
Müller-Breslau-Str. 8, 10623 Berlin, Germany
*Phone: *49-30-314-26283, Fax: *49-30-314-21101*
angelo.carnarius@cfd.tu-berlin.de

Summary

Flow separation and transition to turbulence in a smooth axisymmetric diffuser at $Re_{D1} \approx 7800$ were investigated both numerically and experimentally.

The inlet flow is an incompletely developed laminar pipe flow with a typical boundary layer thickness ($\delta_{99}/D_1 \approx 0.3$). The smooth diffuser contour causes a pressure-induced laminar separation. Due to the inflection point within the shear layer, instabilities cause a transition of the separated laminar flow. Further downstream, the flow reattaches turbulent and recovers slowly to a turbulent equilibrium boundary layer. Periodic disturbances are introduced upstream of the separation point in order to control the breakdown of the separated shear layer. In the present study, two different perturbation modes are tested experimentally and compared in detail with numerical investigations.

1 Introduction

In many aerodynamical applications, e.g. laminar wing design or optimization of high-lift devices, laminar-turbulent transition and separation play an important role. The mechanisms contributing to transitional separation bubbles are not yet fully understood.

Laminar separation bubbles usually develop in laminar boundary layer flows with a sufficiently adverse pressure gradient. According to the linear stability theory[3], the separated laminar flow is unstable with respect to small disturbances due to the inflection point of the velocity profile within the separation region. In the separation bubble, so-called Tollmien-Schlichting waves (TS-instability) of the boundary layer interact with the shear-layer instability[5] and nonlinear interactions of the instabilities lead to transition[1]. The onset of transition causes an increased mixing and momentum transfer within the shear layer, which is associated with a massive increase in the boundary layer thickness. Finally, the momentum transfer towards the wall causes the turbulent flow to reattach. Further downstream, the flow recovers slowly to a turbulent equilibrium boundary layer.

For all presented results, the velocity profile at the diffuser inlet is that of an incompletely developed laminar pipe flow, i. e. the flow has a laminar boundary layer shape ($\delta_{99}/D_1 \approx 0.3$, $Re_{D_1} \approx 7800$). The smooth diffuser contour with a maximum opening angle of approximately 9 and an expansion ratio D_2/D_1 of 1.6 generates a widespread pressure-induced laminar separation bubble close to $X/D_1 = 0$ (see Fig. 1), which covers the entire test-section under investigation. Instabilities grow within the shear layer causing transition several inlet diameters downstream.

Artificial disturbances were introduced at $X/D_1 = -3$ using a membrane actuator in order to study the influence of different perturbation modes. These perturbations increase the instabilities within the separated shear layer, leading to a controlled transition and a turbulent reattachment.

2 Setup

2.1 Experimental setup

The experimental investigations were carried out in a closed-circuit cylindrical pipe facility at the ILR. The inlet diameter is $D_1 = 0.05\,\text{m}$ and the outlet diameter $D_2 = 0.08\,\text{m}$. The test section wall, as well as the pipes and the diffuser, are made of Plexiglas allowing full optical access. The working fluid is a specific oil, whose refraction index matches that of the Plexiglas at a temperature of T=23°C. This allows for interference-free measurements using laser optical methods. Non-intrusive Laser Doppler Velocimetry (LDV) and Particle Image Velocimetry (PIV) were used for the measurements. The two-component LDV was mounted on a three-way traverse unit which allowed precise positioning of the LDV for detailed point-measurements. In contrast, the PIV-system enables velocity field measurements in the light sheet plane. Additionally, wall shear stress fluctuation levels were measured using surface hot-film and surface hot-wire anemometry.

Due to the limited efficiency of the slit actuator developed earlier [4], a new membrane actuator was designed to generate the perturbations, see Fig. 2 a. This actuator is built up of a flush mounted rubber membrane and four separate air chambers in the circumferential direction. The oscillation of the rubber membrane is realized pneumatically using pressure tubes and over- and under-pressure pumps. The increased efficiency causes a reduction of the perturbation momentum and prevents the unwanted forcing of multiple modes. Recently, the actuator has been improved as can be seen in Fig. 2 b. The highly elastic rubber membrane is now driven by four powerful solenoids via a connecting rod. This setup allows for the realization of several perturbation modes and amplitudes, including azimuthal phase shifts. Therefore, this actuator has been used for most of the experiments.

To estimate the perturbation parameter C_μ, the membrane deflection was measured at two cuts using a laser vibrometer. The membrane movement was then approximated in a computer simulation to obtain the change of volume with respect to time. With this, the mean perturbation parameter was calculated as follows

$$C_\mu = \frac{A_p u_p^2}{A_{in} u_b^2} = 1.2 \cdot 10^{-4} \tag{1}$$

In the above equation, A_p is the membrane-surface in undeflected position, A_{in} the inlet face, u_p the maximum perturbation velocity and u_b the average inlet velocity. It should be noted, that the calculated C_μ might differ slightly from the real value, as the perturbation velocity of the membrane is difficult to determine.

2.2 Numerical setup

Large Eddy Simulation was used for the numerical computations. With this technique the large turbulent scales are solved directly, whereas the small scales (or subgrid scales), which contribute only weakly to the turbulent spectrum and are of more universal character, are modeled. In the current study, all computations have been performed with the dynamic one-equation subgrid scale model [2].

The block-structured computational mesh, which has approximately 1.2 million grid points. In the y-z-plane it consists of an octagonal inner block which is surrounded by an outer block that reproduces the diffuser contour. In the y-z-plane the grid is equidistant. Previous investigations have shown that when using a velocity profile, which was approximated by an analytical function or was taken from the experiment, a certain inlet length was needed for the flow to adopt to the behavior of the numerical scheme. To avoid this, the inlet velocity profile was extracted from the simulation of a steady laminar pipe inlet flow.

The periodic perturbations with a frequency of $f = 30\,\text{Hz}$ (see Section 3.2) have been realized by prescribing a wall-normal velocity component at $X/D_1 = -3$. Only results for the anti-phase perturbation are presented here. For this case, the wall-normal velocity component v_r was given as

$$\frac{v_r(t,\phi)}{v_{r,0}} = \sin(2\pi f t)\sin(2\phi)\,. \tag{2}$$

The first term on the right hand side represents the periodicity in time, while the second term gives the phase shift in the circumferential direction ϕ. In test computations, the actuator membranes used in the experiment were modeled by moving boundaries, but the results did not differ significantly. Obviously, the *quantity* of introduced momentum is more important than the *device* used to generate the perturbations.

At the remaining walls the no-slip condition was used and a convective outflow boundary condition was employed at the outlet. Every perturbation period was resolved with approximately 330 time steps resulting in a time step of $\Delta t = 1 \cdot 10^{-4}$ s.

3 Results

3.1 Baseflow

All measurements were performed at a Reynolds number $Re_{D1} \approx 7800$ based on the inlet diameter $D_1 = 0.05\,\mathrm{m}$ and the bulk velocity u_{bulk}. At $X/D_1 = 0$, the flow is laminar. The diverging diffuser contour generates a pressure-induced laminar separation close to the leading edge of the diffuser.

Comparing numerical and experimental results of the baseflow, Fig. 5 shows a very good agreement of axial velocity. The development of the extremely extended separation bubble is clearly reproduced. Initially, the core flow remains almost unaffected. Further downstream of the diffuser, instabilities grow at the edge of the shear layer. These instabilities cause a massive increase in the rms-values of the velocity fluctuations along the shear layer. Fast, high-energy core fluid is moved laterally towards the wall, as low speed fluid originating from the boundary-layer is fed to the center. Finally, the momentum transfer leads to a turbulent break-down of the separated laminar boundary layer, followed by a turbulent reattachment.

3.2 Perturbed flow

To study the influence of controlled perturbations on separation and transition, investigations using two different actuation regimes have been performed. After an experimental analysis of the baseflow, the instability range of the Tollmien-Schlichting (TS) waves was identified (Fig. 3, notice some background noise). Therefore, the perturbation frequency chosen was well within the instability band according to a Strouhal-number $St = \frac{fD_1}{u_b} \approx 0.35$. The artificial disturbances were introduced at $X/D_1 = -3$ by means of the membrane actuator. For all investigations, the perturbation amplitude of the actuator was fixed at a dimensionless perturbation parameter of $C_\mu = 1.2 \cdot 10^{-4}$. In the following, results are shown for in-phase perturbation (all membranes are deflected to the same direction at a time) as well as for anti-phase perturbation (opposing membranes are deflected to the same direction).

In-phase perturbation: For the first case presented (in-phase mode $m = 0$), only experimental data are available. The disturbances forced into the laminar boundary layer cause a growth of the TS instabilities. Fig. 4 shows the amplitude profiles of the rms-values of the axial velocity fluctuations u_{rms} close to the separation point, which have the typical shape including two maxima and one minimum divided by a phase drop. The amplified instability waves grow in the disturbed laminar boundary layer in the downstream direction. The pattern of the resulting wave trains of the growing instabilities are depicted in Fig. 6. As the transition develops, the maximum amplification increases, leading to an increase and spreading of the peak in the rms-profiles.

Anti-phase perturbation: In the upper part of Fig. 7, measurements of the axial velocity for the anti-phase perturbation are compared to numerical results. Up to a position of $X/D_1 \approx 4$, the results agree very well. The separation and the beginning development of the separation bubble are well reproduced by the numerical simulation. Further downstream, the profiles start to diverge slightly, which leads to the assumption that the reattachment points do not match exactly.

A more substantial discrepancy becomes apparent in the rms-values of the axial velocity fluctuations, presented in the lower part of Fig. 7. In the inlet region, the u_{rms}-level of the experiment exceeds that of the numerical computation. This is a consequence of the higher turbulence level in the base flow during the measurements. The development of the rms-peak also differs. In the simulation, a dominating peak can already be observed at $X/D_1 = 0$. In the measurements, this is probably masked by the higher background turbulence level of the flow. At $X/D_1 = 4$ and $X/D_1 = 5$, the location of the dominating peak matches quite well. However, the peak has already started to move towards the axis in the last section of the simulation, while it stays nearly stationary in the measurements. This indicates a more rapid development of transition in the numerical simulation. This might be caused by a slightly differing perturbation parameter C_μ, as this quantity is very difficult to accurately determine from the experiment (see Section 2.1). A solution to this problem needs to be found in subsequent studies.

Comparison: In order to identify differences between both actuation regimes, sections of the axial velocity component u and the u_{rms}-values obtained from LDV measurements are presented in Fig. 8. Comparing the in-phase and anti-phase mode, it can be observed that the recirculation area for the anti-phase case is more extended. The mean flow profiles at $X/D_1 = 6$ still show a small reverse flow zone, whereas the flow is already attached for the in-phase mode. Consequently, the maximum velocity at the center is lower for the in-phase mode. The outer maximum close to the wall grows more rapidly and the high-fluctuation flow already reaches the center at $X/D_1 = 5$. At the farthest downstream section, the fluctuation maximum for the in-phase mode exceeds that of the anti-phase mode in the high-shear zone as well as at the center.

4 Conclusion

In the present paper, the results of a joint experimental and numerical investigation of the transition of a pipe inlet flow in an axisymmetric diffuser have been presented. In order to control the transition over the laminar separation bubble, artificial perturbations were introduced upstream of the diffuser. The base flow, as well as two different perturbation modes were analysed in detail for the present study.

The base flow exhibits an extended laminar separation bubble and an almost unaffected core flow. In this case, the transition to turbulence occurs far downstream of the diffuser. Excellent agreement between measurements and LES were achieved for this configuration. From the measurements of the base flow, an instability band at $f = 30...40\,\text{Hz}$ could be identified. The perturbation frequency for the transition control was therefore set to $f = 30\,\text{Hz}$.

Typical wave trains could be seen in the pressure measurements for the perturbed flow. The rms-values of the axial velocity fluctuations show the typical form close to the separation point and spread to a wedge-shaped distribution further downstream. This is caused by the amplification of the instabilities in the boundary layer.

Of the two investigated perturbation modes, the in-phase disturbances proved to be more effective in triggering transition, although the differences between both modes are small. A comparison of numerical and experimental results showed a good agreement among the velocity profiles but some deviations in the rms-values. Difficulties in determining the perturbation parameter C_μ were probably the reason for this discrepancy. Further investigations will have to be carried out in order to clarify this.

Acknowledgement

The research described herein was funded by the German National Science Foundation (Deutsche Forschungsgemeinschaft, DFG) under the umbrella of the Priority Program (Schwerpunktprogramm, SPP "Transition", Themenkreis V, TKV, "Strömungsphysikalische Modellbildung") at the Technical University Berlin. All simulations were carried out under a grant from the Norddeutscher Verbund für Hoch- und Höchstleistungsrechnen (HLRN) on an IBM pSeries 690 system.

References

[1] Alam, M. and Sandham, N.D., "Direct numerical simulation of 'short' laminar separation bubbles with turbulent reattachment", *Journal of Fluid Mechanics*, Vol. 403, pp. 223-250, 2000.

[2] Davidson, L., "LES of Recirculation Flow Without Any Homogeneous Direction: A Dynamic One-Equation Subgrid Model", *2nd Symposium on Turbulent Heat and Mass Transfer*, K. Hanjalić and T. Peeters (Hrsg.), Delft University Press, pp. 481-490, 1997

[3] Dovgal, A.V., Kozlov, V.V. and Michalke, A. "Laminar boundary layer separation: instability and associated phenomena", *Prog. Aerospace Sci.*, Vol. 30, pp. 61-94, 1994

[4] Martin, A., Thiele, F., Hoefener, L. and Nitsche, W. "Investigation of Flow Separation and Transition to Turbulence in an Axisymmetric Diffuser", *AIAA 2004-1102*, 2004

[5] Maucher, U., Rist, U. and Wagner, S. "Transitional structures in a laminar separation bubble", *Notes on Numerical Fluid Dynamics*, Vol. 72, pp. 307-314, Vieweg-Verlag, 1999

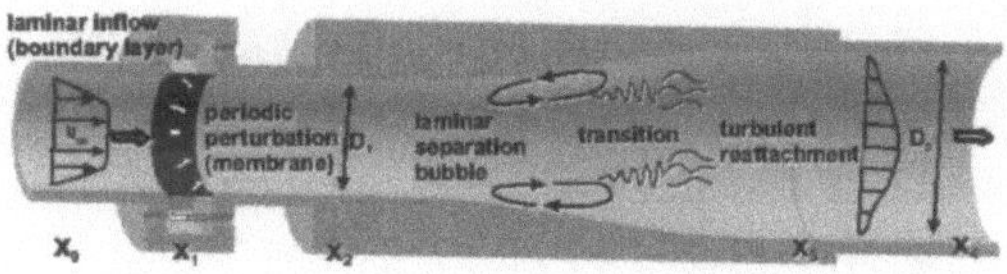

Figure 1 Schematic depiction of the model geometry

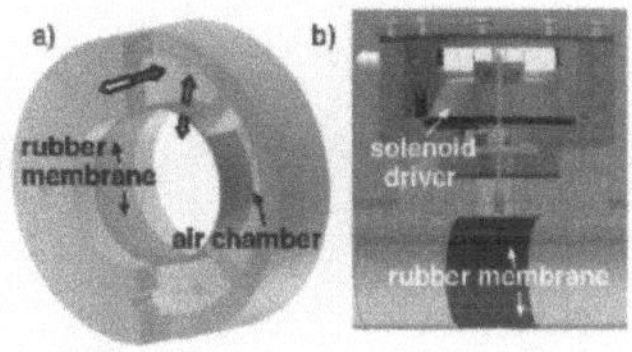

Figure 2 Actuator with rubber membrane: a) pneumatically driven, b) solenoid driven

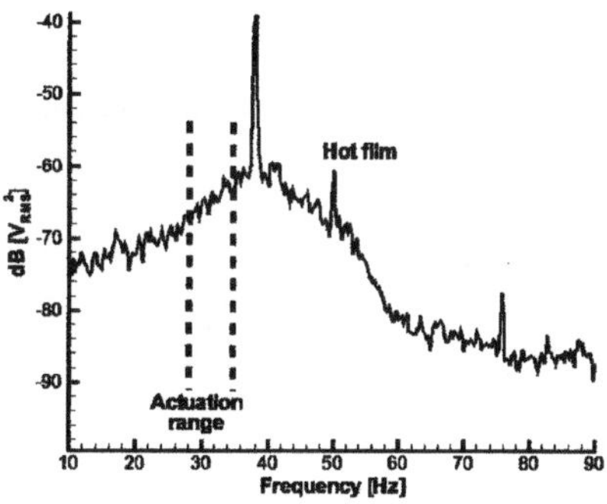

Figure 3 Energy spectrum, hot film upstream of diffuser

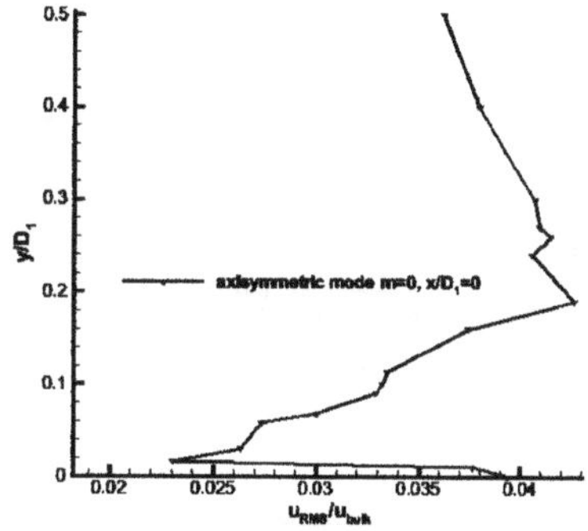

Figure 4 Amplitude profiles: rms-values of axial velocity fluctuations u_{rms}, experiment

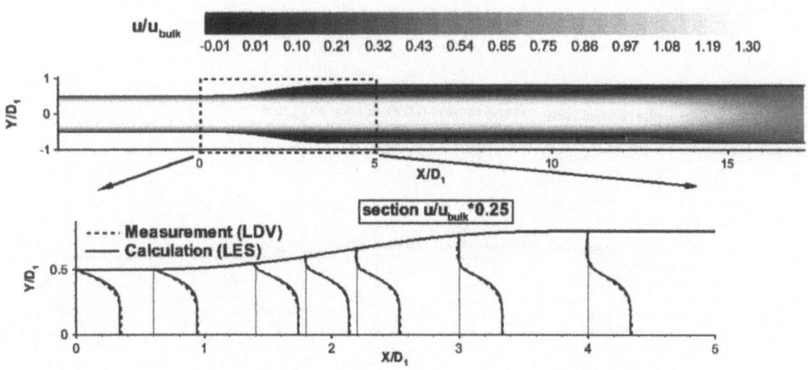

Figure 5 Baseflow: contour plot and sections of the axial velocity, comparison of experiment and LES

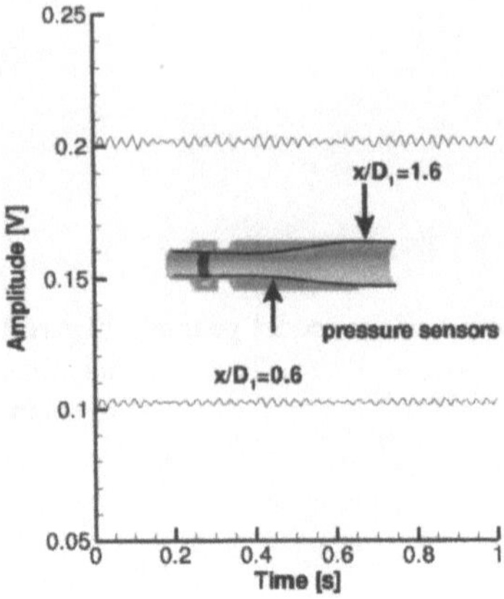

Figure 6 Time traces of pressure sensors at $X/D_1 = 0.6, 1.6$, typical TS wave train

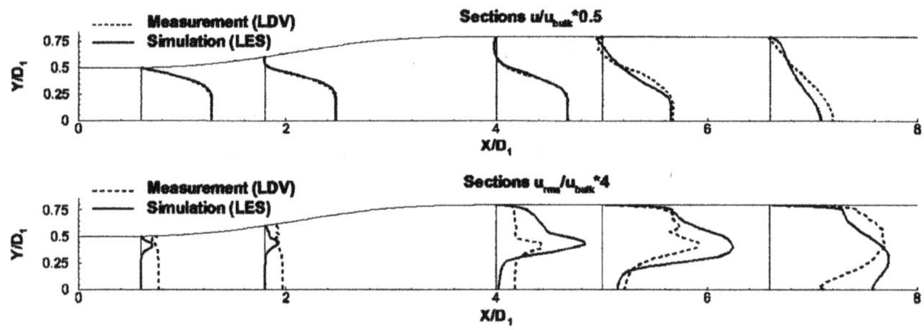

Figure 7 Profiles of normalized axial velocity and normalized u_{rms}-values for the anti-phase perturbation

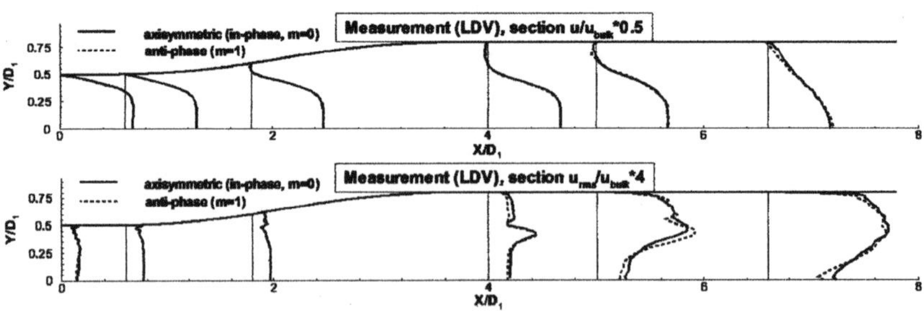

Figure 8 Comparison of two perturbation modes: sections of axial velocity and u_{rms}-values

In-Flight and Wind Tunnel Investigations of Instabilities on a Laminar Wing Glove

I. Peltzer and W. Nitsche

Technical University of Berlin, Institute of Aeronautics and Astronautics
Marchstr. 12, Sekr. F2, D-10587 Berlin, Germany
Inken.Peltzer@tu-berlin.de
http://Aero.ILR.TU-Berlin.DE

Summary

Measurements of the temporal and spatial development of natural and controlled boundary layer instabilities in the linear and weakly nonlinear stage of transition were carried out using a laminar wing glove in-flight and in a wind tunnel. An 74 piezo-sensor-array and a spanwise hot-wire array together with several independent point sources for the controlled experiments were used. Depending on the excitation case, typical structures which characterize fundamental and the oblique breakdown were observed. The results of natural transition show two-dimensional Tollmien-Schlichting waves in the linear stage as well as three-dimensional wave packets in the beginning of nonlinear stage, and were observed during flight and in the wind tunnel. Comparing in-flight and wind tunnel measurements, the disturbance structures are qualitatively similar, but the amplification of 3D instabilities occurs later and more rapidly in the wind tunnel.

1 Introduction

The experimental investigations described in this paper deal with the laminar turbulent boundary layer transition on an airfoil. Transition occurs due to the boundary layer instability. Even minimal disturbances can be frequency-selectively amplified, in accordance to the external boundary conditions (pressure gradient, temperature gradient, etc.), thus leading to transition. This situation is very different in a wind tunnel or during flight. Therefore, comparative in-flight and wind tunnel measurements of the temporal and spatial propagation of instabilities on a laminar wing glove were carried out. Arrays of different types of high-resolution surface sensors (piezo sensors, surface hot-wires) were employed.

Based on the experience gained in the Deutsche Forschungsgemeinschaft -funded university group research project (see e.g. [1]) measurements on controlled transition with several spanwise distributed point sources, as well as on natural transition, were carried out. The subsequent explanations focus on the measurements to investigate the spanwise distribution of instabilities from their early linear to their weakly nonlinear amplification stage.

2 Experimental Set-up

2.1 Laminar wing glove

A laminar wing glove was used for the in-flight measurements. This tool was developed and certified for the Grob G103 TWIN II two-seater sailplane at the Berlin Institute for Aeronautics and Astronautics. Figure 1 shows the sailplane equipped with the measuring glove. The measuring glove has a $1.0\,m$ span and a chord length of $1.22\,m$. A Prandtl tube, which is attached underneath the glove, was used to obtain the freestream velocity. A thermocouple and a manometer are used to measure the air temperature and absolute pressure, respectively. These parameters are continuously recorded to obtain the freestream boundary conditions. The in-flight tests were carried out at flow velocities of 23 m/s to 28 m/s ($Re_c \approx 2 \cdot 10^6 - 3 \cdot 10^6$). The wind tunnel experiments with the same glove were conducted in the laminar wind tunnel at Stuttgart University . This wind tunnel has a very low turbulence level ($Tu = 1.2 \cdot 10^{-4} - 5 \cdot 10^{-4}$) and is therefore suitable for investigations on boundary layer transition [2].

2.2 Sensors and Sensor Positions

For these experiments the glove was equipped with a 74 piezo-sensor-array (fig. 1) and with 16 surface hot-wire sensors arranged spanwise at 33 % of the chord length (not shown). The spanwise distance between the piezo-sensors and the surface hot-wire sensors was seven millimeters. In addition, a spanwise row of seven harmonic point sources was mounted at 20 % of the chord length (spanwise distance of 24 mm) for investigations on controlled transition (fig. 5b).

The surface hot-wire is a highly sensitive wall sensor that was developed for measuring the weakest of wall shear stress fluctuations. A hot wire (5μm diameter) is welded flush to the surface above a thin slot (0.075 mm to 0.1 mm) on a printed circuit board (copper-laminated). Modularly arranged hot-wire bridges were designed especially for in-flight measurements and are able to operate a high number of surface hot-wire sensors during flight in constant-temperature mode (CTA) [3].

The piezo sensor array is made from a 28 μm thin polarized plastic foil (Polyvinylidenfluoride, PVDF), with a metallic layer on both sides. The sensor material produces small electric charges due to pressure (piezo effect) and temperature (pyro effect) fluctuations. The signal-to-noise ratio can be enhanced using the pyro effect. A tiny temperature gradient between the glove and the flow was generated by a heating layer underneath the sensors. In the experiments, only the fluctuations (temperature, pressure) were measured. Detailed sketches of the sensors, the sensor positions and of the harmonic point sources are shown in [4]. A miniaturized measuring system, which was specially developed for in-flight measurements, was employed for the experiments. The system used is a multi-channel data acquisition system described by Suttan [5].

2.3 Boundary conditions

Basic investigations were performed in order to compare the wind tunnel and the in-flight measurements. The velocity distributions over the glove were measured for different freestream velocities in flight, as there is only one particular distribution for each freestream velocity. Then the velocity distribution with the appropriate freestream velocity of the in-flight measurement was reproduced in the wind tunnel. Fig. 2 shows the distributions of the in-flight and the wind tunnel measurement for one freestream velocity ($u = 23.6\,m/s$), which was the one used for most experiments. It can be seen, that the results are really similar, which allows for comparison of in-flight and wind tunnel measurements. Furthermore this agrees well with a calculation of the distribution using a common shared programm (XFoil) [6].

3 Controlled transition

Extensive investigations on controlled transition were carried out. For these experiments the point sources were arranged in a spanwise manner on the wing glove. Loudspeaker were positioned closely underneath the glove surface and introduced controlled disturbances via a circle of holes (each with a 0.2 mm diameter) into the boundary layer. The loudspeakers are controlled individually by a signal processor, which is capable of generating any type of disturbance waves. A point source generates disturbances consisting of a 2D plane wave and an infinite number of oblique wave trains (3D modes). For future investigations dealing with artificially generated disturbances similar to 'natural' disturbances, a point source should be used.

In the most simple case, all disturbance sources were activated in phase with a frequency of 600 Hz. Previous experiments showed that the maximally amplified instabilities were at this frequency. Subsequently, a mixture of signals consisting of two, three and five frequency components were generated. The adjacent point sources were driven with equiphase and with antiphase.

Results from tests using controlled monofrequency-equiphase and antiphase disturbances are shown in fig. 3 for a section of signals from surface hot-wires at 33% cord length, and for a section of the piezo sensor signals. Three sources are located upstream from where the shown signals were measured. Spanwise contiguous 2D-dominated wave trains are observed in the case of equiphase excitation (see contour plots of time traces of hot-wires signals, fig. 3a+b), upper row). The slight waviness of the wave trains is due to the excitation with the harmonic point sources. In contrast, oblique wave trains and the evolution of individual cells downstream of the disturbance sources can be observed in the antiphase case. These are typical structures and were observed in all cases investigated. Looking at the RMS-values of the piezo signals (fig. 3c+d), lower row), in the equiphase case there are relatively small areas with high RMS-values (meaning highly amplified instabilities) downstream of the point sources. On the other side in the antiphase case there are areas of high RMS-values between the upstream point sources.

These observations were also made for the cases of multifrequency excitation. For example, fig. 4 shows the contour plots obtained from the piezo sensor measurement for several cases of disturbances generated by multiple frequencies with opposite phase. Unlike the other figures, fig. 4 illustrates the total piezo sensor array. Basically, the same distribution of RMS-values can be seen. Areas of highest fluctuations are between of the spanwise-arranged, upstream located point sources. The difference between the plots is characterized by the transition moving downstream. According to previous investigations, and to the stability analysis from Stemmer [7] the amplification is highest of a frequency of 600 Hz. Increasing the number of frequencies that participate the generated signal, the amplification of instabilities occurs later, hence the transition moves downstream, as can be seen in fig. 5a).

In general, it has been found that equiphase-generated disturbances lead to fundamental breakdown. In the antiphase case, structures which are typical of oblique breakdown have formed. This is confirmed by comparing a numerical study from Stemmer and measurements using the glove with Stereo-Particle-Image-Velocimetry according to Schröder [8].

4 Natural instabilities measured in-flight and in the wind tunnel

Next, results obtained from natural transition experiments during flight and in the wind tunnel are discussed. Fig. 6 illustrates contour plots of the signals of each spanwise piezo sensor row for in-flight and wind tunnel measurements. The amplification of instabilities can be seen in both cases by comparing subsequent rows. Furthermore, the first row ($x/c = 0.474$) shows similar amplification of almost 2D-Tollmien-Schlichting(TS) wave trains, by indicating the equal phase of the spanwise-adjacent sensors.

Locally limited wave trains with small oblique angles appear in the second row only during in-flight measurements. The 2D structures dominate the second row, as well as the first row during wind tunnel experiments. Obviously, the boundary layer is more receptive for 3D modes during flight than in the wind tunnel. In the wind tunnel, 2D waves still dominate in the third row, but the amplitudes of the fluctuations are higher than in flight. The angle of the oblique wave trains, as well as the number of individual spatial independent wave packets, rise with increasing chord length. The dimension of an individual wave packet decreases as expected, however. In the last row only individual wave packets occur in the in-flight and in the wind tunnel measurements, which means that the 3D-portions dominate the local disturbance structures. This illustrates the beginning nonlinear stage of amplification. The amplitudes of the fluctuations in the wind tunnel measurement are more uniformly distributed, however. Consequently, the RMS-values in downstream direction are smaller in the wind tunnel and they increase later and more rapidly than in the in-flight measurement. Never the less, the fluctuations reach the highest RMS-values at the same chord length in both cases (fig. 7).

Due to the wind tunnel walls, the filter pads, the screens and the huge contraction rate (100:1) in the laminar wind tunnel in Stuttgart, it seems that the flow in this wind tunnel is more parallel, two-dimensional and has fewer freestream disturbances than the in-flight flow. In the free atmospheric flow (at altitudes from 200 m to 2000 m) temporary, locally limited natural disturbances occur which could not observed in the wind tunnel under laboratory conditions. Fig. 8 shows three charts of time traces from surface hot-wire sensors at 33% chord lenght. There are examples of natural disturbances due to temporary atmospheric fluctuations.

5 Conclusions

Temporal and spatial high-resolution measurements of natural and artificially excited instabilities over a laminar wing glove during flight and in the wind tunnel were carried out. For controlled transition, a spanwise row of independent point sources introduced mono- and multifrequency disturbance signals into the boundary layer. The adjacent point sources were activated in phase and with opposite phase. The results show that typical disturbance structures occurred due to excitation. An excitation with mono- and multifrequency equiphase signals leads to a fundamental breakdown, and the antiphase excitation generates structures which are characteristic of an oblique breakdown.

The measurements of natural transition show the development of TS waves from their early linear to the beginning of the nonlinear amplification stage. Comparing in-flight and wind tunnel measurements shows qualitatively-similar disturbance structures. The amplification of the instabilities occurs later and more rapidly in wind tunnel, however. Furthermore the flow in the laminar wind tunnel of Stuttgart seems to be more parallel due to the wind tunnel walls and screens. During flight, 3D instabilities were amplified earlier than in the wind tunnel experiments due to numerous temporary atmospheric disturbances at the altitudes where the in-flight experiments were made.

The results presented in this paper, bring this project, which was funded by the DFG, to it's conclusion. This research topic should, of course, be further investigated to answer open questions.

Acknowledgements

The financial support on this research project by the Deutsche Forschungsgemeinschaft (DFG), reference Ni 282/12-2, is gratefully acknowledged. Special thanks also go to the Institute of Aerodynamics and Gasdynamics, Stuttgart University, where the wind tunnel measurements were carried out.

References

[1] Nitsche, W., Suttan, J., Becker, S., Erb, P., Kloker, M., Stemmer, C.: Experimental and Numerical Investigations of Controlled Transition in Low-Speed Free Flight. Aerosp. Sci. Technol. **5** (2001) 245–255

[2] Herr, S., Würz, W., Wörner, A., Rist, U., Wagner, S., Ivanov, A., Kachanov, Y.: Systematic investigations of 3D acoustic receptivity with respect to steady and unsteady disturbances. Experiment and DNS. In S. Wagner, M. Kloker, U. Rist, ed.: Recent Results in Laminar-Turbulent Transition - Selected Numerical and Experimental Contributions from the DFG-Schwerpunktprogramm "Transition" in Germany (NNFM 86), Springer Verlag (2003) 75–90

[3] Baumann, M.: Aktive Dämpfung von Tollmien-Schlichting Wellen in einer Flügelgrenzschicht. In: Fortschritt-Berichte VDI,Reihe7,Nr.372, Düsseldorf, VDI Verlag (1999) Dissertation an der Technischen Universität Berlin.

[4] Peltzer, I.: Flug- und Windkanalexperimente zur räumlichen Entwicklung von Tollmien-Schlichting-Instabilitäten in einer Flügelgrenzschicht. Mensch und Buch Verlag, Berlin (2004) Dissertation an der Technischen Universität Berlin.

[5] Suttan, J.: Entwicklung und Einsatz von Multisensor–Piezofolienarrays an Laminarflügeln zur zeitlich und flächig hochauflösenden Messung von dynamischen Wandkräften in der Transition. In: Fortschritt–Berichte VDI, Reihe 8, Düsseldorf, VDI Verlag (1999) Dissertation an der Technischen Universität Berlin.
[6] Drela, M.: Xfoil - User Guide. MIT Aero&Astro Harold Youngren. 6.94 edn. (2001) http://raphael.mit.edu/xfoil.
[7] Stemmer, C., Kloker, M.: Interference of Wave Trains with Varying Phase Relations in a Decelerated Two-Dimensional Boundary Layer. In S. Wagner, U. Rist, H.J. Heinemann, R. Hilbig, ed.: New Results in Numerical and Experimental Fluid Mechanics III (NNFM 77), Springer Verlag (2002) 239–246
[8] Schröder, A., Kompenhans, J.: Investigation of Transitional Structures in Artificially Excited Boundary Layer Flows by Means of Stereo and Multi-plane PIV. In S. Wagner, M. Kloker, U. Rist, ed.: Recent Results in Laminar-Turbulent Transition - Selected Numerical and Experimental Contributions from the DFG-Schwerpunktprogramm "Transition" in Germany (NNFM 86), Springer Verlag (2003) 255–268

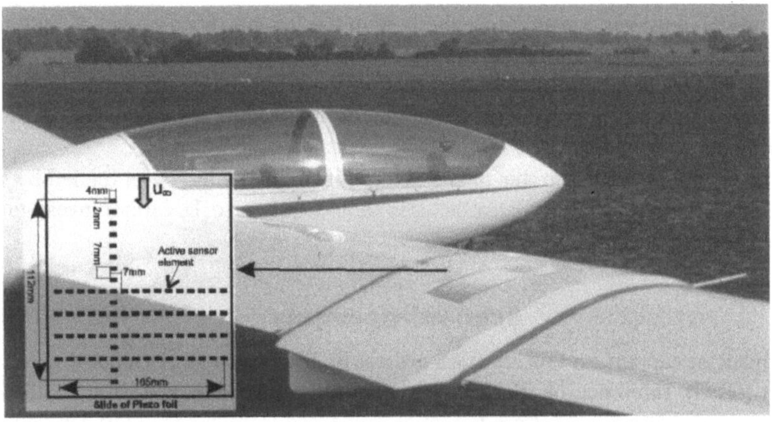

Figure 1 Sailplane with glove and sketch of the piezo-sensor-array

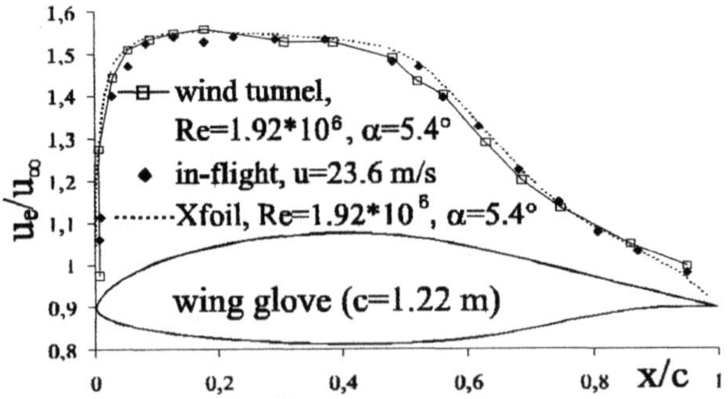

Figure 2 Velocity-distribution from in-flight and wind tunnel measurements, as well as calculated with XFOIL ($u_\infty = 23.6\,m/s$)

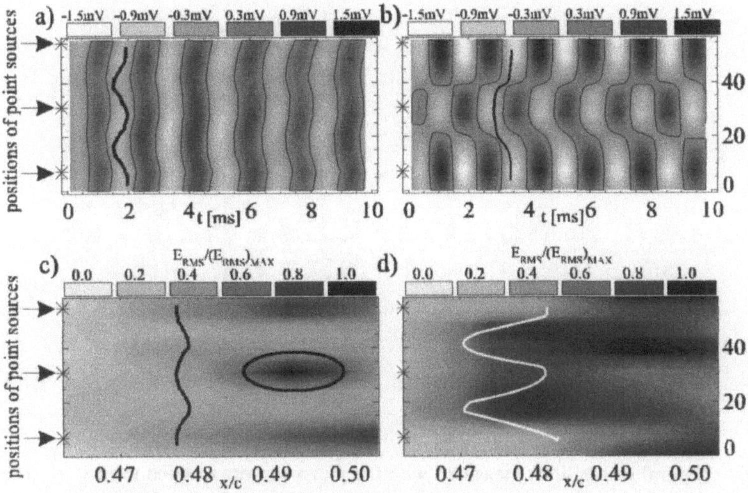

Figure 3 Contour plots of the time traces from the surface hot-wire sensors (a and b, upper row) and of the RMS-values from piezo sensor signals (c and d, lower row) with monofrequency excitation a+c) equiphase and b+d) antiphase (wind tunnel, $u_\infty = 23.6\,m/s$)

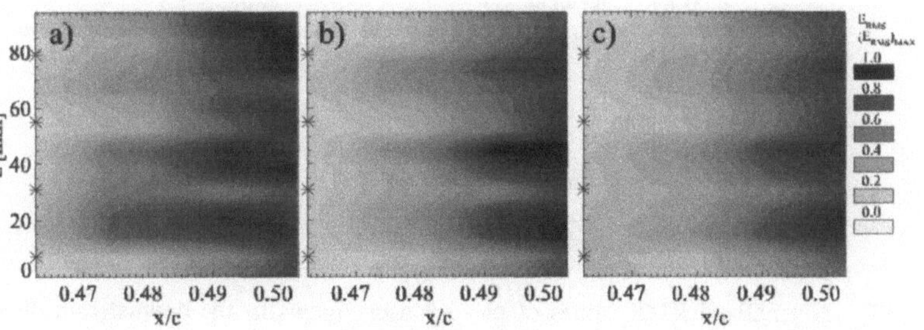

Figure 4 Normalized RMS-values of piezo sensor signals, antiphase a) $(6+7)*100\,Hz$, b) $(6+7+8)*100\,Hz$, c) $(4+5+6+7+8)*100\,Hz$ (wind tunnel, $23.6\,m/s$)

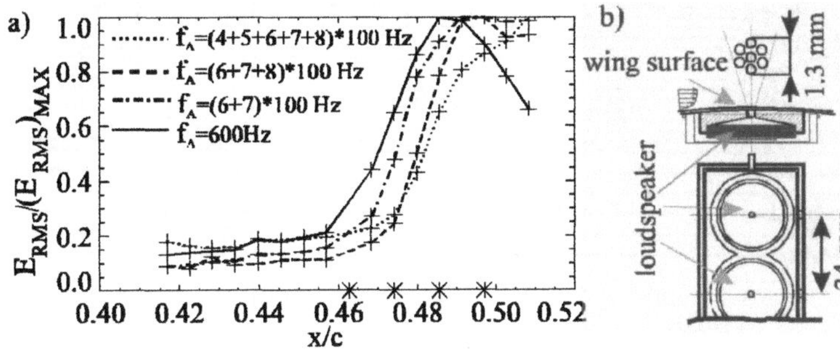

Figure 5 a) Normalized RMS-values of piezo sensor signals in downstream direction, antiphase ($z = 35\,mm$, wind tunnel, $u_\infty = 23.6\,m/s$), b) Sketch of point sources

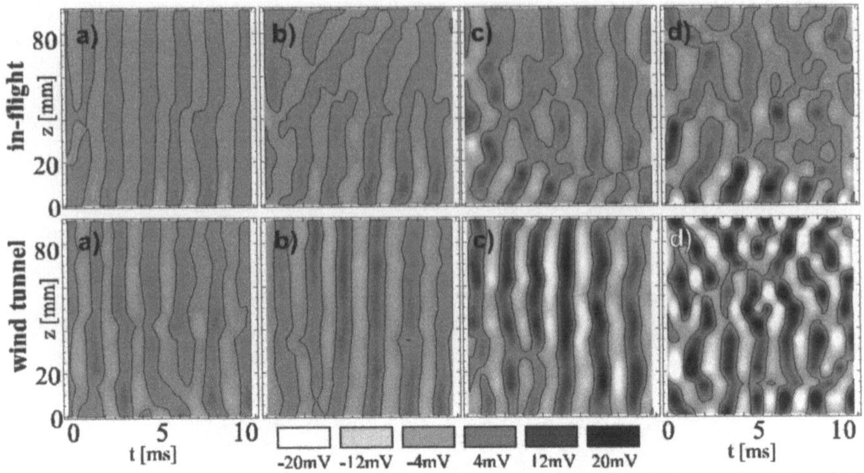

Figure 6 Natural instabilities measured with the piezo sensors in wind tunnel ($23.6\,m/s$) a) row1 ($x/c = 46.3\%$), b) row2 (47.4%), c) row3 (48.6%), d) row4 (49.7%)

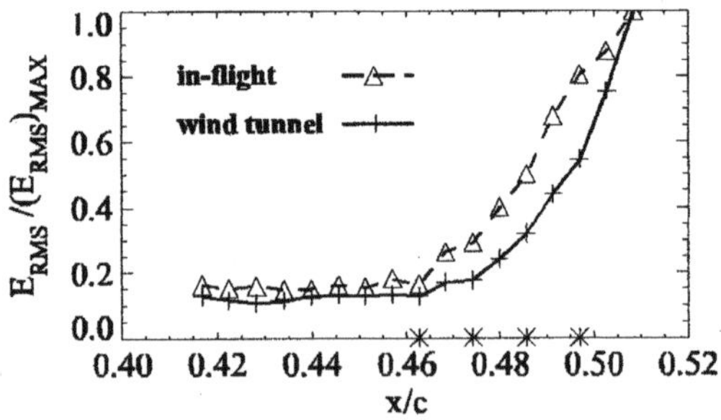

Figure 7 Normalized RMS-values of piezo sensor signals in the downstream direction, natural transition ($z = 35\,mm$, wind tunnel, $u_\infty = 23.6\,m/s$)

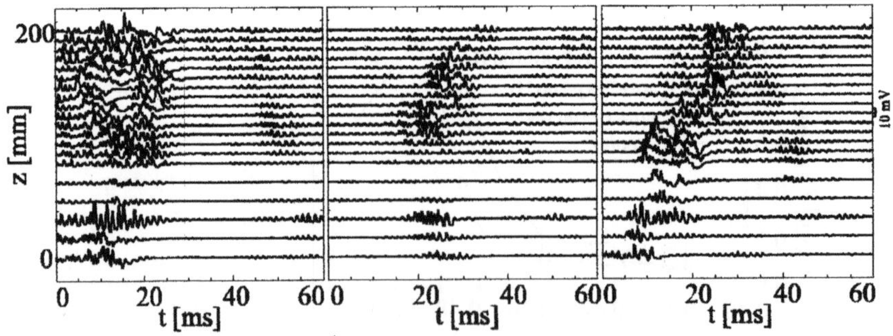

Figure 8 Examples of natural disturbances due to temporary atmospheric fluctuations during flight ($x/c = 33\,\%, u = 22.2\,m/s$)

Steady three-dimensional Streaks and their Optimal Growth in a Laminar Separation Bubble

O. Marxen[1], U. Rist[1], and D. Henningson[2]

[1] Universität Stuttgart, Institut für Aerodynamik und Gasdynamik (IAG)
Pfaffenwaldring 21, D-70550 Stuttgart, Germany,
[2] Department of Mechanics, Royal Institute of Technology (KTH)
SE-100 44 Stockholm, Sweden
olaf.marxen@iag.uni-stuttgart.de

Summary

A laminar separation bubble is formed in a region of adverse pressure gradient on a flat plate by a separating boundary layer that undergoes transition, finally leading to a reattached turbulent boundary layer. Linear amplification of steady three-dimensional disturbances in the flow before separation and in the laminar part of such a separation bubble is studied by means of direct numerical simulation and an adjoint-based optimization technique suited to study spatial optimal transient growth. The steady disturbances develop as streaks following their excitation in the region of favorable pressure gradient. At separation and inside the bubble, numerical and experimental results show good agreement with theoretical predictions for the optimal disturbance. The growth rate of the steady disturbance is seen to possess a maximum around the spanwise wave length that was found to be a preferred one in the corresponding experiment.

1 Introduction

Transition to turbulence in a two-dimensional separated boundary layer often leads to reattachment of the turbulent boundary layer and the formation of a laminar separation bubble (LSB). In many cases, the transition process is solely governed by a strong amplification of *fluctuating* disturbances. However, for environments with higher free-stream disturbance levels or if a strong favorable pressure gradient precedes the adverse pressure gradient, *steady* 3-d disturbances are sometimes observed inside the LSB. Research of bypass transition in zero pressure-gradient boundary layers revealed the possibility of transient growth of such disturbances that are often referred to as streaks [1, 2]. Despite the notation "bypass transition", amplification of these perturbations can still be a linear process [3].

While it is now commonly accepted that the process of formation of 3-d streaks in flat-plate boundary layers can be caused by transient growth, the appearance of 3-d perturbations in separated flows was in the past frequently attributed to a Görtler instability[4, 5]. Amplification of corresponding streamwise vortices is a result of streamline curvature around the separation location. Presence of streaks in conjunction with separated flows was experimentally observed in [6], but not related to transient growth. Their development in LSBs and their relation to transient growth has only recently been studied experimentally and theoretically [7].

This work is a continuation and extension of a DNS study on 3-d steady disturbances that was published in [8]. Therefore, a brief summary of the findings of [8] shall be given here. Four different ways of exciting a steady perturbation either directly via blowing/suction at the wall or by non-linear generation were applied in that study. Even though all cases showed the same finally disturbance shape and growth rate inside the LSB, overall best agreement with a corresponding experiment could be achieved by forcing a pair of oblique waves in the region of favorable pressure gradient. In that case, an initial streamwise vortex relaxed into a streak downstream after a long region of transient behavior.

2 Description of the Flow Field

The reference case is chosen according to a set-up that has been studied extensively by means of numerical and experimental methods[9, 10]. A flat plate is mounted in the free stream of the test section of a laminar water tunnel. A streamwise pressure gradient is imposed locally on the flat-plate boundary layer by a displacement body, inducing a region of favorable pressure gradient followed by a pressure rise (Fig. 1). In the region of adverse pressure gradient (APG, $\breve{x} > 0m$), a laminar separation bubble develops. A rough estimate of the pressure gradient parameter (see [11]) for the APG-region gives $P{=}\breve{\delta}_2/\breve{\nu} \cdot \partial\breve{u}_{slip}/\partial\breve{x}|_{Separation} \approx -0.23$. The transition experiment is performed with controlled disturbance input. Perturbations are forced at $x{=}{-}0.23$ by an oscillating wire with regularly placed 3-d roughness elements (spacers) underneath the wire.

Non-dimensionalization is achieved by a reference velocity $\breve{U}_{ref}{=}0.15\,\text{m/s} \approx 1.2 \cdot \breve{U}_\infty$ and a reference length $\breve{L}_{ref}{=}{}^2/_3\,m$ ($\approx$ length of the body $\breve{L}_{DB}^{Exp}{=}0.69m$), resulting in a Reynolds number $Re_{global}{=}10^5$ in water. At the streamwise position of the inflow boundary $x_{ifl}{=}{-}0.6$, the observed boundary-layer profile can be approximated by a Falkner-Skan similarity solution with $Re_{\delta_1}{=}900$ and $\beta_H{=}1.03$.

To obtain a base flow for subsequent stability calculations (of theoretical nature or based on the Navier-Stokes equations in disturbance formulation), a DNS with controlled disturbance input had to be carried out to generate a laminar separation bubble close in shape to the experimental one. General physical parameters of the flow were chosen to match the set-up described above as accurately as possible. Details of this precursor computation can be found in [12].

The resulting unsteady flow field was averaged in time and subsequently used as a base flow with a height $y_{max}{=}0.12$. Fig. 2 shows the streamwise evolution of some boundary-layer parameters to provide an impression on the flow field. As argued in [8, 12], for the laminar part of the LSB the flow field can be well assumed to be a sufficiently accurate solution to the steady Navier-Stokes equations.

3 Theoretical and Numerical Methods

3.1 Theoretical Method to Compute Optimal Disturbances

The present paper puts its focus on physical aspects regarding the disturbance development. Therefore, the applied methods are described only very briefly. For theoretical investigations, an iterative adjoint-based optimization algorithm is applied. The method is based on the linearized boundary-layer equations. It serves to maximize the kinetic disturbance energy at the downstream position x_1 for a given initial energy at the inflow x_0. Details of the applied method can be found in [13, 14].

In wall-normal direction a spectral method based on Chebychev polynomials with 106 collocation points was used, while the second-order marching procedure in streamwise direction involved 173 steps for the longest domain (reaching from x_0=−0.6 to x_1=0.33). Five different streamwise stations were used as inflow position (x_0=−0.6, −0.45,−0.3,−0.15,0, respectively), while the outflow was kept fixed at x_1=0.33.

If the downstream location for the optimization procedure, i.e. for the theoretical method, is extended beyond $x \approx 0.33$, convergence can not be achieved anymore. This seems to be related to the reverse-flow of the base flow profile, since artificially setting this reverse flow to zero allows to compute through the entire separation bubble *without* giving a visible change in the results up to x=0.33.

3.2 Numerical Method for Direct Numerical Simulations

Spatial direct numerical simulation of the three-dimensional unsteady incompressible Navier-Stokes equations in disturbance formulation serves to compute the disturbance development in the flow field described above. The method uses finite differences of fourth/sixth-order accuracy on a Cartesian grid for downstream (N=1794) and wall-normal (M=241) discretization [15]. Grid stretching in wall-normal direction allows to cluster grid points near the wall. In spanwise direction, a spectral ansatz is applied (K=5). To reduce computational effort, spanwise symmetry is assumed for calculations. An explicit fourth-order Runge-Kutta scheme is used for time integration. The domain for direct simulations covers the streamwise interval $x \in [-0.6, 0.6182]$. Upstream of the outflow boundary a buffer domain starting at $x \approx 0.45$ smoothly returns the flow to a steady laminar state.

Disturbances are forced via blowing and suction at the wall through a disturbance strip $x \in [-0.4268, -0.3398]$. Similarly to case B_{11} of [8], a pair of unsteady oblique 3-d perturbations $A_v(1, \pm 1)$=$1.164 \cdot 10^{-2}$ is forced in the disturbance strip. The fundamental frequency was β_0=30.7 and the fundamental spanwise wavenumber γ_0=72.0. As demonstrated in [8], such forcing results in an immediately decoupled steady mode $(0, 2)$ that is then linearly amplified. As for a linearized Navier-Stokes calculation, no steady-state solution can be obtained anymore for an integration domain where the useful region extends further than $x \approx 0.45$.

4 Results

A double Fourier transform in time and spanwise direction of data sets from measurements or computations yields disturbance amplitudes and phases. Below, the notation (h, k) will be used to specify the modes, with h and k denoting wavenumber coefficients in time and spanwise direction, respectively.

The issue of optimal growth was not explicitly addressed in [8], but results presented there already allow a short discussion of this matter. If we focus on the disturbance evolution in the four small-amplitude cases B_{xx} of [8] and consider only the (decoupled) linear evolution of the steady disturbance slightly downstream of the disturbance strip, we observe (Fig. 3, left) that the perturbation for case B_{11} possess the lowest initial amplitude for the streamwise velocity $\hat{u}'$ – note that the forcing amplitudes were adjusted so that they reach the same final state. Even if we would consider as a criterion the (kinetic) energy instead of the $\hat{u}'$-amplitude, still case B_{11} would give the optimal growing perturbation, since in all cases $\hat{v}', \hat{w}' \ll \hat{u}'$. For that reason, below we will deal only with the case of an excitation by an oblique pair of waves as described above. In sec. 4.1, a comparison with theoretical results is discussed while sec. 4.2 investigates the influence of the spanwise wave length.

4.1 Optimal Growth: DNS and Theory

Results of the theoretical optimization procedure for all the considered inflow positions x_0 are compared with numerical results in Fig. 3 (left). Good agreement of theoretical and DNS results is observed for the streamwise disturbance velocity component, if the inflow position for theoretical calculations is chosen upstream of the disturbance strip in the DNS. Inside the separation bubble, DNS as well as theoretical results become independent on the initial (in x) condition. However, for $x > 0.225$, i.e. inside the LSB, the theoretical prediction gives slightly larger growth rates compared to the DNS. Wall-normal and spanwise velocity components from theory also agree well with DNS (Fig. 3, right), even though the largest differences can be seen in the wall-normal component $\hat{v}'$.

The reason for the mentioned differences are due to different shapes of the wall-normal amplitude functions. These are given in Fig. 5 for three different streamwise locations (from left to right) and all three velocity components (from top to bottom). Similar to transient growth in a Blasius boundary-layer, an initial streamwise vortex (Fig. 5, x=−0.15) relaxes into a streak inside the LSB (x=0.3). Even though theoretical and DNS results possess a slightly different amplitude function initially, inside the separation bubble an almost perfect matching is observable.

As pointed out before (sec. 3.1), downstream of $x \approx 0.33$ theoretical computations fail to converge, but DNS results still show good agreement with the experimental data (Fig. 3, right and Fig. 5 for x=0.405). This hints at the possibility of elliptic effects, not captured by the boundary-layer equations, to set in. However, additional research is necessary for further clarification.

4.2 Influence of the Spanwise Wave Length

A free parameter neglected in the study [8] shall be investigated here: the spanwise wave number. Motivation of such a study comes from the question why in the experiment – of all spanwise wave lengths – the one corresponding to mode $(0, 2)$ in the present nomenclature appears to be a preferred wave length of the set-up. In the experiment, the wave length was chosen by 'trial-and-error', i.e. the length of, and gap between, the spacers (see Fig. 1) was varied until the most regular vortical structures appeared [16, 9, 10]. These showed only a weak deviation from a sinusoidal shape in spanwise direction around the separation location.

Fig. 4 shows results for the entire investigated parameter space. Focus is put on the x-positions that are located in the region of adverse pressure gradient (x=0.15) and at separation (x=0.225). As for the theoretical results, a local maximum in amplification rate (Fig. 4, left) occurs for a wave number close to mode $(0, 2)$, while the DNS results possess such a maximum only for x=0.3 (not shown). Results for very small spanwise wave numbers ($\gamma < 72$) become inaccurate due to too low integration domains in wall-normal direction.

Despite the amplification rates from DNS being largest for small spanwise wave numbers (Fig. 4, left) in the present scenario, it becomes clear that in particular these perturbations experience a strong penalty in absolute amplitude (Fig. 4, right). The reason for this lies in the consideration of transient growth along x (and in case of DNS also receptivity matters) that is taken into account only in the latter case.

5 Conclusion

The present study sheds some light on the origin and linear evolution of steady, spanwise-harmonic perturbations in a laminar separation bubble. All calculations discussed are based on a case for which a profound experimental data base exists.

Good agreement of DNS, theory, and measurements suggests that inside the separation bubble, indeed the optimal steady three-dimensional disturbance could be observed in the present case. Furthermore, the experimentally determined spanwise wave number was confirmed to be the most preferred one of the whole set-up.

Acknowledgments: Financial support of this research by the Deutsche Forschungsgemeinschaft DFG under grant Wa 424/19–1 and Ri 680/10–1 and by the Deutsche Akademische Austauschdienst („Projektbezogener Personenaustausch Schweden") is gratefully acknowledged. Olaf Marxen appreciates the hospitality at the Department of Mechanics (KTH Stockholm) during his visit there and thanks Ori Levin for giving him an introduction to his code used in this paper for the theoretical optimization. We thank Matthias Lang for providing detailed experimental results.

References

[1] Henningson, D.S.: Bypass transition and linear growth mechanisms. In Benzi, R., ed.: Advances in Turbulence V, Kluwer Academic Publishers, Dordrecht, Boston, London (1995) 190–204

[2] Reshotko, E.: Transient growth: A factor in bypass transition. Phys. Fluids **13** (2001) 1067–1075
[3] Henningson, D.S., Reddy, S.C.: On the role of linear mechanisms in transition to turbulence. Phys. Fluids **6** (1994) 1396–1398
[4] Inger, G.: Three-dimensional heat- and mass transfer effects across high-speed reattaching flows. AIAA Journal **15** (1977) 383–389
[5] Wilson, P.G., Pauley, L.L.: Two- and three-dimensional large-eddy simulations of a transitional separation bubble. Phys. Fluids **10** (1998) 2932–2940
[6] Watmuff, J.H.: Evolution of a wave packet into vortex loops in a laminar separation bubble. J. Fluid Mech. **397** (1999) 119–169
[7] Boiko, A.V.: Development of a stationary streak in a local separation bubble. Technical Report IB 224–2002 A04, German Aerospace Center (DLR), Institute for Fluid Mechanics, Göttingen (2002)
[8] Marxen, O., Rist, U., Wagner, S.: Effect of Spanwise-Modulated Disturbances on Transition in a Separated Boundary Layer. AIAA Journal **42** (2004) 937–944
[9] Lang, M., Rist, U., Wagner, S.: Investigations on controlled transition development in a laminar separation bubble by means of LDA and PIV. Experiments in Fluids **36** (2003) 43–52
[10] Lang, M.: Experimentelle Untersuchungen zur Transition in einer laminaren Ablöseblase mit Hilfe der Laser-Doppler-Anemometrie und der Particle Image Velocimetry. Dissertation, Universität Stuttgart (2005)
[11] Gaster, M.: The structure and behaviour of laminar separation bubbles. AGARD CP–4 (1966) 813–854
[12] Marxen, O.: Numerical Studies of Physical Effects Related to the Controlled Transition Process in Laminar Separation Bubbles. Dissertation, Universität Stuttgart (2005)
[13] Andersson, P., Berggren, M., Henningson, D.S.: Optimal disturbances and bypass transition in boundary layers. Phys. Fluids **11** (1999) 134–150
[14] Levin, O., Henningson, D.S.: Exponential vs algebraic growth and transition prediction in boundary layer flow. Flow, Turbulence and Combustion **70** (2003) 183–210
[15] Kloker, M.: A robust high-resolution split-type compact FD scheme for spatial direct numerical simulation of boundary-layer transition. Appl. Sci. Res. **59** (1998) 353–377
[16] Lang, M., Marxen, O., Rist, U., Wagner, S.: Experimental and Numerical Investigations on Transition in a Laminar Separation Bubble. In Wagner, S., Rist, U., Heinemann, H.J., Hilbig, R., eds.: New Results in Numerical and Experimental Fluid Mechanics III. Volume 77 of NNFM., Springer, Heidelberg (2001) 207–214

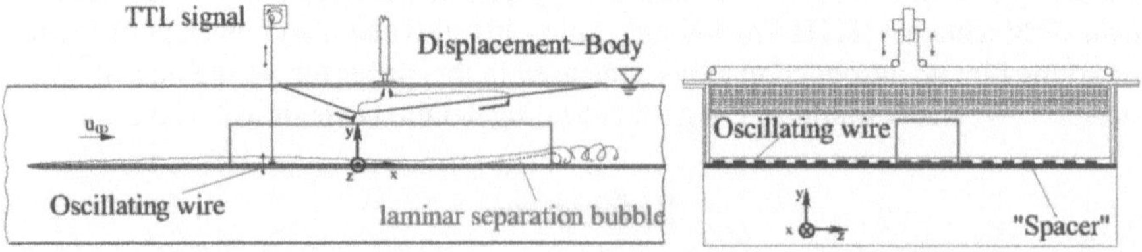

Figure 1 Configuration for the experiment by Lang et al. [9, 10]. The integration domain used for DNS and theory is indicated by a box (not to scale).

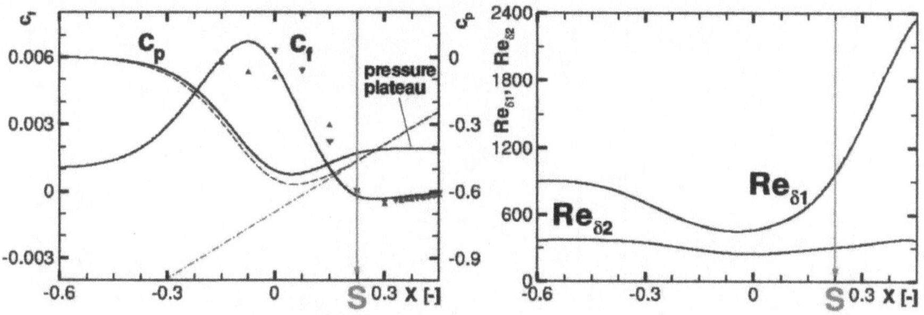

Figure 2 Coefficients for surface pressure c_p, skin friction c_f (left), and Reynolds numbers Re_{δ_1} and Re_{δ_2} (right). Comparison of DNS (solid lines), slip-flow results $c_{p,slip}$ (dashed line), and measurements (symbols).

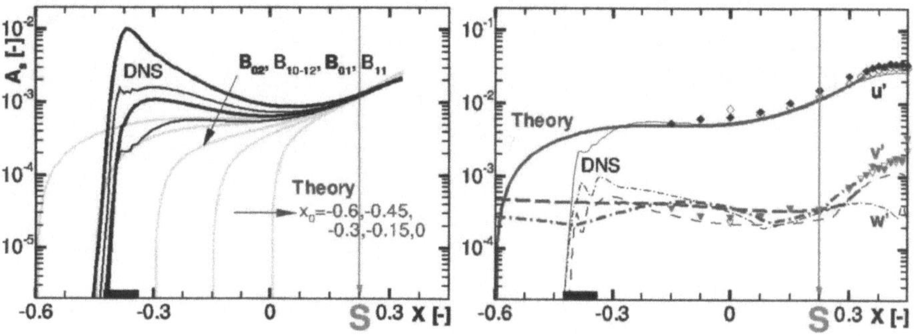

Figure 3 Amplification curves for the maximum (in y) streamwise (solid), wall-normal (dashed), and spanwise (dash-dotted) velocity fluctuation $|\hat{s}'^{(0,2)}_{max}|$. Left: all cases B_{xx} of [8] (dark lines) together with theoretical results for x_0=$-0.6, -0.45, -0.3, -0.15, 0$ (light lines). Right: DNS (thin lines) and theoretical results for x_0=-0.6 (thick lines) together with two different measurements (symbols).

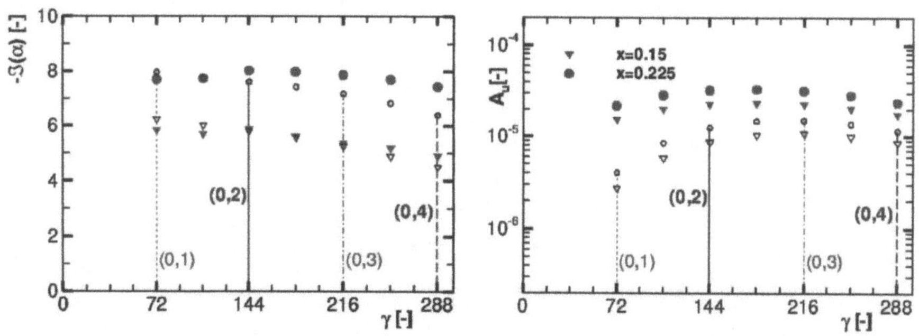

Figure 4 Left: Amplification rate $\Im(\alpha)$ vs. spanwise wave number γ for x=0.15 (triangles), 0.225 (circles) from DNS (open symbols) and theory (x_0=-0.6, filled symbols). Right: Maximum (in y) absolute amplitude $|\hat{u}'_{max}|$ vs. wave number γ at x=0.15 (triangles), 0.225 (circles) for equal forcing in DNS (strip interval $x \in [-0.4268, -0.3398]$, $A_v^{(1,k)}$=$5 \cdot 10^{-3}$) or for equal initial kinetic energy $\int(\hat{v}'^2 + \hat{w}'^2)dy$=$0.625 \cdot 10^{-8}$ in the theory.

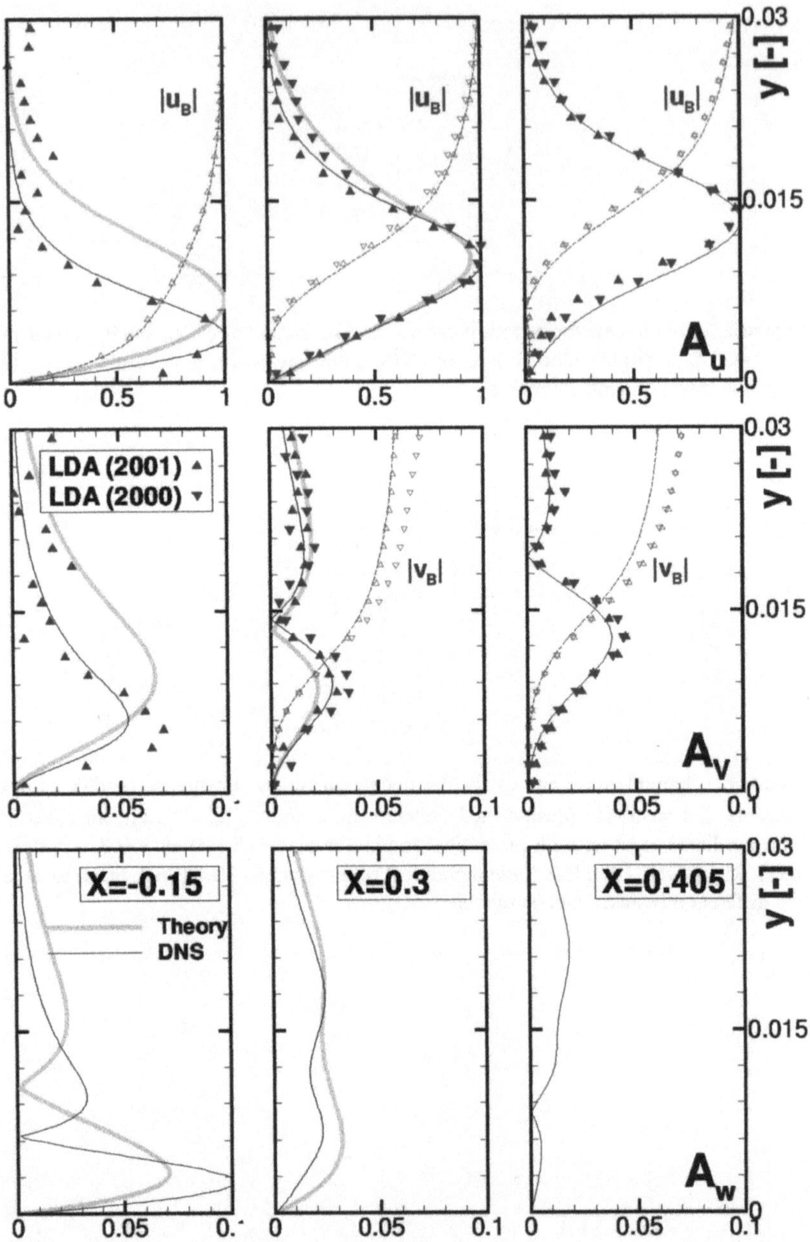

Figure 5 Velocity amplitudes $|\hat{u}'|$, $|\hat{v}'|$, $|\hat{w}'|$ (from top to bottom) of mode $(0, 2)$, normalized by their respective $|\hat{u}'_{max}|$. Comparison of results from DNS (thin line) and theory (x_0=−0.6, thick line) with two different measurements (symbols, as far as available) for x=−0.15, 0.3, 0.405 (from left to right). Base-flow quantities u_B, v_B are also shown.

Applicability and quality of linear stability theory and linear PSE in swept laminar separation bubbles

Tilman Hetsch and Ulrich Rist

Institut für Aerodynamik und Gasdynamik, Universität Stuttgart,
Pfaffenwaldring 21, 70550 Stuttgart, Germany.
hetsch@iag.uni-stuttgart.de

Summary

A family of laminar separation bubbles (LSB) in a swept boundary layer flow - hereafter referred to as "swept LSB"- is used to study the effect of sweep and of the propagation direction of disturbance waves on the quality of linear stability theory (LST) and solutions of the parabolized stability equations (PSE). To this end spatial LST and linear PSE solutions are qualitatively and quantitatively compared to highly resolved results of direct numerical simulations (DNS). The sweep angle of the base flow is systematically varied between 0° and 45° and a variety of Tollmien-Schlichting waves as well as the most amplified stationary cross-flow mode are investigated. It turns out, that even though LST works satisfactory in the presence of sweep, flow separation and back flow, PSE is clearly preferable in terms of accuracy.

1 Introduction

Since the nineties linear PSE-methods are increasingly used for similar tasks as the traditional spatial LST. In aircraft industry their most important field of application lies in the semi-empirical e^N-method for transition prediction, where nowadays both methods are utilised, as for example in [5]. In research, either method provides an inexpensive a priori analysis of single modes or the whole flow field in form of stability diagrams, if base flow is steady. Thus purposeful disturbance scenarios can be determined in advance of computationally more demanding methods like DNS ([6], p. 60). For unsteady flows at least a posteriori analysis of the time averaged DNS-flow field is possible, as done in [3], p. 140. Finally, both methods are frequently used for the validation of codes and measurements.

All these applications depend on the ability of LST and PSE to model the propagation of low amplitude disturbances realistically. Laminar separation bubbles (LSB) represent a demanding base flow for both methods. One might expect problems with non-parallel effects due to curved streamlines and the rapid growth of the boundary layer thickness over the bubble in the case of LST and with the back flow inside the bubble for the streamwise marching procedure on which the efficiency of PSE rests on. But for *unswept* LSB this poses no problem: Over the last decade LST has been very successfully applied to LSB in 2D-base flows (see e.g. [4], [3] and references therein). For PSE growth rates good agreement was also achieved by Hein in [1] (see figure 1 in [1]) compared to DNS results of the unswept version of the same leading edge bubble investigated here. Still unknown on the other hand is the general applicability and overall quality of LST and PSE in laminar separation bubbles in a swept, three-dimensional flow. This is investigated in detail for Tollmien-Schlichting modes in section 3 with emphasis on the effect of sweep and for the linearly most amplified cross-flow mode in section 4. Throughout the paper PSE always refers to linear PSE and LST/DNS to spatial LST/DNS.

2 Base flow and numerical methods

The unswept prototype of the present leading edge bubble was extensively studied by Rist in [4] by means of DNS and LST. Its extension to swept flows, the verification of the base flow by step size tests, as well as the effect of sweep on the base-flow profiles and a first LST analysis were reported by Hetsch & Rist in [2]. The DNS calculations are split in a DNS of the steady laminar base flow and a succeeding DNS for the disturbance propagation. These two DNS-codes solve the complete three-dimensional Navier-Stokes equations for unsteady, incompressible flows in a vorticity-velocity formulation. For an in-depth description of the DNS algorithms see [6]. The LST-code used here is the same as in all three references above. The PSE-results were obtained by the code "nolot" of the DLR-Göttingen, described in [1]. All PSE-calculations were started with local solutions obtained from LST.

The most important base flow parameters described in [2] are repeated here: All quantities in the paper are non-dimensionalized by the reference length $\overline{L} = 0.05\,\mathrm{m}$ and the chordwise free-stream velocity $\overline{U}_\infty = 30\,\frac{\mathrm{m}}{\mathrm{s}}$, which is held constant for all cases. The x- and z-direction are taken normal and parallel to the leading edge with U and W being the associated base flow velocity components, respectively. Periodicity is assumed in spanwise direction only, resulting in a quasi-2D base flow with $(\frac{\partial}{\partial z} \equiv 0)$, but $W(x,y) \neq 0$. The calculation domain itself consists of an infinite flat plate subjected to an adverse pressure gradient. It is introduced to the system by prescribing the chordwise potential flow velocity $U_e(x)$ shown in Figure 1 at the upper boundary of the domain. Different sweep angles Ψ are realized by varying the spanwise free stream velocity $W_\infty = U_\infty \tan(\Psi)$ and setting $W_e(x) \equiv W_\infty$. At the inflow located at $x_o = 0.37$ Falkner-Scan-Cooke profiles are prescribed. With a kinematic viscosity of $\overline{\nu} = 15 \cdot 10^{-6} \frac{\mathrm{m}^2}{\mathrm{s}}$ the flow can be characterised by $Re_{\delta_1} = \overline{U}_\infty \overline{\delta}_1(x_o)/\overline{\nu} = 331$, based on the displacement thickness. The wall-normal coordinate y ranges from 0 to $y = 0.059 = 18 \cdot \delta_1(x_o)$ with $y = \overline{y}/\overline{L}$.

Thus, a family of swept LSB with arbitrary sweep angle is obtained, which –in agreement with the independence principle of infinite swept geometries, see [2]– have identical separation and reattachment positions at $x_s = 1.75$ and $x_r = 2.13$, respectively. The steady calculation of the separation bubble is justified by its small size (e.g. the maximum back flow is $0.3\,\%\,\overline{U}_\infty$) and experience with the unswept case in [4]. It is converged to steady state, if the differences in V and the vorticity components Ω_x and Ω_z of two consecutive time levels are smaller than 10^{-10}.

3 Tollmien-Schlichting waves

For the investigation a packet of Tollmien-Schlichting (TS) waves with varying spanwise wavenumber $\gamma \in [-40, -30, -20, 0, 20, 30, 40]$ was excited by a disturbance strip displayed in Figure 1. They all share the most amplified frequency $\omega = 18$ of the 0°-base flow. In the following the notation (ω/γ) is adopted for discrete modes in the frequency-spanwise wavenumber spectrum. Small initial amplitudes of $A_v = 10^{-10}$ grant a purely linear development throughout the domain. Figure 2 shows two examples of amplification curves $u'_{\omega,\gamma}(x) = \max_y(\widehat{A}^u_{\omega,\gamma}(x,y))$, where $\widehat{A}^u_{\omega,\gamma}$ denotes the amplitude of a double Fourier analysis of the disturbance velocity component u' in time and span. Local DNS-amplification rates may be obtained by $\alpha_i[DNS] := -d(\ln u'_{\omega,\gamma})/dx$. For quantitative comparison local amplification rates α_i obtained by LST are integrated to calculate the amplification curve $A_o\, e^{-\int \alpha_i\, dx}$. Its initial amplitude A_o is fitted to match the DNS results around the neutral point $x_{neut} = 0.95$ (see Fig. 2), after which most modes are amplified for the first time. The relative error $r_{rel} := |DNS - LST|/DNS$ was then evaluated at $x_{DNS-max}$, the position of the peak of the DNS-amplification curve shown in Figure 2. Here the greatest N-factor, defined as $N(x) := \int_{x_{neut}}^{x} -\alpha_i\, dx$, is achieved. All findings were compiled into table 1. Analogous results for PSE are found as a second entry in same table. Table 2 gives the propagation angle $\Psi = \arctan(\gamma/\alpha_r)$ of all modes in a body-fitted and a streamline orientated coordinate system, α_r representing the chordwise wavenumber. For most modes Ψ changes only little in the present base flow, so those values can be regarded as typical.

The effect of sweep: The relative error of the LST-amplification curves is quite high. On the average it was found to be 40% for the 0°-base flow and about 50% for the sweep angles $\Psi_\infty = 30°$ and 45°. All in all there is a tendency towards higher errors for increasing sweep angles Ψ_∞. Note that 45°-mode $(18/-40)$ displays a slightly atypical behaviour in table 1, as it is nearly neutrally stable. LST was applicable for the whole range of $\Psi_\infty \in [0°, 45°]$, but the modes $(18/\pm 40)$ showed first, still negligible convergence problems in their damped regions for $\Psi_\infty = 45°$. The attempt to calculate a complete stability diagram by LST in a 60° bubble failed due to heavy convergence problems already in the amplified regions of the flow. PSE on the other hand was able to predict the disturbance development starting at the neutral point with a relative error of only 6% in the mean, nearly unaffected by Ψ_∞. In the 45°-case beginning convergence problems made it necessary to double the step size for the modes (18/30) and (18/40). This coarse discretisation led to higher errors compared to other PSE-results.

The effect of the propagation direction: Table 2 reveals that the propagation direction Ψ grows monotonically with the spanwise wavenumber γ. The general trend for LST "larger errors for more oblique modes with larger $|\gamma|$ and therefore larger $|\Psi|$" is much more pronounced than the effect of a rising sweep angle. The smallest relative errors of $16\% - 19\%$ are found for modes around $\gamma = 0$. For very oblique modes within the same base flow the errors are up to 4 times higher. Note that there is *no* trend "larger errors for modes with larger $|\gamma_s|$" with respect to the angle Ψ_s relative to the direction of the potential stream line. Obviously, in the 45°-case the TS-wave (18/30) with the smallest angle in the stream-line orientated coordinate system does not exhibit the minimal error. PSE shows also the tendency for higher errors for more oblique modes, but much less pronounced than for LST.

LST for local quantities and qualitative comparison: The high relative errors listed in table 1 should not discourage the use of LST in swept LSBs. Fitting the curves at the neutral point is necessary in the context of N-factor calculation, but shows the *integrated* error from there up to the point of comparison. The direct results of LST are *local* ones –growth rates, wavelengths, propagation directions– for which only the error at the place of comparison is taken into account. Also, many applications of LST mainly require *qualitative* comparisons of curves. This yields improved results, because the curves are fitted at an arbitrary point, for which the error is equally distributed over the whole length of the domain. Furthermore comparison with linear theories are only meaningful until one mode reaches the critical amplitude value of about $1\% U_e$ where nonlinear effects should not be neglected any longer. Therefore, the interval of comparison will typically be smaller than analysed here. An example of a nonlinear scenario is shown in Figure 3, where the LST of the dominating mode (18/0) shows excellent agreement up to the point of saturation and even the mode (18/40) with the extremely high relative error of 78% (at its peak at $x \approx 2.25$) compares more or less satisfactory within the linear regime.

4 Stationary cross-flow modes

In the 45° base flow the strength of the crossflow (CF) velocity W_s, the spanwise base flow velocity in a streamline orientated coordinate system, reaches a value of about $W_{s,max}/U_{s,max} = 8\%$ relative to its streamwise counterpart U_s. Furthermore, W_s-profiles exhibit an inflection point indicating the influence of an inviscid crossflow instability. LST was used to determine the most amplified stationary CF-mode. The maximal growth rates α_i of stationary CF-modes inside the LSB were found to be roughly one-third of those of the most amplified TS-waves. The strongest amplification was exhibited by the modes (0/40) and (0/50), which showed nearly identical amplification curves. The (0/40) was chosen for further comparison. As displayed in Figure 4, the LST- and PSE-amplification curves were individually fitted to the DNS-result in order to achieve the best overall match. After a short transient phase PSE yields excellent agreement with the DNS solution, whereas LST looks even qualitatively less convincing than for the TS-waves from section 3. On the other hand the y-scale is much larger here. Other authors also indicate that LST fails to give good quantitative results in the case of CF-modes. Wassermann and Kloker for example examined several CF-modes in an accelerated 45°-boundary layer without separation in [6]. They report that the LST systematically underpredicts the DNS-growth rates, a trend which is also observable in Figure

2 for the TS-waves investigated here. In their study amplification curves of DNS and LST differed at the middle of the domain already by a factor of $3-4$.

5 Conclusion

The applicability and accuracy of linear stability theory and linear PSE was investigated for a generic family of small laminar separation bubbles. LST and PSE were found to be applicable in the whole sweep angle range of $\Psi_\infty \in [0°; 45°]$. As both methods showed beginning convergence problems in the 45°-base flow increasing difficulties can be expected for higher sweep angles. A packet of TS-waves with systematically varying spanwise wave number γ was compared to DNS-results for all sweep angles. In terms of accuracy PSE is clearly superior to LST, which systematically underpredicts the DNS-growth rates. Compared to DNS-amplification curves the mean error over all modes and sweep angles yielded 6% for PSE compared to 46% for LST. An error of 64% already results in an amplitude factor difference of 2.8 between LST and DNS, which would correspond to a $\Delta N = 1$ in a N-factor prediction. On the other hand the amplification in LSBs is so extreme compared to attached boundary-layer flows –the maximum growth rate of the present LSB is 16 times higher than that of the same inflow without pressure gradient– that even such an error might result in only a small Δx-shift of the predicted transition location.

LST was found to work best for 0°-base flows. The errors increased with rising sweep angle, in the mean by a factor of 1.25 from $\Psi_\infty = 0°$ to 45°. More pronounced is the dependency of LST from the propagation direction Ψ of the analysed mode. It exhibited the general trend "larger errors for modes with larger propagation angles". In the present investigation the errors of the most oblique waves were up to 4 times higher than errors for two-dimensional waves. The accuracy of *PSE*-results on the other hand was rather independent of the sweep angle. But oblique waves with the largest propagation angle also showed the maximum errors. In the mean they differed from the minimum error of the mode $(18/-20)$ by a factor of 1.75.

In application LST is very robust and easy to handle and automate because of its local character. It is well suited to get an overview, for the qualitative comparison of curves or when a great number of modes has to be calculated, as for stability diagrams. Due to its step-size restriction PSE requires more attention per run. PSE comes into play when greater accuracy is desired and is clearly superior for CF-modes. Note that the present base flow was a flat-plate boundary layer. In curved geometries, PSE has the additional advantage of the inclusion of curvature terms.

Acknowledgements

The authors would like to thank the DLR-Göttingen and especially Dr. Hein for the possibility to use the linear "nolot"-PSE-code. The financial support by the Deutsche Forschungsgemeinschaft (DFG) under contract number RI 680/12 is gratefully acknowledged.

References

[1] S. Hein: "Linear and nonlinear nonlocal instability analyses for two-dimensional laminar separation bubbles". Proc. IUTAM Symp. on Laminar-Turbulent Transition, Sedona, USA, Sep. 1999, Springer, p.681-686.

[2] T. Hetsch, U. Rist: "On the Structure and Stability of Three-Dimensional Laminar Separation Bubbles on a Swept Plate". In: Breitsamter, Laschka et al. (ed.): 13th STAB Symposium 02, Munich. NNFM **87**, Springer 2004, pp. 302-310.

[3] O. Marxen, M. Lang, U. Rist, S. Wagner: "A Combined Experimental/Numerical Study of Unsteady Phenomena in a Laminar Separation Bubble". Flow, Tub. and Comb. **71**, Kluwer Acad. Pub., 2003, pp.133-146.

[4] U. Rist: "Zur Instabilität und Transition in laminaren Ablöseblasen". Habilitation, Universität Stuttgart, Shaker 1999.

[5] G. Schrauf: "Industrial View on Transition Prediction". In: S. Wagner et al. (ed.): "Recent Results in Laminar-Turbulent Transition". NNFM **86**, Springer 2003, pp. 111-122.

[6] P. Wassermann, M. Kloker: "Mechanisms and passive control of crossflow-vortex-induced transition in a three-dimensional boundary layer". J. Fluid Mech. **456**, Camb. Univ. Press 2002, pp.49-84.

Table 1 First number: Relative error $r_{rel} := |DNS - LST|/DNS$ of LST results in per cent with respect to DNS amplification curves, evaluated at X-position of greatest amplification of the DNS-curve (see Fig. 2). Its dependency on various modes (ω/γ) (columns) and sweep angles Ψ_∞ of the base flow (rows). Second number: Same findings for PSE.

Ψ_∞	(18/-40)	(18/-30)	(18/-20)	(18/0)	(18/20)	(18/30)	(18/40)	mean
0°	58 / 8	55 / 6	16 / 5	16 / 4	16 / 5	55 / 6	58 / 8	39 / 6
30°	51 / 6	58 / 4	58 / 6	19 / 5	15 / 1	62 / 6	73 / 2	48 / 4
45°	25 / 6	51 / 2	59 / 2	19 / 5	39 / 9	81 / 10	78 / 11	50 / 6
mean	45 / 7	55 / 4	44 / 4	18 / 5	23 / 5	66 / 7	70 / 7	46 / 6

Table 2 Propagation angle $\Psi = \arctan(\gamma_r/\alpha_r)$ in degrees with respect to the X-axis of modes in table 1 according to LST at $x = 1.4$. Second number: Propagation angle $\Psi_s = \arctan(\gamma_r/\alpha_{r_s})$ in streamline orientated coordinate system, displayed in Figure 1.

Ψ_∞	(18/-40)	(18/-30)	(18/-20)	(18/0)	(18/20)	(18/30)	(18/40)
0°	−41	−32	−22	0	22	32	41
30°	$-30/-61$	$-25/-56$	$-19/-49$	$0/-31$	$27/-4$	$43/12$	$58/28$
45°	$-25/-71$	$-22/-67$	$-17/-62$	$0/-46$	$32/-14$	$54/8$	$79/33$

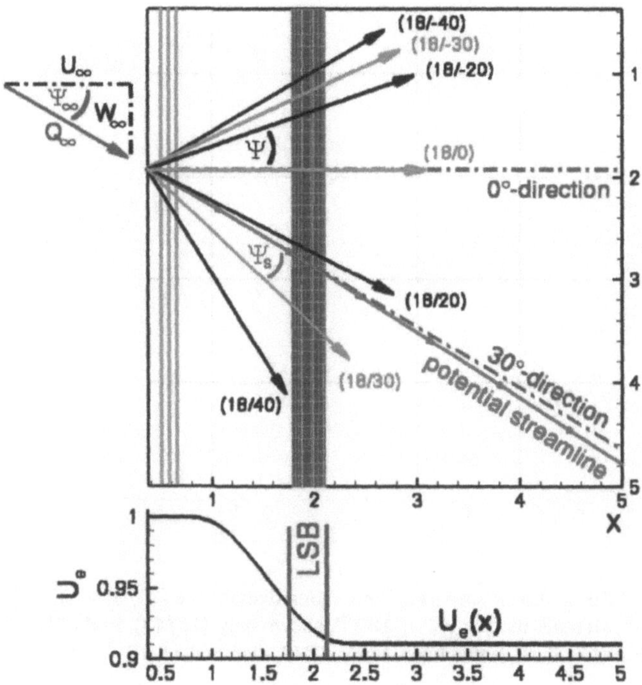

Figure 1 30°-base flow: Topview. Disturbance strip $x \in [0.5; 0.64]$, LSB between $x_s = 1.75$ & $x_r = 2.13$, potential streamline. Propagation direction Ψ of investigated TS-waves (ω/γ) according to LST at $x = 1.4$. Smaller Figure: potential chordwise velocity $U_e(x)$.

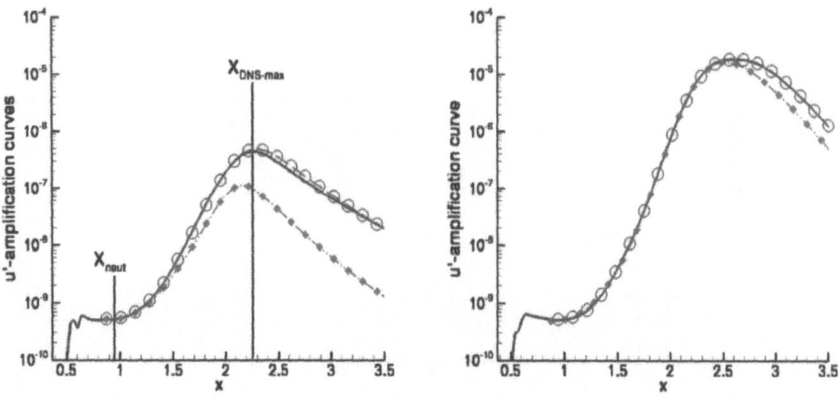

Figure 2 Amplification curves. Left: (18/40) in 45°-base flow with x-positions of error evaluation in section 3 (example of high relative error compared to other TS-waves). Right: (18/20) in 30°-base flow (lowest relative error). Lines: DNS, diamonds: LST, circles: PSE.

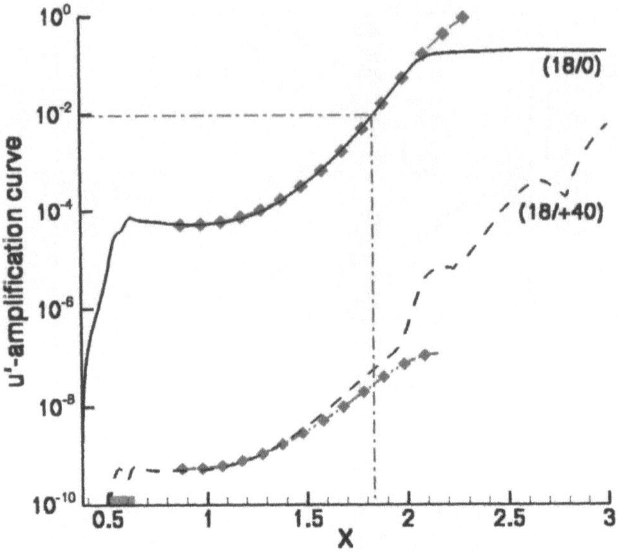

Figure 3 45°-LSB nonlinear scenario: Dominating disturbance (18/0) with initial amplitude $A_v = 10^{-5}$, all other modes: $A_v = 10^{-10}$, shown only (18/40). End of linear regime at $x \approx 1.8$ indicated by dash-dotted lines. Lines: DNS, diamonds: LST.

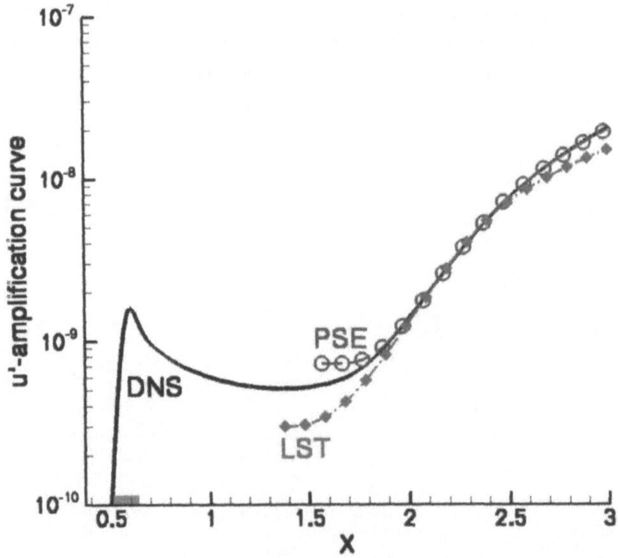

Figure 4 45°-LSB: Linearly most amplified stationary CF-mode (0/40) with disturbance strip. Initial amplitude: $A_v = 10^{-10}$. Lines: DNS, diamonds: LST, circles: PSE.

Drag Reduction of an Ahmed Car Model by Means of Active Separation Control at the Rear Vehicle Slant

A. BRUNN and W. NITSCHE

Institute of Aeronautics and Astronautics
Technical University Berlin, Marchstr. 12, 10587 Berlin, Germany
andre.brunn@tu-berlin.de

Summary

The experimental investigations described in the present paper deal with the reduction of the total aerodynamic drag of a generic car model (Ahmed-Body) by means of periodic forcing. The experiments carried out in this study focus on a unique approach to separation control using fundamental frequencies for local forcing of the shear layer separated from the rear end of the car model. The excitation of large scale vortex structures by periodic forcing intensifies the primary momentum transfer between the separation region and the outer flow, resulting in a substantial reduction of the separation length. A total drag reduction of 27 % was achieved using the flow control method described in this study.

1 Introduction

Flow control over a bluff body for drag and noise reduction purposes is considered to be one of the major issues in aerodynamics. The pressure difference between the front and the rear end of bluff bodies is the driving factor for flow separation at the rear end of the car (Fig. 1, left). Hence, the pressure drag controls the skin friction and the flow field behind the body is determined by a strong wake [3, 9]. The flow situation in the wake is highly three dimensional and unsteady and depends strongly on the rear slant angle φ [7]: Longitudinal vortices occur at slant angles up to 30°, leading to a dramatic increase in pressure drag (Fig. 1, right). With increasing slant angles these horseshoe vortices burst and spanwise vortices dominate the wake region, causing a reduction in drag.

Both active and passive methods of flow control can be applied to avoid or reduce this type of separation-induced performance loss [8, 2] etc. Several investigations on active separation control in diffusers (e.g. [2]) or on simple bluff body geometries ([10, 4] etc.) were successfully conducted using forcing frequencies in the range of the shear layer instabilities observed. In these experiments wall actuators were used to generate periodic perturbations, significantly reducing the separation length.

Nevertheless, practical applications are almost too complicated for an accurate analysis of these typical phenomena. Hence, generic models with the most important boundary conditions are frequently used. The more complicated the configuration is, the more complicated the flow structures become. This is demonstrated, for instance, by Brunn [2] in the transition from plane diffusors to axisymmetric configurations. Here, the spanwise vortex structures, which have been shed from the separation region, lose their initial two-dimensional nature in their early stages of development. The present experimental study focusses on the active separation control methods for the flow behind a generic car model - the Ahmed-Body [1]. The flow control on this car model is an interesting application due to the possibility of increasing the efficiency of land vehicles: Reducing this separation will result in a strongly decreased total drag. The Ahmed-Body combines the essential geometric parameters determining shape, length and position of the separation, and it is used as a reference for numerical and experimental investigations e.g. [5].

2 Experimental Set-Up

The experimental investigations were conducted in a closed water channel with an optically fully accessible water test section using Particle Image Velocimetry (PIV) and digital flow visualization methods (Fig. 2). The PIV-system consists of a frequency-doubled Nd:YAG laser ($\lambda = 532\,nm$, 25 mJ per pulse and 5 ns duration), two CCD-Cross-Correlation-Cameras (12 bit, cooled with $1280 \times 1024\,pxl^2$) and a Synchronization Unit.

The Ahmed-Body, which stretched across the whole width of the test section, was mounted on the channel wall. The slant angle was $\alpha = 35°$, as observations on the fully three-dimensional model by [1, 7] etc., show that the flow field of the slant region under these conditions is dominated by two-dimensional spanwise vortex structures. This kind of vortex structure is expected, even though the aspect ratio of the model is $B/H \approx 4$. All other geometrical parameters of the model are based on the data given by Ahmed [1]. The Reynolds number based on the inflow velocity and the model height was $Re_H = 1.1 \cdot 10^5$. The flow upstream of the car model was fully turbulent.

The periodic pressure perturbations were generated by a water pump connected to a rotating valve (Fig. 3, left) and injected into a cavity-slit-system, resulting in an oscillating synthetic jet with zero net mass flux. The perturbations during the forcing should preferably amplify the vortex structures in the separated shear layer in order to increase the growth of the vortices and intensify the entrainment process.

3 Results

First of all, the unforced base flow was investigated with the time-averaged flow field in terms of streamlines shown in Fig. 4. Here, an ensemble of around 200 instantaneous PIV images in the symmetry plane is depicted. No significant differences in the velocity fields at different spanwise positions could be observed. The flow structures typically occurring behind bluff bodies are visible (e.g. [6]): Two counter rotating vortex structures (time-averaged) and a settling point that closes the separation region at a position of $x \approx 395\ mm$. Spanwise flow structures appear in the near wake region of the Ahmed-Body . These structures are comparable with other, much more simple configurations [2, 10, 4] etc., which is why the experiments focus on an unique approach to drag reduction in terms of active separation control.

Two free shear layers (in the two dimensional point of view) collide and enclose a separation bubble in the wake behind bluff bodies. While the upper shear layer separates from the slant and is consequently driven by the slant configuration, the flow at the bottom is like a backward facing step flow. The snapshot of the instantaneous velocity field in Fig. 5 shows discrete structures of spanwise vortices. Typical wavelengths can be assigned to these structures and they scale with a half of the body height H. Hence, it is the obvious solution to force the flow and amplify these structures with frequencies according to the shear layer instabilities. For this reason, a frequency range according to the Strouhal numbers $0.1 \leq St_H \leq 0.7$ was chosen. Here, the lowest value corresponds to the vortex-shedding frequency estimated with the method of [4] and was confirmed through digital flow visualization documented by [2]. The initial shear layer instability depends primarily on the boundary layer conditions upstream of the separation point [6] and was calculated to be around $St_H = 0.7$. All control experiments in the present study were performed with a forcing intensity of $c_\mu = 3 \cdot 10^{-3}$, where

$$c_\mu = \frac{A_S}{A_0} \cdot \frac{c_S'^2}{\bar{u}_0^2} \tag{1}$$

with $\bar{u}_0$ as the average velocity at the inflow, c_S' as the perturbation velocity at the slit exhaust, the cross section at the inflow A_0 and the active actuator area A_S.

A significant reduction of the car wake separation length was achieved in all forcing cases, with a noticeable difference in the resulting (mean) flow field (Fig. 6). The excitation with the vortex shedding instability ($St_H = 0.2$, Fig. top, right) shows a drastic reduction of the recirculation area compared to the base flow (Fig. 6 top, left). However the forcing effectiveness is consistently reduced with increasing forcing frequency, and in the case of an excitation in the range of the shear layer instability ($St_H = 0.7$, Fig. bottom, right) the reduction is comparatively low.

The distribution of velocity fluctuations is a reliable indicator of an enhanced momentum transfer by means of local forcing. This distribution is depicted in Fig. 7 for forcing cases $0.2 \leq St_H \leq 0.7$, while the picture on the top left represents the flow without forcing. At frequencies close to that of the vortex-shedding, the RMS-values by far exceed the base flow values. At $St_H = 0.7$, the distribution shows no significant enhancement. A closer look at Fig. 7 (top, middle) does not only show increased fluctuations in the slant region, but also demonstrates that the momentum transfer in the near wake directly at the blunt end of the car has much more intensity than at other forcing frequencies. As a result, the entrainment process starts much earlier and the forcing effect is stronger in the near wake region behind the car model.

An analytical method to determine the vehicle drag and the drag coefficient c_D is the use of conservation of momentum with the velocity data of one upstream and at least one downstream position in the integration area far downstream [3]. The profiles of the streamwise velocity component at the downstream position $x = 420\ mm$, depicted in Fig. 8 (left), already show a qualitative drag reduction. The velocity $u(y)$ in the near wake, normalized with the free stream velocity u_∞, is significantly increased due to the local forcing. At this downstream position, an integral value of the velocity deficit was calculated. This one is equivalent to the profile drag coefficient. All three velocity components have to be measured in the near wake in order to fulfill the conservation of momentum. Hence, a calculated drag coefficient with only $(u, v) = f(x, y)$ is incorrect. A velocity deficit in the measurement plane of $c_{Def} = 0.35$ was calculated for the unforced base flow, and it represents the profile drag of a two dimensional configuration. Figure 8 (right) summarizes the drag reductions in terms of a reduced velocity deficit achieved with the active separation control. The values for different forcing frequencies (Strouhal numbers) shown in this figure are normalized with the value of the unforced base flow. An excitation with the frequency of vortex-shedding leads to a dramatic drag reduction in the excitation range $0.2 \leq St_H \leq 0.4$ (approximately 20 %). The minimum drag ($c_{Def} = 0.25$) could be observed at $St_H = 0.2$. The effect decreases with higher forcing frequencies due to the mechanisms explained earlier. Measurements with a force balance in addition to three dimensional far wake velocity analysis will be carried out to verify these results.

4 Conclusions

This study presents experimental investigations on active separation control used to reduce the total drag of an Ahmed-Body car model by means of large scale vortex structure excitation and amplification. Actuators generating periodic perturbations to the flow were used to excite the shear layer separating from the rear slant of the car model. Forcing frequencies in the range of the initial shear layer instability and the vortex-shedding were used to test the receptivity of the flow. The excitation in terms of periodic perturbations at the slant edge lead to increased velocity fluctuations in the shear layers, while the momentum transfer between the recirculation region and the outer flow was significantly intensified due to forcing at the vortex-shedding frequencies. The most effective frequency for drag control was observed for the corresponding Strouhal number $St_H = 0.2$, based on the model height. A total drag reduction of 27 % was achieved for this forcing case. The amplified large scale vortices connected with the vortex-shedding process are the key to reduce separation and consequently the pressure drag of the Ahmed-Body. This result has become the basis for flow control investigations on highly three-dimensional flow configurations.

Acknowledgements

This research was funded by the German Science Foundation (DFG) within the scope of the Collaborative Research Center SFB 557. This support is gratefully acknowledged by the authors.

References

[1] AHMED S.R., RAMM R. & FALTIN G. (1984). Some Salient Features of the Time-averaged Ground Vehicle Wake. *SAE-Techn. Paper Series* **840300**

[2] BRUNN A. (2003). *Aktive Beeinflussung abgelöster turbulenter Scherschichten in überkritischen Diffusoren mit Hilfe periodischer Anregung*, TU Berlin, PhD Thesis, Mensch und Buch Verlag

[3] HUCHO W.-H (2002). Aerodynamik der stumpfen Körper - Physikalische Grundlagen und Anwendung in der Praxis, Vieweg-Verlag

[4] KIYA M., SHIMIZU M. & MOCHIZUKI O. (1997). Sinusoidal Forcing of a Turbulent Separation Bubble. *In: J. Fluid Mech.* **342**, 119–139

[5] KRAJNOVIC S. & DAVIDSON L. (2002). A Test Case for Large-Eddy Simulation in Vehicle Aerodynamics. In: RODI W. and FUEYO N. (Eds.) *Engineering Turbulence Modelling and Experiments* **5**, Elsevier Science Ltd., 647–657

[6] LEDER A (1992). Abgelöste Strömungen - Physikalische Grundlagen, Vieweg-Verlag

[7] LIENHART H., STOOTS C. & BECKER S. (2002). Flow and Turbulence Structures in the Wake of a Simplefied Car Model (Ahmed model). *In: Notes on Numerical Fluid Mechanics* **Bd. 77**, Springer, 323–330

[8] LIN J.C., HOWARD F.G., BUSHNELL D.M. & SELBY B.V. (1990). Comparative Study of Control Techniques for Two-Dimensional Low-Speed Turbulent Flow Separation. In: KOZLOV A.V (Ed.): *Separated Flows and Jets*, Springer, Berlin, 429–474

[9] MOREL T. (1978). The Effect of Base Slant Angle on the Flow Pattern and Drag of Three-Dimensional Bodies with Blunt Ends. *In: Proc. of Symp. Aerod. Drag Mechanisms of Bluff Bodies and Road Vehicles.* Plenum Press, New York, 191–226

[10] SIGURDSON L.W. (1995). The Structure and Control of a Turbulent Reattaching Flow. *In: J. Fluid Mech.* **298**, 139–165

Figures

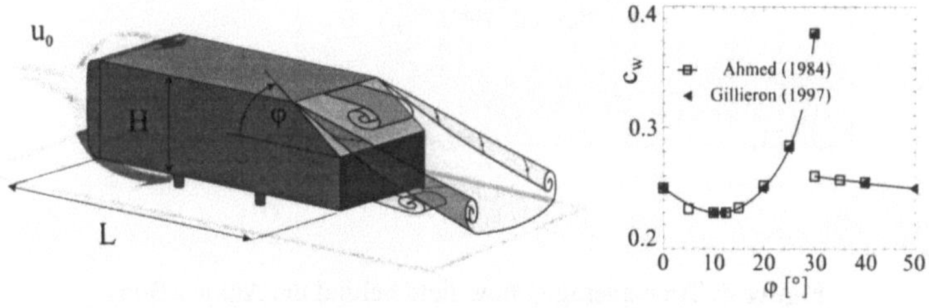

Figure 1: Ahmed Car Model with illustrated vortex structures behind the rear end (left) and the total drag coefficient for different slant angles (right)

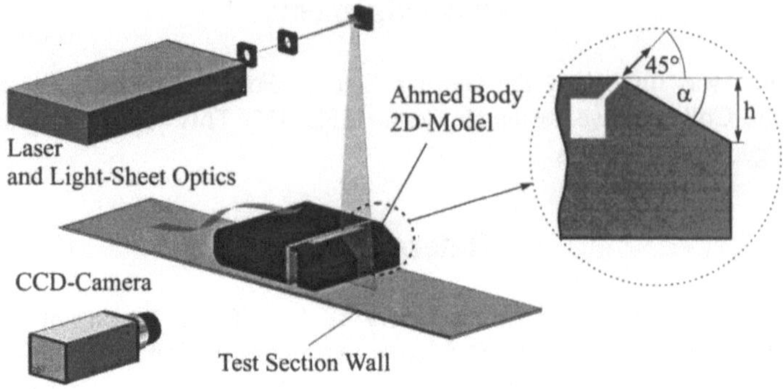

Figure 2: Experimental Set-Up of the Ahmed-Body investigations with Particle Image Velocimetry (PIV)

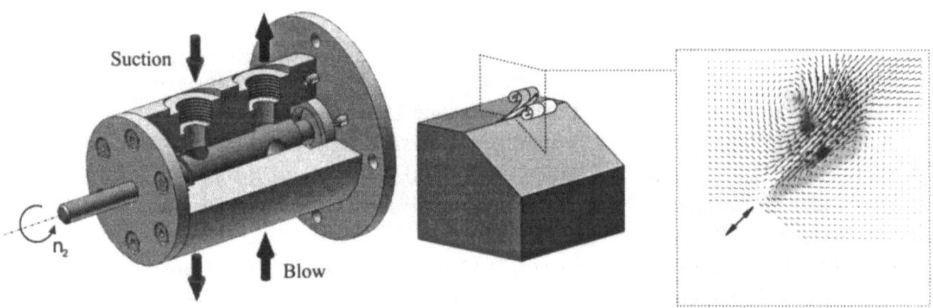

Figure 3: Rotating valve used to provide periodic perturbations in terms of spanwise vorticity (right) at the rear slant edge

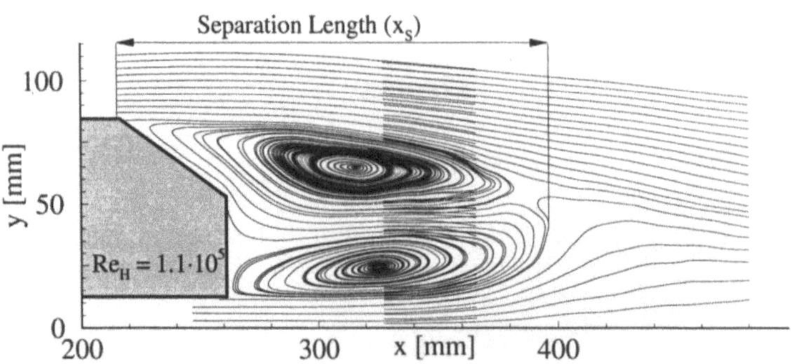

Figure 4: Time averaged flow field behind the Ahmed-Body

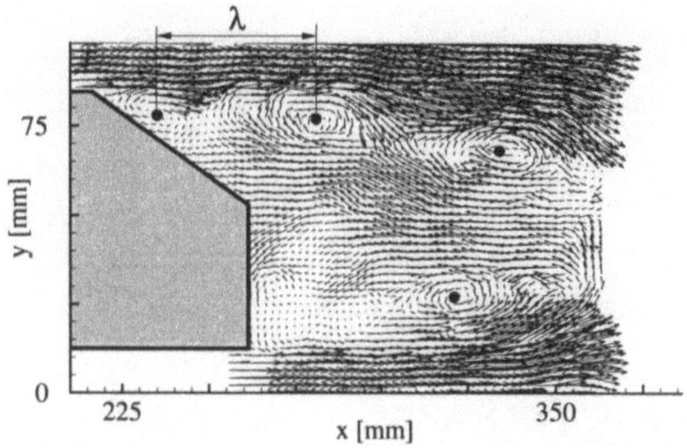

Figure 5: Instantaneous velocity field behind the Ahmed-Body with the characteristic wavelength of spanwise vortices

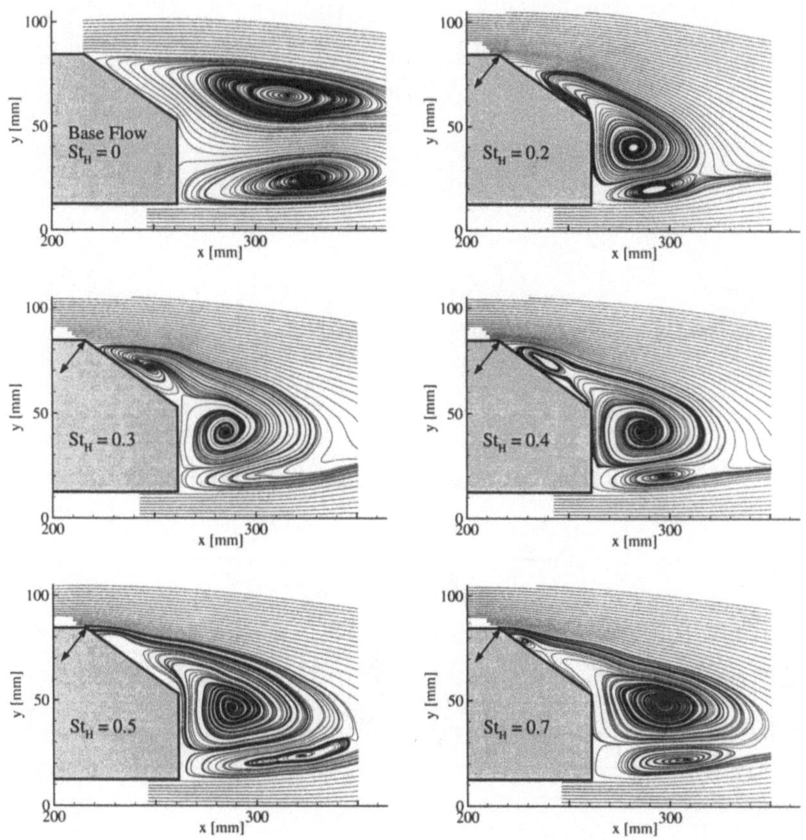

Figure 6: Time averaged flow fields behind an Ahmed-Body at different Strouhal numbers

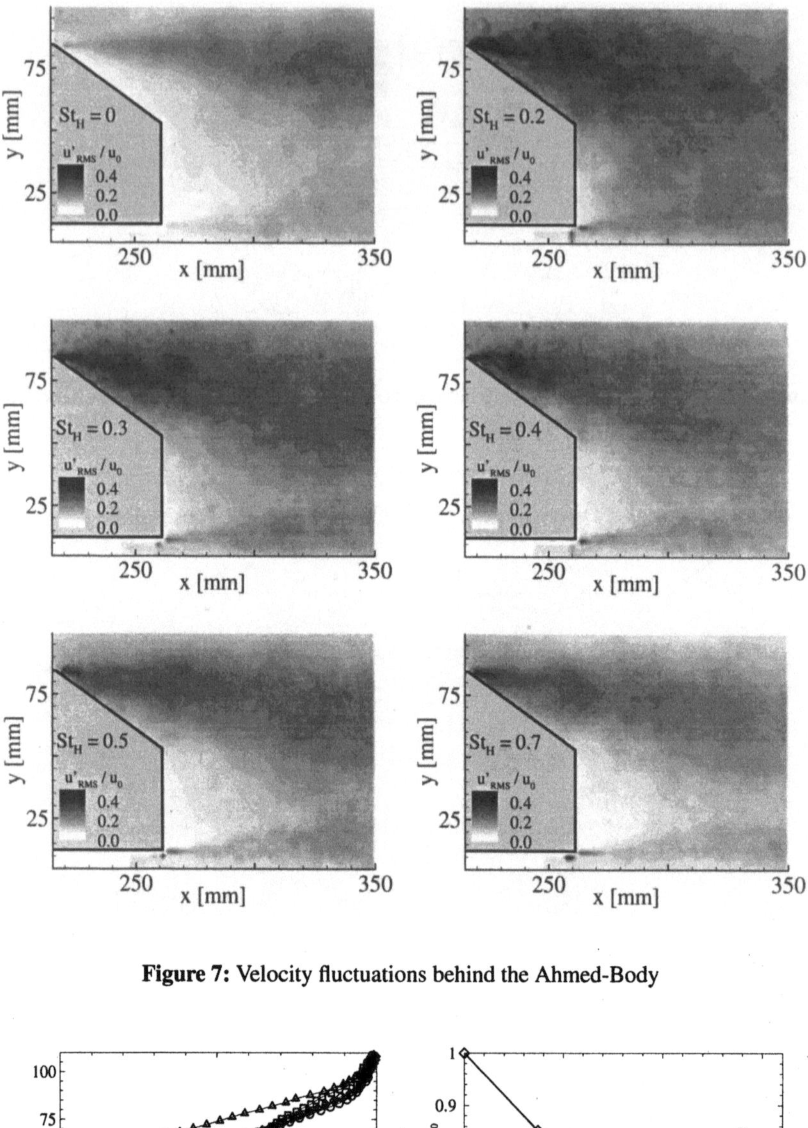

Figure 7: Velocity fluctuations behind the Ahmed-Body

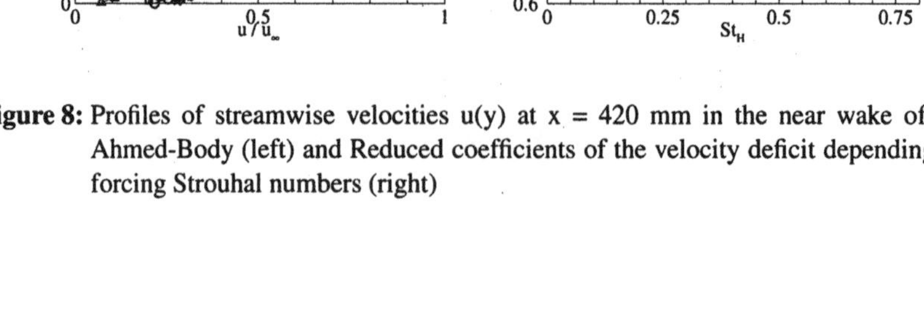

Figure 8: Profiles of streamwise velocities u(y) at x = 420 mm in the near wake of the Ahmed-Body (left) and Reduced coefficients of the velocity deficit depending on forcing Strouhal numbers (right)

Increasing Lift by Means of Active Flow Control on the Flap of a Generic High-Lift Configuration

R. Petz[1], W. Nitsche[1] and M. Schatz[2], F. Thiele[2]

[1] Institute of Aeronautics and Astronautics
Marchstr. 12, 10587 Berlin, Germany,
Ralf.Petz@tu-berlin.de,

[2] Hermann-Föttinger-Institute of Fluid Mechanics
Müller-Breslau-Str. 8, 10623 Berlin, Germany
Markus.Schatz@tu-berlin.de

Summary

This paper demonstrates the use of active separation control on a high-lift configuration in order to enhance the aerodynamic performance in terms of lift and drag. The aim is to delay boundary layer separation on the flap's upper surface by periodic excitation using a pulsed wall jet. The experimental and numerical results show a massive improvement of almost all aerodynamic coefficients over a wide range of angles of attack and flap deflection angles. By actuating with correct excitation parameters, the jet formed by the single slotted flap can be reattached or kept attached, depending on the conditions, to the surface. Lift is increased by up to 12% while drag is reduced by the same amount. As a result, the lift to drag ratio defining the aerodynamic quality is improved by up to 25%. A numerical calculation on the basis of unsteady Reynolds-averaged Navier Stokes equations is also presented to determine the influence of different excitation parameters.

1 Introduction

Active flow control plays an ever growing part in the aerodynamic research area. To control flow with low-power consumption devices enables new approaches to the design or modification of bodies exposed to a moving fluid. Controlling the boundary layer transition is one field of great interest due to its ability to reduce drag [1][2]. The other field of interest which is presented in this paper deals with the delay or suppression of boundary layer separation as it limits angle of attack and flap deflection of high lift devices [3]. The use of active separation control is demonstrated experimentally and numerically on a two dimensional generic high-lift configuration. The aim was to delay separation on the single slotted flap in order to increase lift and flap deflection. Previous investigations have revealed that periodic excitation is by far more efficient than steady blowing or suction [4] [5]. This is due to the fact that periodic excitation uses instabilities such as vortex shedding in the separated shear layer. The excitation is performed at the leading edge of the flap right where the flow field separates from the surface (fig. 1). The flow field around a trailing edge device is very complex due to the fact that there are different mixing layers (separated shear layers, airfoil wake) and therefore a dominant instability has not yet been found. Earlier investigations undergoing phase averaged hot wire measurements led to the assumption that the flapping motion of the separated jet, which is generated by the slotted trailing edge device, is amplified by periodic excitation until it reattaches to the flap's surface [6]. Free wall jets were studied extensively in the 70's and 80's and a flapping motion instability caused by the shedding of large vortices was recognized during this time.[7].

2 Experimental setup

2.1 Wind Tunnel Model

The test model (figure 2) consists of a NACA 4412 main airfoil with a single slotted NACA 4415 shape flap. The flap has a chord length of $0.4 \cdot c_{main}$. The setup is completely two dimensional with a wing span of $1.55m$ which yields an aspect ratio of 3.1. Both angle of attack α and flap deflection η can be automatically adjusted within a wide range in order to allow different configurations to be tested under the influence of excitation [8]. The main wing and flap are equipped with a pressure measurement system in order to acquire complete pressure distributions. To account for the unsteady flow around the flap when flow separation occurs, the flap is equipped with 26 pressure transducers that collect unsteady wall pressure data. The main measurement equipment used in this investigation is a six-component wind tunnel balance which is installed underneath the test section. The balance allows simultaneous acquisition of all three forces (lift, drag and side force) and moments (pitch, roll and yaw). Besides pressure and force measurements a movable hot wire was used to measure time series in the separated shear layer in order to detect possible instability frequencies.

2.2 Excitation system

The actuator development and assembly was governed by four main requirements.

- The actuator has to reach previously defined frequencies and amplitudes.
- The actuator has to fit inside the flap which has a maximum thickness of $30mm$ and is only about $15mm$ thick at the actuator position.
- The velocity distribution of the pulsed jet should be homogenous along the span.
- It has to be very robust and reliable during operation.

To combine all major attributes a system of compressed air and eleven fast switching small solenoid valves was finally chosen. Eleven of those valves were placed inside the flap side by side along the entire span. Each valve produces a very thin and $114mm$ wide (spanwise direction) pulsed wall jet oriented perpendicular to the surface. All valves are connected to the same compressed air line in such a way that a change in pressure changes the amplitudes of all the valves (see fig. 2).

Each valve's opening and closing times are individually controlled by a fast programmable micro controller situated in the main wing. The spanwise segmentation, together with individual control of each valve, allows for the possibility not only to change the excitation frequencies, but also to generate arbitrary opening and closing times for each valve or to use only a selected number of valves along the span for excitation. Besides the geometric parameters α and η, the actuator itself has the four adjustable governing parameters (frequency, amplitude, opening/closing times and number of active valves) mentioned above. Because of the variety of parameters, only the results for a two dimensional excitation are presented, where all valves are active and open and close at the same times. The actuator system was calibrated using a hot wire and fast pressure transducers attached to each of the eleven actuator segments.

3 Numerical Approach

The computational investigations are carried out using ELAN, the numerical code of TU Berlin (Herrmann Föttinger Institute). It is based on a three-dimensional incompressible Finite-Volume scheme to solve the Navier-Stokes equations. The method is fully implicit and of second order in space and time. Based on the SIMPLE pressure correction algorithm, a co-located storage arrangement for all quantities is applied. Advection is approximated by third-order TVD-MUSCL schemes.

The Unsteady Reynolds-averaged Navier-Stokes computations presented here are based on the LLR k-ω model by Rung [9] which in the past exhibited the best performance for steady and unsteady airfoil flows with large separation. It represents an improved two-equation eddy-viscosity model formulated with special consideration of the realizability conditions. The computational mesh is of c-type and provides $45,000$ cells in total. Both profiles are modeled as having bluff trailing edges. The non-dimensional wall-dis-tance of the first cell center remains below $y^+ = 1$ over the complete surfaces. Additional simulations using a finer mesh showed that the results are mesh independent. All simulations are based on a non-dimensional time-stepping of $\Delta t = 0.001\frac{c}{u_\infty}$, which gives a resolution of 400 time steps per oscillation cycle for a typical oscillating frequency.

To model the excitation apparatus, a suction/blowing type boundary condition is used. The perturbation to the flow field is introduced through the inlet velocity to the small chamber modeled as a group of 20 × 20 cells representing the excitation slot:

$$u_{exc}(t) = u_\infty \sqrt{\frac{c_{main}}{H} C_\mu} \cdot cos\left(t \cdot 2\pi \frac{u_\infty}{c_{flap}} St\right) \tag{1}$$

where c_μ is the non-dimensional momentum blowing coefficient defined as $c_\mu = 2 \cdot \frac{H_{slot}}{c_{main}} \cdot \left(\frac{u'_{jet}}{u_\infty}\right)$ and H is the slot width ($H = 0.004\, c_{flap}$). The excitation frequency is given as a nondimensional Strouhal number based on the flap's chord length $St = \frac{f \cdot c_{flap}}{u_\infty}$ where f denotes the excitation frequency.

4 Results

The results presented herein demonstrate the influence of active flow control by yielding lift increase and drag reduction. Tests at different Reynolds numbers were conducted (experimentally and numerically) reaching, a maximum Reynolds number in the experiment of $1 \cdot 10^6$ based on the main airfoils chord.

Numerical Results The numerical investigation shows the effect of different excitation parameters (fig. 3). In the beginning, the basic flow without excitation was computed for $\alpha = 3°$ and $\eta = 37°$ (tripped transition) and compared to the experiments. The obtained solution was later used as initial flow condition for all further simulations. For the low Reynolds number case the simulation yielded a lift coefficient of $c_L = 2.4$. The effect of periodic excitation with $St = 0.62$ and $C_\mu = 50 \cdot 10^{-5}$ is comparable to that obtained by the corresponding experiment (not shown here). Separation can be delayed and the lift coefficient is enhanced remarkably. The unsteady flow field shows qualitatively similar behavior in the experiments and in the numerics. In the simulation however, separation on the flap occurs delayed but a completely attached boundary layer is not predicted.

Further numerical simulations are targeting the effect of the excitation parameters on the flow field and its behavior for higher Reynolds numbers. Experimental results show that the excitation frequency St is of minor importance (in a given range) for the mean lift coefficient whereas the simulation shows a decrease in lift at higher St-numbers. In the low Reynolds number case with $Re = 160,000$ and an excitation intensity of $C_\mu = 50 \cdot 10^{-5}$ up to 14% extra lift can be obtained (fig. 3, left). The excitation frequency -if varied in a useful range- seems to be less important than the excitation intensity. The excitation intensity C_μ however, shows a stronger effect: The mean separation position on the flap can successfully delayed by periodic excitation for the whole investigated range of Reynolds numbers. At $Re = 160,000$ and $Re = 2,000,000$ a certain level of C_μ is required in order to keep the boundary layer attached along the first 15% to 20% of the flap chord (fig. 3, right). Further enhancement does not result in more improvements. At $Re = 1,000,000$ however, suction/blowing initially shows only a small effect on the separation position which continuously grows with increasing C_μ. Mean separation position delay is not captured in the experiment which gives only information on the global flow condition.

Experimental Results Results of force measurements are displayed in figure 4. The diagram on the left shows the influence of excitation (regarding lift and drag) compared to the unexcited case during a complete angle of attack sweep with fixed flap deflection. In the unexcited case the flow on the flap separates at $\alpha = 1°$, marked by a sudden loss of lift. As α is increased further, the flow around the main wing remains attached until it separates at $\alpha = 7°$. The separation process goes along with an increase in drag, which is shown in the drag polar. With the excitation turned on, the same configuration was measured and it was found that separation is successfully suppressed while maximum lift is increased by up to 12% and drag

is lowered by the same amount. In comparison the diagram on the right shows lift and drag during a flap deflection sweep. As can be seen, the onset of separation is delayed, occurring at 39° instead of 30°, an increase of 9°. These two results show that periodic excitation has a dramatic effect on lift and drag.
Figure 5 displays the lift depending on complete α and η sweeps for the non-excited flow. The resulting plot reveals the two separation mechanisms for the high-lift configuration. With increasing flap deflection, separation on the flap starts to occur at angles of about 32°. By increasing α further, flap separation starts earlier and earlier until separation occurs on the main airfoil. The white line in figure 5 indicating the onset of separation divides the plot in attached and separated region. As already shown separation on the flap can successfully be suppressed by local excitation. Figure 6 compares the angles at which the onset of separation is first recognized for the unexcited and excited case. To test for repeatability, the onset of separation was retrieved from two different data sets, namely from angle of attack (fixed flap angle) and a flap deflection sweeps (fixed angle of attack). As long as the separation occurs on the flap, the excitation keeps the flow attached with a maximum flap deflection angle of about 45°. The influence of excitation, on the other hand is limited if the separation occurs on the main airfoil.
After discussing the positive effects of active flow control in terms of delaying separation for different angles of attack and giving an example of how much gain in lift is possible, the next figure gives a quantitative answer to the question of how much gain is achieved at given angles. Figure 7 shows the percentage gain in lift due to excitation. The data was retrieved by first measuring lift in the unexcited case (as shown in fig. 5), and afterwards for the exited case with fixed excitation parameters. For same angles, the gain was calculated giving a two dimensional contour plot. There is a substantial gain in lift, up to 10% over a wide range of flap deflection angles and angles of attack. As mentioned earlier, the gain in lift is accompanied by a decrease in drag. The gain in aerodynamic performance denoted by the lift to drag ratio of the complete configuration is plotted on the right side diagram of figure 7. The percentage gain of the excited case is plotted versus both geometric angles. Again there is an increase of up to 25% for the lift to drag ratio, where the average level of gain is about 20% for a wide range of angles.

5 Conclusion

Experimental as well as numerical investigations concerning local periodic excitation on a generic high-lift configuration were performed with the aim of increasing lift. The results show a significant improvement, resulting in an increase of up to 25% for the lift to drag ratio. The large increase is measurable over a wide range of α and η settings. The beneficial effect is obtainable by an excitation intensity that is higher than a certain limit of c_μ whereas the excitation frequency plays a minor role. In general, the computation together with the experiments show that considerable improvement can be achieved up to a threshold, beyond that only a small improvement is possible.

Acknowledgements

The work was financially supported by the German Science Foundation (DFG) in the framework of the Collaborative Research Center 557 ***Control of complex turbulent shear flows.***

References

[1] Sturzebecher, D., Nitsche, W.: Active cancellation of Tollmien-Schlichting instabilities on a wing using multi-channel sensor actuator systems. International Journal of Heat and Fluid Flow **Vol. 24** (2003) pp. 572–583

[2] Sturzebecher, D., Nitsche, W.: Active Control of Tollmien-Schlichting Instabilities by Means of the Multiple Arrangement of Sensor Actuator Systems. In: Notes on Numerical FluidMechanics. Volume Vol. 77. Springer-Verlag (2002) pp. 375–382

[3] Melton, L.P., Yao, C.S., Seifert, A.: Application of excitation from multiple locations on a simplified high-lift system. AIAA 2004-2324 (2004)

[4] Wygnanski, I.: The variables affecting the control separation by periodic excitation. AIAA Paper 2004-2505 (2004) 2nd AIAA Flow Control Conference, Portland.

[5] Seifert, A., Darabi, A., Wygnanski, I.: Delay of airfoil stall by periodic excitation. Journal of Aircraft **Vol. 33** (1996) pp. 691–698

[6] Tinapp, F., Nitsche, W.: Separation Control on a High-Lift Configuration by Periodic Excitation. In: Notes on Numerical Fluid Mechanics. Volume Vol. 77. Springer Verlag (2002) pp. 42–49

[7] Thomas, F.O., Goldschmidt, V.W.: Structural characteristics of developing turbulent plane jets. Journal of Fluid Mechanics **Vol. 163** (1986) pp. 227–256

[8] Petz, R., Nitsche, W.: Active separation control on a high-lift configuration by a periodically pulsating jet. ICAS 2004-118 (2004)

[9] Rung, T., Thiele, F.: Computational modelling of complex boundary-layer flows. In: 9th Int. Symp. on Transport Phenomena in Thermal-Fluid Engineering, Singapore. (1996)

[10] Schatz, M., Thiele, F., Petz, R., Nitsche, W.: Separation control by periodic excitation and its application to a high lift configuration. AIAA Paper 2004-2507 (2004)

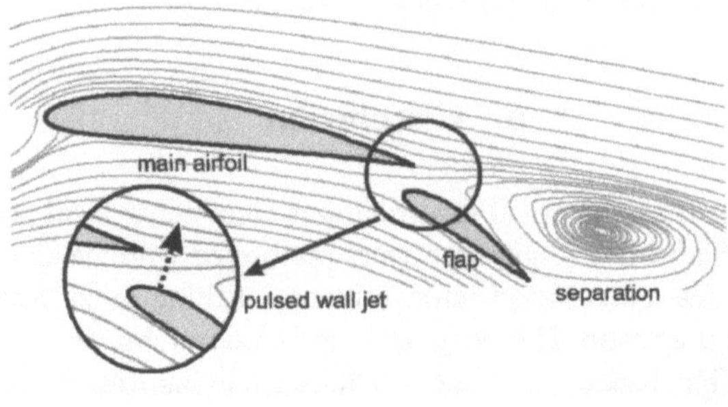

Figure 1 Basic high-lift setup with excitation at the flap's upper surface

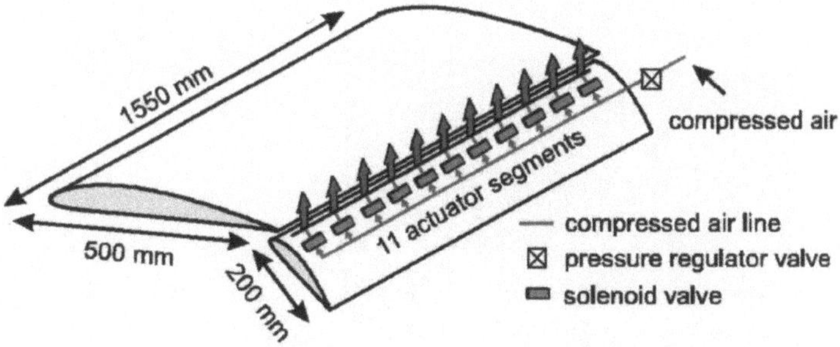

Figure 2 Wind tunnel model with the integrated excitation system.

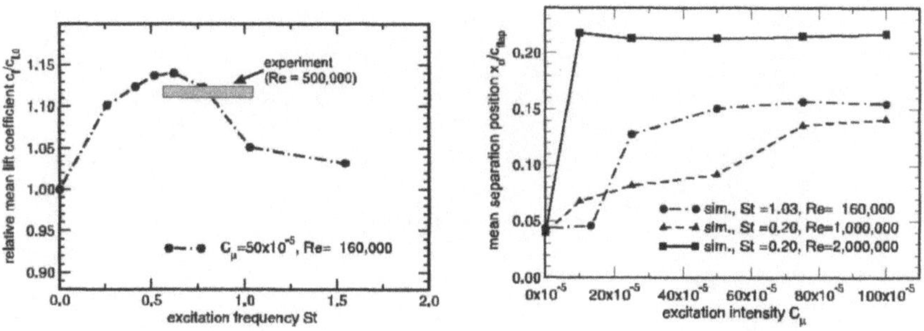

Figure 3 Left: Relative gain in lift for varying excitation frequency; right: Dependency of mean separation position on excitation intensity for different Reynolds numbers.

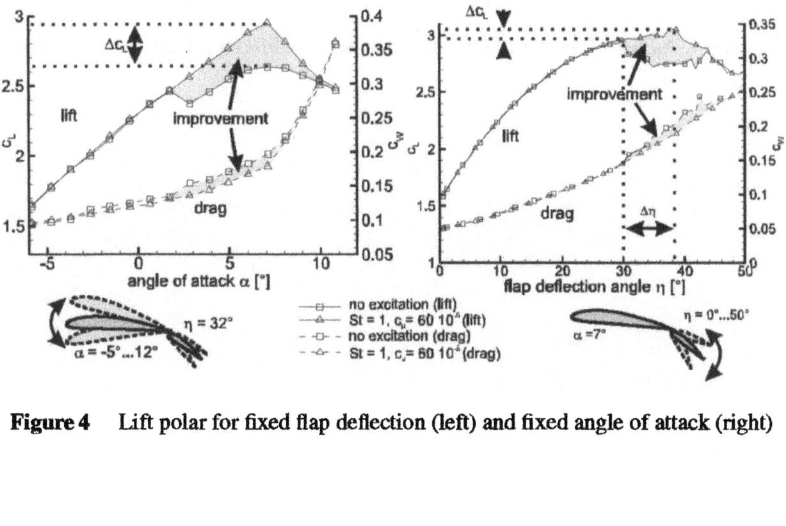

Figure 4 Lift polar for fixed flap deflection (left) and fixed angle of attack (right)

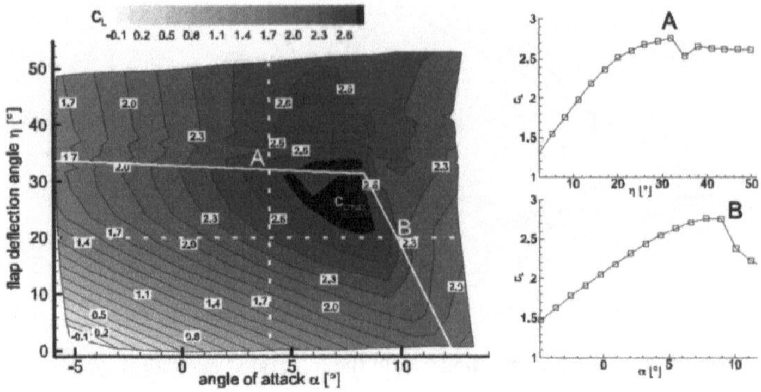

Figure 5 Lift polar for a variety of different angles of attack and flap deflections (white line indicates onset of separation).

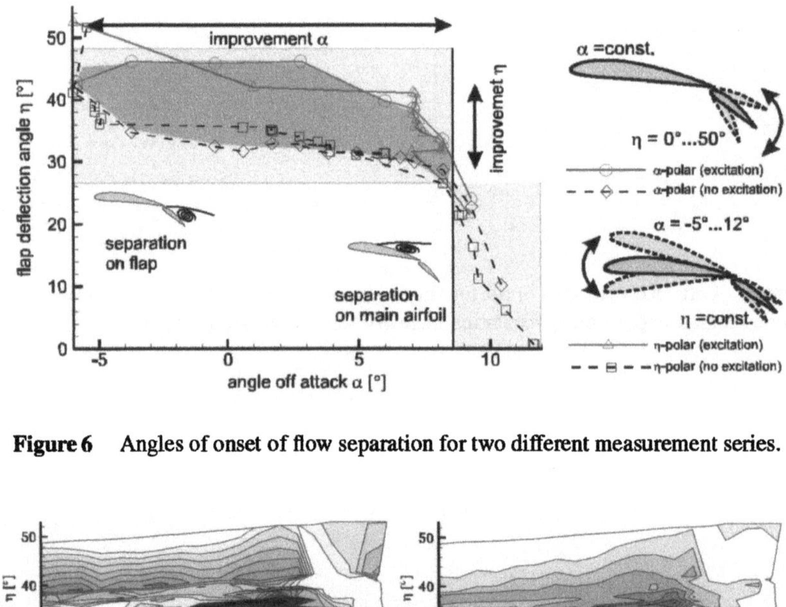

Figure 6 Angles of onset of flow separation for two different measurement series.

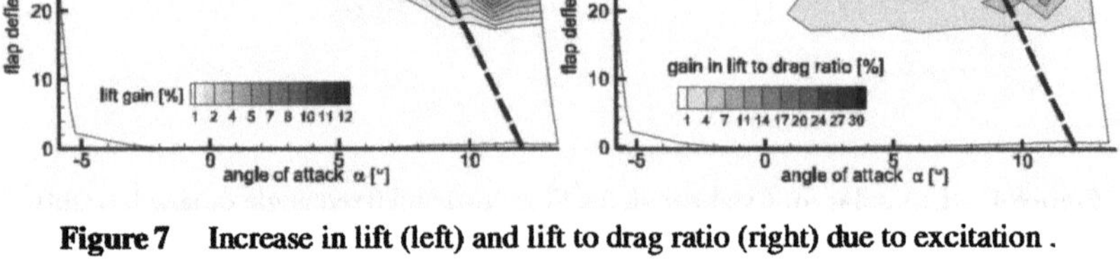

Figure 7 Increase in lift (left) and lift to drag ratio (right) due to excitation .

Influencing the Mixing Process in a Turbulent Boundary Layer by Pulsed Jet Actuators

P. Scholz, J. Ortmanns, C.J. Kähler and R. Radespiel

Technische Universität Braunschweig, Institut für Strömungsmechanik,
Bienroder Weg 3, 38106 Braunschweig, Germany
P.Scholz@tu-bs.de

Summary

Pulsed jet actuators are studied in a low-speed wind tunnel by means of phase-locked stereoscopic particle image velocimetry (PIV) to examine the interaction of the jet with a turbulent boundary layer flow along a flat plate. The aim is the transport of high-momentum fluid from the outer part of the boundary layer flow towards the wall to actively delay or avoid flow separation. It is shown that a properly arranged actuator jet produces a strong streamwise vortex, which is well suited to enhance the desired mixing process. The strength and position of this streamwise vortex is of primary importance for the efficiency of the actuator concept. Different jet-exit-hole geometries, their impact on the vortex-structure and their ability to suppress or delay separation are discussed.

1 Introduction

Since quite a while active control of flow separation is being studied in many varieties [1] because of the strong industrial interest. The most promising way is the stimulation of vortices in the boundary layer by using jet actuators. Such devices can be driven with a continuous ("steady vortex generator jets") or periodic mass flow ("pulsed jet actuators"). Periodic blowing is believed to be superior, as the required energy is smaller, whereas the effectiveness in suppressing stall is enhanced [2, 3]. Basically there are two control strategies:

1. Control by stimulation of natural *instabilities* [6].
2. Control by enhancing the turbulent *mixing*.

In this contribution only the mixing will be considered, as that method seems to work in a more general fashion. However, many details of the pulsing procedure are not very well understood.

Recently it has been shown quantitatively, that a symmetric actuator, e.g. a circular hole drilled perpendicular into the model surface (Fig. 1[(1)]) or a non-skewed slot, is not well suited to enhance turbulent mixing. The blockage of the base flow tends to promote flow separation because of the retardation of the near-wall flow downstream of the jet-axis [2, 5]. Asymmetric blowing with pitched holes (e.g. Fig. 1(2)) or skewed slots on the other hand is efficient, because a strong, streamwise vortex is generated, which transfers high-momentum fluid from the outer parts of the boundary layer towards the wall, provided that the jet axis is properly oriented with respect to the base flow [4, 5].

The goal of the present contribution is to further increase the mixing process by determining the most effective and efficient orifice shape for a pulsed jet actuator. In order to suppress stall, the actuator has to be able to decrease the loss of momentum due to the friction within the boundary layer. Therefore, the analysis has to focus particularly on the impact of the actuator-jet on the region of the boundary layer near the surface. It addresses the following questions:

1. Which kind of structures inside the boundary layer lead to an increase of effectiveness for stall control?
2. How can these structures be created most efficiently?

2 Experimental Setup

The experiments were performed on a 470 mm long, 248 mm wide flat plate, which was mounted horizontally between two end-plates in a Göttingen-type wind tunnel with a 940 mm long, open test section (Ø= 0.5 m, Tu = 0.85%, see [7]). The actuator orifice was a circular 1 mm hole, located 320 mm behind the leading edge, which results in a local Reynolds number of $\mathrm{Re}_x \approx 2.85 \cdot 10^5$ at a freestream-velocity of 13.6 m/s. The thickness of the undisturbed turbulent boundary layer was measured to be $\delta \approx 20$ mm with $\mathrm{Re}_{\delta^*} \approx 2750$. The dynamic actuation was realized with an electromechanical fast-switching valve which ran with 100 Hz at 50% duty cycle. It was supplied with a pressure of $p_V = 1.5$ bar. The jet's exit-velocity was measured with a hot-wire 1 mm above the outlet. Peak exit-velocity during the open part of the cycle was $V_j \approx 70$ m/s, giving a velocity-ratio of $V_R \approx 5.1$ [5].

In the following x denotes the direction of the bulk flow and z indicates the wall-normal direction. The origin coincides with the position of the orifice. The angle between the surface and the blowing axis is called the pitch angle α, the skew angle β is defined as the angle between the blowing axis and the bulk flow direction. Figure 1 gives an overview over the shape of some different actuator-orifices.

The flow field was explored phase-locked using a stereoscopic PIV system, which is capable to capture all three velocity components in the light-sheet plane (3C2D). The camera release was properly synchronized to the actuator process. Eight phase angles were investigated in steps of $\Delta\varphi = 45°$, whereby $\varphi = 0$ is defined as the moment, when the jet reaches 90% of it's maximum exit velocity. See [5] for a more detailed description of the experimental setup. The measurement results below $z \approx 1.5$ mm are affected by spurious vectors (see the lowest row of vectors in Figure 2 for $x = 25\,\mathrm{mm}$) due to reflections on the model surface. The analysis will therefore focus on effects happening above $z \approx 1.5\,\mathrm{mm}$.

All flow fields represent a phase-locked average of 100 cycles. For the visualization the velocity field of the undisturbed boundary layer was subtracted to display the fluctuations Δu, Δv and Δw due to the actuation. If an actuator is able to enforce an area of overspeed (positive Δu) close to the surface, then it becomes likely that separation can be delayed, as the momentum loss due to friction is being compensated. The greater the overspeed gets, the more effective is the control mechanism.

3 Design Studies

Figure 2 displays the results for an unsymmetrical orifice with $\alpha = 45°$ and $\beta = 90°$ for different distances behind the jet ($x = 25\,\mathrm{mm}$ to $60\,\mathrm{mm}$). The flow field is shown at a phase angle of $\varphi = 225°$. As will be shown later, this phasing gives a representative impression of the actuator's flow field. Clearly visible is a counter-rotating vortex pair (CRVP), which is slightly unsymmetrical due to the oblique jet-direction and the right, clockwise vortex features a little more peak-vorticity and is somewhat closer to the wall. As a result, on the right hand side of the CRVP a downwash transfers high-momentum fluid towards the wall such that the streamwise velocity in the near-wall region is increased up to $\Delta u_{\max} \approx 4.5\,\mathrm{m/s}$ or 45% of the local velocity. In contrast, on the left hand side of the CRVP only little overspeed is induced near the wall. This implies, that the left vortex is less efficient for flow control.

A time-series of the streamwise velocity fluctuation Δu for the plane $x = 40\,\mathrm{mm}$ is given in Figure 3. At $\varphi = 0°$ some disturbances of the previous cycle are noticeable, but it takes up to $\varphi = 135°$ until the vortex-system reaches the displayed plane (40 mm behind the orifice) by convection. Although the valve is opened only 50% of the period, each of the subfigures shows an overspeed-area close to the wall of at least $1.0\,\mathrm{m/s}$. This demonstrates clearly, that the benefit of the jet can be obtained throughout the complete cycle with – compared to a steadily blowing jet – reduced mass-flux.

The phase-angles $\varphi = 315°$ to $90°$ (the "passive part" of the period) actually feature the greater overspeed, although the CRVP itself is pronounced during the other phase-angles. The rationale for this behavior is given in Figure 4. At first it can be seen, that the counterclockwise vortex always features less peak vorticity than the clockwise one, thus giving the reason, why the overspeed is always greater on the right side of the vortex system. But peak vorticity of the clockwise vortex is greater during the active part of the period ($\varphi = 135° \sim 270°$), hence still lacking an explanation, why the overspeed is greater during the passive part. Utilizing the "natural" understanding, that the impact of a vortex is increased, if it is located closer, the following values should be noted: Overspeed is greater at $\varphi = 90°$, where $\omega_{min} = -1200$ 1/s at $z = 5.4$ mm, than for $\varphi = 270°$, where $\omega_{min} \approx -2150$ 1/s at $z \approx 14.7$ mm. So the greater overspeed during the passive part seems not to be a result of increased peak vorticity, but of closer distance to the area of impact. Thus, to design an efficient actuator, a combination of peak vorticity and position of the vortex has to be noted and not peak vorticity alone.

As pitched actuators are favorable for the generation of strong vortices the question of the optimum skew angle β comes up. In Figure 5 a comparison is shown for a pitched orifice $\alpha = 45°$ with different skew angles. $\beta = 120°$ means, that the jet has a component *in* bulk flow direction (coflow, Fig. $1^{(2)}$), whereas the orifice with $\beta = 60°$ is partly blowing in countercurrent with the oncoming flow (counterflow). The coflow actuator does not perform very well. Neither can the momentum of the jet be utilized to accelerate the boundary layer near the wall, nor can the CRVP shift high-momentum fluid towards the wall. The counterflow configuration on the other hand does perform better. Compared to the perpendicular configuration peak vorticity is of the same magnitude, but the vortex core is located closer to the wall. As expected, a slight increase in peak overspeed near the wall is discoverable.

Beside the geometry optimization of a single actuator, efficiency can further be increased by utilizing the interactions of multiple actuators. Experiments have shown that the use of two diverging actuators, as illustrated in Figure 6, is very promising. The spacing of the two orifices is 10 mm, both were pitched to $\alpha = 45°$ and skewed to $\beta = 90°$, but in opposite directions (Fig. $1^{(3)}$). Peak vorticity of the two inner vortices is greater than for an isolated orifice. Additionally, the position of the two CRVPs is much closer to the wall, because the two vortices induce a mutual downwind, helping to keep the vortex cores close to the surface. The two downwash areas are being combined, thus leading to a maximum velocity fluctuation near the wall of $\Delta u_{max} \approx 5.5$ m/s, which equals 55% of the local velocity of the oncoming boundary layer. Beside the peak overspeed, the area, which is being accelerated, is larger than for the single actuator. Not shown here is the increase of peak velocity during the passive part of the period, it is even greater for the double-hole actuator than for the single, pitched orifice.

4 Conclusion

Different pulsed jet actuators were examined experimentally using PIV-techniques, to study their efficiency for active flow control. Blockage, displacement and the evolution of a counter-rotating vortex pair were discovered. The latter enhances the turbulent mixing by transferring high-momentum fluid towards the wall and hence increases the streamwise velocity near the surface. The vortex strength – namely peak vorticity – does have an influence on the amount of overspeed, but additionally the distance of the vortex core relative to the wall turned out to be an important parameter for efficient flow control.

From several orifice geometries that were investigated, a 45° pitched hole either blowing perpendicular or with a slight component in countercurrent to the oncoming bulk flow turned out to be capable. Regarding simple actuator arrays it could be shown that an increase of the efficiency is also possible by utilizing the interactions of two properly arranged actuators.

In the future a more refined study has to follow, to find the actual optimum of pitch and skew angle. The relationship between downwash and additional streamwise velocity, depending on the parameters of the oncoming boundary layer, and furthermore the influence of pulsing frequency f, duty cycle Δ, supply pressure p_V and velocity-ratio V_R respectively has to be evaluated for single orifices as well as for arrays of actuators.

Acknowledgements

The project is part of BMWA's Lufo III "Innovative Hochauftriebskonfigurationen" and is being dealt with as participant of the joint research project "Dynamische 3D Strömungskontrolle".

References

[1] Fiedler, H.E.; Fernholz, H.-H.: "On Management and Control of Turbulent Shear Flows", Prog. Aerospace Sci., Vol. 27, Pergamon Press, 1990
[2] Johari, H.; Rixon, G.S. "Effects of Pulsing on a Vortex Generator Jet", AIAA Journal, Vol. 41, No. 12, 2003
[3] McManus, K.R.; Joshi, P.B.; Legner, H.H.; Davis, S.J. "Active Control of Aerodynamic Stall using Pulsed Jet Actuators", AIAA Paper No. 95-2187, 1995
[4] Milanovic, I.M.; Zaman, K.B.M.Q. "Fluid Dynamics of Highly Pitched and Yawed Jets in Crossflow", AIAA Journal, Vol. 42, No. 5, 2004
[5] Ortmanns,J.; Kähler,C.J. "Investigation of Pulsed Actuators for Active Flow Control Using Phase Locked Stereoscopic Particle Image Velocimetry", presented at the 12th Intl. Symposium on Application of Laser Techniques to Fluid Mechanics, Lisbon, 2004
[6] Tinapp, F.: "Aktive Kontrolle der Strömungsablösung an einer Hochauftriebskonfiguration", Dissertation an der Fakultät V: Verkehrs und Maschinensysteme, Institut für Luft- und Raumfahrt, Technische Universität Berlin, 2001
[7] http://www.tu-braunschweig.de/ism/institut/wkanlagen/kwb

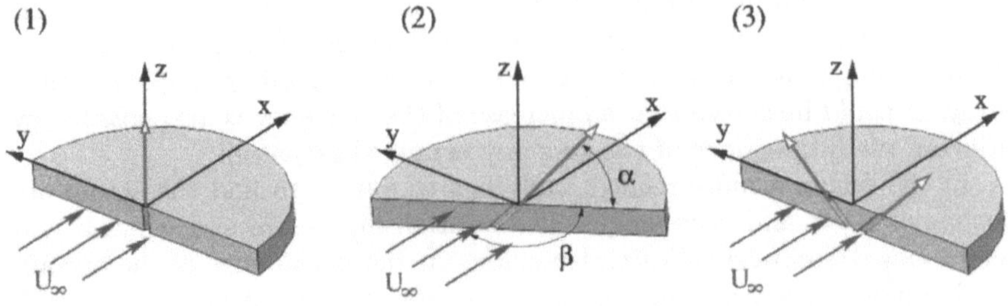

Figure 1 Different actuator-orifices: (1) symmetric actuator, $\alpha = 90°$, β not defined; (2) asymmetric actuator, $\alpha = 45°$, $\beta = 120°$ (coflow); (3) simple array with two orifices, $\alpha = 45°$, $\beta = \pm 90°$

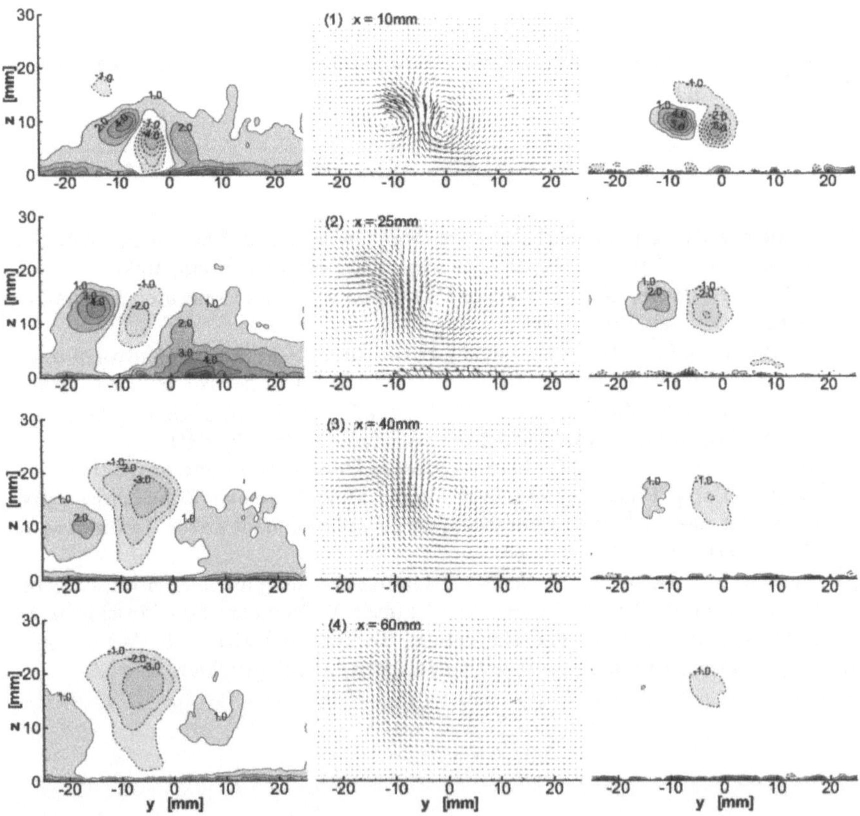

Figure 2 Flow field due to actuation with orifice $\alpha = 45°$ $\beta = 90°$; phasing: $\varphi = 225°$; left: velocity fluctuation Δu [m/s]; middle: velocity fluctuations Δv and Δw [m/s]; right: vorticity ω_x [10^3/s].

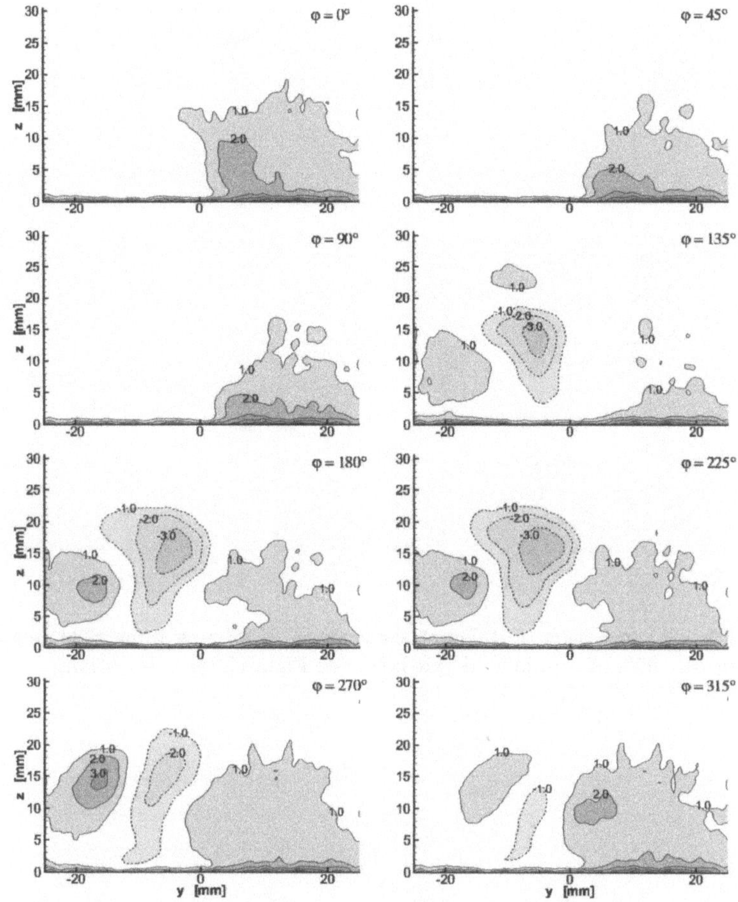

Figure 3 Time series of an actuation with orifice $\alpha = 45°$, $\beta = 90°$; streamwise velocity fluctuation Δu [m/s], 40 mm behind exit-hole.

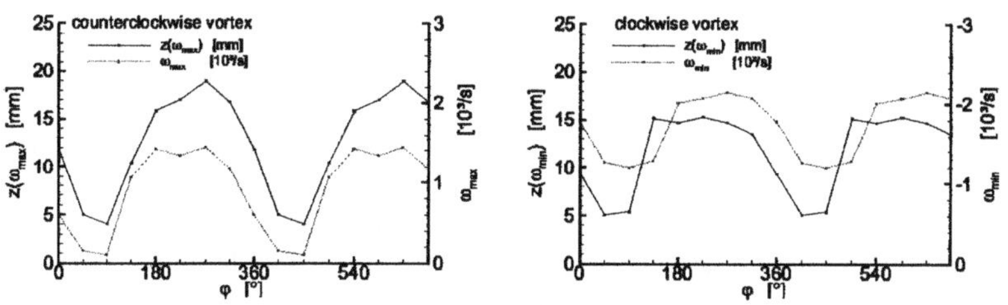

Figure 4 Temporal development of position and peak-vorticity for the counter-rotating vortex pair; orifice $\alpha = 45°$, $\beta = 90°$, 40 mm behind exit-hole.

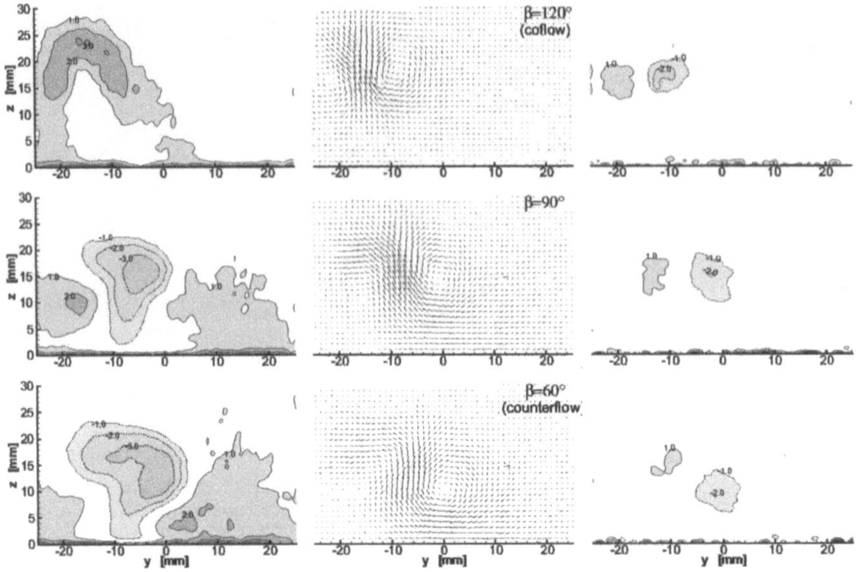

Figure 5 Comparison of different skew angles β for unsymmetrical single-hole actuator $\alpha = 45°$; 40 mm behind exit-hole; See Figure 2 for description.

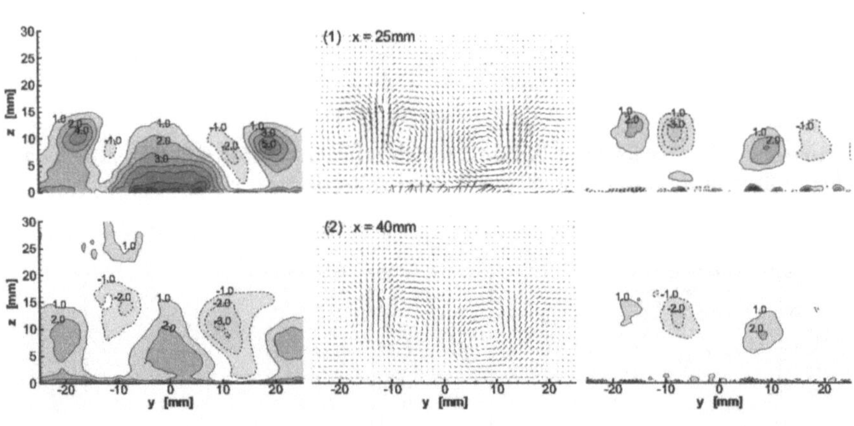

Figure 6 Flow field due to actuation with diverging double-hole actuator $\alpha = 45°$, $\beta = 90°$; phasing: $\varphi = 225°$; 25 mm behind exit-hole; See Figure 2 for description.

Active Flow Control by Surface Smooth Plasma Actuators

B. GÖKSEL[1], I. RECHENBERG[1]
[1] Technische Universität Berlin, Institut für Bionik und Evolutionstechnik, Ackerstr. 71-76, Sekr. ACK1, D-13355 Berlin, Germany

Summary

Surface smooth plasma actuators were used to control leading-edge flow separation on the flying wing airfoil Eppler E338 for angles of attack of up to 12° past stall at low Reynolds numbers. The plasma actuators were operated over a range of free-stream speeds from 2.2 to 6.6 m/s giving chord Reynolds numbers from 26K to 79K. The plasma actuators produced a 2-D wall jet in the flow direction along the surface of the airfoil and thus added momentum to the boundary layer. Each actuator consisted of two metal electrodes separated by a dielectric layer which was part of the airfoil surface. At all five free-stream speeds from 2.2 to 6.6 m/s, the maximum lift coefficient could be reached and the stall regime relaxed. The power to achieve this was approximately 17 watts per meter or 8.6 watts per actuator over the span width. It was found that the application of low power plasma actuators could simplify the design of mini and micro air vehicles (MAVs) by calculating with the maximum possible lift coefficients obtained from simplified fluiddynamic equations.

1 Introduction

At low Reynolds numbers (less than 200K) the flow phenomena are more complicated than those at high Reynolds numbers because of some peculiar features, namely:

- appearance of leading edge laminar flow separation
- nonlinear lift/drag characteristics caused by laminar separation bubble
- l.ift/drag hysteresis at static conditions.

In this regard, the design of mini and micro air vehicles (MAVs) flying at low speeds below 25 mph is mainly based on empirical and intuitive work. Computational analysis is difficult because many of the simplifications of the fluiddynamic equations valid for large Reynolds numbers are not valid for MAV flight regimes [4]. Active flow control can help to overcome the basic design problems of MAVs [2], [3]. Because of the weight and space limitations in small fixed wing construction only unconventional, surface-integrated flow control methods like piezo and plasma actuators are considered [1], [3], [5], [6]. The plasma technique used in these experiments is described in the following.

The plasma actuator consists of two metal electrodes separated by a dielectric layer which is part of the airfoil surface. Sufficiently high voltages (at low radio frequencies in the kHz-range) supplied to the actuator causes the air to weakly ionize at the edges of the upper electrodes. These are regions of high electric field potential. In asymmetric configuration, the plasma is only generated at one edge, Figure 1.

The plasma moves to regions of increasing electric field gradients and induces a neutral air flow by momentum transfer due to Lorentzian collisions. This causes a plasma induced pressure gradient named "electrostatic body force" by Roth [7] which in one-dimension is formulated as follows:

$$F_E = \frac{d}{dx}\left(\frac{1}{2}\varepsilon_0 E^2\right). \tag{1}$$

In this work, plasma induced downstream orientied wall-jets were used to increase the aerodynamic efficiency of a flying wing airfoil by leading-edge separation control at low Reynolds numbers and large angles of attack.

2 Experimental Setup

The experiments to separation flow control by surface smooth plasma actuators were conducted in the small free-stream wind-tunnel in the Institute of Bionics and Evolutiontechnique at Technical University of Berlin. The tunnel has a free-stream cross-section of 0.60 m diameter and a maximum speed of 7.0 m/s. A sensitive two component balance was used to measure the lift and drag forces at five free-stream speeds of 2.2, 3.3, 4.4, 5.5 and 6.6 m/s.

2.1 Airfoil

The airfoil used in the experiment was an Eppler E338, Figure 2. This airfoil had been previously used in flow control experiments with high voltage (10 – 20 kV) charged corona discharge wires [2], [3] and was originally chosen for a mini flying wing design.

The airfoil has a 17.8 cm chord and a 50.0 cm spanwise length. For the five free-stream speeds this give a range of Reynolds numbers from 26K to 79K. End plates from plexiglass with 22 cm diameter were used to reduce end effects, Figure 3.

2.2 Plasma Actuator

The plasma actuator consists of 12 pairs of tinned copper electrodes in asymmetric arrangement separated by a 0.5 mm thick, flexible and self-adhesive Teflon layer which is bonded directly to the surface of the Eppler E338 airfoil and spans its width. The electrodes are made from 0.070 mm thick tinned copper foil tape and have a distance to each other of 9 mm. In these experiments, only two electrodes on the leading edge were supplied with high voltage of 5.8 kV p-p at the operating frequency of 11 kHz, Figure 4. The power consumption was about 17 Watts per meter and one-fourth of the value used in the experiments by Post and Corke with 11 kV p-p [5], [6]. Data acquisition from the polyphase high voltage power supply was done using a self-programmed LabView 8-channel oscilloscope, Figure5.

3 Results

The results presented here relate to laminar leading-edge separation flow control experiments at low Reynolds numbers from 26K to 79K and large angles of attack up to 25 degrees using the Eppler E338 flying wing airfoil.

At Reynolds numbers lower than 60K the flow already separates at small angles of attack below 10 degrees. So the maximum lift coefficient is between 0.6 to 0.8. With plasma actuation the flow separation can be delayed to higher angles of attack up to 16 degrees giving maximum lift coefficients between 1.1 and 1.2. Furthermore, the stall regime is more relaxed up to 24 degrees, Figures 6 and 7.

At Reynolds numbers lower than 30K the lift coefficient drops dramatically even at zero angle of attack but can be recovered after activating the plasma actuators. The lift characteristics is more nonlinear, Figures 8 and 9.

At Reynolds numbers higher than 60K the flow separates at an angle of attack of 20 degrees. Plasma actuation gives a more relaxed stall regime, Figures 8 and 9.

At a free-stream speed of 3.3 m/s, giving Re = 39750, the flow reattached produced by the actuator increases lift by up to 129% for $\alpha = 16°$, Figure 6.

At a free-stream speed of 4.4 m/s, giving Re = 53000, the flow reattached produced by the actuator increases lift by up to 78% for $\alpha = 12°$, Figure 7.

At a free-stream speed of 2.2 m/s, giving Re = 26500, plasma actuation increases lift by 57% for $\alpha = 0°$ and at maximum by 103% for $\alpha = 10°$, Figures 8 and 9.

At a free-stream speed of 6.6 m/s, giving Re = 79500, plasma actuation relaxes the stall regime and increases lift by 62% for $\alpha = 23°$, Figures 8 and 9.

4 Conclusion

Weakly ionized plasma actuators were used for leading-edge flow separation over the flying wing airfoil Eppler E338 for angles of attack of up to 12° past stall at low Reynolds numbers. The effectiveness of the plasma actuator was demonstrated for a range of velocities from 2.2 to 6.6 m/s, corresponding with chord Reynolds numbers from 26K to 79K.

The plasma actuator was arranged in an asymmetric configuration to produce a quasi-steady 2-D wall jet in the flow direction and thus add momentum to the boundary layer. Each of the 12 actuators over the chord length consists of two metal electrodes separated by a dielectric layer which is part of the airfoil surface. In the experiments, only two actuators with 180° phase shift were activated at the leading edge.

At all five free-stream speeds from 2.2 to 6.6 m/s, giving a Reynolds number range from 26K to 79K, the maximum lift coefficient could be reached and the stall regime relaxed. The power to achieve this was approximately 17 watts per meter or 8.5 watts per actuator over the span width.

The application of low power plasma actuators can simplify the design of mini and micro air vehicles by calculating with the maximum possible lift coefficients obtained from simplified fluiddynamic equations. This is of specific interest for the design of bionic flying wing models, currently under development at the Institute of Bionics and Evolutiontechnique, Figure 10. The Reynolds numbers in the outer wing are in the range from 26K to 97K at speeds up to 11 m/s (25 mph).

Acknowledgement

This work was supported by the Bundesministerium für Bildung und Forschung (bmb+f), Germany, under contract No. 0311984.

References

[1] B. Göksel. "Applications of Electric Fields in Aerodynamics". Thesis in Aerospace Engineering, Technical University of Berlin, 1997, p. 120 (in German).

[2] B. Göksel. "Improvement of Aerodynamic Efficiency and Safety of Micro Aerial Vehicles by Separation Flow Control in Weakly Ionized Air". In Proceedings of the German Aeronautics Congress, DGLR-JT2000-203, 2000, pp. 1317-1331 (in German).

[3] B. Göksel, I. Rechenberg. "Wind-Tunnel Experiments to Active Flow Control at Low Re Numbers". In Proceedings of the First European Micro Air Vehicle Conference and Flight Competition, EMAV 2004.

[4] T. J. Mueller (ed.). "Fixed and Flapping Wing Aerodynamics for Micro Air Vehicle Applications". In Progress in Aeronautics and Astronautics, Volume 195, 2001, pp. 115-142.

[5] M. L. Post and T. C. Corke. "Separation Control on High Angle of Attack Airfoil Using Plasma Actuators". AIAA 2003-1024, 2003.

[6] M. L. Post and T. C. Corke. "Separation Control Using Plasma Actuators – Dynamic Stall Control on an Oscillating Airfoil". AIAA 2004-2517, 2004.

[7] J. R. Roth, D. Sherman and S. Wilkinson. "Boundary Layer Flow Control with One Atmosphere Uniform Glow Discharge Surface Plasma". AIAA 1998-0328, 1998.

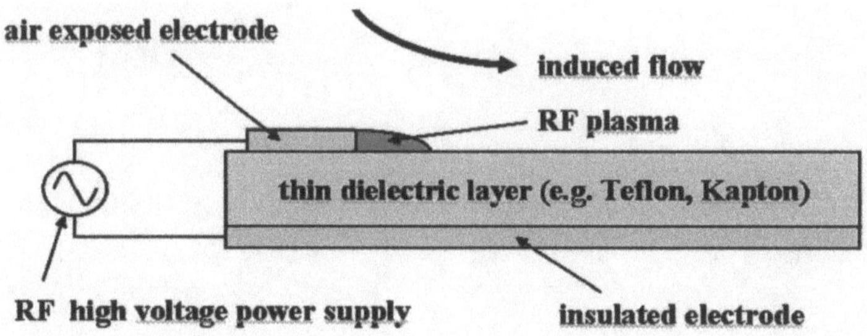

Figure 1 Schematic of asymmetric plasma actuators used in experiments

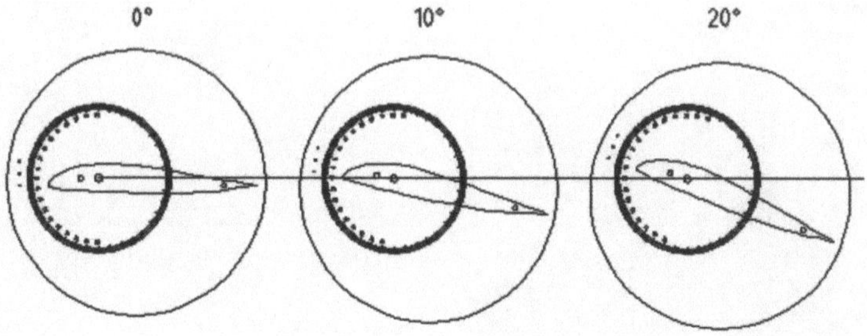

Figure 2 2-D wing with Eppler E338 airfoil and plexiglass end plates

Figure 3 Eppler E338 wing with plasma actuators in the small free-stream wind-tunnel

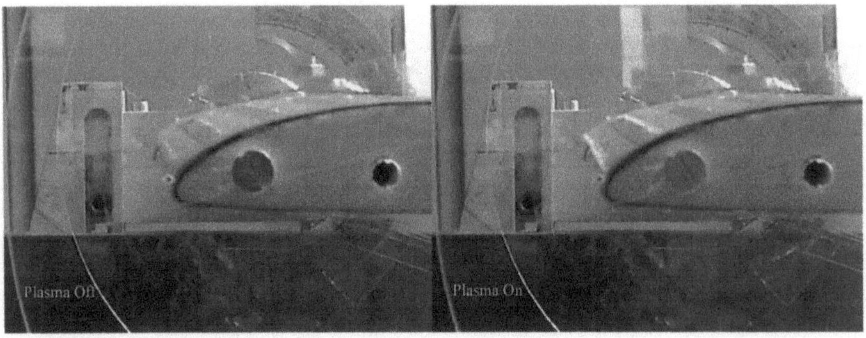

Figure 4 Two activated plasma actuators with 180° phase shift on the leading edge

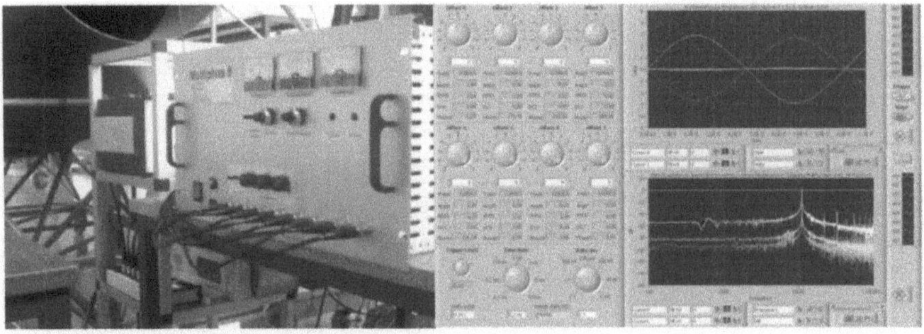

Figure 5 Polyphase high voltage power supply with LabView 8-channel PC oscilloscope

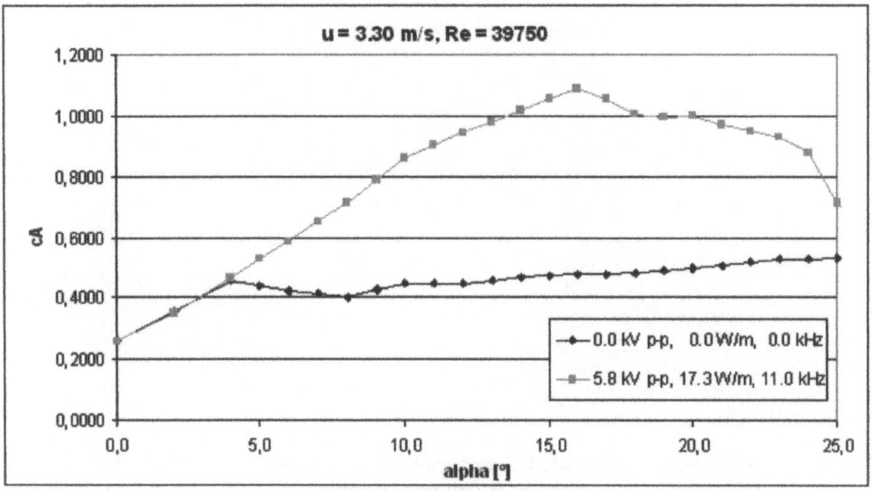

Figure 6 Comparison of lift coefficient versus angle of attack at Re = 39750 without and with RF plasma actuation

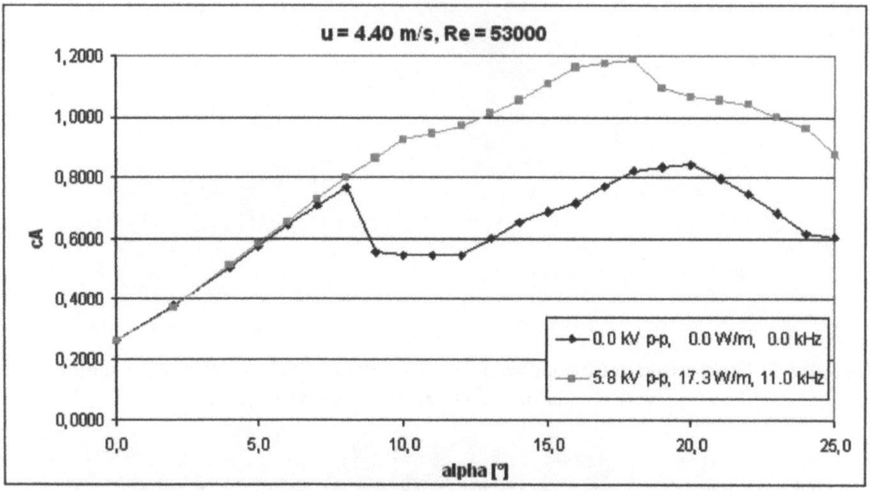

Figure 7 Comparison of lift coefficient versus angle of attack at Re = 53000 without and with RF plasma actuation

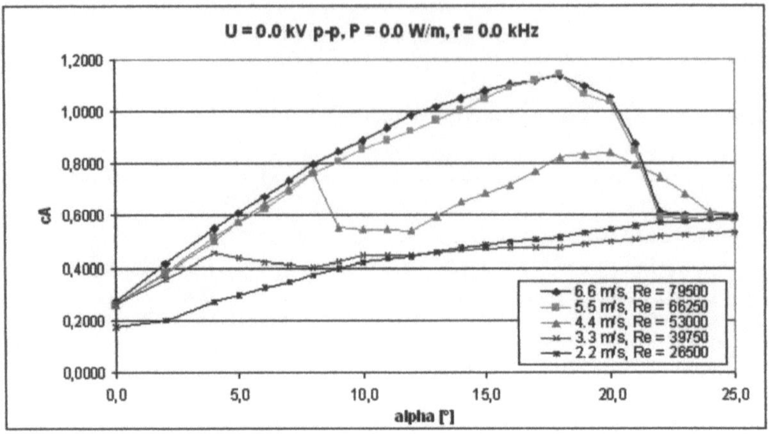

Figure 8 Comparison of lift coefficient versus angle of attack at different Re numbers without RF plasma actuation

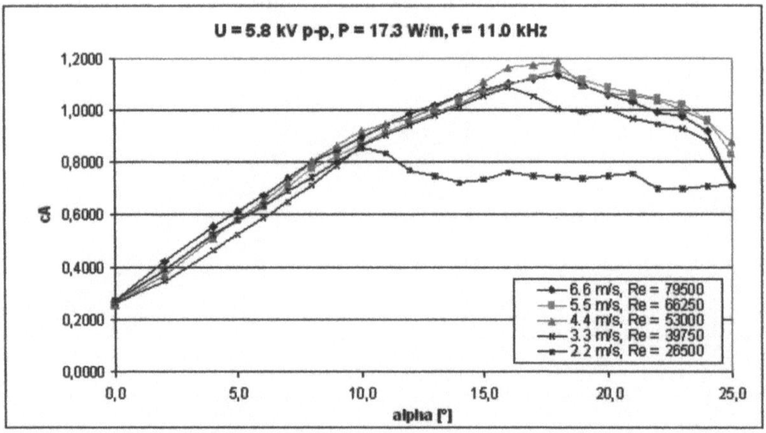

Figure 9 Comparison of lift coefficient versus angle of attack at different Re numbers with RF plasma actuation at the leading edge

Figure 10 Bionic flying wing model with Eppler E338 airfoil

Boundary Layer Influence on Supersonic Jet/Cross-Flow Interaction in Hypersonic Flow

M. Havermann and F. Seiler

French-German Research Institute of Saint-Louis (ISL)
5, Rue du Général Cassagnou, 68301 Saint-Louis, France
havermann@isl.tm.fr

Summary

Supersonic lateral jets are a convenient method for aerodynamic steering of bodies flying at hypersonic speeds. An accurate prediction of the resulting aerodynamic force is difficult, because the flowfield around such a jet in a hypersonic cross-flow is quite complex. In addition to the jet thrust a considerable pressure load on the body can be generated by the jet/cross-flow interaction. Several flow and configuration parameters have an effect on the flow interaction pressure force. In this study, the influence of the body's boundary layer upstream of the side jet is studied experimentally. A shock tunnel is used to generate a Mach-6-flow around a cone-cylinder model with a side jet. By means of flow visualization and unsteady pressure measurements the flowfield around the side jet is characterized for different conditions.

1 Introduction

A promising and convenient way to steer high-speed bodies flying in the earth's atmosphere is the use of supersonic lateral jets. Compared to rudders or flaps, no aerothermal loads are present and no drag is induced when the side jet is inactive. Additionally, at high flight altitudes with low stagnation pressures, side jets offer the only possibility to exert sufficient aerodynamic forces for flight control. A recent example of such an application is given by the flight of SpaceShipOne, which took place on October 4th, 2004 [3]. After being released from the carrier plane at an altitude of 15 km, the pilot continued climbing and accelerated to a Mach number of about 3. Shortly after, a rolling motion started, which could no longer be controlled by the flaps. A side-jet system was then triggered to stabilize the flight path. As known, the mission was successful and Starship One was awarded with the Ansari X-prize.

A disadvantage of lateral control jets, however, is the complex flow pattern around the jet: the jet is bent downstream and a bow shock appears in front of the jet. This shock interacts with the boundary layer leading to a separation shock. Downstream of the jet, a wake with a recirculation zone is formed. Therefore, in addition to the jet thrust, an aerodynamic force resulting from the flow interaction around the jet acts on the body, see Fig. 1. The worst case would be a cancellation of the jet thrust or even a thrust reversal. Therefore, for a correct design of a control system it is necessary to know the effects of the different parameters on the interaction force. In this study, the influence of the lateral jet pressure ratio and the boundary layer is studied experimentally with a cone-cylinder model mounted in the measurement section of a shock-tunnel facility.

2 Experimental arrangements

2.1 The ISL shock tunnel STA

The experiments were conducted in the ISL shock-tunnel facility STA, which allows to duplicate real atmospheric flight conditions for high Mach numbers. The facility consists of a conventional shock tube with an inner diameter of 100 mm connected to a supersonic nozzle and a 10 m^3 dump tank (Fig. 2). The shock tunnel is operated in the shock-reflection mode with tailored interface conditions resulting in testing times of about 2 milliseconds. A single, stainless-steel diaphragm separates the shock tube into a 2.7-m-long driver tube and a 18.4-m-long driven tube. A parallel-flow Laval nozzle with an exit diameter of 219 mm and a design Mach number of 6 is used to expand the flow into the measurement section. For the experiments of this study, two atmospheric flight conditions with altitudes of 15 and 21 km were adjusted by varying the initial pressures in the shock tube. The corresponding freestream flow properties for nitrogen test gas are listed in Table 1.

2.2 The cone-cylinder model

The model consists of a conical nose with a half-angle of 7.1 ° and a circular cylinder with a diameter of 50 mm. The total length is about 507 mm and the exit axis of the side jet is positioned 58 mm behind the cone shoulder, see Fig. 3. The side-jet is generated by expanding pressurized nitrogen gas through a supersonic nozzle (exit diameter: 8 mm; Mach number: 3.3). Before the experiment is started, the stagnation pressure of the jet is adjusted by pressurizing high-pressure bottles, which are connected to the stagnation chamber of the jet. An appropriate triggering of a valve assures that the lateral jet is expanded to a steady state before the shock-tunnel flow starts.

2.3 Flow visualization and pressure measurement

The variation of the density gradient in the compressible flow around the model was visualized by a differential interferometer (Fig. 4). In the set-up used here, the interferometer was adjusted to an infinite fringe spacing so that the interferograms look similar to schlieren pictures. The illumination path is built up with an optical Z-configuration using two identical spherical mirrors (diameter of 250 mm, focal length of 1.5 m). A single-picture CCD camera with a resolution of 1280 x 1024 pixels (PCO PixelFly S) in combination with a continuous light source captures the visualized flowfield during the testing time. The camera's shutter time was set to $100\,\mu s$.

Up to 18 piezoresistive pressure transducers (Kulite XCL-080) were mounted in the model to measure the surface pressure around the lateral jet. The transducers were positioned in an array behind small holes (1 mm in diameter and depth, see Fig. 5), which resulted in a sufficiently good time response for the measurement. A PC-based transient memory with 12 bit A/D-resolution digitized and stored the amplified signals at a sampling rate of 2 Msamples/s. The measurement error is estimated to $\pm 8\%$.

3 Experimental results

The experimental results consist of differential interferograms and pressure distributions for the two altitude conditions and three corresponding jet pressure ratios. The stagnation pressures of the lateral jet were set to values from 8 to 25 bar for each altitude. This resulted in jet pressure ratios (defined as the jet stagnation pressure to the freestream pressure) from 63 to 440.

The differential interferograms of each experiment are shown in Figures 6 and 7 for the 15 and 21 km altitude conditions, respectively. It can be seen that if the pressure ratio is increased, the jet extends more into the external flow. Additionally, the separation shock region in front of the jet also extends more upstream. To quantify this effect, the extension was measured on the pictures and normalized by the jet exit diameter of 8 mm. The results are shown in Fig. 8, which clearly shows that the upstream extension of the separation shock region depends on the pressure ratio, but in a different manner for the two altitudes. If a similar jet pressure ratio of about 150 is compared, the influence of the altitude condition becomes evident: the normalized upstream extension of the separation shock is nearly twice as large for the 21 km altitude condition (image 1 in Fig. 7) as for the 15 km altitude condition (image 2 in Fig. 6).

For each experimental run, the pressure measurement was obtained simultaneously. It is easier to interpret the pressure data, if only the pressures measured directly up- and downstream of the lateral jet are presented. The static wall pressures normalized by the corresponding freestream pressure are plotted against the jet pressure ratio in Figures 9 and 10 for the 15 and 21 km altitude conditions, respectively. Both the jet centerline pressures and the averaged pressures from the first left and right row are shown. Additionally, the pressures for the no-jet case are depicted to give an idea of measurement accuracy. Again, a dependence of the jet pressure ratio on the pressure distribution is visible. However, a clearer estimation of the intensity of the jet/cross-flow interaction can be obtained by relating the upstream pressures to the downstream pressures, which is an estimation of the intensity of the interaction. In Fig. 11 it can be seen that the ratio of upstream to downstream pressure is proportional to the jet pressure ratio. This time, however, the 15 km altitude condition seems to have a stronger effect compared to the results obtained from the flow visualization. For a similar pressure ratio of about 150, the static pressure ratio is about 30% higher for the 15 km than for the 21 km altitude condition.

4 Discussion

The experimental results have shown that the jet interaction zone depends qualitatively and quantitatively on the jet pressure ratio. Additionally, a difference for the extension and the strength of the jet interaction zone was found between the two altitude conditions. This difference can be explained with the boundary layer state upstream of the jet, which depends on the Reynolds number. The unit freestream Reynolds numbers given in Table 1 were multiplied by a characteristic length of 256 mm, which is the distance from the cone tip to the lateral jet orifice (Fig. 3). The corresponding Reynolds numbers are $6 \cdot 10^6$ and $2.4 \cdot 10^6$ for the 15 km and the 21 km altitude condition, respectively. Using a prediction correlation for the transition Reynolds number in hypersonic flow by Bowcutt and Anderson [1], the transition Reynolds number is calculated to $3 \cdot 10^6$. A measurement conducted by Jack [2] with a a cone-cylinder model at a Mach number of 3.12 yielded a transition Reynolds number of $2.4 \cdot 10^6$ at a similar position behind the cone shoulder as in our case. It can therefore be concluded that the boundary layer state at the lateral jet exit is turbulent for the 15 km altitude condition whereas it is laminar or in the transition region for the 21 km altitude condition.

5 Conclusion

The interaction of a supersonic side jet with a hypersonic cross-flow was studied experimentally on a cone-cylinder model in the ISL shock-tunnel facility. The flow visualization showed that an increased pressure ratio leads to an extended separation shock region upstream of the lateral jet with a different correlation for the two altitude conditions. The pressure measurements indicated that the intensity of the flow interaction not only scaled with the jet pressure ratio, but also depended on the altitude conditions. For a similar jet pressure ratio a significant difference was found between the two altitude conditions indicating a stronger interaction for the lower altitude. This effect can be explained with different boundary layer states on the body in front of the lateral jet. For quantitative estimations of the aerodynamic interaction force it is therefore important to know if the boundary layer in front of the lateral jet is laminar or turbulent.

References

[1] J.D. Anderson Jr.: "Hypersonic and High Temperature Gas Dynamics", McGraw-Hill, 1989, pp. 271.
[2] J. R. Jack: "Aerodynamics of Slender Bodies at Mach Numbers of 3.12 and Reynolds Numbers From $2x10^6$ to $15x10^6$, III-Boundary Layer and Force Measurements on a Slender Cone-Cylinder Body of Revilution", NACA RM E53B03, 1953.
[3] Scaled Composites: www.scaled.com, 2004.

Table 1 Flow conditions used for the experiments.

Altitude [km]	M_∞ [-]	u_∞ [m/s]	p_∞ [kPa]	T_∞ [K]	ρ_∞ [kg/m^3]	Re_∞ [m^{-1}]
15	5.94	1782	11.9	217	0.185	$23.5 \cdot 10^6$
21	5.94	1782	4.8	217	0.075	$9.5 \cdot 10^6$

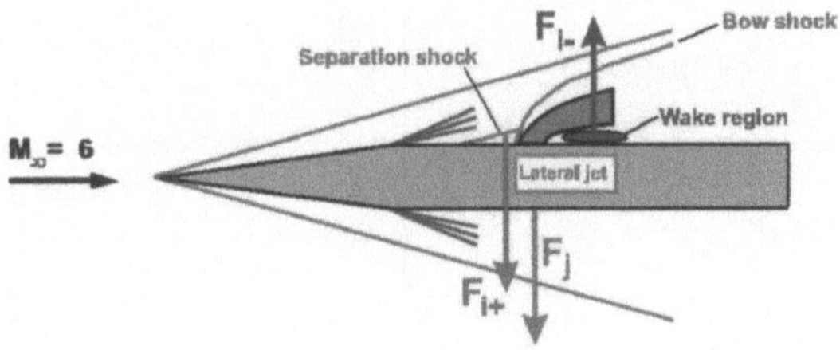

Figure 1 Flowfield around a side jet in a hypersonic cross-flow.

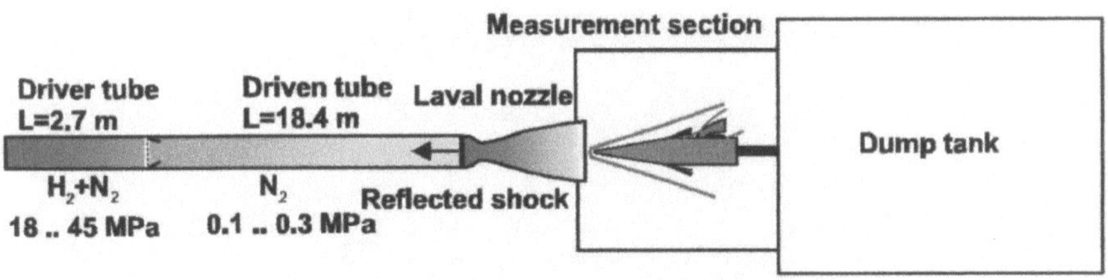

Figure 2 The ISL shock tunnel STA.

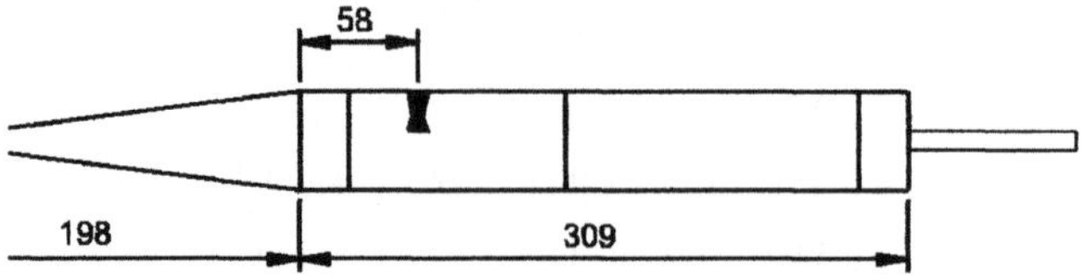

Figure 3 Geometry of the cone-cylinder model.

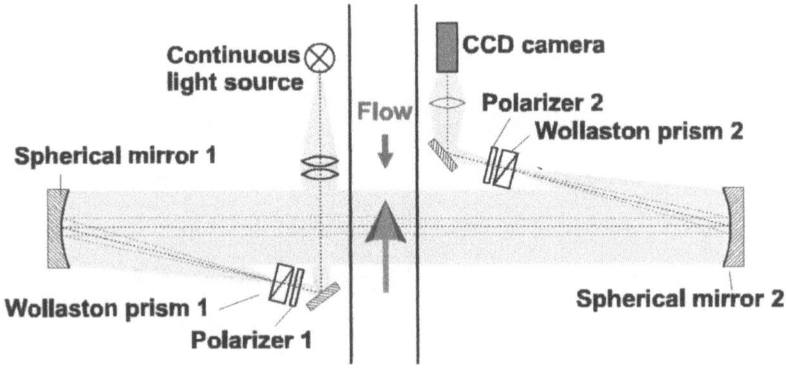

Figure 4 The differential interferometer.

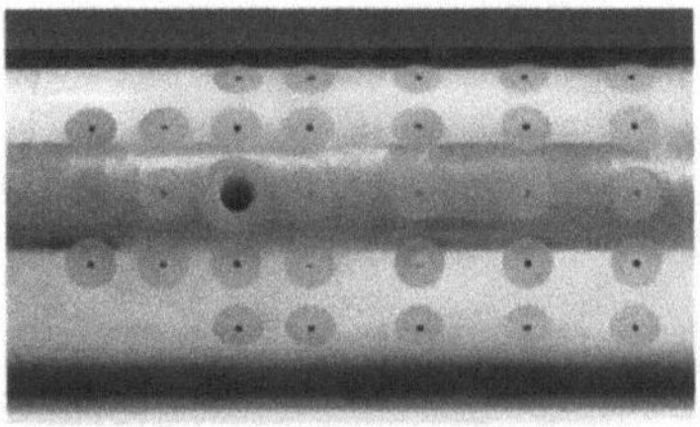

Figure 5 Array of pressure transducers grouped around the lateral jet.

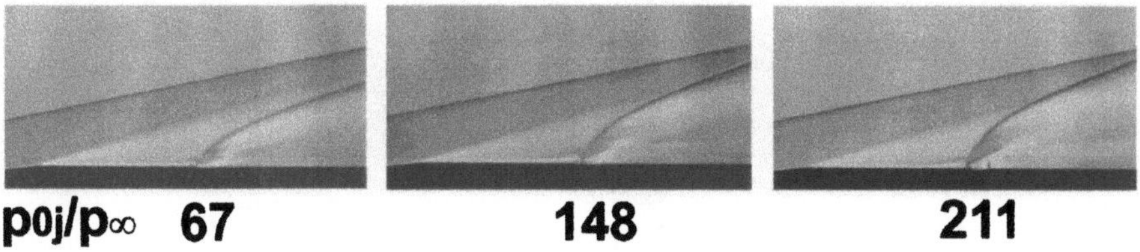

Figure 6 Differential interferograms for three jet pressure ratios at the 15 km altitude condition.

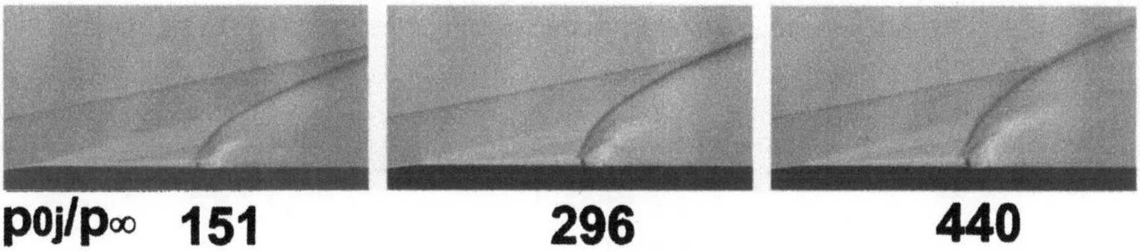

Figure 7 Differential interferograms for three jet pressure ratios at the 21 km altitude condition.

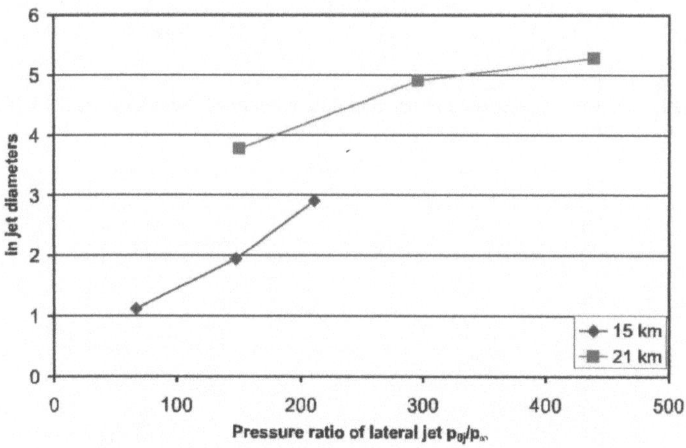

Figure 8 Normalized extension of the separation shock upstream of the lateral jet.

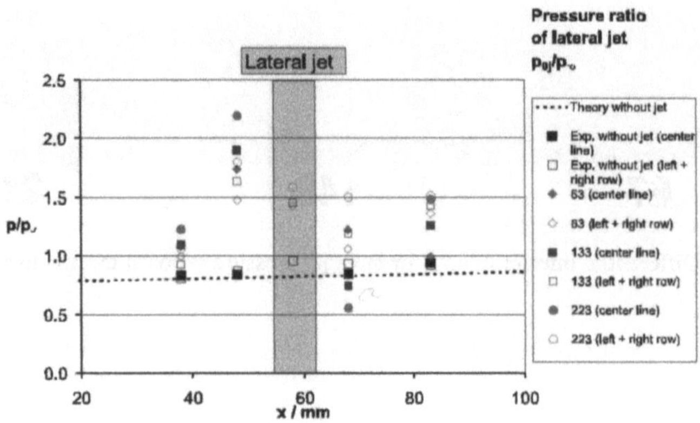

Figure 9 Pressure measurement up- and downstream of the lateral jet (15 km altitude).

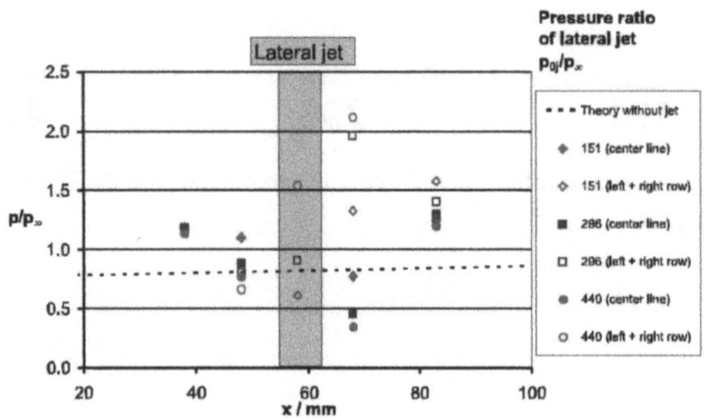

Figure 10 Pressure measurement up- and downstream of the lateral jet (21 km altitude).

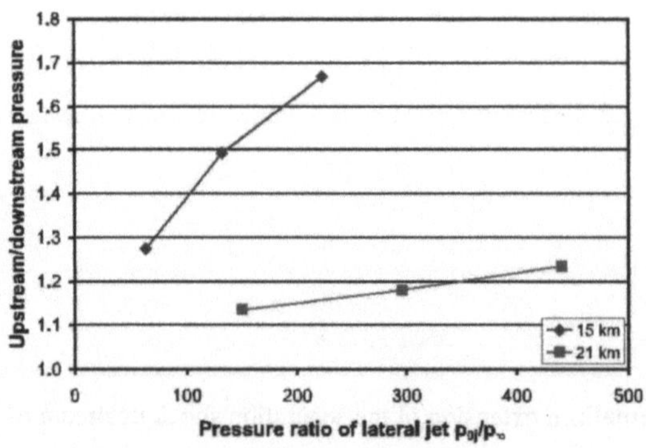

Figure 11 Ratio of the upstream to the downstream pressures relative to the lateral jet.

Numerical Rebuilding of a Generic Body-Flap Model in an Arc Heated Facility

A. MACK[1], M. EMRAN[1], R. SCHÄFER[2]
[1]DLR, Institute of Aerodynamics and Flow Technology,
Lilienthalplatz 7, 38108 Braunschweig, Germany, andreas.mack@dlr.de
[2]DLR, Institute of Structures and Design,
Pfaffenwaldring 38-40, 70569 Stuttgart, Germany

Summary

In the present study a numerical coupling tool for fluid-thermal structure interaction is validated in terms of a generic body-flap model with strong radiation effects on the flap leeside. Coupling effects such as flow topology changes on the flap leeside occur. Special emphasis is placed on the numerical rebuilding of the flow condition entering the test section which has a large impact on the coupling effects. The numerical results found are in good agreement with the experimental data gained in the arc heated facility.

1. Introduction

Efficiency and safety of future space vehicles strongly depend on the design of the thermal protection system. Therefore accurate and reliable predictions of heat flux and thermal loads on the surface are essential for the design. During the German national project IMENS (Integrated Multi-disciplinary dEsigN of hot Structures) three different generic models have been tested experimentally in the DLR arc heated facility L3K [1] and numerically rebuilt [2], [3]. Particularly the numerical results for a generic body flap model presented here are focussing on the radiation effects between the body flap leeside and the opposite model surface. The wind tunnel model is designed resembling a realistic space vehicle body flap configuration. It consists of two flaps separated by a central gap. The actuators that drive the flaps in a real vehicle are also simulated. The model includes optional side plates that close the flap box in spanwise direction.

With this model, the validation of the coupling environment for the prediction of thermal loads, consisting of the DLR's finite volume CFD-solver TAU [4] and a commercial finite element structural solver (ANSYS) linked by an interpolation routine will be done.

2. Numerical Method

The CFD solutions are achieved with the unstructured DLR TAU-code that has been validated in the past for different configurations at super- and hypersonic flow conditions [4]. The time accurate Navier-Stokes equations are marched for steady or unsteady conditions by a three stage Runge-Kutta scheme. The upwind scheme employed is AUSMDV [5]. Available thermochemical models are perfect gas, equilibrium air and chemical non equilibrium. State of the art acceleration techniques such as local time stepping, residual smoothing and multigrid are implemented. The numerical solutions for the thermal analysis of the structure have been carried out with structural finite element solver ANSYS. Quasi structured meshes have been generated to increase the accuracy of the thermal solution. Surface radiation effects are calculated by the structural solver. View factors of the different surface panels are calculated and reflections are taken into account, leading to a degraded surface radiation cooling in cavity areas. The material properties of the C/C-SiC material are nonlinear, they are based on the analysis of the material used during the experimental tests. The coupling algorithm follows a loose thermal coupling and uses Dirichlet-Neumann boundary conditions. Surface heat flux and temperature are calculated by the solvers and exchanged as boundary conditions. The surface values are interpolated from the fluid to the FEM mesh and vice versa with the commercial interpolation routine MpCCI. This routine allows an efficient neighborhood search also for large numbers of surface grid points.

3. Results

The study is carried out for conditions of DLR's arc heated facility L3K (H_0 = 11.5 MJ/kg, Ma = 7.4, Re = 12.000, α = 10°) assuming laminar flow.

3.1 Flow condition

Special attention is given to the flow condition inside the test chamber. Since measuring the hot hypersonic flow core is quite challenging, numerical rebuilding of the wind tunnel has to be done. During the experiments, pressure in the arc heater's reservoir and the mass flow are measured which allows to calculate the other flow condition there. Density and mass fractions of the flow are calculated employing a model for thermochemical equilibrium there. Due to the high temperature in the arc heater the oxygen is almost completely dissociated. This leads to the following reservoir condition, see Table 1.

Table 1 Reservoir conditions

reservoir pressure	p_0	= 4700 hPa	$\alpha(N_2)$	= 0.7253
" temperature	T_0	= 5649 K	$\alpha(O_2)$	= 0.0031
" enthalpy	H_0	= 11.56 MJ/kg	$\alpha(NO)$	= 0.0304
" massflow	$\dot{m}$	= 142 g/s	$\alpha(N)$	= 0.0275
" density	ρ	= 0.2364 kg/m³	$\alpha(O)$	= 0.2136

The flow expands through a conical nozzle with a half angle of 12° until it reaches the test chamber. Inside the test chamber, the flow further expands and reaches the model. In the past, a simple approach with an extrapolated nozzle that includes the test section was chosen in order to calculate the flow condition at the model position (x = 939 mm). By this, the free shear layer of the nozzle is neglected and the background pressure inside the test chamber is not taken into account. Fig. 1 shows the extrapolated nozzle with a schematic model. The characteristic starting at the junction between nozzle throat radius and conical nozzle passes through the flowfield and doesn't reach the model itself. From the axis-symmetric nonequilibrium calculations, the flow condition at the model position are extracted and shown in Table 2. A slight recombination of Nitrogen appears.

Table 2 Flow conditions at model position

Ma_∞	$= 7.36$	$\alpha(N_2)$	$= 0.7621$
p_∞	$= 58.9$ Pa	$\alpha(O_2)$	$= 0.0063$
T_∞	$= 552$ K	$\alpha(NO)$	$= 0.0106$
ρ_∞	$= 0.000308$ kg/m^3	$\alpha(N)$	$= 10^{-5}$
U_∞	$= 3866$ m/s	$\alpha(O)$	$= 0.2210$

The flowfield shows almost no real gas effects further down the throat, the flow is almost completely frozen. As can be seen from the figures, the flowfield displays flow gradients along the nozzle axis. Pressure is reduced by a factor of 2 along the 300 mm model length, see Fig. 4, left. Therefore, an extensive study of the test chamber flow has been performed. First axis-symmetric calculations of the nozzle including the test chamber and the diffuser inlet indicated a strong dependency on the pressure level at the diffuser due to the fact that the flow core is hypersonic and extends only half of the diffuser diameter. Changing the pressure there influences the test chamber pressure and this controlles the state of the nozzle: overexpanded, matched, underexpanded or separated. Due to this problem the whole facility was modeled, including the nozzle, the test chamber (with and without model) and both stages of the diffuser, which leads to a supersonic outflow at the 2nd stage of the diffuser at Mach 2.5 (Fig. 2). The calculation of the flowfield in the test chamber without model showed a shear layer separating the flow core from the low Mach number reverse flow. Inside the shear layer, there is a strong (here 99.9 % total enthalpy) loss of total enthalpy. The diameter of the flow core with constant total enthalpy line follows approximately the Ma = 7.2 line. A second characteristic originating from the nozzle exit propagates through the flowfield and crosses the model area (Fig. 3). Behind this characteristic, the total enthalpy is constant, but the flow condition changes, especially the Mach number decreases. This characteristic limits the homogenous flow core. Inside the core an ideal conical flow is present, between core and shear layer the total enthalpy is constant, but flow conditions including the direction of the flow, change, see Fig. 4, right. However at the model position the flow condition and its gradients do not change because it is well inside the homogenous flow core.

In order to estimate the influence of the model itself, further axis-symmetric calculations have been performed including a generic flap model in the test chamber with a flap deflection angle of 15° and varying the angle of attack. By this, the pressure in the test chamber adjusts itself from the geometry of the flow core and changes the position of the shear layer. Due to the increases background pressure the shear layer moves closer to the model. For 20° angle of attack no steady state solution could be achieved. The model blocked the tunnel flow, just as was observed during the experiments in the past. The pressure in the test chamber increases until a separation in the nozzle appears which interacts with the whole flowfield.

The shear layer has to be investigated particularly weather turbulence has an impact on the topology. Therefore, a turbulent calculation with the Spallart-Almaras turbulence model has been performed. Due to the very low Reynolds-number (12.000) the effects in the flow core can be neglected, the shear layer position and size do not change significantly.

It has to be kept in mind that the axis-symmetric calculation leads to an overprediction of movement of the shear layer position. Due to an increased angle of attack, the cross section area of the axis-symmetric configuration increases and blocks the flow faster than a 3D model. In reality, 10° angle of attack and 30° flap deflection has been performed without blocking. Nevertheless, the calculations without model indicate the maximum size of the usable flow core.

With the help of this study the flow field inside the test chamber can be described more accurately. It turns out that the extrapolated nozzle can be applied to estimate the flow condition and its gradients at the model position. The size of the core diameter indicates that the models used in the following 3D studies are not affected by the shear layer, but are partially outside the homogenous flow core due to their size.

3.2 Generic body-flap model

According to the strategy of the past initial 3D computations were performed with flow conditions determined in the extrapolated nozzle at model position without any flow gradients in x-direction (parallel flow). A hybrid grid of $1.1 \cdot 10^6$ points contains the flow domain which is solved for laminar flow with the TAU-code (Fig. 5, left). The thermal solution inside the structure is treated with a quasi structured mesh of approximately 70.000 points, the radiation and visibility effects particularly on the flap leeside are also considered with the ANSYS code (Fig. 5, right).

The CFD radiation equilibrium solution of the surface temperature and flow topology is shown in Fig. 6, left. Large heat peaks at the side edges are visible, below the flap a large separation region appears. Taking the structural heat conductivity into account, the heat peaks vanish, the lower flap side reaches almost the upper surface temperature. Due to radiation exchange, the lower flap side and the model surface interact, which increases the temperature in the former cold separation area significantly (Fig. 6, right). The separation itself disappears, which is an effect of the structure. The Mach number below the flap is mainly subsonic, tending the flowfield to be non-compressible. Therefore, increasing the surface temperature increases mainly the pressure which creates an artificial pressure gradient, forcing the separation to disappear (Figs. 6 and 7).

According to the previous investigations a generic farfield boundary condition was employed keeping the estimated flow condition and gradient at model position. In this case, due to the pressure gradient in the surrounding flow, the previous effects cannot be observed in Fig. 7, right. As shown in Fig. 8, a large pressure gradient is present along the model, therefore no separation is present for the CFD solution.

The coupling effects on the flowfield therefore are reduced just to increase the pressure gradient further on but not to change the flow topology significantly, see Fig. 7.

Comparisons between numerical and experimental data at the flap middle section are shown in Fig. 8, right, for a fully catalytic surface. Good agreement between the surface infrared data as well as the thermocouple measurement on the model surface and the numerical data can be shown. A crosswise section is shown in Fig. 9, left, which also indicates a good numerical rebuilding of the flowfield for the numerical solution. Further comparisons are made in [6]. Near the symmetry plane, the discrepancies are slightly higher. This might be due to a different surface catalicity or due to the previously described phenomena of the homogenous flow core. For an asymmetrically deflected model, Fig. 9, right, shows the position of the homogenous flow core and the boundary of the shear layer. Only the flap edge is inside the shear layer, which was also observed in the experimental data. But also main parts of the model are outside the homogenous flow core, which means that the surface loads may differ from the numerical ones. Due to the non-ideal flow, forced by the large model compared to the flow core, the overall heat transfer during the experiment is lower than the numerical one. This might explain the lower experimental temperature near the symmetry plane, where the surface temperature is mainly driven by the radiation exchange with the flaps.

4. Conclusions

The numerical rebuilding of the wind tunnel flow in the L3K contributed to get the right understanding of the flow effects around the generic body-flap model. The knowledge of the flow gradient is essential to get the right thermal surface loads and flow topology. Good agreement between the experimental results could be shown. In order to study the effects of the shear layer for larger angles of attack or body-flap deflections, fully 3D computations of the whole wind tunnel domain including the model is necessary which would increase the numerical effort by at least an order of magnitude. In the future, the results of the present study can be introduced to preliminary design of new models, taking into account the effects of flow core sizes and flow gradients in the test chamber region.

Acknowledgement

The authors would like to thank Dr. A. Gülhan and Dr. B. Esser of DLR Cologne for providing and discussing the experimental data and Dr. J.M.A. Longo for many discussions during the study.

References

[1] Gülhan, A., Esser, B., Koch, U.: Experimental Investigation of Gap Flows on a Flap Model in the arc heated facility L3K, DLR-IB-39113-99C01, 1999

[2] Schäfer, R., Mack, A., Esser, B., Gülhan, A.: Fluid Structure Interaction on a Generic Model of a Reentry Vehicle Noesecap. 5th Inter. Congress on Thermal Stresses, Blacksburg, VA., 2003

[3] Mack, A., Schäfer, R.: Flowfield topology changes due to fluid-structure interaction in hypersonic flow using ANSYS and TAU. New Results in Numerical and Experimental Fluid Mechanics IV, p. 196-203, Chistian Breitsamer et al, Springer Verlag Berlin, 2002
[4] Mack, A., Hannemann, V.: Validation of the unstructured DLR-TAU-Code for Hypersonic Flows, AIAA 2002-3111, 2002
[5] Wada, Y., Liou, M.-S., A flux splitting scheme with high-resolution and robustness for discontinuities, AIAA 94-0083, 1994
[6] Mack, A., Schäfer, R., Esser, B., Gülhan, A.: Fluid Structure Interaction of a Hypersonic Generic Body-Flap Model. Proceedings of ICCFD3, Toronto, July 12-16, 2004, to be published in: Computational Fluid Dynamics, Springer Verlag, Berlin (2004)

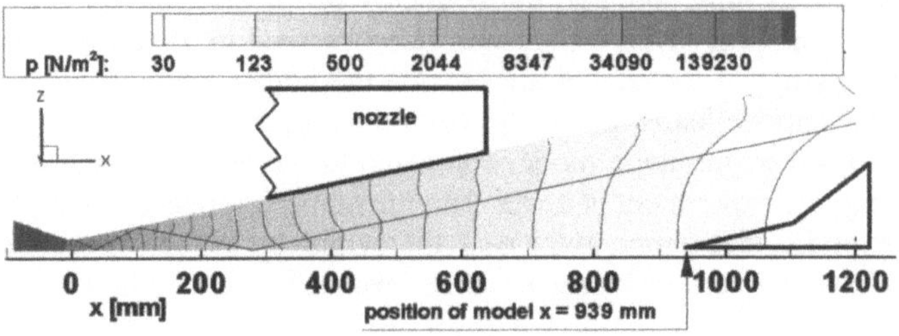

Fig. 1: Extrapolated nozzle: flow topology with characteristic

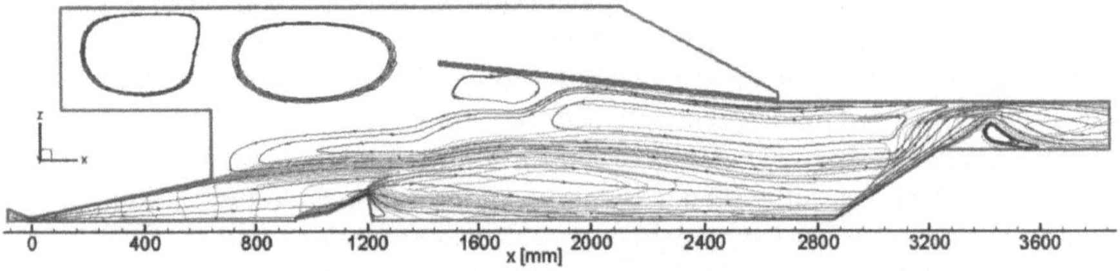

Fig. 2: Wind tunnel including model: geometry, streamlines and Mach number isolines

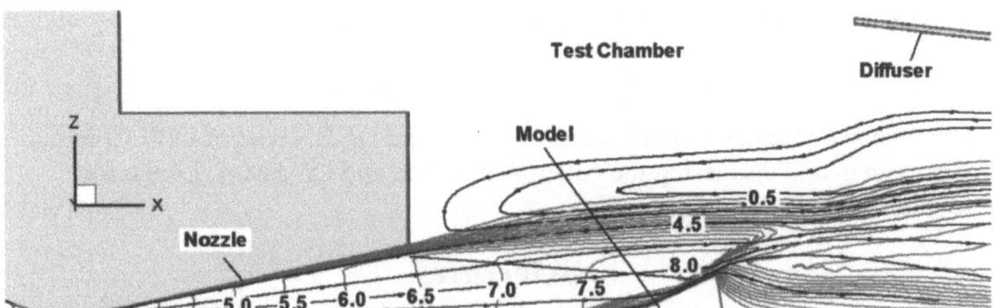

Fig. 3: Wind tunnel including model: detailed view on test chamber, Mach number isolines, characteristic

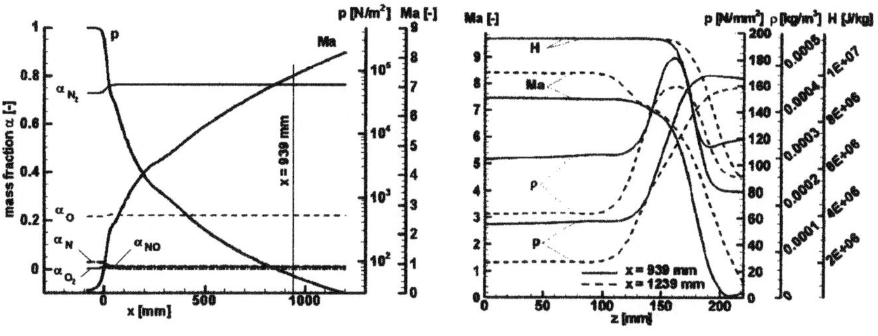

Fig. 4: Flow condition along nozzle axis (left) and across nozzle (right)

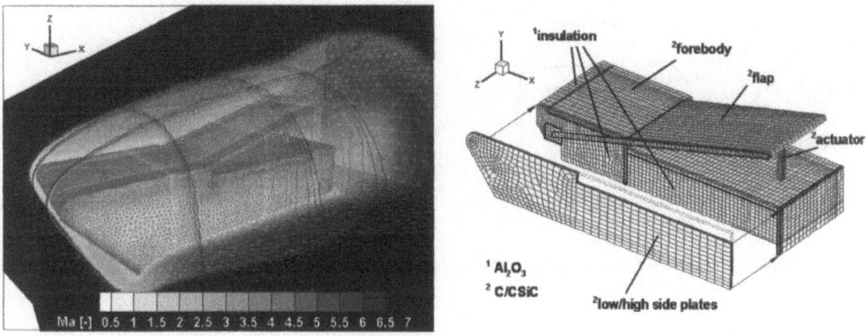

Fig. 5: Body-flap model: CFD-discretisation and flow topology (left) FEM model (right)

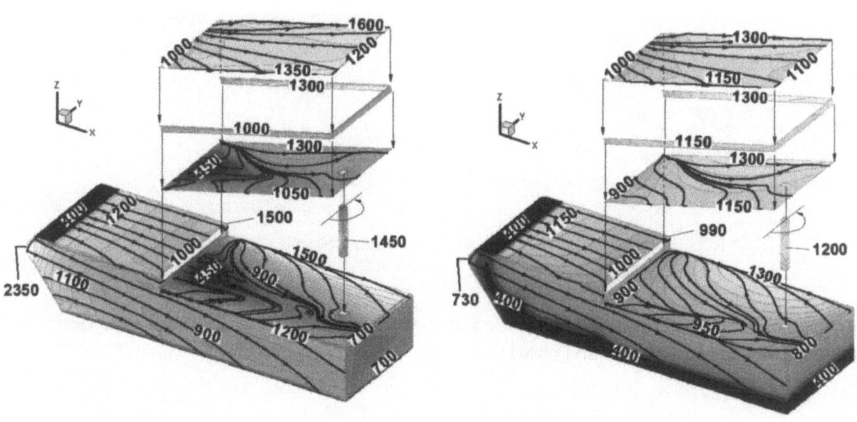

Fig. 6: Body-flap model in parallel flow: CFD-solution (left), coupled solution (right)

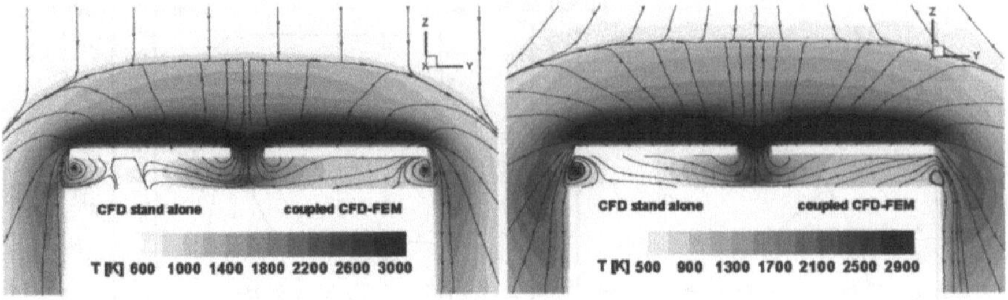

Fig. 7: Cut at x = 160 mm: parallel flow (left), diverging flow (right)

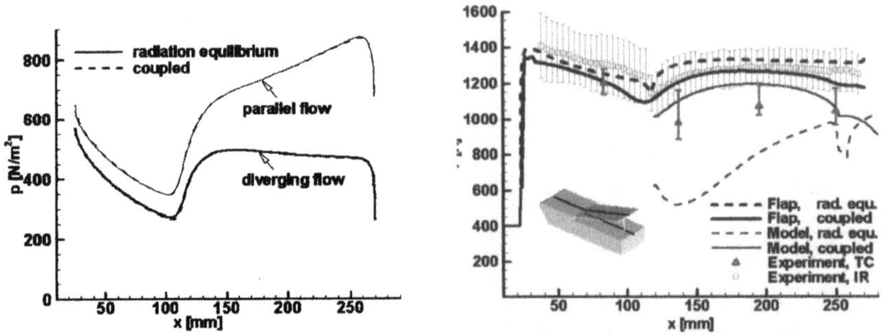

Fig. 8: Pressure level (left), numerical and experimental data at y = 50 mm (right)

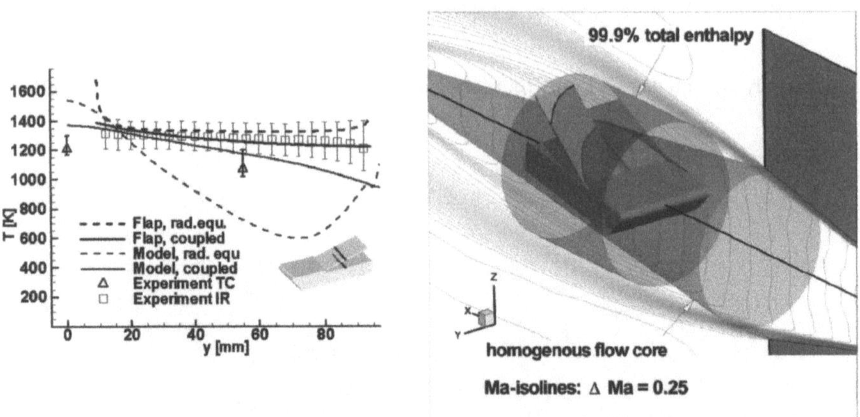

Fig. 9: Numerical and experimental data at x = 190 mm (left), asymmetrically deflected model in wind tunnel flowfield (right)

Experimental and Numerical Investigations of Shock/Turbulent Boundary-Layer Interaction on a Double Ramp Configuration

U. Gaisbauer, H. Knauss
Institut für Aero- und Gasdynamik (IAG), Pfaffenwaldring 21,
70550 Stuttgart, Germany
uwe.gaisbauer@iag.uni-stuttgart.de

N.N. Fedorova, Y.V. Kharlamova
Institute of Theoretical and Applied Mechanics (ITAM), Instututskaya 4/1,
630090 Novosibirsk, Russia
nfed@itam.nsc.ru

Summary

In the presented work the influence of the new defined parameter $d = D/\delta$ on the shock system of a double ramp configuration was investigated, with D describing the distance between the two ramp kinks. On the basis of the experimental and numerical investigations the functional dependency of the most important parameter of the Free Interaction Concept, the upstream interaction length l_0 on the parameter d, could clearly be shown. Therefore an enlargement of the free interaction concept is suggested with respect to the double ramp parameter d. Moreover, through the investigations of the flat plate/double ramp configuration a critical minimum of the first ramp length was found for the investigated R/m-ranges, to realise the wished two shock compression.

1 Introduction

For future, reusable two-stage-to-orbit space transportation systems, the flight in the hypersonic velocity regime by using an air breathing propulsion system is the main problem to be solved concerning the design and the overall vehicle conception. In this context only the use of a scramjet-propulsion system meets all the aerodynamic and gasdynamic requirements. In regard to the highly sophisticated scramjet technology and its requirements to the preconditioning of the incoming flow the purpose of this work was to investigate an highly integrated intake with two combined 2D-ramps under the condition of a not peeled off incoming turbulent boundary-layer to achieve the needed compression with less pressure loss compared with the single ramp case. The projected surface in flow direction of these compression surfaces

should also be minimized to reduce total pressure drag of the vehicle.
In contrast to the 2D-vehicle design with long ramps, like the NASA Hyper-X 43A, here the formed shocks are coupled and can only be treated as a common shock-system with a strong mutual interaction. The combination of two shocks is a very complicate and sensitive structure with respect to small changes in the design point. This concur of a multiple shock structure with the incoming turbulent boundary-layer leads to the problem of the so-called shock boundary-layer interaction, focusing a double ramp configuration (shock/shock) with short as possible length and moderate ramp angles. The semi-empirical theory describing the shock boundary-layer interaction is called the Free Interaction Concept [1]. Here, the most important normalising factor is given by the upstream interaction length L_0, describing the upstream influence of the shock on the boundary-layer. L_0 is used as a function of $L_0(M_\infty, Re_{\delta_0}, \alpha_i, \delta_0)$, usually normalised by the boundary-layer thickness of the undisturbed incoming flow as $L_0/\delta_0 = l_0$. Up to now the theory is limited to the single ramp situation. So, the question is to analyse the influence on the parameter l_0 due to the geometrical double ramp configuration, specified by the two-ramp angles α_1 and α_2 and the distance D between the two ramp kinks, normalised by the boundary layer thickness at the reference position (position of the first ramp kink) $D/\delta_0 = d$. This extension of parameters leads to the advanced functional dependency of the upstream interaction length l_0 at the ramp kink: $l_0(M_\infty, Re_{\delta_0}, \alpha_1, \alpha_2, d)$.
In this field experimental (IAG) and numerical (ITAM) investigations were carried out to clarify the dependencies on the new parameter d with fixed ramp angles $\alpha_1 = 11°$ and $\alpha_2 = 9°$.

2 Experimental facilities

The experimental work was carried out in two different facilities at IAG, Stuttgart University. Visualisation of the flow by the Schlieren method and pressure measurements at a ramp/ramp-model, installed on the wind tunnel wall due to the design of the tunnel, were carried out in the middle sized supersonic wind tunnel HMMS at $M = 2.5$ and 3.0, [3], [7].
The investigations concerning measurements on flat plate models with integrated ramp/ramp-models were carried out at $M = 2.53$ in the so-called shock wind tunnel, SWK [6], at different unit Reynolds numbers.

3 Wind tunnel models

Due to the different requirements of the used wind tunnels and the various intentions concerning the measurements, it was necessary to built different wind tunnel models.

For the tests in the HMMS a ramp/ramp-model was constructed with a movable upper ramp ($\alpha_2 = 9°$) on the first ($\alpha_1 = 11°$) to realize the variation of the distance between the two ramp kinks. The model could be fixed directly on the wind tunnel wall using the turbulent boundary-layer as a kind of base flow in the simulation of shock boundary-layer interactions. For the determination of the boundary conditions, a traversable Pitot-probe was installed on the wind tunnel wall to measure the boundary layer profiles at the reference position.
For the investigation in the SWK several flat plate models with different double ramp configurations were manufactured. In all cases the basic model was a flat plate with a sharp leading edge. Due to the manufacturing the surface was polished to reduce roughness. To simulate a real 2D-flow over the flat plate, the model was completed with trapezoidal side wings, designed with supersonic edges for M=2.5. The model was mounted on a special support to place the whole configuration outside the tunnel sidewall boundary-layer. The first two flat plate models were used to classify the incoming boundary layer. The third used model is of a modular design and gives the possibility of a variable sensor installation, like different kind of hot films, the new developed Atomic Layer Thermo Pile (ALTP), [5], [3] or an insert with static pressure ports.
Moreover, a 11°-ramp/plate insert was used, which allowed to measure the static pressure distribution in front of the first kink and on the ramp simultaneously with a very high spatial resolution, 1 mm in the vicinity of the kink, without mechanical perturbation at the changeover between plate and first ramp. The boundary layer profiles were determined by the traversable Pitot-probe. A more detailed description is given in [3].

4 Numerical model

The computations were performed on the basis of Favre-averaged Navier-Stokes equations closed by a two-equation k-ω turbulence model by Wilcox. A regular grid condensed to the rigid walls was used in calculations. Typically, the grid consists of $100 \div 200$ nodes in y-direction and $200 \div 400$ nodes in x-direction. About 50 % of the grid nodes lay inside the boundary layer developed along the rigid wall. The four-step implicit finite-difference scheme of splitting according to the space directions realized by scalar sweeps was used for time approximation. The TVD scheme of Flux Vector Splitting by van Leer of the third order of accuracy has been used for the approximation of convective terms. The viscous terms have been approximated with the central finite-difference relations of second order of accuracy. The computation domain is restricted by inlet and outlet sections chosen far enough from the interaction zone, solid surface on the bottom and free surface on the top boundary. No-slip condition for velocity and adiabatic condition for temperature are set on solid surfaces. So called simple wave conditions are used on the upper boundary. Profiles of all gasdynamic and turbulence parameters obtained by using a boundary layer approach have been specified at the inlet section. At the outlet section soft conditions have been used. A more detailed description is given in [2], [4].

5 Measurements and results

Boundary conditions

Of main interest for the experimental and numerical research in the field of shock/shock-phenomenon is to specify the incoming flow field and the resulting boundary-layer on the flat plate as well as on the HMMS tunnel floor flow. Therefore in both cases profiles of Pitot-pressure, Mach number, velocity and dimensionless velocity $u^+(y^+)$ as well as the transition Reynolds number in the case of the flat plate in the SWK were determined. In all investigated cases the incoming boundary-layer could be shown as turbulent and two-dimensional at the position of the first ramp, even under the special conditions in the HMMS where the starting point of the turbulent boundary-layer can not be fixed. An overview of the most important incoming flow parameters is given tab. 1.

Measurements in the field of shock boundary-layer interaction

The measurements in the SWK were carried out for $R/m \times 10^{-6} = 9.82$ and 12.41 using a flat plate model with a double ramp configuration installed. It must be emphasized that the first ramp was placed in the already well-developed turbulent boundary-layer and not at the end of transition. For both relevant R/m-ranges the variation of D started at $D = 0\,mm$, equal to a single ramp situation with $\alpha_1 = 20°$, and went up to $D = 40\,mm$, normalised with the respective boundary-layer thickness.

For all values of d in fig. 1 the normalised pressure distributions are plotted versus x/δ_0, in the plate's cartesian coordinate system, where $x/\delta_0 = 0$ indicates the position of the first ramp.

Beginning from values of $d = 3.6$ for $R/m \times 10^{-6} = 9.82$ and of $d = 3.9$ for $R/m \times 10^{-6} = 12.41$ the gradient in the normalised pressure distribution becomes smaller and a constant pressure level on the first ramp is detectable. The desired two-shock system is reattached and the second shock is now separated from the first. Concerning the overall compression, the ratio of the second shock is now independent from the first.

To get a much clearer understanding of the transition process between separated one-shock and attached two-shock system the normalised upstream interaction length l_0, according to [8], is plotted versus d in fig. 3, left. In both relevant R/m-cases a constant value of l_{01} can be defined for the first ramp for $d \geq 3.2$, independent of the double ramp parameter d. At this

critical value of d again the second shock arises from fan-like isentropic compression waves as a single gasdynamic phenomenon. From this point a status of flow is reached where the desired two-shock system develops.
Continuing from $d \approx 4$ a second interaction length l_{02} can be well-defined and for all values $d \geq 5$, l_{02} is only hardly dependent on d. Compared with the results of fig. 1, this second critical value of $d \geq 5$ marks the beginning of an almost constant pressure level on the first ramp, i.e. the boundary-layer starts to redevelop and the whole intake flow attains the working point, so that a compression takes place in two steps.
For the two analysed Reynolds numbers Re_{δ_0} due to the Mach number reduction and the increase of pressure again the values of l_{01} are smaller than the values of l_{02}. As far as the Reynolds numbers are concerned, according to the Free Interaction Concept under the condition of a constant Mach number and fixed ramp-angles for $Re_{\delta_0} < 10^5$ with an increase of Re_{δ_0} a decrease of $l_{0,i}$ is detectable, as it could be shown in fig. 3, left. Concluding, in the case of a constant Mach number a non-negative influence on the theory exists, moreover by adding the parameter d the theory is enlarged and adapted to the case of a ramp/ramp-configuration.

To get a first picture how D effects the coupled shock system under the condition of constant Reynolds number Re_δ and different Mach number initial measurements were conducted in the HMMS at $M = 2.55 \, \text{and} \, 3.0$. In fig. 2 the normalised static pressure distributions are plotted versus x/δ_0 for a range of $20.2\,mm \leq D \leq 40.2\,mm$. Beginning from values of $d = 3.1$ resp. $d = 2.8$ the gradient in the normalised distribution becomes smaller. For a further increase of d only in the case of $M = 2.55$ the pressure distribution reaches a constant level.
To get a much clearer understanding of the transition process between separated one-shock and attached two-shock system, like for the flat plate situation in the SWK, in fig. 3, right, the normalised upstream interaction length l_0 is plotted versus the new parameter d. For $M = 2.55$ starting from a value of $d = 3.8$ a constant distribution of l_{01} is detectable. For smaller values than $d < 2.7$ it is not possible to define an upstream interaction length at the first ramp. One can expect that beginning with this critical value of $d = 3.8$ the second shock arises from fan-like isentropic compression waves as a single gasdynamic phenomenon and there exists no longer an influence on the upstream interaction parameter l_{01}. At the second ramp from a value of $d > 4.1$ the determination of the upstream interaction length l_{02} is possible. But in general the l_{02}-values underlie noticeable fluctuations due to the not yet fully developed boundary-layer on the first ramp. Accordingly, for $M = 3.0$ a constant behaviour of l_{01} for values of $d \geq 2.7$ and of l_{02} for values of $d \geq 3.8$ is detectable but with the difference that in this case no fluctuations for l_{02} can be seen. In both cases the values of l_{01} are qualitatively smaller then the corresponding values l_{02}. With regard to the Free Interaction Concept

again the boundary conditions must be compared. According to the values in tab. 1 it must be pointed out that the Reynolds number based on the boundary-layer thickness is equal for both Mach numbers. So, the functional dependency of l_0 is reduced to only $l_0(M_0, \alpha_1, \alpha_2, d)$. The theory predicts for fixed ramp angles and constant Re_{δ_0} a reduction of l_0 with an increasing Mach number, exactly what can be derived by analysing the presented investigations. Consequently, again the theory is not affected in a negative way by adding the new parameter d.
Additionally, a more detailed insight into the mentioned transition process is given by the analysis of the Schlieren pictures, [3]. The behaviour of $L_{0,i}$ is equivalent to the results derived from the pressure measurements, so that the results of these measurements can be confirmed. As a result, under the special boundary conditions of the HMMS, the behaviour of $l_{0,\,i}$ is in accordance with the Free Interaction Concept, proved by pressure measurements and optical methods.

Parallel numerical investigations concerning the shock/shock problem were performed for all measurements by using the experimentaly defined boundary conditions. For the measurements in the SWK with the above mentioned flat plate/double ramp configuration for 5 different values of d, in fig. 4, left, the numerical results are compared with the experiment, exemplarily for $R/m = 12.41 \times 10^6$. The agreement in the prediction of the pressure distribution is quite good, even in the inflection points due to the ramp/ramp-configuration. Beginning from a ratio of $d = 3.9$ the gradient in the normalised pressure distribution becomes smaller and at $d = 4.9$ the constant pressure plateau is detectable.
For the measurements in the HMMS, representatively for $M = 3.0$ in fig. 4, right, a comparison between a Schlieren picture and the computed density field is given. Here, the case for $d = 3.3$ is shown. The comparison indicates a very good agreement between the experiment and the numerical simulation. The first and the second shock can be identified clearly. Also the structure of the flow field between the two shocks and downstream of the second is well determined by the used numerical method.

6 Conclusion

The presented experimental investigations, pressure measurements and optical visualisation, show clearly the dependency of l_0 as a function of $(M_\infty, Re_{\delta_0}, \alpha_1, \alpha_2)$ and especially of the new defined parameter d. Moreover the good agreement between the experimental results and the numerical simulations demonstrate the applicability of the used numerical code for computations in the field of shock/turbulent boundary-layer interaction.

Acknowledgements

The authors would like to thank the DFG for grants within the already finished SFB 259 established at Stuttgart University in the partial project TPC 5 under which the major part of this project was supported.
The authors would also like to thank the RFBR (Russian Foundation of Basic Research) for the grant of the numerical investigations at ITAM.

References

[1] Delery, J., Marvin, J.: *Shock-Wave Boundary-Layer Interactions.* AGARDograph 280, pp. 51-85, 1986.

[2] Fedorova, N.N., Fedorchenko, I.A., Shuelein, A.: *Experimental and Numerical Study of Oblique Shock Wave / Turbulent Boundary-Layer Interaction at M=5.* Computational Fluid Dynamics, Vol. 10, No. 3, pp. 376-381, 2001.

[3] Gaisbauer, U.: *Untersuchungen zur Stoß-Grenzschicht-Wechselwirkung an einer Doppelrampe unter verschiedenen Randbedingungen.* Dissertation, Universität Stuttgart, 2004.

[4] Gaisbauer, U., Knauss, H., Wagner, S., Kharlamova, Y.V., Fedorova, N.N.: *Shock/Turbulent Boundary-Layer Interaction on a Double Ramp Configuration - Experiments and Computations.* Proceedings of XI International Conference on the Methods of Aerophysical Research, ICMAR, Vol. 3, pp. 56-62, 2002.

[5] Gaisbauer, U., Knauss, H., Weiss, J., Wagner, S.: *Measurement Techniques for Detection of Low Disturbance and Transition Localization in a Short Duration Wind Tunnel.* Proceedings of X International Conference on the Methods of Aerophysical Research, ICMAR, Vol. 2, pp. 54-67, 2000.

[6] Knauss, H., Riedel, R., Wagner, S.: *The Shock Wind Tunnel of Stuttgart University - A Facility for Testing Hypersonic Vehicles.* AIAA-Paper 99-4959, 1999.

[7] Schlaich, F.: *Experimentelle Untersuchungen zur Wechselwirkung zwischen Stoß und turbulenter Grenzschicht an einer Doppelrampe im Überschall.* Dissertation, Universität Stuttgart, 1996.

[8] Settles, G.S.: *An experimental study of compressible turbulent boundary-layer separation at high Reynoldsnumber.* Ph.D. Thesis, Princeton University, 1975.

Tab. 1: Main parameters of the incoming flow in SWK and HMMS.

	R/m	Re_δ	M_∞	c_f
	$\times 10^{-6}$	$\times 10^{-4}$	[/]	$\times 10^{3}$
SWK	9.82 ± 0.13	5.06 ± 0.11	2.513 ± 0.05	1.98 ± 0.08
	12.41 ± 0.17	6.29 ± 0.14	2.543 ± 0.05	1.81 ± 0.07
HMMS	9.17 ± 0.06	6.99 ± 0.05	2.546 ± 0.03	1.84 ± 0.05
	7.49 ± 0.05	7.00 ± 0.05	2.995 ± 0.04	1.53 ± 0.04

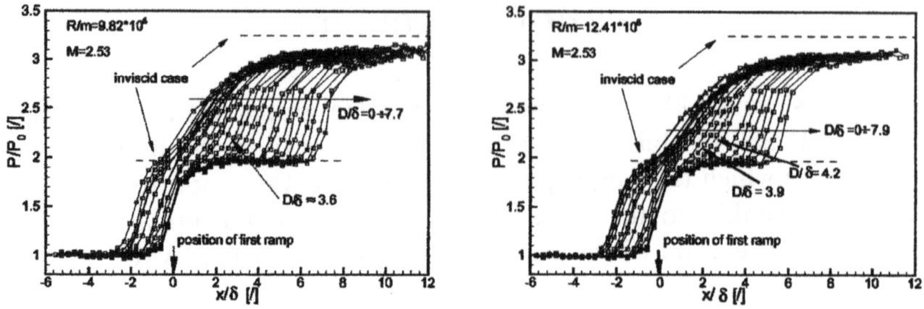

Fig. 1: Normalised static pressure distributions for $R/m \times 10^{-6} = 9.82$ and 12.41.

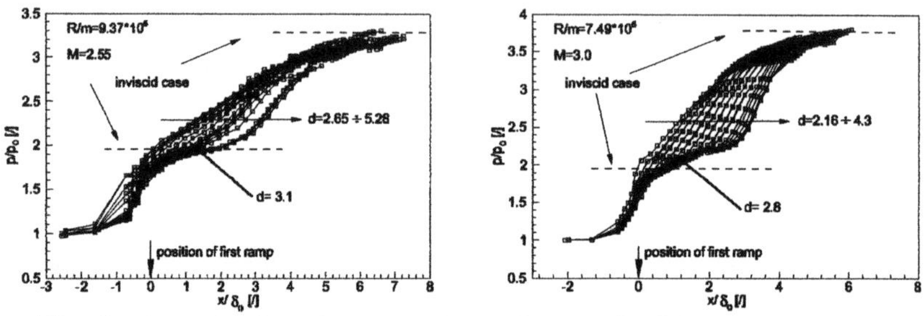

Fig. 2: Normalised static pressure distributions for $M_\infty = 2.55$ and 3.0.

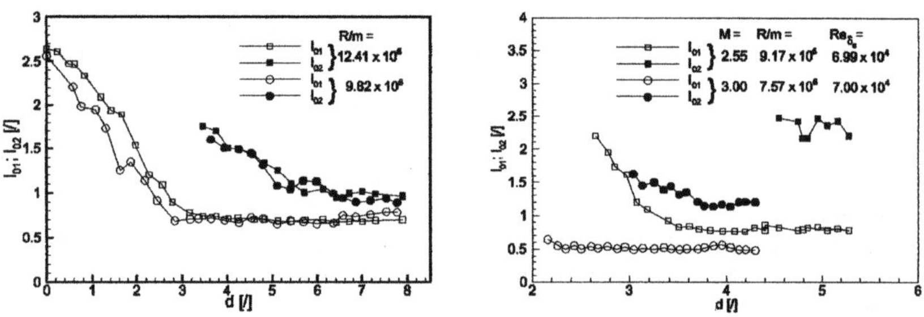

Fig. 3: Dependency of the normalised upstream interaction length l_0 on the parameter d for $M = 2.53 = const.$ (left) and $Re_\delta \approx 7 \times 10^4 = const.$ (right).

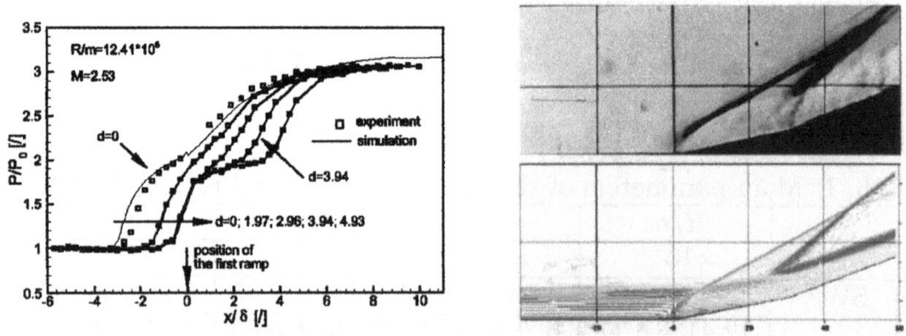

Fig. 4: Comparisson between measured and computed results; left: static pressure distribution, right: Schlieren picture and computed density field.

Time Accurate Simulation of Turbulent Nozzle Flow by the DLR TAU-code

H. Lüdeke and A. Filimon

DLR, Institute of Aerodynamics and Flow Technology,
Lilienthalplatz 7, D-38108 Braunschweig, Germany
Heinrich.Luedeke@dlr.de, Alexander.Filimon@dlr.de

Summary

Fundamental requirements for future launcher technologies are a cost-efficient access to space as well as the improvement of safety and reliability. In the presented contribution an unsteady turbulent flow simulation of the Ariane-5 launcher during ascent will be carried out by detached eddy simulation (DES). In a first step the nature of separation shocks in turbulent over-expanded axisymmetric nozzles is simulated. Different implementations of detached eddy simulation models are investigated for a compressible wake flow as a validation case. Finally steady and unsteady Ariane-5 simulations are carried out at Mach 0.8 wind tunnel conditions including jet flow and the wake of the boosters.

1 Introduction

Substantial requirements for future rocket technologies are the cost-efficient access to orbit as well as the increase in the system reliability. Concerning these requirements the engine is one of the most important parts and a deeper insight into the unsteady phenomena during the start and early flight phase of modern launchers like the Ariane-5 is essential. These requirements have also lead to a greater demand for high expansion ratio nozzles, which ensures a bigger thrust in vacuum, makes a detailed study of rocket nozzle concepts necessary. Especially unsteady side-loads, induced by the interaction of flow separation downstream of the nozzle throat and the wake of the launcher play an important role for the design of future main stage propulsion systems. This so called buffeting coupling, which is generated by an acoustic coupling process in the operational flight case, is one of the main challenges during ascent. In the last decade various experimental and numerical studies of the Ariane-5 launcher ascent have been carried out. However until now it is not possible to predict the unsteady turbulent flow field of the whole launcher configuration numerically. The purpose of this study is to investigate especially such flowfields with recent turbulence models, namely the Detached Eddy Simulation approach, using the unstructured DLR-Tau code. To validate the implemented model a typical compressible test case with a detailed experimental data base of the turbulent quantities is investigated by different variations of the turbulence model. For this validation the axisymmetric base flow behind a blunt cylinder, in comparison with the experiments of Herrin and Dutton is chosen. Unsteady simulations of nozzle flow under harmonic variation of the external pressure field are carried out by standard RANS approach to get basic information of generic cases.

2 Numerical algorithm

The numerical investigations are carried out by the DLR Navier-Stokes Code TAU [2]. The basic solver uses a node based dual mesh with an edge based data structure, created in a preprocessing step. The governing equations are discretised with a Finite Volume method on unstructured hybrid meshes. An explicit iterative scheme with central or upwind space discretisation is the basic solution algorithm. Time accurate simulations are carried out by a dual time stepping scheme. Various one- and two equation turbulence models are implemented for steady simulations. Furthermore in the latest version detached eddy simulation with one- and two equation modeling is possible. Convergence acceleration is achieved with local time stepping, residual smoothing , and an agglomeration based multigrid technique.

3 Axisymmetric base flow

For basic studies of the described turbulence models a coarse hexahedral grid with axisymmetric character, shown at the top of Figure 1, is generated. The cell distribution is inspired by the work of Forsythe et al. [3] and Spalart [4]. For grid convergence studies also a hybrid unstructured grid is in use (Figure 1 right). While the hexahedral grid has only 360.000 cells, to study the accuracy of the numerical schemes, the hybrid grid with about 2.000.000 cells provides a significantly improved resolution in the shear layer and wake.
For both grids the standard DES model, suggested by [4] with a slightly modified filterlength is used. Instead of the maximum cell diameter the third root of the cell volume, typically used for LES simulations, is chosen in the DES-region, defined by the standard procedure. The advantage of this choice is a better treatment of highly stretched cells in hexahedral grids.
The test conditions for the compressible base flow and the experimental data are taken from [3]. These conditions are: $M_\infty = 2.46$, $Re = 45 \cdot 10^6/m$, $U_\infty = 593.8\frac{m}{s}$ and $31.75mm$ cylinder radius.
The averaged Mach number distribution in the cylinder wake agrees well with the experimental data by Herrin and Dutton [1], as shown in Figure 3. Also higher statistical moments, like the resolved turbulent kinetic energy $k = 0.5 \cdot (\sigma_u^2 + \sigma_v^2 + \sigma_w^2)$ are compared, with σ as the standard deviation over the time of the respective velocity component. The result for the standard DES model is shown in Figure 4 with the same k-distribution in the shear layer. The results could be improved by using the refined unstructured grid, as shown in Figure 5. With the improved resolution of the free shear layer it is possible to get a better agreement of the k-distribution in that region. Comparisons of the pressure distribution at the base for RANS- and DES-calculations (Figure 7) show the significant potential advantages of the DES

modelling.
To reduce the number of required grid cells for the very time consuming DES simulations, also a modification of the standard model, proposed by [5] is tested. This algorithm redefines the wall functions of the SA model in the LES region in a way that guarantees a proper LES behaviour of the model in this area. The advantage of this formulation is visible in Figure 6, where computations on the hexahedral grid with both the standard and modified DES methods are carried out. The finer resolution of the turbulent structures in the vorticity is clearly visible.

4 Unsteady nozzle flow simulations

In order to simlify the the complex flow in the launch vehicle base region and to concentrate on the reaction of the separated internal flow to external pressure fluctuations the nozzle is considered at first in isolation. For these nozzle calculations hybrid unstructured meshes including tetrahedral and prismatic cells are used. Concerning the internal flow unstructured cells allow a faster mesh generation and a better adaptation. Nevertheless the resolution of the viscous flow in the boundary layer requires structured cells in this area. The mesh density is adapted locally corresponding to the current separation and shock position (Figure 8).
The main issue of the steady computations is the validation of numerical results by comparison with the experiments by Torngren [6] in the modified FFA HYP500 wind tunnel facility at FOI. These experiments, sponsored by ESA/ESTEC, were carried out with a Truncated Ideal Contour (TIC) sub scale nozzle called Volvo S6 short nozzle. The following computations were done with perfect gas assumption which is appropriate for the experimental setup. It has to be noted, that the main engine of the Ariane-5 is driven by a hydrogen/oxygen system which one has to take into account for simulations of the real operational case. For the experiment preheated air at a temperature of Tc=400K and a stagnation pressure up to $p_c = 25bar$ was expanded in the nozzle. An unsteady external flow was generated by using the plume aspiration effect to draw air through a valve consisting of a static and a rotating slotted disk. By this technique pressure fluctuations between 0 and 900Hz can be generated around the nozzle exit. The steady numerical simulations show the dependency of separation and shock position on the ratio between chamber and ambient pressure (Pc/Pa). The Mach number distribution in Figure 9 shows the well resolved Mach disk for the overexpanded Volvo S6 nozzle. Figure 10 illustrates the wall pressure distribution for pressure ratios between 14.5 and 43.4. It demonstrates the downstream motion of the separation position with increasing pressure ratio. As visible the numerical calculations are completely confirmed by the experimental results.
In former investigations by Schwane et al. [7] the unsteady shock system in the nozzle has shown essentially the behaviour of a damped oscillator. Further examinations of these characteristics are performed by a sudden decrease of the chamber pressure. By this technique the shock is forced to settle at a new position according to the new pressure ratio. However, if the behaviour of a damped oscillator is appropriate, the shock should overshoot this position, performing a damped oscillation. This behaviour is indeed displayed in Figure 11 where the pressure dis-

tribution on the centreline of the nozzle is plotted. It was formally reported by Schwane et al. [7]. Every vertical pressure peak describes the shock position at a different time. This is just the behaviour of a sudden excited damped oscillator. In the experimental investigations further unsteady test runs are included [6], which describes the variations of the ambient pressure with respect to harmonic pressure fluctuations. Several unsteady pressure transducers, located inside and outside of the nozzle, realized time dependent pressure fluctuations. Figure 12 shows the measured pressure at transducer P2 corresponding to x/l = 0.767 in comparison with the external pressure at transducer P10 (see Figure 8). As P2 is fixed for the set-up several pressure ratios are prescribed to investigate different flow conditions at the same position. For a ratio Pc/Pa=20.2 P2 is located inside the separation at all times. Neither a significant phase shift nor a change in the amplitude is visible. at Pc/Pa=24.4 the separation moves forward and backward across the transducer. The resulting signal shows a higher amplitude with an approximated phase shift of 30° w.r.t. the outer pressure. For attached flow at P2 nearly no pressure signal can be measured. This behaviour agrees well with the simulations at a frequency of 40 Hz. Instead of the pressure ratio the positions for the sampling of the wall pressure are varied between x/l=0.804 and 0.577 in this case. The behaviour is the same as in the experiments. For numerical reasons the inflow conditions instead of the ambient pressure is varied to re-sample the wind tunnel data at a frequency of 150 Hz. As the separation and shock position is mostly dependent on the pressure ratio p_c/p_a the results are similar with the exception of x/l=0.534. At this position inside of the separation the pressure is influenced by the inflow conditions and consequently an oscillation is visible in contrast to the experimental case. For this simulation a phase shift between inner and outer pressure of 28° is found. As a conclusion the characteristics of the shock similar to a linear oscillator could be shown by different techniques.

5 DES simulations of the flow past the Ariane-5 launcher

For the calculations of the Ariane-5 nozzle-flowfield, wind tunnel conditions of $M = 0.8$, $Re = 11 \cdot 10^6/m$ are chosen. For these conditions detailed experimental studies of NLR and ESA are carried out. The computational Ariane-5 grid is shown in Figure 13. To reduce computational costs for DES simulations, the relatively coarse hexahedral grid, provided by ESA is used. In this investigation steady RANS as well as unsteady DES simulations for the nozzle part of the launcher are performed. For these investigations the forebody part of the launcher is computed stationary, and the connecting plane of both grid-parts is used as a steady inflow-plane for the transient simulation. The density distribution for the steady RANS-result, (Figure 14) compares well with the experimental results, published in [8]. Preliminary DES results show the complex interaction between the different nozzle plumes of boosters and central nozzle, which is visible especially for the vorticity contours (Figure 15). A cut-out of the flowfield between the three nozzles shows the system of separations and vortices resulting from the interaction of the separated nozzle flow and the jet coming out of the gap between boosters and main stage (Figure 16). This interaction is also responsible for the main aerodynamic frequencies dominating the pressure distribution.

6 Conclusion

In the presented study basic investigations for the simulation of unsteady buffeting coupling are carried out. Isolated axisymmetric nozzle configurations as well as time accurate simulations of an axisymmetric cylinder wake by Detached Eddy Simulation have shown very good comparability with experimental data and encourage the application of these techniques for trans- and supersonic flows of entire launcher configurations. Preliminary investigations of the unsteady phenomenon of buffeting coupling at the Ariane-5 launcher have shown reasonable results and will allow comparisons with detailed experimental data for this case.

References

[1] J.L. Herrin, J.C. Dutton: Supersonic Base flow Experiments in the Near Wake of a Cylindrical Afterbody, AIAA Journal, Vol 32, No. 1, January 1994.

[2] T. Gerhold, O. Friedrichs, J. Evans, M. Galle: Calculation of Complex three-dimensional configurations employing the DLR TAU Code, AIAA-97-0167, 1997

[3] J.R. Forsythe, K.A. Hoffmann, K.D. Squires: Detached-Eddy Simulation with compressibility Corrections Applied to a Supersonic Axisymmetric Base Flow, AIAA 02-0586, 2002.

[4] P.R. Spalart: Young-Person's Guide to Detached-Eddy Simulation Grids, NASA/CR-2001-211032, 2001.

[5] M. Breuer, N. Jovicic, K. Mazaev: Comparison of DES, RANS and LES for the separated flow around a flat plate at high incidence, Int. J. Numer. Meth. Fluids, 41: pp 357-388, 2003

[6] L. Torngren: Correlation between Outer Flow and Internal Nozzle Pressure Fluctuations, Proceedings of the 4th European Symposium on Aerothermodynamics for Space Vehicles, Capua, Italy 15-18 October, 2001

[7] R. Schwane, D. Perigo, Y. Xia: Unsteady Numerical Simulation of the Flow in a Truncated Ideal Contour Nozzle under Free Shock Separation Conditions, (AIAA 2003-3676), 23-26 Jun. 2003/Orlando

[8] J. Muylaert, W. Berry: Aerodynamics for Space Vehicles- ESA's Activities and the Challenges, ESA bulletin 96 (November 1998)

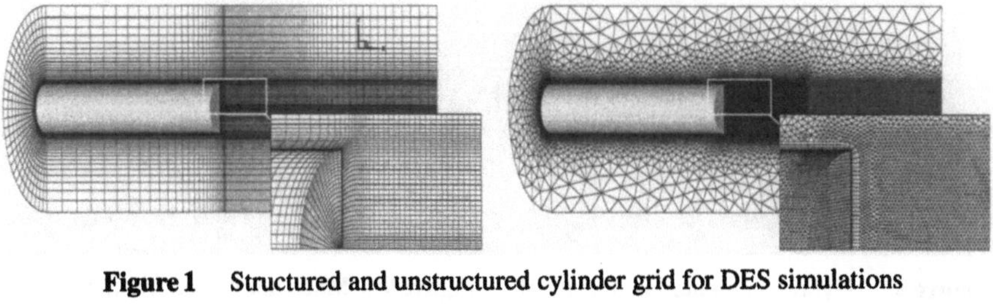

Figure 1 Structured and unstructured cylinder grid for DES simulations

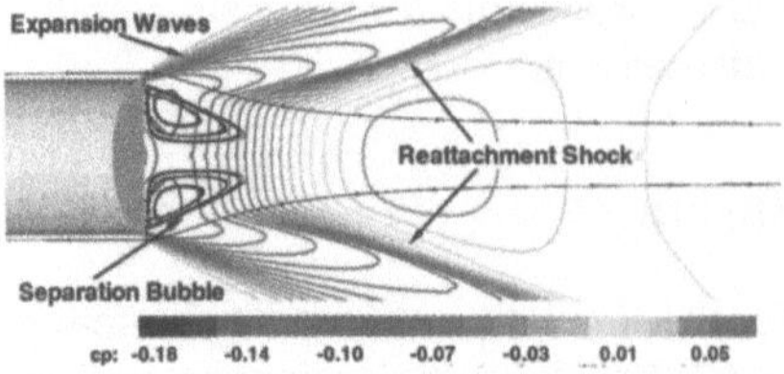

Figure 2 Flow topology and C_p contours of the axisymmetric cylinder base flow at $M_\infty = 2.46$

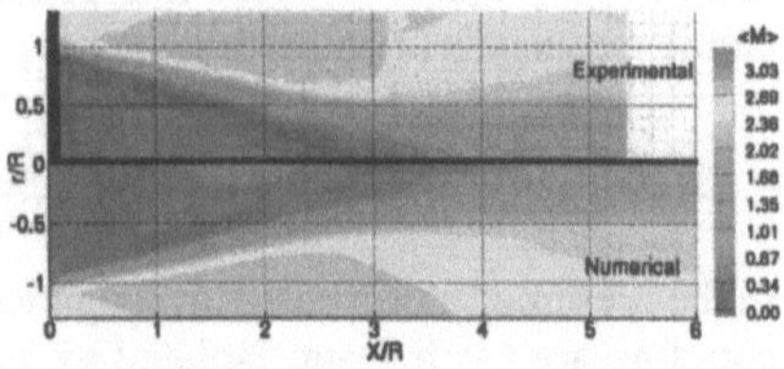

Figure 3 Averaged Mach contours behind cylinder base: DES on hexahedral grid. Experimental part from [3]

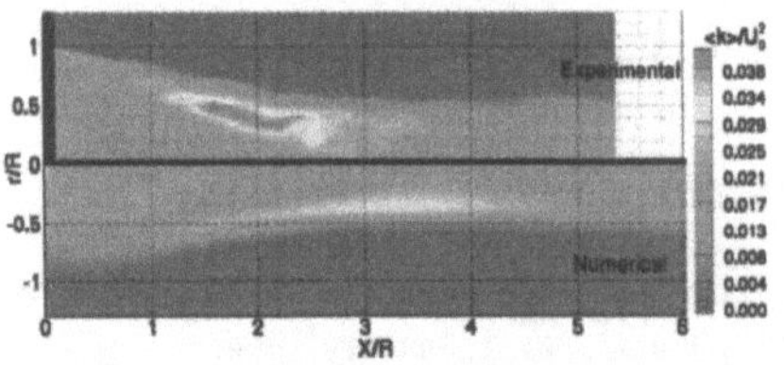

Figure 4 Resolved turbulent kinetic energy behind cylinder base: DES on hexahedral grid. Experimental part from [3]

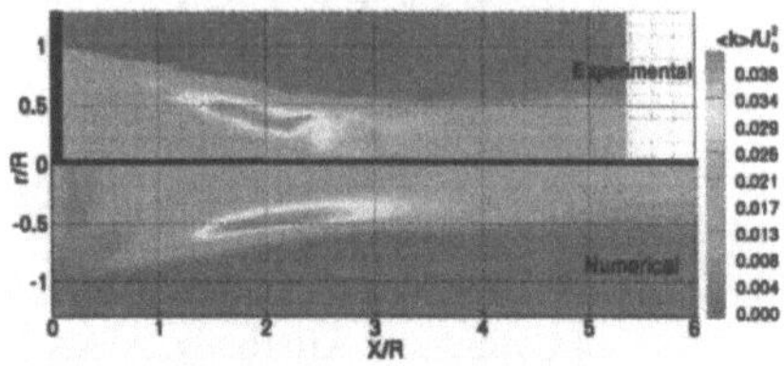

Figure 5 Resolved turbulent kinetic energy behind cylinder base: DES on tetrahedral grid. Experimental part from [3]

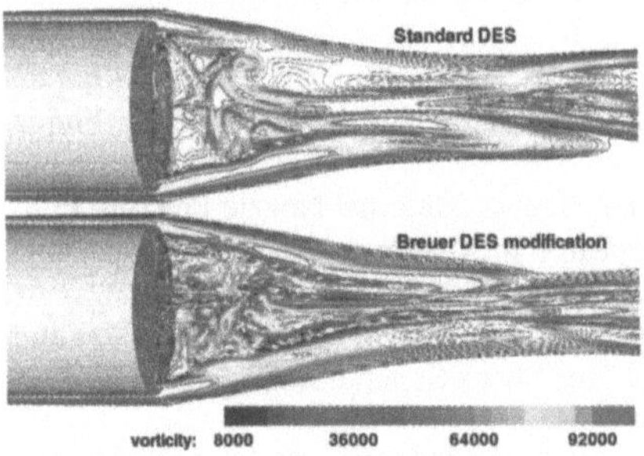

Figure 6 Vorticity in the cylinder wake for different DES modifications on hexahedral grid

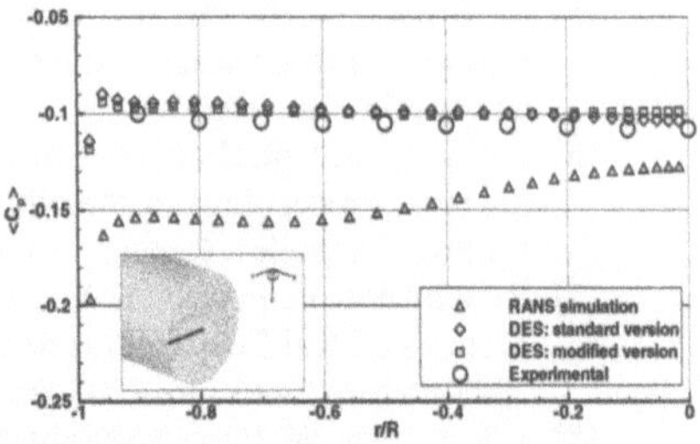

Figure 7 Pressure along the base for RANS and DES models compared with experimental data from [1]

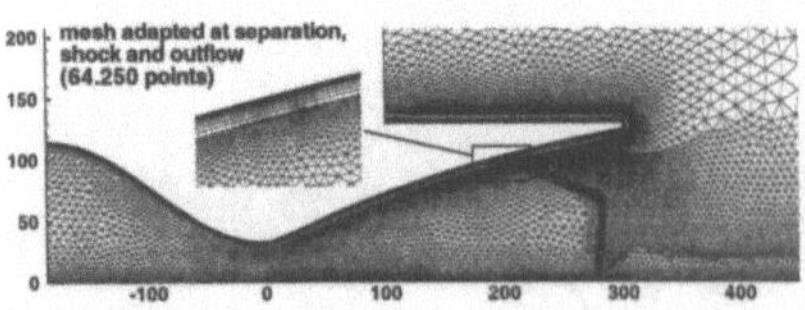

Figure 8 Hybrid computational grid of the Volvo S6 short nozzle after adaptation

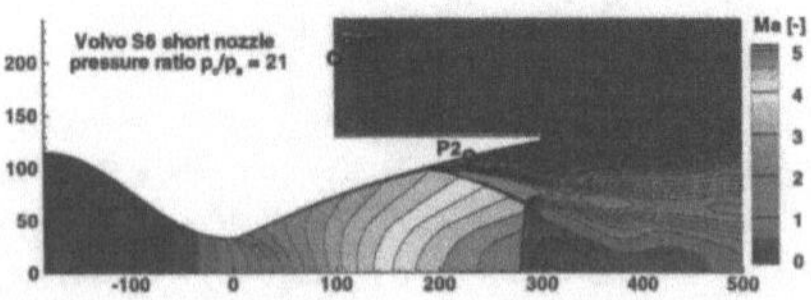

Figure 9 Mach number distribution of the Volvo S6 short nozzle

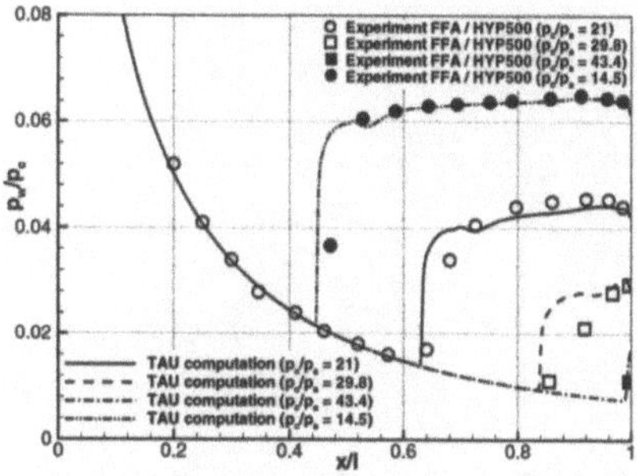

Figure 10 Steady wall pressure at different pressure ratios. Reference length l=300 mm. Experimental data from [6]

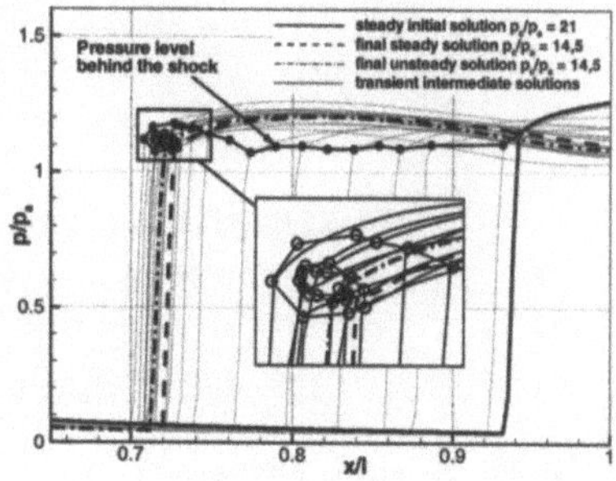

Figure 11 Unsteady shock position after sudden decrease of pressure ratio p_c/p_a from 21 to 14.5

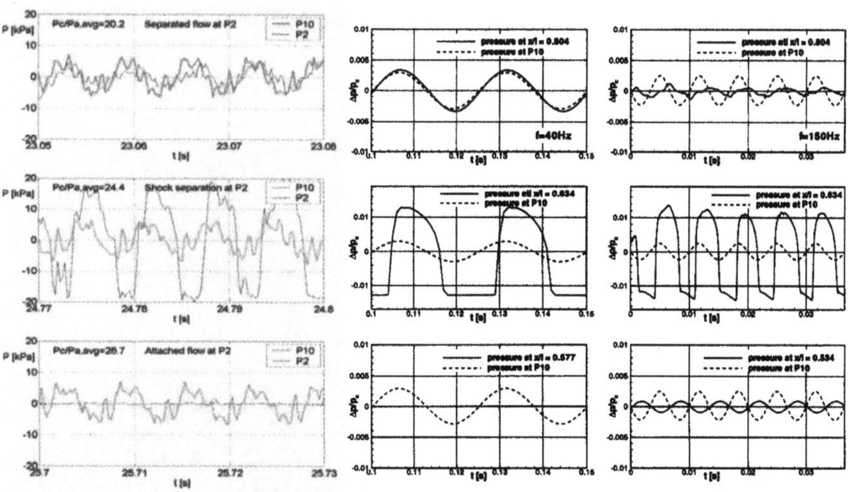

Figure 12 Time dependent pressures at transducer P2 (x/l = 0.767) and P10 (ambient) at different frequencies. Left: L. Torngren [6], middle: CFD 40 Hz, right: CFD 150 Hz

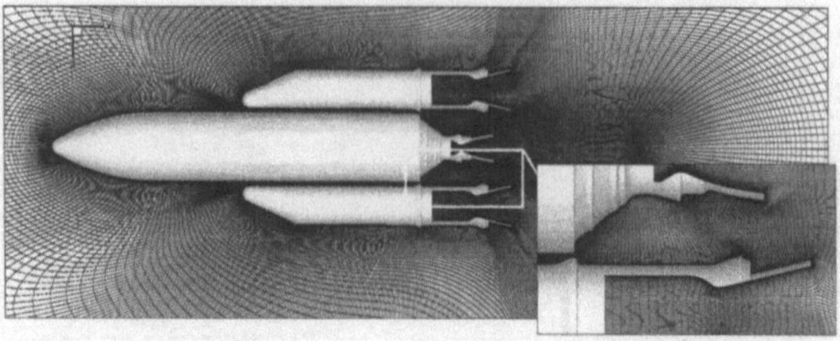

Figure 13 Symmetry plane of the Ariane-5 hexahedral grid, fore- and afterbody

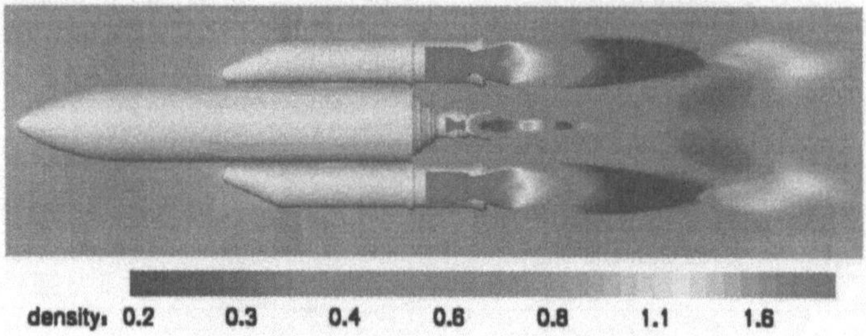

Figure 14 Density contours in the plume of the Ariane-5 launcher. Steady RANS simulation

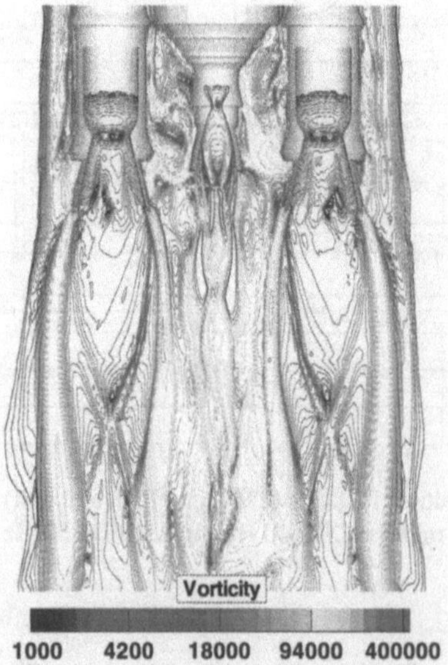

Figure 15 Ariane-5 DES simulation: snapshot of vorticity in the symmetry plane

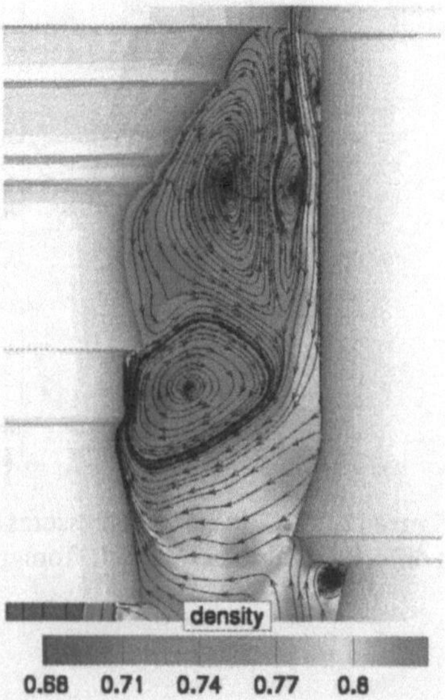

Figure 16 Ariane-5 DES simulation: snapshot of density and flow topology between nozzles

Quantitative Comparison of Measured and Numerically Simulated Erosion Rates of SiC Based Heat Shield Materials

Stefan Löhle, Markus Fertig, and Monika Auweter-Kurtz

Institut für Raumfahrtsysteme, Pfaffenwaldring 31, 70550 Stuttgart, Germany
WWW home page: `http://www.irs.uni-stuttgart.de`
loehle@irs.uni-stuttgart.de

Summary

The main mission critical part of planned reusable re-entry vehicles is its thermal protection system. The surface material has to withstand high heat loads and a chemically aggressive environment in the upper atmosphere. In case of re-usability the candidate material has to withstand these loads several times. Therefore, the mass loss during one re-entry mission has to be as small as possible. In order to predict the mass loss, the material is investigated in plasma wind tunnels and meanwhile numerical simulation of surface processes is possible at Institut für Raumfahrtsysteme (IRS). This paper describes for the first time a quantitative comparison of the specific mass loss estimated on the one hand by plasma wind tunnel experiments and on the other hand by numerical simulation. Results for pressureless sintered silicon carbide (SSiC) and realistic C/C-SiC material are presented. Finally, an attempt is made to interpret the occurring differences.

1 Introduction

Thermal protection systems for future reusable re-entry vehicles have to withstand temperatures of about 2000 K. The up to date favored materials are silicon-based ceramics. The chemical reactions of SiC with oxygen lead to the formation of SiO, SiO_2, CO and some other compounds that can be neglected at elevated temperatures [1]. Since SiO and CO are gaseous at those temperatures erosion and thus material loss is the consequence. SiO_2 in contrast is liquid or solid and forms a layer on top of the SiC so that it acts as a diffusion barrier. This effect is often named a self protection mechanism, because the oxygen flux to the surface is hindered and hence, the mass loss is lowered. The process is called *passive oxidation.* At higher temperatures and lower oxygen pressures, a possible protection layer will be removed and SiO formation becomes easily possible. This is called *active oxidation* and obviously is a state that has to be avoided. But, as can be seen in Fig. 1(a), on planned re-entry missions like EXPERT and X-38 active oxidation may occur. In Fig. 1(a), theoretically calculated reference trajectories are plotted together with a regime known as passive-to-active transition calculated by Lou and Heuer [2]. In this regime, active oxidation is possible and as can be clearly seen, during both missions transition conditions are reached. The limits shown for the regimes (passive-transition-active) are calculated assuming thermal equilibrium. However, the boundary layer around such vehicles during flight in the higher atmosphere is characterised by thermochemical non-equilibrium. At the Institut für Raumfahrtsysteme research in the field of re-entry mechanics is done since the mid eighties. During the HERMES project several plasma wind tunnels (PWT) were qualified to be used for material research for re-entry missions [3]. The purpose of plasma wind tunnels is to rebuild the chemical and thermal boundary layer and consequently to simulate experimentally the chemical behaviour of candidate materials [1], i.e. total pressure, surface temperature, and heat flux resulting from high enthalpy flows. The group for plasma modelling and simulation at the institute uses numerical methods to simulate high-enthalpy re-entry situations, characterised by hypersonic flow conditions including as a consequence thereof high temperature and chemical non-equilibrium effects. The present paper describes the work of comparing the erosion rate, which is possible by numerical modelling of the catalytic and oxidative processes [4].

2 Numerical Simulation

The flow around the hot SiC probe as well as a cooled copper probe is simulated with the 2D/axisymmetric URANUS (Upwind Relaxation Algorithm for Nonequilibrium Flows of the University of Stuttgart) non-equilibrium Navier-Stokes code [5]. The URANUS code has been developed within the framework of the collaborative research center 259 "High Temperature Problems of Re-usable Space Transportation Systems" in order to simulate the flow around re-entry vehicles and the associated surface loads. The 18 governing equations in finite volume formulation are solved fully coupled and fully implicit. The conservation vector

$$Q = \left(\rho_i, \rho u, \rho v, \rho e_{tot}, \rho_k e_{vib,k}, \sum_j \rho_j e_{rot,j}, \rho_e e_e\right)^T$$
$$(i = 1, ..., 10, k = 1, 2, 3, j = 1, 2, 3, 6, 7, 8) \qquad (1)$$

consist of 10 species balance equations for N_2, O_2, NO, N, O, N_2^+, O_2^+, NO^+, N^+ and O^+, two momentum conservation equations, the total energy equation as well as three vibrational, one rotational and the electron energy balance equations. The inviscid fluxes are discretized in physical space with the approximate Riemann solver of Roe [6]. Second order accuracy for the inviscid fluxes is maintained by limited WENO extrapolation [7]. The viscous fluxes are determined in computational space by central differences, as usual. The advanced CVCV multiple temperature gas-phase model is employed to account for the coupling between internal degrees of freedom and chemistry in thermochemical non-equilibrium [8]. The transport coefficients are determined according to the Chapman-Enskog method for multi component mixtures in thermal non-equilibrium [9]. For the simulations of the flow around the probes, thermal and chemical equilibrium at the inflow boundary is assumed.

Under this assumption, the inflow conditions given in Table 1 follow from the specific total enthalpy of $6.2\mathrm{MJ/kg}$, the ambient pressure of 60Pa, the Mach Number of 4 and a minimum species mole fraction of 10^{-12}. Due to the inflow velocity of $3115\mathrm{m/s}$ approximately $4.85\mathrm{MJ/kg}$ arise from kinetic energy such that possible errors based on thermal or chemical non-equilibrium are assumed to be negligible. Nevertheless, significant differences may arise due to changes in radial direction, which have not been determined in experiment so far. The surface boundary conditions based on flux balances account for chemical non-equilibrium, temperature as well as velocity slip. The cooled copper probe surface temperature is set to $300\mathrm{K}$. To account for finite catalycity of copper, recombination coefficients of 0.1 for the atomic species and for NO were applied, the recombination coefficients of the ionized species have been set to 1. For the simulation of the hot SiC probe the surface reaction model presented in [10] has been employed assuming radiation equilibrium.

3 Experimental Setup

A plasma wind tunnel basically consists of three major parts. A vacuum vessel with about 6 m in length and 2 m in diameter is connected on one side to the second part, the vacuum system [1]. The third part is the plasma generator integrated in the front lid. The entire plasma flow can be investigated with optical methods using windows on both sides of the vacuum chamber from the generator exit on. The vacuum system is a roots pumping system with four stages reaching $250000\,\mathrm{m}^3/\mathrm{h}$ at a minimum pressure of 10 Pa at the intake. In the vessel the pressure can be varied between 50 Pa and 1000 hPa. With different generator principles, a wide range of re-entry trajectories can be covered concernig total pressure and enthalpy [11]. Figure 1(b) depicts the operational envelopes for the three different plasma generators used at IRS. Beside their restrictions in total pressure and enthalpy, which is mainly a problem of the maximum heating chamber pressure, the injection of the gas has to be adapted. The thermal plasma generator RB3 used for the present investigation is in use to simulate a segment on the re-entry trajectory, when total pressure is relatively high and the velocities are still high (see fig. 1(b)). Heat fluxes are in the range of $0.1\,\mathrm{MW/m}^2$ to $3\,\mathrm{MW/m}^2$ [12]. The generator is a coaxial thermal arc-jet as can be seen in figure 2(a). The plasma is mainly generated by the nitrogen part of the air passing along the cathode. The oxygen is injected downstream at the end of the anode which is forming the arc chamber. The mixed plasma is expanded into the vacuum chamber. To avoid possible arc attachment a coil is used generating an axial magnetic field such that the arc rapidly spins around.

For a comparison of the experimental and numerical results, the free–stream conditions have to be similar and an appropriate grid for the numerical simulation has to be generated. The condition is defined according to the initial conditions for the numerical calculation which are Mach number and enthalpy (see chapter 2). In thermal arc-jet facilities the mean enthalpy of the gas can be evaluated by balancing the electrical input power P_{el} less the cooling losses Q_w per mass flow $\dot{m}$ [1,

3]. The Mach number is calculated using the measured pressure ratio of total pressure to chamber pressure employing the Rayleigh-Pitot-equation for supersonic flow conditions

$$\frac{p_{tot}}{p_a} = \left(\frac{\kappa+1}{2} Ma^2\right)^{\frac{\kappa}{\kappa-1}} \left(\frac{2\kappa}{\kappa+1} Ma^2 - \frac{\kappa-1}{\kappa+1}\right)^{\frac{1}{\kappa-1}}, \tag{2}$$

where κ was set to 1.3. It should be mentioned that the coefficient κ ranges in between 1.1 and 1.67 depending on molecular or atomic particles as well as internal degrees of freedom and chemical behaviour. For the estimation of the used Mach numbers the influence is smaller than 5%. In the present case, the enthalpy is estimated to 6.2 MJ/kg, the Mach number to 4 and the chamber pressure is 60 Pa. The total pressure of 1250 Pa is measured using the pitot pressure port of the double probe as shown in figure 2(b). Total pressure, surface temperature, and heat flux are naturally also influenced by the distance of the probe from the generator exit, adjusted here to 90 mm. A computer controlled probe holder is installed in the vacuum chamber and protected by copper shields from the hot plasma. The installed probe can be moved in the three directions in space and additionally in one rotational axis in order to use both probe heads in the same experiment. The meanwhile fifteen different probes can be mounted in all plasma facilities at the institute. For erosion rate measurements, a double probe is used. On one side of this double probe, specimen can be mounted while on the other side a pressure port is installed. The sample holder itself is decoupled from the cooled probe base using a canopy made of SiC that holds the sample which is isolated from the cooled parts by a ceramic insulator made of aluminium oxide. All probes have an outer diameter of 50 mm and a sample diameter of 26.5 mm. An experiment is performed as follows: After ignition of the plasma using only nitrogen as working gas, the flow condition is adjusted. When all parameters are in steady state, the probe is moved from its position outside of the plasma flow to the centre axis. The final position is reached after roughly 85 s. This, of course, depends on the plasma condition. After the specified exposure time, the probe is moved out of the plasma axis and the generator is switched off. Two different materials were experimentally investigated. First experiments were performed with commercial pressure-less sintered silicon carbide (SSiC) as it is in use for basic material research. The second specimen is a carbon fibre silicon carbide ceramic provided by the German Aerospace Centre in Stuttgart (DLR-SiC) [13]. The carbon fibres are infiltrated with Silicon to a composite material. Finally, the sample is SiC–coated by chemical vapor deposition processing. The thickness of this layer is 40 μm. The erosion rate is defined as $\Theta_{tot} = \Delta m / A\ t$, where Δm is the mass loss/gain, A the area of the specimen, and t the time [14]. The mass change Δm of the probe is estimated by simple weighing before and after the test. The balance at IRS has an accuracy of 0.01 mg. For accurate measurement the sample is weighed 5 times and average is used [15]. The inserted time t in the above formula is the time period during which the probe remains at its specified position, where pressure, heat flux and temperature are defined. As readily can be imagined, the erosion process starts from the time on when the probe is in contact to plasma. Therefore, the time at the specified position is chosen to be long in comparison to the overall time of the probe in the plasma, for the present test case 1200 s.

4 Results

A comparison of erosion rates is possible when the boundary layers are similar. This is the case for the present study [16]. Measured and calculated total values at the surface are summarised in Table 2.

The temperature derived from numerical calculation is close to the experiment using the carbon fibre reinforced SiC provided by the German Aerospace Centre. The sintered SiC reaches a temperature about 100 K lower compared to the DLR-SiC. Hence, the calculated heat fluxes assuming adiabatic wall behave in the same way. The numerically evaluated particle densities are plotted in figure 3 together with experimentally gained particle densities using laser-induced fluorescence (LIF) measurements. The differences in density measurements compared to numerical calculation depend on differences in the plasma generation processes and in local enthalpy. Measurements of radial profiles of total pressure and heat fluxes of up to 850 $\mathrm{kW/m^2}$ onto a cooled copper probe allow to estimate a maximum in total local enthalpy of about 16 $\mathrm{MJ/kg}$. The comparison actually performed is based on the mean enthalpy of 6.2 $\mathrm{MJ/kg}$ since atomic oxygen number density measured by LIF measurements as well as measured surface temperature agree much better with the numerical results.

However, the boundary layers are generally spoken similar and the erosion rates can be compared. The resulting erosion rates are depicted in table 2.

First, it can be seen that the erosion rate is very small. The difference between the measured SSiC erosion rate and the numerical result is a factor 4000. Taking into account that the thickness Δs increases during the instationary warm-up procedure, it must be noted that the mass of the sample increases, such that the true difference in the stationary mass loss rate is even higher. The experimental investigation of the start-up procedure led to an increase of sample mass of $3 \cdot 10^{-7} kg$. Hence, the overall mass loss rate is $1.7 \cdot 10^{-3}\mathrm{kg/m^2\,h}$, such that the relation of measured to calculated mass loss rate becomes $3.5 \cdot 10^5$. In order to interpret this discrepancy, additional numerical calculations have been performed. Applying the radial enthalpy profile determined from heat flux measurements on a cooled copper probe, the mass loss rate at the stagnation point obtained numerically was $1.3 \cdot 10^{-3}\ \mathrm{kg/m^2\,h}$, which complies with a factor of 1.3. This numerical simulation, however, leads to a stagnation pressure of 1290 Pa and a surface temperature of 1936 K, which corresponds to a surface heat flux close to 700 $\mathrm{kW/m^2}$, i.e. 3.5 times the measured value onto SSiC. The second test with DLR-SiC material shows a similar result. Since the layer on the probe is made of chemical vapour deposited SiC, the same numerical result applies. The experimental erosion rate for this material is about one order of magnitude higher than the mean value for SSiC. Note, that the instationary warm up process is unknown for the DLR-SiC material, such that it is possible that the mass increase found for SSiC does not apply. However, the higher erosion rate corresponds to the higher surface temperature as well. After the test, a part on the surface can be identified where the layer seems to be thin compared to the rest of the probe, although all parameters of the experimental conditions are as specified. Hence, another conclusion that can be made is the following: the material had a higher material loss due to erosion effects arising from an inhomogeneous SiO_2-layer as can be seen in Fig. 4. Additionally, the manufacturing process implies that a higher silicon concentration at the surface occurs which leads to a faster erosion process due to faster chemical reaction processes with the plasma flow, especially during the instationary formation of the SiO_2-layer.

5 Conclusion

For the first time, the erosion rates of SiC ceramic probes measured in a plasma wind tunnel and numerically calculated are compared. This is possible because similar on-stream conditions were found and the chemical boundary layer is proved to be comparable. The erosion rates show rather big differences. One possible reason is the estimation procedure for the determination of the mass loss rate in the plasma wind tunnel. Another possibility is the uncertainty in plasma conditions, due to the fact that the plasma composition departs from equilibrium by a significantly higher dissociation degree of nitrogen. Hence, a final conclusion on local total specific enthalpy requires thoroughly investigation.

The numerical determination of erosion rates under steady state conditions has been newly developed and is unique in the world. Since meaningful and consistent experimental data is lacking, the present model is expected to depart from realistic results of mass loss rates by one to two orders of magnitude, especially if mass loss rate is low.

Although the present results showed rather big differences in measured and numerically determined mass loss rates, they give rise to dedicated numerical an experimental work in future. Since the comparison is the first ever made and due to the complexity of the problem, the current results are considered to be very good.

Acknowledgements

The authors thank all the laboratory people for their help during the experimental campaign and the German Aerospace Centre for the material support.

References

[1] HERDRICH, G. ; LÖHLE, S. ; AUWETER-KURTZ, M. ; ENDLICH, P. ; FERTIG, M. ; PIDAN, S. ; SCHREIBER, E.: IRS Ground-Testing Facilities: Thermal Protection System Development, Code Validation and Flight Experiment Development. In: *37th AIAA Thermophysics Conference*, 2004

[2] HEUER, A.H. ; LOU, V.L.K.: Volatility Diagrams for Silica, Silicon Nitride, and Silicon Carbide and Their Application to High-Temperature Decomposition and Oxidation. In: *Journal of the American Ceramic Society*. 73 (1990), Nr. 10, S. 2789–2803

[3] LAURE, S.: *Experimentelle Simulation der Staupunktströmung wiedereintretender Raumflugkörper und deren Charakterisierung mittels mechanischer Sonden*. Institut für Raumfahrtsysteme, Universität Stuttgart, Diplomarbeit, 1998

[4] FERTIG, M. ; FRÜHAUF, H.-H. ; AUWETER-KURTZ, M.: Modelling of Reactive Processes at SiC Surfaces in Rarefied Nonequilibrium Airflows. In: *8th AIAA Joint Thermophysics and Heat Transfer Conference*. St. Louis, Missouri, USA, 2002

[5] FRÜHAUF, H.-H. ; FERTIG, M. ; OLAWSKY, F. ; BÖNISCH, T.: Upwind Relaxation Algorithm for Reentry Nonequilibrium Flows. In: *High Performance Computing in Science and Engineering 99*. Springer, 2000, S. 365–378

[6] ROE, P.L: Approximate Riemann Solvers, Parameter Vectors and Difference Scheme. In: *Journal of Computational Physics*. 43 (1981), S. 357–372

[7] RÖDIGER, T.: *Analyse limitierter Extrapolationsverfahren zur Rekonstruktion von Hyperschallströmungen im thermochemischen Nichtgleichgewicht*, Institut für Raumfahrtsysteme, Universtität Stuttgart, Diplomarbeit, 2004

[8] KANNE, S. ; FRÜHAUF, H.-H. ; MESSERSCHMID, E.W.: Thermochemical Relaxation through Collisions and Radiation. In: *Journal of Thermophysics and Heat Transfer*. 14 (2000), Oktober, Nr. 4, S. 464–470

[9] FERTIG, M. ; DOHR, A. ; FRÜHAUF, H.-H.: Transport Coefficients for High Temperature Nonequilibrium Air Flows. In: *AIAA Journal of Thermophysics and Heat Transfer*. 15 (2001), April, Nr. 2, S. 148–156

[10] FERTIG, M. ; FRÜHAUF, H.-H. ; AUWETER-KURTZ, M. *Modelling of Reactive Processes at SiC Surfaces in Rarefied Nonequilibrium Airflows*. AIAA-Paper 2002-3102, 8th AIAA Joint Thermophysics and Heat Transfer Conference, St. Louis, Missouri, USA. 2002

[11] AUWETER-KURTZ, M. ; KURTZ, H. ; LAURE, S.: Plasma Generators for Re-Entry Simulations. In: *Journal of Propulsion and Power*. 12 (1996), Nr. 6, S. 1053–1061

[12] HERDRICH, G. ; AUWETER-KURTZ, M. ; ENDLICH, P. ; KURTZ, H. ; LAUX, T. ; LÖHLE, S. ; NAZINA, N. ; PIDAN, S. ; SCHREIBER, E. ; WEGMANN, T. ; WINTER, M.: Atmospheric Entry Simulation Capabilities at IRS. In: *3rd International Symposium on Atmospheric Reentry vehicles and systems*. Arcachon, France, March 2003

[13] HALD, H.: *Faserkeramiken für heiße Strukturen von Wiedereintrittsfahrzeugen - Simulation, Test und Vergleich mit experimentellen Flugdaten*. Institut für Bauweisen- und Konstruktionsforschung, Deutsches Zentrum für Luft- und Raumfahrt, Dissertation, 2001

[14] LAUX, T.: *Untersuchungen zur Hochtemperaturoxidation von Siliziumkarbid in Plasmaströmungen*. Institut für Raumfahrsysteme, Universität Stuttgart, Diplomarbeit, 2004

[15] HILFER, G.: *Experimentelle und theoretische Beiträge zur Plasma–Wand–Wechselwirkung keramischer Hitzeschutzmaterialien unter Wiedereintrittsbedingungen*. Institut für Raumfahrtsysteme, Universität Stuttgart, Dissertation, 1999

[16] LÖHLE, S. ; INFED, F. ; AUWETER-KURTZ, M.: Experimental and Numerical Rebuilding of Re-Entry Plasma Flows. In: *37th AIAA Thermophysics Conference*. Portland,OR, 2004

[17] WALPOT, L. ; OTTENS, H.: FESART/EXPERT Aerodynamic and Aerothermodynamic Analysis of the REV and KHEOPS Configurations. Noordwijk, The Netherlands, 2003 (TOS-MPA/2718/LW). – Forschungsbericht. Technical Report

Table 1 Inflow conditions for the numerical simulations. The minimum mole fraction was set to 10^{-12}.

quantity	value	quantity	value
N_2 partial density $\left(\frac{kg}{m^3}\right)$	$1.0617 \cdot 10^{-4}$	N_2^+ partial density $\left(\frac{kg}{m^3}\right)$	$1.3452 \cdot 10^{-16}$
O_2 partial density $\left(\frac{kg}{m^3}\right)$	$3.2158 \cdot 10^{-5}$	O_2^+ partial density $\left(\frac{kg}{m^3}\right)$	$1.5366 \cdot 10^{-16}$
NO partial density $\left(\frac{kg}{m^3}\right)$	$2.0066 \cdot 10^{-7}$	NO^+ partial density $\left(\frac{kg}{m^3}\right)$	$1.4409 \cdot 10^{-16}$
N partial density $\left(\frac{kg}{m^3}\right)$	$8.6633 \cdot 10^{-17}$	N^+ partial density $\left(\frac{kg}{m^3}\right)$	$6.7258 \cdot 10^{-17}$
O partial density $\left(\frac{kg}{m^3}\right)$	$3.2738 \cdot 10^{-4}$	N^+ partial density $\left(\frac{kg}{m^3}\right)$	$7.6827 \cdot 10^{-16}$
temperature (K)	1502.75	pressure (Pa)	60
Mach Number (-)	4		

Table 2 Resulting surface values from experiment and numerical calculation

	URANUS	Experiment
Total pressure	1253.7 Pa	1250 Pa
Surface temperature	1532.8 K	1420 K (SSiC) 1532 K (DLR-SiC)
Heat Flux	272.29 kW/m^2	195.97 kW/m^2
Erosion rate SSiC	$4.81 \cdot 10^{-9}$ kg/m^{2}h	$7.571 \cdot 10^{-5}$ kg/m^{2}h
Erosion rate DLR-SiC	-	$3.073 \cdot 10^{-4}$ kg/m^{2}h

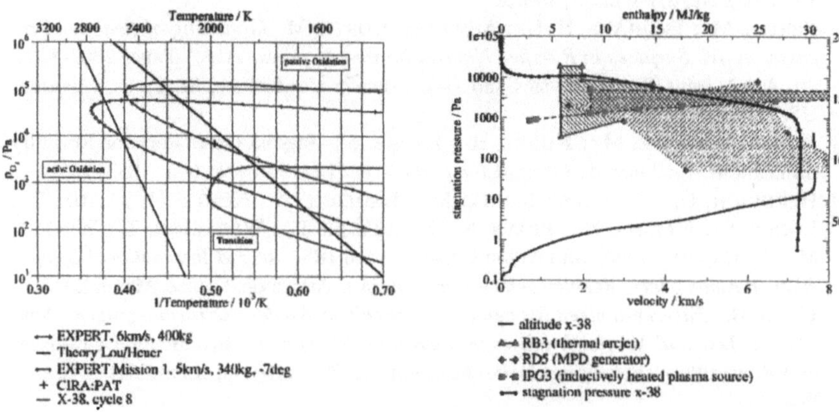

Figure 1 (left) Flight trajectories of EXPERT (nominal) [17] and X-38 (Cycle-8) in relation to active, passive and transitional oxidation regimes based on the model of Heuer and Lou [2] (right) Operational envelope for IRS plasma wind tunnels together with an X-38 flight trajectory

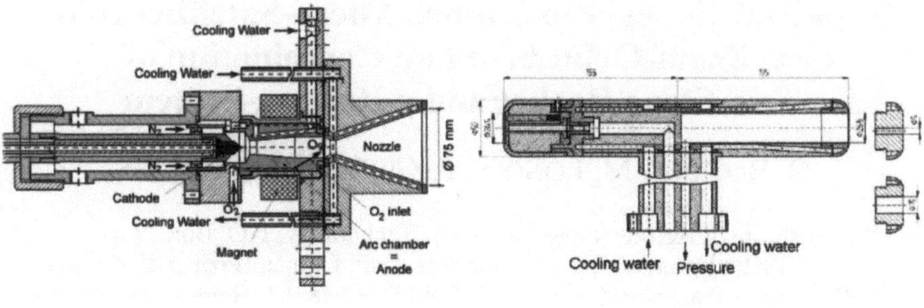

Figure 2 (a) Thermal arcjet generator RB3 (b) Double probe: Material sample holder and pressure port

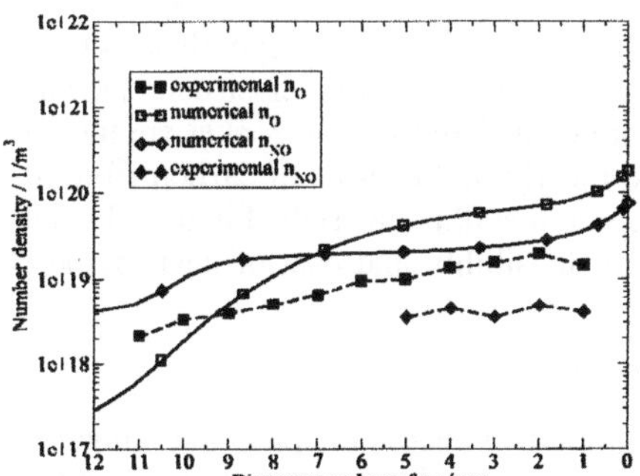

Figure 3 Boundary layer in front of SiC specimen: Experimentally evaluated from laser-induced fluorescence measurements and numerically calculated [16]

Figure 4 DLR-SiC material after test

High-End Concept to Launch Micro-Satellites into Low-Earth-Orbit based on Combination of a Two-Stage Rocket and a Railgun-System

O. BOŽIĆ[1], J. M. LONGO[2], P. GIESE[3], J. BEHRENS[4]

[1,2,3] DLR, Institute of Aerodynamics and Flow Technology, P.O. Box 32 67, D-38108 Braunschweig, Germany, e-mail: Ognjan.Bozic@dlr.de
[4] EADS- Space Transportation, P.O.B. 286156, D-28361, Bremen, Germany e-mail: Joerg.Behrens@space.eads.net

Summary

The electromagnetic railgun technology appears to be an interesting alternative to launch small payloads into Low Earth Orbit (LEO), as this may introduce lower launch costs. A high-end solution, based upon present state of the art technology, has been investigated to derive the technical boundary conditions for the application of such a new system. This paper presents the main concept and the design aspects of such propelled projectiles with special emphasis on flight mechanics, aero-/thermodynamics, materials and propulsion characteristics. Launch angles and trajectory optimisation analyses are carried out by means of 3 degree of freedom simulations (3DOF). The aerodynamic form of the projectile is optimised to provoke minimum drag and low heat loads. The surface temperature distribution for critical zones is calculated with DLR developed Navier-Stokes codes TAU, HOTSOSE, whereas the engineering tool HF3T is used for time dependent calculations of heat loads and temperatures on project surface and inner structures. Furthermore, competing propulsions systems are considered for the rocket engines of both stages. The structural mass is analysed mostly on the basis of carbon fibre reinforced materials as well as classical aerospace metallic materials. Finally, this paper gives a critical overview of the technical feasibility and cost of small rockets for such missions.

1. Introduction

As a first step in the development of this alternative access to space the launch of scientific experiments into suborbital altitudes is planned, permitting to test out and validate all basic features of this new system in a smaller scale. The realization and validation of such a propulsion system is considered as a stepping-stone for a future envisaged launch of small satellites into Low Earth Orbit. One potential market considered is the launch of small payloads like pico-, nano- and micro-satellites[1], which are already today used for several different applications, i.e. education, communication, scientific microgravity experiments, optical or radar spectra imaging or just as low cost technology demonstrators.

[1] pico-satellite (0.1kg -1kg); nano-satellite (1kg-10kg); micro-satellite (10kg-50kg)

A first concept was developed in the frame of an ESA study with the delivery of a 5kg payload into LEO with a one-stage rocket launched by means of an electromagnetic railgun. This goal should be reached by two steps: At first scientific payloads will be launched into suborbital altitudes (~115km) for the research of the upper atmosphere. The railgun needed for this purpose will be 22m long and powered by 32MJ capacitor bank (Fig. 1). The 1m long and 2.2kg heavy projectile will be accelerated within 21msec at 13000g to hypersonic speed (Ma =6.2) and will reach the target altitude in a purely ballistic flight, i.e. it is non-propelled [2]. Secondly, the launch of projectiles (~45kg) with a velocity of 6km/sec at an acceleration level of 10,000 g is foreseen [1], [3]. In this particular case the railgun length will be 180m and the acceleration time is approximately 51 msec. However, to inject the payload into LEO, the total velocity must be larger than 10.6 km/sec to overcome gravitational and aerodynamic drag losses. The difference between the required velocity and the railgun start velocity must be provided by an additional rocket engine, which could be for instance a hybrid rocket motor (HRM). The corresponding projectile design is shown in Fig. 2.

2. High-End Railgun

To determine the technical limits based upon today's state of art technology, a high-end railgun solution to launch small satellites into LEO is derived. The concept consists of a railgun catapult (Fig. 4), the power sources, and systems necessary for flight control and operations. The performance data of the railgun launcher and the propelled rocket are summarised in Table 1. The catapult length limitation imposed by the rail inductance can be circumvented by subdividing the railgun into 70 segments. Each segment will be fed by an independent local energy source.

Table 1: Properties of High-End Railgun System

ROCKET LEO LAUNCHER			HIGH-END RAILGUN		
Payload	(kg)	32.	Total rail length	(m)	$420^{(1)}$ - $740^{(2)}$
Total Mass	(kg)	793.	Muzzle velocity V_{gun}	(m/s)	3700
Sabot - Masse	(kg)	70.	Energy input - facility	[GJ]	16.0 - 17.6
Diameter	(mm)	320.	Maximal armature current	(MA)	5.3
Length	(m)	9.0	Start acceleration, max	(g)	$2000^{(1)}$ - $920^{(2)}$
Mach number at start	(-)	10.83	Resident time of projectile within gun	(msec)	$230^{(1)}$ - $400^{(2)}$
Kinetic energy at start	(GJ)	up 5.2			
Flight time to 42 km	(sec)	24.0	$^{(1)}$ hybrid-railgun		
Apogee point/Orbit	(km)	367.	$^{(2)}$ segmented railgun		
Launch angle	(grad)	29.-39.			

The railgun ramp is installed inside a tube with semi-circled intersection. The tube is evacuated (air pressure from 100 to 200 Pa) to prevent drag and heat loads on the projectile surface during the acceleration phase. In this railgun construction the only moving element is the armature for the acceleration of the rocket launcher (Fig. 3), whereas the sliding contacts of the armature should be able to transfer huge currents up to 5.5 MA. The armature itself is integrated with a wedge, which transfers the momentum from the armature and pushes the projectile with an acceleration of max. 2000g. This solution does not limit the projectile's span, which is given by the associated fins, necessary for aerodynamic stabilization during its flight in the atmosphere.

The power supply must be able to deliver huge energies with huge currents in a short period of time. In general, there are several solutions available: today capacitors are used exclusively, which have sufficient power density but a limited energy density (≤5kJ/kg). A real disadvantage is the very high cost. Alternative solutions are batteries, compulsators (high-speed, compensated pulsed alternators) and homopolar generators (HPG). The batteries have high energy density (≤100kJ/kg), but low power density [4]. Homopolar generators provide just low voltage outputs but they can be used to energize a large storage inductor. At the Australian National University in Canberra a 550MJ homopolar generators have been build (June 1962) and tested, which have operated at 380MJ. Using this technology, 3g projectile was accelerated up to 5,9km/s [5]. Therefore, a battery of HPGs or capacitors seems to be the best solution for a high end railgun

3. Two-Stage Rocket Launcher

In the second design loop the properties of the entire rocket launcher, the propulsion system elements and the proportions of the subsystems were determined, using the DLR code ENGINE-MOD. The proposed configuration of a rocket with a cylindrical body and six fins is shown on the Fig. 5 and Fig. 6. The aerodynamic form of the projectile is optimized to provoke minimum drag and low heat loads. Calculations based on combined measurement data and handbook methods have shown that during flight in the hypersonic velocity range the drag coefficient is lower than 0.1. To obtain low wave drag, the nose cone of the projectile has a spherically rounded tip with a 15 mm radius and a contour designed according to the “power law” with an exponent n = 0.75. A plug nozzle solution is selected because it has a compact form and contributes to the decrease the launcher base drag.

3.1 Propulsion system

The analyses show that the most promising propulsion systems are a semi-cryogen liquid rocket engine (LRE) based on liquid oxygen/hydrocarbon propellants for the marsh stage and a hybrid rocket engine (HRE) based on hydrogen peroxide/wax propellants for the orbital insertion stage. The HRE chosen for the 2nd stage with H_2O_2/wax has a propellant bulk density of 1.3g/cc, which is 30% higher than for LO_2/kerosene (1.03g/cc). Although HRE has a lower specific impulse up to 9% than LRE with LO_2/kerosene, the engine is more compact, simple, cheaper and reliable. Beside this advantage the HRE propulsion system cannot compete with the 1st marsh stage due to its limited thrust, resulting from the low regression rate of the solid fuel. Despite this fact, the selected wax has a better regression rate than all other HRE fuel candidates.

3.2 Flight trajectory

Fig. 7 shows the path for the first 30 seconds of flight, calculated with DLR code TRAJECTORY 3D. Trajectory optimization analysis shows that the optimal projectile start velocity from the railgun is 3.65km/s, i.e. M_a = 10.83. The minimum energy for given start conditions (PT1) are achieved for an elevation launch angle of $\theta = 39^o$. The projectile passes the troposphere after 5.6 sec (PT2), reaches 30000 m after 17 sec (PT3) and will pass the dense atmosphere layer at 42km after just 24 sec (PT4). During the flight with hypersonic velocity the Mach number will be reduced down to M_a=8.4 at point PT4 due to gravitation and aerodynamic drag. The complete mission profile is shown in Fig. 8a. After passing point PT4, the launcher continues to fly without any propulsion for additional 91sec to an altitude of 149 km, where the semi-cryogenic liquid rocket motor of the first stage is fired. Its burnout is 80sec later and the stage will be separated from the vehicle. The launcher continues flying without any propulsion before it reaches LEO (approx. 360km) altitude, where the angle of trajectory inclination becomes almost zero. At this point the launcher ignites the second stage, a hybrid rocket motor, which burns for approximately 40sec providing the launcher with the additional velocity to insert the payload into a circular orbit.

The structural mass is analysed mostly on the basis of carbon-fibre reinforced materials combined with classical light aerospace materials. One more conservative approach to the mass analysis, which takes in account higher fraction of metal components, should lead to rise of total rocket mass up to 793kg (Fig. 8b). That will also have a consequence for moderate increase the start velocity (up to 3.7km/s) and corresponding extension of railgun length (up to 15%).

3.3 Thermal loads and thermal protection system

At the exit of the railgun catapult the rocket will experience the maximum mechanical loads due to the air dynamic pressure. However, these loads will decrease with increasing altitudes due to lower atmospheric density. The maximum thermal loads are achieved after 6sec to 10sec depending on the applied type of TPS protection. The heat loads will decrease rapidly when the launcher passed the densest earth atmosphere after 24sec. The maximum surface temperatures will be established on the nose tip very quickly within the first second, after that it will decrease sharply.

Surface pressure distributions and thermal loads are calculated for a defined geometry configuration with the DLR engineering tool HOTSOSE, based on surface inclination method. The surface temperature distribution is computed by assuming stationary conditions. This is corrected afterwards as time dependent for a trajectory point after 6sec flight (Fig. 9). The areas of high-risk damages due to the high temperatures are the tip of the nose cone (~2100K), the junction between fuselage and fins as well as the leading edges of the fins (~2000K). The cylindrical body is moderately charged (~1600K). These preliminary results are cross-checked for the nose cone only with an additional computer program, the DLR unstructured Navier-Stokes Code TAU for Ma = 10.83. This simulation is also carried out under stationary conditions and then corrected as time dependent for a trajectory point after 6sec flight (Fig. 10). Both simulations, extrapolated under same conditions, show a good correlation of the results with a temperature difference of about 100K.

With the DLR code HF3T (Heat Flux and Temperature as a function of Trajectory And Time) temperatures on the rocket launcher at two reference points were calculated: (a) at the stagnation point at the top of the nose cup (Fig. 11) and (b) at the transition point between the nose cap and the cylindrical body (Fig.12). In the stagnation point a calculation based on "Fay and Riddell-Model" is carried out for a blunted nose cap with a sphere's radius of 15mm. The time dependent temperature is calculated for the nose cup surface and an in-material depth of 8mm for 4 different TPS materials. The best properties show the Re/C-C matrix composite with ULTRA2000 coating, which can withstand for 30sec ultra high temperatures up to 2800K, but even for such an excellent material the surface temperatures are too high for the first 8sec of flight. A possible solution of this problem may be to combine passive protection (high effective TPS material) and controlled ablation within two-layer sandwich structures. Fig. 12 shows the surface temperature in a point representative for the structural design of the entire cylindrical body. The temperature is moderate and it may be easier to manage a thermal protection. In the range up to 1600K more materials do exist, which fulfill those requirements. For the final selection of the material, an optimization with respect of mass and cost has to be made.

4. Conclusion

Investigations conducted in this study demonstrated that an electromagnetic railgun with a length of approx. 485m is in the position to accelerate at the end of railgun launch ramp a projectile to a velocity of 3.7 km/s. In order to achieve orbital motion and to compensate the velocity losses due to gravity and aerodynamic drag, a two-stage rocket is integrated into the projectile's structure. The projectile with a calibre of 320mm itself have a total mass of 793 kg including a payload of 32 kg and is accelerated moderately up to 2000 g's on the railgun ramp.

No major issues have been identified which cannot be implemented. Based upon today's knowledge and the use of already existing potential key technologies, it is possible to complement and compete with conventional atmospheric sounding rockets. The study has shown that the application of railgun as satellite launcher is a realistic goal. Launching up to 800kg mass at 3.7 km/sec is today beyond the demonstrated railgun performances, but there are no physics or technological barriers to prevent the achievement of the goals of this study (high end railgun). The practical limits of the railguns will depend on acceptable cost and service live and therefore the highest emphasis must be put on overall system operability.

References

[1] O. Božić , J. M. Longo , P. Giese, J. Behrens, Design of Propelled Projectile to Provide Escape Velocity for Nano – satellite, 54th Congress of International Astronautic Federation, September 29th – October 3rd. 2003, Bremen, Germany, Paper IAC-03-V.7./ IAA.11.5.02

[2] J. Behrens, O. Božić, Development of Electromagnetic Railgun Technology for the Launch of Payloads into Suborbital and Orbital Altitudes, 55th Congress of International Astronautic Federation, October 3th - 8nd 2004, Vancouver, Canada, Paper IAC -04 -V.5.08

[3] O. Božić, J. M. Longo, P. Giese, New European Concept of Single Staged Rocket to launch nano-satellites in Low Earth Orbit (LEO), International Conference on Recent Advances in Space Technologies (RAST 2003), IEEE, Istanbul, Turkey, November 20th -22nd, 2003

[4] J. L. Brown, K. A. Jamison, N. E. Johnson, G. E. Rolader, J. J. Scanlon III, Earth To Orbit Railgun Launcher, *IEEE Transactions on Magnetics*, Vol. 29. No. 1, January 1993, pp. 373-378

[5] H. D. Fair, Electromagnetic Launch, *International Journal of Impact Engineering*, Vol. 29 , 2003, pp. 247-260

Figure 1. Suborbital railgun - launch facility.

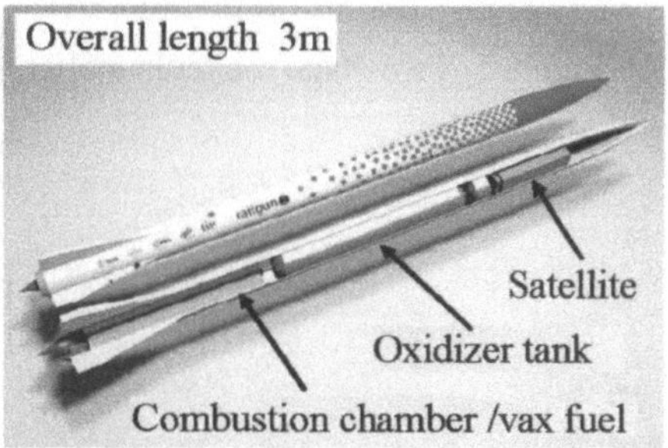

Figure 2. Single-stage LEO launcher for nano-satellites

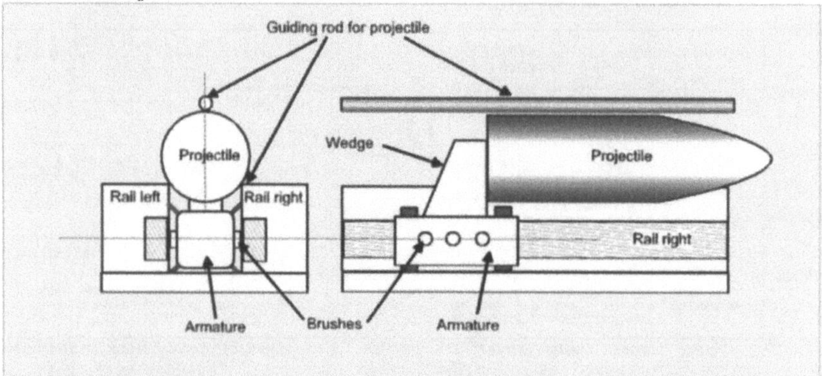

Figure 3. Displaced position of projectile and armature on the railgun ramp

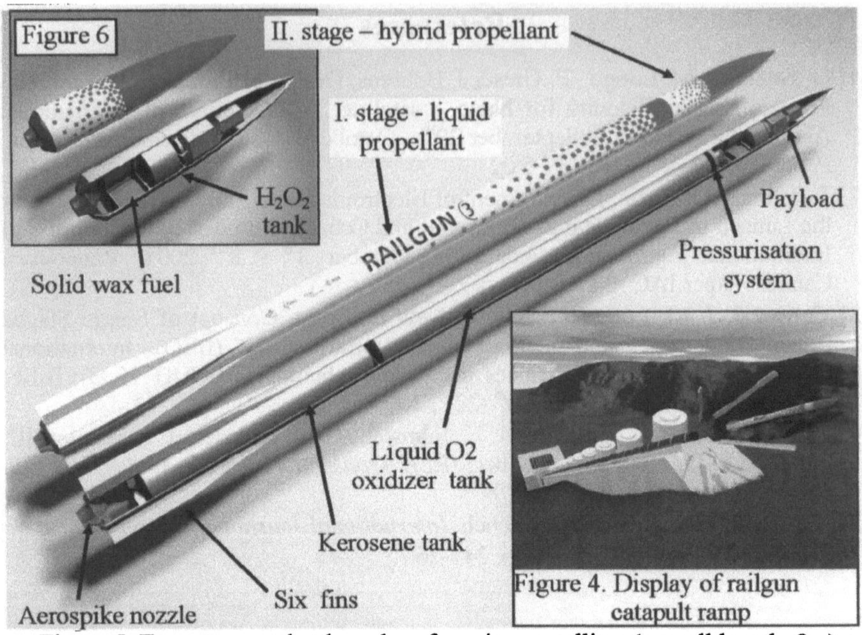

Figure 6

Figure 4. Display of railgun catapult ramp

Figure 5. Two-stage rocket launcher for micro-satellites (overall length 9m)

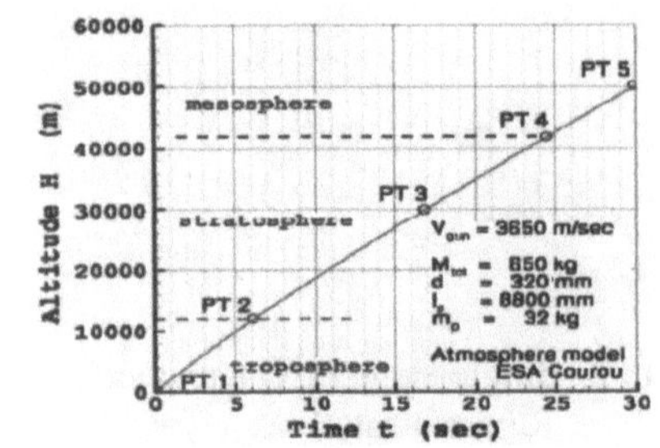

Figure 7. Rocket flight trajectory (first 30 sec)

Figure 8a. Mission profile for the 2-stage rocket (m_{tot} = 650kg)

Figure 8b. Mission profile for the 2-stage rocket (m_{tot} = 793kg)

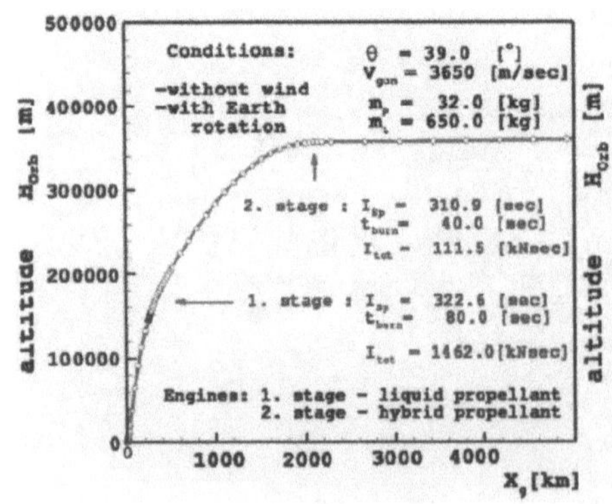

Fig. 8a.

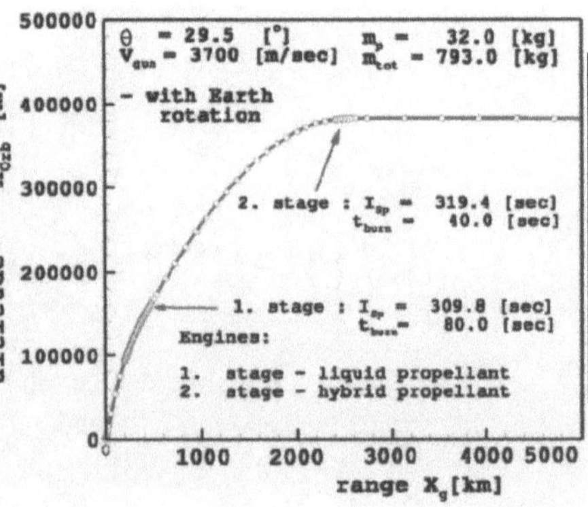

Fig. 8b.

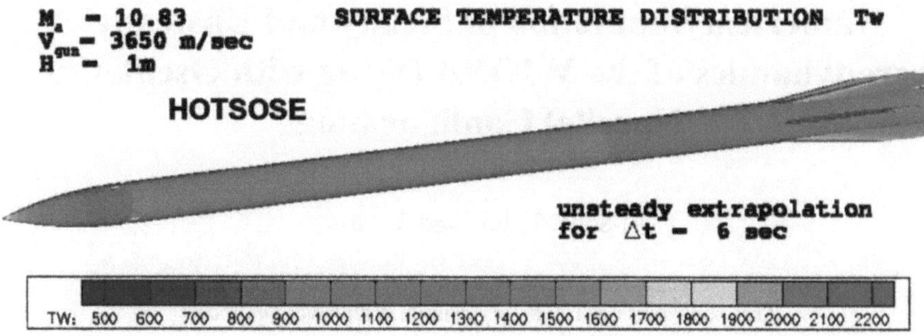

Figure 9. Surface temperature distribution after 6sec of flight (unsteady extrapolation)

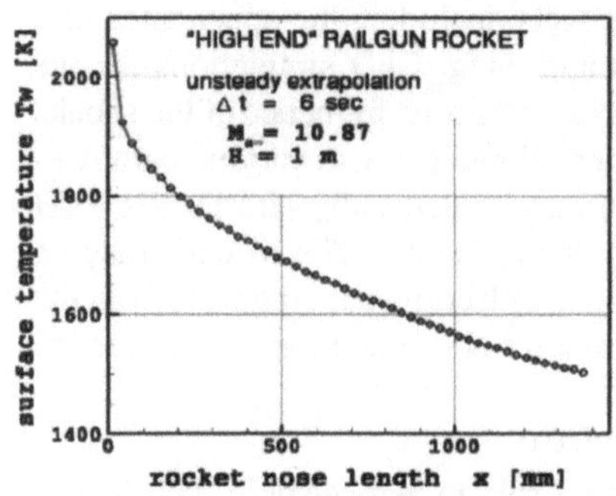

Figure 10. Temperature distribution along the nose cap

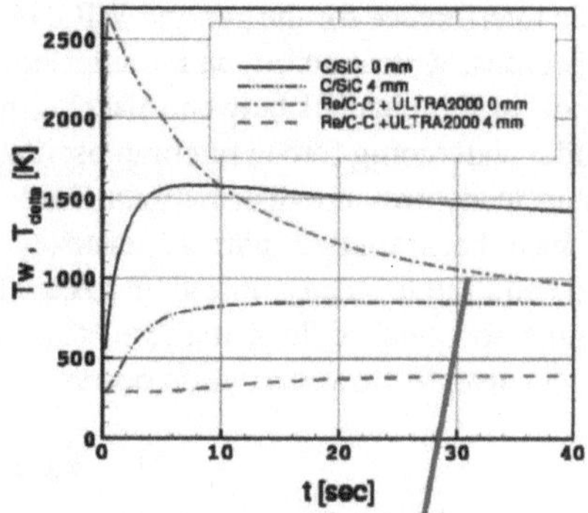

Figure 11. Time dependent temperature on projectile cylindrical body, at surface and 4mm depth for 2 different TPS.

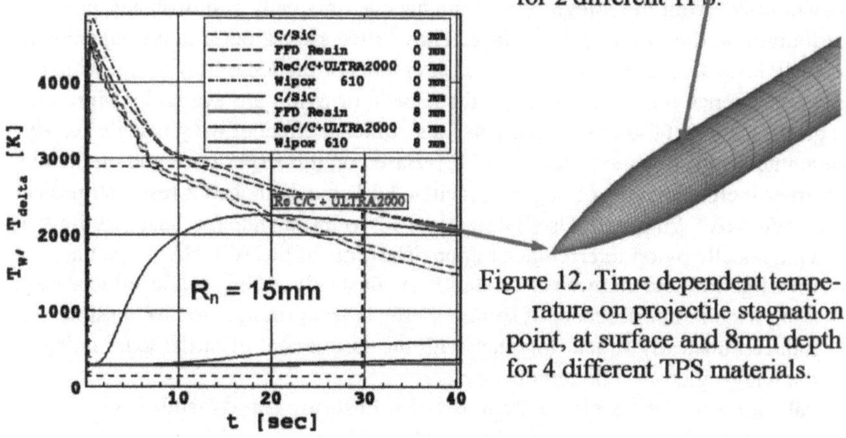

Figure 12. Time dependent temperature on projectile stagnation point, at surface and 8mm depth for 4 different TPS materials.

Numerical Simulation of Steady and Unsteady Aerodynamics of the WIONA (Wing with Oscillating Nacelle) Configuration

A. Soda, T. Tefy and R. Voss

DLR Institute of Aeroelasticity, Bunsenstrasse 10, D-37073 Goettingen, Germany
Ante.soda@dlr.de, Tiana.Tefy@dlr.de, Ralph.Voss@dlr.de

Summary

This paper presents an overview of numerical simulations of unsteady wing nacelle interference for the generic WIONA configuration. The geometry of the model [4] causes, typical aerodynamic interference effects, including the appearance of a shock in the channel between nacelle, pylon and wing. CFD simulations for the model undergoing forced pitching oscillations show a strong influence of the shocks on unsteady pressure distributions. Three numerical models with varying complexity have been applied, namely unsteady RANS and Euler, both with DLR's TAU code and the unsteady time linearized TDLM code. The fast TDLM code may be applied for attached flow, and results are validated with corresponding ones from the Euler solver. If flow separation occurs, the current RANS simulations are necessary.

1 Introduction

Wing nacelle interference can play an important role in aeroelasticity [1]. In former windtunnel tests conducted within the Aeroelastic Model Program (AMP), the impact of unsteady airloads on flutter in transonic flow has been observed for a typical transport aircraft half model configuration [6]. This study clearly indicated the channel effect from the integration of the nacelle on steady and unsteady pressure distribution on the lower wing surface. These effects were not a main objective of the AMP tests, and neither detailed measurements of the local flow in the wing nacelle interference region nor data on the nacelle or pylon are available. Moreover, the global results of flutter tests indicated a significant effect of wing nacelle aerodynamic interference on the flutter speed [6]. Based on this experience and on the fact of further increasing size of engine nacelles, DLR has launched a research project called WIONA (Wing with Oscillating Nacelle) to investigate the unsteady flow in the wing nacelle pylon interference region. The aims of the WIONA project are:

- investigation and numerical modeling of unsteady aerodynamic effects from wing nacelle interference, of spinning effects from the fan and of possible self induced unsteady shock, together with the assessment of buffet originating in the wing-pylon-nacelle region,
- validation of CFD tools for both, detailed unsteady aerodynamic analysis and for a global fast computation of unsteady airloads,
- application of validated tools for flutter computations for complex aircraft configuration and assessment of the impact of engine integration on flutter.

One important part of the project is conducted in cooperation between DLR and ONERA. A joint windtunnel test has been performed in the Transonic Windtunnel Goettingen (TWG). Both partners shared the work of wind tunnel model development and manufacturing and data acquisition, and they have cooperated in codes validation. The current paper presents an overview of DLR numerical work on the WIONA configuration.

2 Methods and results

Steady Navier-Stokes computations

Due to restrictions of the TWG windtunnel (a 1m test section did exclude a half model), a wall-to-wall wing section was chosen and thus excitation on both WT walls was necessary. The existing excitation system did not enable realistic wing sweep so a generic rectangular wing model showing similar flow characteristics at the wing nacelle interference region as real aircraft has been designed using Euler CFD simulations [4]. Navier-Stokes computations were performed with the DLR's TAU code, using a 3-D unstructured grid with 1.8 million cells (hybrid mesh with tetrahedra and hexahedra). Different turbulence models, namely the 1-equation Spalart Allmaras and the 2-equations $k - \omega$ / LEA models have been used for the steady state computations. No significant differences have been observed on the resulting flow field. The basic test case investigated has the following flow parameters: Mach number 0.82, Reynolds number 2.2 million, incidence -0.6 deg. Fig. 1 presents a comparison of numerical results for this case with steady state windtunnel results [2] for the parameters Mach number 0.8269, incidence 0.09 deg. The agreement is good, the only discrepancy is the weaker shock of the test results in the channel between the wing and the nacelle. These results clearly show that the numerical design of WIONA configuration described in [4] was successful, though only Euler solutions were used. A closer inspection of the computed steady state flow field is presented in Fig. 2 for the most interesting position at 47 % wing span, which intersects the channel close to the pylon. There are several local supersonic regions. The most important ones for the current investigation are in the channel flow and on the upper wing surface. Additional small supersonic regions on the lower nacelle surface are not typical for realistic engine nacelles and appear on the WIONA model because the nacelle profile has been chosen to be thick enough for pressure sensors. On the upper wing surface, the shock wave induces a significant increase of boundary layer thickness, but the flow remains attached. In the channel, where the highest local Mach numbers appear, the terminating shock is strong enough to induce flow separation behind the shock. The separation regime on the nacelle extends to the nacelle trailing edge. On the wing's lower surface, after a shock induced separation, the flow attaches again, before a second separated regime close to the lower trailing edges appears further downstream .

Unsteady Navier-Stokes computations

Unsteady RANS computations were performed with the TAU code for a prescribed rigid pitching oscillation on the WIONA configuration around the presented steady flow condition. The unsteady parameters are : pitching axis at 40 % wing chord position, amplitude 1 deg, reduced frequency $k = \omega c/v = 0.2$. The numerical simulations were carried out with 500 time steps per oscillation period, and with 40 inner iterations (pseudo time steps) per physical time step. The simulations were finished after 5 periods with a converged periodic solution, thus resulting in a total number of 2500 time steps. Each period required 68 CPU hours on a cluster of 16 Intel p4 2.0 GHZ processors.

Fig. 3 depicts unsteady results for the oscillating WIONA configuration, in terms of the unsteady pressure coefficient Cp. Three different spanwise positions through the model surface, namely two on the wing, one off the channel (20 % wing span) and one at the channel (47 % wing span), and one on the nacelle (close to the channel) are shown. The lower right graph of each figure presents the different instantaneous pressure distributions for all time steps during the last converged period of motion. Thus covering the shaded region and represent directly the unsteady variation of Cp. The variation of Cp is generally not in phase with the forced motion of the model. For example, on the wing the maxima of extension of supersonic region, of downstream shock position and shock strength on the upper surface and correspondingly of lift are not reached at the time of maximum pitch deflection, but usually later. This phase lag behind the imposed motion depends strongly on the flow field type and on the reduced frequency. In unsteady aerodynamics, the results of a Fourier transform of the periodic time history is usually investigated. The results are expressed as magnitude and phase shift for first and higher harmonics. In the context of classical flutter, only the first harmonic is important, while appearance of higher harmonics during first mode excitation indicates nonlinear aerodynamic effects.

At 20% wing span cut, shocks appear on both upper and lower wing surface. On the upper surface, the shock is close to the trailing edge and of medium strength and the shock motion is nearly harmonical. This is highlighted by the very small value of the second harmonic pressure component. Whereas on the lower surface, a strong shock is oscillating over a distance of 4% wing chord, compared to only 1.5% on the upper surface. Correspondingly, the presure signals are significantly nonlinear, shown by the large value of the second harmonic on the lower surface in the shock region and also downstream of the shock. On the upper surface, the phase lag of the unsteady pressure behind the pitching angle is between 20 and 30 degrees whereas it is larger, namely $140 - 150$ degrees on the lower surface. Note that the phase lag values on the lower surface have positive sign, because they are measured in reference with the upstroke pitching motion. That is, the limiting case of reduced frequency 0 defines a phase lag of 180 degrees instead of 0 degree for the upper surface. Therefore the actual phase lag on the lower surface is about 30-40 degrees.

For the second cut (47%), the results appear to be similar. But on the lower wing surface, the pressure peaks in both the first and second harmonics are higher due to the stronger shock and the width of the peak is wider according to the wider area crossed by the oscillating shock. In addition, the second harmonic contribution is again significant downstream of the shock and both harmonics exhibit a wavy behavior. This may be due to the shock induced separation. The values on the upper

wing surface do not differ much from those at 20% wing cut, with the only exception that the phase lag of the shock motion is different. This reflects the fact that the upper surface pressure is nearly constant along the wing span.
Results on the nacelle differ from those on the wing surface. In the third cut, which represents the lower boundary of the channel, a strong shock with flow separation undergoes heavy oscillations, stronger than in cut 2 (47%). Correspondingly, there are strong pressure peaks for both first and second harmonics. The value at the nacelle trailing edge does not tend to zero and remains at a high level. This is due to the large fluctuations which are excited there by unsteady local flow separation and reattachment between the shock and the trailing edge. In the front part of this cut, downstream to the shock region at $x = 100$, there is a small phase lag of about 20 degrees. Further downstream, entering the shock region the phase angle jumps to $+140$ degrees, preceding the pitch motion and keeping this value for the whole region down to the trailing edge. Thus shock motion, separation and reattachment process are nearly in anti phase with the pitching of the model.

TDLM and Euler unsteady calculations

The development and validation of faster and reduced models for unsteady aerodynamics is of special importance for aeroelasticity. Usually flutter computations need many unsteady aerodynamic calculations (typically for 10 elastic modes, for 10 reduced frequencies per load case : Mach number - lift combination). Thus a flutter analysis applying the unsteady RANS solver would demand an extremely high computational effort which is only adequate for detailed investigations and validation. For classical flutter analysis in aircraft design and development, different strategies have been developed, all of which somehow try to reduce the effort for the CFD generation of unsteady airloads. One of these strategies is the application of time-linearised CFD codes. These model the flow field around an aircraft structure which undergoes forced harmonic oscillations by superposition of the mean flow and an unsteady perturbation flow. For small perturbations as they appear for small amplitudes of the forced motion, the unsteady perturbation flow can be modeled by linear flow equations with varying coefficients coming from the solution of the nonlinear steady equation for the mean flow. This strategy has been realised by the Transonic Doublet Lattice Method (TDLM) [3], solving the unsteady Transonic Small Disturbance Potential Equation. The coefficients may be imported from any steady CFD solution, including Euler or RANS codes. This method provides unsteady CFD results in form of AIC matrices, (all possible elastic oscillation modes are covered), fully compatible with commercial Computational Structural Mechanics (CSM) software like NASTRAN which is in use for industrial flutter computations. The method has been developed for complete aircraft configurations, but it has been validated mainly for clean wing applications. It is tested here for the wing-nacelle problem. Before it is compared with the RANS results for a case with the significant flow separation, it is compared here with inviscid Euler results. Both a steady and an unsteady solution for forced oscillatory motion are generated and compared with Euler results obtained using the TAU code. The steady flow field supplies the coefficients for the time linearised equation, which is then solved by TDLM. TDLM results in form of unsteady first harmonic pressure difference values ΔCp between the lower and upper surfaces are compared with corresponding results from the unsteady CFD Euler run. Fig. 5 depicts a comparison of Euler and TDLM results for real- and imaginary parts of the first harmonic of ΔCp enforced by the same steady and unsteady flow parameters as the RANS test case and presented for the same 3 planes. The TDLM computations have been performed on a structured grid of only 5760 volume cells where wing, nacelle and pylon are geometrically modeled by thin surfaces of 576, 180 and 26 panel cells, respectively. Results are obtained in 15min CPU time on 1 DELL processor in contrast to 100h on 4 processors for the Euler results. The corresponding unsteady Euler results are depicted in Fig.4. The agreement is good between the first harmonics of Euler and TDLM, namely the peaks at the oscillating shock positions are similar for both methods. As for the present amplitude of 1 degree shock motion is quite strong for cuts 2 and 3, the corresponding pressure peaks from unsteady Euler are significantly wider and extend to further downstream positions than the TDLM results. These provide pressure peaks at the position of the steady flow field.

3 Conclusion

Numerical computations, steady as well as unsteady, have been performed for the generic wing nacelle interfering WIONA test configuration. RANS solutions show a strong influence of shocks and flow separations on unsteady airloads. Computations applying the fast TDLM method were able to predict well unsteady airloads and have been validated with corresponding Euler results. Future topic is the validation of TDLM for flows with strong viscous effects, namely flow separation which is of great importance for the interfering flow on the WIONA configuration, as demonstrated by the RANS simulations. Finally, all the current computations need validation by unsteady wind tunnel results which will be available in 2006.

Acknowledgments

The current work has been performed within the DLR project HighPerFlex and the windtunnel test results have been obtained in the DLR-ONERA project WIONA.

References

[1] Förseching, H "New Ultra High Capacity Aircraft (UHCA) : Challenges and problems from an aeroelastic point of view" ZFW 18(1994), pp. 219-231
[2] Dietz, G., Private communication
[3] Lu, S., Voss, R. "A transonic doublet lattice method for 3D potential unsteady transonic flow calculations and it's application to transonic flutter prediction" Proc. Int. Forum on Aeroelasticity and Structure Dynamics, Strasbourg 1993
[4] Soda, A., Tefy, T, "Numerical Investigation of Wing-Nacelle Interference Effects at Transonic Flow Conditions for a Generic Transport Aircraft Configuration", STAB conference 2002 Munich
[5] Weinman, K., et al, "Tau-Code User Guide", Release 2004, DLR, Germany http://www.taurus.dlr.de/taurus_main.htm
[6] Zingel, H, "Measurement of steady and unsteady airloads on a stiffness scaled model of modern transport aircraft wing" Proc. Int. Forum on Aeroelasticity and Structural Dynamics, DGLR 91-06, pp 120-131, 1991

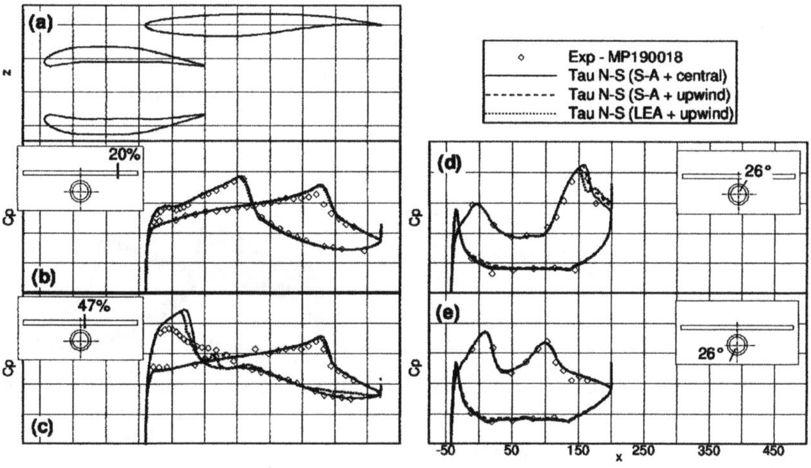

Figure 1 CP distribution for steady viscous flow at 4 cutting planes: 2 on the wing (top) and 2 on the nacelle (bottom).

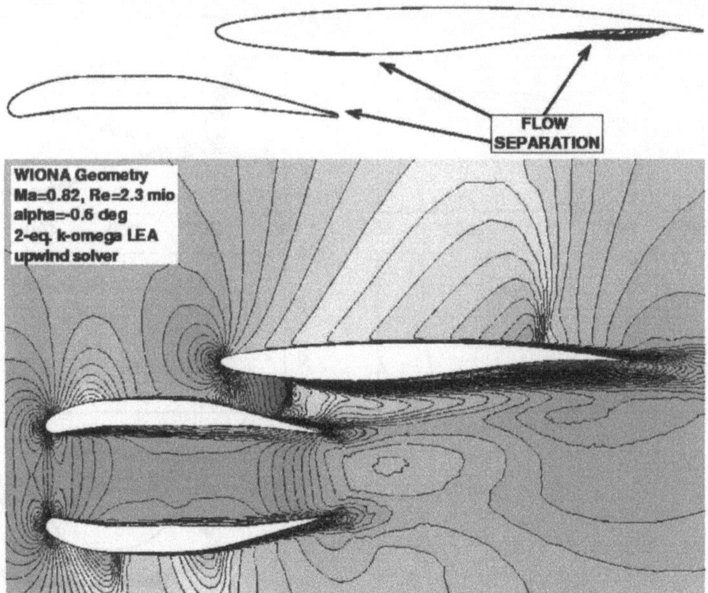

Figure 2 Iso-Mach lines for steady viscous computations.

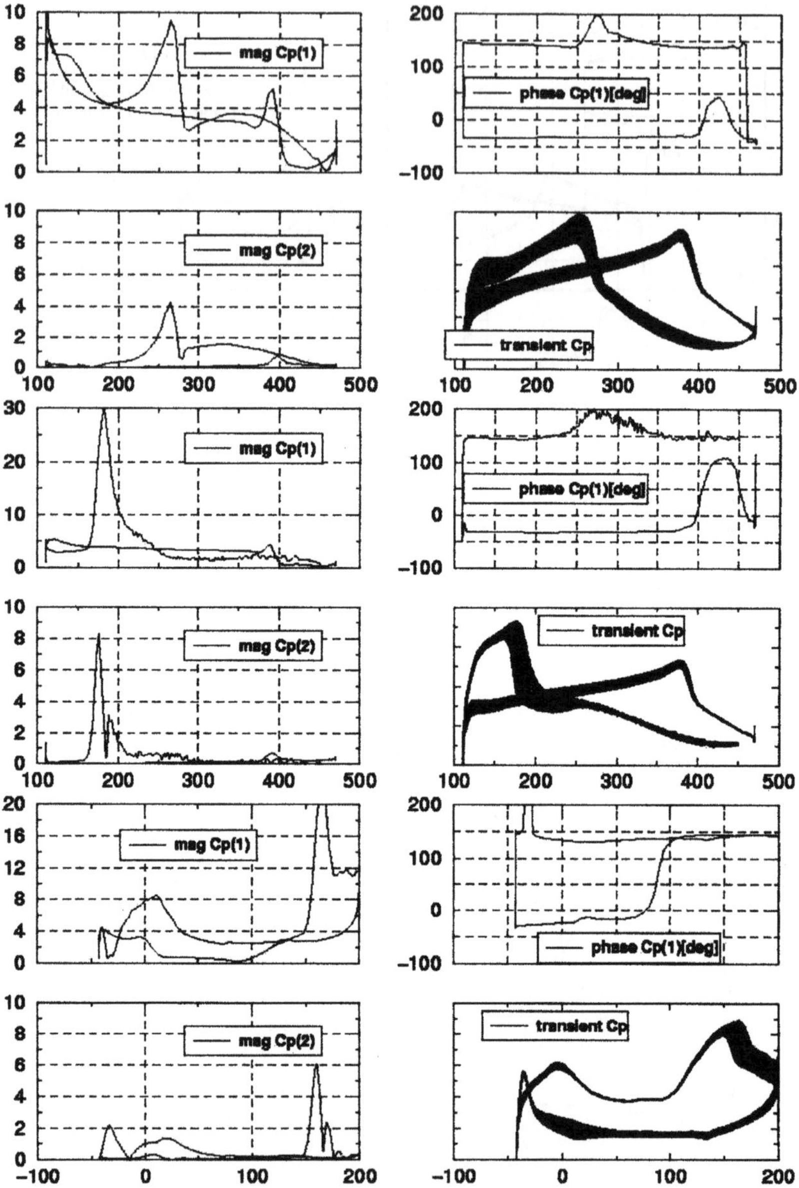

Figure 3 unsteady pressure coefficients of Navier-Stokes solution at 20% wing span (top), at 47% wing span (middle)and at 26% nacelle (bottom). The X-axis is in mm

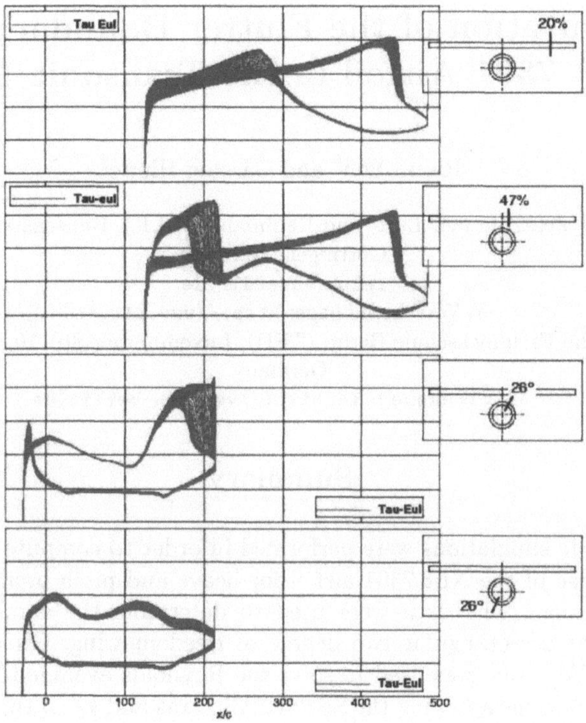

Figure 4 unsteady Cp distribution for Euler calculations.

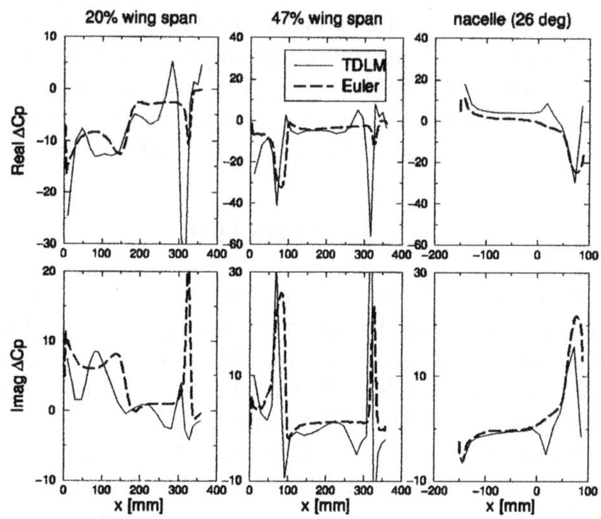

Figure 5 TDLM-Euler comparison, ΔCp at 3 different cut planes.

Computation of the Flutter Boundary of the NLR 7301 Airfoil in the Transonic Range

Ralph Voß[1] and Carsten Hippe[2]

[1] Deutsches Zentrum fuer Luft- und Raumfahrt (DLR), Bunsenstrasse 10, 37073 Göttingen, Germany
ralph.voss@dlr.de,
WWW home page: http://www.dlr.de

[2] Technische Fachhochschule Berlin (TFH), Luxemburger Str. 10, 13353 Berlin, Germany
WWW home page: http://www.tfh-berlin.de

Summary

Aerodynamic simulations were performed in order to compute the unsteady load response of the NLR7301 airfoil for heave and pitch oscillations in the transonic range. The data were used to determine the aeroelastic flutter boundary of a rectangular two degree of freedom wing. On the one hand the DLR TAU code was used to solve the Reynolds averaged Navier-Stokes (RANS) equations applying the Spalart-Allmaras and $k-\omega$ turbulence models. On the other hand computations were made with the DLR TDLM code, which solves the time linearised transonic small disturbance (TSD) equation. The settings were chosen according to experiments by Dietz et al. [1] in the Transonic Wind Tunnel Göttingen (TWG). The comparison between the computed and the experimental results showed that the simulation codes are capable of computing reliable results. Characteristic transonic aeroelastic phenomena were found. The appropriate modeling of viscous effects in transonic flow came out to be an indispensable condition.

1 Introduction

The operational range of airplanes can be limited by the occurrence of flutter. This phenomenon may appear beyond a certain flight speed, which is characteristic for a given airplane configuration. The flutter speed must be higher than the flight speed because loss of dynamical stability may cause unstable wing vibrations and even failure. Efforts are made to treat the problem computationally besides very expensive flight and wind tunnel tests. Test cases for which experimental data are available are used to validate the computational methods. This paper describes the computation of the flutter boundary of the NLR7301 transonic airfoil using different aerodynamic simulation approaches. The structural and aerodynamic settings of the two degree of freedom system were chosen according to those used in wind tunnel tests by Dietz et al. [1]. The experiment provides data for the two dimensional transonic flow around the rectangular wing. The flutter boundary is assumed to be the value of the flutter index Fi for constant free stream conditions in which the wing oscillates harmonically with a constant, small amplitude after an initial excitation. The dimensionless flutter index

$$Fi = \frac{2V_\infty}{\sqrt{\mu} c \omega_\alpha} \; ; \; Fi \propto V_\infty \sqrt{\rho_\infty}$$

which is proportional to freestream velocity V_∞ and the root of air density ρ_∞, describes the influence of the air load on the wing oscillations. It increases with the free stream air speed and the free stream air density. The full chord length c and the eigenfrequency of the pitch degree of freedom ω_α were introduced for normalisation. The mass ratio μ is defined in Table 1. The total damping resulting from structural and aerodynamic effects disappears in the flutter point. The investigated structure was a rigid rectangular wing (span b) with the degrees of freedom heave h and pitch α represented by the displacement vector $\boldsymbol{u}$. The equations of motion state the basic aeroelastic principle - equality between structural (inertia $[M]$, damping $[D]$, stiffness $[K]$) and aerodynamic loads $[A]$).

$$[M]\ddot{\boldsymbol{u}} + [D]\dot{\boldsymbol{u}} + [K]\boldsymbol{u} = [A]$$

Applying the approach $\boldsymbol{u}(t) = \boldsymbol{u}_0 + \boldsymbol{u}_1(t)$ the equations can be split up into a steady and an unsteady problem. The unsteady part leads to an eigenvalue problem assuming simple harmonic vibration, (1). This system of complex equations must be solved to obtain the flutter point Fi_f.

$$\begin{bmatrix} (1+i2\xi_h)\sigma^2 & 0 \\ 0 & (1+i2\xi_\alpha)r_\alpha^2 \end{bmatrix}^{-1} \cdot \left(\begin{bmatrix} 1 & x_\alpha \\ x_\alpha & r_\alpha^2 \end{bmatrix} - \frac{2}{\mu\pi k} \begin{bmatrix} \hat{l}_h & \hat{l}_\alpha \\ \hat{m}_h & \hat{m}_\alpha \end{bmatrix} \right) = \lambda E \quad (1)$$

The problem is presented in a dimensionless form to gain direct comparability with the experimental settings. The dimensionless and physical quantities used to describe the structural wing properties are listed in Tab. 1. The reduced frequency is used in the form $k = \omega c / V_\infty$. The complex coefficients $\hat{l}_\alpha$, $\hat{l}_h$, $\hat{m}_\alpha$ and $\hat{m}_h$ are introduced to describe the unsteady air loads for harmonic oscillations. Their imaginary parts and those of the structural damping approach contained in the stiffness matrix describe the damping behaviour of the aeroelastic system as the flutter problem is solved in the frequency domain. The structural parameters are assumed to be constant. Generally the air loads are depending on the free stream Mach number M_∞, the Reynolds number Re and the reduced frequency of the unsteady motion. In transonic flow the unsteady air loads strongly depend on the steady flow field in contrast to subsonic flow. This is mainly caused by the influence of the local supersonic domains on the phase delay between the displacement and the air load response [5]. A predominant phenomenon of transonic flow is the compression shock that closes the supersonic flow domain on the downstream side. The properties and the unsteady behaviour of the shock systems may cause flow separation that is directly related to the pressure distribution on the airfoil surface and hence the overall air loads. As this is a very sensitive process depending on the conditions within the boundary layer the choice of an appropriate turbulence model is very important. These effects contribute to the so-called transonic dip, which is a transonic aeroelastic phenomenon that occurs in the Mach number range between the first presence of supersonic regions and massive flow separation. The name transonic dip means the special shape of the flutter stability curve in the transonic regime, often showing a pronounced minimum. A reliable prediction is very important for the assessment of the flutter behaviour. The transonic dip has special relevance for supercritical airfoils [2] like the NLR7301 airfoil. The main problem of the presented work was therefore to determine unsteady air loads for simple harmonic heave and pitch motions respecting transonic phenomena. Three ways for obtaining appropriate unsteady air loads were chosen.

1. Experimental data for forced heave and pitch motion by Dietz et al. [1]
2. Data based on the unsteady RANS equations with the DLR TAU code
3. Data based on the TSD equation using the DLR TDLM code

2 Flow Simulations

This section gives an overview of the different aerodynamic data that were used to solve the flutter problem. The DLR TAU code [6] was used to solve the RANS equations. Fig. 1 shows the computational grid that was used in all TAU code simulations. TAU is a 3D code. Symmetric boundary conditions were applied to the side faces to obtain the 2D case. The flow field near the profile surface was discretised with a structured grid of 7488 hexaedra elements. Those were arranged in a way convenient to capture the gradients of the shear stresses, which form the boundary layer. The remaining part of the flow field was approximated with an unstructured grid of 15150 prisms. The turbulence modeling was done following two approaches. The first approach used the one equation turbulence model according to Spalart-Allmaras in its standard formulation (SA) [3]. Further simulations were performed applying the $k - \omega$ Linear Explicit Algebraic model [7]. Tab. 2 shows the parameter settings of the unsteady aerodynamic simulations using the TAU code. The Reynolds number was $Re = 2.30 \cdot 10^6$ in all simulations. The $M_\infty - \alpha_0$ settings in the TDLM simulations were chosen according to the TAU (SA) simulations. The reduced frequency was covered in the range $k = 0.1...0.5$. Experimental data were available for the same Mach number range, $k = 0.1...0.4$, the mean angle of attack $\alpha_0 = 0.33°$ and $Re = 2.21 \cdot 10^6 \pm 9.3\%$. The following section describes the steps of the aerodynamic analysis. Only TAU (SA) results are presented.

2.1 Steady Computations

The first analysis step was meant to determine steady flow fields at the discrete Mach numbers listed in Tab. 2. This was required due to the dependency of the unsteady flow on the steady field in the transonic range, Fig. 2. The pictures show how the airfoil enters the transonic region with increasing Mach number. The shear stress distribution for the highest investigated Mach number shown in Fig. 3 indicates that flow separation occurs already in the steady case. The comparison with an experimental pressure distribution at the semi-span position of the model shows that significant features like shock positions and pressure level at the trailing edge are captured well. Deviations result from the slight differences in Mach number and angle of attack.

2.2 Unsteady Computations

The unsteady computations were performed in the second analysis step. Several cycles of harmonic oscillations had to be simulated for both degrees of freedom to obtain converged conditions of the response curves. This was done for all combinations of the Mach number and the reduced frequency, Tab. 2. Special care was taken of an appropriate choice of numerical parameters of the flow solver. The proper choice of the number of time steps per oscillation period (SPP) and the number of inner iterations per time step (UIPS) is important to insure stable and converged unsteady results with reasonable computational effort. Typical parameters were $SPP = 90$ and $UIPS = 50$ for simulations adopting the SA turbulence model. The simulations were terminated after five oscillation periods. The pitch about the quarter axis $x/c = 0.25$ and heave amplitudes were chosen as $\alpha_1 = 0.5°$ and $h_1/c = 0.005$. These values are sufficiently in order to satisfy the linear approach made in the flutter equations as an amplitude survey proved. The dependency of the unsteady load behaviour on the reduced frequency becomes clear observing the magnitude of the unsteady load loops and phase lag regarding the forcing oscillation. The deviation of the curve shapes from an ellipse indicates nonlinear effects caused by the presence of the supersonic domains and onset of flow separation. It becomes clear from Fig. 4 that the unsteadiness of the air loads (here magnitude and phase of the first harmonic part of the lift coefficient for forcing pitch oscillation) strongly depends on the Mach number, correspondingly the supersonic region and shock position as well. Further data sets were computed for both the real and imaginary parts of the aerodynamic lift and moment components in the flutter equations, (1). Overall 126 unsteady simulation runs were performed ($2dof \cdot 9M_\infty \cdot 7k = 126$). Fig. 5 provides additional information regarding the unsteady load behaviour. There is a significant region of up- and downstream motion of the shock position near $M_\infty = 0.735$. The steady flow in Fig. 2 is nearly shock-free in this Mach number regime, which changes significantly for off-design conditions. The NLR7301 tends to form a strong shock or two separated small supersonic regimes when α is

increased or decreased. Besides the results presented above analogue computations were performed using the $k-\omega$ LEA [7] turbulence model.
The second simulation system used in this analysis was the DLR TDLM code in its 2D version [4]. It is based on the potential theory and solves the transonic small disturbance equation (TSD) assuming a flat plate rectangular wing. In contrast to purely subsonic approaches like the Doublet Lattice Method (DLM) it models the influence of the supersonic flow. The transonic effects on the unsteady loading, namely the phase shift of signal spreading due to the presence and shape of supersonic regions, are computed by solving the time linearised unsteady transonic potential equation. This is based on steady transonic flow field data appearing as locally varying coefficients in the equation. In this case the steady fields, which were computed with the DLR TAU code (SA), were used, Fig. 2. The TDLM system provides two sets of results. One is valid for purely subsonic flow conditions neglecting all transonic effects from the steady flow. The other one respected the steady transonic flow, Tab. 2. Besides the computational results two experimental data sets by Dietz et al. [1] were available. One is based on the pressure distributions at mid span (P), the other one on an overall load balance measurement (B).

3 Solution of the Flutter Problem

The results of the aeroelastic stability computations solving equations (1) are presented in this section. The structural settings of the two degrees of freedom wing system was listed in Tab. 2. An iterative solution algorithm was used to find the points of undamped wing oscillation. The curves in Fig. 6 represent the final results of the investigation. The diagram shows the flutter boundaries based on the different numerical aerodynamic approaches and the experimental results. All critical values of the flutter index Fi_f show good agreement in the subsonic range. The experimental results and those based on the unsteady TAU code simulations clearly describe the transonic dip near $M_\infty = 0.735$. This location corresponds to the pronounced behavior of the Mach number dependency of the unsteady air loads, Fig. 4. The flutter results mainly differ regarding the depth of the dip. The $k-\omega$ LEA turbulence modeling predicts a slightly greater unstable region. This proves the strong dependency of the unsteady air loads and hence the aeroelastic stability on the flow separation process. The divergence of the two TDLM based curves results from the growing influence of transonic effects as the Mach number increases, Fig. 2. The part of the transonic dip that is characterised by the drop of the stability curve is clearly captured using the transonic correction. The values do not find the minimum in contrast to the TAU code and experimental solutions, they keep on falling. This effect may be explained by the onset of unsteady flow separation, which is not captured in an appropriate way. The simulation results agree quite well with the experimental data based on the pressure measurement at semi-span. This measuring method approximates the ideal 2D case much better than the measurement of the overall wing loads. The latter are disturbed by three-dimensional effects. Such effects may be caused by the wind tunnel side walls for example. Discrepancies between the experimental pressure based and the TAU code results may result from measuring problems like unsteady wind tunnel wall interference.

4 Conclusion

The presented investigations strongly underline the presence of viscous phenomena in transonic flow and their influence on aeroelastic stability problems. The DLR TAU code delivers reasonable results simulating unsteady wing oscillations in the transonic range. The transonic dip was found using two turbulence models. A strong influence of the modeling of viscous effects on the flutter stability was found in the region of the dip where unsteady flow separation occurs. The DLR TDLM code can be used to simulate unsteady wing oscillations in transonic flow as long as there is no strong unsteady flow separation. It must be noted that the computational cost of the TDLM code is much less (about a factor 1000) than that required for performing transient analyses solving the RANS. One unsteady simulation for a fixed Mach number, reduced frequency and oscillation mode costs 8000 (RANS) versus 10 (TDLM) CPU seconds on a 2GHz Dell processor. The results could be validated using experimental data.

References

[1] Dietz G. et al.; Experiments on heave/pitch limit-cycle oscillations of a supercritical airfoil close to the transonic dip; Journal of Fluids and Structures 19 1-16, 2004

[2] Nixon D. (ed.); Unsteady Transonic Aerodynamics; Progress in Astronautics and Aeronautics Series, Vol. 120; AIAA, Washington D.C., 1989

[3] Spalart P. R., Allmaras S. R.; A One-Equation Turbulence Model for Aerodynamic Flows; AIAA-92-0439, 1992

[4] Voß, R., Geißler, W.; Investigation of the Unsteady Airloads with Oscillating Control in Sub- and Transonic Flows; Proceedings DGLR International Symposium on Aeroelasticity 1981, Nuremberg. DGLR-Bericht 82-01, pp. 65-80 (1982)

[5] Voß, R.; Über die Ausbreitung akustischer Störungen in transsonischen Strömungsfeldern von Tragflügeln; Göttingen, 1988

[6] Technical Documentation of the DLR TAU code, DLR 2003

[7] Wilcox, D.C.; Turbulence Modeling for CFD, 2nd ed. 1998, DCW Industries, La Canada, California

Table 1 Dimensionless and physical Quantities

$x_\alpha = \frac{S_\alpha}{mc}$	(Static Moment)	m	(Wing Mass)
$r_\alpha = \sqrt{\frac{I_\alpha}{mc^2}}$	(Radius of Gyration)	$I_{c/4}$	(Wing Moment of Inertia)
$\sigma = \frac{\omega_h}{\omega_\alpha}$	(Frequency Ratio)	k_h	(Heav Stiffness)
$\eta^2 = \lambda = \frac{\omega_\alpha^2}{\omega^2}$	(Eigenvalue)	k_α	(Pitch Stiffness)
$\mu = \frac{m}{1/4\pi\rho_\infty c^2 b}$	(Mass Ratio)	S_α	(Static Unbalance)
$\omega_\alpha = \sqrt{\frac{k_\alpha}{I_\alpha}}$	(Pitch Eigenfrequency)	ξ_h	(Lehr's Heave Damping)
$\omega_h = \sqrt{\frac{k_h}{m}}$	(Pitch Eigenfrequency)	ξ_α	(Lehr's Pitch Damping)

Table 2 Settings in unsteady aerodynamic simulations with the TAU code

SA			$k-\omega$		
M_∞	α_0	k	M_∞	α_0	k
0.550	0.46	0.05	0.552	0.33	0.2
0.653	0.46	0.1	0.653	0.32	0.3
0.682	0.46	0.2	0.684	0.32	0.4
0.715	0.46	0.3	0.714	0.31	0.5
0.735	0.46	0.5	0.734	0.31	0.6
0.759	0.46	0.75	0.755	0.30	
0.774	0.46	1.0	0.776	0.31	
	1.5				
	2.5				

Table 3 Structural parameters of the wing model

$m = 26.268kg$	$\xi_h = 0.43\%$
$I_{c/4} = 7.900 \cdot 10^{-2} kgm^2$	$\xi_\alpha = 0.15\%$
$S_\alpha = 3.310 \cdot 10^{-1} kgm$	$x_\alpha = 0.0420$
$k_h = 1.078 \cdot 10^6 N/m$	$r_\alpha = 0.1828$
$k_\alpha = 1.078 \cdot 10^6 Nm/rad$	$\sigma = 0.6991$

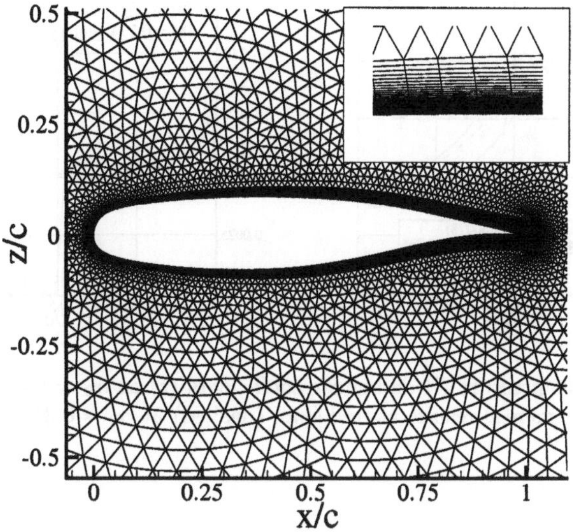

Figure 1 Computational grid of NLR7301 airfoil used for TAU code simulations

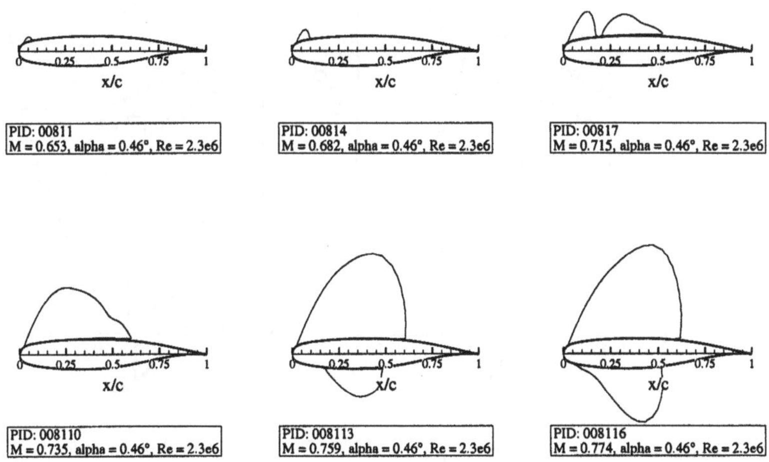

Figure 2 Supersonic regions with increasing Mach number, $\alpha_0 = 0.46°$

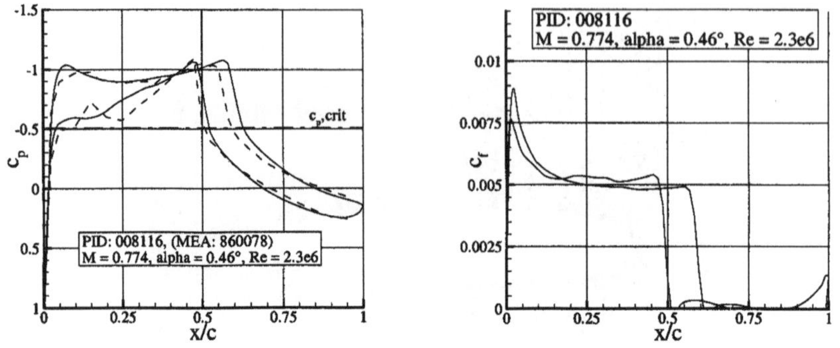

Figure 3 Steady pressure (l) and shear coefficients (r) at $\alpha_0 = 0.46°$; Pressure compared to experimental results at mid span (dashed) at $\alpha_0 = 0.31°$

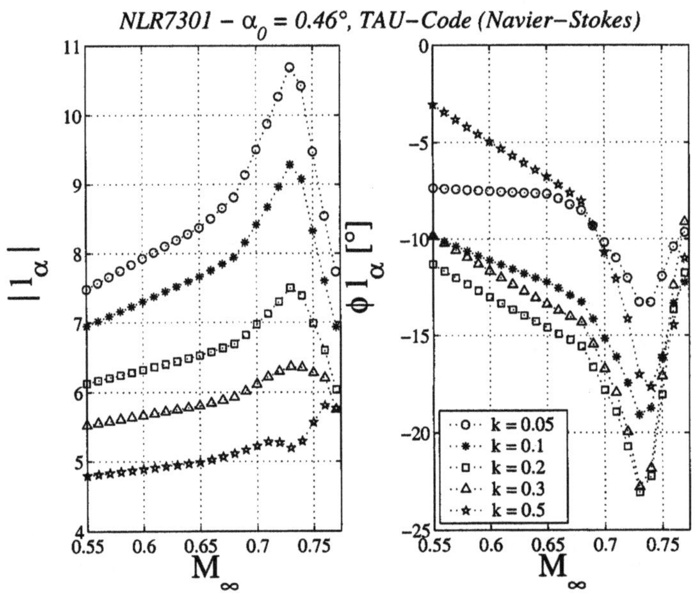

Figure 4 Magnitude and phase shift of unsteay lift due to forced pitching

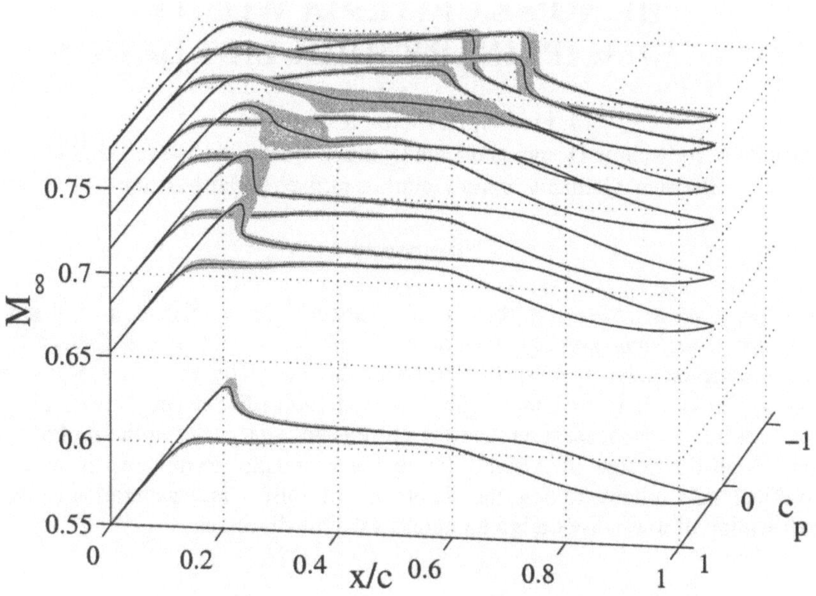

Figure 5 Unsteady pressure distribution during one pitch cycle at $k = 0.3$

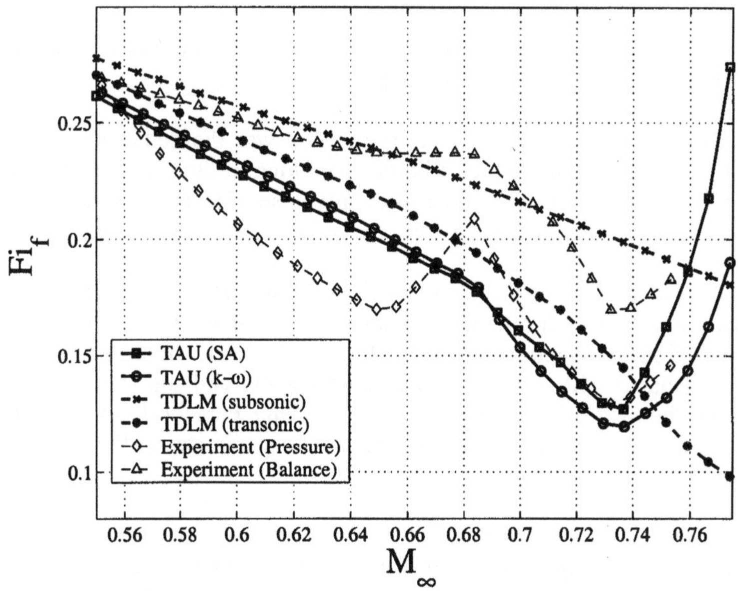

Figure 6 Stability curve of the NLR7301 airfoil

THE NOISE CRITERIA WITHIN MULTIDISCIPLINARY HIGH-LIFT DESIGN

M. Fischer, H. Bieler, R. Emunds
Airbus, Aerodynamic Design and Data Domain, Huenefeldstrasse 1-5, 28183 Bremen, Germany; e-mail: markus.m.fischer@airbus.com

Summary

This paper gives a survey of Airbus' strategy about the low noise high-lift design. The different *airframe noise* sources on a commercial aeroplane are discussed. It will be completed by a physical description of especially those noise sources which typically occur on wings with deployed high-lift devices. Moreover the paper presents the concept to embed the noise criteria basically in the multidisciplinary high-lift design of Airbus. Finally, experimental techniques as well as theoretical tools suitable to describe the process of noise source generation and the identification of low-noise design parameters will be discussed.

1 Introduction

The noise sources of commercial aeroplanes might be split in two categories: The first one covers all the sources generated in connection with the propulsion, the second one describes the noise sources associated with the airframe. In the 2nd category the high-lift devices and the landing gear are the main contributors to airframe noise during take-off and landing.

Up to the sixties of last century the propulsion was the dominant noise source through the entire flight regime. With the beginning of the seventies the introduction of high-bypass engines and successful noise abatement measures on fan, turbine and jet led to a significant noise reduction of more than 10 dB.

As a result of this, the relative part of the airframe noise on overall noise level increases. Especially during approach and landing, airframe noise and propulsion noise are of same magnitude, [4]. This fact in mind and the challenging requirements associated with environmental commitments and the exponential traffic growth in the next 15 years (*Vision 2020*, halving Perceived Noise until 2020) reveal clearly the necessity to embed noise as a parameter in the design of airframe components.

The long-term strategy of Airbus is the design of low-noise airframe components compatible to the design constraints driven by performance or economy. Airbus is addressing this goal through intensive collaborations with Research Establishments, Universities and on European level (EC projects) with other aircraft companies.

2 Airbus' participation in German national and European Research Programmes

Due to the necessity to consider aeroacoustic constraints more intensive in the design of airframe components than it has been done in the past, the Airbus participation in Research Programmes has been increased in the last few years.

On a German national basis (LuFoI, LuFoII) as well as on EC level (particularly in the project RAIN, 4th EC-Framework-Programme) studies with respect to noise generation and prediction on high-lift devices were conducted in close collaboration with Research Establishments and Universities, [3]. The activities in RAIN provided on one hand improved noise prediction methods and on the other hand precise noise reduction measures. Many of these proposed modifications were consequently further investigated with respect to flight-operational applicability in the following EC project SILENCER (5th EC-Framework-Programme). Some of these solutions, which have rather the character of prototypes than certified components, were flight tested on A340. The activities in SILENCER will be continued up to 2006.

There are complementary activities together with Research partners on a German national basis in the two LuFo III-projects FREQUENZ and IHK. FREQUENZ stands for the investigation of noise source generation of commercial aeroplanes and the development of measures to reduce it by means of experiments and numerics. IHK is focusing on the low-noise design of components. The work packages of both projects are well tuned to each other.

3 Low noise high-lift design

This paper gives a survey of Airbus' strategy to use experimental data as well as numerical results to understand the physical mechanisms behind the noise generation process and to include this knowledge into the multidisciplinary high-lift design.

The high-lift design process at Airbus can be circumscribed as follows: The wing mastergeometry and the Top Level Aircraft Requirements (TLAR) represent the requirements with respect to the flight performance and the flight mission to be realized. Further requirements originate from Aviation Rules and Standards (FAR/JAR). Moreover there are requirements coming from the project plan and from economical and operational aspects like the portfolio of airports to be served and in context with this the required runway lengths, airport infrastructure, noise limits (Quota Count levels, night restriction classification scheme).

Even with respect to the QC-levels at take-off and landing the high-lift design is additionally subjected to constraints coming from low-noise procedures operation. This issue was and is presently, with Airbus contribution, scope of the EC programmes SOURDINE I und II and AWIATOR.

The acoustical requirements are based on the Noise Certification Standards of ICAO Annex 16, Volume1. The noise limits and procedures described herein are directly linked to the Aircraft Standards, especially FAR/JAR Part 36. Further acoustical requirements will be taken into account based on development goals for an environmentally-friendly future aircraft outlined in the Vision 2020.

The geometrical design is performed by means of CAD and KBE-tools (knowledge based engineering). These tools provide 3-dimensional surfaces which take fully into account the given constraints from structures and systems (e.g. positioning and kinematics of the moveable surfaces). For the aerodynamic layout coupled theoretical methods (potential/boundary-layer codes) as well as Navier-Stokes codes are applied and from experimental side extensive wind tunnel investigations. The aeroacoustical issues are examined by theoretical and experimental methods which will be described in the following chapters.

It has been shown in a number of investigations that many aerodynamic design parameters may influence directly the noise generation process and the resulting sound pressure field, e.g. the strong dependency of Sound Pressure Level (SPL) from Mach-number. Without consideration of design issues from structures and systems, the aeroacoustical design parameters are a part of the group of aerodynamic parameters. The parameters which are identified of being crucial for low-noise design may be organized into three classes of layout variations:

- Variation of airfoil geometry (e.g. local curvatures, surface irregularities)
- Variation of setting (e.g. deployment angles, gaps, overlaps)
- Variation of flow conditions (e.g. flow velocity, angle-of-attack, flight attitude)

Beside the identification of further crucial parameters it is a major task to define the sensitivity of each of the identified parameters with respect to low-noise design.

3.1 Airframe noise sources on a wing in high-lift configuration

The most important airframe noise sources of a high-lift wing appear at: Slats, wing/fuselage junction, wingtip and flaps. Further noise sources may be generated by gaps or cavities like fuel overpressure holes.

These noise sources are the result of the interaction of flow fluctuations with the surface (hard walls) and there especially with surface irregularities (rapid change of boundary conditions). The character of the generated noise source, mean directivity and spectral distribution of sound pressure, is strongly influenced by the character of the flow (boundary layer status, Re-, St-, Ma-number). The following flow types when interacting with hard walls are of main interest:

- Turbulent boundary layers passing sharp trailing edges
- Laminar or transitional boundary layers coming along with periodic variation of the mean flow by hydrodynamic instabilities passing sharp trailing edges
- Separated flows passing sharp trailing edges
- Attached flows passing blunt trailing edges
- Flow around side edges (development of side edge vortices)

To understand the noise generation on a real high-lift configuration it is not sufficient to assume that the interaction of the flow with the surface happens in an isolated manner, moreover aspects like the turbulence-level of the incident flow or possible wake-interactions between other aircraft components, like landing gear, with the configuration have to be taken into account.

In the end just a small part of the flow energy is transferred into acoustic energy. In the acoustical nearfield, where a strong interaction between hydrodynamics and acoustics is present, the generated noise sources (propagation with speed of sound) are a part of the instantaneous pressure field. The predominant part keeps its hydrodynamic nature and is convected downstream with local flow velocity. With constriction to just the audible range (SPL of 10-130 dB), sound pressures of about 10^{-4} to 10^{2} Pa ($dB=20 \log p_{rms}/(2x10^{5} Pa)$) are present. For comparison, the hydrodynamic pressures of the flow around an airfoil are in the order of standard atmospheric conditions, means 10^{5} Pa. This clarifies the aforementioned small part of acoustic energy in the flow field.

3.2 Theoretical methods for investigation of noise generation

The determination of airframe noise by means of Direct Numerical Simulations (DNS) is due to the present available computational and storage capacity mostly limited to generic test cases. These kind of calculations take place at Universities or Research Establishments. In order to be able to examine real or rather more complex configurations it is necessary to have industrial robust and reliable methods which work cost- and time-efficiently. Thus, Airbus favours the *hybrid approach* to handle the problem: Firstly the flow field is determined by means of established and validated CFD-tools. Secondly the source description as the input for the following CAA-code (*PIANO-code* of DLR, [5]) is based on a description of turbulent quantities either by a direct expression or in terms of correlations or statistical quantities. The CAA-code describes the interaction of turbulent perturbations with hard walls, e.g. a surface, furthermore the generation of acoustic sources and finally the perturbation dynamics. With respect to the last-mentioned point the code itself acts in fact as a perturbation filter. Figure 1 shows the methodology which is presently implemented at Airbus to investigate airframe noise generation particularly with regard to high-lift design. The different ways differ from each other by the extent in modelling:

- *Injection* of a vortex into the flow field. The interaction of the vortex with the given surface allows insights how the vorticity in the flow is transferred into acoustic pressure.
- Examination of sound generation by means of stochastical description of turbulence (based on *SNGR-Ansatz*, [1], using the distribution of kinetic energy and dissipation of the calculated flow field)
- Determination of turbulent sources by means of Large-Eddy Simulation (LES).
- Modelling of the turbulent sources based on data obtained by instantaneous and non-intrusive measurement techniques like PIV

The CAA-code in Figure 1 is based on linearised Euler equations. These equations are derived from the fluidmechanical Euler equations using a perturbation approach of the following form, equation (1):

$$(1)\quad \Phi = \Phi^0 + \Phi'$$

Φ is a variable which may represent pressure, components of the velocity vector or density. This leads to a system of equations for the $'$-quantities which are to be solved and are superimposed to the given stationary flow field (^{0}-quantities). This introduced in the Euler equations, neglecting quadratic terms of the $'$-quantities, a system of equations for density, the sound particle velocity and the sound pressure is created. This system looks in vector notation as follows, equations (2):

$$(2)\quad \frac{\partial U}{\partial t} + A_i \frac{\partial U}{\partial x_i} + HU = S$$

For a 2D-problem U, A_i and H are defined as follows:

$$(3)\quad U = (\rho'; u'; v'; p')^T$$

$$(4)\quad A_i = \begin{pmatrix} u_i^0 & \rho^0\delta_{1i} & \rho^0\delta_{2i} & 0 \\ 0 & u_i^0 & 0 & \delta_{1i}/\rho^0 \\ 0 & 0 & u_i^0 & \delta_{2i}/\rho^0 \\ 0 & \gamma p^0\delta_{1i} & \gamma p^0\delta_{2i} & u_i^0 \end{pmatrix}$$

$$(5)\quad H = \begin{pmatrix} \frac{\partial u_i^0}{\partial x_i} & \frac{\partial \rho^0}{\partial x_1} & \frac{\partial \rho^0}{\partial x_2} & 0 \\ -\frac{1}{(\rho^0)^2}\frac{\partial p^0}{\partial x_1} & \frac{\partial u_1^0}{\partial x_1} & \frac{\partial u_1^0}{\partial x_2} & 0 \\ -\frac{1}{(\rho^0)^2}\frac{\partial p^0}{\partial x_2} & \frac{\partial u_2^0}{\partial x_1} & \frac{\partial u_2^0}{\partial x_2} & 0 \\ 0 & \frac{\partial p^0}{\partial x_1} & \frac{\partial p^0}{\partial x_2} & \gamma\frac{\partial u_i^0}{\partial x_i} \end{pmatrix}$$

whereas δ_{ij} stands for the Kronecker-Symbol and γ is the ratio of the specific heats. The vector S on the right-hand side represents a source term.

The vortex-injection method is presently for Airbus the start of the theoretical aeroacoustic investigations because with this method already essential statements can be obtained how vorticity in the flow is transferred into acoustic pressure. Furthermore issues about the preferred propagation directivities close to the surface of the airfoil can be gained. For the theoretical assessment of the high-lift design in the acoustical farfield empirical methods, integral methods or acoustic analogies are commonly applied.

Figure 2 shows for a 2-dimensional high-lift airfoil with deployed slat and retracted flap an injected vortex into the flow close to the leading edge of the slat at t=0. As the result of the interaction of the vortex with the airfoil, for two times ($t=t_1$ und $t=t_2>t_1$) the generated sound pressure fields can be seen. Presented is also the obtained directivity in the vicinity of the airfoil as a function of Re.

3.3 Experimental methods for investigation of noise generation

In-flight measurements by means of single microphones or microphone arrays allow for a given aircraft configuration the determination of its acoustic signature on the ground. It is possible to obtain the real, absolute sound pressures as a function of the directivity. Furthermore noise sources may be identified and ranked to each other. Nevertheless, it is difficult or even impossible to gather information about the physics behind noise source generation.

In wind tunnel tests the possibility is given to do parametric studies via configuration changes under well defined fluidmechanical and acoustical conditions. Noise relevant parameters may be identified and their impact on acoustics and aerodynamics are determined (*sensitivity study*) simultaneously. The data of these kind of experiments is suitable as a validation basis for theoretical investigations and enhances a deeper understanding of noise source generation.

Pre-design studies or validation experiments on simpler configurations take usually place in small anechoic wind tunnels, whereas combined aerodynamic and acoustic measurements on realistic larger scale models are conducted in large facilities, like the DNW (*Deutsch Niederländischer Windkanal*). The DNW has a closed test section as well as an anechoic open one.

Noise source localisation can be done either by the microphone-array technique, [6], or by means of an acoustic mirror, [2]. The microphone-array technique was still quite recently not equal to the acoustic mirror with respect to spatial resolution, frequency range and signal-to-noise ratio. The remarkable progress in data processing time and data storage capacity reduce more and more the aforementioned system disadvantages. The microphone-array technique takes full advantage of its ability to acquire data in a 2D-field instead of doing point investigations, avoiding time-consuming traversing and adjustment work. Moreover, new algorithms in data processing allow to apply microphone arrays successfully also in wind tunnels with acoustic hard walls. Figure 3 shows an example for this: Within the framework of AWIATOR combined aerodynamic and aeroacoustic tests took place in the closed test section of DNW. On the pressure side of the right wing a wake-vortex device was installed which was before the tests to be assessed to have a potential acoustic impact. On the left side of Figure 3 the upper noise source plot (10 kHz 1/3 Octave band frequency, NLR data) represents the noise source situation without device installed and the lower one with device installed. It can be clearly seen that the device appears in the source plots in terms of two additional noise sources.

For the purpose of complete characterisation of the noise sources the aforementioned measurements are usually accompanied by farfield investigations in the open, anechoic test section of DNW, [7]. Single microphones, which are mostly arranged on a traversable support, determine the directivity as well as the sound pressure as a function of space, time and frequency.

4 Conclusion

The aforementioned goals of the Vision 2020 are not only from aeroacoustical point of view extremely challenging for the high-lift design of an environmentaly friendly aircraft. The first results with the hybrid approach look promising. There is a variety of experimental techniques to validate numerical data. It is rather essential to understand the physical mechanisms of noise generation and finally the implementation of this knowledge in the multidisciplinary high-lift design.

5 References

[1] C. Bailly, D. Juvé. "A stochastic approach to compute subsonic noise using Linearized Euler's Equations". In: AIAA 1999-1872.

[2] F.-R. Grosche, H. Stiewitt, B. Binder. "Acoustic Wind Tunnel Measurements with a highly Directional Microphone". In: AIAA Journal, Vol.15, No.11, pp. 1590-1596, 1977.

[3] K. Mau. "High-Lift Devices Noise Reduction Study". In: RAIN Technical Report TN 3973, 2001.

[4] L.C. Chow, K. Mau, H. Remy. "Landing Gears and High-Lift Devices Airframe Noise Research". In: AIAA 2002-2408.

[5] R. Ewert, J.W. Delfs, M. Lummer. "Perturbation Approaches to the Simulation of Airframe Noise". In: Proceedings of EUROMECH 449, Chamonix, 2003.

[6] S. Oerlemans, P. Sijtsma. "Acoustic Array Measurements of a 1:10.6 scaled Airbus A340 Model". In: AIAA 2004-2924.

[7] W. Dobrzynski. "Quantified Noise Characteristics of the Aircraft with Control Devices". In: AWIATOR Technical Report D1.3.2-3, 2004.

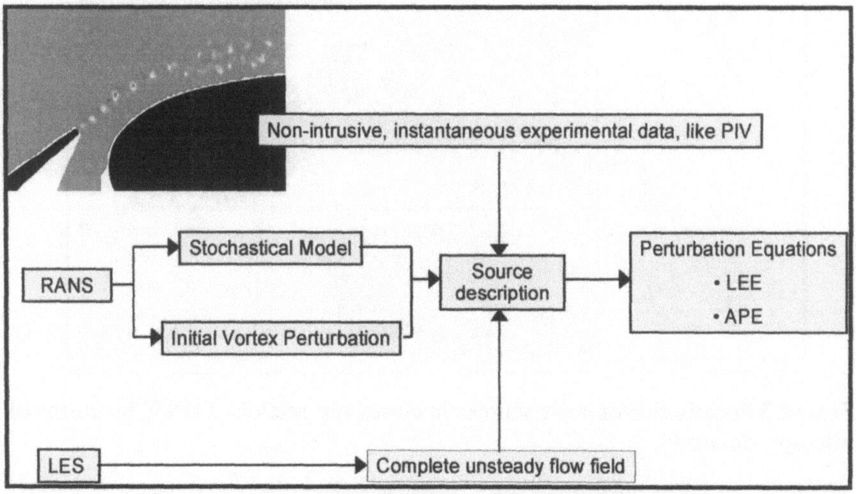

Figure 1 Methodology to investigate the generation of airframe noise.

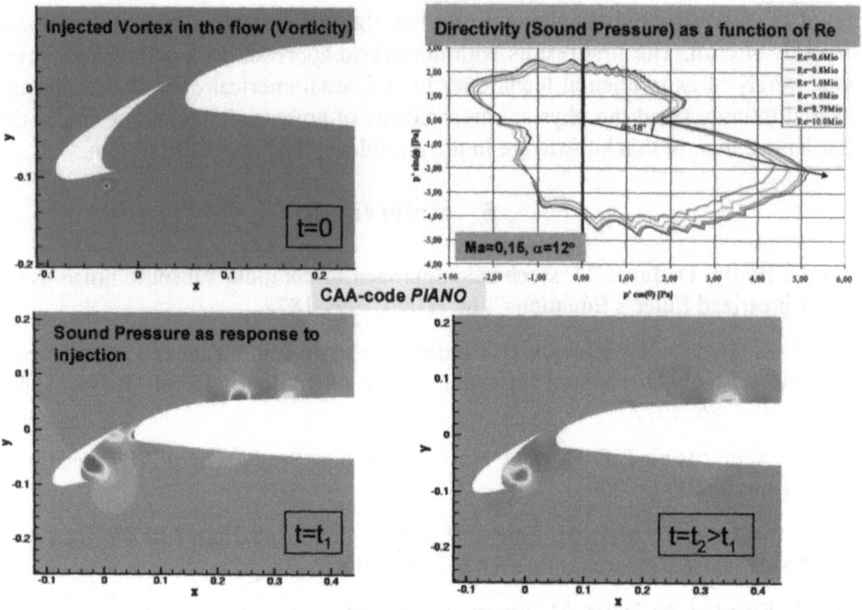

Figure 2 Injected vortex at t=0 close to the slat leading edge and the generated sound pressure at $t=t_1$ and $t=t_2>t_1$. Furthermore the obtained directivity in the vicinity of the profile as a function of Re.

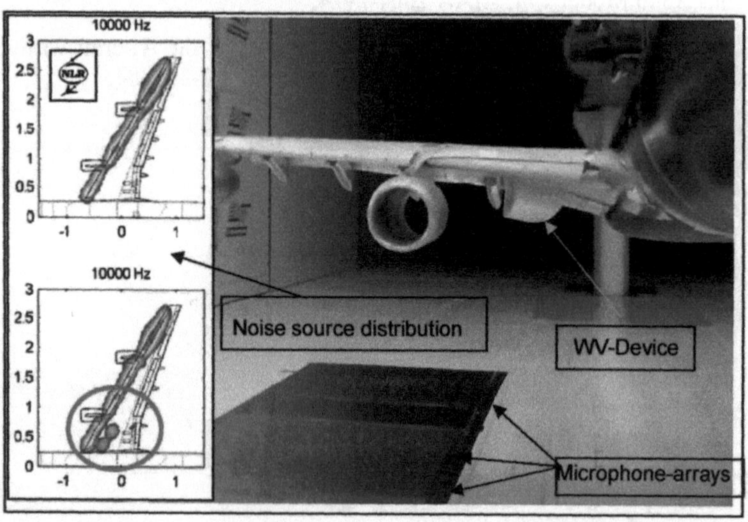

Figure 3 Localisation of noise sources in closed test section of DNW by means of microphone-arrays.

Effect of Noise Reducing Modifications of the Slat on Aerodynamic Properties of the High-Lift System

Judith Ortmann, Jochen Wild

German Aerospace Center (DLR), Institute of Aerodynamics and Flow Technology,
Lilienthalplatz 7, D-38108 Braunschweig, Germany,
Judith.Ortmann@dlr.de

Summary

To minimize slat noise according to the guidelines of "A Vision for 2020" brush-like devices were installed on a slat trailing edge to "soften" it. In experiments slat noise was considerably reduced by such devices. But their effect on aerodynamics is yet unknown.

This study shows that the investigated brush-like devices on slat trailing edges influence the aerodynamic properties of high-lift systems adversely. Because of increasing boundary layer thickness with length and diameter of the brushes the suction peaks are reduced so that C_L decreases.

These are preliminary results. An optimization of slat chord length and brush length should be conducted to minimize C_L reduction.

1 Introduction

During approach the slat trailing edge of large commercial aircraft generates high noise levels. The goal of "A Vision for 2020" [1] is the reduction of airframe noise levels by 6 dB. Experiments have shown that brush-like devices installed on slat and flap tips reduce the noise levels significantly [6].

An extension of this method is the installation of brushes on the entire slat trailing edge to reduce its noise. Acoustic wind tunnel tests of such brush-like devices have already been performed using a flat plate [3]. Mau and Dobrzynski hold a patent on such noise reducing devices for commercial aircraft [5]. Sharp edges produce a high noise level. The brushes "soften" the trailing edge. But until now only acoustical investigations have been performed.

Thus the objective of this study is the assessment of the aerodynamic properties of high-lift systems with brush devices.

2 Grid Generation and Flow Solver

The original geometry is a small wing segment of a swept wing of infinite span whose profile has been deduced from a modern commercial aircraft wing in high-lift configuration with slat and flap (fig. 1(a)). Periodic boundaries are defined to simulate a swept wing with infinite span.

To mount brushes on the slat trailing edge the conventional slat is reduced by the length of the brushes. Cylinders are installed on this (thicker) trailing edge as models for stiff brush hairs (fig. 1(c)) which are arranged parallel to the incoming flow.

A structured grid with roughly 2.5 million nodes is generated with MegaCads [2]. Slat, main wing and flap formed the main grid and the brushes are added in a Chimera block (fig. 2). This way, the discretisation in the main grid stays the same and one obtains flexibility in brush spacing and sizing.

Using the DLR FLOWer code [4] viscous flow is simulated. Fully turbulent calculations are done using the Spalart-Allmaras turbulence model [7] with central spatial discretisation with Jameson dissipation ($k2 = 2$, $k4 = 64$), implicit residual smoothing and three multigrid levels. For time stepping the 5 step Runge-Kutta scheme is used.

The flow conditions are related to those of a model in a low speed wind tunnel: $\alpha = 16°$, $Ma_\infty = 0.22$, $Re = 2.91E{+}06$, $T_\infty = 288.15°$ K and $p_\infty = 101325$ Pa. The conventional configuration features a chord length of 0.56804 m, blunt trailing edges and deployed slat and flap. The modifications are made by shortening the slat trailing edge by the length of the installed brushes. There are three brush parameter: length, diameter and distance. In the following chart the distances of the different devices for the brush length and diameter are listed. In one case no brushes are installed on the shortened trailing edge to obtain the influence of the trailing edge cutback.

Table 1 list of the model parameters

Distance	Diameter (mm)			
Length (mm)	0.2	0.3	0.4	cutback
1.37	0.05 mm 0.01 mm			√
3.0	0.05 mm 0.01 mm	0.05 mm 0.015 mm		
3.5			0.05 mm	

Expecting very small differences between global coefficients of the configurations grid convergence studies are performed for the conventional and one modified configuration.

3 Grid Convergence Study

To minimize the discretisation error a grid convergence study has been carried out for the conventional wing and one modified configuration (length = 1.37 mm, diameter = 0.2 mm, distance = 0.05 mm) using the coarser multigrid levels (fig. 3). The difference between the results on the extrapolated and the fine grid are very small. For the conventional configuration $C_{L,\infty}/C_{L,fine}$ is about 1.26 % and

$C_{D,\infty}/C_{D,fine}$ is 1.42 %. For the modified configuration $C_{L,\infty}/C_{L,fine}$ is 1.65 % and $C_{D,\infty}/C_{D,fine}$ is 1.24 %. Thus, the results of the fine grid are of sufficient accuracy. Therefore, in the following the results of the fine grid will be discussed.

4 Results

As the first result one can see the complex flow phenomena near the brushes (fig. 4 at $\alpha = 16°$). On pressure and suction side an attachment line is generated and on the sides of the brushes separation lines are visible (fig. 4 (c)). These lines are typical for swept cylindrical bodies and are mainly caused by the flow between the brush hairs.

The flow through the brush hairs influence the pressure distribution of all three high-lift elements (fig. 5 and 6). The suction peaks are reduced with increasing length and diameter of the brush hairs, where the diameter dominates the influence on suction peaks of slat and flap (e.g. fig. 5 (a)). The influence of brush devices is also visible on the complete suction side of all three high-lift elements (fig. 6 (a)+(b)). The pressure on almost the complete suction side increases with length and diameter of the brush hairs, whereas near the flap trailing edge the pressure decreases. This is reflected in a thickening of the boundary layer with increasing brush sizes.

Figure 7 and 8 show the effect of the devices on the global aerodynamic coefficients. The C_L-α behavior is altered adversely with brush devices. The reduction of C_L increases with α and depends on length and diameter of the brush hairs. However, the effect of diameter is dominating (fig. 7 (a)). Figure 8 shows that drag decreases at low C_L and increases at high C_L.

ΔC_L (%) and ΔC_D (%) are plotted in figure 9 at $\alpha = 16°$ for a brush distance of 0.05 mm. The slat modification has an increased negative effect on lift of main wing and flap. Furthermore, the slat produces more drag, whereas the drag on main wing and flap decreases. The friction contribution of drag increases on the slat by 5 % to 12 %. But the pressure contribution is dominant for the total drag.

The effect of slat chord reduction on global coefficients is insignificant compared to the modified trailing edge with brushes of the same length. Thus differences were mainly caused by the brush hairs.

5 Conclusions

The flow visualisation shows a complex flow topology with separation and attachment lines in the brush area. This is reflected in the aerodynamic properties of the brushes. All devices altered the pressure distributions on the suction sides of all three high-lift elements adversely. But there is no significant influence on the pressure distribution of the pressure side. With increasing length and diameter of the brushes, the pressure on the suction side increases and the suction peak is reduced.

For the same angle of attack C_L is reduced for all investigated brush-like devices. In addition, C_D increases at high C_L. Shortening the slat cord length has no significant influence on C_L-α and C_L-C_D behavior. Differences in global coefficients are caused by the brush-like devices.

Future studies will contain installed brush-like devices on original slat trailing edges (original slat chord length but thicker trailing edge) and the optimization of slat chord length and brush length to minimize the aerodynamic effects regarding acoustic guidelines. Also, C_D studies at take off conditions will be performed. Additionally the Reynolds number effect up to flight conditions will be investigated.

References

[1] Argüelles, P. et al. (Group of Personalities): "European Aeronautics: A Vision for 2020, Meeting Societys Needs and Winning Global Leadership". Report of the Group of Personalities, the Advisory Council for Aeronautics Research in Europe, Office for Official Publications of the European Communities, Luxembourg, Jan.2001.

[2] Brodersen, O. et al.: "The parametric grid generation system MegaCads". 5th International Conference on Numerical Grid Generation in Computational Field Simulations, Mississippi, 1996, NFS, B.K., pp. 353-362.

[3] Herr, M. and Dobrzynski, W.: "Experimental Investigation in Low Noise Trailing Edge Design". AIAA-2004-2804, 2004.

[4] Kroll, N.; Radespiel, R. and Rossow, C.C.: "Structured Grid Solver I: Accurate and Efficient Flow Solvers for 3D Applications on Structured Meshes". AGARD Report 807, 1995.

[5] Mau, K. and Dobrzynski, W.: "Anordnung zur Minderung des aerodynamischen Lrms an einem Vorflgel eines Verkehrsflugzeuges". Europisches Patent 101 57 849.0.

[6] Phelan, M.: "Airbus combs airflow in quest for aircraft". Flight International 25.Febr.-3.March 2003, p. 24.

[7] Spalart, P.R. and Allmaras, S.R.: "One-Equation Turbulence Model for Aerodynamic Flow". AIAA-92-439, Reno, NV., 1992.

Figures

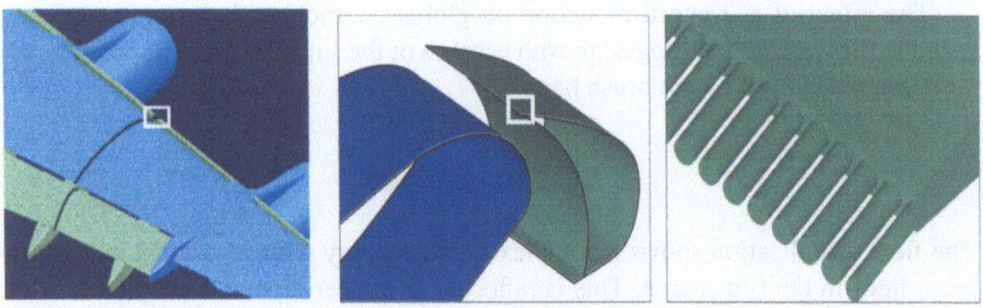

(a) (b) (c)

Figure 1 (a) Wing of a large modern aircraft with cross section used for RANS simulation; (b) closeup of slat; (c) slat trailing edge with brushes.

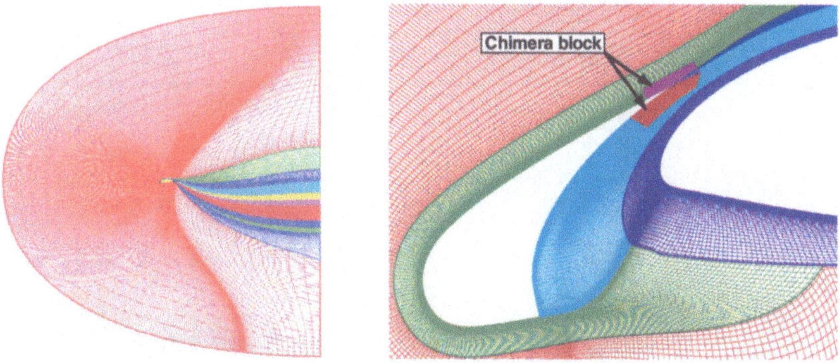

(a) (b)

Figure 2 Structured grid generated with MegaCads. (a) farfield; (b) zoom in of slat trailing edge with Chimera block.

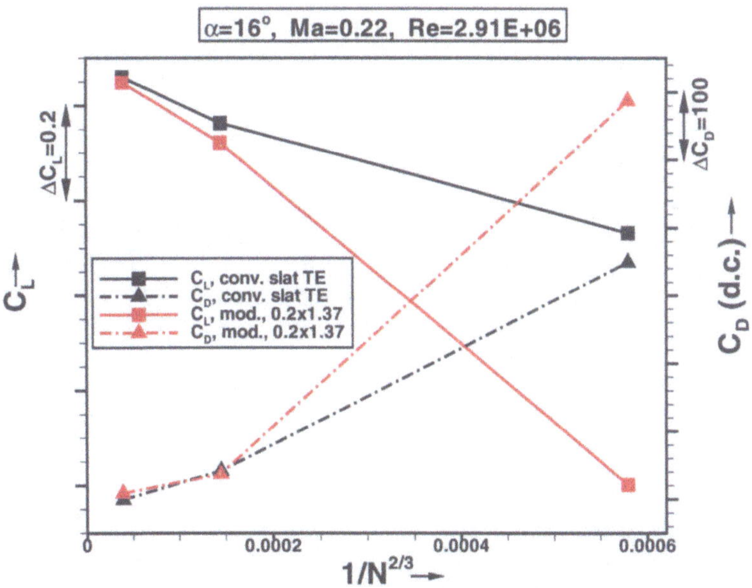

Figure 3 Grid convergence study for conventional wing and the first brush model with length = 1.37 mm, diameter = 0.2 mm, distance = 0.05 mm (1d.c. (drag count)= 0.0001).

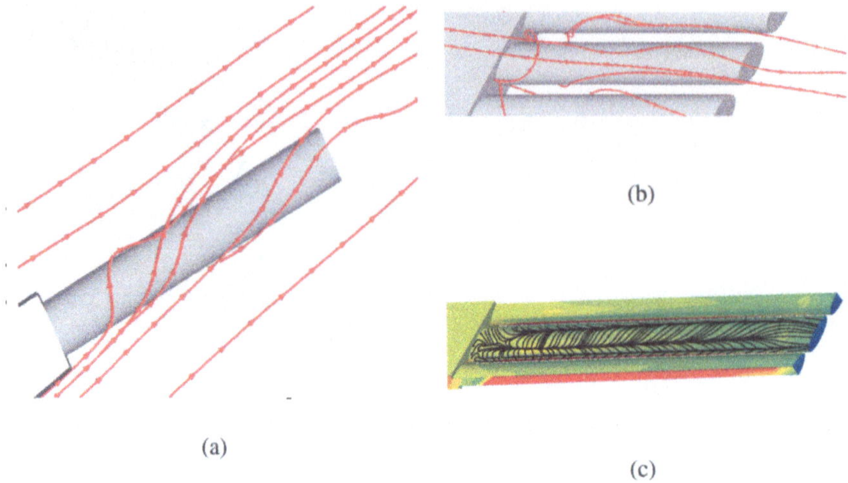

Figure 4 Complex flow topology in brush area. (a) side view; (b) top view; (c) surface pressure distribution and surface streamlines.

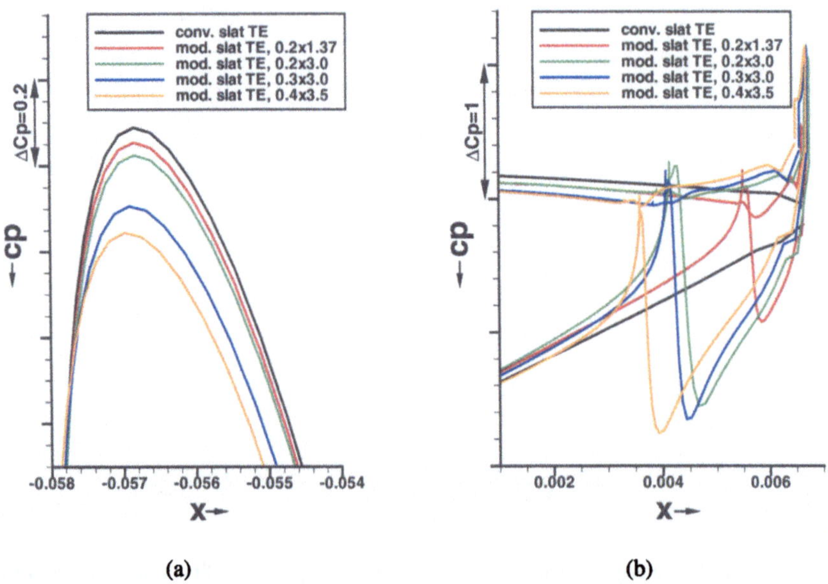

Figure 5 Influence of brushes on pressure distribution. (a) slat suction peak; (b) slat trailing edge;

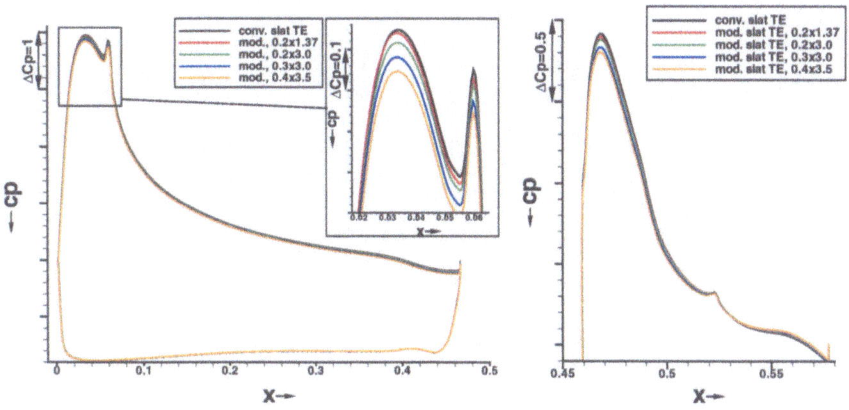

(a) (b)

Figure 6 Influence of brushes on pressure distribution. (a) main wing; (b) flap.

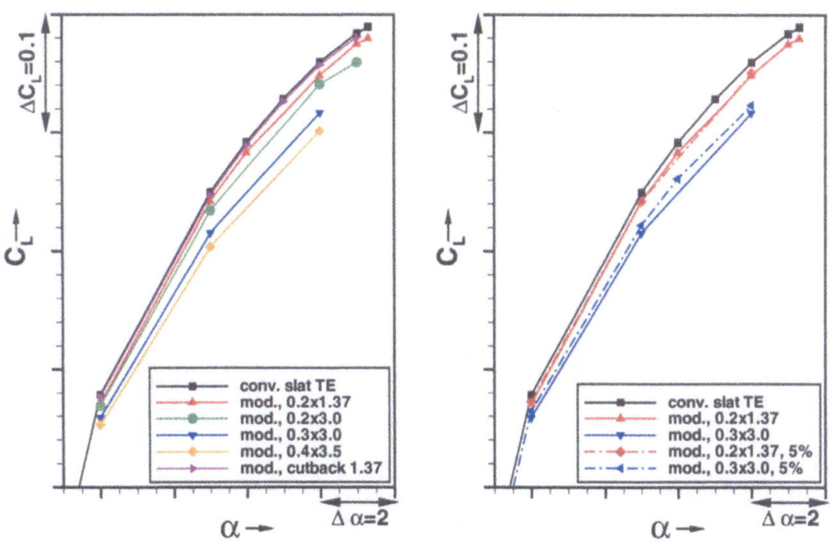

(a) (b)

Figure 7 C_{Lmax} studies. (a) effect of length and thickness of the brushes on C_L-α behavior; (b) effect of brush spacing on C_L-α behavior.

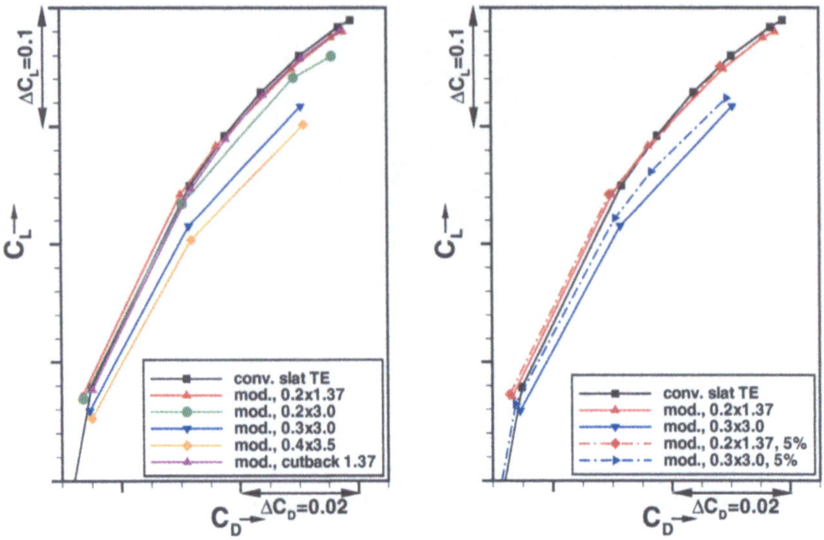

(a) (b)

Figure 8 Polars. (a) effect of length and thickness of the brushes on C_L-C_D behavior; (b) effect of brush spacing on C_L-C_D behavior.

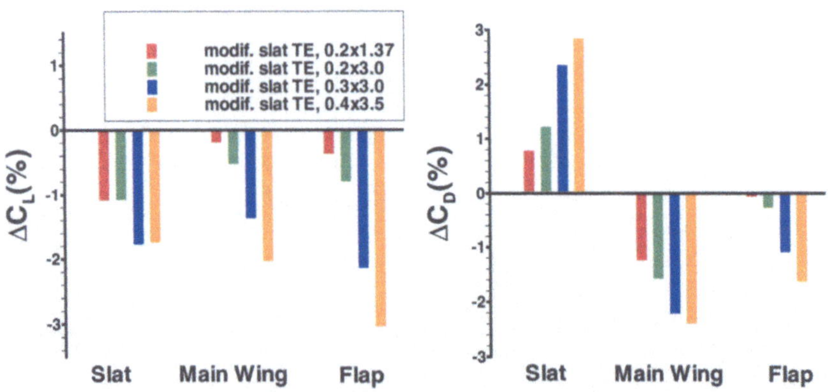

Figure 9 Forces of all three elements (in relation to reference length) for $\alpha = 16°$.

Experimental Study on Noise Reduction through Trailing Edge Brushes

M. HERR
Deutsches Zentrum für Luft- und Raumfahrt e. V. (DLR),
Institut für Aerodynamik und Strömungstechnik, Lilienthalplatz 7,
D-38108 Braunschweig, Germany, Michaela.Herr@dlr.de

Summary

Within an experimental trailing edge noise reduction study in the Aeroacoustic Wind Tunnel Braunschweig (AWB) both acoustic and aerodynamic effects of trailing edge brush devices were examined. Directional microphone and hot wire measurements were undertaken on a zero-lift generic plate model ($Re = 2.1$ to 7.9×10^6). Various brush concepts were tested to clarify the functional relationship between design parameters and the ensuing aeroacoustic properties. First results of this ongoing work indicate a significant source noise reduction in excess of 10 dB, thereby revealing two relevant noise reduction mechanisms. In addition to broadband turbulent boundary layer trailing edge noise also narrow band contributions due to vortex shedding from the edge were alleviated.

1 Introduction

In order to cope with the challenging noise reduction goals as defined in the European strategic paper "A Vision for 2020" [1] a reinforced design effort has to be made towards a significant reduction of airframe noise sources. Since the trailing edges of lifting surfaces (such as slats or flaps of current aircraft) were identified as major noise contributors during landing approach a parametric study was initiated aiming at a phenomenological understanding of trailing edge noise generation and reduction processes. The particular interest within this ongoing research is the reduction of turbulent boundary layer trailing edge noise which is closely related to turbulent boundary layer vortices shed off the trailing edge into the wake. Current attempts to correlate trailing edge noise signatures with measurable flow quantities mainly rely on Howe's approach [9], providing a cardioid directivity and a (velocity)^5-dependence of the sound intensity which is emitted from the solid trailing edge of a semi-infinite plate of negligible thickness. The following equation yields the mean-square farfield pressure as generated by all turbulent boundary layer vortices within a frequency averaged correlation scale l parallel to the edge of wetted span L :

$$\langle p^2 \rangle \approx \rho_0^2 u_0^2 V^2 M_\upsilon \left(L l / R^2 \right) \sin\alpha \sin^2(\theta/2) \cos^3\beta \quad (1)$$

where V is a characteristic mean flow velocity, u_0 the rms turbulence velocity, M_υ the turbulence convection Mach number, and ρ_0 the ambient density. R (distance between source and observer), and θ (observation angle) are the farfield observer coordinates for the retarded source position at reception time, α the angle between the observer direction and the edge and β the trailing edge sweep angle. Based on the finding that trailing edge noise inclines with cosine cubed of sweep angle comb- (or brush-) type trailing edge extensions with fibres aligned with the mean flow direction can be considered an extreme case of serrations with infinitely sharp tip angles. As was demonstrated in various publications [4, 13] serrated trailing edges are means for trailing edge noise reduction, however in some cases a distortion of the flow streamlines was observed leading to less noise reduction than theoretically predicted. A number of studies [3, 8] also applied noise-abating porous edge extensions. During more recent wind tunnel experiments in the framework of the European research program RAIN (Reduction of Airframe and Installation Noise) trailing edge brush extensions [11] were successfully tested. However, within this first turn of experiments neither design rules nor scaling laws for a future technical application of such devices could be deduced.

The current work intends to combine the benefits of flow permeability and a theoretical trailing edge sweep angle of $\beta = \pi/2$ focusing at a prospective application of brush devices as retrofit solutions for current aircraft components.

2 Experimental Set-Up

Both acoustic and aerodynamic measurements were performed in DLR's Aeroacoustic Wind Tunnel Braunschweig (AWB) which is an open jet anechoic test facility with a rectangular 0.8 m by 1.2 m nozzle exit. Directional microphone data were acquired utilizing an elliptical mirror system (with a ¼''-condenser microphone in one focus). A detailed description of the applied mirror system is provided in [7]. Additionally, single and cross-wire measurements served to determine steady and unsteady characteristics of the turbulent boundary layer flow. All documented data were taken in the midspan trailing edge region of a tripped zero-lift plate model where two-dimensionality of the flow was ensured.

The acoustic test set-up is displayed in **Figure 1**. Due to a modular plate design the model exhibits four different chord lengths (0.8, 1.2, 1.6 and 2.0 m) and an exchangeable midspan trailing edge section (with a constant taper of 5°). One insert (trailing edge thickness $h = 1$ mm) served as solid reference configuration. An identical trailing edge section, however, containing a 0.5 mm slit in the downstream face, was used to install different brush extensions which were made of one single row of (flexible) polypropylene fibres. Test parameters included brush design characteristics such as fibre length l_f (15, 30 and 60 mm) and fibre diameter h_f (0.3, 0.4 and 0.5 mm) and Reynolds number (both flow velocity *and* chord length). With regard to future full-scale applications the present study focuses on relatively high Reynolds numbers (2.1 to 7.9 x 10^6) corresponding to flow velocities from 40 to 60 m/s. As to the pertinent literature,

e. g. Blake [2] a solid trailing edge with a "bluntness parameter" of $h/\delta^* < 0.3$ (δ^* being the boundary layer displacement thickness) is considered "sharp", i.e. vortex shedding due to a finite trailing edge thickness is not likely to occur. Within the range of tested chord lengths the ratio h/δ^* covers values from 0.2 to 0.6 for the 4 solid reference trailing edge configurations (with boundary layer thicknesses δ ranging from 14.6 to 33.4 mm). Therefore, tonal noise generation due to Kármán-type vortex shedding from blunt trailing edges (bluntness noise) was expected to be of some relevance only for the short plate versions. Since boundary layer thicknesses were not measured for all the test configurations δ was also calculated by means of flat plate theory, in fact leading to a very good agreement with available test data.

3 Results

3.1 Solid Reference Trailing Edge

Trailing edge noise spectra as obtained for different Reynolds numbers revealed two distinct trailing edge noise contributors. These were (i) broadband turbulent boundary layer trailing edge noise and (ii) narrow band vortex shedding noise (bluntness noise) due to finite trailing edge thickness. In contrast to current knowledge from literature, bluntness noise was identified as major noise component even for small "bluntness parameters" $h/\delta^* < 0.3$, i.e. the above-cited generalizing threshold for "sharp" trailing edges could not be confirmed in the current experiment. In this context, steady CFD-RANS calculations coupled with CAA non-linear perturbation simulations of the vortex shedding phenomenon [12] indicate that the pressure amplitudes are substantially dependent on geometric properties of the edge (e.g. trailing edge taper). As can be seen in **Figure 2** aerodynamic pressure amplitudes as calculated for the plate geometry (5° taper) were five times higher than for a tripped NACA 0012 profile (ca. 15° taper) although the Reynolds- and Mach number as well as the bluntness parameter were identical for both cases ($h/\delta^* = 0.6$). The NACA 0012 profile was chosen with reference to the experimental data from Brooks & Hodgson [5, 6] revealing low pressure levels for small bluntness parameters.

Note that the trailing edge noise spectra as measured in the present experiment show an excellent qualitative agreement with the respective hot-wire u'- and v'-fluctuations spectra in the boundary layer close to the trailing edge (1 mm from the surface, 1 mm upstream of the edge in **Figure 3**).

Figure 4 documents the observed bluntness noise contributions for the 4 solid reference trailing edge configurations and in addition for another 2 trailing edge thicknesses ($h = 5$ mm, implemented by removing the whole exchangeable trailing edge section, and $h = 0.5$ mm by installing a solid 15 mm plate extension in the mounting gap instead of the brushes). Bluntness noise discrete frequency spectral peaks were found to follow a well defined Strouhal-relationship of $St = fh/u_\infty = 0.1$ (see below) which is in good agreement with the results in [5, 6].

As depicted in Figure 4 bluntness noise peaks can be considered tonal for $h = 5$ mm, but broaden and diminish with growing frequency (decreasing h). The effect of broadening with decreasing h/δ^* as reported by Brooks & Hodgson was not clearly observed with regard to the 4 reference configurations ($h = 1$ mm, various δ^*) and is therefore assumed to be rather the result of a decrease in turbulent pressure coherence for high frequencies. As can be seen in the one-third octave band spectra (**Figures 5, 6**) trailing edge farfield noise (as emitted from the source area) generally increases according to a u_∞^5-power law. It should be emphasized that in the present case turbulent boundary layer trailing edge noise levels and frequencies scale with a roughly *constant* correlation length l (referring to Equation 1) despite the variation of chord length. For the sake of simplicity again the trailing edge thickness h was taken as scaling length (Figure 5). In contrast to the commonly applied normalization procedure (choosing δ^* as the relevant correlation scale l) it was found that δ^* indeed does *not* represent the ultimate scaling parameter for trailing edge noise; as presented in Figure 6 such a data presentation leads to a complete mismatch of the correspondingly normalized spectra. This observation may be due to Reynolds number effects on boundary layer turbulence, since in most of the well known trailing edge noise experiments δ^* was varied through testing at different speeds or angles-of-attack, while keeping a constant chord length.

3.2 Brush Trailing Edge

Flow permeable trailing edge brush extensions provide a *significant broad-band noise reduction* (of up to 10 dB) covering a wide frequency range of relevance in terms of human perception. Both stated trailing edge noise contributions, namely turbulent boundary layer trailing edge noise and bluntness noise were alleviated. The maximum noise reduction (up to 14 dB, depending on the configuration) was always achieved in the frequency band containing the bluntness noise spectral peak (at $St = 0.1$). According to the results of near wake hot-wire measurements (not shown here) the removal of bluntness noise spectral peaks can be explained by a suppression of vortex shedding from the edge. The same effect was already found by Khorrami et al. [10] within a computational study on porous slat trailing edge treatment. **Figure 7** documents the test results for the brush configuration ($h_f = 0.4$ mm, $l_f = 30$ mm) which provided the largest peak noise attenuation. In general, the achieved noise reduction potential was found to depend on the frequency and the particular configuration under review. As can be seen in Figure 7 the same scaling laws as formulated for the solid reference configurations (u_∞^5-dependence, constant correlation scale l) pertain for the brush trailing edge configurations. In **Figure 8** the respectively achieved noise reduction potential (in terms of one-third octave band level differences from measurements without and with brush edge extensions installed) is presented for different chord lengths. Correspondingly, except for the high frequency range, boundary layer thickness (plate length) and also flow velocity show almost no effect on the achieved noise reduction potential. The comparison of results from all test cases revealed always the same ranking (in terms of achieved noise reduction) of brush devices, independent on the plate length: According to **Figure 9** (displaying the results for 5 selected brush configurations installed at the 0.8 m plate) long brush devices (l_f= 30 mm, l_f= 60 mm) were found to be more effective than short brushes (l_f= 15 mm) and also thicker fibres (h_f= 0.4 mm, h_f= 0.5 mm) provided a higher noise reduction potential than thinner ones (h_f= 0.3 mm). Since the optimum brush design is considered a function of boundary layer turbulence scales (and therefore of the required correlation scale l) the finding of a minor influence of the plate length is in agreement with the experimental results on the solid reference configurations (revealing a *constant* correlation scale for different chord lengths). As a consequence, all plate configurations are expected to require a similar optimum trailing edge brush design for maximum noise reduction. In summary, configurations with a fibre diameter h_f= 0.4 mm (and l_f= 30 mm, 60 mm) achieved a broadband attenuation in the whole audible frequency range and also the highest peak noise reduction, while brushes with h_f= 0.5 mm (and l_f= 30 mm) achieved the highest trailing edge noise reduction in the low frequency range, however, showing broadband excess noise contributions at high frequencies. Since the latter appear close to the non-audible range, where absolute noise levels are comparatively low, all examined brush devices are considered effective noise reduction means with respect to overall noise levels.

4 Conclusions

Within a parametric wind tunnel study brush edge extensions were proven to be effective means for trailing edge noise reduction. A broadband attenuation of turbulent boundary layer trailing edge source noise in the order of 2 to 10 dB (depending on configuration and frequency) was achieved. This reduction is expected to primarily result from a viscous damping of unsteady turbulent velocities in the trailing edge brush area. Additionally, a strong narrow band contribution related to high-amplitude vortex shedding from the edge (bluntness noise) was alleviated. Bluntness noise contributions in case of the solid and thin reference trailing edge (thin compared to boundary layer displacement thickness) were observed in contrast to the pertinent literature. In this regard the model geometry, particularly the trailing edge shape is considered an essential parameter for bluntness noise amplitudes. Both solid reference trailing edge and brush trailing edge noise spectra show the theoretically expected (velocity)^5-dependence and scale with a roughly constant correlation length within the chosen Reynolds number range of 2.1 to 7.9 x 10^6. Based on this finding, which is in contrast to the widely applied scaling procedure, the boundary layer displacement thickness does not seem to represent the universal scaling parameter for boundary layer trailing edge noise.

Acknowledgements

I wish to thank my colleagues at DLR Braunschweig, namely Julian Alvarez-Gonzalez for his support in data acquisition and processing, Murat Sabanca for the CFD/ CAA computational work, Werner Dobrzynski and Jan Delfs for many helpful discussions. Special thanks are due to Bernd Junker for excellent craftsman's work.

References

[1] P. Argüelles, M. Bischoff, P. Busquin et al. "European Aeronautics: A Vision for 2002, Meeting Society's Needs and Winning Global Leadership". Report of the Group of Personalities, the Advisory Council for Aeronautics Research in Europe, Office for Official Publications of the European Communities, Luxembourg, Jan. 2001.

[2] W. K. Blake. "Mechanics of Flow-Induced Sound and Vibration, Complex Flow-Structure Interactions". In Applied Mathematics and Mechanics, Volume 17-II, Academic Press, Inc., Orlando, June 1986, pp. 756-767.

[3] A. J. Bohn. "Edge Noise Attenuation by Porous-Edge Extensions". AIAA Paper 76-0080, Jan. 1976.

[4] K. A. Braun, A. Gordner, N. J. C. M. v. d. Borg et al. "Serrated Trailing Edge Noise (STENO)". Publishable Final Report of the EU-Project JOR3-CT95-0073, Apr. 1998.

[5] T. F. Brooks and T. H. Hodgson. "Prediction and Comparison of Trailing Edge Noise Using Measured Surface Pressures". AIAA Paper 80-0977, June 1980.

[6] T. F. Brooks, D. S. Pope and M. A. Marcolini. "Airfoil Self-Noise and Prediction". NASA Reference Publication 1218, July 1989.

[7] W. Dobrzynski, K. Nagakura, B. Gehlhar et al. "Airframe Noise on Wings with Deployed High-Lift Devices". AIAA Paper 98-2337, June 1998.

[8] M. R Fink and D. A. Bailey. "Model Tests of Airframe Noise Reduction Concepts". AIAA Paper 80-0979, June 1980.

[9] M. S. Howe. "A Review of the Theory of Trailing Edge Noise". In Journal of Sound and Vibration, Volume 61, No. 3, 1978, pp. 437-465.

[10] M. Khorrami and M. M. Choudhari. "Application of Passive Porous Treatment to Slat Trailing Edge Noise". NASA/TM-2003-212416, May 2003.

[11] K. Mau and W. Dobrzynski. "Anordnung zur Minderung des aerodynamischen Lärms an einem Vorflügel eines Verkehrsflugzeuges". European Patent 101 57 849.0.

[12] M. Sabanca. Private communication.

[13] S. Wagner, G. Guidati, K. Braun et al. "Design and Testing of Aeroacoustically Optimized Airfoils for Wind Turbines (DATA)". Publishable Final Report of the EU-Project JOR3-CT98-0248, Sept. 2000.

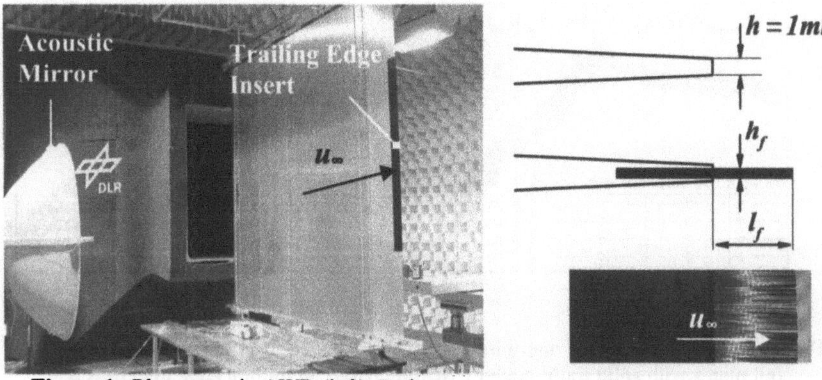

Figure 1 Plate setup in AWB (left). Reference and brush trailing edge inserts (right).

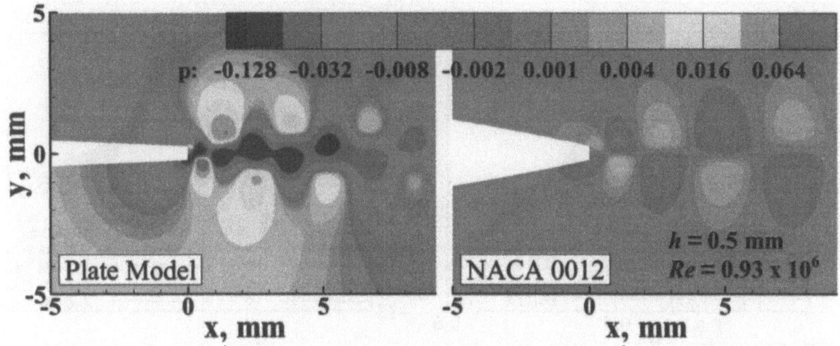

Figure 2 Results of CFD/ CAA computations using DLR codes FLOWer/ PIANO.

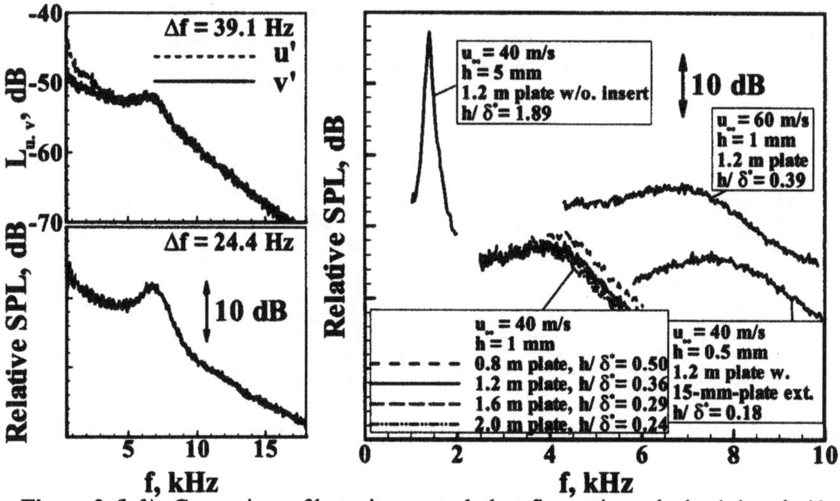

Figure 3 *(left)* Comparison of hot-wire rms turbulent fluctuating velocity (u'- and v'-) narrow band spectra in the very trailing edge region with the corresponding acoustic result, $u_\infty = 60$ m/s, 2.0 m reference. $L_{u,v} = 20 \log (u', v'/ u, v)$. **Figure 4** *(right)* Narrow band bluntness noise contributions for various solid trailing edge configurations, $\Delta f = 24.4$ Hz.

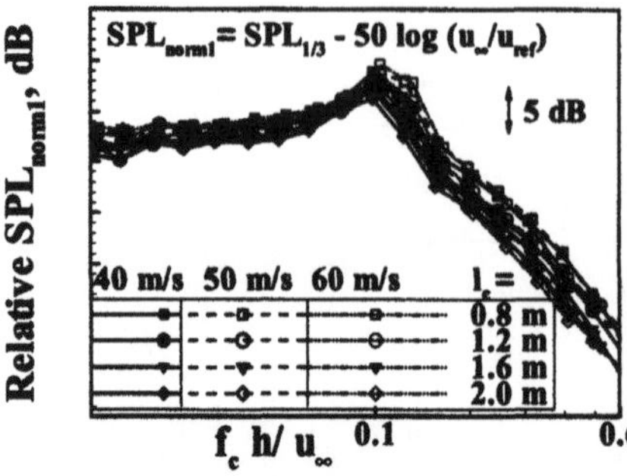

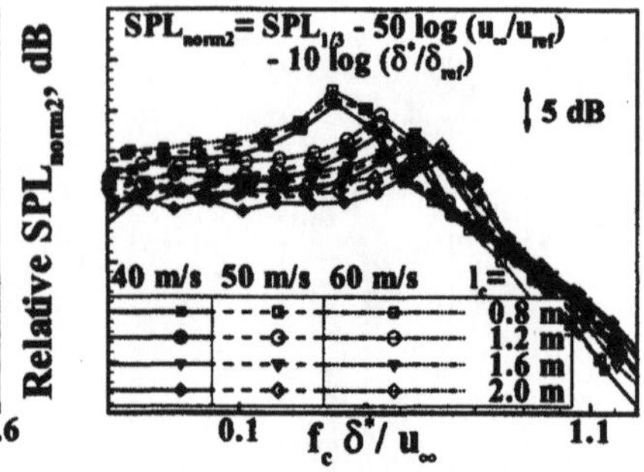

Figure 5 *(left)* Normalized one-third octave band spectra for different chord lengths l_c (all plates with solid trailing edge). Scaling based on trailing edge thickness *h*.
Figure 6 *(right)* Scaling based on boundary layer displacement thickness δ^*, respectively.

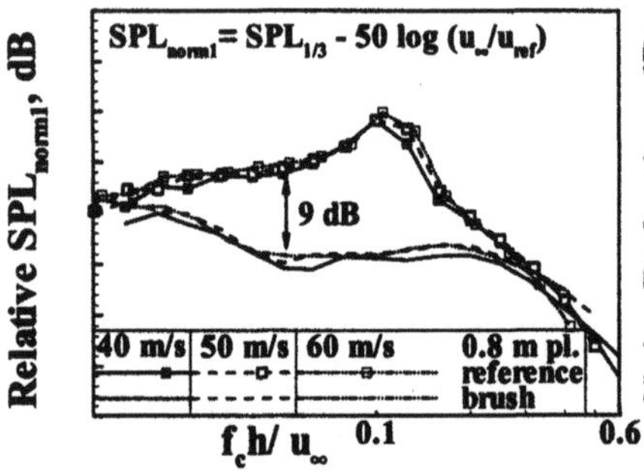

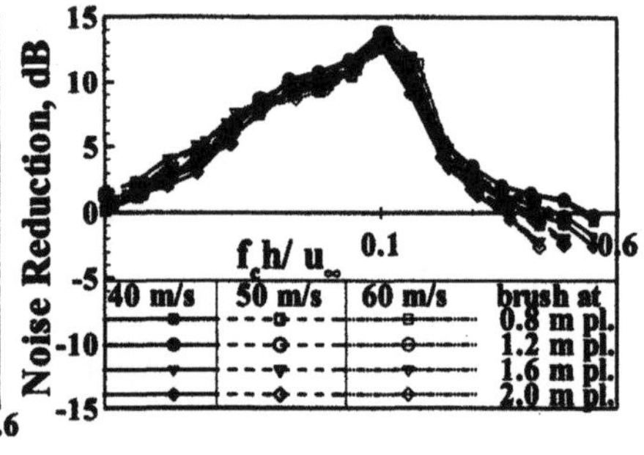

Figure 7 *(left)* Trailing edge noise reduction by means of brush edges (fibre diameter: 0.4 mm, length: 30 mm), comparison with solid trailing edge reference configuration.
Figure 8 *(right)* Influence of plate chord length and flow velocity on the noise reduction potential (same brush configuration as in Figure 7).

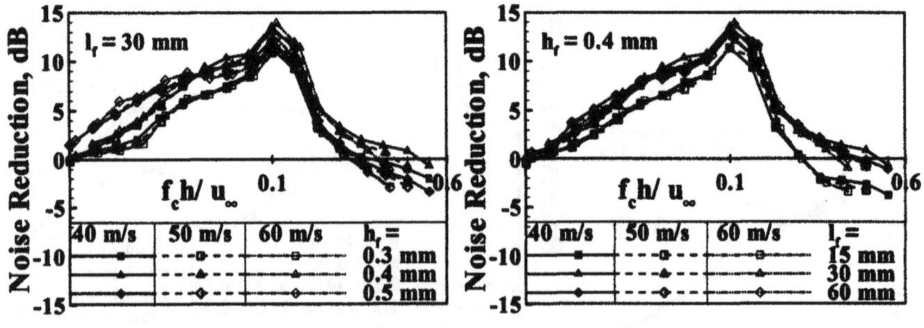

Figure 9 Effect of brush design parameters on trailing edge noise reduction (0.8 m plate).

A Study on Trailing Edge Noise Sources Using High-Speed Particle Image Velocimetry

A. SCHRÖDER, M. HERR°, T. LAUKE°, AND U. DIERKSHEIDE*
Deutsches Zentrum für Luft- und Raumfahrt e.V. (DLR)
Institut für Aerodynamik und Strömungstechnik
Bunsenstrasse 10, 37073 Göttingen, Germany
°Lilienthalplatz 7, 38108 Braunschweig, Germany
email: Andreas.Schroeder@dlr.de
*LaVision GmbH, Anna-Vandenhoeck-Ring 19
37081 Göttingen, Germany

Summary

The noise generation of turbulent flows near the edges of airplanes and automobiles is a general design problem and its importance increases in times of growing traffic. Turbulent boundary layers being convected past the trailing edge into the wake are known to generate an intense, broadband scattering noise. In this feasibility study the high-speed PIV technique was applied to a generic trailing edge noise experiment as performed on a flat plate model in the Aeroacoustic Wind Tunnel Braunschweig (AWB). Trailing edge noise sources have been measured simultaneously with instantaneous velocity vector fields to relate the generated sound to the ensuing aeroacoustic source quantities. The first step towards a new procedure for trailing edge noise prediction, combining numerical methods with the high-speed PIV measurement, is presented.

1 Introduction

Airframe noise is essentially due to the interaction of unsteady, mostly turbulent flow with the structure, particularly caused by vortical flows around edges or over open cavities. A classical problem in this field is the trailing edge (TE) noise, which involves different noise generating mechanisms. Extensive investigations were made on airfoil- and on flat plate trailing edges. Brooks and Hodgson [1] described an experiment on a tripped NACA 0012 model with different TE thicknesses. They observed a spectral "hump" in the acoustic measurements and were able to relate it to blunt TE vortex shedding for relatively thick edges, which vanishes with a decrease in TE thickness. For all test cases under consideration they found a low frequency broadband maximum in the acoustic power density spectrum, caused by the turbulent boundary layer (TBL) convected past the TE, at about 1 kHz. A predictable correlation of surface pressure measurements in the region close to the TE and the far field sound pressure was found.

Howe [6] distinguishes TE noise theories of three different types: i) Those based on Lighthill's acoustic analogy, like from Ffowcs Williams and Hall [4],

ii) those based on special solutions of the linearized hydroacoustic equations and iii) those creating ad hoc models with postulated source distributions. It was pointed out that all theoretical models lead to the same (velocity)^5-scaling of the radiated sound intensity. The TBL flow structures causing the surface pressure fluctuations at the TE were assumed to have characteristic length scales. According to the distribution of the different length scales of the TBL flow structures, the frequency contribution of the respective pressure fluctuation can be recognized in the power density spectrum of the far field noise. Howe's approach, following Ffowcs Williams and Hall [4] provides a scheme how to solve the inhomogeneous wave equation from Lighthill's acoustic analogy for a flat plate of negligible thickness, yielding the mean-square farfield pressure for a fixed observer at a distance R and an observation angle θ:

$$\langle p^2 \rangle \approx \rho_0^2 u_0^2 V^2 M_\upsilon \left(L\,l / R^2 \right) \sin\alpha \sin^2(\theta / 2) \cos^3\beta \qquad (1)$$

with V as the convection velocity, u_0 the rms velocity, M_v the turbulence convection Mach number, ρ_o the ambient density, β the TE sweep angle, L the spanwise edge length and l a characteristic length scale of the flow, which has been often identified with the displacement thickness of the TBL. With the three velocity terms in this formula the scaling of the sound pressure goes with (velocity)^5, consistent with other relevant approaches. An essential result is also the obtained $\sin^2(\theta/2)$-directivity.

Within this feasibility study it is expected that highly time resolved PIV data, as obtained in the region directly up- and downstream of the TE, allow an identification of noise sources and their correlation with the flow structure movement. These velocity field measurements in the source area are a novel attempt to obtain the required data for Howe's approach (Equation 1) but also for the solution of the inhomogeneous wave equation by applying a suitable flat plate Green's function on the measured source quantities. According to Howe [6], Powell [8] and Möhring [7] the major noise contribution is provided by the z-component of vorticity, the corresponding dipole $\varpi \times u$ ("principal edge noise dipole") being perpendicular to the plane of the plate. In this paper the first results of a numerical simulation of trailing edge noise, based on HS-PIV input data are presented.

2 Test Set-Up and Procedure

A flat plate (chord based Re = 5.3×10^6 and 6.6×10^6) with profiled leading and trailing edges was mounted vertically in the Aeroacoustic Wind Tunnel Braunschweig (AWB), which is an open jet anechoic test facility (Figure 1). The flat plate boundary layer was tripped at the leading edge, reaching a thickness of $\delta \approx 0.03$ m on each side of the TE, corresponding to free stream velocities of $U_\infty = 40$ m/s to 50 m/s and a 2 m chord length. Towards the trailing edge the plate is slightly and symmetrically convergent (5° taper), but no flow separation occurs. A full description of the experimental set-up without the HS-PIV system is provided in [5].

The PIV measurement volume was located at the trailing edge in a x-y-plane within the turbulent boundary layer (Figure 1, left) in order to track the flow structures with a spatial resolution of 256 pixels in y- and 1024 pixels (corresponding to 135 mm) in x-direction.

The used high-speed PIV system consists of a *New Wave Pegasus* PIV, dual cavity Nd:YLF laser with an output beam wavelength of 527 nm, a pulse length of 135 ns and 2×10 mJ at 1 kHz and approximately 2×7 mJ at 4 kHz for each resonator, optics to produce a light sheet and a *HighSpeedStar4 (HSS4)* CMOS camera with a spatial resolution of max. 1024×1024 Pixel at 2 kHz frame rate. In this application, a double frame rate of 4 kHz was achieved, thus yielding images at 8 kHz with a spatial resolution of 1024×256 pixel and a 10 bit greyscale dynamics (for details see [10]). 2.6 GB camera memory inside the camera housing allowed to capture 4096 double-images per run. The camera lens was a Nikon 105 mm with an aperture of $f_{\#} = 1.8$. The evaluation of the particle images was performed with a cross correlation scheme using standard FFT with 4× multi-pass with image deformation, interrogation window shift and a final window size of 32×32 pixel, with 75% overlap, corresponding to a resolution down to 3 mm in both directions. The Whittaker reconstruction was used for the deformation scheme and peak detection was achieved by a three point Gaussian fit. For post-processing, a median filter was used to remove vector outliers.

The fluctuation velocity (u', v') vector fields for each run were assessed by subtracting the ensemble average of the 4096 instantaneous velocity vector fields from each instantaneous one. The latter were then split into the single component scalar fields. Additionally, all instantaneous z-vorticity fields, the rms-fields, probability density functions and space-time-correlations were calculated.

The far field sound pressure, as emitted from the source area, was measured simultaneously with a directional microphone (acoustic mirror) at $\theta = \pi/2$ and $R = 1.15$ m, focused at the TE (Figure 1, right). Acoustic data were recorded at a sampling rate of 80 kHz. An average run time difference from the TE noise sources to the microphone of $T_r \approx 6.05$ ms has to be taken into account for comparison of the time resolved flow fields with the respective sound pressure time history. Because of the local focusing of the microphone a frequency cut off at about 1 kHz for the long wave-lengths side has to be considered. Figure 2 shows an example of an instantaneous velocity vector field from a series of time resolved HS-PIV measurements and the corresponding acoustic data.

A "straight forward" method which should result in a direct calculation of the source terms and therefore a reconstruction of the whole sound field was applied: As the major vortex source term, the perturbed Lamb vector $L' = \varpi \times u - \varpi \times u$, i. e. the source term of the acoustic analogies of Powell [8], Howe [6] and Möhring [7], was directly computed from the measured HS-PIV velocity field quantities (namely the instantaneous velocity, vorticity and the mean flow). After linear interpolation onto the body-fitted block-structured grid for the trailing edge, these source term values were fed into the subsequent computational aeroacoustic (CAA) simulations. Assuming a mean flow at rest ($Ma = 0$) the computations were performed by the DLR acoustic code PIANO (**P**erturbation

Investigation of Aerodynamic Noise), which in this case solves the acoustic perturbation equations (APE, [2, 3]). PIANO is a high resolution, high-order finite difference code advancing the numerical solution in time by the well known 4-stage Runge-Kutta scheme. Spatial derivatives are discretized by the 4th order dispersion relation preserving (DRP) scheme of Tam and Webb [9]. A 6th/8th-order digital filter is used to eliminate spurious waves.

3 Results

As a selection of the different HS-PIV runs the results of two test cases at $U_\infty = 40$ m/s and $U_\infty = 50$ m/s are presented in the following.

Figure 3 (top) shows the rms velocity field as measured for $U_\infty = 50$ m/s. The color coding makes visible that the highest values can be found directly downstream of the TE and in the shear layers between the wake and the "free" TBLs, which is approximately the extrapolated position of the highest rms values in the TBLs before they convect past the TE, at about 15 % of the boundary layer thickness. Consequently, the shear layers on both sides of the wake show also the highest contribution to the average z-vorticity (Figure 3, bottom), which is mostly represented by shear vorticity in this case. The spatial extent of high vorticity flow is narrow at the TE, as the transition between wake and TBL is of small scale, and with mixing further downstream the shear smoothens out.

In Figure 4 space-time correlation functions, resulting from one-point space correlations, of the z-vorticity fluctuation fields are shown for a wake position in the very vicinity of the TE ($U_\infty = 40$ m/s). The size of the region of peak correlation is considered directly related to the average size of the coherent flow structure in x- and y- direction, revealing a characteristic vortex size of 6 mm in the current. Also, the space-time correlation provides information about the corresponding average convection velocity and the coherence of the wake itself.

The negative correlations on the upper and lower side are due to local shear vorticity with an inverse sign. The convection velocity in the wake region can be determined to 24 m/s at $U_\infty = 40$ m/s free stream velocity. Together with the average size of the vortex structures of about 6 mm a characteristic frequency of 4 kHz is obtained which corresponds to an appropriate "hump" in the acoustic spectrum. Figure 5 (left) shows the respective narrow band spectra for U_∞= 40 m/s, 50 m/s and 60 m/s as obtained with the directional microphone ($\Delta f = 37$ Hz). Herein, the known (velocity)^5-dependence of the noise intensity is confirmed. According to Herr and Dobrzynski [5] the dominant spectral peak is related to blunt TE vortex shedding following a Strouhal-dependence of $St = fh/U_\infty \approx 0.1$ (h is the TE thickness). Because a Kármán-type vortex shedding is present, a relatively extended coherence length was obtained for the wake. For correlation coefficients down to 0.4 a length of ±0.03 m was determined.

In Figure 5 (right) a snapshot of the numerically calculated propagation of pressure sound waves as emitted from the TE is shown. This reconstruction is a first result which was obtained by applying CAA, based on HS-PIV input data ($U_\infty = 40$ m/s). In this regard, further investigations (in particular, the comparison with the acoustic measurement data) have to show whether the new method is successful.

4 Conclusions and Outlook

The work presented in this paper is one of the first applications of highly time-resolved PIV to a classical aeroacoustic problem at industrially relevant Reynolds numbers. A new method for the prediction of trailing edge noise was suggested with the future aim to compute the noise field and directivity by using PIV data, namely the aeroacoustic source quantities, as input for a CAA calculation.

Both high-speed PIV and acoustic experiments (for a later validation of the suggested method) were performed on a flat plate model in the Aeroacoustic Wind Tunnel Braunschweig (AWB). The PIV data-set was captured at a double-frame rate of 4 kHz with a sufficiently large field of view and enough spatial resolution to resolve all main features of the sound generating flow. Time-space-correlations of the velocity- and vorticity-fluctuations provided information of the coherent structure sizes within the trailing edge flow and their specific convection velocities.

The obtained data-base will be used for further investigations concerning the development of prediction methods. Time resolved PIV allows the non-intrusive quantification of the relevant flow parameters and helps to investigate vortex-structure interactions. In terms of the aeroacoustic optimization of existing aircraft components such an “optical” detection of aeroacoustic source terms will be beneficial, since a huge amount of (at least low speed) problems could be investigated at lower costs without the need of quiet test facilities.

Acknowledgement

The authors would like to thank Prof. J. Delfs from the Institute of Aerodynamics and Flow Technology, Department of Technical Acoustics, DLR, Braunschweig, for the idea of applying HS-PIV on the generic trailing edge noise experiment in the AWB and Dr. Roland Ewert from the same department for the essential idea of using the measured source quantities as input data for a CAA simulation [3].

References

[1] T.F. Brooks and T.H. Hodgson. “Trailing Edge Noise Prediction from Measured Surface Pressures”. In Journal of Sound and Vibration, Volume 78, No. 1, 1981, pp. 69 -117

[2] R. Ewert and W. Schröder. “Acoustic Perturbation Equations based on Flow Decomposition via Source Filtering“. In Journal of Computational Physics, Volume 188, 2003, pp. 365-398

[3] R. Ewert and W. Schröder. "On the Simulation of Trailing Edge Noise with a Hybrid LES/APE Method". In Journal of Sound and Vibration, Volume 270, 2004, pp. 509-524

[4] J.E. Ffowcs Williams and L.H. Hall. "Aerodynamic Sound Generation by Turbulent Flow in the Vicinity of a Scattering Half Plane". In J. Fluid Mech., Vol. 40, 1970, pp. 657-670.

[5] M. Herr and W. Dobrzynski. "Experimental Investigations in Low Noise Trailing Edge Design". 10th AIAA/CEAS Aeroacoustics Conference, 10-12 May 2004, Manchester, UK, Paper 2004-2804

[6] M.S. Howe. "A Review of the Theory of Trailing Edge Noise". In Journal of Sound and Vibration, Volume 61, No. 3, 1978, pp. 437-465

[7] W. Möhring. "Modelling Low Mach Number Noise". In Mechanics of Sound Generation in Flows, edited by E. A. Müller, Springer, 1979

[8] A. Powell. "Theory of Vortex Sound". In Journal of the Acoustical Society of America, Volume 36, No 1, 1964, pp. 177-195

[9] C. Tam and J. Webb. "Dispersion-Relation-Preserving Finite Difference Schemes for Computational Acoustics". In J. Comp. Phys., Volume 107, 1993, pp. 262-281

[10] http://www.lavision.de

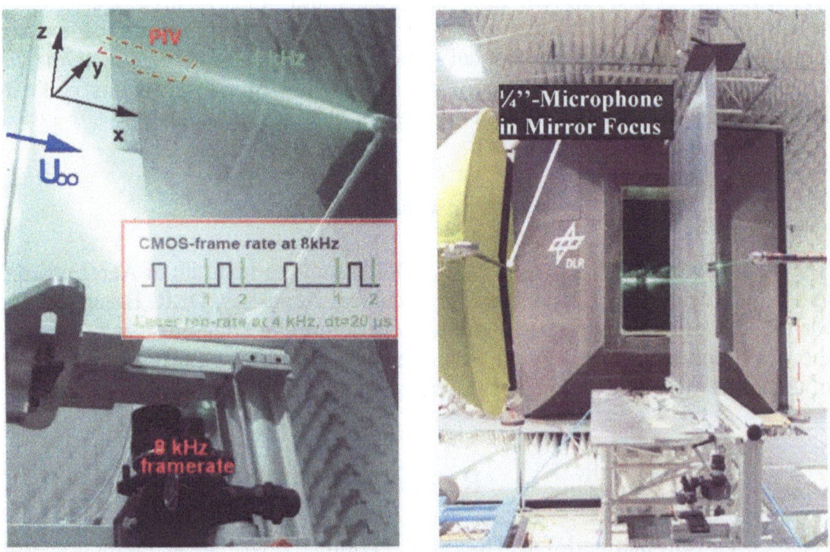

Figure 1: Set-up and trigger scheme of the high-speed PIV system at the flat plate trailing edge (left), view against flow direction with the acoustic mirror (right) in the AWB, DLR-Braunschweig

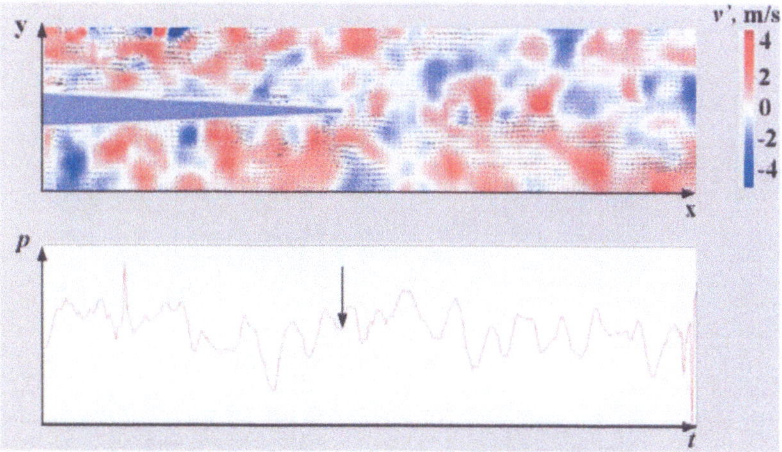

Figure 2: Instantaneous velocity vector field (v' scalar field color coded) out of a 4 kHz run in the TE region (top) and corresponding directional microphone signal (actual value at the arrow position).

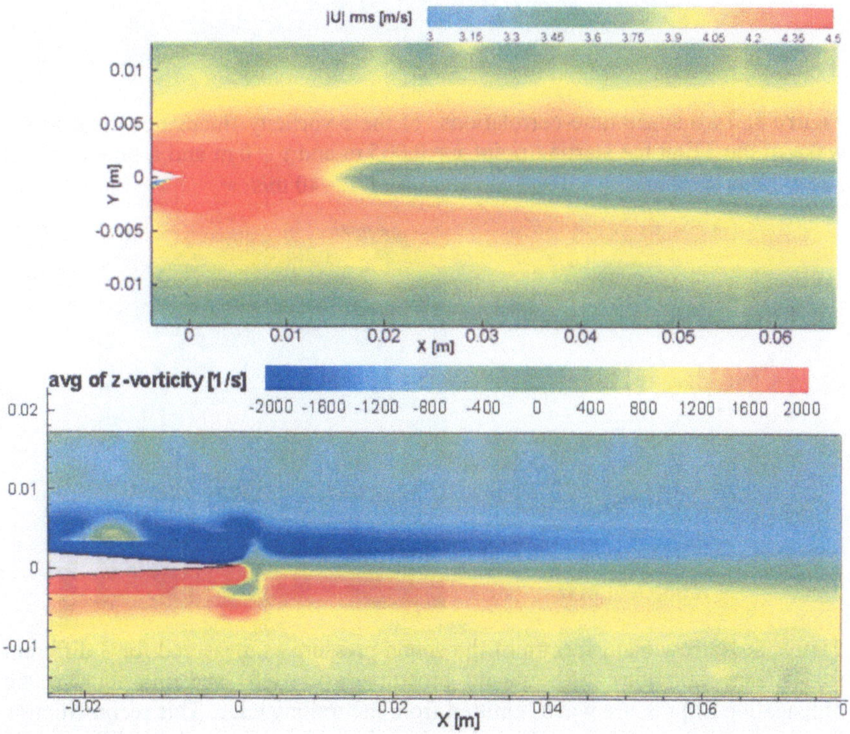

Figure 3: Rms velocity field at $U_\infty = 50$ m/s with high values at the shear layers of the wake region (top) and corresponding field of average z-vorticity (bottom)

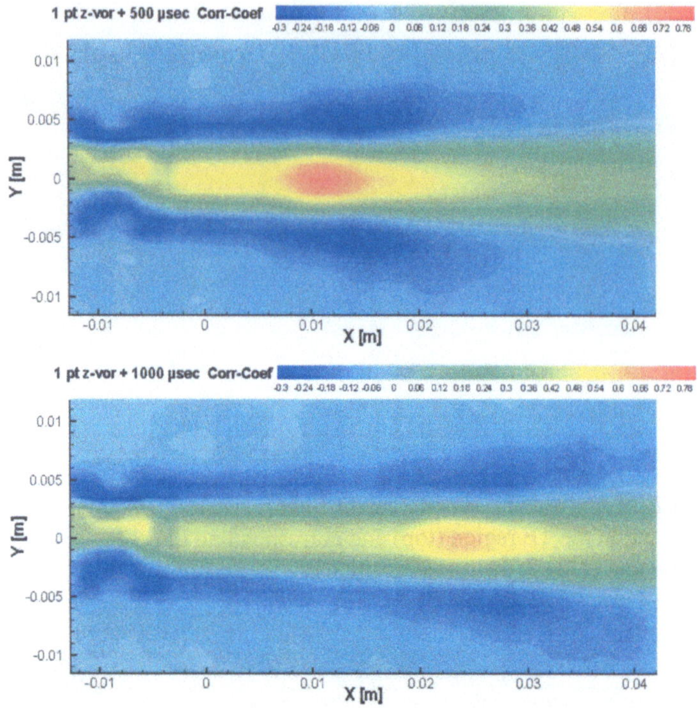

Figure 4: Two space-time-correlations of the *z*-vorticity fluctuation field in the wake of the TE at the starting point $x = 0.005$ m and $y = 0$ m and a time delay of $\Delta t = 500$ µs (top) and $\Delta t = 1000$ µs (bottom), $U_\infty = 40$ m/s

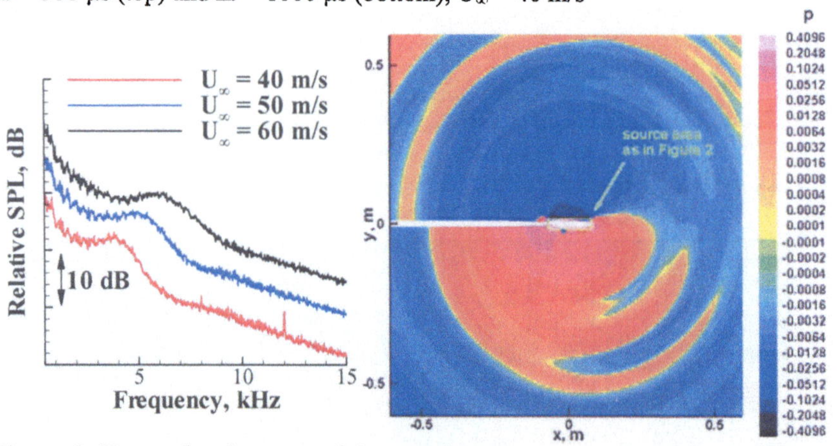

Figure 5: Narrow band spectra of the sound pressure as measured for 3 different free stream velocities, SPL: sound pressure level (left) and snapshot of the propagation of pressure waves emitted from the trailing edge. This reconstruction is performed by a CAA code and is based on the measured HS-PIV 4 kHz velocity vector fields (right)

Towards the Applicability of the Modified von Kármán Spectrum to Predict Trailing Edge Noise

Marcus Bauer[1] and Andreas Zeibig[2]

[1] Institute of Acoustics and Speech Communication,
Dresden University of Technology, 01062 Dresden, Germany

Present address: DLR (German Aerospace Center), Institute of Aerodynamics and Flow Technology, Lilienthalplatz 7, 38108 Braunschweig, Germany
Marcus.Bauer@dlr.de

[2] Institute of Acoustics and Speech Communication,
Dresden University of Technology, 01062 Dresden, Germany
Andreas.Zeibig@ias.et.tu-dresden.de

Summary

Trailing edge noise can be predicted with the help of a synthetic turbulent velocity field. Assuming isotropic conditions this field may be generated via an energy spectrum of turbulence like the modified von Kármán spectrum (MVKS). In this work one-dimensional wavenumber spectra of turbulence obtained from hot-wire measurements at the trailing edges of a thin, flat plate and a NACA0012 airfoil were compared to the respective spectra extracted from the MVKS. Good agreement at all measuring positions is possible with a modified form of the MVKS. The remaining discrepancies can be attributed to the anisotropy of boundary layer turbulence.

1 Introduction

Airframe noise is an important part of an aircraft's landing approach noise. In order to design quieter aircraft there is a strong demand for tools that allow a computational optimisation of its respective components, e.g. landing gears, slats and flaps. Unfortunately this is hardly possible with present computer resources via a Direct Numerical Simulation (DNS) solving the full governing flow equations (Navier Stokes Equations, NSE). This has given rise to the application of a hybrid CFD[1]/ CAA[2] approach based on simplified forms of the NSE and on a synthetic stochastic isotropic turbulent velocity field [7]. With this means broadband trailing edge noise, a significant contribution to

[1] Computational Fluid Dynamics
[2] Computational Aeroacoustics

airframe noise, has successfully been computed [2, 5]. The synthetic velocity field is based on the assumption of the modified von Kármán spectrum (MVKS) as the prevailing energy spectrum of turbulence [3, 9, 10, 11].

The aim of the present investigations was to find out to what extent the MVKS is appropriate to describe turbulent trailing edge flows. Therefore one-dimensional turbulent wavenumber spectra obtained from hot-wire measurements at the trailing edges of two test objects were compared to the respective spectra extracted from the MVKS. Please note that the topic of this paper is *not* to present trailing edge noise results computed with the hybrid approach (please refer to references [2, 5] in case of interest). Consequently the approach will only be described briefly in so far as the MVKS is concerned.

2 Procedure of the Studies

2.1 Coordinate System and Measurements

Figure 1 shows the coordinate system. Test objects were a NACA0012 airfoil (dimensions $L_x = 120\,\text{mm}, L_z = 140\,\text{mm}$) and a thin, flat plate ($L_x = 200\,\text{mm}, L_y = 0.3\,\text{mm}, L_z = 100\,\text{mm}$). In both cases the angle of attack was zero degrees and the incoming flow velocity $v_{0x} = 37.5$ m/s, corresponding to a Mach-number $Ma = 0.11$. The measurements were carried out in the aeroacoustic wind-tunnel of the Institute of Acoustics and Speech Communication at Dresden University of Technology using a triple sensor hot-wire probe. It was positioned with a lightweight traverse system of accuracy ± 0.01 mm in every coordinate direction. At both objects time responses of the velocity vector components v_i $(i = x, y, z)$ were recorded in approximately 40 y-coordinates at $z = 0$ as close as possible to the trailing edge in x-direction ($x \approx 1\,\text{mm}$). $16,384 = 2^{14}$ samples were recorded at a sampling frequency of 25 kHz (plate) and 250 kHz (airfoil) using anti-aliasing low-pass filters with cut-off frequencies of 10 kHz and 100 kHz, respectively. The boundary layer was tripped with a tape of height 0.3 mm and length 1.2 mm on both sides of the objects. At the flat plate this tape was attached 6 mm downstream of the leading edge and at the NACA0012 airfoil at the x-coordinate of maximum profile thickness, respectively.

2.2 The MVKS in the Framework of the Hybrid Approach

The hybrid approach used in [2, 5] to compute broadband trailing edge noise is based on splitting the field variables into a time averaged and a generic fluctuating part. In the first step the mean flow field is calculated with DLR's CFD-Code FLOWer as a solution of the RANS[3]-equations using the standard Wilcox $k - \omega$ turbulence model [12], k denoting the turbulent kinetic energy

[3] Reynolds Averaged Navier Stokes

and ω its specific dissipation rate, respectively. In the second step a modified form of the LEE[4] with a source term is solved with DLR's CAA-Code PIANO to compute the time-dependent variables. The source term is calculated from a frozen synthetic turbulent velocity field $\boldsymbol{v}_{st}$ which is given by a sum of discrete Fourier modes [7].

Generating $\boldsymbol{v}_{st}$ requires a turbulent energy spectrum $E(\alpha)$, $\alpha = \sqrt{\alpha_x^2 + \alpha_y^2 + \alpha_z^2}$ denoting the wavenumber. At present $E(\alpha)$ is modelled by the MVKS:

$$E(\alpha) = C_1 \frac{2/3k}{\alpha_e} \frac{(\frac{\alpha}{\alpha_e})^4}{(1+(\frac{\alpha}{\alpha_e})^2)^{17/6}} \mathrm{e}^{-C_2 (\frac{\alpha}{\alpha_\nu})^2} \ . \tag{1}$$

It has its maximum at wave number α_e and reaches up to the Kolmogorov wave number α_ν. Both α_e and α_ν, and thus the whole spectrum, can be calculated from the RANS-results of k and ω. The parameter C_2 in the exponent is $C_2 = 2$ and the scaling factor C_1 ensures that

$$\int_0^\infty E(\alpha)\mathrm{d}\alpha = k. \tag{2}$$

Here C_1 is (for every set of input parameters k and ω) calculated as the ratio of k and the integral over $E(\alpha)$, which is computed numerically with the trapezoidal rule (for $C_1 = 1$). Note that turbulence can only be described by an energy spectrum $E(\alpha)$ if isotropy is assumed.

2.3 Characterising the Applicability of the MVKS

One-dimensional wavenumber spectra $\Phi_{xx}(\alpha_x)$, $\Phi_{yy}(\alpha_x)$ and $\Phi_{zz}(\alpha_x)$ obtained from measurement as well as from (1) were compared to each other, see Figure 2. $E(\alpha)$ given by the MVKS was transformed into these longitudinal (Φ_{xx}) and lateral (Φ_{yy}, Φ_{zz}) one-dimensional spectra via [1]:

$$\Phi_{xx}(\alpha_x) = \frac{1}{2} \int_{\alpha_x}^{\infty} \frac{E(\alpha)}{\alpha} \left(1 - \frac{\alpha_x^2}{\alpha^2}\right) \mathrm{d}\alpha \ , \tag{3}$$

$$\Phi_{yy/zz}(\alpha_x) = \frac{1}{4} \int_{\alpha_x}^{\infty} \frac{E(\alpha)}{\alpha} \left(1 + \frac{\alpha_x^2}{\alpha^2}\right) \mathrm{d}\alpha \ . \tag{4}$$

2.4 Generation of the MVKS

The parameter C_2 governs the decrease of the MVKS at high wavenumbers. To provide for a better agreement with the one-dimensional wavenumber

[4] Linearized Euler Equations

spectra $\Phi_{ii}(\alpha_x)$ from measurement modified forms of the MVKS with $C_2 \neq 2$ were used in this study, too, see below.

Furthermore the input parameters k and ω were not taken from the RANS computation here. Actually, k was calculated from the measured $v_i(\boldsymbol{x}, t)$ at coordinate $\boldsymbol{x} = [x, y, z]^{\mathrm{T}}$ via

$$k = \frac{1}{2}\left(\overline{v_x(\boldsymbol{x},t)^2} + \overline{v_y(\boldsymbol{x},t)^2} + \overline{v_z(\boldsymbol{x},t)^2}\right) \quad , \tag{5}$$

the overline denoting the time-average. The value of ω was estimated to provide for a best possible coincidence of the $\Phi_{ii}(\alpha_x)$ from measurement and theory. If k and ω would have been taken from the RANS-solution, the discrepancy of these spectra would have increased due to RANS inaccuracies, see below. However, the main goal of this study was not to assess the applicability of the input parameters k and ω rather than that of the MVKS itself.

3 Results and Discussion

3.1 Wavenumber Spectra of Trailing Edge Turbulence

Figure 3 compares one-dimensional wavenumber spectra $\Phi_{ii}(\alpha_x)$ obtained from the MVKS and from a measurement at the NACA0012 airfoil at $y \approx -1.6$ mm. Due to the limited signal-to-noise ratio of the hot-wire measurement technique no values of $\Phi_{ii}(\alpha_x)$ below approximately $10^{-5}\,\mathrm{m}^3/\mathrm{s}^2$ could be resolved. Using the generic form of the MVKS with $C_2 = 2$ ($\Rightarrow C_1 = 1.69$ for the employed values of k and ω) there is a good agreement at wavenumbers $\alpha_x < 2.0 \cdot 10^3/\,\mathrm{m}$ but a rather poor agreement at higher wavenumbers. A modified form with $C_2 = 50$ ($C_1 = 2.24$) provides for a good coincidence throughout all α_x. The remaining deviations can be attributed to the anisotropy of boundary layer turbulence [4, 6, 8]. They can not be overcome by another $E(\alpha)$ spectrum as for example in Figure 3 $\Phi_{xx}(\alpha_x)$ from measurement falls below $\Phi_{xx}(\alpha_x)$ from the MVKS while at the same time $\Phi_{yy}(\alpha_x)$ and $\Phi_{zz}(\alpha_x)$ from measurement exceed the respective spectra extracted from the MVKS.

Figure 4 shows results from the flat plate at $y \approx -0.7$ mm. There is a satisfying agreement of the onedimensional wavenumber spectra from measurement and the generic MVKS throughout all wavenumbers. The remaining discrepancies are again due to anisotropy. The upper cut-off frequency of the measurement underlying Figure 4 is lower than in Figure 3 because as mentioned above the sampling frequency at the flat plate was only 25 kHz instead of 250 kHz as in the NACA0012 measurement. The spurious peaks in the measured spectra at wavenumbers $\alpha_x < 100/\mathrm{m}$ are most problably due to low-frequency electromagnetic disturbance (e.g. PC-monitor) of the hot-wire sensor signals.

At both objects the measured spectra confirm that the share of the small eddies increases when the coordinate $y = 0$ is approached [6], i.e., the $\Phi_{ii}(\alpha_x)$ spectra decay slower towards higher wavenumbers as $y \to 0$. Considering similar y-coordinates, however, the share of the small eddies at the flat plate is always higher than at the NACA0012 airfoil. This leads to the following consequences:

- **Flat plate:** There is a good agreement of the $\Phi_{ii}(\alpha_x)$ from measurement and from the generic MVKS for larger values of y, cp. Figure 4, but as $y \to 0$ the $\Phi_{ii}(\alpha_x)$ from measurement exceed those from the MVKS at high wavenumbers.
- **NACA0012:** The $\Phi_{ii}(\alpha_x)$ spectra from measurement fall below those of the generic MVKS at high wavenumbers in case of large y, cp. Figure 3, but on the other hand there is a good agreement as $y \to 0$.

3.2 Trailing Edge RANS Results

Next to a correct formulation of the turbulent energy spectrum $E(\alpha)$ itself reliable values of k and ω from the RANS-computation are also crucial to generate a proper $E(\alpha)$. In this context Figure 5 illustrates $k(y)$ and $\omega(y)$ from two RANS mean flow solutions for the flat plate. In the first computation its small thickness of only $L_y = 0.3$ mm was neglected, while in the second it was taken into account. Obviously this results in significant differences in k and ω as $y \to 0$. Regarding the finite thickness clearly provides for a better agreement with measured data.

However concerning the agreement of the onedimensional wavenumber spectra from measurement and the MVKS, the remaining RANS inaccuracies may be as important as the choice of the energy spectrum, see the dashed line in Figure 4 where k and ω were taken from the RANS solution of the flat plate to obtain the MVKS.

4 Conclusion

At some coordinates the one-dimensional wavenumber spectra computed from the generic MVKS are in good agreement with respective spectra obtained from hot-wire measurements at the trailing edges of a thin, flat plate and a NACA0012 airfoil. However, in general, modified forms of the MVKS provide for a significant improvement resulting in a good agreement with measured spectra throughout all positions at the trailing edges of both objects. The remaining discrepancies are attributed to the anisotropy of the boundary layer turbulence.

Thus it seems worthwhile to investigate the influence of enhanced turbulence spectra and anisotropy, respectively, on the computed trailing edge noise. Also the influence of the fidelity of the RANS results underlying an energy spectrum of turbulence should be subject of future interest.

Acknowledgements

The authors wish to thank the DLR and the DFG (German Research Foundation) for kind funding of this work under grant number KO1242/6 in the framework of the *SWING+* project (Simulation of Wing Flow Noise Generation).

References

[1] G. K. Batchelor. *The Theory of Homogeneous Turbulence.* Cambridge at the University Press, 1956.
[2] M. Bauer. Berechnung der Schallabstrahlung überströmter Hinterkanten. Diploma thesis, Institute of Acoustics and Speech Communication, Dresden University of Technology, 2002.
[3] W. Béchara, C. Bailly, P. Lafon, and S. Candel. Stochastic Approach to Noise Modeling for Free Turbulent Flows. *AIAA Journal*, 32(3):455–463, March 1994.
[4] D. C. Dunn and J. F. Morrison. Anisotropy and Energy Flux in Wall Turbulence. *Journal of Fluid Mechanics*, 491:353–378, 2003.
[5] R. Ewert and M. Bauer. Towards the Prediction of Trailing Edge Noise via Stochastic Surface Sources. AIAA-paper 2004-2861, American Institute for Aeronautics and Astronautics, 2004.
[6] J. O. Hinze. *Turbulence.* Mc Graw Hill Book Company, Inc., 1959.
[7] R. H. Kraichnan. Diffusion by a Random Velocity Field. *Phys. Fluids*, 13(1):22–31, 1970.
[8] J. M. Österlund and A. V. Johansson. Turbulence Statistics of Zero Pressure Gradient Turbulent Boundary Layers. (to be submitted), Dept. of Mechanics, KTH, SE-100 44 Stockholm, Sweden.
[9] Y. H. Pao. Structure of Turbulent Velocity and Scalar Fields at Large Wave Numbers. *Phys. Fluids*, 8(6):1063–1075, 1965.
[10] P. G. Saffman. On the Fine-Scale Structure of Vector Fields Convected by a Turbulent Fluid. *Journal of Fluid Mechanics*, 16:545–572, 1963.
[11] T. von Kármán. Progress in the Statistical Theory of Turbulence. *Proc. Natl. Acad. Sci. U.S.*, 34:530–539, 1948.
[12] D. C. Wilcox. *Turbulence Modeling for CFD.* DCW Industries, Inc., 1993.

Figure 1 Coordinate system; the origin was centered at the trailing edge in the y and z-direction, respectively

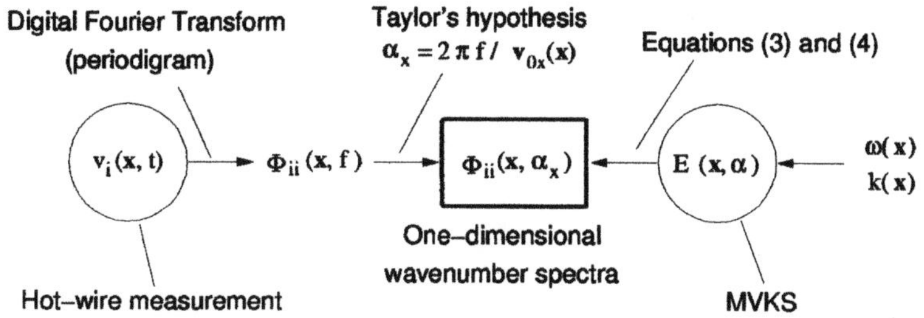

Figure 2 Illustration of the procedure to compare data from the hot-wire measurement to the MVKS; $i = x, y, z$

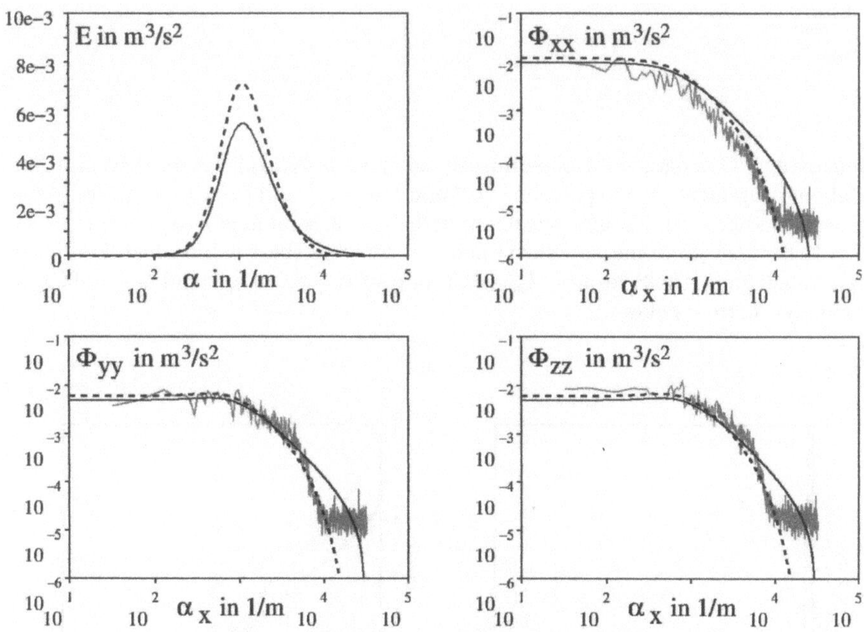

Figure 3 Comparison of one-dimensional spectra $\Phi_{ii}(\alpha_x)$ obtained from a NACA0012 measurement at $\boldsymbol{x} = [1.0\,\mathrm{mm}, -1.6\,\mathrm{mm}, 0.0\,\mathrm{mm}]^T$ and from $E(\alpha)$ given by the modified von Kármán spectrum. $E(\alpha)$ was calculated for $k = 19.0\,\mathrm{m}^2/\mathrm{s}^2$ (from measurement) and $\omega = 25000\,/\mathrm{s}$ (best fit); solid line: $C_2 = 2$, dashed line: $C_2 = 50$

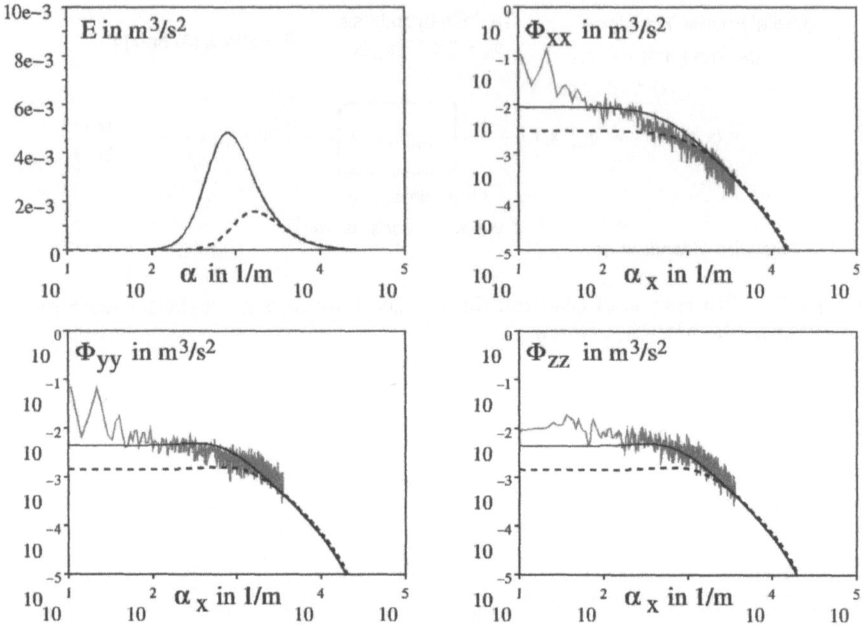

Figure 4 Comparison of one-dimensional spectra $\Phi_{ii}(\alpha_x)$ obtained from a flat plate measurement at $\boldsymbol{x} = [1.0\,\mathrm{mm}, -0.7\,\mathrm{mm}, 0.0\,\mathrm{mm}]^{\mathrm{T}}$ and from $E(\alpha)$ given by the generic modified von Kármán spectrum with $C_2 = 2$; solid line: $E(\alpha)$ calculated for $k = 12.0\,\mathrm{m}^2/\mathrm{s}^2$ (from measurement) and $\omega = 15000\,/\mathrm{s}$ (best fit); dashed line: $E(\alpha)$ calculated from RANS results ($L_y = 0.3$ mm) of $k = 8.0\,\mathrm{m}^2/\mathrm{s}^2$ and $\omega = 25000\,/\mathrm{s}$, compare Figure 5 below

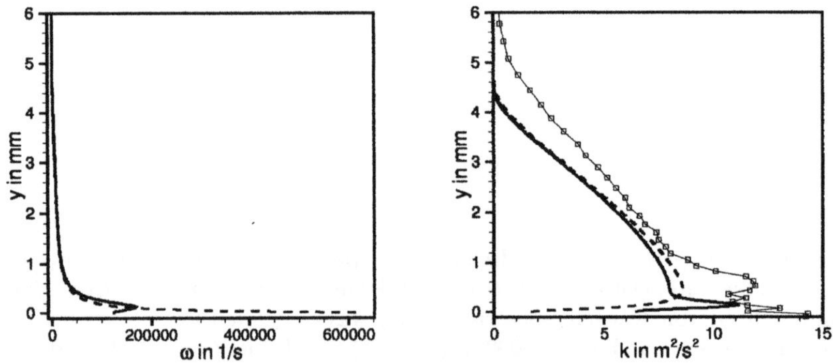

Figure 5 RANS-solutions of $\omega(y)$ and $k(y)$ from the thin, flat plate at $x \approx 1$ mm for $L_y = 0.0$ mm (dashed lines) and $L_y = 0.3$ mm (solid lines) plus the respective $k(y)$ from the hot-wire measurement (symbols)

Numerical simulation of combustion noise using acoustic perturbation equations

T. Ph. Bui, W. Schröder and M. Meinke

Institute of Aerodynamics, RWTH Aachen University,
Wüllnerstraße zw. 5 und 7,
52062 Aachen, Germany,
phong@aia.rwth-aachen.de,
WWW home page: http://www.aia.rwth-aachen.de

Summary

Combustion noise of unconfined turbulent flames has been investigated using a hybrid CFD/CAA Method. A large-eddy simulation (LES) of a turbulent non-premixed flame is used to determine the source terms for the computational aeroacoustics (CAA) simulation. The governing CAA equations, namely the Acoustic Perturbation Equations (APE), have been extended to take into account noise generated by reacting flow effects. The right-hand side of the pressure-density relation within the APE system shows that the major source term, the heat release per unit volume, is encoded in the density fluctuation. Therefore the total time derivative of the density is used as source term to simulate combustion noise.

1 Introduction

The analysis of combustion noise which already has been of major interest in the seventies and eighties, has been revived over the last couple of years. On the one hand, today's available computation power allows the simulation of the acoustic output of combustion systems, while on the other hand, a solution of the unsteady Navier-Stokes equations by direct numerical simulations (DNS) is still infeasible for real technical applications. Especially the simulation of reacting flows is very expensive as the complete chemical reactions have to be captured additionally. An alternative to a DNS approach concerning combustion noise is a hybrid LES/CAA method, which will be presented here.

This research project is part of the Research Unit FOR 486 "Combustion Noise", which is supported by the German Research Council (DFG). The objective of the Institute of Aerodynamics of the RWTH Aachen University is to investigate the origin of combustion noise and its mechanisms. The numerical approach is a two-step method, the first of which is based on an LES, which is performed by the Insitute for Energy and Powerplant Technology from Darmstadt University of Technology, followed by the CAA simulation to compute the acoustical field. Due to the large difference in the characteristic fluid mechanical and acoustical length scales it is reasonable to restrict the LES to the source region only, whereas the CAA simulation is performed on a much coarser grid covering the larger acoustical field. This hybrid LES/CAA approach is similar to that in [2]. However, in this study the Acoustic Perturbation Equations are extended to reacting flows. In flows, where chemical reactions have to be considered, the application of such an approach is essential as the disparity of the characteristic fluid mechanical and acoustical length scales is even more pronounced than in the non-reacting case.

It is well known from the literature, e.g. [3,4], that noise generated by combustion in low Mach number flows is dominated by heat release effects, whereas in jet or airframe noise problems the major noise contribution originates from the Lamb vector ($\boldsymbol{L}' = (\boldsymbol{\omega} \times \boldsymbol{u})'$), which can be interpreted as a vortex force [1,5]. In principle it is possible to treat this task by extending Lighthill's Acoustic Analogy to reacting flows as was done in the past [3,4]. This, however, leads to an inhomogeneous wave equations with an ordinary wave operator e.g. [4,6], which is valid for homogeneous mean flow only. Therefore, this approach is restricted to the acoustic far field. The APE approach remedies this drawback. It is valid in non-uniform mean flow and takes into account convection and refraction effects, unlike the linearized Euler equations [1].

2 Governing equations

To derive the extended APE system the governing equations of mass, momentum, and energy for reacting flows are rearranged such that the left-hand side describes the APE-1 system [1], whereas the right-hand side (RHS) consists of all non-linear flow effects including the sources related to chemical reactions.

$$\frac{\partial \rho'}{\partial t} + \nabla \cdot (\rho' \bar{\boldsymbol{u}} + \bar{\rho} \boldsymbol{u}') = q_c \tag{1}$$

$$\frac{\partial \boldsymbol{u}'}{\partial t} + \nabla (\bar{\boldsymbol{u}} \cdot \boldsymbol{u}') + \nabla \left(\frac{p'}{\bar{\rho}} \right) = \boldsymbol{q_m} \tag{2}$$

$$\frac{\partial p'}{\partial t} - \bar{c}^2 \frac{\partial \rho'}{\partial t} = q_e \tag{3}$$

As was mentioned before, the heat release effect dominates the generation of combustion noise. Therefore the investigations have been performed using q_e only, i.e. assuming $q_c = 0$ and $\boldsymbol{q_m} = 0$.

2.1 Thermoacoustic Source Terms

In the proposed APE system the source term containing heat release effects appears on the RHS of the pressure-density relation, i.e. q_e. This term vanishes when only isentropic flow is considered. However, due to the unsteady heat release in a flame the isentropic pressure-density relation is no longer valid in the combustion area. Nevertheless, it is this effect, which defines the major source term in comparison to the sources (q_c, $\boldsymbol{q_m}$) in the mass and momentum equations within the APE system.

Concerning the other source mechanisms, which lead to an acoustic multipole behavior though it can be conjectured that they are of minor importance in the far field. Using the energy equation for reacting flows the pressure-density relation becomes:

$$
\begin{aligned}
\frac{\partial p'}{\partial t} - \bar{c}^2 \frac{\partial \rho'}{\partial t} &= -\bar{c}^2 \cdot \frac{\partial \rho_e}{\partial t} \\
&= \bar{c}^2 \left[\frac{\bar{\rho}}{\rho} \cdot \frac{\alpha}{c_p} \cdot \left(\sum_{n=1}^{N} \left. \frac{\partial h}{\partial Y_n} \right|_{\rho,p,Y_m} \rho \frac{DY_n}{Dt} + \nabla \cdot \boldsymbol{q} - \frac{\partial u_i}{\partial x_j} \tau_{ij} \right) \right. \\
&\quad - \nabla (\boldsymbol{u} \rho_e) - \frac{1}{\bar{c}^2} \left[\left(1 - \frac{\bar{\rho}\bar{c}^2}{\rho c^2} \right) \cdot \frac{Dp}{Dt} - \frac{p - \bar{p}}{\rho} \cdot \frac{D\rho}{Dt} \right] \\
&\quad \left. + \left[-\frac{\gamma - 1}{\gamma} \boldsymbol{u} \cdot \nabla \bar{p} - \frac{p}{\bar{c}^2} \cdot \boldsymbol{u} \left(\frac{\nabla \bar{p}}{\bar{p}} - \frac{\nabla \bar{\rho}}{\bar{\rho}} \right) \right] \right]
\end{aligned}
\tag{4}
$$

where ρ_e is defined as

$$
\rho_e = (\rho - \bar{\rho}) - \frac{p - \bar{p}}{\bar{c}^2} \tag{5}
$$

Perturbation and time averaged quantities are denoted by a prime and a bar, respectively. The volumetric expansion coefficient is given by α and c_p is the specific heat capacity at constant pressure. For an ideal gas the equation $\alpha / c_p = (\gamma - 1)/c^2$ holds. The quantity Y_n is the mass fraction of the nth species, h the enthalpy and $\boldsymbol{q}$ the heat flux.

2.2 Evaluation of the thermoacoustic source terms

The investigations have been performed by considering q_e only. Reformulating the energy equation for a gas with N species [4] leads to

$$
\frac{D\rho}{Dt} = \left[\frac{1}{c^2} \frac{Dp}{Dt} + \frac{\alpha}{c_p} \cdot \left(\sum_{n=1}^{N} \left. \frac{\partial h}{\partial Y_n} \right|_{\rho,p,Y_m} \rho \frac{DY_n}{Dt} + \nabla \cdot \boldsymbol{q} - \frac{\partial u_i}{\partial x_j} \tau_{ij} \right) \right] \tag{6}
$$

Since the combustion takes place at ambient pressure and the pressure variations due to hydrodynamic flow effects are of low order, the whole combustion process can be assumed to be at constant pressure. From our analysis [5] and from literature [4] it is known that combustion noise is dominated by heat release effects and that all other source mechanisms are of minor importance. Assuming combustion at constant pressure and neglecting all mean flow effects q_e reduces to sources, which are related to heat release effects, non-isomolar combustion, heat flux and viscous effects. Adding up all these sources under the aforementioned restrictions the RHS of the pressure-density relation can be substituted by the total time derivative of the

Concerning the other source mechanisms, which lead to an acoustic multipole behavior though it can be conjectured that they are of minor importance in the far field. Using the energy equation for reacting flows the pressure-density relation becomes:

$$\begin{aligned}\frac{\partial p'}{\partial t} - \bar{c}^2 \frac{\partial \rho'}{\partial t} &= -\bar{c}^2 \cdot \frac{\partial \rho_e}{\partial t} \\ &= \bar{c}^2 \left[\frac{\bar{\rho}}{\rho} \cdot \frac{\alpha}{c_p} \cdot \left(\sum_{n=1}^{N} \left.\frac{\partial h}{\partial Y_n}\right|_{\rho,p,Y_m} \rho \frac{DY_n}{Dt} + \nabla \cdot \boldsymbol{q} - \frac{\partial u_i}{\partial x_j} \tau_{ij} \right) \right. \\ &\quad -\nabla (\boldsymbol{u} \rho_e) - \frac{1}{\bar{c}^2} \left[\left(1 - \frac{\bar{\rho} \bar{c}^2}{\rho c^2} \right) \cdot \frac{Dp}{Dt} - \frac{p - \bar{p}}{\rho} \cdot \frac{D\rho}{Dt} \right] \\ &\quad \left. + \left[-\frac{\gamma - 1}{\gamma} \boldsymbol{u} \cdot \nabla \bar{p} - \frac{p}{\bar{c}^2} \cdot \boldsymbol{u} \left(\frac{\nabla \bar{p}}{\bar{p}} - \frac{\nabla \bar{\rho}}{\bar{\rho}} \right) \right] \right] \end{aligned} \tag{4}$$

where ρ_e is defined as

$$\rho_e = (\rho - \bar{\rho}) - \frac{p - \bar{p}}{\bar{c}^2} \tag{5}$$

Perturbation and time averaged quantities are denoted by a prime and a bar, respectively. The volumetric expansion coefficient is given by α and c_p is the specific heat capacity at constant pressure. For an ideal gas the equation $\alpha / c_p = (\gamma - 1)/c^2$ holds. The quantity Y_n is the mass fraction of the nth species, h the enthalpy and $\boldsymbol{q}$ the heat flux.

2.2 Evaluation of the thermoacoustic source terms

The investigations have been performed by considering q_e only. Reformulating the energy equation for a gas with N species [4] leads to

$$\frac{D\rho}{Dt} = \left[\frac{1}{c^2} \frac{Dp}{Dt} + \frac{\alpha}{c_p} \cdot \left(\sum_{n=1}^{N} \left.\frac{\partial h}{\partial Y_n}\right|_{\rho,p,Y_m} \rho \frac{DY_n}{Dt} + \nabla \cdot \boldsymbol{q} - \frac{\partial u_i}{\partial x_j} \tau_{ij} \right) \right] \tag{6}$$

Since the combustion takes place at ambient pressure and the pressure variations due to hydrodynamic flow effects are of low order, the whole combustion process can be assumed to be at constant pressure. From our analysis [5] and from literature [4] it is known that combustion noise is dominated by heat release effects and that all other source mechanisms are of minor importance. Assuming combustion at constant pressure and neglecting all mean flow effects q_e reduces to sources, which are related to heat release effects, non-isomolar combustion, heat flux and viscous effects. Adding up all these sources under the aforementioned restrictions the RHS of the pressure-density relation can be substituted by the total time derivative of the

3.4 CAA computation

For the CAA computation this proposed APE-System has been implemented into the PIANO (Perturbation Investigation of Aeroacoustic Noise) Code from the DLR (Deutsches Zentrum für Luft- und Raumfahrt e.V.).
The source terms on the right-hand side of the APE system has to be interpolated in time during the CAA computation. Using a quadratic interpolation method at least 25 points per period are required to achieve a sufficiently accurate distribution. Hence, the maximal resolvable frequency is $f_{max} = 1/(25\Delta t) = 800Hz$ since the LES solution comes with a time increment of $\Delta t = 5 \cdot 10^{-5}s$ [7]. This frequency is much smaller than the Nyquist frequency. The CAA code is based on the fourth-order DRP scheme of Tam and Webb [8] for the spatial discretization and the alternating LDDRK-5/6 Runge-Kutta scheme for the temporal integration [9]. At the far field boundaries a sponge-layer technique is used to avoid unphysical reflections into the computational domain.

Solving the APE system means to solve five equations (3D) for the perturbation quantities ρ', u', v', w' and p' per grid point and time level. No extra equations for viscous terms and chemical reaction need to be considered since these terms can be found on the RHS of the APE system and are provided by the LES within the source region. On the other hand the time step within the CAA computation can be chosen much higher than in the LES. This means, using a rough estimation, that the ratio of the computation times between LES and CAA is approximately $t_{LES}/t_{CAA} \approx 4/1$.

4 Results

Fig. 2 shows a snapshot of the acoustic pressure field in the streamwise center plane at the dimensionless time $t = 100$. The source region is evidenced by the dashed box. This computation was done on a 27-block domain using approximately 4×10^6 grid points, where the arrangement of the blocks is arbitrary provided that one block contains all acoustical sources.

The acoustic directivity patterns (Fig. 3) are computed for different frequencies on a circle in the $z = 0$ plane with a radius $R/D = 17$ whose center point is at $x = (10, 0, 0)$. The jet exit diameter is denoted by D. From 150° to 210° the directivity data is not available since this part of the circle is outside of the computational domain. In general an acoustic monopole behaviour with a small directivity can be observed since this circle is placed in the acoustic near field.

5 Conclusion

The APE system has been extended to compute noise generated by reacting flow effects. The heat release per unit volume, which is expressed in the total time derivative of the density, represents the major source term in the APE system when combustion noise is analyzed. The main combustion noise characteristic, i.e., the monopole nature caused by the unsteady heat release, could be verified. In the present work we have demonstrated that the extended APE System in conjunction with a hybrid LES/CAA approach and with the assumptions made, is capable of simulating an acoustic field of a reacting flow, i.e., of a non-premixed turbulent flame.

Acknowledgements

The authors would like to thank the Institute for Energy and Powerplant Technology from Darmstadt University of Technology for providing the LES data of the non-premixed flame.

References

[1] R. Ewert and W. Schröder. Acoustic perturbation equations based on flow decomposition via source filtering. J. Comput. Phys. 188 (2003), pp. 365-398

[2] R. Ewert and W. Schröder. On the Simulation of Trailling Edge Noise with a Hybrid LES/APE Method. J. Sound and Vibration (2004) 270, pp. 509-524

[3] W. C. Strahle. Some results in combustion generated noise. J. Sound and Vibration (1972) 23 (1), pp. 113-125

[4] D.G. Crighton, A.P. Dowling, J.E. Ffowcs Williams, M. Heckl and F.G. Leppington, Modern Methods in analytical acoustics, Lecture Notes, Springer-Verlag (1996)

[5] T. Ph. Bui, M. Meinke, W. Schröder, A Hybrid Approach to Analyze the Acoustic Field Based on Aerothermodynamics Effects. Proceedings of the joint congress CFA/DAGA '04, Strasbourg, March 22-25, 2004

[6] S. Kotake. On Combustion Noise related to chemical reactions. J. Sound and Vibration Vol. 42 (3), pp. 399-410

[7] M. Düsing, A. Kempf, F. Flemming, A. Sadiki and J. Janicka. Combustion LES for Premixed and Diffusion Flames. VDI-Berichte Nr. 1750, 21. Deutscher Flammentag, pp. 745 - 750, Cottbus, 2003

[8] C. K. W. Tam and J. C. Webb Dispersion-Relation-Preserving Finite Difference Schemes for Computational Acoustics. Journal of Comp. Physics 107 (1993), pp. 262-181

[9] F. Q. Hu, M. Y. Hussaini and J. L. Manthey. Low-dissipation and low-dispersion Runge-Kutta schemes for computational acoustics. J. Comput. Phys. 124 (1996), pp. 177-191

[10] M. Germano, U. Piomelli, P. Moin, and W. H. Cabot. A dynamic subgrid-scale viscosity model. Phys. Fluids A, Vol. 3, No. 7, 1991, pp. 1760–1765.

[11] N. P. Waterson. Development of a bounded higher-order convection scheme for general industrial applications. Project Report 1994-33, von Karman Institute, June 1994.

[12] M. Klein, A. Sadiki and J. Janicka. A digital filter based generation of inflow data for spatially developing direct numerical or large eddy simulations. J. Comput. Physics, Vol. 186, 2003, pp. 652–665.

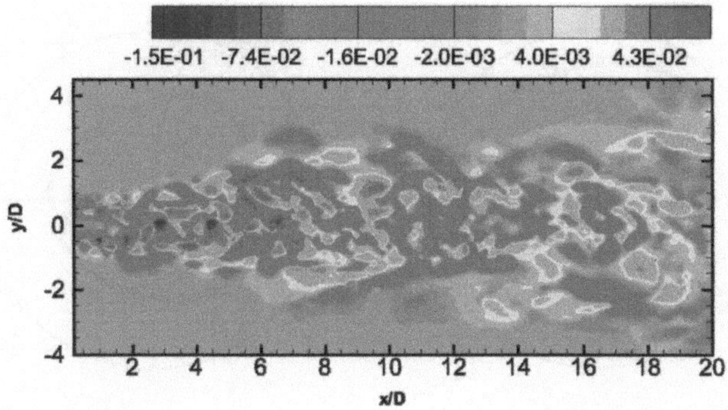

Figure 1 Contours of the total time derivative of the density $(D\rho/Dt)$ at $t = 100$ in the streamwise center plane

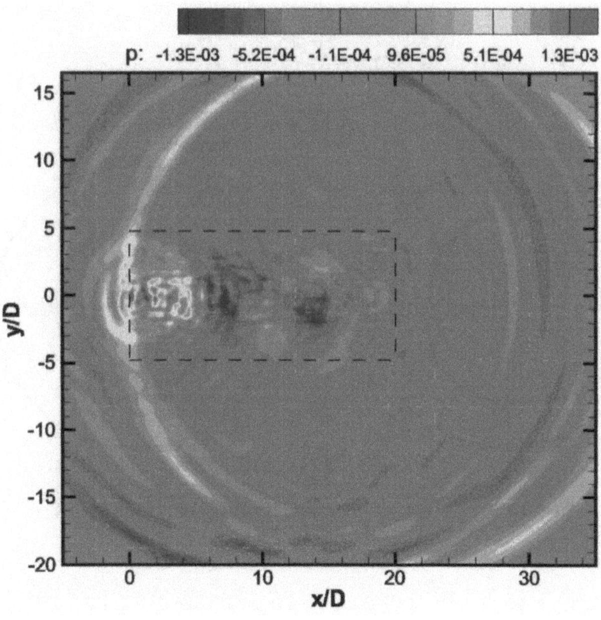

Figure 2 Pressure contours of the APE solution at $t = 100$ in the streamwise center plane

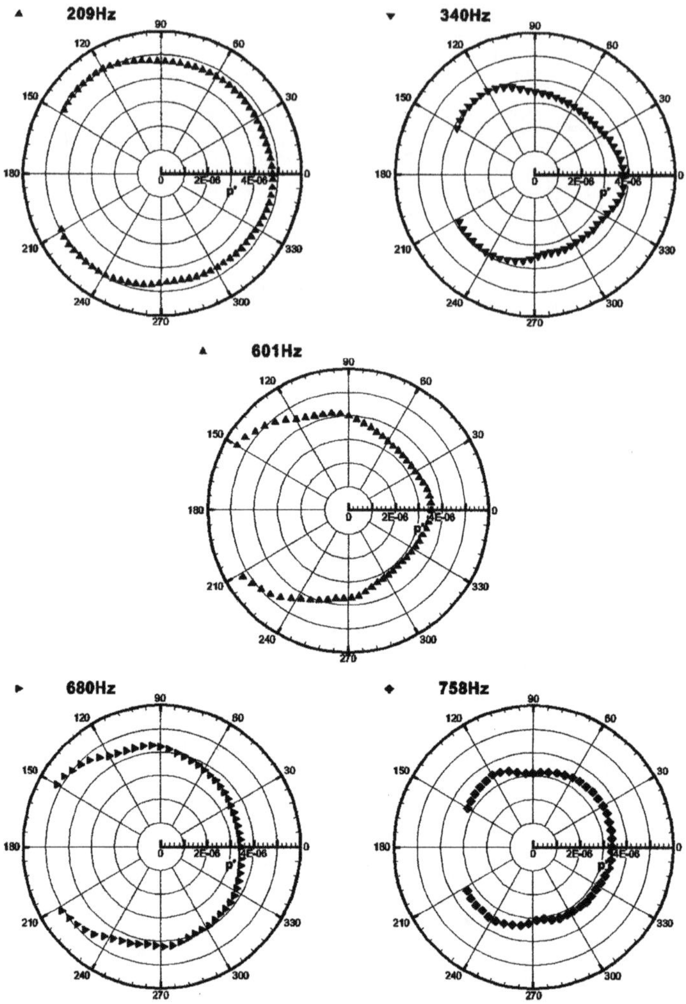

Figure 3 Directivity patterns for different frequencies

Toward efficient solution of the compressible Navier-Stokes equations

C.-C. Rossow
DLR, Institut für Aerodynamik und Strömungstechnik
Lilienthalplatz 7, D-38108 Braunschweig, Germany
cord.rossow@dlr.de

Summary

A fully implicit relaxation technique is developed to accelerate convergence of current multigrid solvers. The flux Jacobians are efficiently recomputed during iteration. The implicit operator allows increasing CFL numbers of the basic explicit scheme to O(100), and it properly addresses the stiffness in the discrete equations associated with highly stretched meshes. Compared to a well tuned, standard reference code, computation times are more than halved.

1. Introduction

Fast convergence is a prerequisite for efficient use of CFD in the design process, and multigrid has become a widespread acceleration technique. Methods as devised by Jameson et al. [5], where explicit Runge-Kutta time integration is combined with implicit residual smoothing and multigrid, represent a well balanced compromise of simplicity in the explicit relaxation scheme and implicit consideration of the cell aspect ratio. However, solving the Navier-Stokes equations at high Reynolds numbers, the resolution of thin boundary layers requires highly stretched meshes with very high cell aspect ratios. The large disparity in the corresponding spectral radii of the associated coordinate directions results in high stiffness of the discrete equations, thus severely deteriorating convergence of the numerical method.

In the present work, the conventional implicit residual smoothing is converted into a fully implicit operator with symmetric Gauss-Seidel iteration to address the stiffness problem. The associated memory overhead usually renders such methods prohibitive for large scale computations. To address this challenge, a new formulation of the necessary flux Jacobians is derived which allows efficient on-the-fly computation at every smoothing step, thus significantly reducing storage requirements. Algebraic expressions are simplified by exploiting a Mach number expansion of flux Jacobians. Transformation to primitive variables leads to economic evaluation of flux Jacobians on cell faces, and a fully implicit operator is constructed for implicitly smoothing residuals in the framework of a Runge-Kutta time stepping scheme. The implications of this technique are then assessed for computation of viscous, turbulent airfoil flow. Comparison with a standard approach for implicit smoothing in the same baseline code allows for direct estimation of the potential of the present method for convergence and efficiency improvement. To assess the applicability to high Reynolds number flows, a variation of Reynolds number by more than an order of magnitude is carried out.

2. Governing Equations

We consider the two-dimensional compressible Navier-Stokes equations. For a control volume fixed in time and space, the system of partial differential equations in integral form is given by:

$$\iint_{Vol} \frac{\partial \boldsymbol{W}}{\partial t} dV + \int_{S} \boldsymbol{F} \cdot \boldsymbol{n} \, dS = 0 \ , \tag{1}$$

where $\boldsymbol{W} = [\rho, \rho u, \rho v, \rho E]^T$ represents the vector of conservative variables, $\boldsymbol{F}$ is the flux-density tensor, and *Vol*, *S*, and $\boldsymbol{n}$ denote volume, surface, and outward facing normal of the control volume. The flux density tensor $\boldsymbol{F}$ may be split into an inviscid, convective part $\boldsymbol{F}_c$ and a viscous part $\boldsymbol{F}_v$:

$$\boldsymbol{F} = \boldsymbol{F}_c + \boldsymbol{F}_v \ , \tag{2}$$

where $\boldsymbol{F}_c$ and $\boldsymbol{F}_v$ are given by

$$\boldsymbol{F}_c = \begin{bmatrix} \rho \boldsymbol{q} \\ \rho u \boldsymbol{q} + p \boldsymbol{i}_x \\ \rho v \boldsymbol{q} + p \boldsymbol{i}_y \\ \rho H \boldsymbol{q} \end{bmatrix} \ , \ \boldsymbol{F}_v = \begin{bmatrix} 0 \\ \tau_{xx} \boldsymbol{i}_x + \tau_{xy} \boldsymbol{i}_y \\ \tau_{xy} \boldsymbol{i}_x + \tau_{yy} \boldsymbol{i}_y \\ \left(u\tau_{xx} + v\tau_{xy} + k\frac{\partial T}{\partial x} \right) \boldsymbol{i}_x + \left(u\tau_{xy} + v\tau_{yy} + k\frac{\partial T}{\partial y} \right) \boldsymbol{i}_y \end{bmatrix} \ , \tag{3}$$

$\boldsymbol{q}$ is the velocity vector with Cartesian components u, v, and $\boldsymbol{i}_x$, $\boldsymbol{i}_y$ denote the unit vectors in direction of the Cartesian coordinates x and y. ρ, p, H, T represent density, pressure, total specific enthalpy, and temperature, k is the coefficient of thermal heat conductivity, and $\tau_{xx}, \tau_{yy}, \tau_{xy}$ are the viscous stress tensor components.

In order to close the system given by eq. (1), the equation of state

$$\frac{p}{\rho} = R \cdot T \tag{4}$$

is used with R as specific gas constant.

3. Basic Solution Scheme

The basic solution scheme is a cell centered, finite volume method as employed in Ref. [9]. A semi-discrete form of the governing equations (1) may be written as:

$$\frac{\partial \boldsymbol{W}}{\partial t} + \frac{1}{Vol} \sum_{k=1}^{K} \boldsymbol{F}_{\perp,k} \, S_k = 0 \ , \tag{5}$$

where $F_{\perp}$ is the flux density vector corresponding to the direction normal to the cell face, and K represents the maximum number of cell faces of the corresponding control volume with k as running index. Using Flux Difference Splitting (FDS) [8], the convective part of the flux density vector F_c normal to a cell interface reads:

$$F_c = \frac{1}{2}\left(F^L + F^R\right) - \frac{1}{2}|A| \cdot \Delta W \quad , \tag{6}$$

where F^L and F^R are the left and right states of the inviscid flux density vector normal to the cell interface, and A is the corresponding flux Jacobian. The expression ΔW denotes differences in conservative variables on the left side L and right side R of a cell interface, giving $\Delta W = W^R - W^L$. Similarly to Ref. [9], in the present work an implementation is used where the expression $|A| \cdot \Delta W$ is expanded in terms of the interface normal Mach number M_0, with M_0 defined as [9]:

$$M_o = \min\left(|\mathrm{M}|, 1\right) \cdot sign\,(M) \quad . \tag{7}$$

Discretization of the viscous terms F_v is performed by central difference operators, and for implementation details the reader is referred to Refs. [7], [9].

Time integration of eq. (5) is achieved with an explicit 5-stage Runge-Kutta scheme. To accelerate convergence toward steady state, the explicit time stepping is augmented by implicit residual smoothing, which allows an increase of the CFL number of the basic scheme by a factor of 2-3 [6]. Time integration may be further enhanced by employing multigrid following the ideas of Jameson [5]. The influence of turbulence is modeled according to Baldwin and Lomax [2]. The basic outline of the framework for time integration may be found in Ref. [7].

4. Derivation of Present Approach

A fully implicit formulation of eq. (5) may be written as:

$$\frac{\partial W}{\partial t} + \frac{1}{Vol}\sum_{k=1}^{K} F_{\perp,k}^{(n+1)}\, S_k = 0 \quad , \tag{8}$$

where $F_{\perp}^{(n+1)}$ is evaluated at the new time level (n+1). Linearizing $F_{\perp}^{(n+1)}$ about the current time level (n) leads to

$$F_{\perp}^{(n+1)} = F_{\perp}^{(n)} + \left(\frac{\partial F_{\perp}}{\partial W}\right)\delta W = F_{\perp}^{(n)} + A_{\perp}\delta W \quad , \tag{9}$$

with δW defining the time difference of conservative variables and A_n the flux Jacobian in normal direction:

$$\delta W = W^{(n+1)} - W^{(n)} \quad , \quad A_{\perp} = \frac{\partial F_{\perp}}{\partial W} \quad . \tag{10}$$

Expressing the time derivative in eq. (8) by the forward finite difference, substituting eq. (9) into eq. (8), and rearranging leads to

$$\left(1 + \frac{\delta t}{Vol}\sum_{k=1}^{K} A_{\perp,k} S_k\right)\delta W = -\frac{\delta t}{Vol}\sum_{k=1}^{K} F_{\perp,k}^{(n)} S_k = R^{(n)} \quad , \tag{11}$$

where $R^{(n)}$ has been introduced to denote the residual at current time level (n), and δt is the local time step. The matrix $A_{\perp}$ has real eigenvalues and may be split into two matrices $A_{\perp}^{+}$ and $A_{\perp}^{-}$, where

$$A_{\perp}^{+} = 0.5\left(A_{\perp} + |A_{\perp}|\right) \quad , \qquad A_{\perp}^{-} = 0.5\left(A_{\perp} - |A_{\perp}|\right) \quad . \tag{12}$$

With definition (12), eq. (11) may be rewritten as a point implicit scheme:

$$\left(1 + \frac{\delta t_{i,j}}{Vol_{i,j}}\sum_{k=1}^{K} A_{\perp,k}^{+} S_k\right)\delta W_{i,j} = R_{i,j}^{(n)} - \frac{\delta t_{i,j}}{Vol_{i,j}}\sum_{k=1}^{K} A_{\perp,k}^{-} S_k \, \delta W_{NB} \quad , \tag{13}$$

where indices i,j denote the current cell and NB are all direct neighbors of cell i,j.

The challenge of eq. (13) is the memory requirement for storing the matrices $A_{\perp}^{+}$ and $A_{\perp}^{-}$. For solution of a 2D (3D) problem, these are 4x4 (5x5 in 3D) matrices, requiring storage of 32 (50) variables per cell face. Instead of storing these values, in the present work the matrix components will therefore be recomputed whenever needed. For efficient evaluation of matrices $A_{\perp}^{+}$ and $A_{\perp}^{-}$, the formalism of the Mach number expansion for the absolute Jacobian according Ref. [9] will employed. Eq. (11) is first transformed to the set of primitive variables $U = [\rho, p, u, v]^T$:

$$\frac{\delta U}{\delta t} + \frac{1}{Vol}\sum_{k=1}^{K} P_{\perp}\delta U \, S_k = -\frac{\partial U}{\partial W}\frac{1}{Vol}\sum_{k=1}^{K} F_{\perp}^{(n)} \, S_k \quad , \tag{14}$$

with $P_{\perp}$ as the analog of the normal flux Jacobian in primitive variables:

$$P_{\perp} = \frac{\partial U}{\partial W} A_{\perp} \frac{\partial W}{\partial U} = \frac{\partial U}{\partial W}\left(A_{\perp}^{+} + A_{\perp}^{-}\right)\frac{\partial W}{\partial U} = P_{\perp}^{+} + P_{\perp}^{-} \quad . \tag{15}$$

Using the definitions of eq. (12) and the formalism of Mach number expansion [9], performing the multiplications with the transformation Jacobians of eq. (15), multiplying with time changes in primitive variables δU, and sorting terms gives

$$P_\perp^+ \cdot \delta U = \begin{bmatrix} (q_\perp + |q_\perp|)\delta\rho & + & \frac{1}{c}(1-|M_0|)\delta p & + & \rho(1+M_0)\delta q_\perp \\ (q_\perp + |q_\perp|)\delta p & + & \frac{h^*}{c}(1-|M_0|)\delta p & + & \gamma p(1+M_0)\delta q_\perp \\ (q_\perp + |q_\perp|)\delta u & + & n_x\frac{1}{\rho}(1+M_0)\delta p & + & n_x c(1-|M_0|)\delta q_\perp \\ (q_\perp + |q_\perp|)\delta v & + & n_y\frac{1}{\rho}(1+M_0)\delta p & + & n_y c(1-|M_0|)\delta q_\perp \end{bmatrix}, \quad P_\perp^- \cdot \delta U = \begin{bmatrix} (q_\perp - |q_\perp|)\delta\rho & - & \frac{1}{c}(1-|M_0|)\delta p & + & \rho(1-M_0)\delta q_\perp \\ (q_\perp - |q_\perp|)\delta p & - & \frac{h^*}{c}(1-|M_0|)\delta p & + & \gamma p(1-M_0)\delta q_\perp \\ (q_\perp - |q_\perp|)\delta u & + & n_x\frac{1}{\rho}(1-M_0)\delta p & - & n_x c(1-|M_0|)\delta q_\perp \\ (q_\perp - |q_\perp|)\delta v & + & n_x\frac{1}{\rho}(1-M_0)\delta p & - & n_y c(1-|M_0|)\delta q_\perp \end{bmatrix} \tag{16}$$

where γ, h, c denote the ratio of specific heats, specific enthalpy, and speed of sound, respectively, and $\delta q_\perp$ and h^* are defined as

$$\delta q_\perp = n_x \delta u + n_y \delta v \quad , \quad h^* = (\gamma - 1)h \tag{17}$$

with n_x and n_y as the components of the outward facing normal $\boldsymbol{n}$.

The definitions for $P_\perp^+ \delta U$ and $P_\perp^- \delta U$ by eq. (16) allow an implementation of the implicit scheme of eq. (13) with a minimum of computational effort: only the expressions $|q_\perp|, M_0, (1-|M_0|)$ are pre-computed and stored for each cell face, all other components can efficiently be re-computed whenever necessary. The contributions of the viscous flux Jacobians can easily be incorporated with the definitions outlined in Ref. [1] for primitive variables.

The implicit scheme of eq. (13) is implemented into the framework of the basic code analogously to implicit residual smoothing: in each Runge-Kutta stage, prior to updating conservative variables, the explicitly evaluated residuals are transformed to residuals in primitive variables to form the right hand side of eq. (13). Solution of eq. (13) with three symmetric Gauss-Seidel (SGS) sweeps through the computational domain yields new expressions for residuals, which are then transformed to conservative residuals and used to update conservative variables.

5. Computational Results

The proposed method was used to compute viscous, turbulent flow around the RAE 2822 airfoil. From the investigations of Cook, McDonald and Firmin [3], the subsonic Case 1 $(M_\infty = 0.679, \alpha = 1.98°, \mathrm{Re} = 5.700.000)$ and the transonic Case 9 $(M_\infty = 0.73, \alpha = 2.31°, \mathrm{Re} = 6.500.000)$ were selected. Structured C-meshes were used, where for the multigrid algorithm coarse meshes were created by successively omitting every second grid line. Computations were performed on a SGI Octane workstation using a single 360 MHz processor.

This investigation is focused on enhancement of time integration, therefore always an identical space discretization is employed. For fair comparison, all computations were started from free stream on the finest mesh. Pressure distributions obtained on a reference mesh with 256 cells along the airfoil, 32 cells in the wake region, and 64 cells in normal direction are shown in Figs. 1 and 2. For the investigated Cases 1 and 9, a fairly well agreement with experimental results was achieved.

Figs. 3 and 4 show convergence histories for two different time integration strategies employed on the reference mesh. In all cases, a 5-stage Runge-Kutta scheme with stage coefficients for a second order upwind discretization was used. The first strategy employs the standard implicit residual smoothing of Ref. [6], denoted by rk5-s, with admissible CFL number of 7.5. For the second strategy as proposed in this study, denoted by rk5-i, with smoothing according to eq. (14), the CFL number was set to 16 for the first 8 multigrid cycles, and then fixed to 160. The scheme achieved a convergence improvement by a factor of 9 compared to the standard scheme, and CPU time was reduced by a factor of about 2.5-2.8. The rate of convergence, expressed by the ratio of residuals at end and begin of iteration to the power of the reciprocal number of iterations, improved from 0.98 for the smoothing of Ref. [6] to about 0.85 for the present method.

To investigate whether the proposed method alleviates stiffness associated with high aspect ratio cells, a variation of Reynolds-number by more than an order of magnitude was carried out. The subsonic test case was used, since here it was possible to use the second order space discretization without any limiter functions, which otherwise may lead to stalling after convergence by 3 or 4 orders of magnitude. Computational meshes were taken from Ref. [4], where the effect of Reynolds number variation was investigated for turbulence modeling. All meshes have a C-topology with 312 cells along the airfoil, 56 cells in the wake region, and 88 cells in normal direction. The meshes were adapted to the corresponding Reynolds number, leading to cell aspect ratios varying from about 3.000 to over 50.000. As can be seen from the convergence histories in Fig. 5, the effect of cell aspect ratio when varying the Reynolds number from 5.700.000 to 100.000.000 is only moderate: the number of multigrid cycles required for convergence remains at O(100) even for the highest Reynolds-number, only increasing by a factor of 2 with Reynolds-number. In Ref. [4], where similar computations were made using a standard scheme, the number of cycles is already O(1000) and then more than doubles with increasing Reynolds-number.

6. Conclusion

A computational approach was derived which addresses the stiffness problem associated with high aspect ratio cells for viscous flow computations by introducing a fully implicit operator for smoothing of residuals in the framework of Runge-Kutta time stepping. To avoid memory overhead usually associated with such methods, the components of the Jacobians are recomputed during iteration. Efficient evaluation of the corresponding terms is achieved by exploiting a Mach number expansion of the flux Jacobians as outlined in Ref. [9]. To solve the implicit equation, symmetric Gauss-Seidel iteration with three sweeps through the computational domain was employed, and in combination with Runge-Kutta time stepping CFL numbers of O(100) could be used. The method was applied to solve for compressible, turbulent airfoil flow, and convergence rates from 0.8 to 0.85 were achieved. In comparison to well tuned common multigrid methods with Runge-Kutta time stepping and implicit residual smoothing, reductions in CPU time by more than a factor of 2.5 were realized. Variation of Reynolds number by more than an order of magnitude showed only moderate influence on convergence: the number of multigrid cycles required for convergence increased only by a factor of 2. This confirms that the stiffness in the discrete equations associated with very high cell aspect ratios is properly addressed by the proposed method. It is worth noting that in contrast to commonly employed implicit smoothing techniques, the present method is not restricted to structured meshes, since the fully implicit operator does not rely on information obtained from curvilinear coordinates.

References

1. Abarbanel, S. and Gottlieb, D., "Optimal Splitting for Two and Three Dimensional Navier-Stokes Equations with Mixed Derivatives", ICASE Report No. 80-6, Feb. 18th, 1980.

2. Baldwin, B. and Lomax, H., "Thin Layer Approximation and Algebraic Turbulence Model for Separated Turbulent Flows", AIAA-Paper 78-257, 1987.

3. Cook, P. H., McDonald, M. A., and Firmin, M. C. P., "Aerofoil RAE2822 Pressure Distributions and Boundary Layer and Wake Measurements", AGARD-AR-138, 1979.

4. Faßbender, J., "Improved Robustness of Numerical Simulation of Turbulent Flows around Civil Transport Aircraft at Flight Reynolds Numbers", Ph.D. Thesis, Technical University of Braunschweig, http://opus.tu-bs.de/opus/volltexte/2004/579, 2004.

5. Jameson, A., "Multigrid Algorithms for Compressible Flow Calculations", in *Second European Conference on Multigrid Methods, Cologne, 1985*, ed. W. Hackbusch and U. Trottenberg, [Lecture Notes in Mathematics, Vol. 1228], Springer Verlag, Berlin, 1986.

6. Martinelli, L., "Calculations of Viscous Flow with a Multigrid Method", Ph.D. Thesis, Princeton University, 1987.

7. Radespiel, R., Rossow, C.-C., and Swanson, R.C., "An Efficient Cell-Vertex Multigrid Scheme for the Three-Dimensional Navier-Stokes Equations", *AIAA J.*, Vol. 28, No.8, 1998, pp.423-459.

8. Roe, P.L., "Approximate Riemann Solvers, Parameter Vectors and Difference Schemes", *J. Comput. Phys.*, Vol. 43, 1981, pp.357-372.

9. Rossow, C.-C., "A Flux Splitting Scheme for Compressible and Incompressible Flows", *J. Comput. Phys.*, Vol. 164, 2000, pp.104-122.

Figures

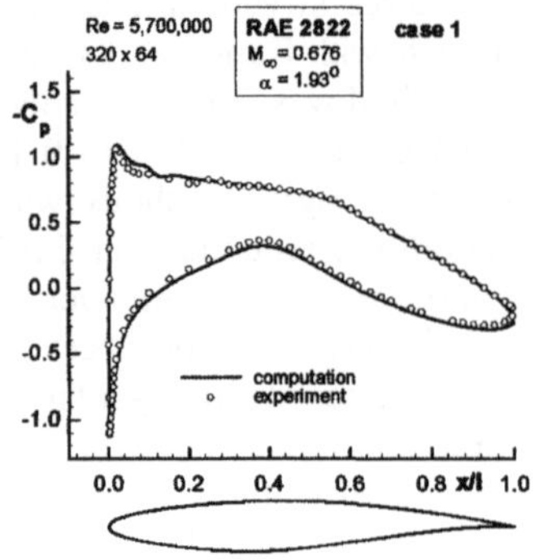

Fig. 1: Pressure distribution for Case 1

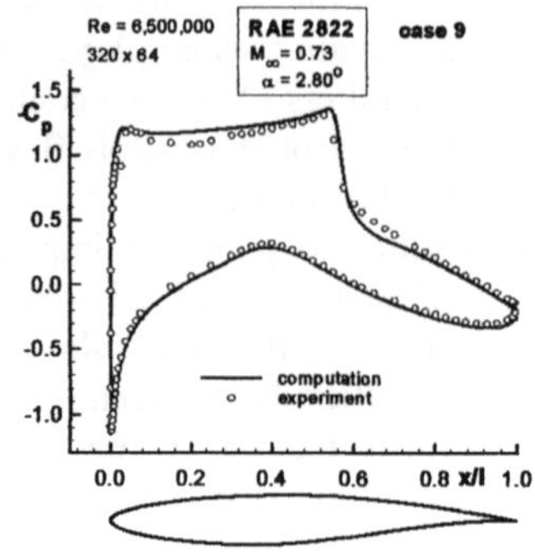

Fig. 2: Pressure distribution for Case 9

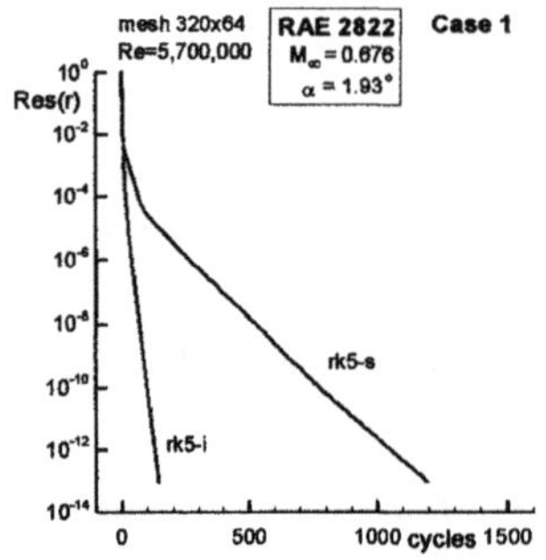

Fig. 3: Convergence histories for Case 1

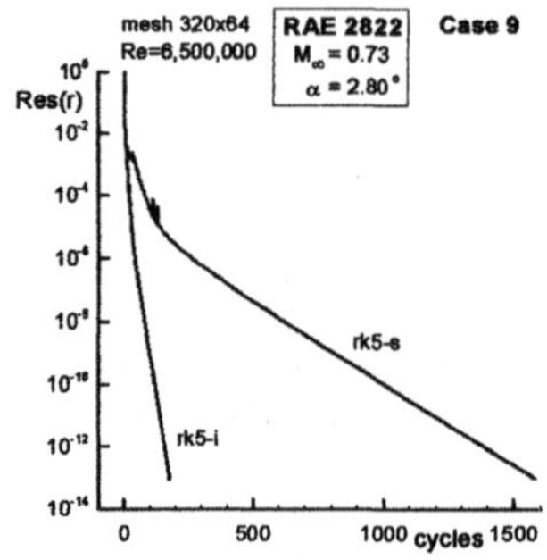

Fig. 4: Convergence histories for Case 9

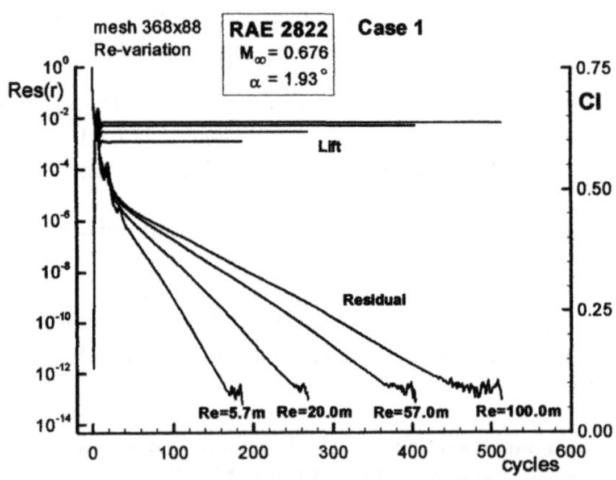

Fig. 5: Convergence histories for Reynolds number variation

Direct Numerical Simulation of a Turbulent Flow Using a Spectral/hp Element Method

Andrei Shishkin and Claus Wagner

DLR - Institute for Aerodynamics and Flow Technology
Bunsenstr.10, D-37073 Göttingen, Germany
e-mail: Andrei.Shishkin@dlr.de, Claus.Wagner@dlr.de

Summary

Direct numerical simulation (DNS) of incompressible turbulent pipe flow was carried out on unstructured grids for a Reynolds number based on the friction velocity and the pipe diameter of $Re_\tau = 360$ using the spectral/hp element method (SEM) by Karniadakis and Sherwin [3]. The main objective was to investigate the computational aspects of this DNS with respect to accuracy, CPU time and memory requirements. DNS results of evaluated statistical moments of up to fourth order agree well with data from the literature. A conducted performance study reveals the computational requirements for DNS of turbulent flows using this SEM.

1 Introduction

Numerical simulations of turbulent flows by means of Direct Numerical Simulation (DNS) and Large Eddy Simulation (LES) have been the subject of intensive research during the last years. These techniques require accurate numerical methods which do not produce significant artificial viscosity. Spectral methods which provide accurate spatial discretization have been used successfully in the past mainly for simple geometries. In these methods the flow variables are expanded in terms of smooth (infinitely differentiable), mostly orthogonal trial functions, which are defined over the whole computational domain. Typical trial functions are the Fourier series components (domain with periodic boundary conditions) or Chebyshev and Legendre polynomials (nonperiodic boundary condition). Principally, DNS and LES of turbulent flows in complex geometries can be conducted using the spectral/hp element method (SEM) by Karniadakis and Sherwin [3]. In this method high-order polynomials are used to approximate the flow fields within finite elements to provide spectral accuracy on unstructured meshes. The advantage is that this SEM inherits the geometrical flexibility of finite element methods and the exponentially accurate polynomial resolution of spectral methods.

The main objective of this work is to use the SEM for a full 3D DNS of a turbulent flow in a computational domain with curved boundaries (which is a challenge for SEM) on an unstructured mesh and to compare the obtained results with those from the literature. The turbulent pipe flow was selected as a test case since it meets the above requirements although the geometry is fairly simple for the SEM. The other advantage is that turbulent pipe flows have been intensively investigated in numerical and experimental studies. Some of the results are collected in the AGARD Advisory report No. 345 [1, Ch.5].

The structure of the paper is the following. The computational domain and parameters are presented in section 2. Section 3 contains a short description of the used SEM, i.e the $\mathcal{N}\varepsilon\kappa\mathcal{T}\alpha r$ code by Karniadakis and Sherwin [3]. In Section 4 CPU time and storage requirements of the SEM are compared with those of the second order accurate finite volume method (FVM) by Wagner and Friedrich [5]. In this comparison the requirements of FVM serve only as a reference since the method is only applicable in simple computational domains due to the use of Fourier transformations in streamwise and azimuthal directions. Finally, in Section 5 the mean streamwise velocity $\langle u_z \rangle$, the rms velocity fluctuations $u_{\alpha,rms}$, the skewness S_α and flatness F_α factors ($\alpha = z, \varphi, r$) are compared to turbulent pipe flow data by Loulou *et al* [1, 4] who used a B-spline spectral method (B-spline SM) for their DNS and by Wagner and Friedrich [5]. Further, experimental data from Laser Doppler Anemometry (LDA) and Particle Image Velocimetry (PIV) measurements by Durst *et al* [2] and by Westerweel *et al* [8], respectively, are included.

2 Computational details

We consider a turbulent pipe flow for a Reynolds number, based on friction velocity $\hat{u}_\tau$, pipe diameter $\hat{D}$ and kinematic viscosity $\hat{\nu}$, of $Re_\tau = \frac{\hat{u}_\tau \hat{D}}{\hat{\nu}} = 360$, where $(\hat{\cdot})$ stands for dimensional quantities. The computational domain Ω is a pipe section of diameter D and length $L = 5D$. For an incompressible Newtonian fluid the flow is governed by the dimensionless Navier-Stokes equations

$$\frac{\partial \boldsymbol{u}}{\partial t} + (\boldsymbol{u} \cdot \nabla)\boldsymbol{u} = -\nabla p + \frac{1}{Re_\tau}\nabla^2 \boldsymbol{u} + \boldsymbol{F}, \qquad \nabla \cdot \boldsymbol{u} = 0, \tag{1}$$

where $\boldsymbol{u}$ denotes the velocity vector field, p - the pressure field, $\boldsymbol{F}$ - the forcing vector. For a fully developed turbulent pipe flow the driving force $\boldsymbol{F}$ corresponds to the mean pressure gradient in streamwise z-direction, $\langle \frac{\partial p}{\partial z} \rangle = -4$. At the solid wall $\mathbf{u}|_{\partial\Omega} = 0$ enforces the impermeability and no-slip boundary conditions. Furthermore, periodical boundary conditions are applied in streamwise direction. The simulations were started from an initial turbulent field which was obtained in a DNS using a second order accurate FVM.

3 Spectral/hp element method for incompressible flows

The $\mathcal{N}\varepsilon\kappa\mathcal{T}\alpha r$ code exploits structured and unstructured conforming meshes or a combination of both. Considering the generation of 3D meshes, this admits any combination of hexagonal, tetrahedral and prismatic elements provided that the corresponding faces of contiguous elements coincide.

The velocity and pressure fields are sought in terms of series expansions. The velocity field, for example, is expressed as

$$u(\boldsymbol{x}, t) = \sum_i \hat{u}_i(t)\varphi_i(\boldsymbol{x}), \tag{2}$$

where $\varphi_i(\boldsymbol{x})$ denotes an i-th mode (base function) and $\hat{u}_i(t)$ – the corresponding expansion coefficient. Each mode $\varphi_i(\boldsymbol{x})$ is a piecewise polynomial function supported by one element of the mesh. The system $\{\varphi_i(\boldsymbol{x})\}$ is constructed using the tensor product of the polynomials, orthogonal over a simplex, and the classical Jacobi polynomials (for details see [3, Ch.3]).

The solution of Eq. (1) is obtained using the splitting method by Karniadakis and Sherwin [3, p.247] which combines semi-implicit time integration using a forward Adams scheme [3, p.193] with the enforcement of the continuity equation. This leads to a Poisson equation for the pressure field and a Helmholtz equation for the velocity field which are solved with a preconditioned conjugate gradient method using the Schur complement (static condensation) technique.

4 Performance study

We construct an unstructured mesh consisting of 79860 prisms presented in Fig. 1. The elements are uniformly distributed along the axis and non-uniformly along the pipe radius according to a quadratic law to provide higher spatial resolution in the vicinity of the wall. The highest order of the expansion polynomials is taken equal to 3 for both components of the tensor product, resulting in 40 modes within each element.

In the spectral/hp element approach the combination of two parameters, i.e. the characteristic size h of the elements and the order p of the polynomial modes, defines the mesh resolution. In order to compare the spatial resolution achieved in SEM and FVM simulations it seems reasonable to aggregate both h and p characteristics as follows. Consider the number M_{df} of degrees of freedom of the expansions (2), i.e. the number of independent global expansion modes, to calculate the average value $M_{df/el}(e)$ of degrees of freedom for each element e. The ratio $\delta(e) = \frac{M_{df/el}(e)}{v(e)}$, where $v(e)$ is the volume of the element e, serves as a measure of the spatial resolution of the element e. Finally, we define a piecewise constant function $\delta(\boldsymbol{x}) = \delta(e)$, $\boldsymbol{x} \in e$, for each point $\boldsymbol{x}$ of the computational domain Ω, which indicates the quality of spatial resolution. Considering a cylindrical domain, the averaged functions $\delta(r,\varphi) = \frac{1}{L}\int \delta(\boldsymbol{x})dz$ and $\delta(r) = \frac{1}{2\pi L}\int\int \delta(\boldsymbol{x})dzd\varphi$, provide the appropriate spatial resolution characteristics of the mesh.

The well resolved DNS by Wagner and Friedrich [5] who used a non-equidistant cylindrical structured grid with $N_z = 256$, $N_\varphi = 128$, $N_r = 70$ grid points in z, φ and r-directions, respectively, serves as reference case. For this mesh and the FVM the function $\delta(r,\varphi)$ can be determined taking $M_{df/el} = 1$.

In Fig. 2 profiles of the mesh resolution functions $\delta(r,\varphi)$ of the SEM and the above mentioned FVM are presented. It is observed that the FVM mesh has a significantly higher resolution near the centerline due to its singularity at the axis. The unstructured mesh has a comparable resolution beyond the center zone and even higher resolution than the FVM mesh in the viscous sublayer near the rigid wall. Therefore, it is concluded that the constructed SEM mesh is fine enough to resolve all relevant turbulent scales.

The spectral element simulations were conducted in a parallel mode using 10 processors of the 32-bit Linux Cluster machine. The main characteristics of the meshes used and the computational requirements are summarized in Table 1. It is observed that the SEM needs 80 times more CPU time and 60 times more Random Access Memory (RAM) than the FVM. It must be noted that in a similar performance study the second order accurate FVM for curvilinear staggered grids by Wesseling *et al* [7] required 358 times more CPU time and 32 times more RAM than the FVM for structured cylindrical grid (see Wagner and Dallmann[6]).

5 Numerical results

In Fig. 3-6 the turbulence statistics obtained in the DNS with the SEM (SEM DNS) are compared with results of different authors.

In Fig. 3 profiles of the mean axial velocity obtained in the SEM DNS are depicted together with those obtained in the FVM DNS by Wagner and Friedrich [5], in the B-spline SM DNS by Loulou *et al* [4] and in the PIV measurements by Westerweel *et al* [8]. The comparison reveals a good agreement to the reference data. Fig. 4 displays the root-mean-square (rms) values of the fluctuating velocity components for the SEM DNS in comparison with the FVM DNS by Wagner and Friedrich [5], the B-spline SM DNS by Loulou *et al* [4], the PIV measurements by Westerweel *et al* [8] and the LDA measurements by Durst *et al* [2]. Again, an exellent agreement of all the data is observed. The skewness and the flatness factors (Figures 6,5) are compared with the numerical data by Wagner and Friedrich [5] and by Loulou *et al* [4] and the measurements by Durst *et al* [2]. There is a certain variance in the behavior of the skewness S_r and the flatness F_r factors of the radial velocity component. Contrary to the low-order FVM DNS, the skewness factors S_r of the SEM DNS and the B-spline SM DNS change their sign near the wall in an agreement with the measurements. Fig. 6 exhibits a discrepancy in the flatness factor F_r obtained in the numerical simulations and in the experiments. The measurements provide lower values of F_r than the simulations near the wall. Presently this contradiction is an open problem and needs to be investigated precisely in the future. We note also that the SEM DNS data present a lower near-wall value of F_r than other considered numerical methods.

6 Conclusion

In the paper we investigated the computational aspects of DNS of a turbulent pipe flow using the spectral/hp element method. To this end, we generated an unstructured mesh consisting of 79860 prisms and conducted the simulation using the SEM code $\mathcal{N}\varepsilon\kappa\mathcal{T}\alpha r$ [3].

The statistical results obtained in the SEM DNS are in a good agreement with other numerical and experimental data. Particularly, the skewness and the flatness factors of SEM DNS agree better with the experimental data than those of FVM DNS. As compared with the FVM for simple geometries the SEM needs substantially more CPU time and RAM. Note that the considered version of SEM does not exploit the symmetry properties of the computational domain. Therefore, the SEM can be used for turbulent flow simulations in complex geometries with about the same computational expenses.

Acknowledgments

We are grateful to Dr. S.Sherwin for providing the 3D version of $\mathcal{N}\varepsilon\kappa\mathcal{T}\alpha r$.

References

[1] AGARD Advisory Report No. 345. *"A Selection of Test Cases for the Validation of Large-Eddy Simulations of Turbulent Flows"*, 1998

[2] F. Durst, J. Jovanovic, J. Sender. "Detailed measurements of the near wall region of turbulent pipe flows". Proc. 9th Symp. on Turbulent Shear Flows, Kyoto, Japan, August 16-18, 2/2/1–2/2/6 (1993)

[3] G.E. Karniadakis and S.J. Sherwin. *"Spectral/hp Element Method for CFD"*. Oxford University Press, 1999.

[4] P. Loulou, R. Moser, N. Mansour, B. Cantwell. *"Direct simulation of incompressible pipe flow using a b-spline spectral method"*. Technical Report TM 110436, NASA, 1997.

[5] C. Wagner, R. Friedrich. "On the turbulence structure in solid and permeable pipes". International Journal of Heat and Fluid Flow, **19**(1998), 459-469.

[6] C. Wagner, U. Ch. Dallmann. "A direct Navier-Stokes solver for turbulent flows over round steps", in: W. Nitsche, R. Hilbig (Eds.), 1999, *New Results in Numerical and Experimental Fluid Mechanics, Contributions to the 11th AG-STAB/DGLR Symposium, Berlin, Germany, 1998*, Vieweg-Verlag

[7] P, Wesseling, A. Segal, J.J.I.M. van Kan, C.W. Oosterlee, C.G.M. Kassels. "Finite volume discretization of the incompressible Navier-Stokes equations in general coordinates on staggered grids". Comp. Fluid Dynamics Journal, **16**(1992), 27-33

[8] J. Westerweel, R.J. Adrian, J.G.M. Eggels, F.T.M. Nieuwstadt. "Measurements with particle image velocimetry on fully developed turbulent pipe flow at low Reynolds number". Proc. of the 6th Int. Symp. on Applications of Laser Tech. to Fluid Mechanics, Lisbon, Portugal, July 20-23, 1993

Table 1 Computational requirements of a spectral/hp element method (SEM) and a finite volume method (FVM) for structured cylindrical grid in DNS of turbulent pipe flow.

	N_{el}	M_{df}	number of proc.	memory usage	memory/ N_{el}	memory/ M_{df}	CPU-time/M_{df} per time step
SEM	79860	1045770	10	8030 Mb	103 Kb	7.86 Kb	$9.8 \cdot 10^{-4}$ sec
FVM	2293760	2293760	1	131.25 Mb	60 byte	60 byte	$1.22 \cdot 10^{-5}$ sec

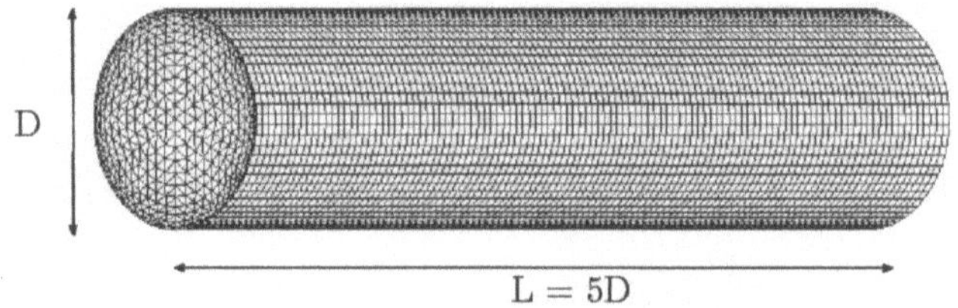

Figure 1 Unstructured mesh for the spectral/hp element method.

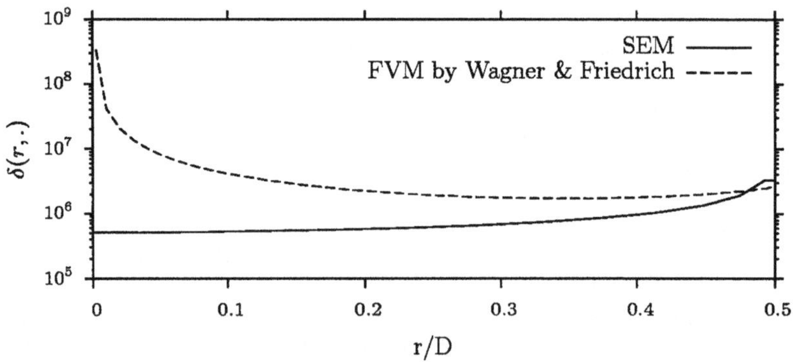

Figure 2 The profiles of the mesh resolution function $\delta(r,\varphi)$ (see Section 4) for the spectral/hp mesh (SEM) and the structured cylindrical non-equidistant grid ($N_z = 256$, $N_\varphi = 128$, $N_r = 70$)(Wagner and Friedrich [5]).

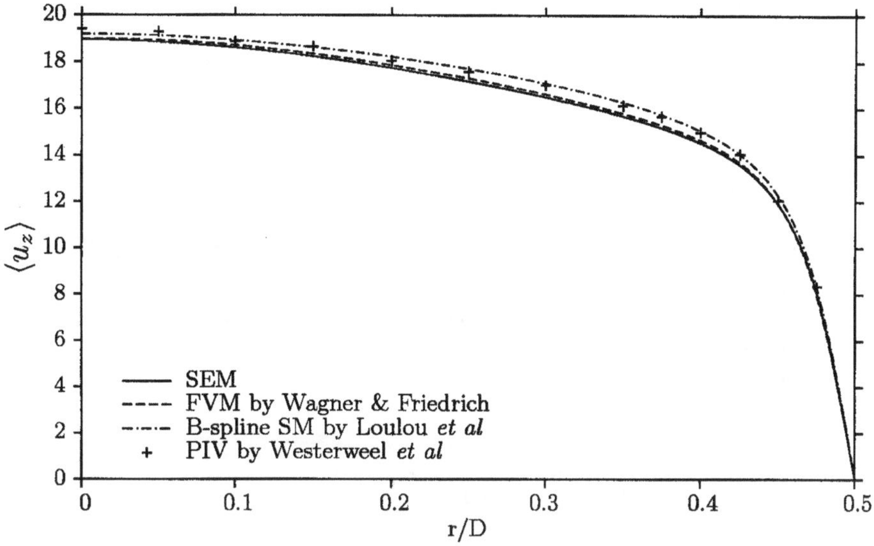

Figure 3 Mean streamwise velocity $\langle \bar{u}_z \rangle$ profiles obtained in the DNS of turbulent pipe flow using the spectral/hp element method, the FVM DNS by Wagner and Friedrich [5], the B-spline SM DNS by Loulou *et al* [4] and the PIV measurements by Westerweel *et al* [8].

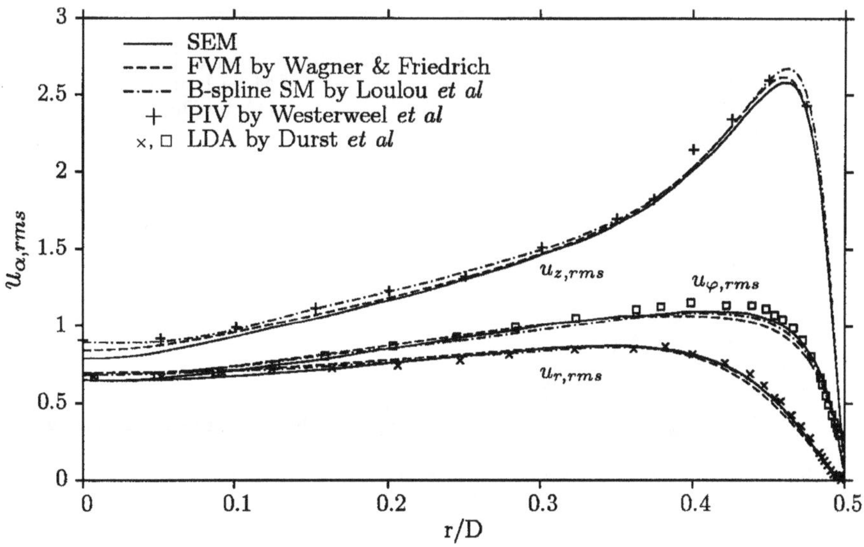

Figure 4 rms-velocities obtained in the DNS of turbulent pipe flow using the spectral element method, the FVM DNS by Wagner and Friedrich [5], the B-spline SM DNS by Loulou *et al* [4], the PIV measurements by Westerweel *et al* [8] and the LDA measurements by Durst *et al* [2].

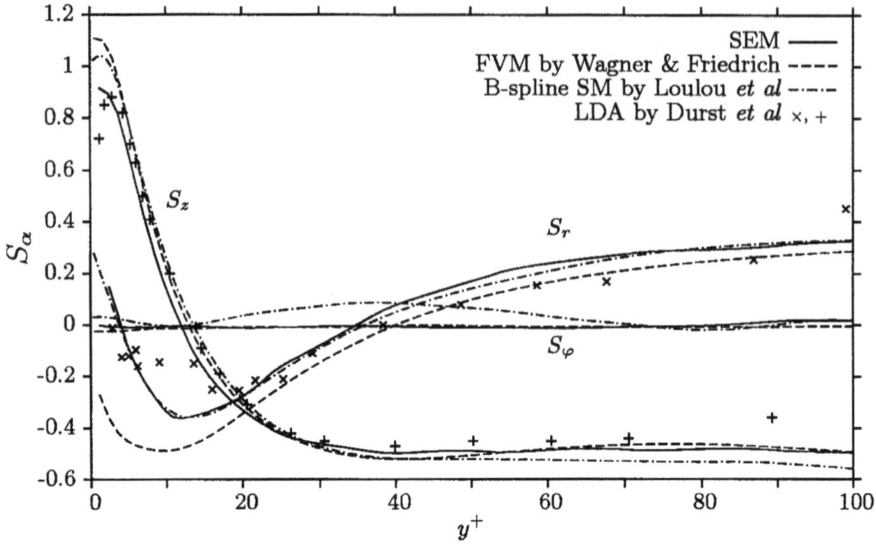

Figure 5 Skewness factors of the velocity fluctuations obtained in the DNS of turbulent pipe flow using the spectral element method, the FVM DNS by Wagner and Friedrich [5], the B-spline SM DNS by Loulou *et al* [4] and the LDA measurements by Durst *et al* [2].

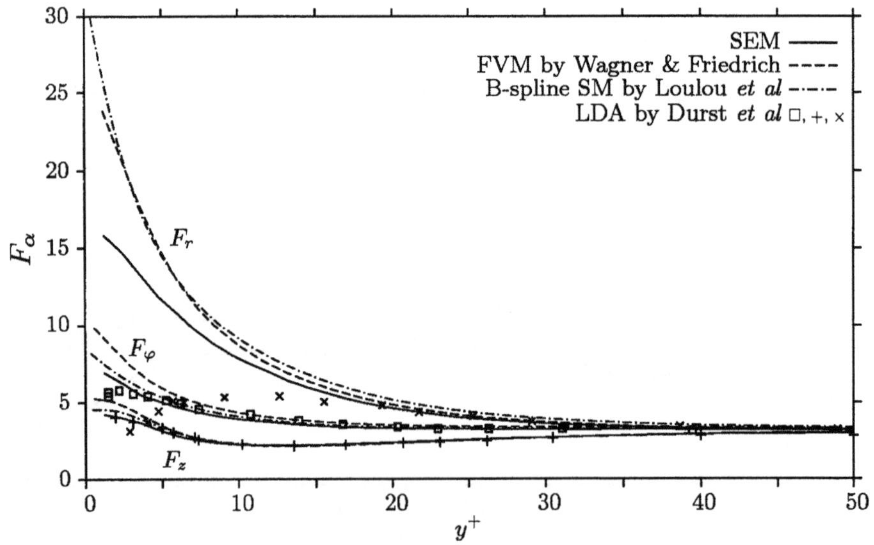

Figure 6 Flatness factors of the velocity fluctuations obtained in the DNS of turbulent pipe flow using the spectral element method, the FVM DNS by Wagner and Friedrich [5], the B-spline SM DNS by Loulou *et al* [4] and the LDA measurements by Durst *et al* [2].

Numerical simulation of aerodynamic problems with a Reynolds stress turbulence model

Bernhard Eisfeld

Institute of Aerodynamics and Flow Technology, German Aerospace Center (DLR),
Lilienthalplatz 7, 38108 Braunschweig, Germany,
Bernhard.Eisfeld@dlr.de

Summary

The Speziale-Sarkar-Gatski (SSG) Reynolds stress model is implemented into DLR's Navier-Stokes solver FLOWer blended with the Wilcox stress-ω model in the near wall region. The length scale is supplied by Menter's ω-equation. Results for 2D flows are presented for the transonic flow around the RAE 2822 airfoil, Cases 9 and 10, and the Aérospatiale A airfoil at $\alpha = 13.3^0$. Results for 3D flows are shown for the transonic flow around the ONERA M6 wing and the DLR-ALVAST wing-body configuration. Improvements are achieved with respect to predictions with the Wilcox k-ω model concerning shock positions, trailing edge separation and the pressure distribution near the wing tip due to an improved resolution of the wing tip vortex.

Symbols

b	semi-span, m
C_L	lift coefficient
L	chord length, m
Ma	Mach number
Re	Reynolds number
x	chordwise coordinate, m
y	spanwise coordinate, m
y_n	wall normal coordinate, m
α	incidence, $deg.$
$\eta = y/b$	dimensionless spanwise coordinate
τ_{wall}	wall shear stress, $kg/(m \cdot s^2)$

1 Introduction

The fluid flow investigated in aircraft aerodynamics usually is turbulent. Therefore turbulence modelling is a major issue for accurate predictions especially of shock locations, separation zones or free vortices.

Currently the majority of models in use is based on the Boussinesq hypothesis, assuming the same proportionality between turbulent (Reynolds) stresses and the mean shear tensor as for the Newtonian stresses in laminar flow. The corresponding proportionality factor, the eddy viscosity, depends on the mean flow and is simply added to the molecular viscosity, which is a property of the fluid.

Several models have been developed for providing an eddy viscosity, which are classified according to the equations to be solved. Algebraic models, where the eddy viscosity is an explicit function of the local flow properties, usually rely on Prandtl's mixing length concept [10], e. g. the Baldwin-Lomax model [1]. Alternatively transport equation models provide differential equations describing the transport of turbulent quantities, from which the eddy viscosity can be computed. Popular models for aeronautical applications are the one-equation model of Spalart and Allmaras [12] and two-equation models based on the turbulence kinetic energy k and the length scale ω, e. g. the Wilcox model [15] or the Shear Stress Transport (SST) model of Menter [9].

Nevertheless the Boussinesq hypothesis brings along some disadvantages. First of all it is known that by a scalar eddy viscosity the normal stress anisotropy close to solid walls cannot be resolved. Furthermore, practical computations show, that flow properties, such as shock positions in transonic flows or separation, are difficult to predict accurately with eddy viscosity models. Improvements have been achieved by explicit algebraic Reynolds stress models (EARSM) [11], [14] and additional modifications for free voritices [3]. However, these extensions are naturally limited by the characteristics of the required baseline eddy viscosity model.

A substantial improvement of flow predictions is expected from full differential Reynolds stress models, because they naturally include effects of streamline curvature and secondary motions [16]. In this approach the assumption of an eddy viscosity is dropped. Instead, for each component of the Reynolds stress tensor an individual transport equation is specified, augmented with an equation for the turbulent length scale. In the past this type of models has been mainly applied in academic research, rather than in industrial aeronautical computations.

Therefore one major focus of the EU project FLOMANIA has been to close the gap between academic research and industrial aeronautical application with respect to Reynolds stress turbulence models. Within this project DLR has implemented a Reynolds stress model based on the SSG re-distribution model [13] into its structured flow solver FLOWer [7]. This paper presents first results of its application to aeronautical flow problems.

2 Reynolds stress modelling

The transport equation for the Reynolds stress components R_{ij} to be solved reads for compressible flow

$$\frac{\partial\left(\rho R_{ij}\right)}{\partial t}+\frac{\partial}{\partial x_k}\left(\rho R_{ij}U_k\right)=\rho P_{ij}+\rho\Pi_{ij}-\rho\epsilon_{ij}+\rho D_{ij}+\rho M_{ij}, \qquad (1)$$

where ρ represents the mean density and U_i the components of the mean velocity vector. In this equation the production term

$$\rho P_{ij} = -\rho R_{ik} \frac{\partial U_j}{\partial x_k} - \rho R_{jk} \frac{\partial U_i}{\partial x_k} \tag{2}$$

is given exactly, while all other terms require modelling.

Various models exist for the re-distribution term Π_{ij}, from which in FLOMANIA the SSG-model [13]

$$\begin{aligned} \Pi_{ij} = &-\left(C_1^{(0)}\epsilon + C_1^{(1)} P^{(k)}\right) b_{ij} + C_2 \epsilon \left(b_{ik} b_{kj} - \frac{1}{3} b_{mn} b_{mn} \delta_{ij}\right) \\ &+ \left(C_3 - C_3^* \sqrt{b_{mn} b_{mn}}\right) k S_{ij} + C_4 k \left(b_{ik} S_{jk} + b_{jk} S_{ik} - \frac{2}{3} b_{mn} S_{mn} \delta_{ij}\right) \\ &- C_5 k \left(b_{ik} W_{kj} - W_{ik} b_{kj}\right) \end{aligned} \tag{3}$$

has been selected for implementation. In this equation S_{ij} and W_{ij} represent the symmetric and the antisymmetric parts of the velocity gradient tensor $\partial U_i / \partial x_j$ and $b_{ij} = R_{ij}/2k - \delta_{ij}/3$ the anisotropy tensor. $P^{(k)} = \frac{1}{2} P_{kk}$ is the production of kinetic turbulence energy. Since this model has been devised for homogenous turbulence, it is combined with the stress-ω model of Wilcox [16] near walls, using Menter's F_1 function [9] for blending the C_i, $C_i^{(k)}$ and C_3^* coefficients.

Isotropy is assumed for the destruction term, i. e.

$$\epsilon_{ij} = \frac{2}{3} \epsilon \delta_{ij} \qquad \text{with} \qquad \epsilon = C_\mu \omega k, \tag{4}$$

where ω is provided by the length scale equation of the SST model [9].

Finally, the diffusion term is modelled as

$$\rho D_{ij} = \frac{\partial}{\partial x_k} \left(\mu \frac{\partial R_{ij}}{\partial x_k} + C_s \frac{\rho k}{\epsilon} B_{km} \frac{R_{ij}}{\partial x_m}\right), \tag{5}$$

where $B_{km} = k\delta_{km}$ represents simple (SGDH) and $B_{km} = R_{km}$ represents generalized gradient diffusion (GGDH). In this expression μ denotes the molecular viscosity of the fluid and C_s a model coefficient.

The contributions of the mass flux variations ρM_{ij} are neglected.

3 Numerical method

The above turbulence models are implemented into DLR's structured flow solver FLOWer [7]. This code solves the coupled system of equations by different methods for the mean flow and the turbulence equations. Second order central discretization with artificial dissipation is applied to the mean flow equations, and an explicit hybrid five-stage Runge-Kutta scheme is used for time integration. The convergence of the mean flow solution is accelerated by local time stepping, implicit residual smoothing and a multigrid algorithm. In contrast the turbulence equations are discretized by a first order upwind scheme for the convective terms and integrated by a special implicit scheme on the finest grid level only. This technique has been successfully used in the past with one- and two-equation turbulence models [5].

4 Results

In order to assess the feasibility of Reynolds stress turbulence models for aeronautical applications, a number of simple test cases has been computed with the above SSG-ω model. They are described in the following.

4.1 RAE 2822 airfoil

The RAE 2822 airfoil is a common test case for transonic flow, in particular the so-called case 9, ($Ma = 0.73$, $Re = 6.5 \cdot 10^6$, $\alpha = 2.8^0$), where the shock does not induce separation, and case 10 ($Ma = 0.75$, $Re = 6.2 \cdot 10^6$, $\alpha = 2.8^0$), where the shock is strong enough to induce separation at its foot [2]. There is some doubt on the two-dimensionality of these experiments, nevertheless turbulence models that are able to predict both cases reasonably well usually perform also well for complex industrial applications in transonic flow. A mesh of 736x176 cells has been used for the computations, which has been checked to ensure grid independent solutions.

Figure 1 shows the pressure distributions for cases 9 and 10 obtained with the Wilcox k-ω and the SSG-ω turbulence models. As one can see, the Wilcox k-ω model predicts the shock location quite well for case 9, but for case 10 it yields the shock position rather far downstream of its experimental location. In contrast the SSG-ω model predicts shock positions that are slightly upstream the experimental location for case 9 and only slightly downstream for case 10. Thus the overall prediction quality of the SSG-ω model is considered superior to the Wilcox k-ω model with respect to this transonic airfoil flow. Note that the freestream sensitivity of the Wilcox model appeared to be of minor importance for this test case.

4.2 Aérospatiale A airfoil

In FLOMANIA the flow around the Aérospatiale A airfoil has been computed for $Ma = 0.15$, $Re = 2 \cdot 10^6$ and $\alpha = 13.3^0$, representing high lift conditions. For these conditions a laminar separation bubble occurs on the suction side between 10% and 12% chord, so that a very fine grid of 512x128 cells has been used to resolve this phenomenon. This was the finest grid, where steady solutions could have been obtained. Nevertheless it appeared in FLOMANIA, that the transition and wake region were not resolved optimally for obtaining grid independence [6].

Figure 2 shows details of the pressure distribution near the leading edge and the trailing edge. As one can see, the agreement with the experiment is improved, when using the SSG-ω model instead of the Wilcox k-ω model. This holds even more, when comparing the predicted shear stress distributions with the experiments

(Figure 3). As one can see, the length of the separation bubble at the trailing edge predicted with the SSG-ω model in conjunction with generalized gradient diffusion compares well with the experiment. This good agreement of the SSG-ω model with the experiment is also reflected by the Reynolds stresses at $x/L = 0.96$ depicted in Figure 4 normal to the wall.

4.3 ONERA M6 wing

Like the RAE 2822 airfoil the ONERA M6 wing has been established as a standard test case for transonic flow [2]. In FLOMANIA the conditions $Ma = 0.84$, $Re = 11.72 \cdot 10^6$ and $\alpha = 3.06^0$ have been selected, corresponding to a well behaved flow field with a lambda shock forming on the suction side, but without shock induced separation. Thus this test case is mainly devoted to demonstrate the applicability of Reynolds stress modeling to three-dimensional aeronautical flows. The computational grid contains 240x64x52 cells, where the distance of the wall nearest grid lines, d_1, is in wall units $y_1^+ = d_1 \rho u_\tau / \mu \approx 1$ with $u_\tau = \sqrt{\tau_{wall}/\rho}$ the friction velocity. No further grid refinement study has been performed.

Figure 5 shows the pressure distributions at two different sections, namely at $\eta = 0.65$, where two distinct shocks are present, and at $\eta = 0.99$, where the influence of the tip vortex is already felt. As one can see, at dimensionless span coordinate $\eta = 0.65$ the pressure distributions predicted with the Wilcox k-ω and the SSG-ω model are virtually identical and in good agreement with the measurements. Only the first shock is smoothed out a bit. In contrast, at $\eta = 0.99$ the SSG-ω prediction agrees much better with the measurements than the Wilcox k-ω prediction. This improvement can be attributed to the better resolution of the tip vortex by the SSG-ω model compared to the Wilcox k-ω model, which is documented by the iso-C_p plots in Figure 6. This confirms that, due to the exact production term, Reynolds stress models should be superior to two-equation models in predicting vortex dominated flow fields.

4.4 DLR-ALVAST wing-body

The DLR-ALVAST wing-body model has been tested experimentally by ONERA within the EU-project Enifair [4]. Unfortunately the model heated up during the windtunnel tests and therefore deformed, so that only the mid-span sections, in particular $\eta = 0.62$, are considered meaningful for a comparison with CFD simulations. First computations for that configuration were carried out within the EU-project Airdata, where the grid has been provided by NLR [8]. This grid, which consists of 41 blocks with $1.1 \cdot 10^6$ cells, is used also here without any further grid refinement study. The flow conditions considered are $Ma = 0.75$ and $Re = 4.8 \cdot 10^6$, while the incidence is varied, in order to achieve a given lift coefficient.

Figure 7 shows the computed pressure distributions for $C_L = 0.492$ in the two mid-span sections $\eta = 0.46$ and $\eta = 0.62$ compared to the experiments. As one can see, for $\eta = 0.46$ the Wilcox k-ω and the SSG-ω model both predict the shock position too far upstream. Nevertheless, with the SSG-ω model the prediction is closer to the experiments. For the section at $\eta = 0.62$, which is considered the most meaningful one, the SSG-ω model predicts the shock position in accordance with the measurements, while with the Wilcox k-ω model the shock is still too far upstream. Thus the Reynolds stress model appears superior also for the prediction of the flow field around more complex aircraft type configurations.

5 Conclusion

The SSG-ω Reynolds stress turbulence model has been successfully implemented into DLR's structured flow solver FLOWer [7] and applied to the flow around airfoils and 3D configurations. No special numerical difficulties have been encountered, when solving the complete system of RANS and turbulence equations.

The results obtained show that for the RAE 2822 airfoil the shock locations predicted with the SSG-ω model are in better agreement with the measurements than those predicted with the Wilcox k-ω model. The same holds for the trailing edge separation of the Aérospatiale A airfoil, where the shear stress predicted with the SSG-ω model is in good agreement with the measurements as well as the Reynolds stress distribution normal to the wall. For the ONERA M6 wing the SSG-ω model improves the prediction of the pressure distribution close to the wing tip, because the tip vortex is clearly resolved better with the Reynolds stress model. Target lift computations for the DLR-ALVAST wing-body show an improved agreement of predicted pressure distributions with the measurements, when applying the SSG-ω instead of the Wilcox k-ω model.

As it seems, the SSG-ω model provides accurate results for a wide range of applications. With respect to aeronautical flows the SSG-ω model has clearly demonstrated a greater universality than the Wilcox k-ω model. The memory and the CPU time per iteration required by the SSG-ω model appeares to be about twice as high as the values for the Wilcox k-ω model. The number of iterations needed depends on the test case and is not necessarily higher, when using a Reynolds stress model.

Acknowledgements

This work was mainly supported by research grants from the European Union under the FLOMANIA project (Flow Physics Modelling - An Integrated Approach). The project is administrated by the CEC, Research Directorate-General, Growth Programme, under Contract No. G4RD-CT2001-00613.

References

[1] B. S. Baldwin, H. Lomax: "Thin Layer Approximation and Algebraic Model for Separated Turbulent Flows". AIAA-Paper 78-0257, 1978

[2] J. Barche (Ed.): "Experimental Data Base for Computer Program Assessment", AGARD-Report AGARD-AR-138, 1979

[3] H. S. Dol, J. C. Kok, B. Oskam: "Turbulence Modelling for Leading-Edge Vortex Flows". AIAA-Paper 2002-0843, 2002
[4] A. Dumas, J.-L. Godard, L. Marin: "Analysis of ENIFAIR High Speed Test Results – An Evaluation of Interference Effects". Workshop on European Research on Aerodynamic Engine/Airframe Integration for Transport Aircraft, DLR Braunschweig, Germany, 26-27 September 2000.
[5] J. K. Fassbender: "Improved Robustness for Numerical Simulation of Turbulent Flows around Civil Transport Aircraft at Flight Reynolds Numbers". Dissertation, Technische Universität Braunschweig, DLR Forschungsbericht, DLR–FB 2003–09, 2004
[6] Haase, W., Aupoix, B., Bunge, U., Schwamborn, D. (Eds.): "FLOMANIA - Flow Physics Modelling - An Integrated Approach", Notes on Numerical Fluid Mechanics, Springer to appear
[7] N. Kroll, C. Rossow, D. Schwamborn, K. Becker, G. Heller: "MEGAFLOW – a numerical flow simulation tool for transport aircraft design", ICAS 2002 Congress
[8] M. Laban: "Aircraft Drag and Thrust Analysis (AIRDATA)". Publishable Synthesis Report. NLR-TP-2000-473
[9] F. R. Menter: "Two-Equation Eddy-Viscosity Turbulence Models for Engineering Applications". AIAA Journal **32**, 1994, pp. 1598-1605
[10] L. Prandtl: "Über die ausgebildete Turbulenz". ZAMM, **5**, 1925, pp. 136-139
[11] T. Rung, H. Lübcke, M. Franke, L. Xue, F. Thiele, S. Fu: "Assessment of Explicit Algebraic Stress Models in Transonic Flows". Proc. 4th Int. Symp. Engineering Turbulence Modelling and Measurements, Corsica, France, 1999, pp. 659-668
[12] P. R. Spalart, S. R. Allmaras: "A One-Equation Turbulence Model for Aerodynamic Flows". AIAA-Paper, 92-439, 1992
[13] C. G. Speziale, S., Sarkar, T. B. Gatski: "Modelling the pressure-strain correlation of turbulence: an invariant dynamical systems approach". J. Fluid Mech. **227**, 1991, pp. 245-272
[14] S. Wallin, A. V. Johansson: "An explicit algebraic Reynolds stress model for incompressible and compressible turbulent flows". J. Fluid Mech. **403**, 2000, pp. 89-132
[15] D. C. Wilcox: "Reassessment of the Scale Determining Equation for Advanced Turbulence Models". AIAA Journal **26**, 1988, pp. 1299-1310
[16] D. C. Wilcox: "Turbulence Modeling for CFD". DCW Industries, La Cañada, USA, 2nd ed., 1998

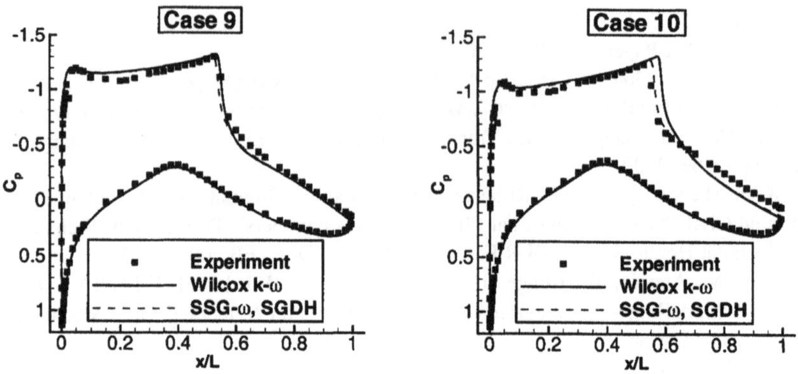

Figure 1 RAE 2822. Pressure distributions for case 9 ($Ma = 0.73$, $Re = 6.5 \cdot 10^6$, $\alpha = 2.8^0$) and case 10 ($Ma = 0.75$, $Re = 6.2 \cdot 10^6$, $\alpha = 2.8^0$).

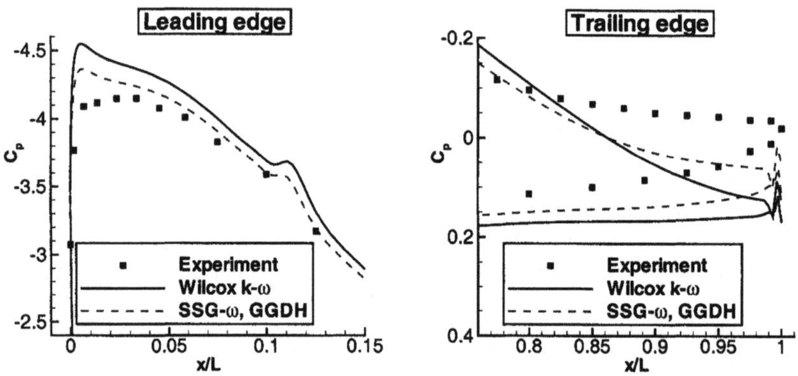

Figure 2 Aérospatiale A, $\alpha = 13.3^0$. Details of pressure distribution.

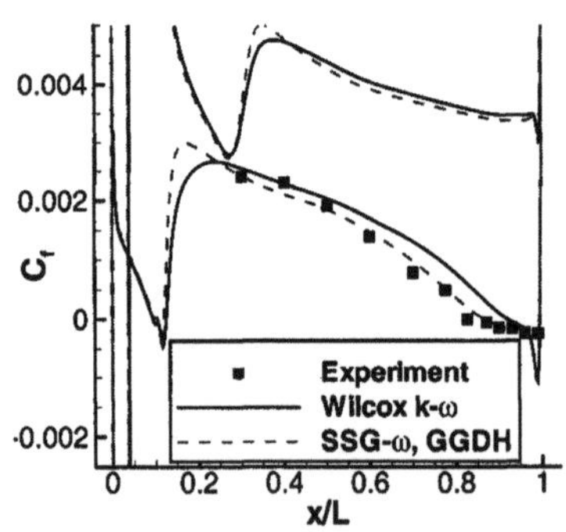

Figure 3 Aérospatiale A, $\alpha = 13.3^0$. Shear stress distribution.

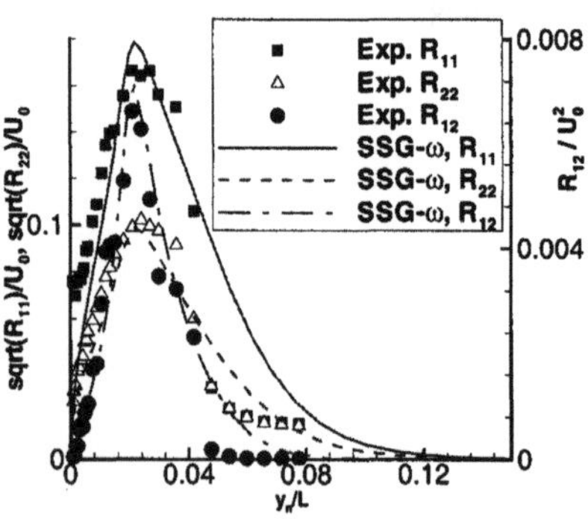

Figure 4 Aérospatiale A, $\alpha = 13.3^0$, $x/L = 0.96$. Reynolds stress distributions.

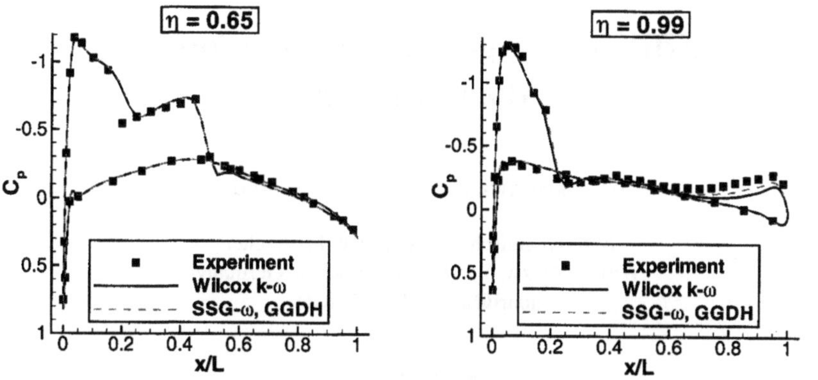

Figure 5 ONERA M6, $\alpha = 3.06^0$. Pressure distributions at $\eta = 0.65$ and $\eta = 0.99$.

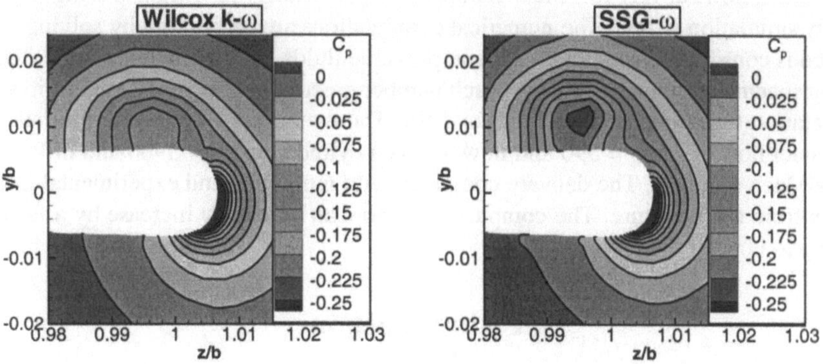

Figure 6 ONERA M6, $\alpha = 3.06^0$. Tip vortex in plane $x/b = 0.90$

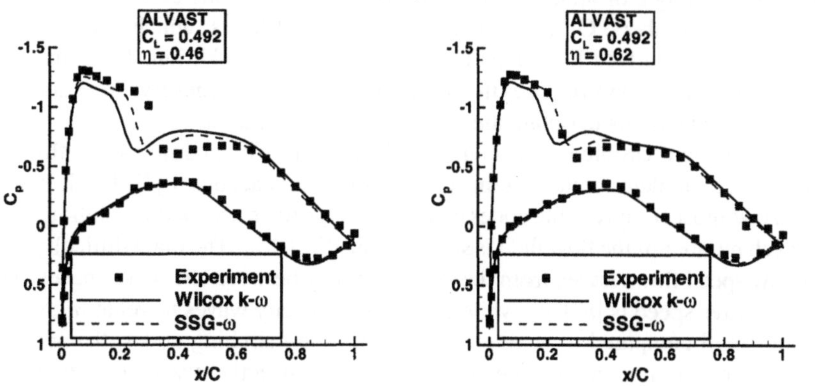

Figure 7 DLR-ALVAST, $C_L = 0.492$. Pressure distributions at $\eta = 0.46$ and $\eta = 0.62$.

Efficient Large-Eddy Simulation of Low Mach Number Flow

N. Alkishriwi, M. Meinke, W. Schröder

Aerodynamisches Institut, RWTH Aachen
Wüllnerstr. zw. 5 u. 7, 52062 Aachen, Germany
nouri@aia.rwth-aachen.de

Summary

Turbulent flows at very low Mach numbers are numerically investigated using large-eddy simulation (LES). The numerical computations are carried out by solving the viscous conservation equations for compressible fluids. An implicit dual time stepping scheme combined with low Mach number preconditioning and a multigrid acceleration technique is developed for LES. The method is validated for turbulent channel flow at $Re_\tau = 590$ and flow across a cylinder at Re = 3900 and different low Mach numbers. The data are compared with numerical and experimental findings from the literature. The computations show an efficiency increase by a factor of up to 60.

1 Introduction

In many engineering problems compressible and nearly incompressible flow regimes occur simultaneously. In aerodynamics, for example, the flow over a multi-element airfoil at high angle of attack, contains transonic flow regions even for small free stream Mach numbers. Other examples are low speed flows, which may be compressible due to surface heat transfer or volumetric heat addition. The numerical simulation of such flows requires the solution of the viscous conservation equations for compressible fluids to obtain the physically correct results.

When a compressible flow solver is applied to a nearly incompressible flow, its performance can deteriorate in terms of both speed and accuracy [1], [2]. It is well known that most compressible codes do not converge to an acceptable solution when the Mach number of the flow field is smaller than $\mathcal{O}(10^{-1})$. The main difficulty with such low speed flows arises from the large disparity between the wave speeds. The acoustic wave speed is $|u \pm c|$, while entropy or vorticity waves travel at $|u|$, which is quite small compared to $|u \pm c|$. In explicit time-marching codes, the acoustic waves define the maximum time step, while the convective waves determine the total number of iterations such that the overall computational time becomes large for small Mach numbers.

Different methods have been proposed to solve such mixed flow problems by modifying the existing compressible flow solvers. One of the most popular approaches is to use low Mach number preconditioning methods for compressible codes [1-4]. The basic idea is to modify the time marching behavior of the system of equations without altering the steady state solution. This is, however, only useful when the steady state solution is sought.

Large-eddy simulations for low Mach number flows, i.e., flows that can be considered almost incompressible, have hardly been discussed in the context of solutions of the compressible NAVIER–STOKES equations. The reason is the strong susceptibility of the small scales to the dissipation level contained in the numerical scheme. The overall artificial dissipation is changed, however, when an efficient and accurate flow solver at low Mach number flows with, e.g., preconditioning is developed. This means, the dilemma is on the one hand, preconditioning is required to satisfy the physics of low compressible flows while on the other hand, the artificial dissipation has to be well below the dissipation of the high wave number vortices, i.e., the small scales.

In the present paper, a highly efficient large-eddy simulation method is presented based on an implicit dual time stepping scheme combined with preconditioning and multigrid.

First, the implementation of preconditioning in the LES context using the dual time stepping technique is described. Then, the time marching solution technique and the discretization within the dual time stepping approach are discussed. Subsequently, the multigrid and the numerical methods are outlined and finally, the numerical and efficiency results are presented and analyzed.

2 Mathematical Formulation

The governing equations are the unsteady three-dimensional compressible NAVIER–STOKES equations written in generalized coordinates $\xi_i, i = 1, 2, 3$

$$\frac{\partial \mathbf{Q}}{\partial t} + \frac{\partial(\mathbf{F}_{c_i} + \mathbf{F}_{v_i})}{\partial \xi_i} = 0 \quad . \tag{1}$$

where the quantity $\mathbf{Q}$ represents the vector of the conservative variables and $\mathbf{F}_{c_i}$, $\mathbf{F}_{v_i}$ are inviscid and viscous flux vectors, respectively.

2.1 Preconditioned Dual-Time Stepping Method

As mentioned before, preconditioning is required to provide an efficient and accurate method of solution of the steady NAVIER–STOKES equations for compressible flow at low Mach numbers. For unsteady flow problems, a dual time stepping technique [5], [6] for time accurate solutions is used. In this approach, the solution at the next physical time step is determined as a sequel of steady state problems to which preconditioning, local time stepping and multigrid are applied.

Introducing of a pseudo-time τ in (1), the unsteady two-dimensional governing equations with preconditioning read

$$\Gamma^{-1}\frac{\partial \mathbf{Q}}{\partial \tau} + \frac{\partial \mathbf{Q}}{\partial t} + \mathbf{R} = 0 \tag{2}$$

where $\mathbf{R}$ represents

$$\mathbf{R} = \left(\frac{\partial \mathbf{E}_c}{\partial \xi} + \frac{\partial \mathbf{F}_c}{\partial \eta} + \frac{\partial \mathbf{E}_v}{\partial \xi} + \frac{\partial \mathbf{F}_v}{\partial \eta} \right) \tag{3}$$

and $\mathbf{\Gamma}^{-1}$ is the preconditioning matrix, which is to be defined such that the new eigenvalues of the preconditioned system of equations are of similar magnitude. A preconditioning technique based on Turkel [1] has been implemented.

It is clear that only the pseudo-time terms in (2) are altered by the preconditioning, while the physical time and space derivatives retain their original form. Convergence of the pseudo-time within each physical time step is necessary for accurate unsteady solutions. This means, the acceleration techniques such as local time stepping and multigrid can be immediately utilized to speed up the convergence within each physical time step to obtain an accurate solution for unsteady flow. The derivatives with respect to the real time t are discretized using a three-point backward difference scheme that results in an implicit scheme, which is second-order accurate in time

$$\frac{\partial \mathbf{Q}}{\partial \tau} = \mathbf{RHS} = -\mathbf{\Gamma} \left(\frac{3\mathbf{Q}^{n+1} - 4\mathbf{Q}^{n} + \mathbf{Q}^{n-1}}{2\Delta t} + \mathbf{R}(\mathbf{Q}^{\mathbf{n+1}}) \right) \quad . \tag{4}$$

Note that at $\tau \rightarrow \infty$ the first term on left-hand side of (2) vanishes such that (1) is recovered. To advance the solution of the inner pseudo-time iteration, a 5-stage Runge-Kutta method in conjunction with local time stepping and multigrid is used. For stability reasons [7] the term $\frac{3\mathbf{Q}^{n+1}}{2\Delta t}$ is treated implicitly within the Runge-Kutta stages

$$\begin{aligned}
\mathbf{Q^0} &= \mathbf{Q^n} \\
&\vdots \\
\mathbf{Q^l} &= \mathbf{Q^0} - \alpha_l \Delta\tau \left[\mathbf{I} + \frac{3\Delta\tau}{2\Delta t} \alpha_l \mathbf{\Gamma} \right]^{-1} \mathbf{RHS} \\
&\vdots \\
\mathbf{Q^{n+1}} &= \mathbf{Q^5} .
\end{aligned} \tag{5}$$

Note, the additional term means that in smooth flows the development in pseudo-time is proportional to the evolution in t.

3 Multigrid Method

The multigrid method [8] is based on a Full Approximation Storage (FAS) scheme. Following Turkel [2], we transfer the residuals to the next grid based on the preconditioned system, since these residuals are more balanced than the non-preconditioned residuals. A 3-level sawtooth multigrid cycle is used in all calculations. To go from the fine to the coarse mesh, which contains every second grid point of the fine mesh, a residual based full weighting restriction operator is used. On the coarsest mesh the same iterative method of solution as on the finest mesh is applied. The corrections are transferred from the coarse to the fine grid by a trilinear prolongation operator. In general, one iteration is performed on the finest and each intermediate grid level, and two iterations on the coarsest mesh before the prolongation.

4 Numerical Method

The governing equations are the NAVIER–STOKES equations filtered by a low-pass filter of width Δ, which corresponds to the local average in each cell volume. The monotone integrated large-eddy simulations (MILES) approach is used to implicitly model the small scale motions through the numerical scheme.

The approximation of the convective terms of the conservation equations is based on a modified second-order accurate AUSM scheme using a centered 5-point low dissipation stencil to compute the pressure derivative in the convective fluxes. This scheme is described in detail in Meinke et al. [9], [17]. The viscous stresses are discretized to second-order accuracy using central differences, i.e., the overall spatial approximation is second-order accurate. The method is formulated for multi-block structured curvilinear grids and implemented on vector and parallel computers.

5 Results

The objective of this section is to show the improvement in the overall accuracy when a preconditioned LES solver for the compressible NAVIER–STOKES equations is applied to flows at almost vanishing compressibility effects. The above described numerical method based on implicit dual-time stepping, multigrid, and preconditioning has been used to perform large-eddy simulations for well-documented internal and external flow problems in the literature. We start by considering turbulent channel flow at a Reynolds $Re_\tau = 590$ based on the friction velocity u_τ and then, turbulent flow past a circular cylinder at a diameter based Reynolds number of $Re_D = 3900$ is investigated. After that, the efficiency increase of the method is concisely discussed.

5.1 Channel Flow

The channel flow computations at $Re_\tau = u_\tau\delta/\nu = 590$ correspond to $Re_{cl} = u_{cl}\delta/\nu = 12448$, where u_{cl} is the maximum velocity on the centerline and δ is the channel half-width. The Mach numbers based on the centerline velocity are set to $Ma = 0.01, 0.05,\ 0.1$. The size of the computational domain in the streamwise x, spanwise z, and normal y direction is given by $L = 2\pi\delta$, $B = \pi\delta$, and $H = 2\delta$. A mesh of 153x145x93 cells is used.

The boundary conditions consist of no-slip and isothermal conditions on the walls and periodic boundary conditions in the spanwise direction. In the streamwise direction the pressure and temperature fluctuations and the mass flow are assumed to be periodic. Using the relation $Re_\tau = 0.1097 Re_{cl}^{0.911}$ from Hussain and Reynolds [10] the pressure difference $\Delta p = \rho u_\tau^2 L/H$ that drives the flow is determined by the maximum velocity. After a fully developed turbulent channel flow has been established the simulations are continued for another 200 dimensionless time units to compute the statistical results. Samples are stored each quarter of a dimensionless time unit. A direct numerical simulation for an incompressible flow at $Re_\tau = 590$ carried out by Moser et al. [11] on a mesh of 384x257x384 cells is used as a reference solution for the LES results. The comparison of the DNS data and the LES findings at $Ma = 0.01$ and $\Delta t = 0.01\delta/u_{cl}$ in Fig.1 show overall a convincing agreement for the mean velocity profile, the turbulence intensity distributions, and the resolved Reynolds stress component profile although the peak value in the streamwise fluctuations is slightly overpredicted. This is due to an insufficient turbulent energy redistribution from the streamwise to the spanwise direction, caused by slightly too weak a coherent near wall structure.

5.2 Cylinder Flow

The flow around a cylinder is performed at a Reynolds number $Re_D = 3900$ based on the diameter D, a free stream Mach number $M_\infty = 0.05$, and a physical time step $\Delta t = 0.02 D/U_\infty$. The computational domain extends $25D$, $15D$, and $1D$ in the streamwise, normal and spanwise direction. The C-type mesh, the near-wall resolution of which is shown in Fig.2, possesses 105x273x49 cells in the circumferential, normal, and spanwise direction resulting in approx. $1.4 \cdot 10^6$ cells. No-slip and isothermal conditions are specified on the walls and periodic boundary conditions are imposed in the spanwise direction. On all inflow and outflow boundaries non-reflecting boundary conditions with pressure relaxation are applied [12].

To show the quality of the results the LES findings are compared with experimental data from Lourenco and Shih [13] and Ong and Wallace [14] and LES results from Beaudan and Moin [15]. To further evidence the impact of the preconditioning on the solution, distributions computed with and without preconditioning are also presented. The statistics is based on a non-dimensional time interval of $\tilde{T} = T u_\infty / D = 170$ plus averaging in the spanwise direction. Fig.2 shows the development of the streamwise velocity component on the centerline in the wake of the cylinder. Unlike the LES distribution computed using a pure explicit scheme without preconditioning the new LES method captures the location and the minimum value in the u-distribution measured by Lourenco and Shih [13]. The comparison of the distribution of the normal distribution of the u-component further downstream in the wake of the cylinder at $X/D = 1.06$ and $X/D = 2.02$ in Fig. 3 with LES and experimental data [15], [13] confirms the convincing quality of the results of the LES method using preconditioning, dual time stepping and multigrid.

6 Conclusion

An efficient large-eddy simulation method for nearly incompressible flows based on solutions of the governing equations of viscous compressible fluids has been introduced. The method uses an implicit time accurate dual time-stepping scheme in conjunction with low Mach number preconditioning and multigrid acceleration. To validate the scheme, large-eddy simulations of turbulent channel flow at $Re_\tau = 590$ and cylinder flow at $Re_D = 3900$ for several Mach numbers and physical time steps have been performed. The results show the scheme to be efficient and to improve the accuracy at low Mach number flows. The efficiency gain from the implicit method is presented in Fig.4 which are based on channel flow solutions. These figures show the convergence of the inner iterations at $Ma = 0.01$ and the impact of the Mach number on the efficiency as a function of the physical time step. A detailed efficiency analysis can be found in [16]. Generally, the new method is 6-60 times faster than the basic explicit 5-stage Runge-Kutta scheme requiring a storage increase by 15 percent.

References

[1] Turkel, E.: "Preconditioning techniques in computational fluid dynamics". Annual Review of Fluid Mechanics 31:385-416,1999.

[2] Turkel, E.:" A review of preconditioning methods for fluid dynamics". Applied Numerical Mathematics 12, 257-285, 1993.

[3] Weiss, J.M.; Smith, W.A.:" Preconditioning applied to variable and constant density flows". AIAA Journal, Vol. 33, No. 11, pp. 2050-2057, 1995.

[4] Choi, Y.H.; Merkle, C.L.: "The application of preconditioning in viscous flows". J. of Computational Physics. 105:207-223, 1993.

[5] Jameson, A.: "Time dependent calculations using multigrid, with applications to unsteady flows past airfoils and wings". AIAA Paper 91-1596, 1991.

[6] Schröder, W.; Keller, H.B.:" Wavy Taylor-Vortex flows via multigrid-continuation method". J. of Computational Physics, Vol. 91, No. 1, pp. 197-227, 1990.

[7] Arnone, A.; Lion, M.S.; Povinelli, L.A.: "Integration of Navier-Stokes equations using dual time stepping and multigrid method". AIAA Journal, Vol. 33, No. 6, pp. 985-990, 1995.

[8] Brandt, A.: "Guide to multigrid development". In Lectures Notes in Mathematics, pp. 220-312, Springer Berlin, 1981.

[9] Meinke, M.; Schröder, W.; Krause, E.; Rister, T.: "A comparison of second- and sixth-order methods for large-eddy simulations". Comp. & Fluids, Vol. 31, pp. 695-718, 2002.

[10] Hussain, A.K.; Reynolds, W.C.: "Measurement in fully developed turbulent channel flow". Trans. ASME, pp. 568-580, 1975.

[11] Moser, R.D.; Kim, J.; Mansour, N.N.: "Direct numerical simulation of turbulent channel flow up to $Re_\tau = 590$". Physics of Fluids, Vol. 11, No. 4, pp. 943-445, 1999.

[12] Poinsot, T.J.; Lele, S.K.: "Boundary conditions for direct simulations of compressible viscous flows". J. of Computational Physics, Vol. 101, pp. 104-129, 1992.

[13] Lourenco, L.M.; Shih, C.: "Characteristics of the plane turbulent near wake of a circular cylinder". A particle image velocimetry study. (Data taken from [15]), 1993.

[14] Ong, L.; Wallace, J.: "The velocity field of the turbulent very near wake of a circular cylinder". Exp. Fluids 20, pp. 441-453, 1996.
[15] Beaudan, P.; Moin, P.: "Numerical experiments on the flow past a circular cylinder at sub-critical Reynolds numbers". Technical Report TF-62, Center Turb. Res., Stanford, 1994.
[16] Alkishriwi, N.; Meinke, M.; Schröder, W.: "A large-eddy simulation method for low Mach number flows using preconditioning and multigrid". Submitted to Computers & Fluids, 2004.
[17] Rütten, F.; Schröder, W.; Meinke, M.: "Large-Eddy Simulation of low frequency oscillations of the Dean vortices in turbulent pipe bend flows". Physics of Fluids 17 (3), 2005.

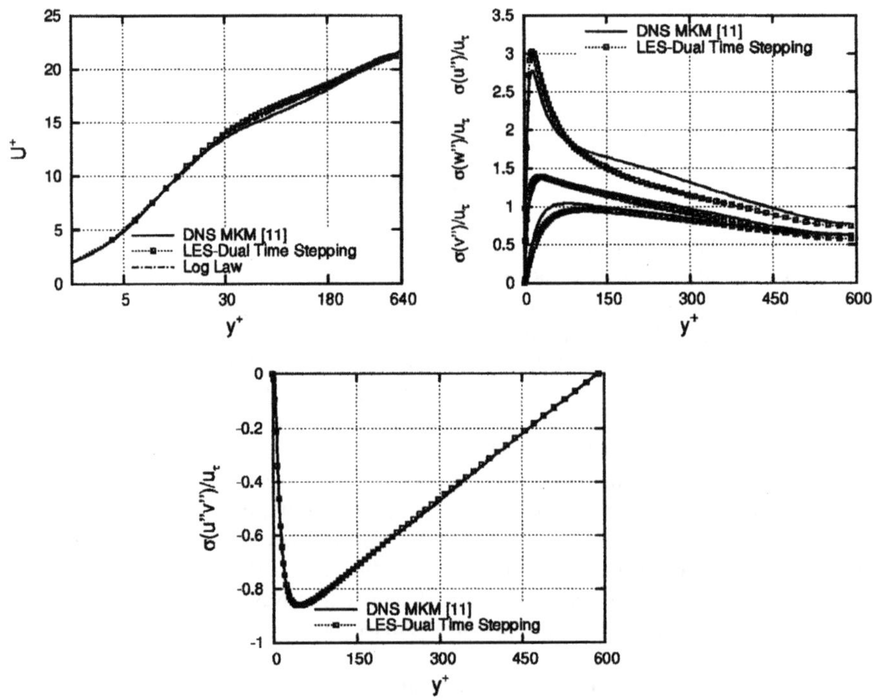

Figure 1 LES of turbulent channel flow at Re_τ=590 and $Ma = 0.01$; mean velocity distribution (top left); distribution of the streamwise, spanwise, and normal Reynolds stresses (top right); distribution of the Reynolds shear stress component $(u''v'')$ (bottom).

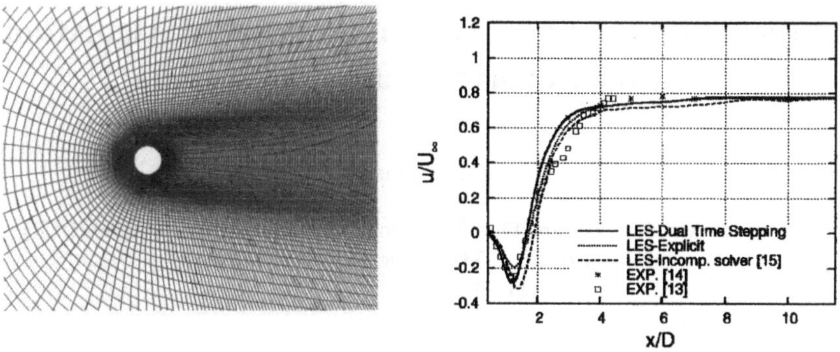

Figure 2 LES of flow across a cylinder at $Re_D = 3900$; C-type computational mesh (left); Streamwise velocity as a function of X/D on the centerline $Y/D = 0$, $Z/D = 0$ in the cylinder wake (right).

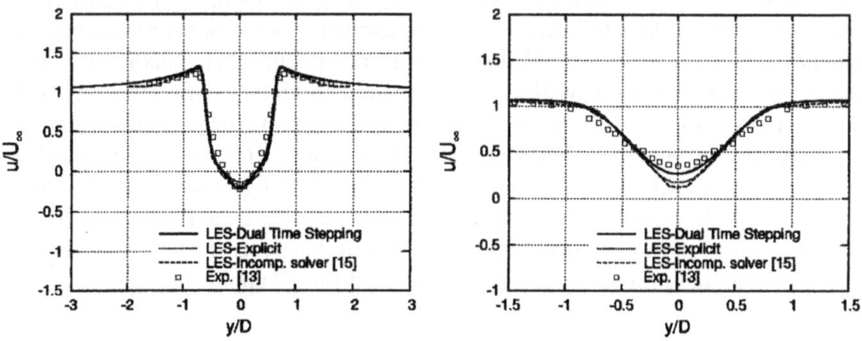

Figure 3 Streamwise velocity as a function of Y/D in the cylinder wake at $X/D = 1.06$ (left); Streamwise velocity as a function of Y/D in the cylinder wake at $X/D = 2.02$ (right).

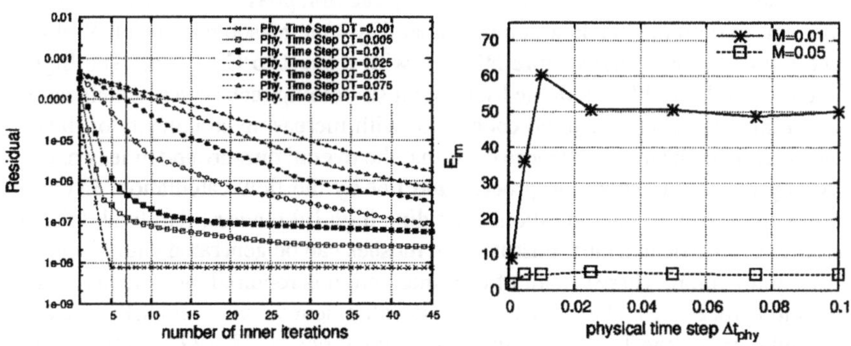

Figure 4 Convergence of the inner iterations at $Ma = 0.01$ (left); Efficiency E_{im} as a function of Δt_{phy} at $Ma = 0.01$ and $Ma = 0.05$ (right).

Implementation and Usage of Structured Algorithms within an Unstructured CFD-Code

Ralf Heinrich

Institute of Aerodynamics and Flow Technology, German Aerospace Center (DLR), Lilienthalplatz 7, 38108 Braunschweig, Germany, Ralf.Heinrich@dlr.de

Summary

In the CFD community structured and unstructured codes are in use and under further development. Both types of codes have their specific advantages. Unstructured codes enable a higher geometric complexity in the simulation, whereas structured codes are usually characterized by better performance properties. From developer and user side it would be desirable to have one code being able to handle structured and unstructured meshes as well as grids containing both types of regions. At DLR a first prototype of a mixed structured / unstructured code has been developed, in order to explore the potential of such a combined procedure. The prototype is based on the unstructured DLR TAU code and includes ingredients from the block-structured FLOWer code like implicit smoothing techniques, which are applied in the structured regions of the computational domain. Within this paper the prototype will be described and first promising results are presented for 2D applications.

1 Introduction

Within the national research project MEGAFLOW [3] two CFD codes dedicated for aerodynamic applications have been developed: The block-structured FLOWer code [3] and the unstructured TAU code [2]. During the first phase of MEGAFLOW emphasis was laid on the block-structured capability. One reason has been that block-structured codes are usually more efficient with respect to CPU time. The higher efficiency is achieved for example by implicit smoothing techniques, which can be easily implemented in structured codes. But with increasing complexity of the applications it became more and more clear that the overall turn-around time is mainly driven by the mesh generation process and not by the flow calculation. Even for very complex aircraft configurations including high lift devices, flap-track fairings and other geometrical details, unstructured meshes can be generated with a high degree of automation, whereas a lot of user interaction is required by structured mesh generation procedures. Therefore the CFD development activities at DLR were concentrated more on the TAU code during the second phase of MEGAFLOW and the follow on project MEGADESIGN.

Both codes have now reached a high level of maturity and are in daily use at DLR, industry and universities. For applications up to medium complexity the

FLOWer code is usually preferred (wing, fuselage, winglet, tails), for higher complexity the TAU code is the first choice. So currently two codes have to be maintained. New features like improved turbulence models have to be implemented twice. The user has to train his skills for two CFD-codes. It becomes clear, that it is desirable to have one CFD-tool which is able to combine the advantages of structured and unstructured technologies. This would help to reduce costs on developer and customer side. Within the DLR-ONERA project MIRACLE [4] an activity has been launched to find an efficient way for combining structured and unstructured technologies. It is not the idea to develop a completely new code. The plan is to adapt the existing TAU code, if possible while keeping the basic concepts and data structures. The desired features of the adapted code are:

- The same or better performance and accuracy on purely unstructured meshes as the existing TAU-code
- The same or better performance and accuracy on a purely block-structured mesh as the existing FLOWer-code
- Simultaneous usage of structured and unstructured technologies in case of meshes including both block-structured and unstructured regions.

Within MIRACLE a first prototype based on TAU has been developed. In order to find an appropriate way for implementing structured algorithms in the prototype, only a small subset of the existing TAU code with reduced functionality has been used. In the following sections the main features of the prototype named TAUijk as well as the ideas behind them are described. First results are presented showing that the code is on a good way to fulfil the desired features mentioned above.

2 The Prototype TAUijk

The prototype TAUijk is based on a cell centred developer version of TAU [6]. To find an appropriate way for implementing structured algorithms in the code, it was sufficient to transfer only a small subset of the functionalities of the complete TAU-system. To enable a simple transfer from TAU to the prototype, TAUijk is written like TAU in the C programming language. The basic data structure is the same. In the following sections the main features transferred from TAU to TAUijk are described. Afterwards focus is set on the changes enabling structured algorithms.

TAUijk solves the three dimensional Euler equations. For the spatial discretization the MAPS+ [5] scheme has been transferred from TAU to TAUijk. To achieve second order accuracy, gradients are used to reconstruct the values of variables at the cell faces. A slip wall and a farfield boundary condition have been implemented. For time integration a Runge-Kutta scheme is used. To accelerate the convergence, local time stepping, explicit residual smoothing and a multigrid method are used. All mesh based data needed by the solver is created during the preprocessing phase. The element types allowed within the mesh are tetrahedral, prism, hexahedra and pyramids. The solver itself does not need any information about the shape of elements or the number of neighbours of an element, because like in TAU a face based data structure is used. This makes the solver very flexible and it is especially very helpful for the implementation of multigrid in an unstructured code. During the agglomeration process, where fine mesh cells are fused together to coarse cells, arbitrarily shaped polyhedra are created.

What is now the difference between TAU and the new prototype, enabling the usage of structured algorithms? TAUijk is able to read in meshes which include block-structured regions. The additional information needed are the number of blocks included in the mesh, the number of cells in i, j and k direction of each block and a list of hexahedral elements belonging to the structured region. During the preprocessing phase, these elements are sorted in such a way, that from a block-number mbl, the loop indices i, j and k, the element address can be easily computed. Let's assume, for simplicity in 2D like in fig. 1, we have a mixed block- structured / unstructured mesh containing 2 blocks and an unstructured region. Block 0 contains 3 cells in i-direction and 2 cells in j-direction, block 1 contains 1x4 cells and the unstructured region 10 additional cells. So the number of cells in block 0 is 6, in block 1 it is 4. The element data (e.g. residual, size of control volume) of block 0 is addressed with the indices 0...5, of block 1 with the indices 6...9 and the unstructured part with the indices 10...19. The element numbers are printed in fig. 1. To compute the element numbers as a function of the indices i,j and the block-number, the following formula is used:

$$elem\_id = eoffset[mbl] + i + imax[mbl] * j \tag{1}$$

$eoffset[mbl]$ is the first element number of a block. In the previous example eoffset[0] = 0 and eoffset[1] = 6. In the code the formula is realized as a macro function. Using this technique, it is very easy to arrange structured loops in i, j, k direction for the structured part of the mesh. All unstructured elements are stored behind the structured part. No additional sorting is used within the unstructured part currently. As a demonstration the implicit residual smoothing with constant coefficients has been implemented in the prototype, which is known to be usually more efficient compared to the explicit residual smoothing available in TAU. In the following a comparison of a structured loop from FLOWer and the realization in TAUijk is shown.

```
! FLOWer like



do j = 2, j2
   do i = 3, i2


      t = dijk(i)
      rr1  = r(i,j,1)
      rr2  = r(i,j,2)
      rr3  = r(i,j,3)
      rr4  = r(i,j,4)

      r(i,j,1)= t*(rr1+am*r(i-1,j,1))
      r(i,j,2)= t*(rr2+am*r(i-1,j,2))
      r(i,j,3)= t*(rr3+am*r(i-1,j,3))
      r(i,j,4)= t*(rr4+am*r(i-1,j,4))
   enddo
enddo
```

```
\\ TAUijk like
#define EID(I,J,IM,EOFF) = \
           (I + (J) * (IM) + EOFF)
for (j = 0; j < jm; j++)
  for (i = 1; i < im; i++)
  { int p0= EID(i,j,im,eoffset);
    int p1= EID((i-1),j,im,eoffset);
    t = dijk[i];
    rr0  = r[p0][0];
    rr1  = r[p0][1];
    rr2  = r[p0][2];
    rr3  = r[p0][3];

    r[p0][0]= t*(rr0+am*r[p1][0]);
    r[p0][1]= t*(rr1+am*r[p1][1]);
    r[p0][2]= t*(rr2+am*r[p1][2]);
    r[p0][3]= t*(rr3+am*r[p1][3]);
  }
```

This loop is taken from the implicit residual smoothing, for simplicity again in 2D. Please note, that in FLOWer, written in FORTRAN, the physical range of cells starts always with 2, in TAUijk with 0. In the loop the cell addresses (i,j) and (i-1,j) are used. In FLOWer, of course, the indices are used directly via multidimensional arrays. In TAUijk the element indices p0 and p1 are computed from (i,j) and (i-1,j). The variable r in the example is the residual of an element after summing up all the fluxes. This example shows that element based loops can be transferred from FLOWer to TAUijk easily without memory overhead.

Fig. 2 shows results of a benchmark of the residual smoothing routine alone from FLOWer and TAUijk. The routine has been called 10 times, the element number has been varied from 20000 up to 10000000. TAUijk and FLOWer have been compiled with Intel compilers and GNU compilers on a LINUX PC with the same optimization level. In a first step the array of the residuals in the FORTRAN routine has been arranged in the same way like in FLOWer (r(imax,jmax,kmax,5)). With respect to the location in memory the variables of equations 1 are far away from the variables of equation 2. Within the inner part of the loop all 5 equations are used, which can result in a performance loss for large arrays due to the limited size of the cache memory of the computer. This becomes visible in fig. 2. Using the Intel FORTRAN compiler the CPU time shows a linear behaviour up to 80000 elements (dashed line filled boxes), then the slope of the curve is increased . Switching the arrays from r(imax,jmax,kmax,5) to r(5,imax,jmax,kmax) the situation is improved. Variables of equation 1 are now direct neighbours of equation 2. For the whole range of element numbers the slope is now constant. The same positive behaviour is found for the smoothing routine of TAUijk. If for both codes the Intel compilers are used, the performance is more or less the same (solid lines, filled box for FLOWer and filled triangle for TAU). For the comparison of the performance of the whole codes TAUijk and FLOWer we concentrate within this paper on small cases, where the cache optimizing does not play a big role. Large cases in 3D are currently under investigation.

Beside the implicit smoothing techniques, the construction of the coarse grids in the frame of multigrid has been transferred from FLOWer to TAUijk. So in the structured area 8 fine cells in 3D or 4 cells in 2D are agglomerated to a coarse cell, see fig. 3. The agglomeration of the structured part is done in a first step. The rest of the computational domain is agglomerated like in TAU as described in [1]. Here it is irrelevant, if the number of cells is divisible by 2, 4, 8 or other numbers, in contrast to the structured part of the mesh.

3 Results

Within this paper we concentrate on inviscid 2D applications. 3D applications are currently under investigation. All meshes used here have been generated with a MEGACADS developer version [7]. Within this version a 2D triangular mesh generator and a 3D tetrahedral mesher is included. This MEGACADS version does not only allow the generation of purely block-structured meshes, but also of mixed structured / unstructured meshes. 3 meshes around the NACA0012 airfoil have been generated. Details of the meshes are shown in fig. 4. For all simulations the inflow Mach number is 0.7 with 0° angle of attack. For time integration 5 Runge-Kutta stages in combination with 3 or 4 level v-cycle multigird has been selected. For a verification of TAUijk, results have been compared to TAU and FLOWer. Fig. 5 shows the comparison of the pressure coefficient on the profile for all three codes. As should be expected for this simple test case the agreement is very good. For a fair comparison of CPU-times and convergence rates between FLOWer and TAUijk, FLOWer has been slightly modified. For example the reconstruction of flow variables needed for 2nd order upwind schemes is now done based on gradients, which are computed using the theorem of Green-Gauss. As next test case mesh II and III have been used. In fig. 7 and fig. 8 different convergence histories (L2-norm of density residual) are shown. Mesh II is a purely structured mesh. Computations have been performed with FLOWer and TAUijk. TAUijk has been used in purely unstructured mode and structured mode. Purely unstructured mode means like the original TAU-code using explicit residual smoothing, structured mode means that the implicit residual smoothing taken from FLOWer has been used. It becomes visible, that FLOWer and TAUijk in structured mode show the same convergence rates. Compared to TAUijk in unstructured mode the convergence is improved, see fig. 7. Although the convergence properties of FLOWer and TAUijk in structured mode are the same, the CPU time required for one multigrid cycle is not the same. For one cycle FLOWer needs for this test case 57 percent more CPU time. If TAUijk is used in purely unstructured mode the CPU time is only 1 percent higher than in structured mode. So the best performance for this test-case is achieved by TAUijk in structured mode, because of better convergence compared to TAUijk in unstructured mode and a better CPU time efficiency compared to FLOWer. The result with respect to the CPU time can additionally be improved by using the mixed structured, unstructured mesh (mesh III). Convergence histories are shown in fig. 8. The area close to the wall is identical to mesh II, but in order to save memory, the number of elements is reduced to 70 percent of mesh II by using triangular elements for the connection to the farfield. In fig. 6 for this test case the total pressure loss on mesh II and III is compared on the profile. On both meshes the same accuracy level has been reached, although mesh III has only 70 percent of the cells compared to mesh II. This shows that a mixed structured / unstructured code can help to improve the performance by a selective usage of the advantages of structured und unstructured technology.

4 Conclusion

Within this paper the prototype TAUijk has been introduced. TAUijk allows the simultaneous usage of unstructured and structured algorithms in one CFD tool. The code is based on the unstructured TAU code. With relatively small effort it was possible to implement for example the implicit residual smoothing technique from the structured FLOWer code without memory overhead. First results show that TAUijk is able to achieve the same convergence rates like FLOWer, but with better CPU-time efficiency on purely block-structured meshes for the selected architecture. A further improvement with respect to memory and CPU-time requirements can be achieved by using mixed structured / unstructured meshes. Block-structured meshes are used close to the walls, unstructured meshes for filling the computational domain between block-structured region and the farfield. In the near future the described method will be tested for 3D viscous appications.

References

[1] Galle, M.: "Ein Verfahren zur numerischen Simulation kompressibler, reibungsbehafteter Strömungen auf hybriden Netzen". DLR-FB, 99-04, 1999

[2] Gerhold, T., Friedrich, O., Evans, J., Galle, M.: "Calculation of Complex Three–Dimensional Confi gurations Employing the DLR–TAU–Code". AIAA-Paper 97–0167, 1997

[3] Kroll, N., Rossow, C.-C., Schwamborn, D., Becker, K., Heller, G.: "MEGAFLOW - A Numerical Flow Simulation Tool For Transport Aircraft Design". ICAS Congress 2002, Toronto (can), 09.-13.09.2002, ICAS, CD-Rom, S. 1.105.1-1.105.20, 2002

[4] Kroll, N., Cambier, L.: "MIRACLE: Merging and Development of Industrial and Research Aerodynamic Environment. Schedule of DLR-ONERA project MIRACLE, 2004

[5] Rossow, C.-C.: "A Flux Splitting Scheme for Compressible and Incompressible Flows". Journal of Computational Physics, Vol. 164, (2000), S. 104-122, 2000

[6] Widhalm, M., Rossow, C.-C.: "Improvement of upwind schemes with the Least Square method in the DLR TAU Code". 13. DGLR-Fach-Symposium AG STAB, München, 12.-14.11.2002, DGLR, New Results in Numerical and Experimental Fluid Mechanics IV, Springer Verlag, 2002

[7] Wild, J.: "Acceleration of Aerodynamic Optimization Based on RANS-Equations by Using Semi-Structured Grids". Sedign Optimization International Conference, March 31- April 2, Athens, Greece, 2004

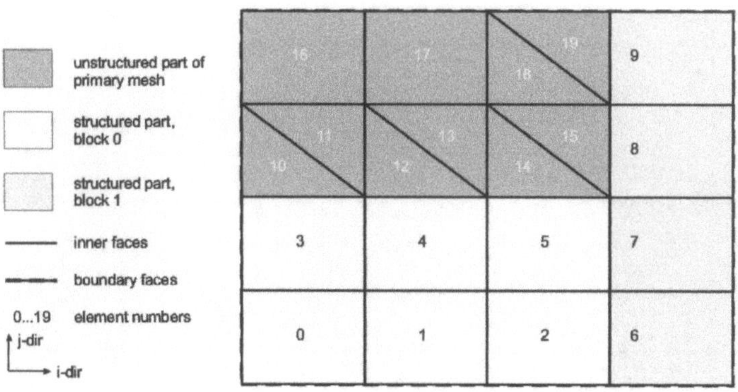

Figure 1 Element numbering in a mixed structured / unstructured mesh

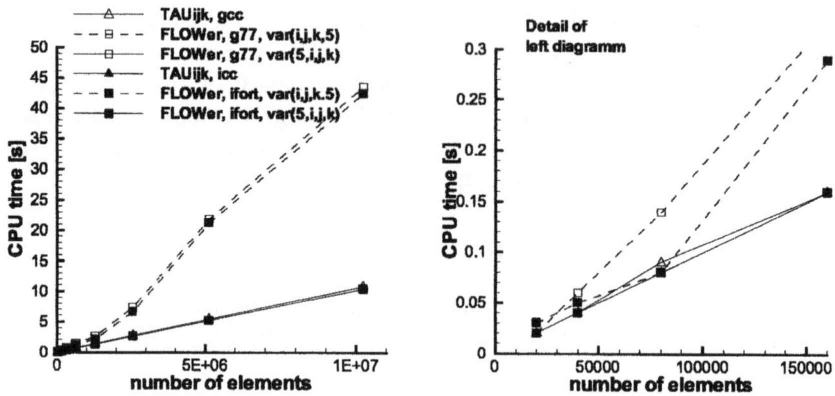

Figure 2 Comparison of CPU-times for TAUijk and FLOWer smoothing routines as function of the number of elements

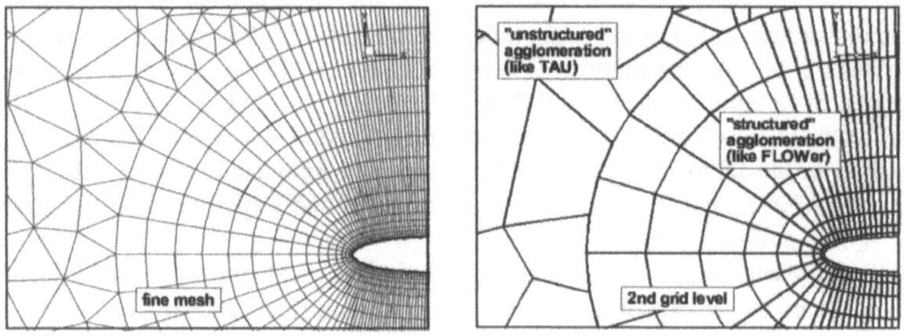

Figure 3 First and second grid level for a mixed structured / unstructured mesh

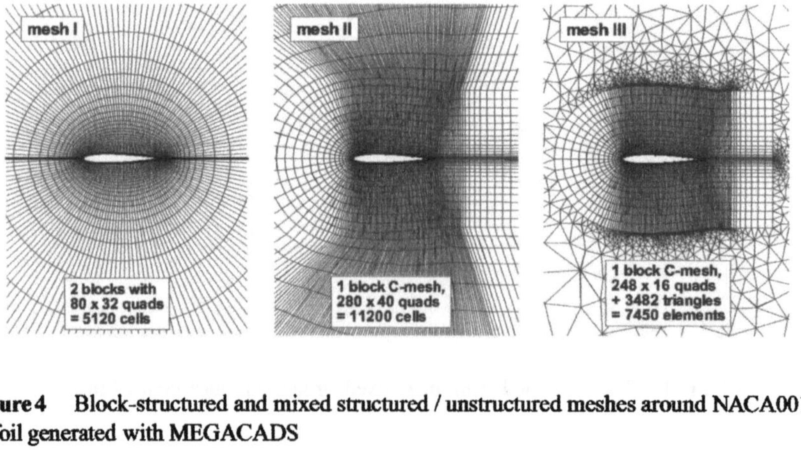

Figure 4 Block-structured and mixed structured / unstructured meshes around NACA0012 airfoil generated with MEGACADS

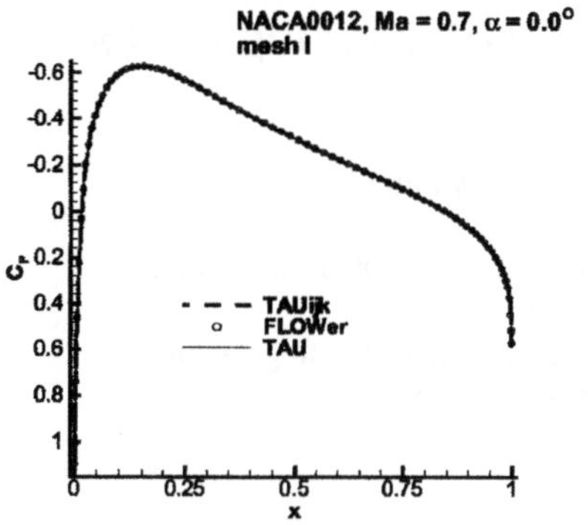

Figure 5 Comparison of c_P distributions of FLOWer, TAUijk and TAU on mesh I

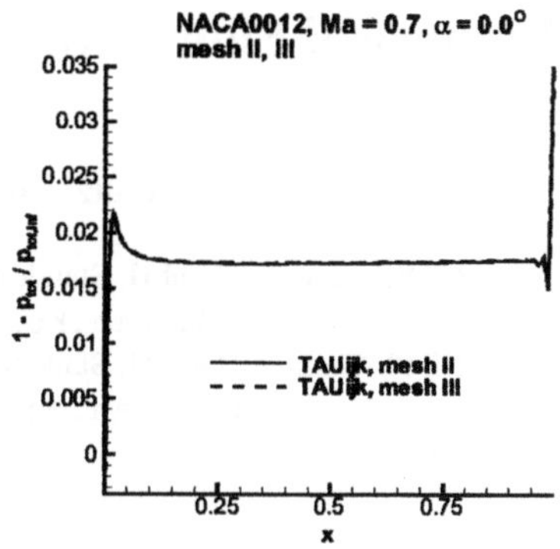

Figure 6 Comparison of total pressure losses on mesh II and III using TAUijk

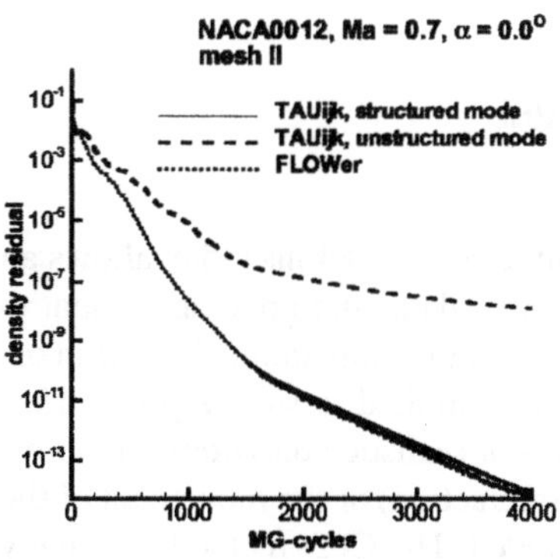

Figure 7 Convergence history of density residual of FLOWer and TAUijk in structured and unstructured mode on mesh II

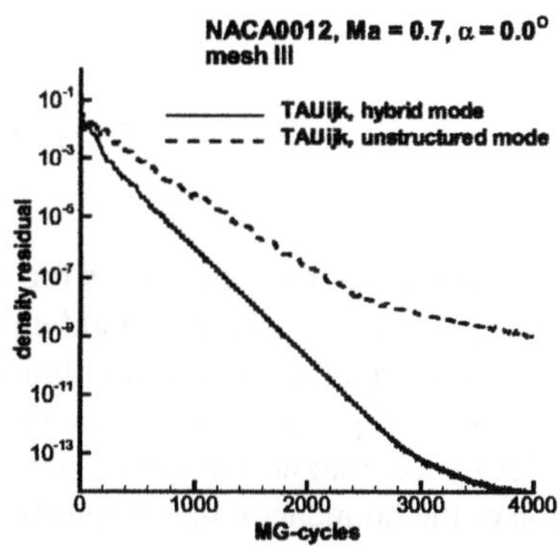

Figure 8 Convergence history of density residual of TAUijk in mixed structured / unstructured and purely unstructured mode on mesh III

Numerical Flow Simulation with Moving Grids

Martin Kuntz[1] and Florian R. Menter[2]

[1] ANSYS Germany GmbH, Staudenfeldweg 12, 83624 Otterfing, Germany
Martin.Kuntz@ansys.com
[2] ANSYS Germany GmbH, Staudenfeldweg 12, 83624 Otterfing, Germany
Florian.Menter@ansys.com

Summary

The numerical analysis of aerodynamic flows is in general limited to steady geometries. Depending on the flow conditions steady or transient flow solutions in the relative frame of the body are computed. In order to take into account the flexibility of the body (e.g. fluttering wing) and the motion of the body (manoeuvre flight), moving computational meshes are required. The CFD method has to take into account meshes with moving nodes and deforming control volumes. The present paper shows computational results of different applications with moving grids, e.g. an oscillating airfoil, a fluttering wing and a guided manoeuvre flight of an airplane.

1 Introduction

The numerical simulation of flows for aerodynamic configurations today allows an accurate prediction of the flow field and performance data with good agreement to experimental data [11]. These simulations require a high computational effort in order to resolve e.g. the boundary layers, but are still limited to steady geometries. In order to take into account more physical effects or realistic conditions the deformation of the boundary contour (fluid-structure interaction) or the movement of the whole geometry (manoeuvre flight) must be included. The CFD method for steady meshes has to be extended with regard to two aspects: First taking into account the motion (non-zero grid velocity) and the deformation of the control volumes (non-constant control volumes) for the descretized governing equations and second the mesh morphing in case that the mesh motion is specified at the boundaries of the computational domain and the location of the remaining nodes has to be determined.

The implementation of a moving grid algorithm for a CFD method has to take into account the space conservation law to avoid loosing conserved quantities during a transient run [4],[9]. An example for mesh morphing in the field of external aerodynamics is the deformation of the geometry, whereas other parts of the grids (e.g. the farfield) are steady. The transient mesh is created using a method based on a diffusion equation.

2 Numerical Method

The numerical method of moving and deforming grids is implemented in CFX-5, the general purpose CFD code of ANSYS Inc. CFX-5 has a finite-volume method and the governing equations for turbulent flow are the Reynolds-Averaged Navier-Stokes (RANS) equations. In CFX-5 these equations are disretized using a finite volume method, which is conservative and time-implicit [12],[14],[3]. The computational hybrid and unstructured mesh can consist of different element types such as hexahedrals, prisms, wedges, and tetrahedrals. A control volume is constructed around each nodal point of the mesh and the fluxes are computed at the integration points located at the subfaces between two control volumes. The discrete equations are evaluated using a bounded high resolution advection scheme similar to that of Barth and Jesperson [1] being second order accurate on regular meshes. The mass flow is evaluated such that a pressure-velocity coupling is achieved by the Rhie and Chow [13] algorithm. The discrete systems of equations are solved by a coupled algebraic multi-grid method developed by Raw [12]. The numerical effort of this method scales close to linearity with the number of grid nodes, which is selected to resolve the computational domain. Steady state applications are computed by time step iteration until a user defined convergence level is reached. For a transient computation an iterative procedure updates the non-linear coefficients within each time step (coefficient loop) while the outer loop advances the solution in time using a first order time discretization.

Moving meshes introduce additional terms into the conservation equations of fluid dynamics. The straightforward discretization of the equations can lead to a loss of the conserved variables due to a first order time error in the volume discretization. This error manifests itself by a change in density (and any other conserved variable), even for trivial constant density, stagnant flow conditions.

The consistent discretization of the moving mesh equations in CFX-5 is described in the present chapter. The conservation equation for a quantity Φ is written as:

$$\frac{\partial}{\partial t}\left(\int \Phi dV\right) + \int \Phi(u_j - v_j)dA_j = \dot{\overline{\Phi}}V + \overline{\Phi}\dot{V} + \int \Phi(u_j - v_j)dA_j = 0 \quad (1)$$

u_j and v_j are the fluid and grid velocity respectively and V is the control volume size. The average quantity is defined as:

$$\overline{\Phi} = \frac{1}{V}\int \Phi dV \quad (2)$$

The time derivative of the control volume can be expressed as:

$$\overline{\Phi}\dot{V} = \int \overline{\Phi} v_j dA_j \quad (3)$$

The conservation equation can therefore be re-written as:

$$\dot{\overline{\Phi}}V + \int \Phi u_j dA_j + \int \left(\overline{\Phi} - \Phi\right) v_j dA_j = 0 \tag{4}$$

The third term on the left hand side includes the moving grid terms. For $\Phi = 1$, this equation satisfies the space conservations requirement for a divergence free velocity field u_j. The solution variables are the velocity components u_j defined in the absolute reference frame.

Mesh morphing computes the mesh shape based on a specified motion of one or multiple boundaries. The location of the nodes in the remaining part of the computational grid is obtained by solving a diffusion equation for the mesh deformation component d_i:

$$\frac{\partial}{\partial x_i}\left(\gamma \frac{\partial}{\partial x_i} d_i\right) = 0 \tag{5}$$

The prescribed mesh motion is the boundary condition for this equation. In order to force the mesh motion also for nodes inside the computational domain, the corresponding nodal equations for the mesh deformation in the equation system are replaced by algebraic equations. The diffusivity γ of equation 5 is used to control the mesh deformation. To preserve the node distribution in certain areas (e.g. boundary layer), a high value for the diffusivity is chosen, resulting in a small gradient of the mesh displacement in this area. The diffusion equation is implemented equivalent to the governing flow and turbulence equations. Therefore features like parallelization of the solution procedure can be applied.

3 Applications

3.1 Oscillating Airfoil

Transient aerodynamics are computed for the flow over an NACA 0012 airfoil pitching around its quarter-chord point. Experimental data is available in the AGARD report R-702 [10], dataset 3. The pitching motion is defined by the mean angle and by the amplitude and frequency of the sinusoidal oscillation $\alpha(t) = \alpha_m sin(\omega t)$, whereby ω is expressed by the reduced frequency $\kappa = (\omega c)/(2U_\infty)$.

O-type grids with a distance from the airfoil to the farfield of about 22 chord lengths are used. Despite the closeness of the farfield no special treatment of the farfield boundary is taken into account. The grid size is 192 x 64 points. The grid points are clustered in the boundary layer resulting in an average y^+ value of 7. The moving grids are obtained by prescribing the airfoil motion, fixing the farfield boundary and computing the remaining nodes with the mesh morphing algorithm. Studies about the time step size are carried out in order to obtain a time-accurate lift and momentum coefficient independent on the time step size. An optimal value of 160 time steps per period with 5 subiterations per timestep was found. All computations are carried out with the BSL turbulence model with an automatic wall treatment [7] and a second order accurate high-resolution advection scheme. Computational results of the CT1 test conditions with Reynolds number of 4.6 million,

Mach number 0.6, asymmetric pitching movement of $\alpha_m = 2.89^o$ and $\alpha_0 = 2.51^o$ and a reduced frequency of 0.0808 are presented here. Figure 1 shows the comparison of the lift and momentum coefficient with experimental data. There is a slight mismatch of the average slope of the lift curve and a shift of the momentum curve. A possible reason for this deviation is a too small distance of the wing to the farfield. Although, similar deviations to the experimental data can also be seen in results of other authors [6]. The hysteresis in the lift and momentum curves is mainly an inviscid effect. Therefore, no strong influence of the turbulence model on the result is expected.

3.2 Wing Flutter

As an example for aeroelastic coupling the AGARD wing 445.6 test case has been analyzed. A detailed description of the experimental set-up is given in the AGARD report R-765 [15]. The wing span is 0.76 m and is swept by 45 degrees at the quarter chord line (see figure 2). The total weight of the wing is 1.8 kg. The wing is disretized with 6300 solid nodes and the computational fluid domain includes about 300.000 nodes. The material of the wing is laminated mahogany. For the present investigations the so-called weakened model (with holes in the wing in order to reduce the material stiffness) is selected. The stiffness along the grain of the wood is taken from the AGARD report with a Young's modulus of $E = 3.25 \cdot 10^9 Pa$, a shear modulus of $G = 0.412 \cdot 10^9 Pa$ and a Poisson ration of $\nu = 0.31$.

As a first step, a steady CFD solution is generated (without coupling) for each operating point using the experimental data for Mach number, inlet velocity and density using the SST turbulence model. A high resolution spatial discretization and a first order time discretization is used. Starting from this flow solution the coupled fluid-structure computation is launched. Sensitivity studies of the time step size (from 0.2 ms to 1.0 ms) showed a minor influence on the flutter frequency. Therefore, a time step of 0.001 s is chosen, which is expected to be sufficient to resolve the wing flutter frequency in the range of 10 to 40 Hz. At each time step, CFX sends the pressure plus viscous stress data to ANSYS and ANSYS sends nodal displacements of the interface to CFX (explicit coupling). The displacements of the remaining nodes are computed with the mesh morphing algorithm.

The wing starts to oscillate with the flutter frequency. The comparison of predicted and experimental flutter frequency as function of the Mach number is shown in Figure 2. The computational results are slightly below the curve of the experimental data in the high Mach number regime. A reason for this deviation is probably a too coarse grid for aerodynamic computations in the transonic and supersonic regime. Also the chosen time step size could be too big for the explicit coupling algorithm. The results are comparable to the studies of other authors summarized in [8].

3.3 Guided Manoeuvre Flight

The moving grid algorithm is applied for the computation of a guided flight manoeuvre. The testcase is a forward swept wing configuration (see figure 3). Grids were generated with ICEM-Hexa in the frame of the European project FLOMANIA [5]. Different grid sizes of 320.000 and 2.5 million nodes were used for the present computations ($y^+ = 2$). Experimental data such as global performance values and velocity contours are provided by Breitsamter [2].

The flow conditions for the computations are a Reynolds number of 0.46 million, an inflow Mach number of 0.118 and a variable angle of attack in the range from 10^0 to 45^0. First, computations are carried out for different fixed angle of attack with steady inflow conditions on non-moving meshes using a time step of 0.5 ms (5 sub-iterations per time step) and the SST turbulence model. Lift and drag coefficients agree quite well with the experimental data, even the coarse mesh values (see Figure 4). For an angle of attack of 20^0 a comparative computation with a translational moving grid (moving opposite to the flow direction) and zero inflow velocity is carried out on the coarse grid. Lift and drag is compared to a steady state result computed with a first order and a second order temporal discretization scheme. The moving grid result agrees well with the steady state results (see figure 4). Figure 5 shows the convergence history (RMS residual) for both computations (steady inflow on fixed grid and moving grid with zero inflow). There is a quite similar convergence behaviour.

The main advantage of the moving grid method is the applicability to unsteady flight conditions. A guided transient motion with an increase in angle of attack from 0^0 to 45^0 and a decrease back to 0^0 is computed ($\alpha = 45^0(0.5(1 - cos(\omega t)))$). The rotation is around the mean aerodynamic center. The time step is chosen again as 0.5 ms. Figure 6 shows the variation of lift and drag coefficient during the motion for different values of rotational frequency ω, whereby ω can be related to the maximum acceleration g of the front part of the aircraft at maximum angle of attack. For both grid densities a similar transient behaviour is obtained. The results for the moderate acceleration $g = 1.5$ is as expected nearer to the steady state results, but all results show a hysteresis effect.

The computational effort for this manoeuvre is about a factor 10 higher than a single steady state computation. Typical computation time for the whole transient run on a Linux cluster with 3GHz processors is about 8 h on the coarse grid using 4 processors and 14 h for the fine grid using 16 processors.

4 Conclusion

Three different moving grid test cases demonstrate the performance of the CFD method for different simulation conditions: an oscillating airfoil with a prescribed motion, a transient fluid-structure interaction of a wing and a simulation of a manoeuvre flight using translating and rotating meshes. All results show the applicability of the method. The transient results are obtained with reasonable computational effort.

Some detailed studies about the influence of different parameters on the computational result are missing in the present paper (like time step and grid size studies for the CFD computations). These studies are planned for the future and will help to judge the accuracy of the predictions. More accurate results are expected also by implementing a second order accurate time integration scheme for moving grid applications.

Acknowledgements

The forward swept wing test case is provided by EADS Military Aircraft in the frame of the project FLOMANIA (Flow Physics Modelling - An Integrated Approach), a collaboration between industry, research centers and universities in Europe. The project (contract No. G4RD-CT2001-00613) is funded by the European Union and administrated by the CEC, Research Directorate-General, Growth Programme.

References

[1] T.J. Barth and D.C. Jesperson. The design and application of upwind schemes on unstructured meshes. AIAA Paper 89-0366, 1989.
[2] Breitsamter. Vortical Flow Field Structures at Forward Swept Wing Configurations. ICAS Proceedings 1998, 21st Congress, Melbourne, Australia, 1998
[3] CFX-5.7 Solver Manual. ANSYS Inc., 2004.
[4] I. Demirdzic and M. Peric. Space conservation law in finite volume calculations of fluid flow. Int. J. Num. Methods in Fluids, 8, pp1037-1050, 1998.
[5] Final report of European project FLOMANIA (Flow Physics Modelling - an Integrated Approach), to be published.
[6] C. Gao, S. Luo, F. Liu and D.M. Schuster. Calculation of Unsteady Transonic Flow by and Euler Method with Small Perturbation Boundary Conditions. AIAA 03-1267, 2003.
[7] H. Grotjans and F.R. Menter. Wall Functions for General Application CFD Codes. Computational Fluid Dynamics, Proceedings of the 4th Computational Fluid Dynamics conference, 7-11 Sept. 1998, Athens, Greece, Vol. 1, Part 2, ECCOMAS, John Wiley & Sons, pp. 1112-1, 1998.
[8] W. Haase, V. Selmin, B. Winzell. Progress in Computational Flow-Structure Interaction. Notes on Numerical Fluid Mechanics and Multidisciplinary Design, Springer, Volume 81, 2003.
[9] I.R. Hawkings and N.S. Wilkes. Moving Grids in Harwell-FLOW3D. AEA, InTec-0608, 1991.
[10] R.H. Landon. NACA0012 Oscillatory and Transient Pitching, Compendium of Unsteady Aerodynamic Measurements, Data Set 3. AGARD Report R-702, Aug. 1982.
[11] R. Langtry, Drag Prediction of Engine-Airframe Interference Effects with CFX-5, AIAA 2004-0391.
[12] M.J. Raw. Robustness of coupled algebraic multigrid for the Navier-Stokes equations. AIAA Paper 96-0297, 1996.
[13] C.M. Rhie and W.L. Chow. Numerical study of the turbulent flow past an airfoil with trailing edge separation. AIAA Journal, 21:1525-1532, 1983.

[14] G.E. Schneider and M.J. Raw. Control volume finite-element method for heat transfer and fluid flow using colocated variables. 1. Computational procedure. Numerical Heat Transfer, 11:363-390, 1987.
[15] E.C. Yates. AGARD Standard Aeroelastic Configuration for Dynamic Response, I, Wing 445.6. AGARD-R-765, 1988.

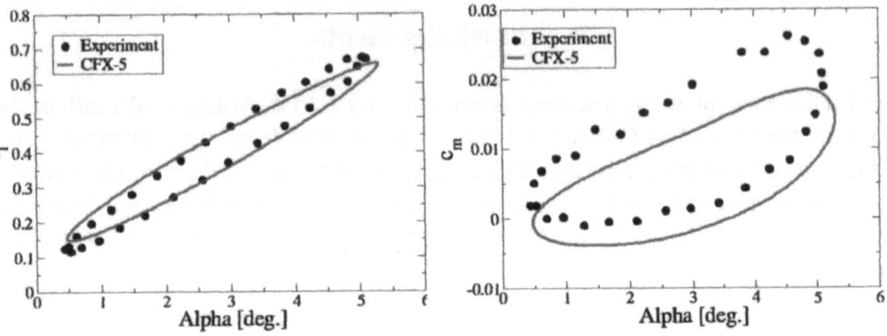

Figure 1 NACA 0012 oscillating airfoil, comparison of lift and moment coefficients for experimental data and CFX-5 simulation ($\alpha_m = 2.89^o$ and $\alpha_0 = 2.51^o$, reduced frequency $k = 0.0808$).

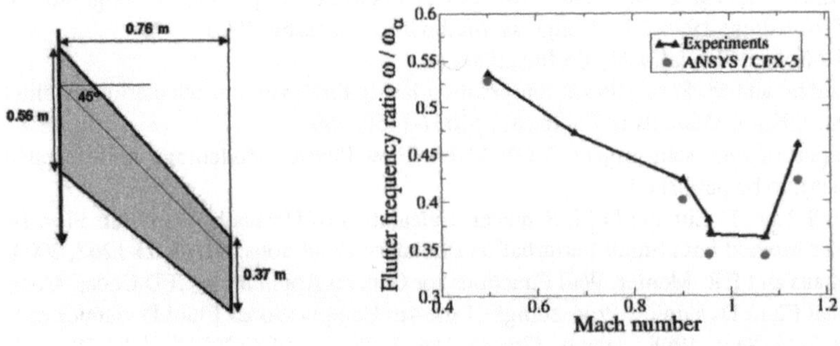

Figure 2 AGARD 445.6 wing: geometry (left) and flutter frequency ratio as function of Mach number.

Figure 3 Forward swept wing: geometry.

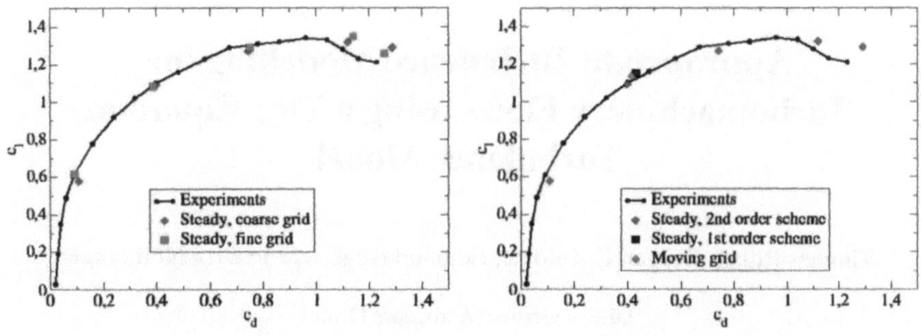

Figure 4 Forward swept wing: drag polar for steady state computations (left) and comparison with moving grid computations (right).

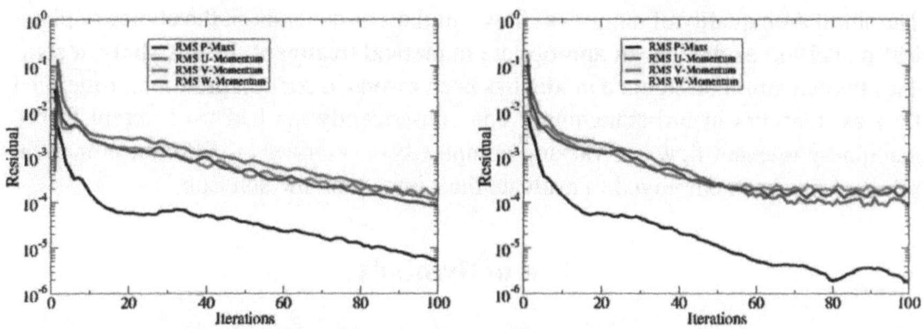

Figure 5 Forward swept wing: convergence history for steady state and moving grid computation.

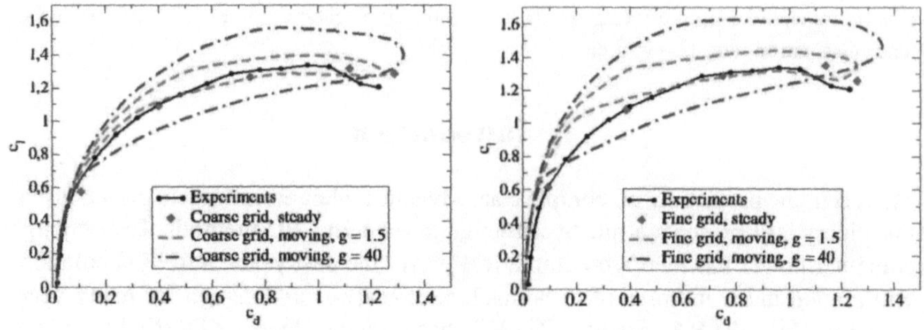

Figure 6 Forward swept wing: drag polar for moving grid computations of guided flight manoeuvre on coarse grid (left) and fine grid (right).

Appropriate Turbulence Modelling for Turbomachinery Flows using a Two-Equation Turbulence Model

Thomas Röber, Dragan Kožulović, Edmund Kügeler, and Dirk Nürnberger

DLR – German Aerospace Center
Institute of Propulsion Technology, 51147 Köln, Germany
thomas.roeber@dlr.de

Summary

The simulation quality of numerical flow simulations depends on the choice of physical modelling as well as an appropriate numerical treatment. In this study, a standard two-equation turbulence model has been extended for compressible, rotational flow as it occurs in turbomachinery and subsequently applied to different turbomachinery relevant flows of varying complexity. A number of different numerical schemes has been employed to evaluate their impact on the solution.

List of Symbols

a – speed of sound, b – half-width of wake, c – chord, k – specific turbulence kinetic energy, p – pitch, S – strain rate, t – time, u – velocity , u_d – velocity defect, U_∞ – free stream velocity, x, y, z – cartesian coordinates, y^+ – dimensionless wall distance,
α, β, σ – closure coefficients, Γ – generic exchange coefficient (e.g. viscosity, heat conductivity), Θ – momentum thickness, $\mu_{(t)}$ – dynamic (eddy) viscosity, $\nu_{(t)}$ – kinematic (eddy) viscosity, ξ, η, ζ – curvilinear coordinates, ρ – density, τ_{ij} – Reynolds stress tensor, φ – generic transported property (e.g. k, ω), ψ – convection scheme limiter, ω – specific turbulence dissipation rate, Ω – vorticity

1 Introduction

The accurate prediction of complex aerodynamic phenomena occuring in turbomachinery still presents a major challenge to modern CFD methods. Even though computer power has increased dramatically over the past years, RANS simulations will remain the workhorse of turbomachinery analysis and design for many years to come. Thus, the RANS solver TRACE has been developed at DLR's Institute of Propulsion Technology in Cologne over the last years and is currently being applied at a number of research institutions and industry companies.

TRACE features parallel, distributed computations of three-dimensional, structured and unstructured multi-block meshes [7]. The fluxes of the RANS equations are discretized using a finite volume scheme and the TVD upwind scheme by Roe

[9], the equations are being solved using an implicit time-stepping algorithm. The turbulence model equations are solved using a similar algorithm, but are decoupled from the mean equations of motion [16].

This study is mainly concerned with the improvement of turbulence modelling within TRACE. To this end, turbulence model extensions aimed at compressible, rotational turbomachinery flow are incorporated into a standard two-equation model and tested in conjunction with various numerical schemes. Even though newer turbulence modelling strategies, e.g. Reynolds stress models or hybrid approaches exist, the majority of turbomachinery calculations are still carried out using two-equation models.

2 Turbulence Modeling

2.1 Turbulence Model Equations

For closure of the Reynolds averaged Navier-Stokes equations, the two-equation k-ω model as presented by Wilcox [14] is being used. One main advanage of choosing a k-ω model over, say, a k-ϵ type model is that the ω-equation can be integrated through the viscous sublayer without using special damping functions, making it easy to implement into a CFD solver. Several extensions of this model exist, furthermore, extensions aimed at other two-equation models are straightforward to implement into this model [16]. A number of adaptions and enhancements have been made in order to improve the model's behaviour in unsteady, compressible flow. The original turbulence model equations are as stated below:

$$\frac{\partial \rho k}{\partial t} + \frac{\partial \rho u_j k}{\partial x_j} = \tau_{ij}\frac{\partial u_i}{\partial x_j} - \beta_k \rho \omega k + \frac{\partial}{\partial x_j}\left[(\mu + \sigma_k \mu_t)\frac{\partial k}{\partial x_j}\right] \tag{1}$$

$$\frac{\partial \rho \omega}{\partial t} + \frac{\partial \rho u_j \omega}{\partial x_j} = \alpha \frac{\omega}{k}\tau_{ij}\frac{\partial u_i}{\partial x_j} - \beta_\omega \rho \omega^2 + \frac{\partial}{\partial x_j}\left[(\mu + \sigma_\omega \mu_t)\frac{\partial \omega}{\partial x_j}\right] \tag{2}$$

The turbulence model production terms have been altered to enhance the model's behaviour in the stagnation point region. Following a proposal by Kato and Launder [5], the strain rate $|S^2|$ has been replaced by $|S\Omega|$, where $|\Omega|$ is the trace of the vorticity tensor. Note that, strictly speaking, this modification violates the Boussinesq assumption and the conservation of energy, while on the other hand it effectively suppresses the unphysical overproduction of turbulent quantities at the stagnation point.

In order to account for pressure dilatation and pressure diffusion, additional source terms depending on the turbulent Mach number $\mathrm{Ma}_\mathrm{t}^2 = 2k/a^2$ are inserted into the k-equation following a proposal by Sarkar [10]:

$$P_\mathrm{d} = \alpha_2 \rho \mathrm{Ma}_\mathrm{t} \tau_{ij}\frac{\partial u_i}{\partial x_j} + \alpha_3 \beta_k \mathrm{Ma}_\mathrm{t}^2 \rho \omega k \text{ , with } \alpha_2 = 0.15 \text{ , } \alpha_3 = 0.2 \tag{3}$$

Furthermore, the turbulence model has been adapted to improve prediction accuracy in compressible turbulent mixing layers by adjusting the destruction terms' closure coefficients [15]

$$\begin{aligned} \beta_{k,c} &= \beta_k \left[1 + \tfrac{3}{2} \max\left(\mathrm{Ma}_t^2 - \tfrac{1}{16}, 0\right)\right] \\ \beta_{\omega,c} &= \beta_\omega - \beta_k \left[\tfrac{3}{2} \max\left(\mathrm{Ma}_t^2 - \tfrac{1}{16}, 0\right)\right] \end{aligned} \tag{4}$$

Finally, a new destruction term has been added to the ω-equation to model the effect of streamline curvature and rotation on the dissipation of turbulent kinetic energy as suggested by Bardina et al. [1]:

$$D_r = -\alpha_4 \rho \omega |\Omega| \quad \text{with:} \quad \alpha_4 = 0.15 \tag{5}$$

A more detailed discussion of these turbulence model modifications can be found in [6].

2.2 Assessment of Turbulence Model Transport Mechanisms

To appropriately model multi-stage turbomachinery flows, both the wall-bounded shear layers and the free shear layers (i.e., wakes) have to be accurately predicted. The wall-bounded shear layers are dominated by the source terms of the governing equations, while the free shear layers are dominated by transport mechanisms, i.e. convection and diffusion. In this chapter, the numerical modelling of transport terms will be analysed.

Numerical Treatment of Transport Terms The popular first-order upwind scheme [4] originally employed to spatially discretize convective fluxes of the turbulence model's transported properties has the advantage of being unconditionally stable. This, however, is gained at the expense of a considerable truncation error of the order $\mathcal{O}(\Delta x)$ and leads to an unphysically large amount of *numerical* homogenization of the transported quantities.

In order to improve this, the cut-off of the discretization has been shifted so that the scheme is now second order accurate. Using a second-order upwind scheme rather than a first order one results in a considerable gain in accuracy at the expense of stability, especially in the transient phase of a computation. Therefore, a limiter of the form (assuming Δx is constant and convection velocity $u > 0$, i.e., backward differencing is applied):

$$\psi = \frac{2r}{r^2 + 1} \quad \text{with:} \quad r = \frac{\Delta\varphi_j^{Bw}}{\Delta\varphi_{j-1}^{Bw}}, \tag{6}$$

$$\frac{\Delta\varphi_j^{Bw}}{\Delta x} = \frac{\varphi_j - \varphi_{j-1}}{\Delta x} \quad \text{and} \quad \frac{\Delta\varphi_{j-1}^{Bw}}{\Delta x} = \frac{\varphi_{j-1} - \varphi_{j-2}}{\Delta x}$$

has been adopted and inserted into the scheme, so that the scheme for the convection of a property φ now takes the form:

$$u\frac{\partial \varphi}{\partial x} \approx u\left\{\frac{\Delta\varphi_j^{Bw}}{\Delta x}+\frac{1}{2}\psi\left(\frac{\Delta\varphi_j^{Bw}-\Delta\varphi_{j+1}^{Bw}}{\Delta x^2}\right)\right\} \quad . \tag{7}$$

Note that only the convective terms of the turbulence model equations are considered in this study, while the convective terms of the Reynolds averaged equations of motion are always modeled by a second order accurate Fromm-scheme and are *not* altered.

The full discretization of the diffusion term will be compared to a variant of the thin-layer approximation, in which only derivatives along the curvilinear direction normal to the wall are taken into account. Practically, in 3D turbomachinery flow, there are several wall-normal directions so that in fact only mixed derivatives that occur when converting the coordinate system from Cartesian (x, y, z) to generalised curvilinear (ξ, η, ζ) coordinates are not taken into account [1]:

$$\begin{aligned}\tfrac{\partial}{\partial x}\left(\Gamma\tfrac{\partial\varphi}{\partial x}\right) = {} & \tfrac{\partial}{\partial\xi}\left[\xi_x\Gamma\left(\varphi_\xi\xi_x+\varphi_\eta\eta_x+\varphi_\zeta\zeta_x\right)\right]+\\ & \tfrac{\partial}{\partial\eta}\left[\eta_x\Gamma\left(\varphi_\xi\xi_x+\varphi_\eta\eta_x+\varphi_\zeta\zeta_x\right)\right]+\\ & \tfrac{\partial}{\partial\zeta}\left[\zeta_x\Gamma\left(\varphi_\xi\xi_x+\varphi_\eta\eta_x+\varphi_\zeta\zeta_x\right)\right]\\ \approx {} & \tfrac{\partial}{\partial\xi}\xi_x\Gamma\varphi_\xi\xi_x+\tfrac{\partial}{\partial\eta}\eta_x\Gamma\varphi_\eta\eta_x+\tfrac{\partial}{\partial\zeta}\zeta_x\Gamma\varphi_\zeta\zeta_x\end{aligned} \tag{8}$$

Both the turbulence model equations and time-mean equations of motion are affected by this change in the discretization scheme. Generally, the computational cost of these changes is quite low, the more accurate convective flux computation takes slightly more time than the fully resolved diffusion terms. For a typical compressor test-case (one-stage TU Darmstadt research compressor, cf. below), Tab. 1 states the increase in computational cost for the different levels of refinement. Note that using both improved schemes together is actually cheaper than the sum of using each scheme separately, as some auxiliary computations are carried out only once by the solver.

Assessment of Numerical Schemes For a meaningful analysis of transported turbulence quantities, three combinations for different transport term modelling, denoted as testcase A, B and C in Tab. 2 are to be tested.

Since the far wake is mainly dominated by transport mechanisms, it is a suitable flow for the current analysis. Also, the far wake posesses known self-similar solutions of mean flow quantities[2] [8]:

$$x\text{-dir.:}\quad \left(\frac{b}{\Theta}\right)^2 = 16\frac{\nu_t}{U_\infty\Theta}\ln 2\left(\frac{x}{\Theta}\right) \quad , \quad \left(\frac{U_\infty}{u_d}\right)^2 = 4\pi\frac{\nu_t}{U_\infty\Theta}\left(\frac{x}{\Theta}\right) \tag{9}$$

[1] An index x, ξ, η and ζ denotes differentiation with respect to x, ξ, η and ζ respectively.

[2] For this analysis, $b = 0.01$ m, $U_\infty = 30$ m/s and $u_d = 1.5$ m/s at the inlet of the computational domain, so that the incompressible momentum thickness $\Theta = 0.001034$, cf. Fig. [1].

$$y\text{-dir.:} \quad \frac{U}{u_d} = \exp\left\{-4\left(\frac{y}{b}\right)^2 \ln 2\right\} \tag{10}$$

As inlet boundary condition for the turbulence model, it is assumed that $\nu_t/U_\infty\Theta = 0.032$ [12], so that $\nu_t/\nu = 66.2$. The latter is presumed to be constant in y-direction. The turbulent kinetic energy $k(y/b)$ also approaches similarity. This distribution is taken from [3] and is plotted in Fig. 2. The resulting far wake is considered as self-similar at the inlet boundary, since the fictive origin lies at a distance of $x/\Theta \approx 1000$ further upstream.

Two grids with ca. 36 000 nodes each were generated in order to assess the different schemes: The cells of the first grid are aligned to the mean flow, while the cells of the second grid are inclined at an angle of 45° with respect to the mean flow. Thus, the aligned grid represents the most favorable and the non-aligned grid the worst case for numerical resolution of transport mechanisms. In practice, during the convection of a wake through a blade passage, a mixture of both cases occurs.

Grid-independence studies (not shown here) suggest a number of at least 20 cells for the width of the wake ($2b$), in order to properly resolve the flow. In the present study, 30 cells have been applied. Assuming that the numerical time integration behaves similarly as the spatial approximation, the time for lateral passing of the wake for a distance of $2b$ should also be resolved with at least 20 nodes in unsteady simulations.

Figure 1 shows the influence of presented modeling configurations on the development of mean flow in x-direction. The agreement of simulations performed on the aligned grid with the theoretical values of $b(x)$ and $u_d(x)$ is quite good. The simulations from the non-aligned grid show higher spreading and stronger homogenization of the flow. The effect of different modeling of diffusion terms on wake spreading is much smaller than the effect caused by the grid. However, the fully resolved approach shows sligthly better predictions than the thin layer approximation. Furthermore, regarding the mean flow properties, the testcases A, B and C show almost no difference on corresponding grids.

In contrast, the influence on the simulated turbulent quantities is substantial. In Fig. 2, k is shown in relation to $\sqrt{u_d d(U_\infty b)/dx}$ [11], as this relation is more suitable for self-similarity considerations than u_d, which is also widely spread in literature. All testcases preserve the similarity of k very well on the aligned grid, while only testcases B and C show satisfactory preservation of similarity on the non-aligned grid.

Primarily responsible for the better results is the higher order discretization scheme for convective fluxes. To a smaller degree, the fully resolved viscous fluxes show better results than the thin layer approximated.

Together with the turbulent kinetic energy k, the eddy viscosity μ_t is also affected by the convection term discretization scheme. This quantity serves as a connection between the turbulence model and the RANS equations in Boussinesq-style turbulence modelling.

2.3 Application to Turbomachinery Flows

Transonic Compressor Cascade Next, the flow through a highly loaded transonic compressor cascade, DLR TSG-97, is considered. This test case was designed at DLR Cologne for validation purposes [13]. The flow is characterized by shock/boundary-layer interaction, separation on the suction side and a large corner separation in the suction side/side wall edge, which makes it particularly suitable to investigate the effect of the numerical scheme on separated flows.

To investigate the test case, an 8-block O/H-grid with 712 754 nodes (49 nodes in lateral direction) representing half the wind tunnel width was used with a symmetry boundary condition. The dimensionless wall distance of the first node was always $y^+ \approx 1$, so that a low-Reynolds number formulation of the boundary conditions could be used.

The pressure distribution at midspan (Fig. 3, left) shows that the pre-shock suction side flow is particularly dependent on the discretization scheme, while behind the shock and on the pressure side the influence is only slight. The laminar separation bubble below the shock was not captured because the computation was carried out without a transition correlation, rather, the flow was assumed to be fully turbulent.

Inspection of the turbulence kinetic energy and axial velocity contours (Fig. 3, right) in the rear corner of the suction side shows that the shape and size of the corner vortex and the separation zone is predicted differently by both discretization approaches. While the fully resolved diffusion terms predict a very round separation zone, the corner vortex computed using the simplified modelling is rather flat and more expanded in pitchwise direction. One entire blade pitch (pitch-to-chord ratio $p/c = 0.7$) and one-half blade span are shown.

This difference in the corner separation shape has an essential impact on the blade loading, and therefore the entire flow is predicted differently using both approaches.

Transonic Compressor Stage As final test-case, a single-stage, transonic research compressor thoroughly investigated at TU Darmstadt [2] was chosen. The relative tip Mach number of this test case is $\text{Ma} = 1.34$ at the aerodynamic design point, so that the use of a turbulence model adapted to compressibility and rotation is appropriate. A study on the effect of the rotational and compressibility extension of the turbulence model on the characteristic map can be found in [6].

The computational mesh of this test case consists of an OCH topology, with an extra H-grid in the tip clearance. Altogether, rotor and stator have been discretized using 250 000 nodes, with a first wall distance of $y^+ \approx 30$ for the rotor and $y^+ \approx 50$ for the stator. Therefore, wall function solid surface boundary conditions have been used. For convective transport, the second order accurate scheme was applied in all cases, and only the numerical scheme for diffusion was altered to assess its influence on the flow behaviour under turbomachinery conditions.

For the design point as shown in Figure 4, the radial distributions of the total pressure ratio and isentropic efficiency behind the stator (not shown here) show good agreement with the experiment for both discretization mechanisms. However, looking at the radial flow angle at rotor exit (Fig. 4, left) , a greater difference between both schemes' predictions can be observed. The radial flow angle coincides very well for both discretization schemes for most of the radius, while the tip vortex is less pronounced in the solution using the thin layer approximation. This can also be observed in Figure 4 (right), in which the different rotor tip vortex shapes look somewhat similar to the different shapes of the corner vortex in the compressor cascade.

This leads to a small underprediction of mass flow for the same back pressure, and thus a slight shift of the compressor performance map. In this particular compressor, however, due to the stator effect, the change in the overall performance is minor and in fact hardly visible, but, judging from the different prediction of the vortices, the effect of the numerical scheme is expected to grow in off-design and multi-stage computations.

3 Conclusion

To improve the accuracy of turbomachinery flow prediction, a good physical model is needed as well as an appropriate numerical scheme. Therefore, a number of extensions have been incorporated into the k-ω model and the transport terms of the model and mean equations of motion discretized using different levels of numerical approximation.

It can be seen that the prediction accuracy of flows in general is very dependent on the type of numerical scheme applied to the turbulence model and equations of motion.

By and large, proper numerical modelling of diffusion is essential for a good simulation of separated flow, as it occurs in end wall regions, e.g. rotor tip vortices or edge vortices. As can be seen from the pressure distribution at midspan of the compressor cascade, it also helps capture the pre-shock flow development, which in turn is necessary for proper transition modelling. On the other hand, in two-dimensional, attached and simple flows, the influence of mixed derivatives on the flow solution is negligible so that the thin layer approximation remains a very useful tool for a proper prediction of these types of flows. In multi-stage, three-dimensional turbomachinery flows, however, it has a considerable impact on the overall simulation accuracy.

The numerical scheme applied to the convective fluxes plays a significant role in free stream and wake regions. As usually a mixing plane model is employed for steady multi-stage computations, the accurate prediction of the convection of turbulent model quantities is particularly important for the unsteady simulation of multi-stage flows or blade-row interaction, e.g. wake induced transition.

References

[1] Bardina, J., Ferziger J.H., Rogallo, R.S.: Effect of Rotation on isotropic Turbulence: Computation and Modelling. J. Fluid Mech. Vol. **154** (1985) 321–336

[2] Blaha, C., Henneke, D.K., Fritsch, G., Hoeger, M., Beversdorff, M.: Laser-2-Focus Measurements in a Transonic Compressor Blisk-Rotor and Comparison with 3D Numerical Simulations. ISABE 97-7069, proc. *13th ISABE*, Chattanooga, Tennessee (1997)
[3] Durbin, P.A., Pettersson Reif, B.A.: *Statistical Theory and Modeling for Turbulent Flows*. John Wiley & Sons Ltd, Chichester (2001).
[4] Fletcher, C.A.J.: *Computational Techniques for Fluid Dynamics*. Volume I Springer-Verlag, Berlin (1990).
[5] Kato, M., Launder, B. The Modelling of Turbulent Flow around Stationary and Vibrating Square Cylinders. Proc. *Ninth Symposium on Turbulent Shear Flows*, Kyoto (1993)
[6] Kožulović, D., Röber, T., Kügeler E., Nürnberger, D.: Modifications of a two-equation Turbulence Model for Turbomachinery Fluid Flows. Deutscher Luft- und Raumfahrtkongress, Dresden (2004)
[7] Kügeler, E., Nürnberger, D.: *TRACE User's Manual*, Version 2.2, DLR IB-325-07-04, Cologne (2004)
[8] Ramaprian, B.R., Patel, V.C., Sastry M.S.: The Symmetric Turbulent Wake of a Flat Plate. AIAA J. Vol. **20**, No. 9 (1982) 1228–1235.
[9] Roe, P.: Approximate Riemann solvers, parameter vector and difference schemes. J. Comp. Phys. Vol. **34** (1981) 357–372
[10] Sarkar, S.: The pressure-dilatation correlation in compressible flows. J. Phys. Fluids, Vol. **4**, No. 12 (1992) 2674–2682
[11] Thomas, F. O., Liu, X.: An experimental investigation of symmetric and asymmetric turbulent wake development in pressure gradient. J. Phys. Fluids. Vol. **16**, No. 5 (2004) 1725–1745.
[12] Townsend, A.A.: *The Structure of Turbulent Shear Flow*. Cambridge University Press, Cambridge, 1956.
[13] Weber, A., Steinert, W., Fuchs, R., Schreiber, H.A.: Testcase: *3D Flow in a Transonic Compressor Cascade (DLR TSG-97)*, DLR IB-325-06-99, Cologne (1999)
[14] Wilcox, D.C.: Reassessment of the Scale-Determining Equation for Advanced Turbulence Models. AIAA J. Vol. **26**, No. 9 (1988) 1299–1310
[15] Wilcox, D.C.: Dilatation-Dissipation corrections for Advanced Turbulence Models. AIAA J. Vol. **30**, No. 11 (1992) 2639–2646
[16] Wilcox, D.C.: *Turbulence Modeling for CFD*, DCW Industries, La Cañada, (1998)

Table 1 Computational cost for different transport term formulations

	thin layer	fully resolved
1st order upwind	100%	106%
2nd order upwind	109%	111%

Table 2 Analysed combinations of transport terms modeling

Testcase	Convective terms of turbulence model	Diffusive terms of turbulence model and RANS eqs.
A	1st order upwind	fully resolved
B	2nd order upwind	fully resolved
C	2nd order upwind	thin layer approximation

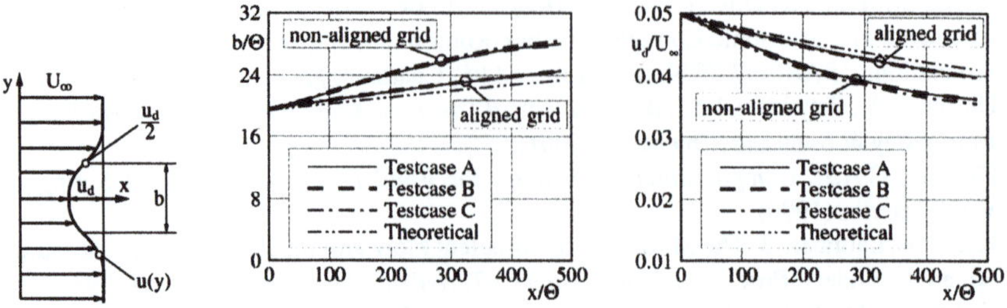

Figure 1 Far wake notation and influence of discretization scheme of convective and diffusive terms at the mean flow of a far wake

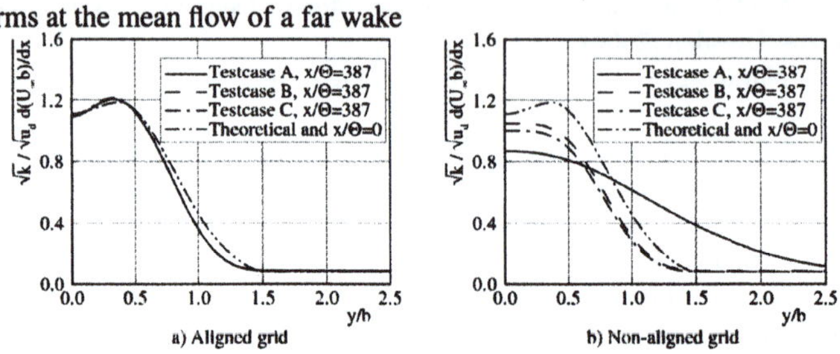

Figure 2 Influence of discretization scheme of transport terms on the turbulence quantities of a far wake

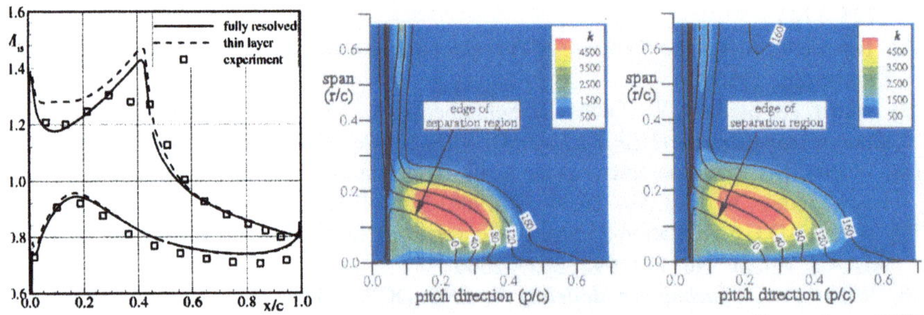

Figure 3 Suction side corner vortex (k and axial velocity iso-lines) at trailing edge of TSG compressor cascade with fully resolved diffusion terms (*l.*) and thin layer approximation (*r.*)

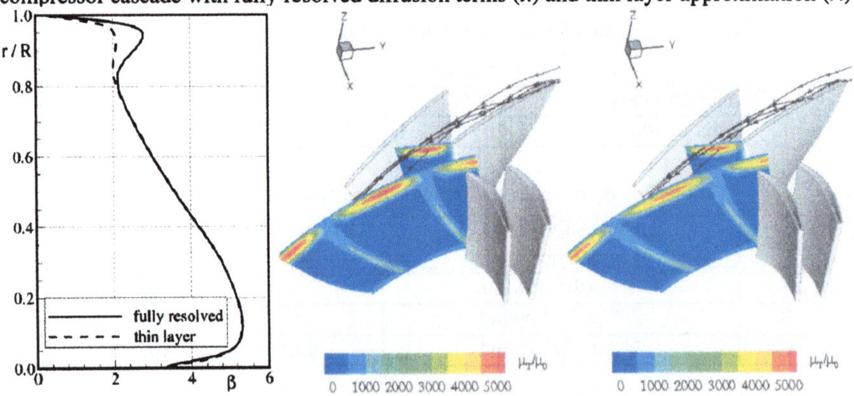

Figure 4 Rotor tip vortex near blade trailing edge and at rotor-stator interface for Darmstadt compressor stage with fully resolved diffusion terms (*l.*) and thin layer approximation (*r.*)

Numerical determination of dynamic derivatives for transport aircraft

A.-R. Hübner

Deutsches Zentrum für Luft- und Raumfahrt (DLR),
Institute of Aerodynamics and Flow Technology,
Lilienthalplatz 7, D-38108 Braunschweig

andreas.huebner@dlr.de

Summary

The represented investigations are concerned with simulations of unsteady aircraft aerodynamics and thus belongs to the research field of computational fluid dynamics (CFD) for aerospace applications. The calculations are based on the surface singularity panel method VSAERO for simulation of quasi-steady motions and on the solution of the Time-dependent Reynolds-averaged Navier-Stokes (TRANS) equations using the finite volume parallel solution algorithm with an unstructured discretization concept (DLR TAU-code). The analysis consists of two parts, in order to obtain a more detailed understanding of the unsteady aerodynamical and flight mechanical behaviour of an airplane. For this purpose systematic investigations have been performed with basic configurations which have NACA0012 profiles (e.g. wing, wing + horizontal tail, wing + vertical tail). The second part of the investigation is the calculation of the dynamic derivatives of a modern transport aircraft configuration (DLR-F12). The objective of this investigation is to adapt the numerical methods for the calculations of dynamic derivatives and to validate the numerical results against wind-tunnel data.

Nomenclature

p	roll rate	$c_{Lq} + c_{L\dot{\alpha}}$	lift due to pitch motion
q	pitching rate	$c_{mq} + c_{m\dot{\alpha}}$	pitchdamping-rerivative
α	angle of attack	c_{lp}	rolldamping-rerivative
$\dot{\alpha}$	angular velocity	c_{np}	yaw moment due to roll motion
F	reference area	c_{Yp}	side force due to roll motion
l_μ	reference chord length	$FWVH$	full confi guration (fuselage, wing,
$q_\infty = \frac{\varrho}{2}V^2$	dynamic pressure		vertical/horizontal tail plane)
$c_L = \frac{L}{q_\infty F}$	lift coeffi cient	FW	confi guration without tails
$c_M = \frac{M}{q_\infty F l_\mu}$	pitching moment coeffi cient		(fuselage, wing)

1 Introduction

The prediction of the flight mechanic behavior of an aircraft in development has to be as precise as possible. Systematic investigations have been performed during several wind tunnel entries from 1998 to 2004 ([1],[4] and [5]). For new configurations the standard prediction tools (half empirical methods) are not as accurate as required. The panel method VSAERO and the higher-order numerical DLR-TAU code are used to determine the dynamic behaviour of aircraft configurations.

2 Numerical Approach

2.1 Panel Method VSAERO

VSAERO is a computer program for calculating the nonlinear aerodynamic characteristics for aircraft of arbitrary configurations in subsonic flow. Nonlinear effects of wake shape are treated in an iterative wake relaxation procedure, while the effects of viscosity are treated in an iterative loop coupling potential flow and integral boundary layer calculation. The simulation of quasi-steady motions about the body-fixed axis is a very important capability of the program [6].

The basis of the program is a surface singularity panel method using quadrilateral panels on which doublet and source singularities are distributed in a piecewise constant form. The panel source values are directly determined by the Neumann boundary condition controlling the normal component of the local flow. The doublet values are solved after requiring the equivalent flow to be irrotational and incompressible. Surface perturbation velocities are obtained from the gradient of the surface potential, while field velocities are obtained by directly summating all singularity panel contributions (further informations can be found in Nathman [8] and Maskew [9]).

2.2 CFD Solver TAU

The calculations are based on the solution of the Time-dependent Reynolds-averaged Navier-Stokes equations. This is accomplished by employing a finite volume flow solution method, the DLR TAU code [7], which is characterized by an unstructured mesh concept. As the data structure is based on the edges of the control volume, the code is independent of the type of grid cells, allowing it to handle unstructured, structured and hybrid grids. The governing equations for three-dimensional compressible flows are solved on a dual background grid, which is determined directly from the primary grid specify the type of the dual grid. The flow variables are calculated at the centers of the dual grid, i.e. at the vertices of the primary grid. The code use an explicit multistage Runge-Kutta time stepping scheme. For steady applications the convergence is accelerated using local time stepping, residual smoothing and multigrid. For time accurate computations the implicit dual time stepping approach has been implemented, to bypass the stability restiction of the explicit baseline method. The inviscid fluxes are calculated either by a Roe- or AUSM-type 2nd-order upwind scheme, or by employing a central scheme with scalar dissipation, which will be used for the present study. The gradients of the flow variables are determined with a Gauss-Green formula. The viscous fluxes are discretized using central differences. For time-accurate computations the well-known dual-time stepping approach for 1st-, 2nd- or 3rd-order discretization in physical time is implemented in TAU.

The turbulence models available in the TAU-code are the Spalart-Allmaras model [11], the Wilkox $k-\omega$ -model [12] and the SST-model according to Menter [10]. Transition can be fixed by assigning a flag (laminar or turbulent) to each surface node. Optimisation for different architectures is achieved by vector- or cache-type coloring of the edges, on which most of the work is done. For parallel computations a domain-decomposition is used providing a subset of dual-grids.

The adaptation module detects regions with insufficient grid resolution by gradient sensors of flow variables and performs local grid refinement by bisection of cells. The initial solution is then interpolated to the adapted grid. In addition, the adaptation also allows the redistribution of the prismatic layers to capture the viscous boundary-layer in Navier-Stokes computations adequately. The solver is part of the MEGAFLOW-project and is presented in more detail e.g. in [2] and [3].

3 Experimental Data

In order to evaluate the accuracy of the experimental and numerical methods used for the prediction of the dynamic derivatives a new lightweight model is used, which was completed in September of 2004. This DLR-F12 model should not only allow the measurement of unsteady forces and moments but also unsteady pressure distributions using pressure taps at specific stations on the wing and horizontal and vertical stabilizer.

In October 2004 the first test campagne was performed in the DNW Low-Speed-Wind-Tunnel Braunschweig (DNW-NWB). The experiment includes measurements under both steady-state and unsteady conditions in order to get the resulting force and moment to determine the dynamic derivatives. These examination are used to validate the numerical codes.

4 Results

Preliminary examinations with simple configurations have been performed. These configurations are: a NACA0012 wing, a wing with an additional horizontal tail plain (htp) (see Fig. 2) and a wing with an additional vertical tail plain (vtp). The size of these configurations (wingspan, lever arms of the horizontal and vertical tail plane) and of a conventional wind tunnel model are similar. Using the panel method VSAERO longitudinal/lateral solutions under steady state and under inviscid and viscous conditions have been calculated. Also quasi-steady movements around all body fixed axes are determined. Finally the Euler-/Navier-Stokes equations are going to be solved for quasi-steady and unsteady movements using the DLR-TAU code.

It is planned to follow the same strategy for the DLR-F12 configuration. Different configurations will be simulated to allow the analysis of the influence of individual components of the aircraft. Comparable to the measurements the calculations are done by model frequencies up to 3 Hz and velocities of 70 m/s.

Applying a quasi-steady pitching motion for different pitching rates q the total lift C_L and the moment C_m are calculated (see Fig. 3). The characteristics are linear for these small angles of attack. The TAU results correspond well with the inviscid VSAERO solutions. Nearly the same results can be found for the pitching moment.

Furthermore Fig. 4 compares the unsteady Euler and Navier-Stokes solutions with the quasi-steady solutions. Shown is the sum of the coefficients C_{Lq} and the C_{mq} for the quasi-steady motion and the derivatives with the additional $\dot{\alpha}$ term ($C_{L\dot{\alpha}}$ and $C_{m\dot{\alpha}}$) for the unsteady solutions. The data evaluation of the hysteresis loop of the unsteady solutions is done by the same program used in the DNW-NWB. For each physical point two periods including a minimum of 50 physical time steps per period are needed.

The quasi-steady solutions calculated with the TAU code are in a good agreement with the VSAERO solutions. The predictions of these derivatives are a bit higher in values, due to the influence of the flow field, which can not be taken into account by VSAERO. But in case of the unsteady Euler result the additional lift is not as high as for the quasi-steady solution. The reason for this is the $C_{L\dot{\alpha}}$ term. Due to the unsteady behaviour of the flow field the lift is reduced. However the pitch damping increases. This combination of the derivatives is not as expected. Further investigations will be to simulate the pure heave oscillation in order to obtain directly the unsteady $C_{L\dot{\alpha}}$ term and thus the percentage of this coefficient to the complete derivative. Furthermore the unsteady Navier-Stokes solutions are shown in these figures. It is to be expected that due to the viscous terms of the equations the absolute values of the derivatives are reduced.

The first experimental data (DLR-F12) of the roll motion in comparison with inviscid solutions calculated with VSAERO are shown in Fig. 5. For small angles of attack the calculations reveal satisfactory agreements with the measurements but the gradients are not similar. The differences between the numerical and experimental results are the influences of frictional losses and due to unsteady flow field phenomena. The same calculations using the NACA0012 wing + htp configuration (see Fig. 6) confirm these conditions in case of the yaw moment and the side force due to the roll motion.

5 Conclusions

In order to improve the accuracy of the experimental and numerical methods used for the prediction of the dynamic derivatives systematic investigations have been performed using different generic models. With the panel method VSAERO quasi-steady motions are calculated and with the DLR-TAU code quasi-steady and unsteady motions are simulated. The results obtained with the panel method match well with the present measurements. The quasi steady solutions calculated with VSAERO and TAU reveal very similar results in the damping derivatives. By the cross coupled derivatives the tendency is good. But it is necessary to solve the unsteady Euler and Navier-Stokes equations in order to consider the influence of the unsteady flow field. Further investigations have to be done in order to understand the unsteady behaviour in more detail. Furthermore due to these calculations viscosity effects are also taken into account. Using the simple generic models which have NACA0012 profiles, basic informations about the unsteady flow field behaviour are investigated. In addition the measurement of the unsteady pressure distributions using the pressure taps of the DLR-F12 full configuration which will be performed in the next test campaigns are needed to validate the DLR-TAU Code accurately

References

[1] *Dornier GmbH (Ed.):* Oszillierende Derivativwaage für die 3m-Unterschall-Windkanäle der BRD / Entwicklungsphase IV Benutzerhandbuch für die Messdatenerfassungsanlage der MOD. Forschungsbericht 76/13 B, 1976.

[2] *Galle, M.:* Ein Verfahren zur numerischen Simulation kompressibler, reibungsbehafteter Strmungen auf hybriden Netzen. DLR-FB 99-04 1999

[3] *Galle, M.; Gerhold, T.; Evans, J.:* Technical Documentation of the DLR TAU-Code DLR-IB 233-97/A43 1997

[4] *A.-R. Hübner, S. Peters:* Ermittlung aerodynamischer Dämpfungsderivativa zweier Airbus-Typen. AG STAB Jahresbericht 1999, pp 37 -38.

[5] *A.-R. Hübner, T. Löser:* Recent Improvements in the Measurement of Aerodynamic Damping Derivatives. 12. DGLR-Fach-Symposium der STAB. Stuttgart, 2000.

[6] *A.-R. Hübner:* Bestimmung der dynamischen Dämpfungsderivativa von Verkehrsflugzeugen mit dem Rechenverfahren VSAERO, IB 129 -2001/1

[7] *Kroll, N.; Rossow, C.-C.;Becker, K.; Thiele, F.:* MEGAFLOW - A Numerical Flow Simulation System.
21st ICAS congress, 1998, Melbourne, 13.09-18.09. 1998, ICAS-98-2.7.4, 1998.

[8] *Nathman, J.K.:* VSAERO a computer program characteristics of arbitrary confi gurations. User's manual, Analytical Methods Inc., Redmond, Washington, Nov. 1997.

[9] *Maskew, B.:* Predicting aerodynamic characteristics of vortical flow on three-dimensional confi gurations using a surface-singularity panel method. Analytical Methods, Inc.

[10] *Menter, F.R.:* Two-Equation Eddy-Viscosity Turbulence Models for Engineering Applications. AIAA-Journal, Vol. 32, No. 8, pp. 1598-1605, 1994.

[11] *Spalart, P. R.; Allmaras, S.R.:* A One-Equation Turbulence Model for aerodynamic Flows. AIAA-papaer 92-0439, 1992

[12] *Wilcox, D.C.:* Reassessment of the Scale Determining Equation for Advanced Turbulence Models. AIAA-Journal, Vol. 26, No. 11, pp. 1299-1310, 1988.

Figures

Figure 1: DLR-F12 model in the DNW-NWB Braunschweig.

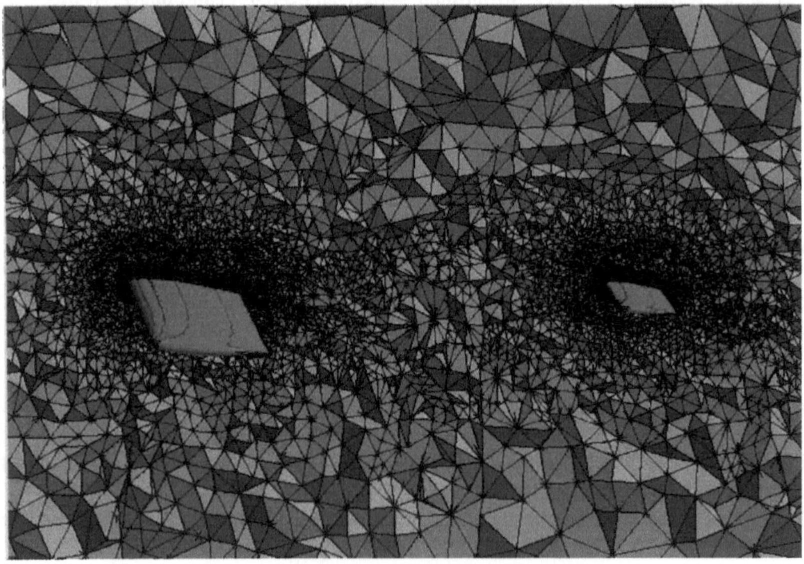

Figure 2 NACA0012 wing + horizontal tail plane

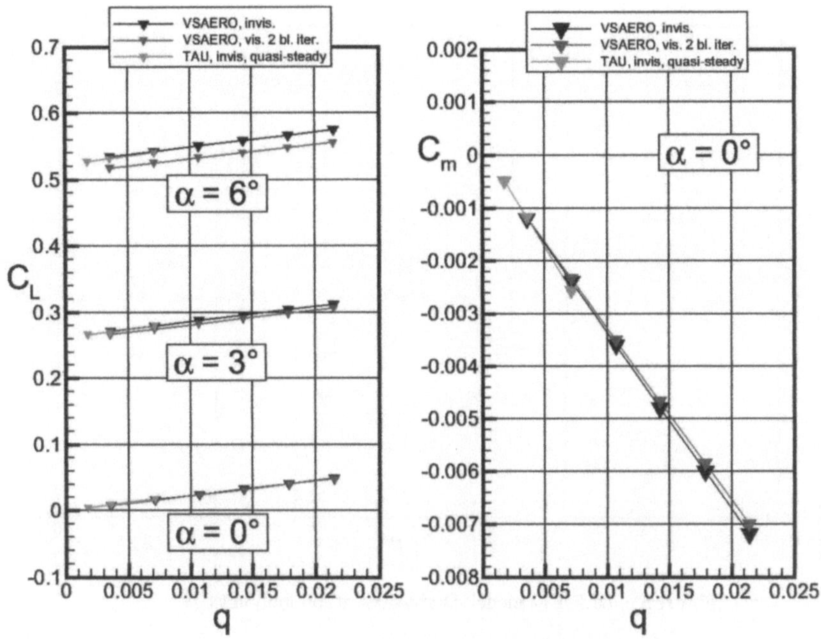

Figure 3 Results of the quasi-steady pitching motion

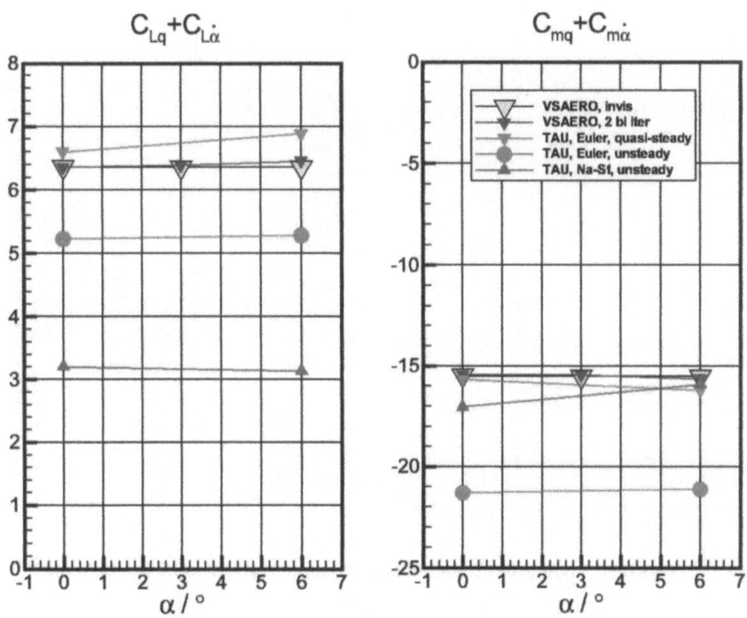

Figure 4 Derivatives of pitching motion vs. α

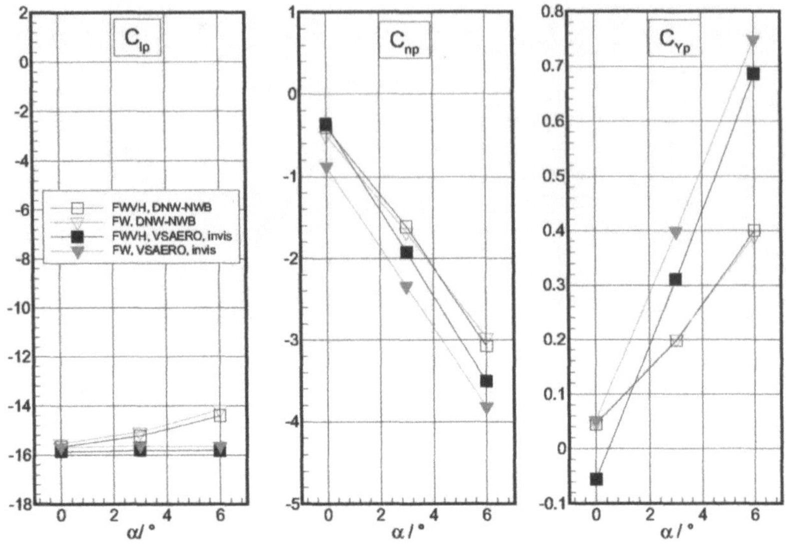

Figure 5 DLR-F12 model. Derivatives of roll motion vs. α

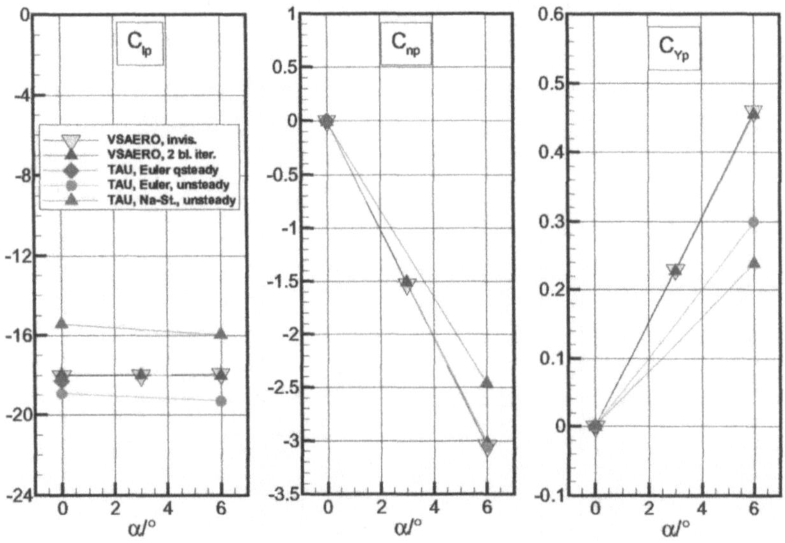

Figure 6 NACA0012 wing + htp, Dynamic derivatives of the roll motion

Detached–eddy simulation of the delta wing of a generic aircraft configuration

A. Gurr[1], H. Rieger[2], C. Breitsamter[3], and F. Thiele[1]

[1] Technical University Berlin, Hermann–Föttinger–Institute for Fluid Mechanics, Sekr. HF1, 10623 Berlin, Germany, andreas.gurr@cfd.tu-berlin.de
[2] EADS Deutschland GmbH, Military Aircraft, Willy-Messerschmitt-Strasse, 85521 Ottobrunn, Germany
[3] Technische Universität München, Lehrstuhl für Fluidmechanik, Boltzmannstr.15, 85748 Garching, Germany

Summary

In this paper, the results of a Detached–Eddy Simulation (DES) of the burst vortex system over a delta wing using a structured, locally refined mesh is described. For evaluation of the simulation quality, averaged velocities are compared directly with the experiment, and an investigation of the instantaneous structure of the spiral burst of the primary vortex is presented. To assess the modelling characteristics, a comparison with an equivalent RANS technique is conducted, in which the topology of the time–averaged surface flow is discussed in relation to the global structure of the vortex system. The simulation of the single wing presented here constitutes the basis of ongoing investigations, and is thereby intended to represent the minimum simulation quality which could be obtained for a complete configuration.

1 Introduction

A particular characteristic of delta wings, compared to wings of large aspect ratio, is the strongly separated and three–dimensional vortex system. At small angles of attack (α), the flow separates from the leading edge forming a vortex system over the suction side of the wing, which remains even after the maximum lift has been exceeded. As well as a nonlinear dependence of the lift on α, a dominant unsteady character and high anisotropy of the turbulence is caused by the phenomenon of vortex burst in the higher α range [3], [5]. These, together with the high curvature of the flow, give rise to special demands on the modelling of this problem, as the physical assumptions inherent in traditional linear eddy viscosity RANS models (such as isotropic eddy viscosity and structural equilibrium of the turbulence) are invalidated. Because of this, it is commonly observed that the simulation quality of such techniques falls with increasing α.

Focussing on the unsteady character mentioned above, the accurate prediction of unsteady aerodynamic loads is of high relevance to certain practical design issues, such as the treatment of fluid–structure interaction or the design of flight control systems. In recent years a novel hybrid modelling strategy known as Detached–Eddy

Simulation (DES) has been introduced [9], which promises improved performance over traditional RANS approaches for separated flows dominated by large–scale unsteadiness. The investigations presented are therefore intended to evaluate the performance and applicability of DES for the simulation of a realistic delta wing aircraft configuration at high angle of attack. In order to assess this, an analysis of the instantaneous structures characterising the vortex burst is presented. This initial investigation of the wing in isolation therefore represents the basis for ongoing investigations, which are to include the evaluation of unsteady field values and integral force coefficients of the full delta–canard configuration. In addition to the DES, the use of different RANS modelling techniques was tested, selected results of which are presented for comparison.

2 Method

Numerical Background. The flow was computed numerically using ELAN3D [12], an in–house finite volume–based code solving the unsteady, and in this case incompressible, Reynolds–averaged Navier–Stokes (RANS) equations in an implicit manner of second order accuracy in space and time. All scalar quantities as well as the cartesian components of tensorial quantities are stored in the cell centres of arbitrarily curvilinear, semi–structured grids that can capture complex geometries and incorporate local refinement. Diffusive terms are approximated with central schemes, whereas convective terms can be treated with central or upwind–based, limited schemes of higher order. The linearized equations are solved sequentially and the pressure is iterated to convergence using a pressure–correction scheme of the SIMPLE type. A generalized Rhie & Chow interpolation is used to avoid an odd–even decoupling of pressure, velocity– and Reynolds–stress components. A variety of turbulence models featuring various degrees of complexity are implemented (algebraic, one or two–equation, and explicit algebraic stress models (EASM)).

Detached–Eddy Simulation. The computations were performed using the hybrid approach introduced as Detached–Eddy Simulation in 1997 [9] which seeks to combine the strengths of RANS and LES [10], where here the SALSA model [8] is employed as the background model. The DES approach exhibits a larger dissipation of eddy viscosity further away from the wall, which reduces the overestimated production and thereby the damping effect of the turbulence. In the LES–mode of this simulation, implicit grid filtering is applied requiring an unsteady, fully 3D computation. The constant $C_{DES} = 0.65$ is calibrated by the simulation of decaying isotropic turbulence [2]. A hybrid blending of central and upwind–based limited schemes of higher order is used, as suggested in [11].

Geometry and Simulation Parameters. The DES is performed for the flow over the delta wing presented in Figure 1. As well as a sweep angle of 50°, a root chord of $c_r = 0.529m$ and a span of $0.38m$, this delta wing is characterised by its three-dimensional form. The simulation parameters are summarised in Table [1]. Since a wide range of periodic physical phenomena are present in the flow in question, the determination of the time step was carried out iteratively. Thus, the chosen time step $\Delta t = 0.002s$ was found at the point when further reduction did not result in a change in the unsteady solution behaviour of the integral forces, Figure 2. It was assumed that with a smaller time step, the influence of the more highly–resolved turbulent scales has no global effect on the flow structure.

The mesh used was based on a 5.2 million point multi–block structured mesh created by EADS–M. In order to reduce the computational burden, this mesh was modified such that a locally–refined region was maintained from the trailing edge, over the suction side up to the region of the stagnation line on the pressure side. Through systematic coarsening of factor 2 in each direction outside this zone, the number of points was reduced to 1.5 million. The height of the fine mesh zone, divided for topological reasons into seven blocks, was selected such that at high angles of attack the vortex system is completely enclosed in the local refinement, Figure 1. In the span–wise direction, the fine region extends from the symmetry plane up to the beginning of the wingtip pod. A refinement of the pod was not conducted, owing to the relatively small influence on the primary flow structure. This local refinement naturally leads to the use of hanging nodes at the relevant block interfaces. Because of the surface–based formulation of the gradients and the unstructured data storage at the block interfaces, the influence and numerical cost of this is small. All computations used six 1.6GHz CPUs connected by a 100Mbit/s ethernet, upon which a full DES computation required $\approx$ 31 days, and a stationary RANS simulation $\approx$ 3 days.

Important for the quality of the DES is the requirement that the switch to LES–mode occur at a sufficient distance from the wall. It was ascertained that in all regions, this criterion was satisfied, such that a negative influence on the boundary layer can be excluded for the mesh used.

Experiment. In the benchmark experiment [1], the flow over the left wing of a delta–canard configuration was measured using hot-wire anemometry in an open-jet wind tunnel of Goettinger design. For a time interval of $6.4s$, the averaged velocities and their statistic characteristics were obtained using a sampling rate of $3kHz$ and a low-pass filter at $1kHz$. Small–scale eddy structures of maximum size $9.5mm$ were captured at 18 planes each having 33×16 lateral and vertical points. Regarding direct comparsion, it must be borne in mind that the measurements were conducted for the complete aircraft, whereas the simulations consider the isolated wing. Therefore, the simulated symmetry–plane at the wing root and the lack of the canard are misrepresentative of the experimental conditions.

3 Results

In the interests of brevity, only the areas of greatest contrast or similarity have been selected for the following comparative discussions. Furthermore, the SALSA model is chosen as representative of all URANS models used, as the variations between these, although existant, are not relevant in the current context. In particular, as shall be described, all URANS models deliver exclusively steady results

Comparison to Experiment. The downstream development of the flow is represented in Figure 3 at a horizontal slice in the lower region of the vortex system. The DES longitudinal velocity exhibits a large negative region in the center of the vortex system, extending upstream to the vortex burst–point. In the experiment by contrast, a much weaker reduction of the longitudinal velocity values is observed, which remain positive throughout the entire vortex system. Furthermore, a stronger thickening of the vortex system is apparent in all velocity components.

The position of the vortex burst is characterised in the experiment by a sudden increase of the longitudinal velocity rms–values, and is most easily found in the simulation due to the accompanying sudden pressure rise. The experimental burst location lies within the range $0.35 < x/c_r < 0.40$, and $0.25 < x/c_r < 0.35$ for the DES.

Comparison to URANS. For the DES, a surface flow pattern dominated by the vortex burst can be seen in Figure 4. The primary attachment line is bent sharply towards the symmetry plane at burst location due to the sudden rise in turbulence intensity and thickening of the vortex system, and is thereafter curved towards the leading edge. Due to the sudden conversion of rotational energy to friction and reduction of vorticity, the suction peak is truncated by the burst point. The secondary vortex is furthermore forced to separate at the leading edge by the enlargement of the rotation core in this region. The primary vortex separates downstream in parts of the rear portion of the wing.

Turning to the URANS, the clearly–defined suction peak continues up to three times longer in the wing depth direction. Furthermore, the primary attachment shows a smoother curvature toward the symmetry plane. Likewise, the distribution of the secondary separation and attachment lines shows a stable, unburst vortex system. Between these, a weakly pronounced tertiary vortex is suggested, with two additional turning points in the streamlines. As no temporal change in the solution took place with the SALSA-URANS simulation, the result corresponds to that of a RANS. In agreement with [6] and [7], the same steady behaviour was found irrespective of RANS-model used ($k-\omega$, $k-\epsilon$ or EASM). The time–independence also remained unaffected by a variation of time step and convection scheme. It is therefore concluded that URANS is not capable of delivering information on the presence of unsteadiness and/or vortex burst, instead delivering the overly–conservative prediction of a stable vortex system continuing up to the trailing edge.

Vortex burst. The type of vortex core visualisation described in [4] offers an impression of the spiral instabilities accompanying the vortex burst, Figure 5. These structures arise from inside the burst vortex system and move in the axial direction with a simultaneous radial enlargement. The spiral shape results from the influence of the incident flow and the circumferential velocity of the primary vortex.

The clear identity of the spiral–type burst process can be further clarified in the instantaneous image of Figure 6, in which regions of strongly reversed flow and low total pressure are represented. The spiral form of the low pressure region shows an avoidance of the primary vortex core, caused by a stability–related change in the upstream vortex system structure. Due to the sudden decrease in, and even reversal of the axial velocity in the core of the primary vortex, this is pushed towards the rotating outer area. The associated expansion of the vortex system causes axial gradients in the circumferential velocity. Thus, the spiral shape of the low pressure region rotates in the opposite direction to the primary vortex.

Conclusions

In this work, the behaviour of the hybrid DES method for such vortical delta wing flows has been investigated. The appearance of clearly identifiable unsteady behaviour and spiral vortex burst is highly encouraging against the backdrop of the universal failure of conventional (U)RANS methods in this respect.

However, the remaining deviations from the experiment still raise questions concerning the simulation parameters and modelling assumptions, and are being addressed in ongoing work. It is important to mention that it is essential for properly–defined DES modelling that fine enough spatial and temporal discretization is used, although precise practical requirements remain rather elusive and highly case--specific. Therefore, a grid sensitivity study and investigation into the interaction between numerically and physically–modelled dissipation forms a further part of future work. Additionally, the use of different RANS models as subgrid–scale models and the optimal integration of these into the DES methodology with respect to their individual modelling assumptions is to be scrutinised. To increase the level of comparability, it is necessary to simulate the flow on the complete delta – canard configuration used in the experiment.

Bearing in mind the failure of industrially–established URANS methods to capture the natural unsteadiness in this case, it is concluded that the DES method is much better suited to near–future requirements in the simulation of such practically–relevant flows for which unsteady effects play a major role.

References

[1] C. Breitsamter: "Turbulente Strömungsstrukturen an Flugzeugkonfigurationen mit Vorderkantenwirbeln". Dissertation, Technische Universit München 1997.

[2] U. Bunge, C. Mocket, F. Thiele: "Calibration of Different Models in the Context of Detached–Eddy Simulation". AG STAB Mitteilung 2003, DGLR, Göttingen.

[3] D. Hummel: "Untersuchungen über das Aufplatzen der Wirbel an schlanken Deltaflügeln". Zeitung für Flugwissenschaften **5**, 1965, pp. 158-168.

[4] J. Jeong and F. Hussain: "On the identification of a vortex". J. Fluid Mech. **285**, 1995, pp. 69-94

[5] A. Mitchell, P. Molton, D. Barberis, J. Delery: "Characterisation of Vortex Breakdown by Flow Field and Surface Measurements". AIAA 2000-0788.

[6] S. Morton, J. R. Forsythe, A. Michell, D. Hajek: "DES and RANS simulation of delta wing vortical flows". AIAA 2002-0587.
[7] S. Morton, M. B. Steenman, R. M. Cummings and J. R. Forsythe: "DES grid resolution issues for vortical flow on a delta wing and an F–18C". AIAA 2003-1103.
[8] T. Rung, U. Bunge, M. Schatz, F. Thiele: "Restatement of the Spalart-Allmaras Eddy-Viscosity Model in Strain-Adaptive Formulation". AIAA Journal **41**, Nr. 7, pp. 1396-1399.
[9] P. R. Spalart, W-H. Jou, M. Strelets, S. R. Allmaras: "Comments on the Feasibility of LES for Wings, and on a Hybrid RANS/LES Approach". Advances in DNS/LES, 1st AFOSR Int. Conf. on DNS/LES, Aug 4-8, 1997, Greyden Press, Columbus Oh.
[10] A. Travin, M. Shur, M. Strelets, P.R. Spalart: "Detached-Eddy Simulation Past a Circular Cylinder". Flow, Turbulence and Combustion **63**, pp. 293-313, 1999.
[11] A. Travin, M. Shur, M. Strelets, P.R. Spalart: "Physical and numerical upgrades in the Detached-Eddy Simulation of complex turbulent flows". Fluid Mechanics and its Applications **65**, pp. 239-254, 2002.
[12] L. Xue: "Entwicklung eines effizienten parallelen Lösungsverfahren zur dreidimensionalen Simulation komplexer turbulenter Strömungen". Dissertation, Technische Universit Berlin 1998.

Table 1 Flow parameter

Inflow velocity	U_∞	$40.0m/s$
Turbulent intensity	Tu	0.4%
Reynolds number	Re_{l_μ}	970000
Reference length	l_μ	$0.36m$
Angle of attack	α	15.0^o
Angle of yaw	β	0.0^o

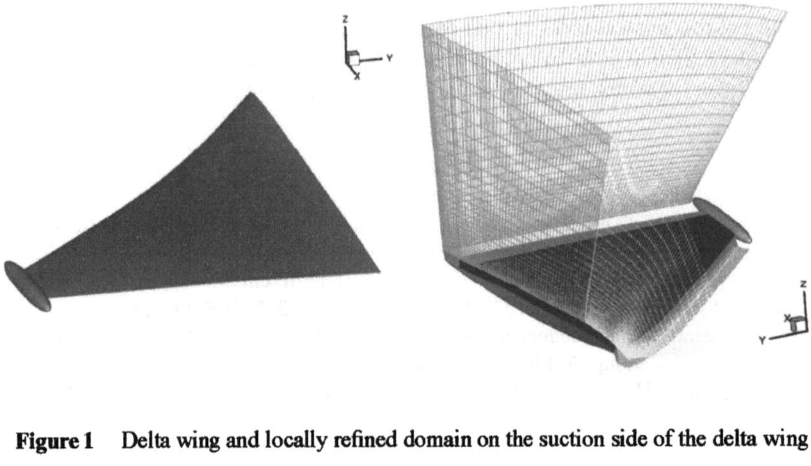

Figure 1 Delta wing and locally refined domain on the suction side of the delta wing

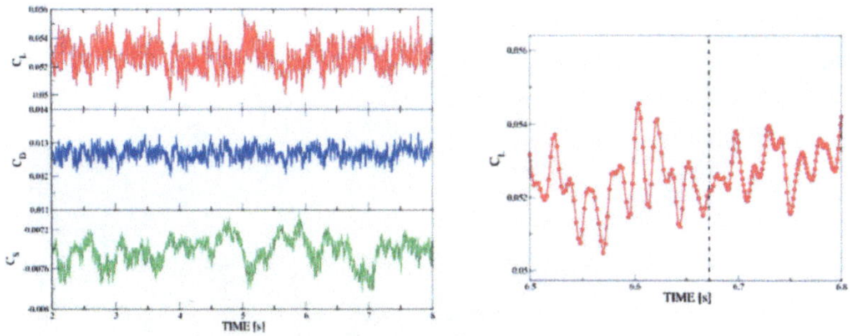

Figure 2 Time series of integral coefficients and effects of a reduction in timestep

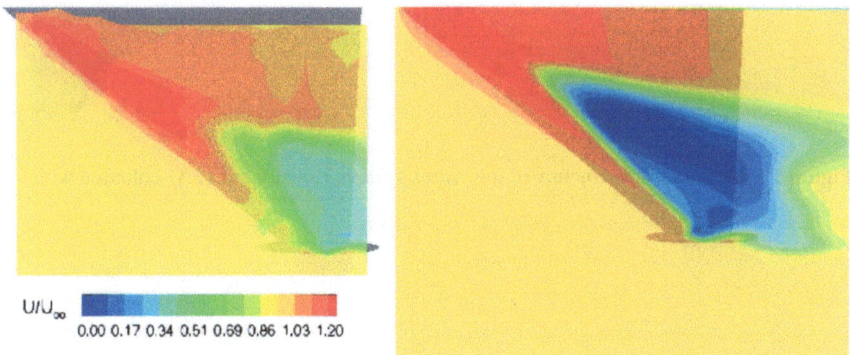

Figure 3 Comparison of U-velocity between experiment (left) and DES (right) in a horizontal cut through the vortex system

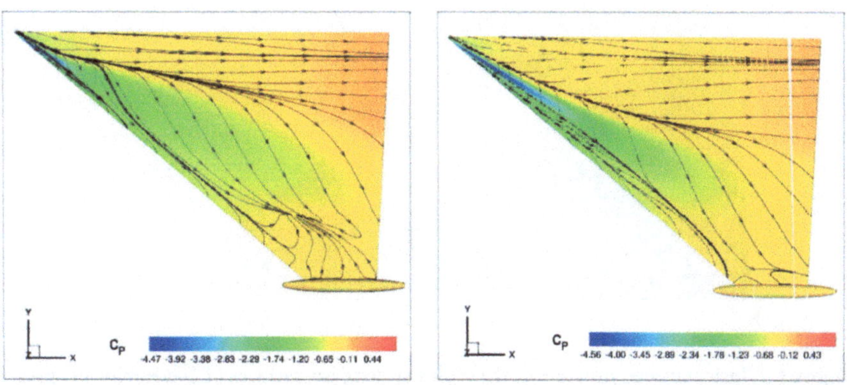

Figure 4 Comparison of the wall bounded flow between and DES (left) and URANS (right)

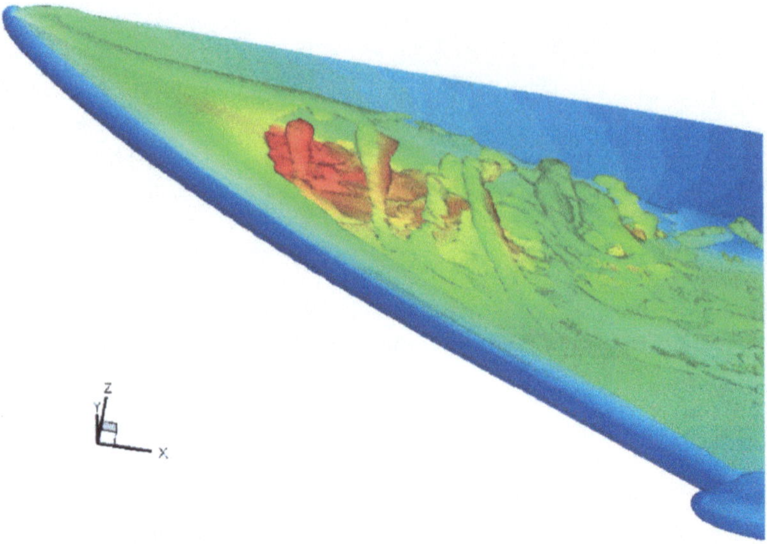

Figure 5 Instantaneous structure of the vortex system, iso-surfaces of λ_2 coloured with ν_t

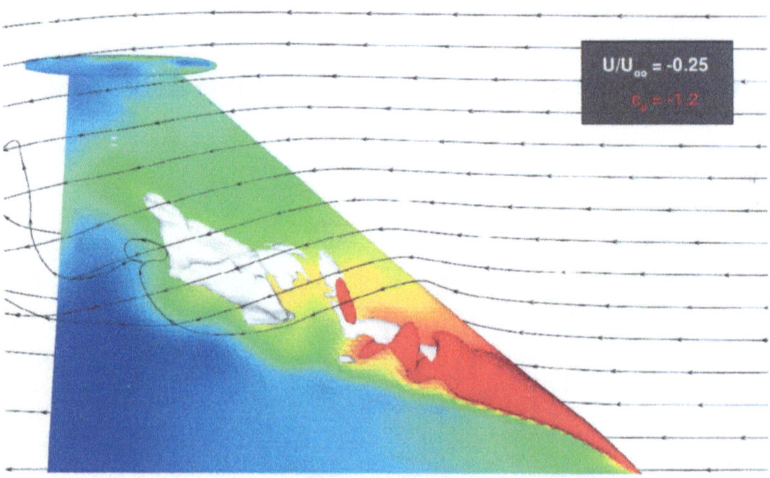

Figure 6 Instantaneous structure of the vortex system, iso-surfaces of U-velocity and pressure, the wing is coloured with pressure

Small Disturbance Navier-Stokes Equations: Application on Transonic Two-dimensional Flows Around Airfoils

Michail Iatrou, Christian Breitsamter, Boris Laschka

Lehrstuhl für Fluidmechanik, Abteilung Aerodynamik, Technische Universität München
85747 Garching, Germany
michail.iatrou@aer.mw.tum.de

Summary

The objective of this numerical investigation is the evaluation of the small disturbance Navier-Stokes method FLM-SD.NS for test cases of two-dimensional transonic flow. For this reason the results of a NLR7301 flap-oscillation and a pitching NACA 64A010 airfoil are compared to experimental data. The influence of viscosity is shown by comparison to results of Euler computations while the time effort is judged by comparing with results of unsteady full Navier-Stokes computations.

1 Introduction

The variety of parameters in aeroelastic investigations like angle of attack, frequencies, Mach number and eigenforms demands a time-efficient and accurate method to calculate the unsteady aerodynamic forces. In order to work towards this goal a small disturbance Euler-code (FLM-SDEu) [4] based on a full Euler-code (FLM-Eu) was developed at the Lehrstuhl für Fluidmechanik of the Technische Universität München. The advantage of a small disturbance method lies in the fast direct calculation of the flow amplitudes compared to their extraction after a time costly unsteady flow calculation. Therefore, calculations of the unsteady aerodynamic forces with respect to aeroelastic problems may be performed better by a factor of ten or more. Because of lack of accuracy of an inviscid approach and the absence of the ability to describe phenomena based on viscous effects like Buzz and Buffeting, a small disturbance Navier-Stokes method (FLM-SD.NS) was developed on basis of the full Navier-Stokes-code (FLM-NS) and the small disturbance Euler-code (FLM-SDEu).

2 The Small Disturbance Navier-Stokes Equations

In this section the transformation of the conservative full Navier-Stokes equations for curvilinear coordinates (1) in a small disturbance formulation is presented.

$$\frac{\partial \mathbf{Q}}{\partial \tau} + \frac{\partial(\mathbf{F} - \mathbf{F_v})}{\partial \xi} + \frac{\partial(\mathbf{G} - \mathbf{G_v})}{\partial \eta} = 0 \qquad (1)$$

$\mathbf{Q}$ denotes the curvilinear state vector, $\mathbf{F}$, $\mathbf{G}$ the convective fluxes and $\mathbf{F_v}$, $\mathbf{G_v}$ the viscous fluxes. The flow quantities indicated by Ψ are splitted to three parts [5], [9]: a steady quantity $\bar{\Psi}$, an unsteady periodic quantity $\tilde{\Psi}$ and an arbitrary turbulent quantity Ψ'

$$\Psi = \bar{\Psi} + \tilde{\Psi} + \Psi'. \tag{2}$$

The periodic flow quantities of the aerodynamic cases we are focusing on here, result exclusively from the periodic motion of the airfoil geometry

$$x(\xi, \eta, \zeta, \tau) = \bar{x}(\xi, \eta, \zeta) + \tilde{x}(\xi, \eta, \zeta)e^{ik\tau}. \tag{3}$$

The steady reference position is denoted by $\bar{x}$ while the amplitude of the harmonic grid oscillation is represented by $\tilde{x}$. The variables ξ, η and ζ denote the curvilinear coordinates while τ represents the dimensionless time and k the dimensionless frequency.
In accordance to the decomposition (2) the curvilinear state vector $\mathbf{Q}$ is splitted to the steady quantity $\bar{\mathbf{Q}}$, the unsteady periodic quantity $\tilde{\mathbf{Q}}$ and the turbulent quantity $\mathbf{Q}'$,

$$\mathbf{Q} = \bar{\mathbf{Q}} + \tilde{\mathbf{Q}} + \mathbf{Q}'. \tag{4}$$

Because of the harmonic behaviour of the airfoils motion, it is assumed, that the unsteady periodic state vector quantity $\tilde{\mathbf{Q}}$ also has a periodic character:

$$\tilde{\mathbf{Q}} = \hat{\mathbf{Q}} \cdot e^{ik\tau}, \tag{5}$$

while the conservative steady state vector amplitude $\hat{\mathbf{Q}}$ is composed on the one hand by the known steady metric $\bar{J}$ and the unknown disturbed flow quantities $\hat{\mathbf{q}}$ marked by the superscript (1) and on the other hand by the given disturbed metrics $\hat{J}$ and the known steady flow quantities $\bar{\mathbf{q}}$ marked by the superscript (2).

$$\hat{\mathbf{Q}} = \hat{\mathbf{Q}}^{(1)} + \hat{\mathbf{Q}}^{(2)}, \quad \hat{\mathbf{Q}}^{(1)} = \bar{J}\hat{\mathbf{q}}, \quad \hat{\mathbf{Q}}^{(2)} = \hat{J}\bar{\mathbf{q}} \tag{6}$$

The steady flow quantities $\bar{\mathbf{q}}$ are taken from the steady-state solution of the full Navier-Stokes method and are therefore known in advance.
Introducing the decomposed flow quantities in the full curvilinear Navier-Stokes equations (1), eliminating the a priori satisfied steady mean terms and the harmonic term $e^{ik\tau}$ and neglecting the second order disturbed terms, finally leads to the small disturbance Navier-Stokes Equations formulation (7).

$$\begin{aligned} &\frac{\partial \hat{\mathbf{Q}}^{(1)}}{\partial \tau} + \hat{\mathbf{Q}}^{(1)} \cdot ik + \frac{\partial(\hat{\mathbf{F}}^{(1)} - \hat{\mathbf{F}}_{\mathbf{V}}^{(1)})}{\partial \xi} + \frac{\partial(\hat{\mathbf{G}}^{(1)} - \hat{\mathbf{G}}_{\mathbf{V}}^{(1)})}{\partial \eta} \\ &= -\hat{\mathbf{Q}}^{(2)} \cdot ik - \frac{\partial(\hat{\mathbf{F}}^{(2)} - \hat{\mathbf{F}}_{\mathbf{V}}^{(2)})}{\partial \xi} - \frac{\partial(\hat{\mathbf{G}}^{(2)} - \hat{\mathbf{G}}_{\mathbf{V}}^{(2)})}{\partial \eta} \end{aligned} \tag{7}$$

$\hat{\mathbf{F}}^{(1)}$, $\hat{\mathbf{G}}^{(1)}$, $\hat{\mathbf{F}}^{(2)}$, $\hat{\mathbf{G}}^{(2)}$ are the perturbed inviscid fluxes and $\hat{\mathbf{F}}_{\mathbf{V}}^{(1)}$, $\hat{\mathbf{G}}_{\mathbf{V}}^{(1)}$, $\hat{\mathbf{F}}_{\mathbf{V}}^{(2)}$, $\hat{\mathbf{G}}_{\mathbf{V}}^{(2)}$ the perturbed viscous fluxes. The arbitrary turbulence quantity $\mathbf{Q}'$ is part of the viscous fluxes in form of a time-averaged mean turbulent viscosity $\bar{\mu}_t$ and a phase-averaged turbulent viscosity $\tilde{\mu}_t$, with respect to the Boussinesq Formulation [9]. A detailed description of the small disturbance Navier-Stokes equation development is presented in [3], [6], [7].

3 Numerical method

The numerical approach that is used for the solution of the small disturbance Navier-Stokes equations is based on a cell-centered finite volume method on structured grids with second order accuracy in time and space. For the time-integration an explicit Runge-Kutta method is used. An upwind flux-difference splitting scheme of Roe is applied for the discretisation of the convective fluxes. An extensive presentation of the discretisation of the perturbed convective fluxes can be found in [4]. In this section the focus is on the discretisation of the perturbed viscous fluxes, particularly on the formulation concerning the spatial derivatives of the perturbed velocities u, v and the temperature T. They are treated in accordance to the spatial derivatives appearing in the viscous fluxes of FLM-NS, which remain cartesian. Hence, a special treatment is needed due to the fact that spatial derivatives expressed in cartesian coordinates, cannot be discretized in a curvilinear space simply by averaging cell-center values at the cell interface of two adjoining cells.

3.1 Discretisation of the spatial derivatives of the perturbed quantities

In order to explain the discretisation of the spatial derivatives of the perturbed primitive quantities, the method of Chakravarthy [1] on the spatial derivatives of the instantaneous primitive quantities will be presented. The primitive quantities u, v and T are denoted by Φ. Further, an auxiliary cell defined by the midpoints of the two neighboured cells is introduced, Fig. 3. The spatial derivative of the primitive quantity Φ is being assumed constant over the whole auxiliary cell and it is assigned to its center. Φ is assumed constant at every face of the auxiliary cell. The values of Φ_W, Φ_E are obvious

$$\Phi_E = \Phi_{k+1,l}, \quad \Phi_W = \Phi_{k,l} , \tag{8}$$

while the values Φ_N and Φ_S are determined by averaging

$$\begin{aligned} \Phi_N &= \frac{1}{4}\left(\Phi_{k,l} + \Phi_{k,l+1} + \Phi_{k+1,l+1} + \Phi_{k+1,l}\right) , \\ \Phi_S &= \frac{1}{4}\left(\Phi_{k,l} + \Phi_{k+1,l} + \Phi_{k+1,l-1} + \Phi_{k,l-1}\right) . \end{aligned} \tag{9}$$

Due to these assumptions, the expression we get by applying the Gauss formula on the auxiliary cell leads to the spatial derivative of the primitive quantity Φ. It is shown here exemplarily for the derivative in x-direction (10).

$$S_{aux}\frac{\partial \Phi}{\partial x} = \int\limits_{B_{aux}} \Phi \, dy \qquad \text{or} \qquad \frac{\partial \Phi}{\partial x} = \frac{1}{S_{aux}} \int\limits_{B_{aux}} \Phi \, dy \tag{10}$$

The substitution of the instantaneous surface area of the auxiliary cell S_{aux}, its boundary B_{aux} and the instantaneous primitive quantities Φ by their mean steady values $\bar{S}_{aux}$, $\bar{B}_{aux}$ and $\bar{\Phi}$, and their perturbed unsteady values $\tilde{S}_{aux}$, $\tilde{B}_{aux}$ and $\tilde{\Phi}$, respectively (11)

$$S_{aux} = \bar{S}_{aux} + \tilde{S}_{aux}, \quad B_{aux} = \bar{B}_{aux} + \tilde{B}_{aux}, \quad \Phi = \bar{\Phi} + \tilde{\Phi} \tag{11}$$

yields after neglecting second order terms and some transformation to the formula for the spatial derivatives of the perturbed primitive quantities as follows

$$\frac{\partial \tilde{\Phi}}{\partial x} = \frac{1}{\bar{S}_{aux}} \left(\int\limits_{\bar{B}_{aux}} \tilde{\Phi}\, dy + \int\limits_{\tilde{B}_{aux}} \bar{\Phi}\, dy - \tilde{S}_{aux} \frac{\partial \bar{\Phi}}{\partial x} \right) . \tag{12}$$

This formula (12) is a necessery part for enabling the discrete formulation of (7). A full description of the numerical treatment of (7) can be studied in [3], [4], [7], [8]. With respect to turbulence modelling the algebraic Baldwin-Lomax model was chosen for both FLM-NS and FLM-SD.NS due to its simplicity.

4 Results

In order to show the advantage of a Navier-Stokes method versus an Euler method, the presence of viscous effects in the flowfield should be noticeable. Therefore, a test case with experimental data related to a shock-induced separation bubble at an NLR7301 airfoil [10] is chosen to show the abilities of the small disturbance Navier-Stokes method. The transonic free-stream conditions are $M_\infty = 0.7$ and $Re_\infty = 2.14 \times 10^6$ and the angle of attack is $\alpha = 3.0^o$. A harmonic flap motion around three quarters of the chord length with a reduced frequency of $k_{red} = 0.142$ ($f = 30Hz$ [10]) and an amplitude of $\Delta\eta = 1.0^o$ generates the unsteadiness of the flow. The shock-induced separation-bubble is steady at steady flow conditions and remains more or less stable during the harmonic flap motion.
Computations have been performed also with the full Navier-Stokes method FLM-NS and the small disturbance Euler method FLM-SDEu. Two C-topology grids with 384×96 points have been used. One with the flap being undeflected and one with deflected flap. On the one hand these grids were interpolated to simulate the actual grid for the unsteady calculations with the full Navier-Stokes method FLM-NS, and on the other hand they were used to calculate the disturbed metrics for the small disturbance methods FLM-SD.NS and FLM-SDEu. The off-body distance for the Navier-Stokes calculations was chosen at 10^{-5}. The density residual reached was for all methods 10^{-4}. The time-integration is performed explicitly with the small disturbance solvers FLM-SD.NS and FLM-SDEu and implicitly by applying the LU-SSOR method for the unsteady calculations with the full Navier-Stokes solver FLM-NS.
The quality of the results can be shown by the distribution of the zeroth and the real and imaginary part of the first harmonic of the pressure coefficient in Fig. 1a. There

is a good agreement with the experimental data outside the shock region for all the methods employed. The lack of accuracy of the FLM-SDEu method to predict the position and the strength of the shock is obvious. The position of the shock is further downstream and the peak too high due to the absence of the boundary layer interaction. It is rather difficult to decide which of the two Navier-Stokes methods FLM-SD.NS and FLM-NS (Fourier-transformed) show a better agreement to the experimental data. The peak is not that strong with the FLM-NS method and its position and the broadness of the shock region varies from that of the FLM-SD.NS. These variations may be caused due to the time development of the separation bubble. By using the full Navier-Stokes Method FLM-NS the separation-bubble is being calculated at every time-step of the flap-motion, while the flow of the bubble is given to the FLM-SD.NS method as an input from the steady state solution during the whole calculation. Generally, the agreement with the experimental data is good. Further comparison of the methods at reduced frequencies of $k_{red} = 0.071$ and $k_{red} = 0.284$ are made. Experimental data at these frequencies exist only for the first harmonic of the lift and pitching moment coefficients. A comparison is being presented in Fig. 2. In most cases the numerical data match with the experimental data. The results of the Euler calculations deviate more significantly from the experimental data, especially for the pitching moment coefficients, because of the inaccuracy in predicting shock position and strength.

The main objective of improving time-efficiency is achieved. At the reduced frequencies of $k_{red} = 0.071, 0.142$ and 0.284 the calculation with FLM-SD.NS was 2.2, 2.5 and 3.3 times faster than the one performed with FLM-NS.

Additionally, a harmonic pitching oscillation of a NACA64A010 airfoil around one quarter of chord length is presented. The free-stream conditions are $M_\infty = 0.8$ and $Re_\infty = 1.26 \times 10^7$ and the angle of attack is $\alpha = 0.0^o$. The reduced frequency is $k_{red} = 0.102$ and the amplitude $\Delta\alpha = 1.0^o$. In this case, the influence of viscosity appears as a shock-boundary layer interaction. The results are shown in Fig.1b. For this test case no flow separation occurs. The shock is predicted a little more downstream with FLM-SDEu than with FLM-SD.NS and FLM-NS, respectively. Because of the boundary layer displacement the flow around the airfoil reaches the speed of sound further upstream compared to the inviscid calculation, so that recompression also takes place further upstream. The comparison with experimental data [10] shows good agreement outside the shock region. It is difficult to judge the accuracy between the Navier-Stokes methods due to the fact, that only one experimental data point lies in the shock region. The calculation with FLM-SD.NS was twice as fast than with FLM-NS.

5 Conclusions and Outlook

The comparison of results of a small disturbance Navier-Stokes method with experimental data and, especially, with the results of a small disturbance Euler method indicate an improvement in the prediction of shock position and strength. Compared to the full Navier-Stokes method a reduction in computational time using the small disturbance approach, is given by a factor of three. The extension of the small disturbance Navier-Stokes method for three-dimensional flows is conducted and test cases of delta wing flap oscillations and wing pylon nacelle oscillations are selected for further detailed investigations.

Acknowledgements

The authors would like to thank Dipl.-Ing. A. Pechloff (Technische Universität München) for his substantial efforts, Dr.-Ing. C. Weishäupl (EADS-MT 242) and Dipl.-Ing. A. Allen (Technische Universität München) for the good cooperation.

References

[1] Chakravarthy, S.R.: "High Resolution Upwind Formulations for the Navier-Stokes Equation." Rockwell Int. Sciense Center, USA, 1988.
[2] Davis, S.S.: "NACA64A010 (NASA AMES MODEL) Oscillatory Pitching". AGARD Report No.702 Compendium of Unsteady Aerodynamic Measurements, August 1982.
[3] Iatrou, M., Weishäupl, C., Laschka, B.: "Entwicklung eines instationären Navier-Stokes-Verfahrens bei kleinen Störungen für aeroelastische Problemstellungen.", Technische Universität München, Arbeitsbericht TUM-FLM-2002/09, Abschlussbericht TUM-FLM-2003/27.
[4] Kreiselmaier, E.: "Berechnung instationärer Tragflügelumströmungen auf der Basis der zeitlinearisierten Eulergleichungen," Dissertation, Lehrstuhl für Fluidmechanik, Technische Universität München, 1998.
[5] Laschka, B.: "Unsteady Flows - Fundamentals and Applications," AGARD Conference Proceedings No. 386, Unsteady Aerodynamics - Fundamentals and Applications to Aircraft Dynamics, Göttingen, FRG, 1985.
[6] Pechloff, A.: "Triple Decomposition of the Two-dimensional Navier-Stokes Equations in Cartesian Coordinates and Linearization for Small Disturbances."Institutsbericht, Technische Universität München, TUM-FLM-2001/4, 2001.
[7] Pechloff, A., Iatrou, M., Weishäupl, C., Laschka, B.: "The Small Disturbance Naviar-Stokes Equations: Development of an Efficient Method for Calculating Unsteady Air Loads" 13. DGLR-STAB-Symposium, 12.-14. Nov. 2002 published in: Notes on Numerical Fluid Mechanics and Multidisciplinary Design, Volume 87, NNFM.
[8] Pechloff, A. Laschka, B.: "Efficient Calculation of Unsteady Air Loads with a Small Disturbance Navier-Stokes Method", ICAS 2004-2.1.3, Yokohama Japan, 29 Aug. - 03 Sept. 2004.
[9] Telionis, P.,D.: Unsteady Viscous Flows. Springer-Verlag, 1981.
[10] Zwaan R.J.: "NLR 7301 Supercritical Airfoil Oscillatory Pitching and Oscillating Flap". AGARD Report No.702 Compendium of Unsteady Aerodynamic Measurements, August 1982.

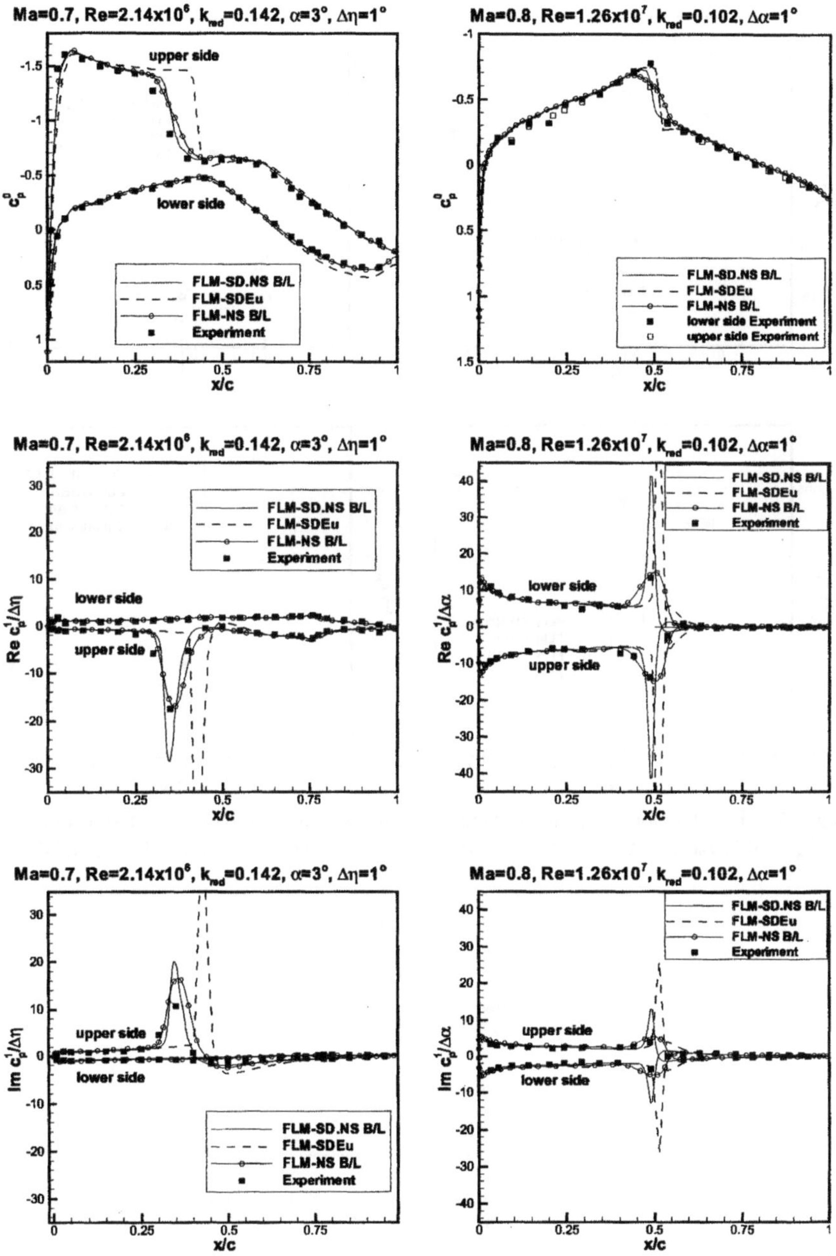

a) NLR7301 flap-oscillation around $x/c = 0.75$

b) NACA64A010 pitching oscillation around $x/c = 0.25$

Figure 1 Zeroth and first harmonic pressure coefficients
B/L: Baldwin-Lomax

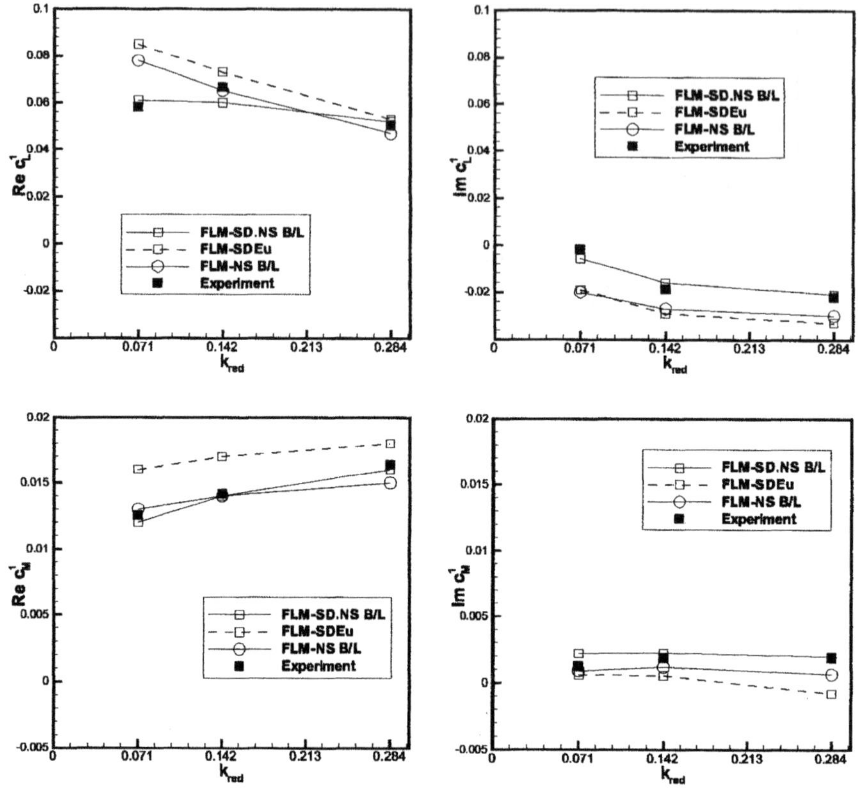

Figure 2 First harmonic lift and pitching moment coefficient for NLR 7301 flap-oscillation at $M_\infty = 0.7$, $Re_\infty = 2.14 \times 10^6$, $\alpha = 3^o$, $\Delta\eta = 1.0^o$ and $k_{red} = 0.071, 0.142$ and 0.284

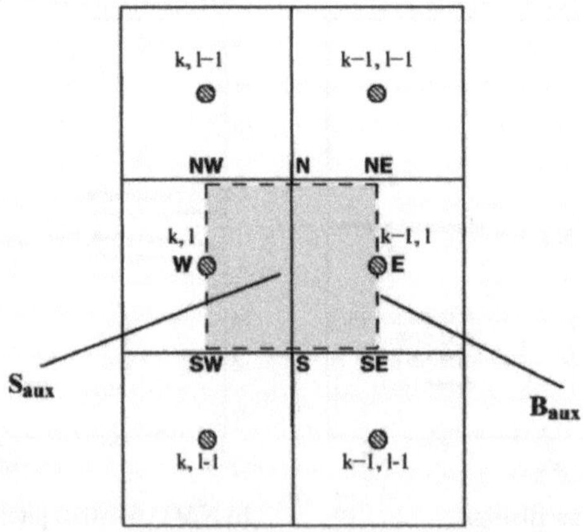

Figure 3 Computational space section

Numerical and experimental investigations of turbulent convection with separation in aircraft cabins

G. GÜNTHER, J. PENNECOT, J. BOSBACH and C. WAGNER

German Aerospace Center, Institute of Aerodynamics and Flow Technology, Bunsenstr. 10, D-37073 Göttingen, Germany
Email: Gero.Günther@dlr.de, julien.pennecot@dlr.de, johannes.bosbach@dlr.de, claus.wagner@dlr.de

Summary

Turbulent mixed convection in two generic configurations, the geometry of which was deduced from a passenger cabin and a sleeping bunk of a modern long distance passenger aircraft, has been investigated numerically by means of Reynolds averaged Navier-Stokes (RANS) computations. In the same configurations Particle Image Velocimetry (PIV) measurements were conducted to validate the numerical results. The performed comparison indicates that results obtained using RANS with low Reynolds number turbulence models agree considerably better with experimental data than those of RANS with high Reynolds number turbulence models.

1 Introduction

Confined mixed convection is of essential importance for a variety of practical applications like air-supply in offices or residential buildings, the cooling of micro-electronic devices, and the air conditioning of vehicle and aircraft cabins. The numerical prediction of such flows by means of Computational Fluid Dynamics (CFD) allows for an optimised design of configurations, while excessive prototyping can be prevented. But systematic comparative studies by Costa et al. [1] and Blay et al. [2] reveal large deviations of RANS results obtained utilizing different turbulence models, when it comes to the quantitative description of mixed convection flows. Especially, flow reversal due to premature separation of a cooling wall jet is of major importance as pointed out by Sinha [3]. Incautious use of inappropriate turbulence models in such configurations might prevent even the correct qualitative prediction of the flow direction.

With this study we want to improve the understanding of airflow in aircraft passenger cabins and sleeping bunks and its effect on thermal comfort as described in the ISO norm [9] and discussed by Silva [8]. To achieve this, numerical simulations and validation measurements have been conducted and are discussed below. Fig. 1 shows a central cross section of a modern megaliner with passenger cabins at two levels and bunks at the level of the cargo bay. The figure also contains the CAD representation of an idealized sleeping bunk (left) and a generic cabin configuration (right), respectively, together with the according mock-up's (below).

2 Methodology

2.1 Numerical method

The commercial code StarCD, which solves the Reynolds averaged Navier Stokes (RANS) equations together with various optional turbulence models, was used for all computations presented in this paper. Three different groups of turbulence models were investigated:

- The low Reynolds number k-ε turbulence models includeing the cubic low Reynolds number k-ε turbulence model by Shih [5] for which the same transport equations are solved in the whole computational domain.
- The high Reynolds number k-ε turbulence models together with wall functions which are used to determine velocity and turbulence quantities close to the wall.
- The two layer k-ε turbulence models for which a low Reynolds number turbulence model is used close to the walls and a high Reynolds number turbulence model in the remaining domain.

2.2 Measurement technique

Particle Image Velocimetry (PIV) of the airflow in the cabin mock-up was conducted using an air cooled Nd:YAG double oscillator (Brilliant Twins, Quantel) with pulse energy of 160 mJ while a water-cooled Nd:YAG double oscillator (Brilliant Twins, Quantel) with a pulse energy of 350 mJ was used for PIV of the flow in the sleeping bunk mock-up. In both measurements a pulse duration of 5 ns was selected for particle illumination. The scattered light of the tracer particles was detected by a CCD-Camera (Sensicam QE, PCO) with an actively cooled CCD-sensor and a spatial resolution of 1378 x 1040. The camera was used in combination with lenses from Zeiss (Planar T* 1.4/85). While the flash lamps of the laser were operated at a repetition rate of 10 Hz, the pockets-cells as well as the camera were operated at a frequency of 2.5 Hz.

Small droplets of Di-Ethyl-Hexyl-Sebacat (DEHS), synthetic oil, served as tracer particles. They were produced with a Laskin nozzle atomizer (Pivtec) together with an impactor. The generated particles have a mean diameter of 1 μm and are stable in air for hours. For more details the reader is referred to Raffel et al. [4]. The light sheet thickness was adjusted for these measurements to 4 mm.

3 Idealized passenger cabin

3.1 Configuration

The rather complex geometry of a realistic passenger cabin has been simplified for this more fundamental investigation. We focused on the flow along the luggage compartment and close to the air inlet. As a consequence, only a part of the real aircraft cabin was modelled by the mock-up of height 1.35 m, length 3.43 m and width 2.0 m presented in Fig. 1. The air enters the mock-up horizontally through three slot-shaped air inlets, which supply in total 40 l/s of air through slits of length 1.01 m and width 10 mm, each. They are positioned 11 mm below the luggage compartment, regularly along the mock-up. The distance between the inlets amounts to 125 mm. The air outlet is located below the air inlet at the ground level of the mock-up. It spans over the whole length of the mock-up and has a width of 50 mm (see Fig. 1).

3.2 Numerical simulation

Simulations were performed for a sub-domain of the whole model in order to save computational resources. Taking into account, that the mock-up is ventilated by three regularly spaced nozzles, the domain can be considered to consist of three equally sized cells, each of which is ventilated by one nozzle. Further, assuming that the flow in the central cell is weakly influenced by the flow in the neighbouring cells, only the central cell was taken into account. Since this central cell is symmetric, it can be further reduced to one half of the cell (i.e. one sixth of the full domain) using symmetry boundary conditions.

For the computations with the low Reynolds number turbulence models we used a hybrid mesh with the first point close to the wall at $y+ = 1$ and with 343199 points, 1036033 volume elements, where y+ denotes the normal distance to the nearest wall in inner coordinates. Simulations with the high Reynolds number turbulence models were performed on a second mesh with 166826 points, 901533 volume elements and the first point at $y+ = 30$.

3.3 Experiment

As depicted in Fig. 2 (right), the wall jet can be resolved using oil droplets as tracer particles. The velocity vectors determined by PIV are always averaged over the area of the interrogation window. The interrogation window size amounted to 32 x 32 pixels, resulting in a spatial resolution of 17.5 mm. Figure 2 (right) shows grey levels of the resulting averaged in plane velocity magnitude as obtained from 480 double images.

Following Bosbach et al. [6], an increase of the interrogation windows has the following two consequences for the PIV representations of the wall jet. First, velocity values close to the wall are biased to zero due to the overlap of the interrogation window with the boundary of the measurement domain. Further away from the wall, the values are too large due to the overlap with the jet peak. This leads to a shift of the maximum to larger distances from the wall. Second, statistical averaging leads to a broadening of the jet profile with increasing window size. Therefore, for comparing wall jet velocity profiles, as presented in Fig. 3, the field of view has been further reduced, so that the wall jet could be resolved with an interrogation window size of only 4 mm (24x24 pixels).

3.4 Comparison between experiment and computations

In Fig. 2 the statistically averaged velocity magnitude fields obtained in RANS with the cubic low Reynolds number k-ε turbulence model (left) and with PIV (right) are presented. It must be noted, that, independently from the turbulence model used, all simulations produced qualitatively the same overall flow structure. Due to the Coanda effect, the jet generated by the nozzle remains attached to the luggage compartment wall until shortly before the ceiling where it separates. Then, the flow follows each sidewall separating and reattaching at every corner, before it leaves partially through the outlet. In the core of the mock-up, the mean velocity is always very small.

A more detailed comparison of the results is presented in Fig. 3. Mean velocity magnitudes obtained using RANS with various turbulence models and from PIV are plotted as a function of the wall distance along the wall normal B and C (as indicated in Fig. 2) in Fig. 3 left and right, respectively. Depending on the turbulence model applied, the results reflect significant differences with respect to the width and height of the predicted wall jet. Obviously, the turbulence models belonging to the low Reynolds number k-ε group produce the smallest jet diffusion, followed by the high Reynolds number turbulence models, while the two layer turbulence models predict the largest beam. From all investigated turbulence models the low Reynolds number k-ε turbulence models compare best to the PIV data.

4 Idealized sleeping bunk

4.1 Configuration

The complete crew rest compartments consists of up to 12 single person sleeping bunks. The flow in such a single bunk is considered below. The experiments were conducted on a simplified box of height 0.60 m to 0.68m, length 1.98 m and width 0.48m to 0.55m. Fresh air at temperature 20°C enters through an inflow slot of size 1.59m x 15 mm and leaves the mock-up through an outflow slot of 1 m x 80 mm.

4.2 Numerical simulation

RANS computations were performed on a hybrid grid of tetras and prisms containing 381614 points and 11954505 volume elements. The surface mesh of the grid is presented in Fig. 4. Again, the smallest wall distance was chosen to achieve a dimensionless wall distance of $y+ < 1$. The simulations were performed with the Shih cubic low Reynolds k-ε turbulence model [5]. An inflow velocity of 0.78 m/s at 15% turbulence intensity with a turbulent length scale of 1.5 mm was selected. Further, the heat release of the sleeping digital human was assumed to be 90 W in total, according to DIN 1946 [7]. This norm defines that at room temperature (23°C) the heat release is transported in nearly equal parts by convection, radiation and transpiration. Additionally, the heat releases due to convection and radiation were set to 27.8 W and 26.95 W, respectively. It must be noted, that the assumption of an isotropic heat release of the naked human body is supported by the ability of the blood to transfer 130 W to the extremities at minimum heartbeat. It was further assumed, that heat radiation from the digital human is transferred isotropically and directly to the walls of the bunk.

Consequently, the heat fluxes due to convection and radiation were modelled assuming constant fluxes of 21.67 W/m^2 at the surface of the digital human and of 5.4 W/m^2 at the bunk walls, respectively. The latent heat transport was not considered directly.

4.3 Experiments

In the experiment the flow was driven by suction to the outlet. The flow rate of 15 l/s was provided by a PIV calibrated tuneable fan at the end of a 2m long nozzle. For the velocity field measurements acryl glass windows in top and front plane were used and the incoming air was seeded with small tracer particles of oil. The mock up was insulated to be nearly adiabatic and pre-heated. Measurements were taken at steady state with respect to the enthalpy balance. In order to receive realistic heat transport inside the bunk the PIV measurements were performed with a human test person, the contours of which are visible in Fig. 6 (right).

4.4 Results

The large scale flow structure obtained in the RANS computation is characterized by a large convection vortex originating at the inflow slot, as shown in Fig. 5. The comparison of a computed flow field in a cross section to PIV measurements is presented in Fig. 6. The simulation predicts an injected jet which separates 0.53 m downstream of the entrance with reattachment at the long sidewall in front. This results in qualitative agreement with the PIV measurements presented in Fig. 6 (right).

Finally, the computed temperature distribution on the surface of the human dummy in Fig. 7, reveals temperature variations between 24.1 C and 36.6°C. It is observed that temperature are high in regions where the human body has many perspiratory glands. Therefore, heat release due to perspiration should be included in a human comfort model. This result is essential for further thermal comfort investigations.

5 Conclusions

Aircraft cabin and bunk airflow have been investigated in a generic cabin/bunk mock-up by means of RANS and PIV. The comparison of the results reveals a satisfactory agreement between RANS and PIV. It turns out, that for simulation of aircraft cabin/bunk flow low Reynolds number turbulence models have to be used instead of high Reynolds number or two layer turbulence models because of their better ability to describe jet diffusion and separation. It is concluded that, with respect to heat release, a more sophisticated model of a digital human will enable us to give even better predictions of thermal comfort of cabins/bunks in transport vehicles.

References

1. J.J. Costa, L.A. Oliveira, D. Blay: *Test of several versions for the k-ε type turbulence modelling of internal mixed convection flows*, Int. Journal of Heat and Mass Transfer **42**, 1999.
2. D. Blay, S. Mergui, C. Niculae: *Confined Turbulent Mixed Convection in the Presence of a Horizontal Buoyant Wall Jet*, Fundamentals of Mixed Convection **213** (1992) 62.4391.
3. S.L. Sinha: *Behaviour of inclined jet on room cooling*, Building and Environment **36** (2001) 569.
4. M. Raffel, C. Willert, J. Kompenhans: *Particle Image Velocimetry*, Springer publishers,1998.
5. T.H Shih, J. Zhu, J.L. Lumley: *A realizable Reynolds stress algebraic equation model*, NASA TM-105993, 1993.
6. J. Bosbach, J. Pennecot, C. Wagner, M. Raffel, T. Lerche, S. Repp: *Numerical and experimental simulations of turbulent ventilation in aircraft cabins*, Proc. of the ASME-ZSIS Internationsal Thermal Science Seminar II, 2004.
7. Recknagel, Sprenger, Schramek, *Taschenbuch für Heizung + Klimatechnik*, Oldenbourg Verlag, München, 2004.
8. G. da Silva: *Measurements of comfort in vehicles*, Measurement Science Technology 13, 2002.
9. International Standards Organisation, ISO 7730 *Moderate Thermal Environments – Determination of the PVM and PPD Indices and the Specifications of the Conditions for Thermal Comfort*, 1993.

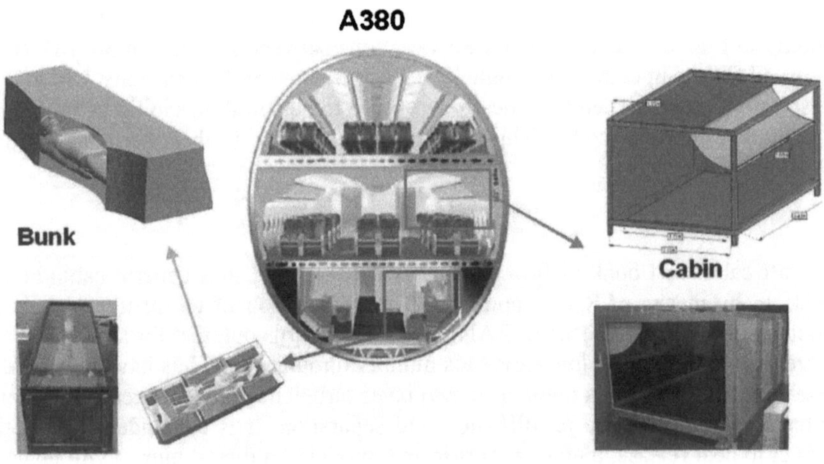

Figure 1: Placement of bunks and cabins in a modern megaliner and deduced bunk and cabin mock-ups

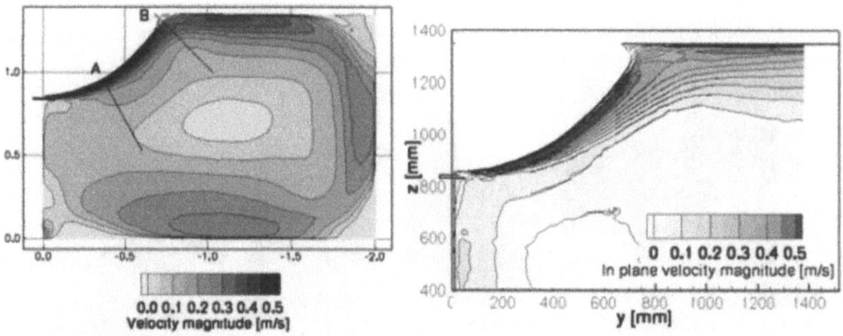

Figure 2: In symmetry plane velocity magnitude obtained in RANS with a low Reynolds number turbulence model (left) compared to statistically averaged PIV results (right)

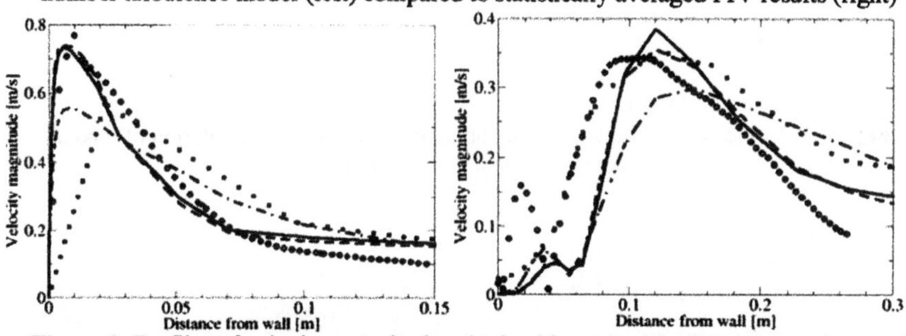

Figure 3: Profiles of velocity magnitudes obtained in various RANS computations and with PIV (◦) along the wall-normal plane A (left) and plane B (right). Solid lines denote Standard Low Reynolds Number, dashed lines Cubic Low Reynolds Number, dotted lines Standard High Reynolds Number and dashed-dotted lines Two Layer RNG turbulence models.

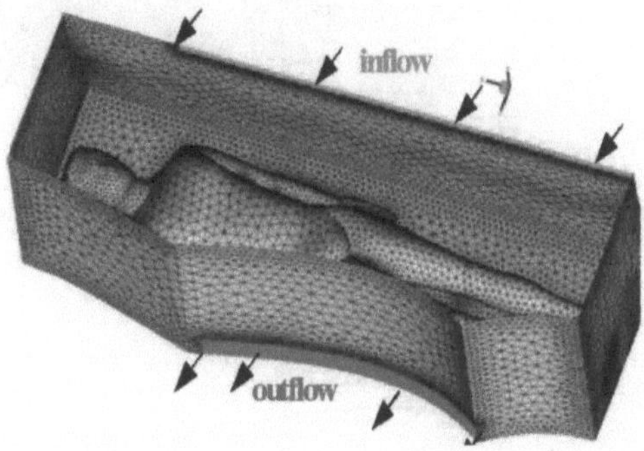

Figure 4: Computational surface mesh of sleeping bunk simulation.

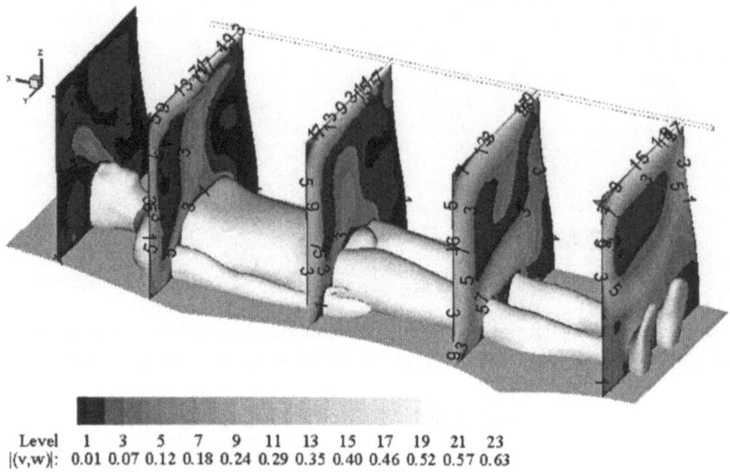

Figure 5: Velocity magnitude contours obtained in a RANS computation of mixed convection in a sleeping bunk.

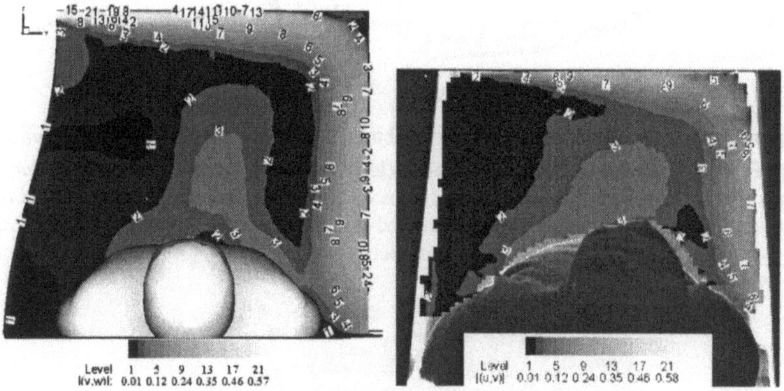

Figure 6: In center plane velocity magnitude predicted by CFD (left) compared to PIV measurments (left).

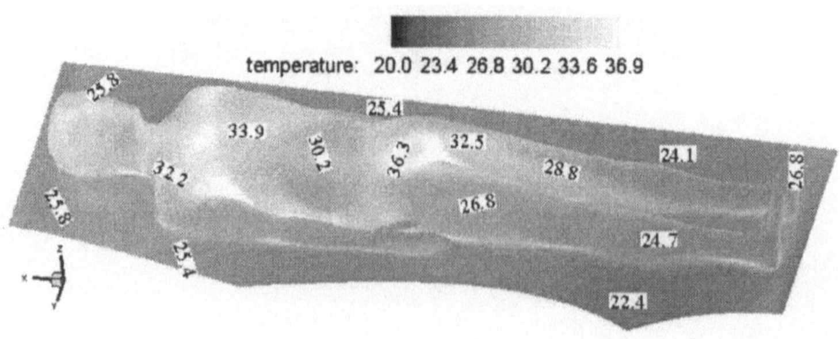

Figure 7: Temperature distribution on the surface of a digital dummy determined in a RANS computation of mixed convection in a sleeping bunk.

Investigation of Influence of Different Modelling Parameters on Calculation of Transonic Buffet Phenomena

Ante Soda and Nicolas Verdon

DLR Institute of Aeroelasticity, Bunsenstrasse 10, D-37073 Goettingen, Germany
E-mail: ante.soda@dlr.de

Summary

This paper is a summary of an ongoing numerical investigation dealing with the influence of modelling parameters on numerical simulation of 2-D unsteady transonic flows. The focus of research lies on a simulation of self-sustained shock oscillation (shock-buffet) since this mechanism can play an important role in the aeroelastic behaviour of modern large-span aircrafts. Three different 2-D profiles have been used in the investigation and the influence of following modelling parameters has been analysed: (a) turbulence modelling, (b) flow solver spatial discretization schemes, and (c) temporal resolution parameters.

1 Introduction

Aerodynamic excitation due to transonic shock oscillation can occur with a stationary geometry (buffet phenomena). In general, shock-buffet is a large-scale flow-induced shock motion that involves alternating separation and reattachment of the boundary layer. It can generate large transient aerodynamic loads in very narrow ranges of transonic Mach number and incidence angle values, and have characteristic frequencies similar to those involved in flutter. Thus, buffet onset boundary is one of the key features that define the upper limit of the cruise flight speed of today's transonic transport aircraft.

The typical flight envelope of modern large-span aircraft, flying in the transonic flow regime, is characterized by the appearance of following features: local supersonic regions closed by shock waves, strong interactions between shock wave and boundary layer, and flow separation. The influence of these flow features on the flight characteristic of an aircraft can be twofold. On one hand they can result in drastic changes of the aerodynamic loads (fluctuations of lift, drag and pitching moment) for both stationary and oscillating wings, and therefore represent the limiting factor in pure aerodynamic sense. On the other hand, the unsteady loads can cause aerodynamic excitation of the elastic structure of the aircraft (transonic flutter, buffeting, buzz and limit cycle oscillation phenomena) and subsequently restrict the flight envelope of an aircraft in an aeroelastic sense.

Periodic shock motions on stationary aerofoils at transonic flow conditions are experimentally well documented phenomena [4, 6, 8], and they have also been detected using various viscous [1, 7] and viscous-inviscid interaction numerical methods [3, 5]. In spite of numerous experimental and computational studies in the last three decades, a reliable and efficient numerical simulation of unsteady, transonic viscous flows still remains an important problem in fluid dynamics.

2 Variation of Modelling Parameters

In the investigation the DLR-Tau code [9], a time-accurate URANS flow solver based on the finite-volume approach, was used. It employs an unstructured spatial flow field discretisation and dual-time stepping scheme for time-accurate (unsteady) calculations. Three different profiles have been used for flow simulations: two conventional symmetrical NACA aerofoils (64A010 and 0012) and a supercritical NLR 7301 aerofoil. Experimental and numerical buffet results can be found in the literature for the second and third profile [6, 8], while there are only steady and pitching measurements for the first profile available [2].

Influence of following modelling parameters has been analyzed: (a) turbulence modelling, (b) flow solver spatial discretization schemes, and (c) temporal resolution parameters. Two linear eddy-viscosity-based turbulence models have been used in the investigation: the 1-equation model of Spalart and Allmaras (S-A) and the linearised explicit algebraic (LEA) 2-equation k-ω model [9]. For the spatial gradients of fluxes, both schemes available in the Tau code, upwind and central differencing, have been tested. In the case of upwind solver, the Van Leer scheme was employed on coarse grids and the AUSMDV scheme on the finest grid level. In order to vary the temporal discretization of time-accurate calculations, two additional numerical parameters have been varied in the dual-time-stepping scheme. These are the number of time-steps per period of oscillation (NTPER) and the number of pseudo-time steps (iterations) per physical time step (NINNER).

3 Results for NACA 64A010 Profile

The first investigated profile was a 10% thick symmetrical NACA 64A010 aerofoil. As a starting point for the buffet investigation, the steady and unsteady calculations were made for the AGARD SSC test case [2]. This case represents a challenge for numerical codes since the strong shock/boundary-layer interference and the shock-induced boundary layer separation are present in the flow. Figure 1 shows general differences between results calculated with two turbulence models and two solver schemes. It can be seen that the upwind scheme (dashed line) predicted more accurate shock location, while the central scheme (solid line) returned higher pressure level at the trailing edge (less severe flow separation). Variation of flow parameters (Figure 2) showed that Mach number values, critical for the onset of shock-buffet (according to buffet-onset criteria in [4]), are: Ma=0.72 (start of shock-induced boundary layer separation), Ma=0.76 (separation reaches trailing edge) and Ma=0.789 (start of trailing edge pressure divergence). Similarly, for constant Mach number (Ma=0.789), the critical α values are 2 deg (start of shock-induced separation) and 3 deg (separation reaches trailing edge, start of trailing edge pressure divergence).

The analysis of results for the AGARD SSC pitching test case (Ma=0.789, Re=12 mio, α_{mean}=4.0 deg, $\Delta\alpha_{max}$=1.0 deg, ω^*=0.4) showed that qualitative aspects of the unsteady flow development were captured with all combinations of turbulence models and solver schemes. The only exception was the 2-equation model combined with the central solver, which predicted a wrong phase shift in the shock wave region. Figure 3 shows the influence of pitching reduced frequency ($\omega^* = \omega\, c/U_\infty$) on surface pressure distribution. Analysis of the phase angle for the 1st harmonic component of unsteady pressure (Figure 3, right) shows an interesting development: with increasing frequency the phase angle at shock location (x/c=0.40-0.55) changed from 120° to 60° (instantaneous pressure leading the profile motion). This phase shift, together with the similar phase shift observed at the trailing edge, suggests the existence of aerodynamic natural (buffet) frequency close to the pitching reduced frequency ω^*=0.4. Another notable feature in Figure 3 is an almost perfect agreement between the measured phase angle for ω^*=0.4 and the computed angle for ω^*=0.5.

In order to investigate the possible existence of shock-buffet on a NACA 64A010 aerofoil, a series of buffet calculations (stationary geometry + dual-time stepping) were performed. The buffet onset was taken as the first incidence angle at which the lift oscillation was persistent (constant amplitude) and the amplitude of oscillation was at least two orders of magnitude larger than any oscillations due to numerical instabilities. Numerical results confirmed the existence of shock-buffet, although at slightly higher incidence angles than expected from the analysis of steady and forced-response results. Furthermore, the shock oscillations could be captured by only one combination of modelling parameters (2-equation model + central solver), while all other combinations returned damped oscillations (no evidence of self-sustained shock-buffet). Figure 4 gives an overview of numerical results in the parametric α-Ma plane and there it can be seen that buffet was found in the Mach number window 0.70<Ma<0.80 and for incidence angles 4°<α<8°. Reduced frequencies of these oscillations were in the range 0.4<ω^*<0.8 (depending on Mach number and incidence). In Figure 5 the time-evolutions of lift coefficient oscillations for selected test points (Ma-α combinations) from Figure 4 are given. For α-sweep at constant Mach number Ma=0.750 (Figure 5, a and b) the harmonic shock-buffet appears between α-values 4 and 5 deg. With increasing incidence the amplitudes are growing but the harmonic nature of oscillations is lost (shock motion still remains periodic though) until, at α=8 deg, the fluctuations become stochastic (break-up of periodic shock-buffet). With increasing Mach numbers the buffet α-window becomes narrower and lift amplitudes smaller (Figure 5 d) until no buffet could be found for Ma>0.80 (damped oscillations converging to steady solutions). In Figure 5, c and d, the influence of different modelling parameters and Δt-values on the buffet results can be seen.

4 Results for NACA 0012 Profile

Several reports can be found in the literature dealing with the influence of turbulence modelling on the calculated buffet onset boundary appearing on a 12% thick NACA

0012 aerofoil [1, 7]. In the current study the influence of one additional parameter, solver discretization scheme, has also been investigated. Figure 6 gives an overview of buffet onset boundaries calculated with the Tau code, and compares them to the measured and computed results found in the literature [1, 3, 5, 6]. It can be seen that, different from the NACA 64A010 buffet results, the shock-buffet oscillations on a NACA 0012 could be predicted by both turbulence models and both solver schemes (bold solid lines). Furthermore, it has to be noted that for Ma>0.77 no buffet could be found with either 1- or 2-equation turbulence model (damped oscillations only). The reduced frequencies of shock oscillations were within the range $0.4<\omega^*<1.0$ and, in general, they were increasing with both Mach number and incidence angle. With respect to the temporal resolution used in these calculations, the time-step used was $\Delta t \approx 1 \cdot 10^{-4}$ s, which means that each cycle of buffet oscillations was resolved with at least 250 time-steps.

For one selected point from the buffet graph (Ma=0.70, α=6 deg) Figure 7 shows time-evolutions of lift coefficient calculated with three combinations of modelling parameters. Both calculations with the central discretisation of fluxes (solid and dashed line) predicted somewhat higher reduced frequency ($\omega^* \approx 0.48$) than calculation using the upwind solver ($\omega^* \approx 0.43$, marked with dotted line). Furthermore, the 2-equation turbulence predicted almost 80% larger amplitude of lift oscillation than the 1-equation model. In all calculations the shock amplitude (peak-to-peak) was between 12.5 and 15% of chord length.

5 Results for NLR 7301 Profile

For a 16.8% thick NLR 7301 supercritical profile a distinct buffet behaviour was found with both turbulence models and both solver schemes (Figure 8). Comparison of the calculated results and the measured buffet onset boundary from [8] shows that for this profile the 1-equation model and the upwind solver were the most successful combination of modelling parameters (dashed line). Shock motion amplitudes were somewhat smaller than amplitudes observed with a NACA 0012 (approximately 10% chord). With respect to the buffet amplitudes, the solver scheme influenced the amplitude of lift fluctuations significantly, with upwind scheme predicting higher values than the central solver. The frequency of shock motion, on the other hand, was less sensitive and the frequencies obtained with upwind solver were only marginally higher. In general, reduced frequencies for the NLR 7301 shock-buffet were similar to those found with symmetrical aerofoils ($0.4<\omega^*<1.0$).

6 Conclusions

In this study three different 2-D profiles have been employed in order to investigate the influence of various modelling parameters on the calculations of unsteady transonic flows, with the emphasis on simulation of self-sustained shock oscillations.

It has been shown that the 2-equation turbulence model was more successful than the 1-equation model in predicting the shock-buffet phenomena on thin symmetrical aerofoils. For a thick supercritical aerofoil, on the other hand, the 1-equation model gave buffet onset boundary closer to the experimental data. With respect to the spatial discretisation scheme, the central solver returned more accurate results for flows with weaker boundary layer separation (thin aerofoils). The upwind solver scheme, while overpredicting the separation effect, was more accurate in capturing the shock wave location and this could be an explanation for a better buffet prediction on a thick NLR aerofoil.

Regarding the analysis of physical mechanism of buffet phenomena, it has been shown that self-sustained shock oscillations on all three aerofoils occur in the Mach number window 0.70<Ma<0.80. The buffet frequency is proportional and the buffet amplitude is inversely proportional to the Mach number (ω^* increases and ΔCl decreases with increasing Ma-value). Furthermore, relatively high incidence angles (α>4 deg) are necessary to provoke the onset of shock-buffet on a thin aerofoil (maximum thickness about 10% chord). With increasing thickness the buffet onset shifts toward lower α-values and, in the case of thick supercritical aerofoil, buffet can occur even at small negative incidence angles. Finally, it has also been shown that buffet reduced frequency is independent of the aerofoil thickness, i.e., for all three profiles the reduced frequency values were in the range $0.4<\omega^*<1.0$. Since three geometrically different aerofoils shared the same chord length dimension (c=1), similarity of calculated reduced frequencies hints at the possible acoustic origin of buffet excitation mechanism. In such a mechanism the crucial role is played by pressure waves propagating through the flow field and spreading the disturbances in both upstream and downstream directions.

References

[1] Barakos, G., and Drikakis, D., 'Numerical simulation of transonic buffet flows using various turbulence closures', Int. Journal of Heat and Fluid Flow 21 (2000), pages 620-626

[2] Davis, S. S., 'NACA 64A010 (NASA AMES Model) Oscillatory Pitching', Compendium of Unsteady Aerodynamic Measurements, AGARD Report No. 702, 1982

[3] Edwards, J. W., 'Transonic Shock Oscillations Calculated With a New Interactive Boundary Layer Coupling Method', AIAA-93-0777, 1993

[4] Finke, K., 'Stoßschwingungen in schallnahen Strömungen', VDI-Forschungsheft 580, Germany, 1977

[5] Girodroux-Lavigne, P., and Le Balleur, J. C., 'Time Consistent Computation of Transonic Buffet Over Airfoils', ICAS Paper No. 88.5-5.2, 1988 (ONERA TP No. 1988-97)

[6] McDevitt, J. B., and Okuno, A. F., 'Static and Dynamic Pressure Measurements on a NACA 0012 Aerofoil in the Ames High Reynolds Number Facility', NASA Technical Paper 2485, USA, 1985

[7] Raghunathan, S., Mitchell, R. D., and Gillan, M. A., 'Transonic shock oscillations on NACA 0012 aerofoil', Shock Waves (1998) 8, pages 191-202, Springer Verlag

[8] Schewe, G., Knipfer, A., Mai, H. and Dietz, G., 'Experimental and Numerical Investigation of Nonlinear Effects in Transonic Flutter', DLR IB 232-2002 J 01, Germany, 2002

[9] Weinman, K., et al, 'Technical Report - Technical Documentation of the DLR Tau-code', DLR, Germany, 2004

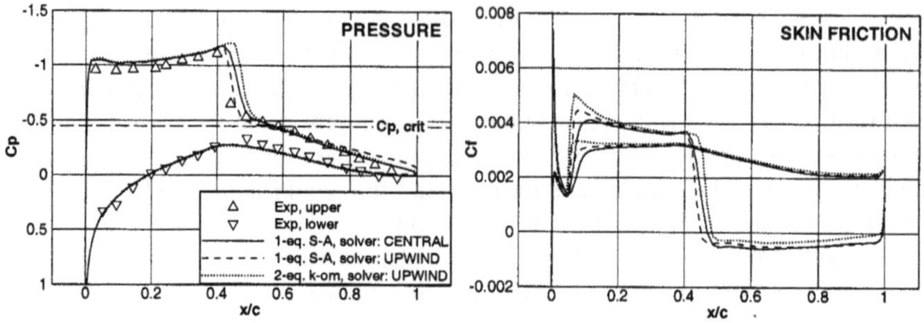

Figure 1 NACA 64A010 - steady AGARD SSC results (Ma=0.789, Re=12 mio, α=4.0 deg) calculated with different combinations of turbulence models and solver schemes. Calculation with the 2-equation model and central solver did not converge.

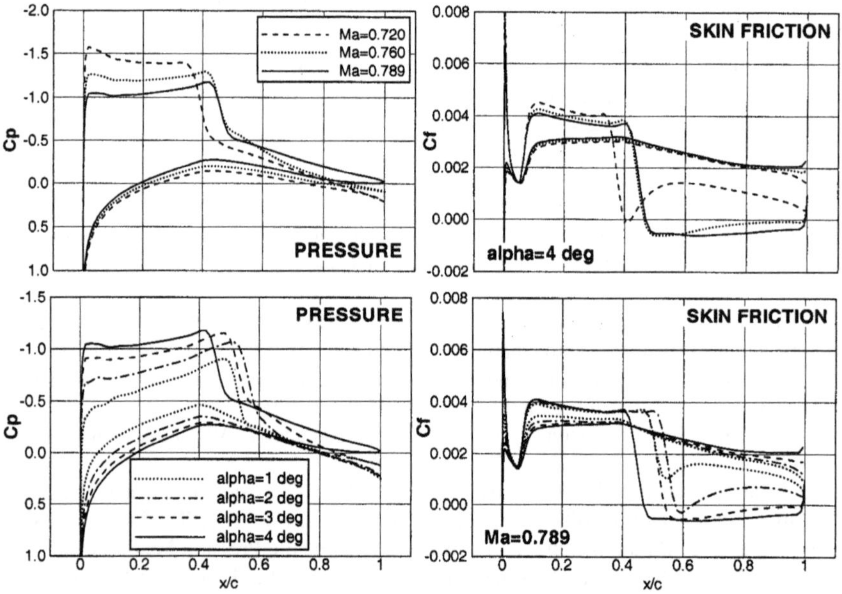

Figure 2 NACA 64A010 - variation of Mach number (top) and incidence angle (bottom). Steady flow results, calculated with the S-A turbulence model and central solver.

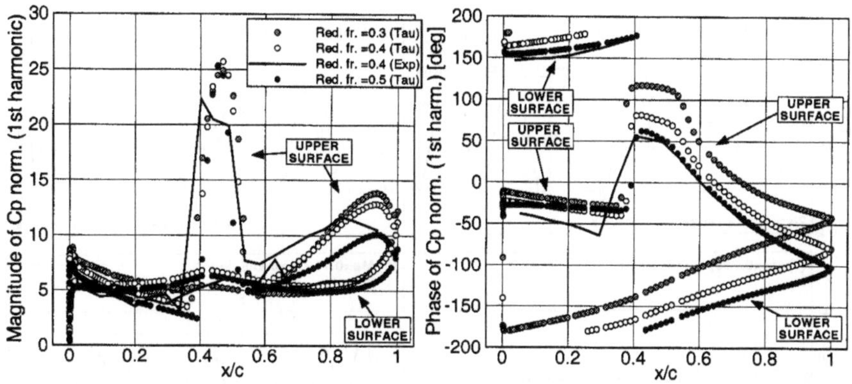

Figure 3 NACA 64A010 - influence of reduced frequency on unsteady pressure distribution (pitching forced oscillations). Calculated with the 1-equation model and upwind solver. Flow parameters: Ma=0.789, Re=12 mio, α_{mean}=4.0 deg, $\Delta\alpha_{max}$=1.0 deg. Temporal parameters: NTPER=250, NINNER=100. Unsteady pressure normalised with the amplitude of oscillation $(2\pi/360 \cdot \Delta\alpha_{max})$.

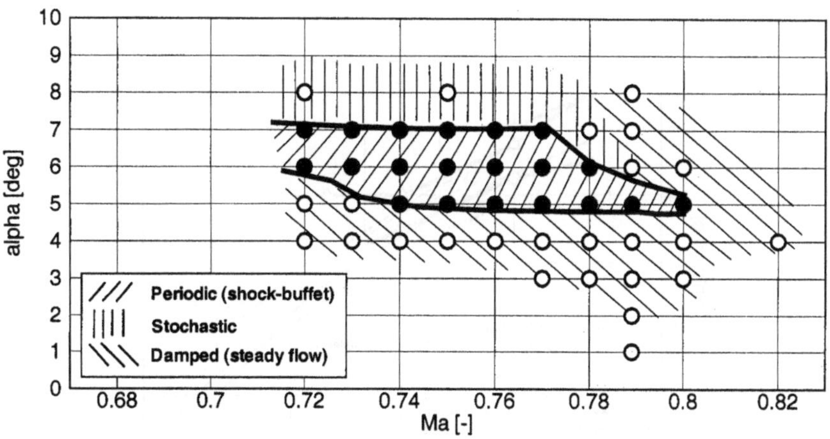

Figure 4 NACA 64A010 - numerical buffet results in parametric α-Ma plane (symbols represent test points where calculations were made). All results obtained with the 2-equation turbulence model and central solver. Other relevant parameters for these calculations: Re=12 mio, NINNER=100, $\Delta t \approx 0.5 \cdot 10^{-4}$ s (giving, for example, NTPER≈500 for $\omega^* \approx 0.8$). For Mach number values smaller than Ma=0.72 no calculations have been made.

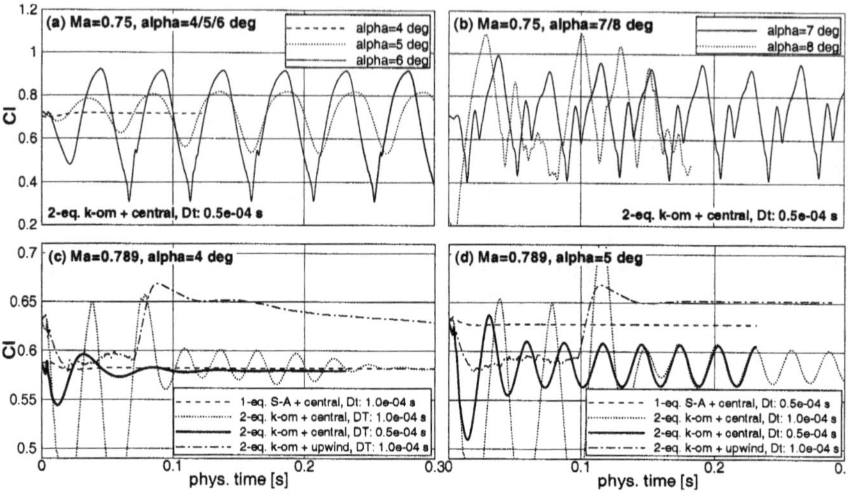

Figure 5 NACA 64A010 - results of buffet simulations at two Mach numbers. Top: for Ma=0.750 large-amplitude shock oscillations were found with the 2-equation model and central scheme. Bottom: for Ma=0.789 buffet was targeted with different combinations of two turbulence models, two solver schemes and different time-step values (for example, $\Delta t \approx 1.0 \cdot 10^{-4}$ s gives NTPER≈500 for $\omega^* \approx 0.4$). For all calculations: NINNER=100.

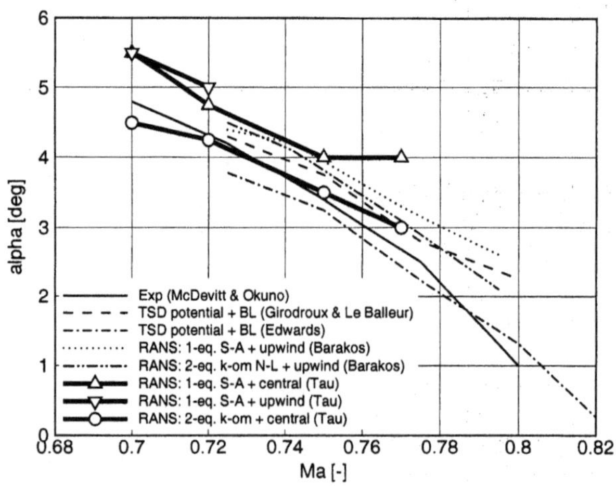

Figure 6 NACA 0012 - buffet onset boundaries calculated with different combinations of modelling parameters (bold solid lines). Relevant parameters for these calculations: Re=10 mio, $\Delta t \approx 1.0 \cdot 10^{-4}$ s (giving temporal resolution of NTPER≈500 for $\omega^* \approx 0.5$), NINNER=200. Tau-code results compared to numerical and experimental results found in literature.

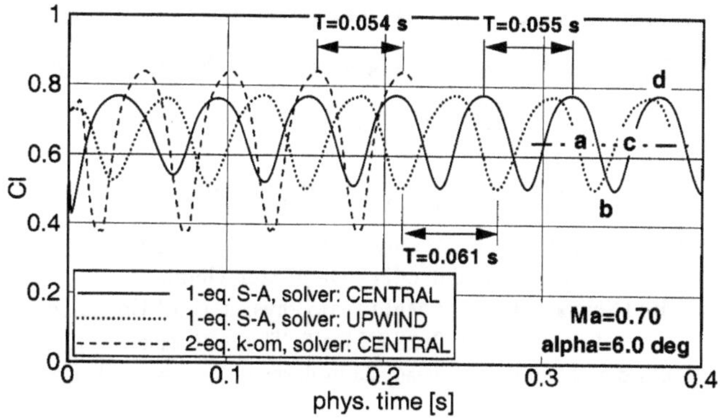

Figure 7 NACA 0012 - buffet results for Ma=0.70 and α=6.0 deg. Time-evolutions of lift coefficient calculated with different combinations of modelling parameters are shown. For period T$\approx$0.055 s the corresponding reduced frequency was $\omega^* \approx 0.48$ and every period was resolved with more than 500 time-steps (NTPER$\approx$550).

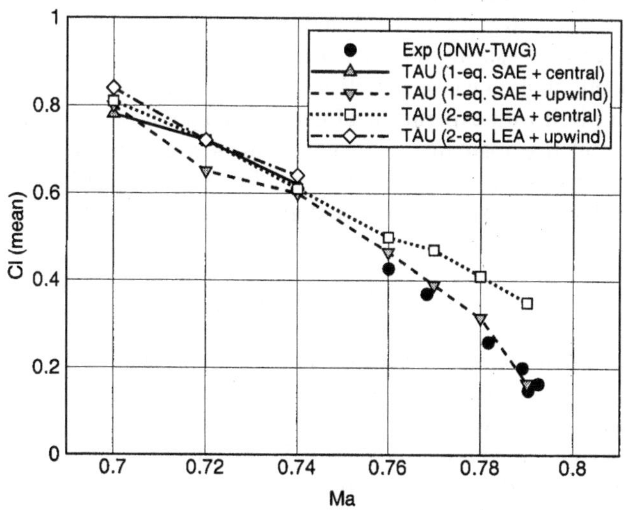

Figure 8 NLR 7301 - comparison of calculated buffet onset boundaries to the experimental data from [8]. Relevant parameters for these calculations: Re=2.3 mio, central solver, $\Delta t \approx 1.0 \cdot 10^{-4}$ s (giving temporal resolution of NTPER$\approx$500 for $\omega^* \approx 0.55$), NINNER=100. NOTE: On y-axis the mean lift coefficient is plotted.

Anisotropy Evolution in Relaminarizing Turbulent Boundary Layers: a DNS-aided Second-Moment Closure Analysis

S. Jakirlić[1], K. Hanjalić[2] and C. Tropea[1]

[1]Fachgebiet Strömungslehre und Aerodynamik, Darmstadt University of Technology
Petersenstr. 30, 64287 Darmstadt, Germany
[2]Thermal and Fluid Sciences Dept. of Multi-scale Physics, Delft University of Technology
Lorentzweg 1, 2628 CJ Delft, The Netherlands
s.jakirlic@sla.tu-darmstadt.de, ctropea@sla.tu-darmstadt.de, hanjalic@ws.tn.tudelft.nl

Summary

Turbulent boundary layers subjected to strong acceleration (favourable pressure gradient - FPG) with reference to the Direct Numerical Simulation (DNS) of Spalart (1986) were examined computationally using a differential, near-wall Second-Moment Closure (SMC) model within the RANS (Reynolds-Averaged Navier-Stokes) framework, accounting separately for viscous and kinematic wall blocking. Besides the three accelerating cases treated by DNS, characterized by the acceleration parameter $K = 1.5 \cdot 10^{-6}$; $2.5 \cdot 10^{-6}$ and $2.75 \cdot 10^{-6}$, the laminarizing turbulent boundary layer was also investigated. The value of $K = 3.2 \cdot 10^{-6}$ was found to be sufficiently high to cause complete relaminarization of the initially turbulent flow, agreeing well with experimental findings. Integral flow parameters, mean velocity and turbulence quantities are computed in close agreement to available DNS database. The analysis of the anisotropy evolution reveals a continuous tendency of the flow turbulence to reach the one-componental isotropic state with an increase of the acceleration intensity.

1 Introduction

Flow acceleration is common place in industrial flow systems, often a consequence of cross-section reduction or strong streamline curvature, as e.g. in flows along curved surfaces. Associated with flow acceleration is a decrease of static pressure, a so-called favourable pressure gradient. This negative pressure gradient, if sufficiently strong, can lead to reverse transition, that is the flow relaminarization[1]. This phenomena is desirable in flow configurations where a higher turbulence level can negatively influence the quality of final products, e.g. in the course of liquid film drying in coating processes. On the other hand, it can cause negative effects in some other applications, e.g. sudden decrease in the heat transfer rate in the blade passages of a gas turbine. From the computational point of view, the flow acceleration

[1] The laminarization phenomena is also observed in oscillating flows, Hanjalić et al. (1995), flows in rotating pipes and channels, Jakirlić et al. (2002) or during the cyclic compression/expansion in a piston-cylinder assembly, Hadžić et al. (2001), etc.

is characterized by a strong departure from the conditions of local equilibrium, thus invalidating the law-of-the-wall-related assumptions. The portion of the boundary layer exhibiting strong viscosity effects is extended, the fact being documented by a Reynolds number decrease. The viscous sublayer grows continuously, causing the production and dissipative regions in the turbulent energy spectra to approach each other. Consequently, the inertial region ($\propto k^{-5/3}$) disappears entirely. The latter is a typical low Reynolds number phenomena. This process is followed by a gradual transition of the velocity profile to a laminar-like one, leading to a decay of the wall shear stress and an increase of the form factor ($H_{12} = \delta_1/\theta$, where δ_1 denotes the displacement thickness and θ the momentum thickness of the boundary layer). The favourable pressure gradient suppresses the wall-normal velocity fluctuations. This leads to an increase in the near-wall anisotropy, expanding the extent of the region where the turbulence is closer to the two-component limit.

Accelerating turbulent boundary layers have been studied by numerous researchers, e.g. Launder (1964), Patel and Head (1968), Jones and Launder (1972), Simpson et al. (1975), Warnack and Fernholz (1998) and others. Spalart (1986) performed direct numerical simulations of self-similar sink flow boundary layers and produced valuable details of turbulence properties. Since the flow is self-similar, the integral parameters retain constant values. For 2-D accelerating boundary layers the most common criterion for laminarization is the acceleration parameter $K = \nu/U_e^2(dU_e/dx)$, with U_e representing the free stream mean velocity. Alternatively, the non-dimensional pressure gradient $P^+ = -\nu/(\rho U_\tau^3)(dP/dx)$ can be used. The computational treatment of the relaminarization process was often ignored in computations of aerodynamic configurations. However, the in-flight measurements due to van Dam et al. (1997) confirmed the existence of laminar regions over all elements of a high-lift wing system. Yip et al. (1993) underlined the importance of accounting for laminarization. They have observed a sudden reduction of the lift coefficient caused by attachment line transition (up to 15%) on a subsonic transport high-lift wing section, which is expected to be partially recovered after onset of flow relaminarization. A sensitive test for a turbulence model is the ability to predict laminarization at the appropriate value of the parameter K. The critical value of K at which a turbulent boundary layers reverts to laminar was experimentally found to be between 2.8 and $3.2 \cdot 10^{-6}$. The corresponding range of critical P^+ is between 0.021 and 0.024. The maximum value of the K parameter in the in-flight experiment of van Dam et al. (at the 40 deg. flap) was around $8 \cdot 10^{-6}$. Spalart provided DNS results for three values of the acceleration parameter $K = 1.5, 2.5$ and $2.75 \cdot 10^{-6}$.

2 Computational method

Numerical framework A parabolized Navier-Stokes solver based on the finite volume method was used to solve the RANS equations for the axial velocity and turbulence quantities. The radial velocity was obtained from the continuity equation. The solution domain has a rectangular form ($L_x \cdot L_y = 4m \cdot 100mm$) covering the entire boundary layer development. The flow acceleration was imposed through the prescription of the acceleration parameter, i.e. velocity/pressure change along the upper boundary. Both diffusion and convective fluxes were discretized by using the 2nd order central differencing scheme. The longitudinal gradients of all variables were treated explicitely. The flows considered were computed also by an elliptic solver in order to check the functionality of the parabolic solver. Such a comparison is displayed in Fig. 1 indicating no noticable difference between two treatments.

Turbulence models In such a transitional flow, only a model scheme accounting for near-wall turbulence can be successfully applied. The best prospect for accurate predictions of turbulent flows with favourable pressure gradients offers the second-moment closure model which can mimic the dynamics of the turbulent stress field and evolution of each stress component, as well as the dynamics of the turbulence scale. The most important prerequisite for capturing the stress anisotropy modification due to strong flow acceleration is to account separately for both viscosity effects (modelled in the conventional manner through the turbulence Reynolds number $Re_t = k^2/(\nu\varepsilon)$) and non-viscous, wall blockage effects (modelled independent of the distance from the wall and its topography in terms of the invariants of the turbulent stress tensor $A_2 = a_{ij}a_{ji}$, $A_3 = a_{ij}a_{jk}a_{ki}$, $A = 1 - 9(A_2 - A_3)/8$ and stress dissipation tensor $E_2 = e_{ij}e_{ji}$, $E_3 = e_{ij}e_{jk}e_{ki}$, $E = 1 - 9(E_2 - E_3)/8$, where $a_{ij} = \overline{u_i u_j}/k - 2\delta_{ij}/3$ and $e_{ij} = \varepsilon_{ij}/\varepsilon - 2\delta_{ij}/3$ are the corresponding anisotropy tensors). Such a differential, near-wall, Reynolds stress model is proposed by Hanjalic and Jakirlic (1998, denoted throughout the work as HJ low-Re RSM). This model was consequently applied for the computations of all cases considered in this work. This model was intensively tested in a number of 2D and some 3D high- and low -Re-number wall flows, including cases of severe acceleration, by-pass and separation-induced laminar-to-turbulent transition, oscillating flows at transitional and high Re numbers, flows subjected to pressure gradients, separating and reattaching flows, rotating and swirling flows as well for flows with mean compression with and without swirl (Hanjalic and Jakirlic, 2002; Jakirlic, 2004). An illustration of the model capability to account for different levels of the mean rate of strain modulation through various pressure gradients (vortex filaments stretching) modifying strongly the Reynolds stress anisotropy is displayed in Fig. 2.

3 Results and discussion

Figs. 3 to 9 illustrate the computational results in a range of K-values considered by DNS. Also, some results will be presented for $K = 3.2 \cdot 10^{-6}$, which has been regarded as sufficiently high to cause flow relaminarization irrespective of the initial turbulence level.

Fig. 3 (left) displays the evolution of the free stream velocity along the flow corresponding to the K-values considered. The vertical lines indicate the so-called sink positions, the locations at which the free stream velocites reach the infinite value. In order to illustrate the model performance in the limit of vanishing Reynolds number of turbulence Re_t and in predicting the laminarization, the evolution of the turbulence field along the flow for two different values of K prior to and around the critical value, 2.75 and $3.2 \cdot 10^{-6}$, illustrated by the plot of maximum values of the turbulence kinetic energy and shear stress and compared with the DNS solution, is presented in Fig. 3 (right). Both solutions were obtained by starting the computations with the Spalart's solutions for $K = 2.75 \cdot 10^{-6}$. Apart from the 'settling' length, where the flow properties 'adjust' to the new set of equations, the agreement is very good. The full recovery is achieved after the profiles, which satisfy the modelled equations, are obtained. For $K = 2.75 \cdot 10^{-6}$, the solution yields the self-similar values very close to the DNS results. Whereas for $K = 3.2 \cdot 10^{-6}$ the turbulence shows a steady decay and the flow laminarizes as indicated by zero values of both the kinetic energy of turbulence and shear stress at the end of the solution domain. This is further illustrated in Fig. 4 by displaying the contours of the kinetic energy of turbulence k for the acceleration parameters mentioned above. Besides the flow laminarization, these figures document also the shifting of the maximum k-values towards the wall.

As already discussed in the introduction, a streamwise favourable pressure gradient thickens the viscous sublayer, thus strengthening the influence of the viscosity affected region on the overall flow development. The best representation of the flow evolution under the conditions of a severe acceleration is given through the behaviour of some integral parameters, such as Reynolds number based on the momentum thicknes Re_θ, form factor H_{12} and friction coefficient C_f, Figs. 5 and 6 (left). Whereas the Reynolds number and friction coefficient show a clear reduction, typical for the growing viscosity influence, the form factor exhibits a continuous increase until the laminar value ≈ 2.1 was reached. The model results are compared with the DNS data for all K-values and with laminar solution for K=3.2×10^{-6}. Again, apart from the 'adjusting' length, the agreement is very good. Fig. 6 (right) represents the mean velocity evolution with increase in the acceleration intensity illustrating clearly the departure of the logarithmic law being manifested through the log-line overshoot. The velocity profile for the laminarizing case $K = 3.2 \cdot 10^{-6}$ (the computation was started from the turbulent profiles for $K = 2.75 \cdot 10^{-6}$) and the laminar solution for the same acceleration parameter coincide completely.

Figs. 7 show a comparison between model results and DNS results for all three normal stress components, documenting very good agreement, particularly close to the wall. The shape of the stress profiles resembles those in a constant-pressure boundary layer or channel flow at corresponding Reynolds numbers, but with an obvious reduction of all three components, particularly in the outer flow region. It is also noticable that the damping due to acceleration affects the spanwise component w^+ the most and the streamwise component u^+ the least, increasing the stress anisotropy in the near wall region (Fig. 8) as compared with a constant-pressure boundary layer. This points out the ability of the turbulent model to predict proper asymptotic behaviour of all turbulence quantities as the wall is approached. Because the fluctuations normal to the wall die out faster than those in the plane parallel to the wall, the turbulence approaches the two-component limit: $A \propto y^2$. Fig. 9 shows typical behaviour of $f(-II, III)$ $(-II = A_2/8;\ III = A_3/24)$ in the case of flow affected by a favourable pressure gradient. Strong acceleration promotes the approach to one-componental isotropic turbulence (left end of the straight line denoting $A = 0$). The evolution of $f(-II, III)$ by increasing the K-value, displaying in Fig. 9 (left) only the results obtained by DNS, clearly follows this tendency. Fig. 9 (right) shows a further increase of the accelerating parameter K compared to the DNS range to values corresponding to the critical one, at which the laminarization occurs. The curve $f(-II, III)$ for $K = 3.2 \cdot 10^{-6}$ was taken at the position $x^+ \approx 13.500$ (Fig. 3 right) before the flow completely laminarizes. The curve $f(-II, III)$ corresponding to the laminar situation can obviously leave the invariant map only at its right corner, denoting one-component isotropic turbulence.

4 Conclusions

A relaminarization of an initially turbulent boundary layer was analyzed computationally by using a low Reynolds number Second-Moment Closure model. Hereby, the DNS database of a series of self-preserving sink flows (Spalart, 1986) was used as a reference. An in-depth analysis of the stress anisotropy evolution under the conditions of severe acceleration yelded the conclusion that the anisotropy maximum was shifted towards the wall, extending the region (in wall units) where turbulence approaches the two-component limit. The final stage of the relaminarization process corresponds to the one-component isotropic state.

References

[1] Hadžić, I., Hanjalić, K. and Laurence, D. (2001): *Physics of Fluids*, Vol. **13(6)**, pp. 1739-1747

[2] Hanjalić, K., Jakirlić, S., and Hadžić, I., (1995): *Turbulent Shear Flows*, Vol. **9**, pp 323-342., Eds. F.Durst et al., Springer Berlin

[3] Hanjalić, K., and Jakirlić, S. (1998): *Computers and Fluids*, **27**, pp. 137-156

[4] Hanjalic, K., and Jakirlić, S. (2002): Second-Moment Turbulence Closure Modelling. In *Closure Strategies for Turbulent and Transitional Flows*, B.E. Launder and N.H. Sandham (Eds.), Cambridge University Press, Cambridge, UK, pp. 47-101

[5] Jakirlić, S., Hanjalić, K., and Tropea, C. (2002): *AIAA J.*, **40(10)**, pp. 1984-1996

[6] Jakirlić, S. (2004): DNS-based scrutiny of RANS-approaches and their potential for predicting turbulent flows. *Habilitation thesis*, Darmstadt University of Technology

[7] Jones, W.P., and Launder, B.E. (1972): The prediction of the laminarization with a two-eqaution model of turbulence. *Int. J. Heat and Mass Transfer*, Vol. **15**, pp. 301-313

[8] Launder, B.E. (1964): Laminarisation of the turbulent boundary layer in a severe acceleration. *ASME J. Appl. Mech.*, Vol. **31**, p. 707

[9] Patel, V.C., and Head, M.R. (1968): Reversion of turbulent to laminar flow. *J. Fluid Mech.*, Vol. **34**, p. 371

[10] Simpson, R.L., and Wallace, D.B. (1975): Laminarescent Turbulent Boundary Layers: Experiments on Sink Flows. TR SMU-1-PU, Southern Methodist University, Dallas

[11] Spalart, P.R. (1986): Numerical study of sink-flow boundary layers. *J. Fluid Mech.*, Vol. **172**, pp. 307-328

[12] van Dam, C.P., Los, S.M., Miley, S.J., Roback, V.E., Yip, L.P., Bertelrud, A., Vijgen, P.M.H.W. (1997): In-Flight Boundary-Layer State Measurements on a High-Lift System: Slat (pp. 748-756) and Main Element and Flap (pp. 757-763), *Journal of Aircraft*, Vol. **34(6)**

[13] Warnack, D., and Fernholz, H.H. (1998): The effects of favourable pressure gradient and of the Reynolds number on an incompressible axisymmetric turbulent boundary layer: the boundary layer with relaminarization. *J. Fluid Mech.*, Vol. **359**, pp. 357-381
[14] Yip, L.P., Vijgens, P.M.H.W., Hardin, J.D., van Dam, C.P. (1993): In-flight pressure-distribution and skin-friction measurements on a subsonic transport high-lift wing section. AGARD Conference Proceedings 515: *High-Lift System Aerodynamics*, 71st Fluid Dynamics Panel Meeting and Symposium, Banff, Alberta, Canada, October 5-8, 1992

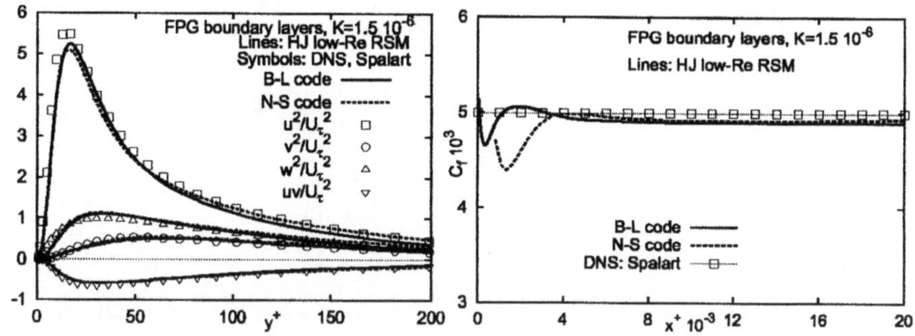

Figure 1: Comparison between turbulent stresses and friction coefficients obtained by the Navier-Stokes solver (N-S) and parabolized N-S (B-L, boundary layer code)

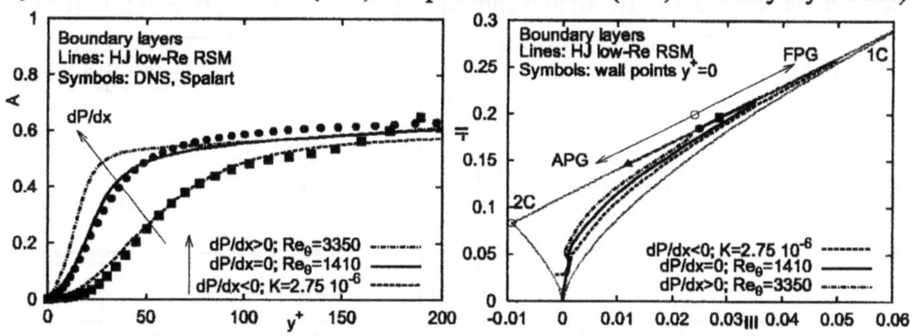

Figure 2: Lumley's two-componentality ("flatness") parameter A and the invariant map for boundary layers at zero, favorable and adverse pressure gradients

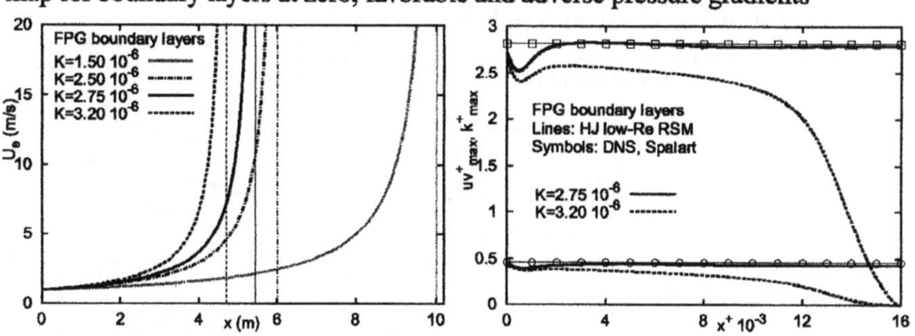

Figure 3: Evolution of the free stream velocity with increasing acceleration parameter K and evolution of the maximum values of the kinetic energy of turbulence and shear stress component along the boundary layer for the cases with strongest acceleration ($K = 2.75 \cdot 10^{-}6$) and with ($K = 3.20 \cdot 10^{-}6$) consequent laminarization

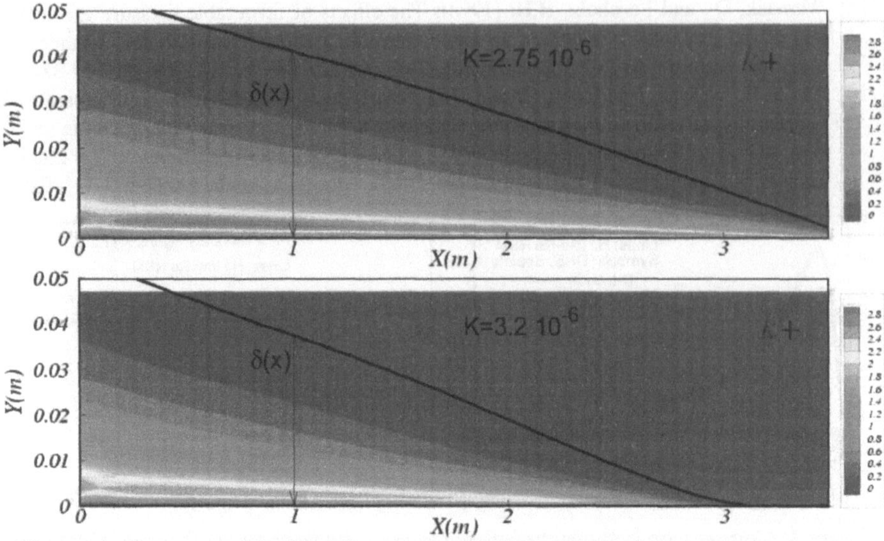

Figure 4: Contours of the kinetic energy of turbulence for the cases with strongest acceleration ($K = 2.75 \cdot 10{-}6$) and consequent laminarization ($K = 3.20 \cdot 10{-}6$)

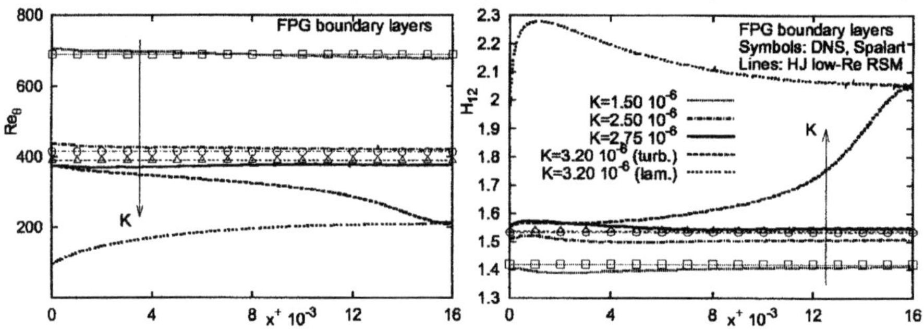

Figure 5: Reynolds number (Re_θ) decrease and form factor (H_{12}) increase under conditions of strong acceleration. Symbols as in Fig. 6

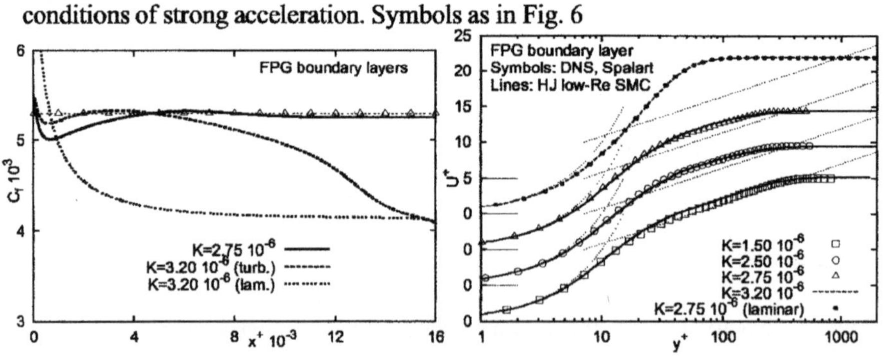

Figure 6: Friction coefficients for the case with strongest acceleration ($K = 2.75 \cdot 10{-}6$) and the laminarizing case ($K = 3.20 \cdot 10{-}6$) and semi-log plots of mean velocity profiles for all sink flow cases. Lines as in Fig. 5

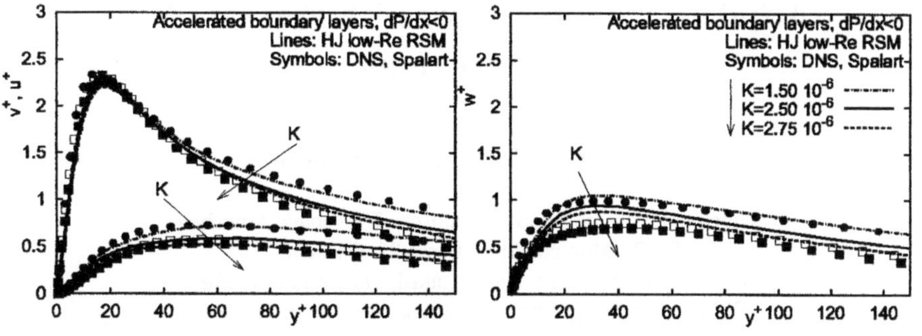

Figure 7: Evolution of the normal Reynolds stress components under the conditions of strong acceleration. Symbols as in Fig. 8

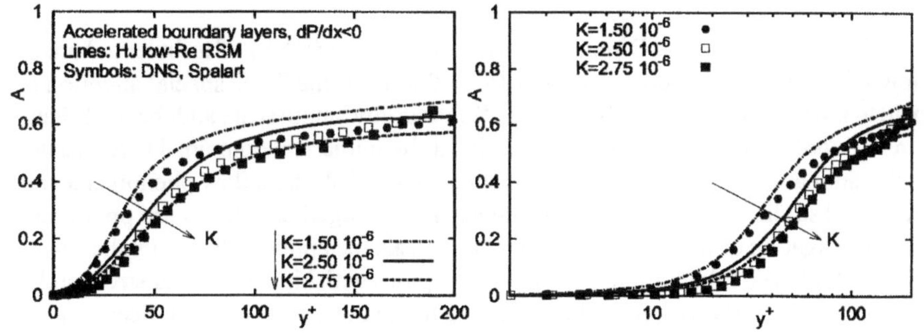

Figure 8: Evolution of the two-componentality parameter of Reynolds stress anisotropy A under conditions of strong acceleration

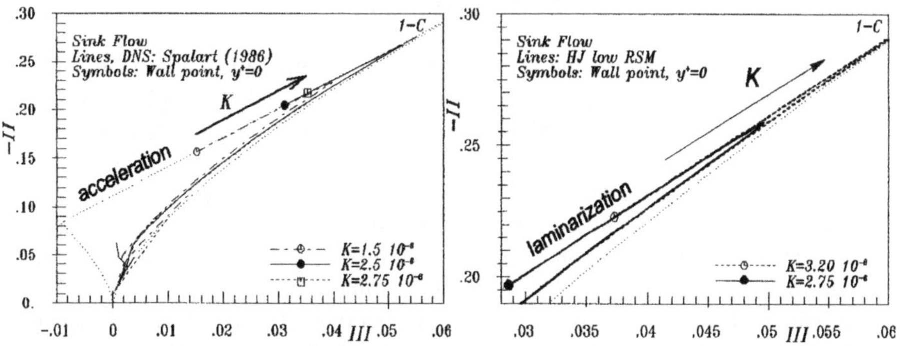

Figure 9: Invariant map of the Reynolds stress anisotropy tensor in the favourable pressure gradient (FPG) boundary layer displaying the process of flow acceleration and consequent laminarization

Turbulent Channel Flow with Periodic Hill Constrictions

Nikolaus Peller[1] and Michael Manhart[1]

Fachgebiet Hydromechanik, Technische Universität München,
80333 Munich, Arcisstrasse 21, Germany,
n.peller@bv.tum.de, m.manhart@bv.tum.de,
WWW home page: http://www.hy.bv.tum.de

Summary

This paper presents a Direct Numerical Simulation (DNS) of turbulent channel flow with periodic hill constrictions at $Re = 2808$. For the DNS, special attention is paid to the grid design by analysis of the Kolmogorov length scale and the wall shear stress in order to be well resolved over the entire numerical domain. Therefore the DNS simulation provides data for a detailed study of physical flow phenomena and near-wall studies. Results for instantaneous and averaged flow fields are presented. An investigation of the near wall behaviour reveals the applicability of explicit wall models for Large Eddy Simulation (LES) in regions with high wall shear stress and a complete failure of conventional wall scaling in separation and reattachment regions where the wall shear stress is small.

1 Introduction

Detached turbulent flows are difficult to predict numerically and an improvement of turbulence models requires a better understanding of the underlying physical processes. Therefore investigations of basic flow phenomena are especially important. In that respect it is reasonable to restrict oneself to simple standard test cases.

An example of such a simplified flow is a channel flow with periodic hill constrictions that has been analysed by Almeida et al. [1] in an experiment. The challenging properties of this flow are separation and reattachment on a smoothly curved geometry, strong pressure gradients and a fluctuating separation point in time. The clean definition of boundary conditions by using periodicity in streamwise and spanwise direction makes it a good test case for numerical simulations such as Reynolds Averaged Navier Stokes (RANS) and LES simulations. Consequently, this flow has then been chosen as a standard test case for various research groups in order to test turbulence simulation strategies. As an example, in the framework of the French German research group "LES of complex flows" this flow has been chosen as test case for benchmarking different LES codes and models. In the present work, we are concerned with the testing, improving, and the development of new wall models for LES. In comparison to the experimental setup of Almeida the numerical setup currently used as benchmark case has been changed. According to the geometry used by Mellen et al. [9], the distance between two hills has been enlarged in order to achieve reattachment of the flow before the slope of the next hill. Temmermann et al. [13] have found that this test case is especially sensitive to near-wall resolution and wall modeling approaches. This is the reason for studying the near-wall behaviour in detail in this work by means of Direct Numerical Simulation (DNS).

The paper proceeds with the numerical setup and test case definitions. It then explains the means of grid design for the Direct Numerical Simulation and presents the results for instantaneous and averaged flow fields. In comparison to a coarse grid simulation differences are pointed out. Consequently the attention is drawn to the near-wall behaviour and plots of instantaneous velocities in inner coordinates. Finally the paper finishes with a conclusion on near-wall modeling.

2 Numerical framework and test case definition

The simulation was carried out with the finite volume code MGLET [5] for the incompressible Navier Stokes equations. The numerical grid is cartesian and non-equidistant with a staggered arrangement of variables. For the discretization in time a “Leapfrog” scheme is used.

$$u^{n+1} = u^{n-1} + 2\Delta t[C(u^n) + D(u^{n-1}) - G(p^{n+1})]$$

The abbreviations are u for the velocity and p for the pressure, C for the convective term, D for the diffusive term, and G for the pressure term. The discretization in space is a second order central approximation scheme. It allows an efficient formulation in combination with the immersed boundary method and makes it expecially simple to parallelize the code. It has been shown by several authors that second order accuracy can be sufficient for DNS of flows provided the grid resolution is sufficient (Eggels et al. [2], Friedrich et al. [3], Manhart and Friedrich [7]). Especially for geometrically complex flows, true higher order is extremely difficult and expensive to achieve, so we prefer to use the efficiency of our 2nd order scheme to obtain high spatial resolution. For the solution of the Poisson equation the “Stone Implicit Procedure” (SIP) is implemented. For more information on the code, please refer to Manhart [8]. For the representation of the hill geometry in the cartesian grid an immersed boundary technique was developed which allows a smooth representation of the body surface by using high order interpolation for the interface cells [14]. For the simulation presented here, a least squares interpolation has been implemented [10]. This method prevents instabilities which are present in high order Lagrange interpolation schemes.

The numerical setup has been introduced by Mellen et al. [9]. Based on hill height h, the channel extends $9.0h$ in streamwise, $4.5h$ in spanwise and $3.035h$ in wall-normal direction. The hills are arranged periodically at the distance $9.0h$. In streamwise and spanwise direction periodic boundary conditions are used. On the hill surface a no-slip condition is applied as well as on the top of the channel wall. A sketch of the geometry can be seen in Figure 1.

3 Grid design and resolution study

In order to find proper grid spacings, we use estimations based on wall shear stress for the wall normal resolution and estimations based on the Kolmogorov scale in the core of the flow. A first evaluation of these criteria has been obtained by a preliminary simulation with a relatively coarse grid. The finally computed values confirmed these resolution estimations. The grid is refined by geometric stretching in z-direction close to the hill surface where strong gradients must be resolved. The stretching factors are kept below 3%. Due to the cartesian grid the refinement is not exactly wall normal at all positions of the geometry. The grid spacings in streamwise and spanwise direction are equidistant. The actual wall shear stresses along the bottom wall from the DNS simulation can be seen in Figure 2. Based on the wall shear stress the grid spacing in z-direction remains below $\Delta z^+ = (\sqrt{\frac{\tau_w}{\rho}})\frac{\Delta z}{\nu} = 1.4$ at the lower wall.

The Kolmogorov length scale for the cross section at the hill crest can be seen in Figure 3 in comparison to the one of a plane channel flow at comparable bulk Reynolds number. Two facts can be pointed out. Firstly, the Kolmogorov length scale for the periodic hill simulation is smaller over the whole cross section in comparison to channel flow. Secondly the length scale is decreasing when the wall ist approached. Therefore a much finer grid is needed in the near-wall region than in plain channel flow which is accounted for by the grid refinement in vertical direction towards the wall. The Kolmogorov length scale is not directly linked to the wall shear stress as Figure 4 demonstrates for two streamwise positions – one at maximal wall shear stress ($x/h = 8.7$) and one at a relatively small shear stress ($x/h = 0.0$).

In Table 1 the grid spacings and resulting number of cells used for the preliminary coarse grid and the final grid are listed. The total number of cells equals 47 million for the fine and 7 millions for the coarse grid. The computation was performed on the super computer Hitachi SR8000 at the Leibniz Computing centre of Munich with our fully MPI parallelized code [5].

A comparison of the time averaged streamwise velocity profile just after separation at $x/h = 0.5$ for both simulations can be seen in Figure 5. The beginning of the separation seems to be predicted very well even by the coarse simulation. More pronounced differences are visible towards the center of the channel. Contrary to the averaged velocity, the profiles of the Reynolds stresses in the coarse grid simulation differ substantially from the fine grid results. Figure 6 and Figure 7 show the profiles of the $< u'u' >$ and $< u'w' >$ Reynolds stress components at the position $x/h = 0.5$. The peak values in the coarse simulation are overpredicted and some details of the profiles are not captured in the coarse simulation as can be seen in Figure 6.

4 Physical flow features

During the flow separation from the smoothly contoured surface of the hills, the boundary layer evolves to a free shear layer that is highly active and leads to strong oscillations of the separation and reattachment points (Figure 8). Because of the length of the domain, the reattachment region is located on the channel bottom before the next hill slope.

The recirculation region of the time averaged flow field can be identified from Figure 9. The white area marks the recirculation bubble. Corresponding to the wall shear stress in Figure 2 the reattachment point can be found at $x/h \approx 5.1$. Fr¨ohlich et al. [4] point out that the reattachment position is strongly dependent on the point of flow detachment. Both positions are well predicted even by the preliminary simulations (Figure 2). From the wall shear stress in Figure 2, a second tiny recirculation zone can be found in the mean flow at $x/h \approx 7.2$. A considerable negative wall shear stress can be recognized at $x/h \approx 3.0$ within the recirculation bubble which shows that a strong backflow develops in the recirculation zone.

The focus of our study lies in investigating the near-wall behaviour of separating and reattaching flows. As already pointed out by Simpson [12] it is extremely difficult to find a wall scaling even for the time averaged flow. When developing wall models for statistical and Large-eddy simulations, however, some knowledge of possible universal near-wall behaviour is required. For LES the situation is even more complicated than for RANS, since this universal behaviour has to be found in the instantaneous near-wall flow fields. A step in this direction is the notion of the importance of instantaneous pressure gradients in a turbulent boundary layer on a flat plate with a marginal separation bubble [6].

In the present simulation several flow regions have been identified which are or are not in agreement with the assumption of a canonical boundary layer as expressed by a lin/log law of the wall as described in Sagaut [11]. Figure 10 is a scatter plot of instantaneous velocity profiles over the wall distance normalized by the instantaneous local wall shear stress. This position at the hill crest ($x/h = 0.002$) is close to the point of detachment where the wall shear stress vanishes. There is a great scatter in these instantaneous profiles which differ substantially even from the linear law of the wall. Further downstream at $x/h = 8.84$ (Figure 11) the instantaneous velocity profiles in inner coordinates are much narrower around the lin/log law of the wall. As can be identified from the wall shear stress plot in Figure 2 the position $x/h = 8.84$ is located in the region of maximum wall shear stress.

Further investigations of the velocity profiles suggest that the flow can be divided into four regions. The first region characterizes the hill top where the flow approaches the detachment region and is under a strong adverse pressure gradient. In this region a very poor agreement with the law of the wall is observed. The second flow region is located within the recirculation bubble. Surprisingly, there is a much better agreement with the law of the wall. It seems that there is a developing boundary layer in the upstream direction within the backflow region. The third region can be recognized by the reattachment of the flow where the agreement of the profiles is again very poor. In the fourth region, which is characterized by the maximum shear stress the law of the wall is again a good approximation for the instantaneous velocity profiles in inner coordinates.

5 Conclusions

A DNS of turbulent flow in a channel with periodic constrictions by smoothly contoured hills has been presented. The DNS has been carefully checked in terms of grid resolution to satisfy criteria based on the wall shear stress and the Kolmogorov length scale. Therefore, it can be regarded as a reliable data base for detailed investigation of the flow.

The flow exhibits two separation regions, the main region leeward of the hill and a second one, which is very small just upstream of the hill. A third recirculation region on the top of the hill as found by Fr¨ohlich [4] at $Re = 10595$ is not present at the considered $Re = 2808$.

Our focus lies on the near-wall behaviour. We have found four regions along the bottom wall of distinct near-wall behaviour. In two of them, the main backflow region and the strongly accelerated region, the law of the wall seems to give a reasonable description of the near-wall profiles. As soon as the wall shear stress is large enough, scaling based on it seems to be sufficient, independent of how complex the surrounding flow situation is. This investigation also confirms the observation from other studies that conventional wall scaling completely fails when the average wall shear stress is small.

Acknowledgements

The authors gratefully acknowledge the financial support by the "Deutsche Forschungsgemeinschaft" under reference number MA 2062/3 and the support of the "Leibniz Rechenzentrum" (LRZ) of the Bavarian Academy of Sciences.

References

[1] G.P. Almeida, D.F.G. Durão and M.V. Heitor, Wake flows behind two-dimensional model hills. *Experim. Therm. Fluid Sci.* , 7,87-101, 1993

[2] J.G.M. Eggels, F. Unger, M.H. Weiss, J. Westerweel, R.J. Adrian, R. Friedrich, and F.T.M. Nieuwstadt. Fully developed turbulent pipe flow: A comparison between direct numerical simulation and experiment. *J. Fluid Mech.*, 268:175–209, 1994.

[3] R. Friedrich, T. Hüttl, M. Manhart, and C. Wagner. Direct numerical simulation of incompressible turbulent flows. *Computers and Fluids*, 30(5):555–579, 2001.

[4] J. Fröhlich, C.P. Mellen, W. Rodi, L. Temmerman and M.A. Leschziner, Highly-resolved large eddy simulation of separated flow in a channel with streamwise periodic constrictions. *J. Fluid Mechanics* , Cambridge University Press, 2005, 256,19-66.

[5] M. Manhart, F. Tremblay, and R. Friedrich, MGLET: a parallel code for effi cient DNS and LES of complex geometries *Parallel Computational Fluid Dynamics 2000,* Elsevier Science B.V., Amsterdam, Jensen et al., 449-456

[6] M. Manhart. Analysing near-wall behaviour in a separating turbulent boundary layer by DNS. In B. Geurts, Hrsg., *Direct and Large-Eddy Simulation IV.* Kluwer Academic Publishers, Dordrecht, 2001.

[7] M. Manhart and R. Friedrich. DNS of a turbulent boundary layer with separation. *International Journal of Heat and Fluid Flow*, 23(5):572–581, 2002.

[8] M. Manhart. A zonal grid algorithm for DNS of turbulent boundary layers. *Computers and Fluid*, 33(3):435–461, March 2004.

[9] C.P. Mellen and J. Fröhlich and Rodi. Large eddy simulation of the flow over periodic hills. *In Proc. IMACS World Congress* ,(ed. M.Deville & R.Owens), Lausanne

[10] N. Peller, Anne Le Duc, Frédéric Tremblay and Michael Manhart, Lagrange versus least-square direct forcing in an immersed boundary method, *In Preparation*

[11] P. Sagaut, Large Eddy Simulation for Incompressible Flows. *Scientific Computation* , Second Edition, Springer, 2002.

[12] R.L. Simpson. Turbulent boundary-layer separation. *Ann. Rev. Fluid Mech.* , 21:205–234, 1989.

[13] L. Temmerman, M.A. Leschziner, C.P. Mellen and J. Fröhlich. Investigation of wall-function approximations and subgrid -scale models in large eddy simulation of separated flow in a channel with streamwise periodic constrictions. *Intern. Journal of Heat and Fluid Flow.* Science Direct, 2003, 24,157-180.

[14] F. Tremblay, M. Manhart and R. Friedrich. LES of flow around a circular cylinder at a high subcritical Reynolds number. *Direct and Large-Eddy Simulation IV* , Kluwer Academic Publishers, Dodrecht, 2001.

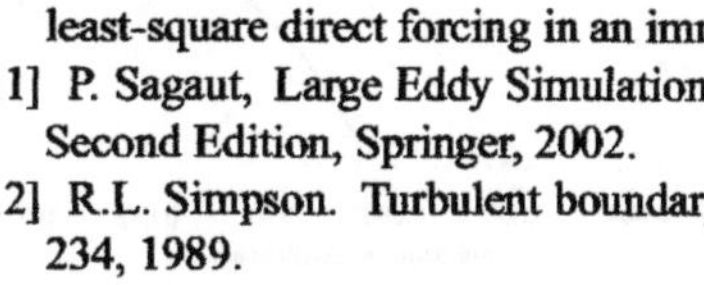

Figure 1 Numerical domain

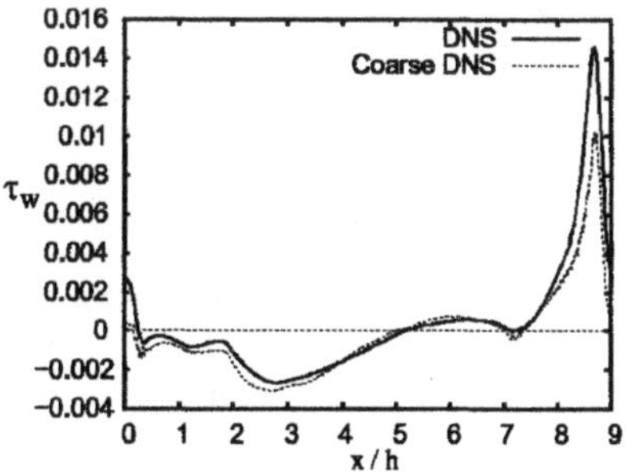

Figure 2 Wall shear stress on bottom wall for both simulations.

Table 1 Number of grid points and grid spacings based on wall shear stress at $x/h = 8.7$; The grid spacing in z-direction is given at the hill crest.

	Points			Grid spacing		
	Nx	Ny	Nz	Δx^+	Δy^+	Δz^+_{max}
Coarse DNS	172	154	289	15	8.4	5.5
DNS	464	304	338	5.5	4.2	1.4

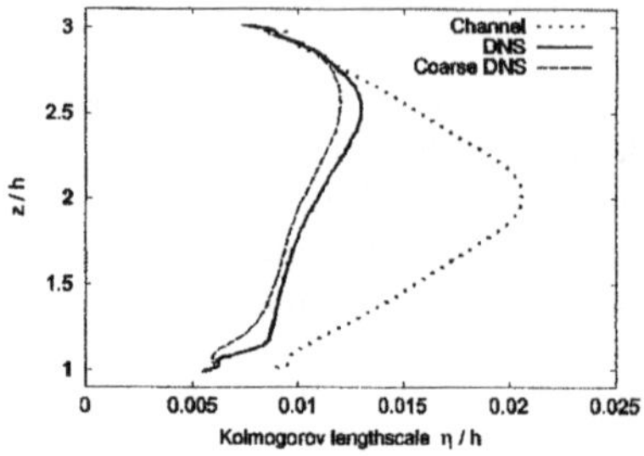

Figure 3 Kolmogorov length scale of plain channel flow and periodic hill simulation at $x/h = 0.0$

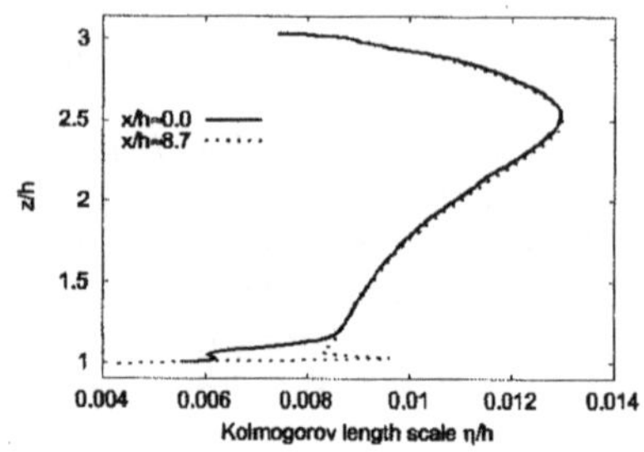

Figure 4 Kolmogorov length of periodic hill simulation at $x/h = 0.0$ and $x/h = 8.7$

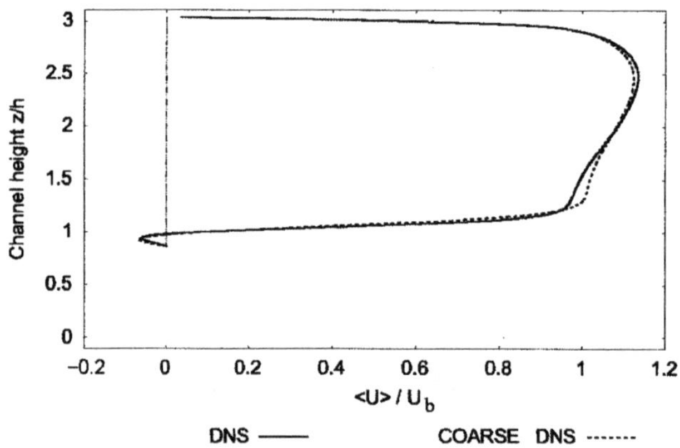

Figure 5 Streamwise velocity profi les at $x/h = 0.5$ for both resolutions;

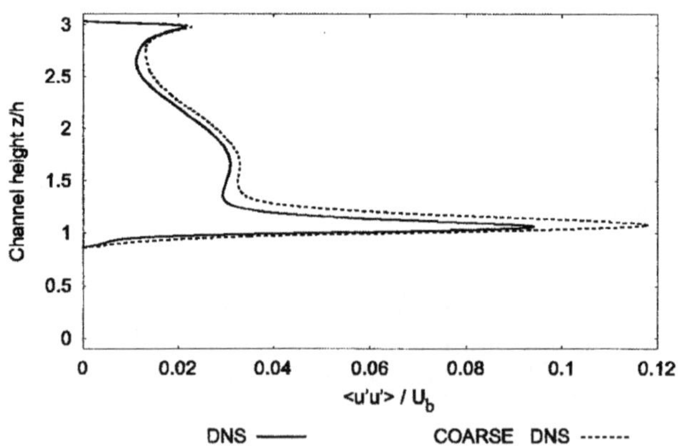

Figure 6 Comparison of DNS with coarse grid DNS at $x/h = 0.5$; Reynolds normal stress $< u'u' >$ for both grid resolutions.

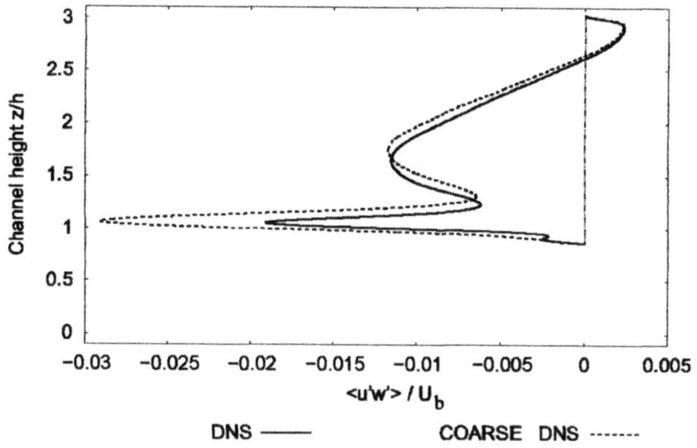

Figure 7 Comparison of DNS with coarse grid DNS at $x/h = 0.5$; Reynolds shear stress $< u'w' >$ for both grid resolutions.

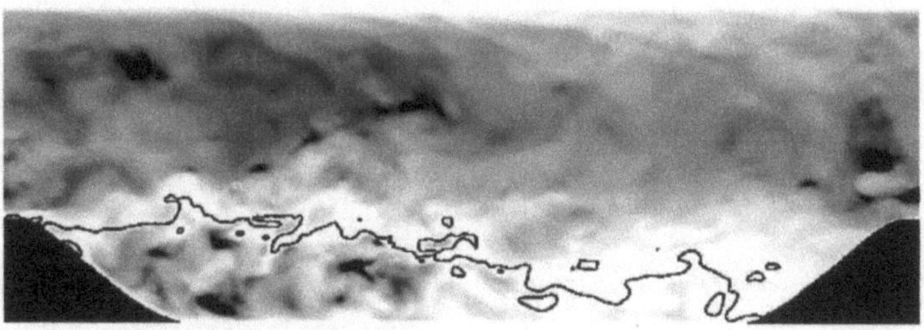

Figure 8 Absolute value of streamwise velocity; High values are colored dark. The recirculation bubble is located below the black line

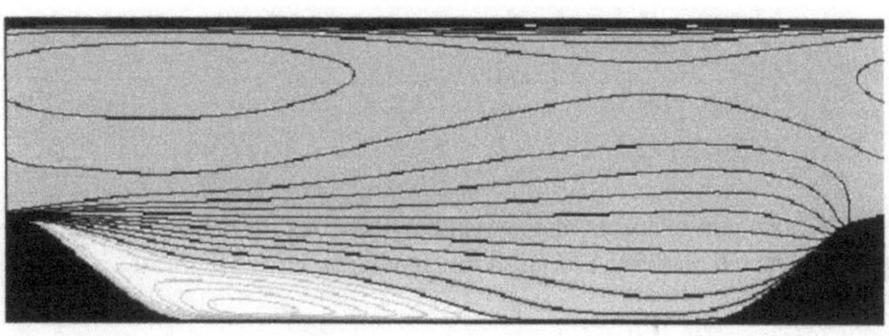

Figure 9 Isolines of streamwise velocity from statistical flow field. The white area marks the recirculation zone. Increments are $0.1u_b$.

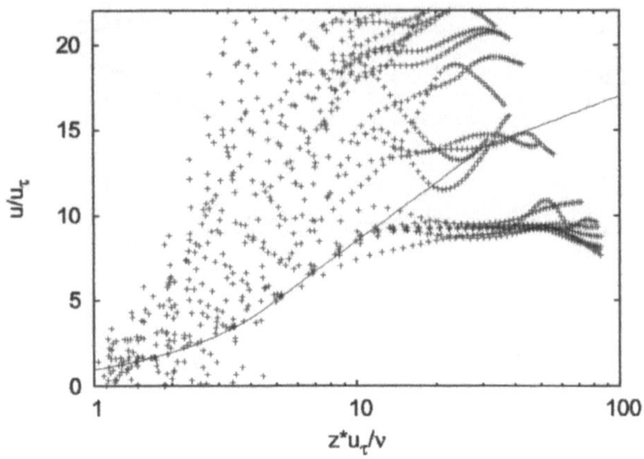

Figure 10 Scatter plot of instantaneous velocity profi les normalized by the instantaneous wall shear stress at $x/h = 0.002$; The solid line marks the law of the wall.

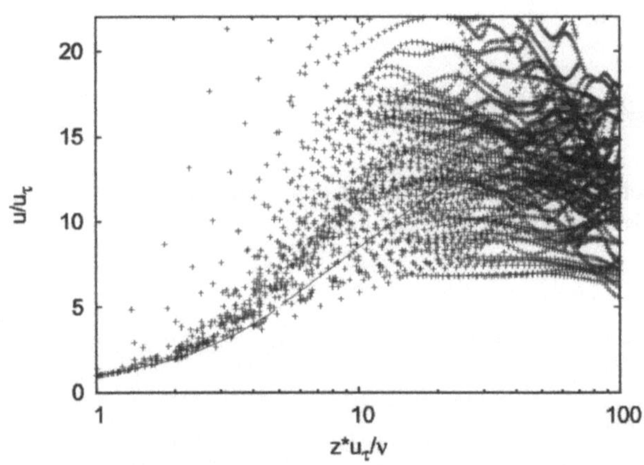

Figure 11 Scatter plot of instantaneous velocity profi les normalized by the instantaneous wall shear stress at $x/h = 8.84$; The solid line marks the law of the wall.

Turbulent Flow Separation Control by Boundary-Layer Forcing: A Computational Study

S. Šarić, S. Jakirlić and C. Tropea

Fachgebiet Strömungslehre und Aerodynamik, Technische Universität Darmstadt
Petersenstr. 30, 64287 Darmstadt, Germany
saric@sla.tu-darmstadt.de, s.jakirlic@sla.tu-darmstadt.de, ctropea@sla.tu-darmstadt,de

Summary

Various methods for unsteady flow computations: LES (Large Eddy Simulation), DES (Detached Eddy Simulation) and URANS (Unsteady Reynolds-Averaged Navier-Stokes) were used to study the effects of boundary-layer forcing on the mean flow and turbulence. Two flow configurations were considered: a periodically perturbed flow over a backward-facing step (with fixed separation point) at a low Reynolds number ($Re_C = 3700$, Yoshioka et al., 2001 [10]) and a high Reynolds number ($Re_c = 9.36 \cdot 10^5$) flow over a wall-mounted hump with steady suction (flow separation at the smooth surface), Rumsey et al., 2004 [5]. Whereas LES and DES reproduced all important flow characteristics observed in the experiments, RANS (S-A and k-ω SST models were employed) method exhibited a weaker sensitivity to the shear layer oscillations i.e. boundary-layer suction. It is shown that DES method, representing a hybrid RANS/LES approach, is capable of reproducing the mean flow and turbulence features of a quality comparable with the LES method, but employing significantly coarser spatial resolution.

1 Introduction

Active flow separation control is a topic of significant interest, being especially challenging for unsteady flow simulation strategies. The phenomena of flow separation is accompanied by large energy loses, limiting the performance and the design of fluid flow devices. The prevention (or at least delay) of separation or reduction of reattachment length can lead to performance gains in numerous engineering applications. Seifert and Pack [6], [7] studied configurations at high Reynolds numbers relevant to the aircraft aerodynamics: flow past an airfoil and flow over a wall-mounted hump. Chung and Sung [1], Yoshioka et al. [10] investigated the flow over a backward-facing step. Turbulent flow over a wall-mounted hump (simulating the upper surface of a Glauert-Goldschmied type airfoil at zero angle of attack) at high Reynolds number ($Re_c = 9.36\text{x}10^5$ based on the hump cord length and the free-stream velocity) served as a test case for computational methods and turbulence models validation at the recent NASA Langley Research Center Workshop [5]. Flow control, i.e. shortening of the recirculation bubble was accomplished by steady suction through a slot (0.004 C wide) located at approximately 65 % cord length (C). The spatially uniform suction velocity (along 0.5842 C span) corresponds to the mass flow rate of 0.01518 kg/s. The oncoming flow is characterized by turbulent boundary-layer thickness being approximately half of the maximum hump height measured at the location about two cord lengths upstream of the hump leading edge. The flow over a backward-facing step ($Re_C = 3700$, based on the step height H and the centerline velocity of the inlet channel U_C), perturbed periodically by an alternate blowing/suction jet through a thin slit situated at the step edge, has been investigated experimentally by Yoshioka et al. [10]. Computational studies pertinent to this flow configuration were recently reported by Šarić et al. [8] and Dejoan et al. [2]. The uniform injection velocity is parameterized by a sinusoidal law: $v_e = V_e \sin \phi$, V_e being the velocity amplitude ($V_e = 0.3U_c$) and ϕ is the phase angle. All perturbation frequencies f_e ($\phi = 2\pi f_e t$) provided experimentally, corresponding to the Strouhal numbers: $St = 0.08, 0.19$ and 0.30 ($St = f_e H/U_c$), were investigated numerically. It was found experimentally that the perturbation frequency corresponding to $St = 0.19$ represents the most effective one, leading to the minimum reattachment length. The purpose of the present work was the computational study of the effects of boundary layer, i.e. shear layer forcing on the mean flow and turbulence using various methods for unsteady flow computations: LES (Large Eddy Simulation), DES (Detached Eddy Simulations) and URANS (Unsteady Reynolds-Averaged Navier-Stokes) aiming also at mutual comparison of their features and performance in such complex flow situations.

2 Numerical Method

All computations were performed with an in-house computer code based on a finite volume numerical method for solving both three-dimensional filtered and Reynolds-Averaged Navier-Stokes equations on block-structured, body-fitted, non-orthogonal meshes. The sub-grid scales were modelled by the most widely used model formulation proposed by Smagorinsky (1963) in the LES framework (including Van Driest damping of the Smagorinsky coefficient $C_s = 0.1$). A one-equation turbulence model by Spalart and Allmaras (S-A, 1994), based on the transport equation for turbulent viscosity, was employed to model influence of the smallest, unresolved scales on the resolved ones in the framework of the DES computational scheme (e.g. Travin et al. [9]). Different statistical turbulence models including the above-mentioned S-A model (it was interesting to see, how the same model performed in two different computational frameworks: RANS and DES) and the k-ω SST model (Menter, 1994) were examined. The convective transport of all variables was discretized by a second-order central differencing scheme. Time discretization was accomplished by applying the (implicit) Crank-Nicolson scheme. The CFL number, representing the time step chosen, was less then unity in largest portion of the solution domain for both flow configurations. The only exception is narrow region around the thin slot at the hump (≈ 2 mm width) / step edge (1 mm width). The solution domain for the hump geometry (dimensions: 6.14Cx0.91Cx0.152C) is meshed with almost 4 Mio. (426x145x64) grid cells for LES computations. The solution domain employed for DES with somewhat larger spanwise dimension (0.2C),

was meshed by approximately 1.7 Mio. (426x145x28) grid cells. RANS computations have not shown significant difference in the solutions obtained if the computational domain was extended further upstream ($6.39C$) as in the experiment. Furthermore, it has been demonstrated experimentally that the flow is insensitive to the upstream boundary conditions. Therefore, in all LES and DES computations available steady profiles (the mean experimental velocity profiles) were imposed at the inlet plane placed at $2.14C$ upstream of the hump leading edge.The focus region, just downstream the slot including the region arround reattachment, was meshed to provide $\Delta x^+ = 80$, $\Delta y^+ = 1 - 80$, $\Delta z^+ = 150$ (maximum values for DES grid). RANS/LES interface arround the hump was at $y^+ = 30 - 90$. Compared to DES, LES resolution in the spanwise direction was finer providing $\Delta z^+ = 50$.
The size of solution domain adopted behind the step was $30H$x$3H$xπH. The inlet data corresponding to a fully developed channel flow were generated by pre-cursor LES and the channel length of H/2 was taken before expansion. Different grids were employed as described in [8], with the final ones comprising 590 000 (LES-220x82x32) and 190 000 (DES-142x82x16) grid cells. No-slip boundary conditions were applied at the walls ($y^+ < 1$), convective outflow at the outlet and periodic boundary conditions along the streamwise direction.

3 Results

The results of backward-facing step simulations will be presented first, followed by the predictions of separated flow over a wall-mounted hump. The time-averaged results have been extracted to provide comparison with the available experimental data.

3.1 Backward-facing step flow

The results of the reference case (without perturbation) are considered first. Figs. 1 and 2 show the mean velocity and Reynolds shear stress profiles. All computations significantly overpredict the measured reattachment length ($X_R = 6.0H$), presumably because of 3D contamination in the experiment due to spanwise confinement (aspect ratio $L_z/H = 12$ is regarded as too short for providing a 2-D flow in the mid-span plane). Nevertheless, the main goal was to study the effects of perturbations on flow characteristics compared to the reference unperturbed case. The additional simulations (not shown here) using the finer grids and larger spanwise computational domain ($8H$) yield the same reattachment length ($X_R = 7.18H$), which is also in accordance with the LES predictions of Dejoan et. al ($X_R = 7H$) [2]. For this reason five specific streamwise locations are normalized by both step height (H) and the reattachment distance of the unperturbed flow (X_{R0}). Agreement between DES and LES solutions and experiments is very close. Due to the lack of space only the results obtained for the optimum frequency $St = 0.19$ (as observed in the experiment) are presented (see ref [8] for further details). Fig 3 shows the profiles of the mean axial velocity and Reynolds shear stresses in all characteristic regions behind the step: within the recirculation zone ($x/H = 2$ and 4), at the reattachment ($x/H = 6$) and in the recovery region ($x/H = 8$ and 10). The main effect of the perturbation is reproduced by LES and DES. A significant increase in the shear stress causes a higher momentum transfer and consequently shortening of the recirculation length. This is clearly illustrated in Fig. 4, where the perturbation effectiveness can be deduced by analysing the mean skin-friction distributions at the bottom wall for all perturbation frequencies. Secondary corner separation bubble is clearly represented. The maximum and minimum levels of skin-friction obtained by LES are not significantly affected by the perturbation, except for the low frequency case ($St = 0.08$). Gross effects of the perturbation are reproduced by both LES and DES as summarized in Fig. 5 which displays the evolution of reattachment length depending on perturbation frequency (normalized by the reattachment length obtained for the unperturbed case - X_{R0}). The closest agreement with the experimentally observed reduction in the reattachment length (28.3 %) is obtained by LES (24.5 %) and DES (35 %), whereas URANS employing S-A and k-ω SST models (these computations were performed by P. Queutey and E. Guilmineau, EC Nant, see Jakirlić et al. [3]) show a weak sensitivity to the perturbation, 5.9 and 12.9 % respectively.

3.2 Flow over a wall-mounted hump

Predictions of the separation and reattachment locations for the computed hump configurations are summarized in Table 1. In spite of premature separation in the baseline case, LES captures clearly the reduction in recirculation bubble in the suction case. The S-A RANS overpredicts significantly reattachment but returns the correct separation location. DES predictions for the baseline case are closest to the experiment, clearly showing the advantage of the DES as a hybrid RANS/LES approach (operating as RANS method within the boundary layer and LES method in the separated region), as far as baseline case is concerned. However, in the suction case DES predictions are poor indicating the importance of grid design, i.e. RANS-LES interface in DES. Fig. 6 displays pressure coefficient distributions. Generally, both LES and DES results agree better with the measurements than S-A RANS, however, in the baseline case, the peak suction pressure is underpredicted irrespective of computational method. This could be explained by possible blockage effects of the wind tunel side walls, not accounted for in the computations. Mean streamwise velocity profiles are shown in Figs. 7 and 8. It is interesting to see that DES predictions for the baseline case are superior to the ones of LES. On the other hand, agreement of LES predictions with the experimental data is excellent in the case with flow control. This could be explained by the fact that the separation point is to a large extent fixed by activating the flow suction (see small differencies in the separation point locations obtained by different methods, Table 1). The foregoing discussion is supported and clearly illustrated by Fig 9 which shows the selected shear stress profiles across the recirculation region at $x = 0.8C$. LES and DES are superior to the S-A RANS which fails to predict the peak shear stress in the separated shear layer. This is crucial for prediction of the main flow features downstream, particularly for capturing the reattachment location.

4 Conclusion

The potential of the methods for unsteady flow computations: LES, DES and U-RANS, to predict flow and turbulence in configurations relevant to the active flow control was investigated. LES and DES of a periodically perturbed backward-facing step flow reproduce all important effects observed in the experiments, whereas RA-NS (S-A and k-ω SST) exhibit a weaker sensitivity to the perturbation frequency. The LES and DES predictions of the main characteristics of separated flow over a wall-mounted hump, obtained on relatively coarse grids for the flow Reynolds number considered ($Re_c = 9.36 \cdot 10^5$), are encouraging, outperforming significantly the S-A RANS. It is interesting to see that DES (1.7 Mio. grid cells) predictions of the mean flow features for the baseline case are superior to the ones of LES (4 Mio. grid cells). On the contrary, in the steady suction case LES is superior to DES, indicating importance of DES grid design which dictates the RANS-LES interface.

Acknowledgements

The authors gratefully acknowledge financial support of the Deutsche Forschungsgemeinschaft through the grants GK "Modelling and numerical description of technical flows" and the research group on "LES of complex flows" (FOR 507/1, JA 941/7-1).

References

[1] Chun, K.B., and Sung, H.J. (1996): Control of turbulent separated flow over a backward-facing step. *Experiments in Fluids*, Vol. **21**, pp. 417-426
[2] Dejoan, A., Jang, Y.-J., and Leschziner, M.A. (2004): LES and unsteady RANS computations for a periodically-perturbed separated flow over a backward-facing step. *ASME Heat Transfer/Fluids Enginering Summer Conference, July 11-15, Charlotte, NC, USA*
[3] Jakirlić, S., Jester-Zürker, R., and Tropea, C. (2002): 9th ERCOFTAC/IAHR/COST Workshop on Refined Turbulence Modelling. *ERCOFTAC Bulletin*, December, No. **55**, pp. 36-43
[4] Obi, S.(2002): private communication.
[5] Rumsey, C., Gatski, T., Sellers, W., Vatsa, V. and Viken, S. (2004): Summary of the 2004 CFD Validation Workshop on Synthetic Jets and Turbulent Separation Control. *AIAA* 2004-2217
[6] Seifert, A., and Pack., L.G. (1999): Oscillatory control of separation at high Reynolds numbers. *AIAA J.*, Vol. **37**, No. **9**, pp. 1062-1071
[7] Seifert, A., and Pack., L.G. (2002): Active flow separation control on wall-mounted hump at high Reynolds numbers. *AIAA J.*, Vol. **40**, No. **7**, pp. 1363-1372
[8] Šarić, S., Jakirlić, S., and Tropea, C.: A Periodically perturbed backward-facing step flow by means of LES, DES and T-RANS: an example of flow separation control. *ASME Heat Transfer/Fluids Enginering Summer Conference, July 11-15, Charlotte, NC, USA*
[9] Travin, A., Shur, M., Strelets, M., and Spalart, P.R. (2002): Physical and numerical upgrades in the Dettached-Eddy Simulation of complex turbulence flows. In *Fluid Mechanics and Its Application*, R. Friedrich and W. Rodi (Eds.), Vol. 65, pp. 239-254
[10] Yoshioka, S., Obi, S., and Masuda, S. (2001): Turbulence statistics of periodically perturbed separated flow over backward-facing step. *Int. J. Heat and Fluid Flow*, Vol. 22, pp. 393-401

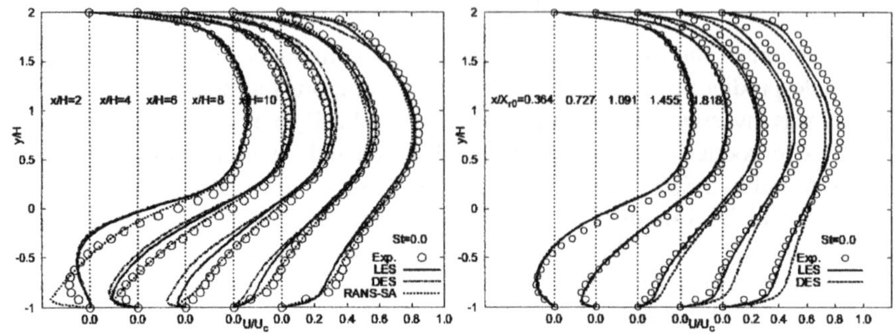

Figure 1 Mean streamwise velocity profiles for $St = 0.0$ normalized by H and X_{R0}

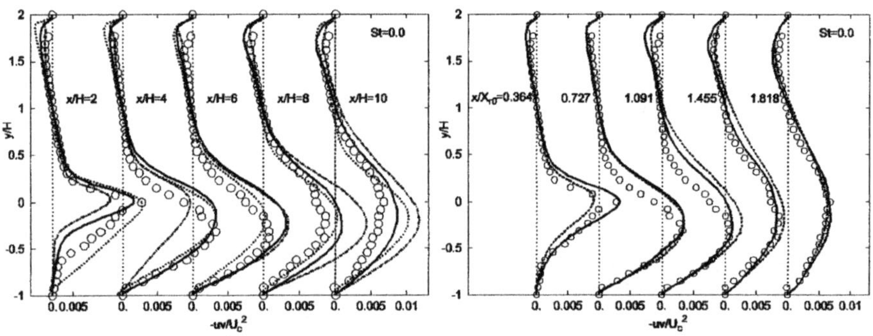

Figure 2 Reynolds shear stress profiles for $St = 0.0$ normalized by H and X_{R0}

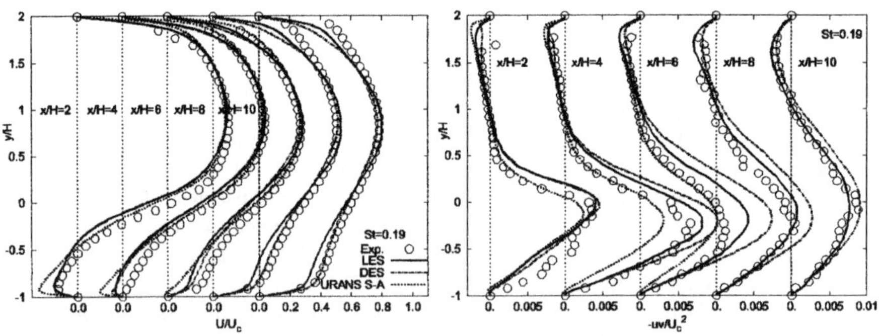

Figure 3 Mean streamwise velocity (left) and shear stress profiles (right) for $St = 0.19$

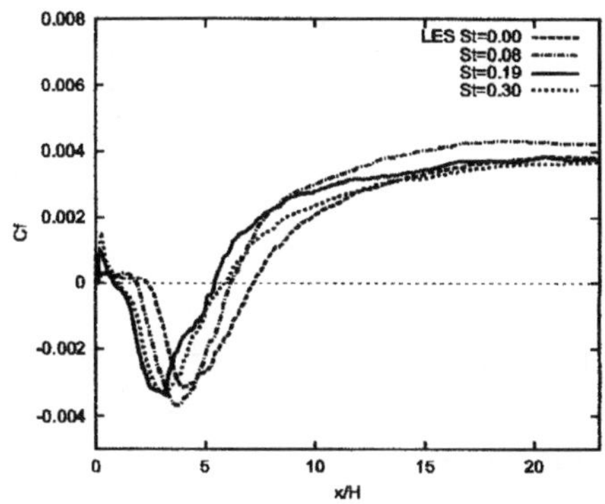

Figure 4 Mean skin-friction coefficient distr. obtained by LES.

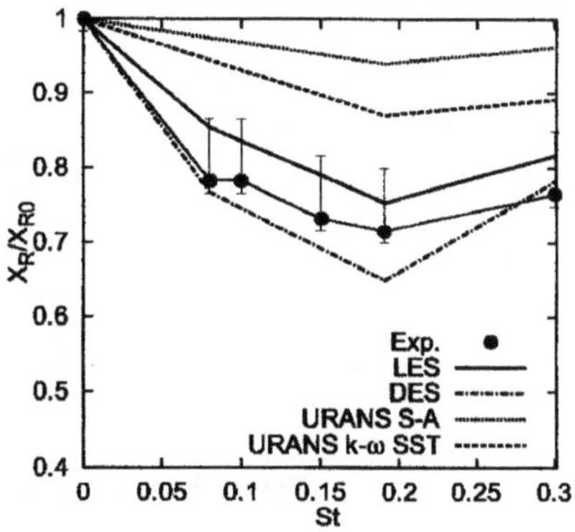

Figure 5 Evolution of rel. reattachment lengths vs. St, Exp.: Obi [4]

Table 1 Separation and reattachment locations

Computation	Grid size	L_Z	sep. (x/C)	reatt. (x/C)
LES	426x145x64	0.152 C	0.645	1.180
DES	426x145x28	0.2 C	0.663	1.121
S-A RANS	426x145	-	0.667	1.259
Exp. (no flow)	-	-	0.673	1.110
LES	426x145x32	0.152 C	0.671	0.947
DES	426x145x28	0.2 C	0.674	1.105
S-A RANS	426x145	-	0.674	1.098
Exp. (suction)	-	-	0.686	0.940

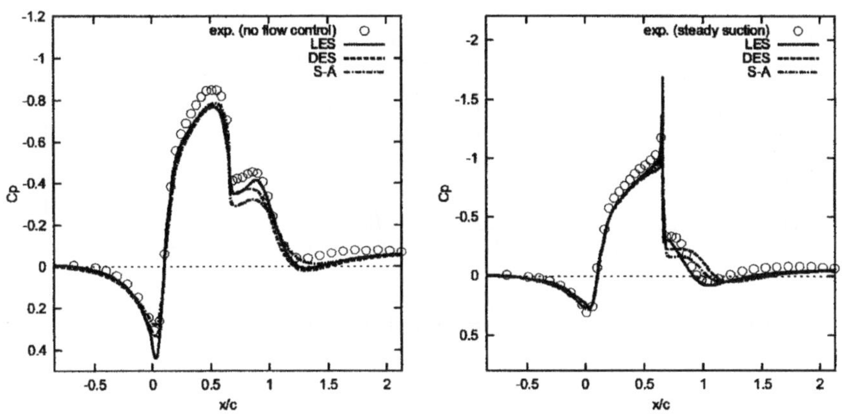

Figure 6 Pressure coefficient predictions, baseline (left) and suction (right) case

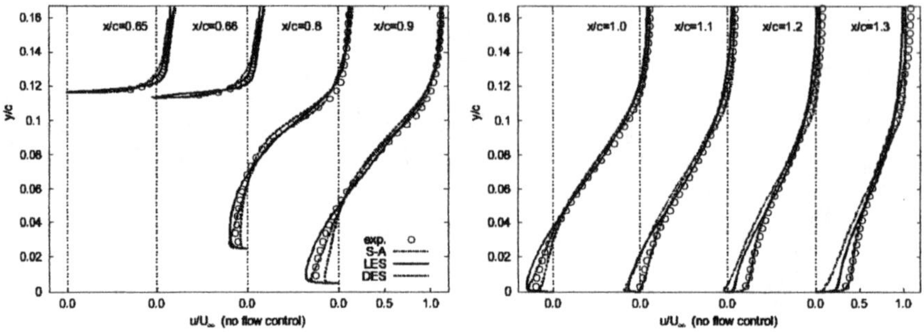

Figure 7 Mean velocity profiles for the baseline case

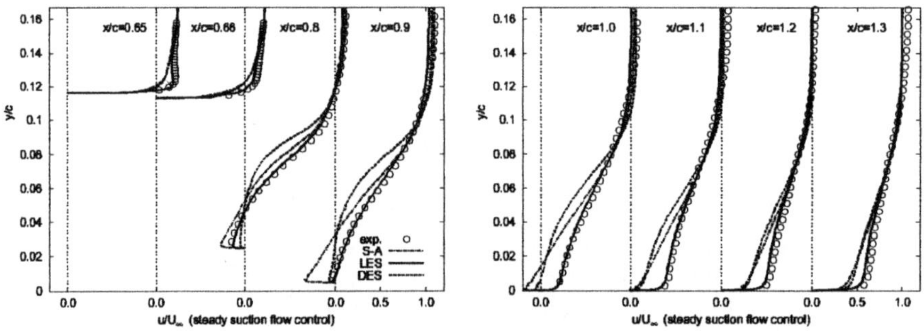

Figure 8 Mean velocity profiles for the case with steady suction flow control

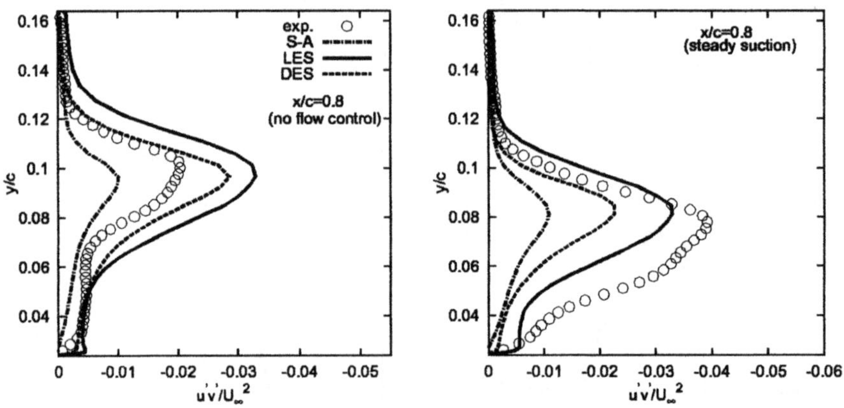

Figure 9 Reynolds shear stress profiles across the recirculation zone at $x = 0.8C$

Implementation of Propeller Simulation Techniques at DNW

I. PHILIPSEN
German Dutch Wind Tunnels, DNW, P.O. Box 175, 8300 AD Emmeloord, The Netherlands, iwan.philipsen@dnw.aero

Summary

Several improved or new techniques for propeller integration testing have been implemented at the German-Dutch Wind Tunnels (DNW). Presented here is an air-return line bridge system along with the applicable correction methodology. The implementation path followed is illustrated by the air-return line bridges systems concept validation and the systems performance validation during wind tunnel operation.

1 Introduction

At present the most commonly used test set-up for a full model with propeller simulation is a sting-mounted model with an internal balance. In order to establish the effects of the simulated propeller slipstream on the model it is necessary to determine the so-called clean aerodynamic forces and moments. As an example Figure 1 shows the deduction of the drag force (thrust bookkeeping) from the measured balance force. The customer required accuracy and repeatability of the aerodynamic coefficients translates into a requirement for a high accuracy for the main balance and the rotating shaft balances (RSB) that are used to determine the propeller forces and moments. This holds in particular for testing in a large low-speed facility. Furthermore, on the one hand the main balance needs to have a large load range due to the model weight and the high thrust simulation conditions. On the other hand the relatively low dynamic pressure, when compared to transonic cruise simulations [4], leads to relatively small aerodynamic forces and moments. This results in a simulation requiring a large (load range) main balance with a high resolution and accuracy. The latter must be an order of magnitude higher than those for cruise simulations.
The propellers are powered either by electro-, hydro- or (compressed) air-motors. Advanced propeller blades call for a relative high torque of the motor at the requested advance ratios. To facilitate this at model scales used in the DNW-LLF only air motors have sufficient torque. To get the compressed air to the motors the air-supply lines have to cross the internal balance. This necessitates an air-supply line bridge system to overcome the parasitic forces and moments introduced by the momentum of the compressed air mass flow.
For a good propeller simulation on a full model it is furthermore essential that the

exhaust flow of the air motor does not disturb the flow around the model. This would lead to an incorrect aerodynamic simulation since the exhaust flow of the air motor does not scale appropriately with the exhaust flow of the real engine. Relaxing the air directly after passing the air motor would mostly results in a too large model engine exhaust flow and momentum, causing interference effects on the wing and high lift devices that are incorrect. This situation can be improved by returning the exhaust flow of the air motor through the model and the model support system. Returning the air motor mass flow through the model implies that mass flow needs to pass the internal balance again. So also an air-return line system is necessary.
All the above mentioned testing and simulation techniques: main internal balance, rotating shaft balances (RSB), air-supply line bridges and air-return line bridges have been implemented at the DNW-LLF for an Airbus A400M model. The available state of the art simulation techniques [1], [4] are not sufficient to obtain the accuracy required at a large low speed facility. So a new small high load range internal balance was put into operation [1]. A new type of RSB was conceived and extensively tested to verify its performance before it was successfully used in a wind tunnel test [2]. Finally also an existing air-supply line bridge system was successfully combined with a newly developed air-return line bridge system. This paper will deal in more detail on how this later system was implemented and focus on some of the lessons learned. This work started back in 1998 and was only recently finished.

2 Implementation of an air-return line bridge system

2.1 Requirements

To achieve a good overall accuracy for aerodynamic coefficients it is necessary to minimize the effects that the air line bridge (ALB) has on the performance of the main balance. To achieve this there are two approaches possible. One, accept the fact that the ALB has considerable parasitic forces and moments and try to correct for them accurately or, second, aim to have a ALB with only small parasitic forces and moments in the first place. The second approach has the advantage that correction of the residual parasitic forces and moments to the same relative accuracy level becomes easier. So the accuracy of the corrective terms could be of less importance for the second approach. The aim therefore was to develop an ALB that has corrective terms that are of the same order of magnitude as the balance accuracy of the balance to which it is mounted.
The new air-return line bridge system had to have a capacity of 12 kg/s at a pressure of 16 Bar and an air temperature of 263 K, and had to fit into a given model fuselage shape. The power and size of the air motors in general dominate the scale of a powered full model leading to a relatively small model given the size of the facility. This implies that space for any air line bridge system is limited.

2.2 Correction Methodology

The following possible influences of the ALB system on the balance performance in terms of attainable accuracy are to be distinguished:

- Change in the general characteristics of the balance on which the ALB system is mounted. Especially the change in sensitivity of the balance for certain loads, due to the additional stiffness introduced by the ALB system.
- Pressure effects. Pressurized air in the ALB system might result in parasitic forces and moments (off-set) or even a change in the sensitivity of the balance (pressurized air in the ALB makes the ALB system stiffer).
- Temperature effects. Temperature differences between the main balance and ALB and gradients in the bridges themselves will lead to a strain (dilatation) in the ALB, possibly resulting in parasitic forces and moments to the balance.
- Momentum effects. The check on the primary goal of the air-supply and air-return line bridge, reduce parasitic forces and moments due to a momentum flow crossing the balance from non-metric to metric part.

Note that in order for the ALB system to function well the main balance should have heat insulation to minimize the temperature effects caused by the heated air-supply lines. It also should have limited deflections to minimize the displacement between metric and non-metric part of the balance that the ALB has to cope with.

The deduction of the corrections is split into two parts. The first part is aimed at obtaining a correction for the change in sensitivity of the balance due to the ALB and establishing preliminary pressure and temperature corrections. This can be done without a wind tunnel model well in advance of the actual wind tunnel test. The second part is the actual checkout of the system prior to a wind tunnel test to obtain final corrections for the pressure, temperature and momentum effects on the balance due to the ALB. Such a checkout is customary for each wind tunnel test.

2.3 Concepts

Two air-return line bridge systems concepts were developed and manufactured.
The first concept designated as the straight pipe ALB is shown in figure 3. This concept comprises two air return lines, one on port and one on starboard side of the model. The ALB-s consist of concentrically aligned pipes, where the gap between the inner and outer pipes provides the same cross sectional area as the inner pipe. In this arrangement, where the pipes are flexibly coupled, the airflow through the inner tubes is directed across the couplings via the outer flow channel (the concentric gap). These couplings are bellow like devices that allow two free rotational movements around axes that are perpendicular to the pipe's axis and also would allow a displacement in axial direction. Using two of these couplings it is possible to get the desired six degrees of freedom. The area of the inner and outer pipe is equal so the couplings should not have any pressure forces working on them. The main advantage of this concept is that it needs very little space in the model fuselage. The second concept designated as RALD 2001 is based on the

principles already established for the air-supply line bridges system (designated RALD 2000) with which the return lines were combined, Figure 2. This ALB is a so-called swan-neck type ALB. In this concept three flexible pipeline connectors (FPC-s) per air line are used. It was possible to design and manufacture a device with a bellows function that is compact, symmetric, stable and has almost negligible hysteresis. Each FPC allows two free rotational movements by means of flexure elements. Both rotations are around axes that are perpendicular to the flow in the air-line. The arrangement of the three FPC-s is such that the ALB has six degrees of freedom. The space in the model only allowed an arrangement of the line in such a way that the mass flow enters the system vertically and leaves the system horizontally, not the desired configuration for momentum decoupling.

2.4 Concept validation

First investigations with the system focused on the performance of the system for axial loading the load component expected to be the most affected. The pipe ALB concept was tested first and was rejected since it did not fulfill the requirement of having only small parasitic influences. Comparing the pipe ALB results with the RALD 2001 results clearly illustrates why the pipe ALB concept was rejected for not complying with the requirements. For the signal for axial load R1 the change in sensitivity of the balance is illustrated in Figure 4. The pipe ALB results in a 0.97% change of the sensitivity compared to only a 0.07% change in sensitivity change for the combined RALD 2000 and RALD 2001 ALB system. Pressure in the pipe ALB generates relatively large off-sets and leads to hysteresis effects compared to the pressure effect of the RALD 2001 ALB as shown in Figure 5. Note that the pipe ALB is only pressurized on one side at the time so the effects need to be doubled for comparison with the RALD 2001 ALB. Temperature effects caused by temperature differences in the pipe ALB are ambiguous and large as shown in Figure 6. The return ALB temperature will in general be below the balance temperature during normal operation. Since the inner pipes of the pipe ALB were not accessible for cooling with nitrogen they were heated by hot air. Finally, the mass flow effect was established for RALD 2001 [1] indicating that the concept may be successfully implemented for use in a wind tunnel model.

2.5 Full calibration and performance verification

After the concept validation the sensitivity change due to the combined ALB systems RALD 2000 and RALD 2001, was calibrated for all load components of the balance. The pressure and mass flow effects were calibrated during a wind tunnel test. Since the temperature effect was negligibly small it was not corrected for. The pressure corrections are also very small but not negligible (about 0.1% of the FS load range of the balance). Note that since the air-supply and air-return line bridges system are now connected via the air motor, special model provisions need to be made for calibration of the pressure effect on the air-supply line (80 Bar) and the air-return line (16 Bar) separately.

The mass flow effect is calibrated by pumping mass flow through the model and the blocked air motors. The blocked air motors, not doing any work, will in this situation require a lower mass flux than during normal operation for the same air return conditions. The obtained mass flow correction has to be extrapolated during normal operation. The mass flow effect of for the complete supply and return ALB system is shown for the axial component in figure 7. Again, the effect is very small, reproducible and therefore correctable. The model provision to block the air motors while pumping mass flow through the model makes it possible to verify performance of the ALB-s (and the application of all corrections) during real wind tunnel operation. Figure 8 shows a comparison of several α-polars (near negative stall) for the same model configuration with 0, 4.1 and 7.3 kg/s mass flow over the combined ALB system. The results repeat within the balance accuracy limits. The accuracy of a balance for a given component depends on the overall relative combined load level of all six load components [2]. The accuracy in Fx for the given test condition is between 0.09% and 1.41% FS (Fx), in this case between ±24 and ±37 drag counts. Finally figure 9 shows cleaned aerodynamic results for repeat polars during normal operation with the air motors and propellers. The repeatability shown here indicates that the ALB performance is at least as good as during the verification.

3 Conclusions

Implementation of new simulation techniques like an air-return line bridge system as required for propeller simulation testing is not a straight forward first time right development. Implementation can be successfully achieved when the correct correction and verification methodology is adopted. For air-line bridge systems used for propulsion simulation testing in low-speed wind tunnels this implies the effects of these systems on the main balance must be minimized before correction. The overall accuracy of the balance will in that case not be affected.

References

[1] Catalanotto, C. and Javis, R. F. "a significant improvement of an air supply /balance cross-over system", AIAA-84-0600

[2] Philipsen., I, Hoeijmakers, H., Alons, H.J., "a new balance and air-return line bridges for DNW-LLF models (b664 / RALD 2001)", 4th International Symposium on Strain-Gauge Balances", San Diego, USA, 10–13 May 2004

[3] Philipsen, I., Hoeijmakers, H., "dynamic checks and temperature correction for six-component rotating shaft balances", 4th International Symposium on Strain-Gauge Balances", San Diego, USA, 10–13 May 2004

[4] Prieur, J., Sechaud, J.F., Francois, G., "wind tunnel testing of propeller driven aircraft models the FLA experience at ONERA", 2nd International Symposium on Strain-Gauge Balances", Bedford, UK, 4–7 May 1999

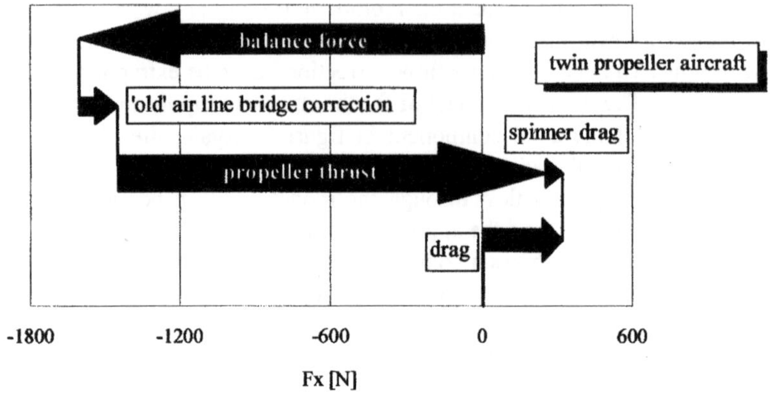

Figure 1 Thrust bookkeeping example

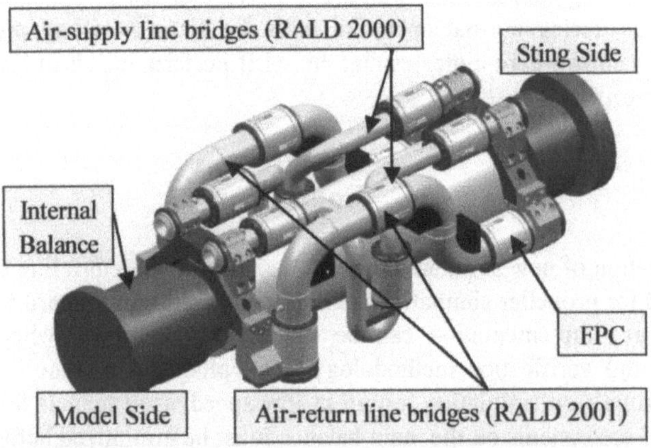

Figure 2 Sketch of internal balance with air-return line bridge systems

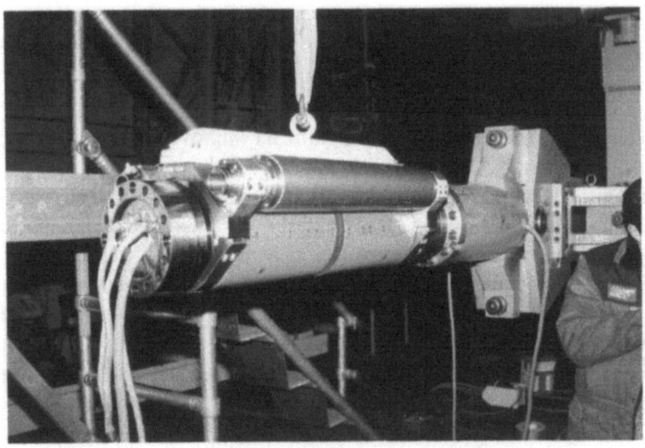

Figure 3 Straight pipe air-return line bridge system

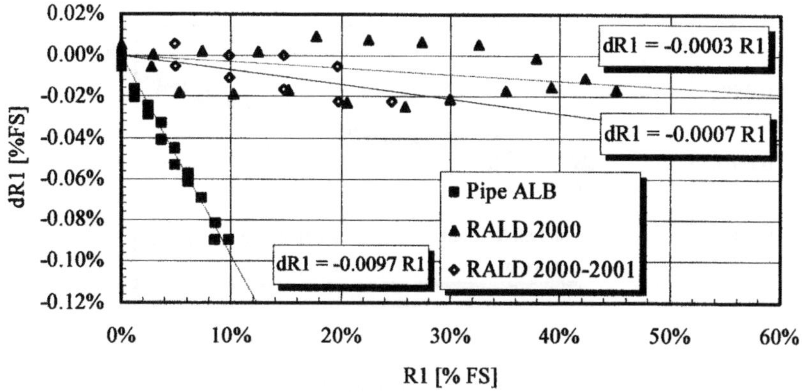

Figure 4 Mechanical stiffening effect for the axial load

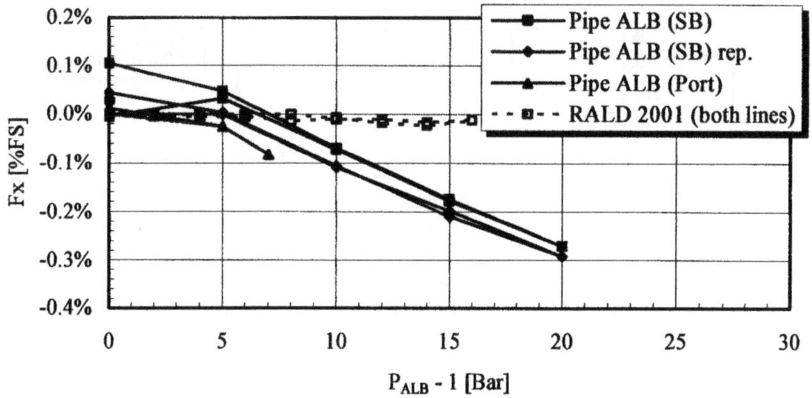

Figure 5 Pressure effect of return-line ALB on Fx

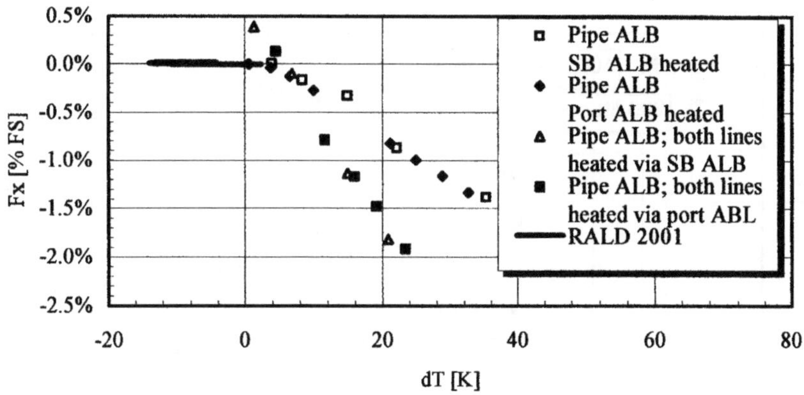

Figure 6 Temperature effect of ALB-s on Fx

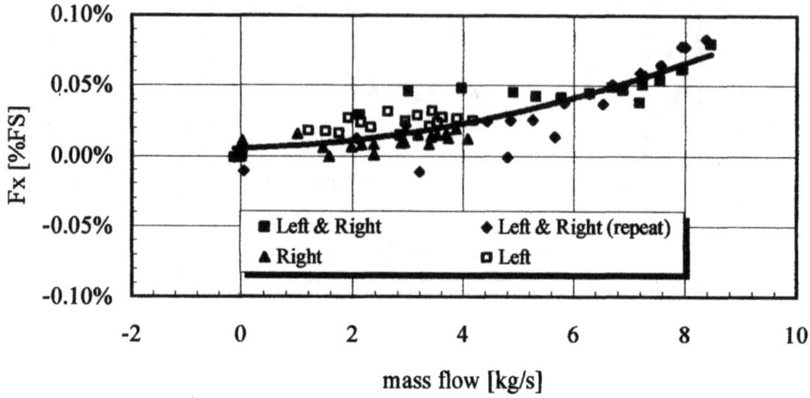

Figure 7 mass flow calibrations of RALD 2000 & RALD 2001 in wind tunnel

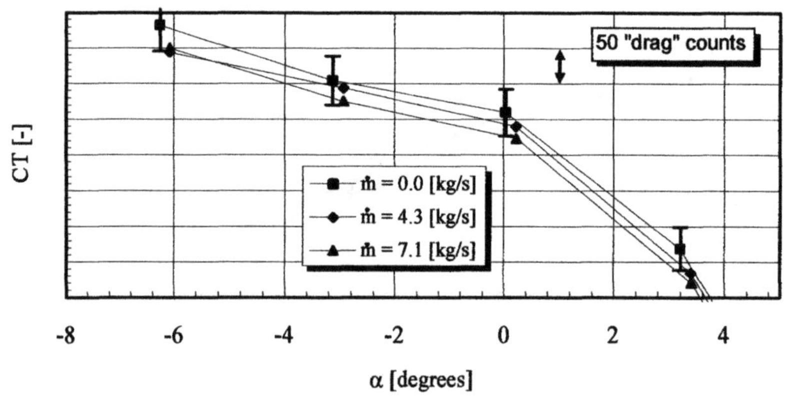

Figure 8 mass flow calibration verification

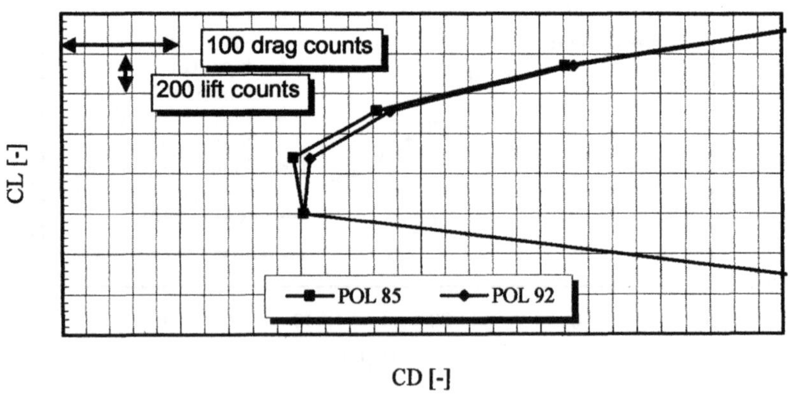

Figure 9 Repeatability for cleaned drag and lift

Notes on Numerical and Fluid Mechanics and Multidisciplinary Design

Available Volumes

Volume 94: W. Haase, B. Aupoix, U. Bunge, D. Schwamborn (eds.): FLOMANIA - A European Initiative on Flow Physics Modelling - Results of the European-Union funded project 2002 - 2004. ISBN 3-540-28786-8

Volume 93: Y. Shokin, M. Resch, N. Danaev, M. Orunkhanov, N. Shokina (eds.): Advances in High Performance Computing and Computational Sciences - The 1th Kazakh-German Advanced Research Workshop, Almaty, Kazakhstan, September 25 to October 1, 2005. ISBN 3-540-33864-0

Volume 92: H.-J. Rath, C. Holze, H.-J. Heinemann, R. Henke, H. Hönlinger (eds.): New Results in Numerical and Experimental Fluid Mechanics V - Contributions to the 14th STAB/DGLR Symposium Bremen, Germany 2004. ISBN 3-540-33286-3

Volume 91: E. Krause, Y. Shokin, M. Resch, N. Shokina (eds.): Computational Science and High Performance Computing II - The 2nd Russian-German Advanced Research Workshop, Stuttgart, Germany, March 14 to 16, 2005. ISBN 3-540-31767-8

Volume 90: K. Fujii, K. Nakahashi, S. Obayashi, S. Komurasaki (eds.): New Developments in Computational Fluid Dynamics - Proceedings of the Sixth International Nobeyama Workshop on the New Century of Computational Fluid Dynamics, Nobeyama, Japan, April 21 to 24, 2003. ISBN 3-540-27407-3

Volume 89: N. Kroll, J.K. Fassbender (eds.): MEGAFLOW - Numerical Flow Simulation for Aircraft Design - Results of the second phase of the German CFD initiative MEGAFLOW, presented during its closing symposium at DLR, Braunschweig, Germany, December 10 and 11, 2002. ISBN 3-540-24383-6

Volume 88: E. Krause, Y.I. Shokin, M. Resch, N. Shokina (eds.): Computational Science and High Performance Computing - Russian-German Advanced Research Workshop, Novosibirsk, Russia, September 30 to October 2, 2003. ISBN 3-540-24120-5

Volume 86: S. Wagner, M. Kloker, U. Rist (eds.): Recent Results in Laminar-Turbulent Transition - Selected numerical and experimental contributions from the DFG priority programme 'Transition' in Germany. ISBN 3-540-40490-2

Volume 82: E.H. Hirschel (ed.): Numerical Flow Simulation III - CNRS-DFG Collaborative Research Programme, Results 2000-2002. ISBN 3-540-44130-1

Volume 81: W. Haase, V. Selmin, B. Winzell (eds.): Progress in Computational Flow-Structure Interaction - Results of the Project UNSI, supported by the European Union 1998-2000. ISBN 3-540-43902-1

Volume 80: E. Stanewsky, J. Délery, J. Fulker, P. de Matteis (eds.): Drag Reduction by Shock and Boundary Layer Control - Results of the Project EUROSHOCK II, supported by the European Union 1996-1999. ISBN 3-540-43317-1

Volume 79: B. Schulte-Werning, R. Grégoire, A. Malfatti, G. Matschke (eds.): TRANSAERO - A European Initiative on Transient Aerodynamics for Railway System Optimisation. ISBN 3-540-43316-3

Volume 78: M. Hafez, K. Morinishi, J. Periaux (eds.): Computational Fluid Dynamics for the 21st Century. Proceedings of a Symposium Honoring Prof. Satofuka on the Occasion of his 60th Birthday, Kyoto, Japan, 15-17 July 2000. ISBN 3-540-42053-3

Volume 77: S. Wagner, U. Rist, H.-J. Heinemann, R. Hilbig (eds.): New Results in Numerical and Experimental Fluid Mechanics III. Contributions to the 12th STAB/DGLR Symposium Stuttgart, Germany 2000. ISBN 3-540-42696-5

Volume 76: P. Thiede (ed.): Aerodynamic Drag Reduction Technologies. Proceedings of the CEAS/DragNet European Drag Reduction Conference, 19-21 June 2000, Potsdam, Germany. ISBN 3-540-41911-X

Volume 75: E.H. Hirschel (ed.): Numerical Flow Simulation II. CNRS-DFG Collaborative Research Programme, Results 1998-2000. ISBN 3-540-41608-0

Volume 66: E.H. Hirschel (ed.): Numerical Flow Simulation I. CNRS-DFG Collaborative Research Programme. Results 1996-1998. ISBN 3-540-41540-8

MIX
Papier aus verantwortungsvollen Quellen
Paper from responsible sources
FSC® C105338

If you have any concerns about our products,
you can contact us on
ProductSafety@springernature.com

In case Publisher is established outside the EU,
the EU authorized representative is:
Springer Nature Customer Service Center GmbH
Europaplatz 3, 69115 Heidelberg, Germany

Printed by Libri Plureos GmbH
in Hamburg, Germany